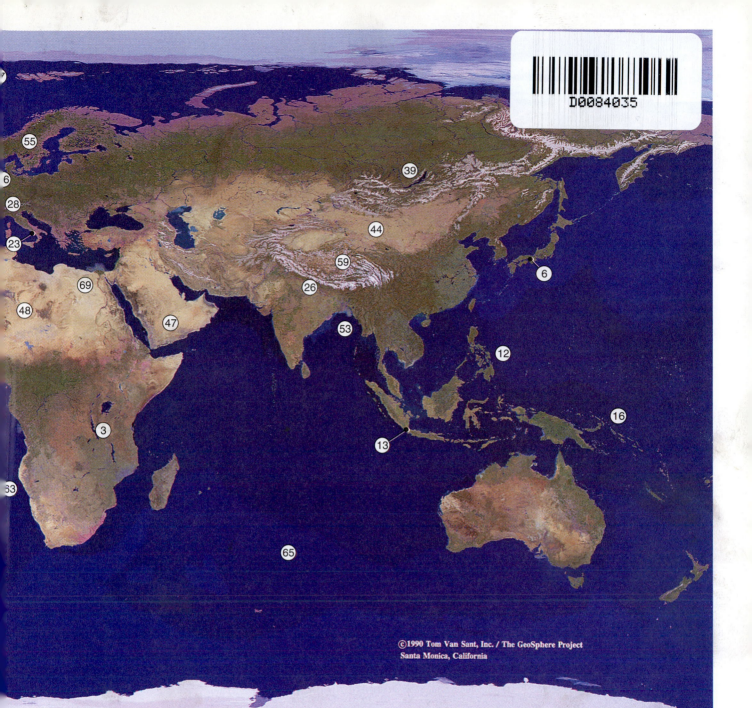

Earth Science

and the
Environment

second edition

Graham R. Thompson, PhD
University of Montana

Jonathan Turk, PhD

BROOKS/COLE
THOMSON LEARNING

Australia • Canada • Mexico • Singapore • Spain
United Kingdom • United States

BROOKS/COLE
THOMSON LEARNING ™

Publisher: John Vondeling
Acquisitions Editor: Jennifer Bortel
Developmental Editors: Susan Dust Pashos, Ed Dodd
Product Manager: Erik Fahlgren
Production Manager: Alicia Jackson
Project Editor: Anne Gibby
Text Designer: Chazz Bjanes
Art Director: Caroline McGowan
Copy Editor: Marne Evans
Illustrator: Rolin Graphics, Inc.
Cover Designer: Larry Didona
Cover Printer: Lehigh Press
Compositor: TechBooks
Printer: Von Hoffman Press

Cover Image: *Top, background:* USA, Alaska Chugach Mountains, aerial view of range and glacier *(RGK Photography/Tony Stone Images).* *Bottom, background:* San Andreas Fault, California, USA, aerial view *(James Balog/Tony Stone Images).* *Inserts,* clockwise from top left: Meadow, Paradise River, Washington *(© 1998 Charles Mauzy/Panoramic Images, Chicago, All rights reserved).* Deep space *(Robert Williams and the Hubble Deep Field Team STScI, and NASA).* Australia, Queensland, Whitsunday Island, Hill Inlet, aerial view *(Robin Smith/Tony Stone Images).* Molten lava *(G. Brad Lewis/Tony Stone Images).*

Frontispiece: Mt. Edith Cavell, Jasper National Park, Alberta, Canada *(Richard During/Tony Stone Images)*

Printed in the United States of America
5 6 7 05 04 03 02

For more information about our products, contact us at:
Thomson Learning Academic Resource Center
1-800-423-0563

For permission to use material from this text, contact us by:
Phone: 1-800-730-2214
Fax: 1-800-730-2215
Web: http://www.thomsonrights.com

Library of Congress Catalog Number: 98-86272
Earth Science and the Environment, second edition
ISBN: 0-03-006048-6

Asia
Thomson Learning
60 Albert Street, #15-01
Albert Complex
Singapore 189969

Australia
Nelson Thomson Learning
102 Dodds Street
South Melbourne, Victoria 3205
Australia

Canada
Nelson Thomson Learning
1120 Birchmount Road
Toronto, Ontario M1K 5G4
Canada

Europe/Middle East/Africa
Thomson Learning
Berkshire House
168-173 High Holborn
London WC1 V7AA
United Kingdom

Latin America
Thomson Learning
Seneca, 53
Colonia Polanco
11560 Mexico D.F.
Mexico

Spain
Paraninfo Thomson Learning
Calle/Magallanes, 25
28015 Madrid, Spain

Fossil ferns and dinosaur bones in Connecticut tell us that this temperate region was once warm and tropical. Millions of years after the dinosaurs became extinct, continental glaciers bulldozed across New England, leaving huge mounds of rock and sediment. Old desert sand dunes lie beneath the fertile plains near Denver, Colorado. Dead, giant cedars stand in a salt marsh in Washington State—but cedars don't grow in salt water. Combining this observation with other evidence, geologists concluded that the trees grew on dry land that dropped into the sea during a massive earthquake. Just about anywhere on our planet, scientists find evidence that the rock, the landscape, the climate, and the living organisms have changed throughout Earth history. In modern times, earthquakes, volcanic eruptions, floods, and violent storms remind us that the Earth continues to change, sometimes dramatically and cataclysmically.

Earth science is a study of the world around us: the rocks beneath our feet, the air we breathe, the water that exists almost everywhere in our environment, and all the living organisms that share our planet with us. Earth science also extends a view into space to look at distant planets, stars, and even faraway galaxies.

Earth scientists study the mechanisms of change. How did the climate in Connecticut change from warm and humid to frigidly cold? What forces drive earthquakes and why do they concentrate in some regions but not in others? As scientists seek to answer these and other related questions, they have found that the Earth is a system composed of smaller, interconnected, and interacting systems. In preparing the second edition of *Earth Science and the Environment*, we have substantially rewritten the text to emphasize the nature of Earth systems and their interactions.

Many of the Earth systems interactions discussed in the text summarize the results of research published within the past five years. Thus, at the same time that we are presenting the most accurate modern picture of planetary change, we also teach that science is a living process and not a set of stale facts to be memorized. We emphasize that scientists propose models, frequently disagree with one another, and then return to the laboratory, the computer screen, or the field to gain more information and to try to answer difficult questions.

The primary purpose of the second edition of *Earth Science and the Environment* is to explain fundamental Earth processes to students who have little or no previous science background. We go back in time and ask how the Earth evolved, how and when water collected to form oceans, how continents rose out of the watery wilderness, and what processes created an atmosphere favorable to life.

New to this edition is Unit VI: Human Interactions with Earth Systems. This unit consists of four chapters: Chapter 18, Climate Change; Chapter 19, Air Pollution; Chapter 20, Water Resources; Chapter 21, Geologic Resources.

Although some of the material in this unit appeared in the first edition, much of it did not. We include these chapters because humans are affected by natural environmental change, and in turn, we are changing the world around us. We believe that one of the purposes of science is to create a better life for all of humankind. In order to achieve this goal, we must look backward at events that formed the Earth's turbulent and changing past, and with this understanding, we must study how our present activities may alter the systems that sustain us.

Teaching Options

Earth science is a broad field covering many separate scientific disciplines. During the development of this book we communicated with many geologists, geographers, meteorologists, oceanographers, and astronomers who teach Earth science. The pedagogic strategy of this book was derived from several themes emphasized by these scientists.

We recognize that not all students in introductory Earth science courses have backgrounds in chemistry and physics. Therefore, this text explains the workings of planet Earth accurately, but in a language and style that is readily understood by students with little or no college-level science or mathematics background.

An exhaustive treatment of all the topics introduced in this course would involve several books, not just one. While recognizing the inherent limitations of a survey course, we offer balanced coverage of *all* topics. Few professors will assign the entire book in a quarter or semester course; there is just too much material. In order to accommodate a wide range of course emphases, each of the seven units offers a holistic introduction to the focus of the unit. Both the units and the chapters themselves are written to be as independent as possible so that the individual professors can pick the topics that are important to their course requirements. Furthermore, students can pursue any subject by independently reading units or chapters that interest them. This book is designed to be a useful reference for all Earth science topics in addition to being a text for a quarter or semester course.

Sequence of Topics

Just as different Earth science courses have many different emphases, a wide variety of logical sequences exist. We have chosen to introduce the Earth's materials and geologic time, and then to start from the Earth's interior and work outward. Thus the book is divided into seven units:

Unit I: Earth Materials and Time
Unit II: Internal Processes
Unit III: Surface Processes
Unit IV: The Oceans
Unit V: The Atmosphere
Unit VI: Human Interactions with Earth Systems
Unit VII: Astronomy

Some instructors may prefer other sequences. The unit structure of this book allows any alternative topic sequence.

SPECIAL FEATURES

Unit Opener Essays   We have introduced **Earth Systems Interactions** essays as a new feature in the second edition of *Earth Science and the Environment*. These essays highlight the central theme of the book: complex system interactions drive change in the structure of the continents, the oceans, climate, and life. For example, rocks and air seem to be separate realms, but the essay introducing Unit V explores the interactions between atmospheric composition and the minerals that comprise certain rocks. Furthermore, it shows that these interactions were profoundly influ-

enced by living organisms. If the seven essays were collected and read in sequence, they would provide a short monograph on the development and future of the Earth's four spheres—rock, water, air, and life—with a final essay on the evolution of the Universe. If the essays are read in their present order before each unit, they introduce concepts to be presented in the unit, add new concepts and principles, and reinforce the systems approach. The seven essays are:

Unit I: Earth Rocks, Earth History, and Mass Extinctions
Unit II: Plate Tectonics and Earth Systems
Unit III: Evolution of the Oceans, the Atmosphere, and Life
Unit IV: Flowers Bloomed on Earth while Venus Boiled and Mars Froze
Unit V: The Origin of Iron Ore and the Evolution of Earth's Atmosphere, Biosphere, and Oceans
Unit VI: Human Population and Alteration of Earth Systems
Unit VII: Birth of the Universe

Special Topics   In a survey course where many subjects are introduced, it is refreshing to read more in-depth discussions of related topics. Therefore, interesting special topics are set aside and highlighted in color as **Focus On** boxes. These topics are not necessary to the sequential development of each chapter, but from our own teaching experience, we have learned that students are drawn into a subject by specific examples. Some *Focus On* boxes cover topics in traditional Earth science such as "The Upper Fringe of the Atmosphere." Others cover environmental topics of current interest such as "Asbestos and Cancer." Numerous **Case Studies** incorporated into the text also highlight important concepts through the use of specific examples.

Chapter Review Material   A short **summary** of important chapter material is provided at the end of each chapter. Important **key words** that are highlighted in bold type in the text are also listed with page numbers at the end of each chapter.

Questions   Two types of end-of-chapter questions are provided. The **review questions** can be answered in a straightforward manner from the material in the text, and allow students to test themselves on how completely thay have learned the material in the chapter. On the other hand, **discussion questions** challenge students to apply what they have learned to an analysis of situations not directly described in the text. These questions often have no absolute correct answers. In sis of situations not directly described in the text. These questions often have no absolute correct answers.

Glossary and Appendix A **glossary** is provided at the end of the book. In addition, **appendices** cover the elements, mineral classification and identification, metric units, rock symbols, and star maps.

ANCILLARIES

This text is accompanied by an extensive set of support materials.

Instructor's Manual with Test Bank

The Instructor's Manual, written by the authors of the text, provides teaching goals, alternate sequences of topics, answers to discussion questions, and a short bibliography. An extensive test bank, written by Christine Seashore, is included in the Instructor's Manual as well. The Test Bank includes multiple choice, true or false, and completion questions for each chapter of the text.

Computerized Test Bank

A computerized version of the test bank is available for Windows, IBM PC, and Macintosh. These versions allow instructors the flexibility to add their own questions or modify existing questions. Easy-to-follow commands make customizing tests and quizzes simple and efficient. Instructors without access to computers may receive customized tests within five working days by calling 1-800-447-9457. FAX service is also available.

Study Guide

The Study Guide, written by James Albanese, provides review and study aids to further enhance the students' understanding of the text. The Study Guide includes chapter objectives, chapter outlines, multiple choice questions that test vocabulary, objective questions that test recall of important factual information, and short answer questions that require students to use and apply important chapter concepts. The student is also asked to think critically about environmental issues or concerns.

Instructor's Resource CD-ROM

A CD-ROM containing images of all of the text's illustrations. This CD-ROM can be used as a presentation tool in its own right or in conjunction with other commercial presentation packages such as PowerPoint™ software. It is available on a dual platform CD-ROM for both Windows and Macintosh platforms.

Web Support

To broaden the material we can offer you, we have built the following areas of Brooks/Cole's Internet site. Please visit us at www.brookscole.com/earthscience_d/. Earth Science Online provides multiple avenues for Earth science investigation, from our Top Ten List of related websites to scientific news groups, resource directories, virtual field trips, educational links, and help with specific searches and net navigation. We also have an area directly geared toward users of *Earth Science and the Environment*, 2/e. Here you will find practice quizzes, updates resulting from new discoveries in Earth science, and other interactive activities. Again, all of this material may be found at www.brookscole.com/earthscience_d/, under the discipline of Earth Science.

Earth Systems Interactive CD-ROM

This visually captivating student CD-ROM presents some of the fundamentals of Earth science through video, photos, and animations. Students are led on a tour of internal and surface processes that enriches their learning experience. This serves as an excellent, affordable supplement to traditional textbook learning.

Overhead Transparencies and 35 mm Slides

Over one hundred carefully chosen color illustrations and photographs from *Earth Science and the Environment, Second Edition* are available for use with overhead projectors or slide projectors for classroom presentation.

If you require additional flexibility, we are also offering you a choice of 100 images from the text to suit your specific needs. Please contact your sales representative for more information about this option for custom overhead transparencies.

Also available for qualified adopters is the **Saunders Earth Science Slide Set**. The set includes five hundred 35 mm slides of illustrations and photographs covering a wide range of topics such as physical and historical geology, oceanography, meteorology, and astronomy, and can be used in the classroom or the laboratory.

Brooks/Cole may provide complimentary instructional aids and supplements or supplement packages to those adopters qualified under our adoption policy. Please contact your sales representative for more information. If as an adopter or potential user you receive supplements you do not need, please return them to your sales representative or send them to:

Attn: Returns Department
Troy Warehouse
465 South Lincoln Drive
Troy, MO 63379

ACKNOWLEDGMENTS

We have not worked alone. The manuscript has been extensively reviewed at several stages and the numerous careful criticisms have helped shape the book and ensure accuracy:

James Albanese, *State University of New York at Oneonta*
Sandra Brake, *Indiana State University*
Mark Evans, *University of Pittsburgh*
Bryan Gregor, *Wright State University*
Clay Harris, *Middle Tennessee State University*
Steve LaDochy, *California State University—Los Angeles*
John Madsen, *University of Delaware*
Joseph Moran, *University of Wisconsin—Green Bay*
Paul Nelson, *St. Louis Community College at Meramec*
Adele Schepige, *Western Oregon State College*
Richard Smosna, *West Virginia University*
Ronald Wasowski, *King's College*

We would also like to thank the reviewers of the first edition:

James Albanese, *State University of New York at Oneonta*
John Alberghini, *Manchester Community College*
Calvin Alexander, *University of Minnesota*
Edmund Benson, *Wayne County Community College*
Robert Brenner, *University of Iowa*
Walter Burke, *Wheelock College*
Wayne Canis, *University of Northern Alabama*
Stan Celestian, *Glendale Community College*
Edward Cook, *Tunxis Community College*
James D'Amario, *Harford Community College*
Joanne Danielson, *Shasta College*
John Ernissee, *Clarion University*
Richard Faflak, *Valley City State University*
Joseph Gould, *St. Petersburg Junior College*
Bryan Gregor, *Wright State University*
Miriam Hill, *Indiana University—Southeast*
John Howe, *Bowling Green State University*
DelRoy Johnson, *Northwestern College*
Alan Kafka, *Boston College*
William Kohland, *Middle Tennessee State University*
Thomas Leavy, *Clarion University*
Doug Levin, *Bryant College*
Jim LoPresto, *Edinboro University of Pennsylvania*
Joseph Moran, *University of Wisconsin—Green Bay*
Glenn Mason, *Indiana University—Southeast*
Alan Morris, *University of Texas—San Antonio*
Jay Pasachoff, *Williams College*
Frank Revetta, *Potsdam College*
Laura Sanders, *Northeastern Illinois University*
Barun K. Sen Gupta, *Louisiana State University*
James Shea, *University of Wisconsin—Parkside*
Kenneth Sheppard, *East Texas State University*
Doug Sherman, *College of Lake County*
Gerry Simila, *California State University—Northridge*
Edward Spinney, *Northern Essex Community College*
James Stewart, *Vincennes University*
Susan Swope, *Plymouth State College*
J. Robert Thompson, *Glendale Community College*
Brooke Towery, *Pensacola Junior College*
Amos Turk, *City University of New York*
Jeff Wagner, *Fireland College*
Tom Williams, *Western Illinois University*
J. Curtis Wright, *University of Dubuque*
William Zinsmeister, *Purdue University*

Earth science is a visual science. We can readily observe landforms and weather patterns in our daily lives. Although we cannot see many processes, such as movements of tectonic plates, collisions of air masses, or the violent interior of a galactic nucleus, these events can be visualized through an artist's eye. George Kelvin and Carlyn Iverson have painted most of the illustrations in this book. It has been a pleasure to work with both of these artists.

We would never have been able to produce this book without professional support both here in Montana and at the offices of Saunders College Publishing. Thanks to Christine Seashore and Eloise Thompson for finding photographs, contributing personal photographs, and for logistic collaboration. Special thanks to John Vondeling, our Publisher. One of us, Jonathan Turk, has worked with John for over twenty years and has developed a long-lasting friendship and a superb professional relationship with him. Thanks to Jen Bortel, who coordinated the project, Susan Pashos, who lifted it off the ground, and Kathy Walker, who completed its development. Many thanks to Ed Dodd and Anne Gibby, who remembered all the details we forgot as the book's production raced to the home stretch. Thanks to the photo research team, led by George Semple, and to art directors Caroline McGowan, Susan Blaker, and Kim Menning, who designed, coordinated, and tracked the entire art and design project with ease. Finally, thanks to Dana Desonie, who reviewed the art, Marne Evans, our copyeditor, and Alexandra Buczek, our ancillary editor. All of these people and many of their associates have worked hard and efficiently to produce the finished product.

Graham R. Thompson
Missoula, Montana

Jonathan Turk
Darby, Montana

July 1998

CONTENTS OVERVIEW

CONTENTS

U N I T

III

Surface Processes   15

UNIT
IV
The Oceans   283

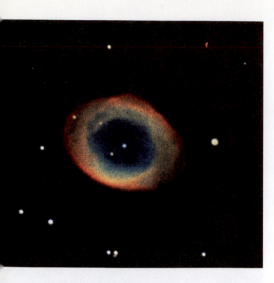

Earth Systems

The Earth's four spheres interact where wind and waves batter a coast, buffeting scattered trees that hang tenaciously to a rocky island in Olympic National Park. *(SharkSong/M. Kazmers/ Dembinsky Photo Associates)*

Imagine walking along a rocky coast as a storm blows in from the sea. Wind whips the ocean into whitecaps, gulls hurtle overhead, and waves crash onto shore. Before you have time to escape, blowing spray has soaked your clothes. A hard rain begins as you scramble over the rocks to your car. During this adventure, you have observed the four major realms of the Earth.

The rocks and soil underfoot are the surface of the **geosphere,** or the solid Earth. The rain and sea are parts of the **hydrosphere,** the watery part of our planet. The wind is the **atmosphere** in motion. Finally, you, the gulls, the beach grasses, and all other forms of life in the sea, on land, and in the air are parts of the **biosphere,** the realm of organisms. ●

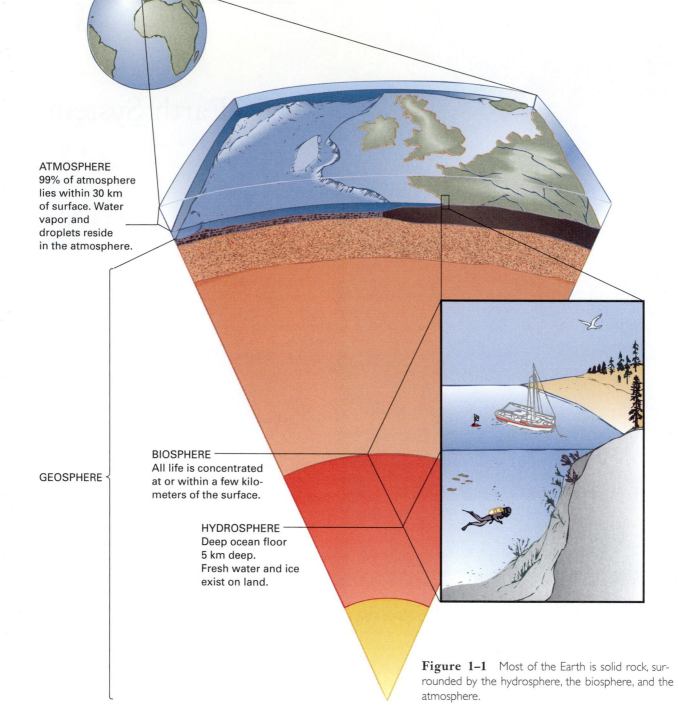

ATMOSPHERE
99% of atmosphere lies within 30 km of surface. Water vapor and droplets reside in the atmosphere.

GEOSPHERE

BIOSPHERE
All life is concentrated at or within a few kilometers of the surface.

HYDROSPHERE
Deep ocean floor 5 km deep. Fresh water and ice exist on land.

Figure 1–1 Most of the Earth is solid rock, surrounded by the hydrosphere, the biosphere, and the atmosphere.

1.1 Earth Science

As you stand on the coast watching the storm, the rocks feel solid and immobile, while the water, the air, and the gulls swirl around you. Then a large wave crashes against the shore, rolling sand and pebbles in the surf. You wonder how long even massive cliffs can withstand the relentless pounding and abrasion.

Earth is a dynamic planet. Even rocks that seem so permanent change slowly. Earth science is the study of our planet and the extraterrestrial bodies. Scientists look at our world and our Universe as it is today, search back into prehistory to study the past, and peer into the future. No person can be an expert in such a broad and complex field, so Earth science is divided into many disciplines: geology, oceanography, climatology, meteorology, physical geography, paleontology, and astronomy.

As you drive on a highway or walk along the seacoast, you experience your surroundings more richly if you understand how the hills and rocks formed and why the sky is stormy or clear. Thus, we study Earth science to satisfy our natural curiosity. In addition, we depend on planetary and extraterrestrial resources. Sunlight warms our world and provides the energy for plant growth. We burn fossil fuels and use mineral resources. Soil supports crops and forests. Weather and climate affect us daily. The oceans regulate climate and provide food and sea lanes for commerce and travel. Clean, fresh water is vital to agriculture, to industry, and for human consumption.

Today, many people have become concerned that humans may endanger our own well-being by altering our environment and depleting Earth resources. In this book we introduce our planetary and extraterrestrial environment, how it evolved, how it is changing today, and how we alter the systems that sustain us.

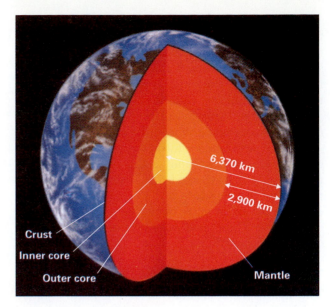

Figure 1–2 The geosphere is divided into three major layers: the crust, mantle, and core.

1.2 The Earth's Four Realms

Figure 1–1 shows that the geosphere is by far the largest of the four realms. The Earth's radius is about 6400 kilometers, 1-1/2 times the distance from New York to Los Angeles. Despite this great size, nearly all of our direct contact with the Earth occurs at or very near its surface. The deepest well penetrates only about 12 kilometers, 1/640 of the total distance to the center. The oceans make up most of the hydrosphere, and the central ocean floor is about 5 kilometers deep.

Nearly all of the atmosphere lies within 30 kilometers of the surface, and the biosphere is a thin shell about 15 kilometers thick.

The Geosphere

The geosphere consists of three major layers (Fig. 1–2). The outermost layer is a thin veneer called the **crust.** Below a layer of soil and beneath the ocean water, the crust is composed almost entirely of solid rock. Even a casual observer sees that rocks of the crust are different from one another: Some are soft, others hard, and they come in many colors (Fig. 1–3).

A

Figure 1–3 The Earth's crust is made up of different kinds of rock. (A) The granite of Baffin Island is gray, hard, and strong. (B) The sandstone and limestone of the Utah desert are red, crumbly, and show horizontal layering.

The **mantle** lies beneath the crust and contains almost 80 percent of the Earth's volume. Although the mantle is mostly solid rock, it is so hot that it contains small pools of liquid rock called **magma.** The chemical composition and physical properties of the mantle are quite different from the crust.

The third and innermost layer is a dense, hot, partly molten **core** composed mainly of iron and nickel.

Each of these layers is further subdivided. The crust consists of continental and oceanic crust. The uppermost mantle is hard, solid rock, like the crust, while the rest of the mantle is hot, weak, and plastic. The core consists of an outer liquid region surrounding a solid center. These subdivisions are described in more detail in Chapter 5.

The Hydrosphere

The hydrosphere includes all of the Earth's water, which circulates among oceans, continents, and the atmosphere. Oceans cover 71 percent of the Earth and contain 97.5 percent of its water. Ocean currents transport heat across vast distances, altering global climate.

About 1.8 percent of the Earth's water is frozen in glaciers. Although glaciers cover about 10 percent of

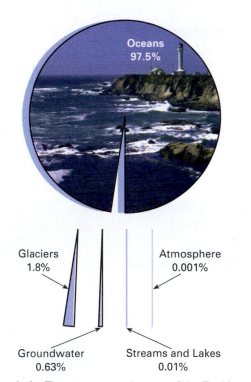

Figure 1–4   The oceans contain most of the Earth's surface water. Most fresh water is frozen into glaciers. Most available fresh water is stored underground as ground water.

● Interactive Question: How would a doubling of the amount of water in the atmosphere affect the oceans? How would it affect weather and climate?

the Earth's land surface today, they covered much greater portions of the globe as recently as 18,000 years ago.

About 0.63 percent of the Earth's water exists as **ground water** saturating rock and soil of the upper few kilometers of the geosphere. Only 0.01 percent of the total forms streams and lakes. A minuscule amount, 0.001 percent, exists in the atmosphere, but this water is so mobile that it profoundly affects both the weather and climate of our planet (Fig. 1–4).

The Atmosphere

The atmosphere is a mixture of gases, mostly nitrogen and oxygen, with smaller amounts of argon, carbon dioxide, and other gases. It is held to the Earth by gravity and thins rapidly with altitude. Ninety-nine percent is concentrated in the first 30 kilometers, but a few traces remain even 10,000 kilometers above the Earth's surface.

The atmosphere supports life because animals need oxygen, and plants need both carbon dioxide and oxygen. In addition, the atmosphere supports life indirectly by regulating climate. Air acts both as a filter and as a blanket, retaining heat at night and shielding us from direct solar radiation during the day. Wind transports heat from the equator toward the poles, cooling equatorial regions and warming temperate and polar zones (Fig. 1–5).

The Earth, Venus, and Mars have approximately the same composition and all share the same general environment in space. Yet, the three planets have radically different atmospheres and climates. The Venusian atmosphere is hot, dense, and rich in carbon dioxide. The surface temperature is 450°C, as hot as the interior of a self-cleaning oven, and sulfuric-acid-rich clouds cover the sky. In contrast, Mars is frigid, with an atmospheric pressure only 0.006 that at the surface of the Earth. Light winds carry dust across cold, barren deserts. At the same time, life thrives on a temperate Earth.

In Unit V we will learn why the Earth's atmospheric composition and temperature are so favorable to life, while conditions are so hostile on neighboring planets.

The Biosphere

The biosphere is the zone inhabited by life. It includes the uppermost geosphere, the hydrosphere, and the lower parts of the atmosphere. Sea life concentrates near the surface, where sunlight is available. Plants also grow on the Earth's surface, with roots penetrating a few meters into the soil. Animals live on the surface, fly a kilometer or two above it, or burrow a few meters underground. Large populations of bacteria live in rock to depths as much as four kilometers, some organisms live on the ocean floor, and a few wind-blown

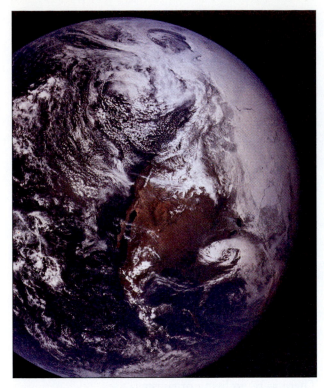

Figure 1–5 Winds often drive clouds into great swirls. At the same time, global winds carry warm air to cooler regions and cool air to equatorial parts of the Earth. The flow of heat makes the climates of both regions more favorable to humans. *(Courtesy NASA)*

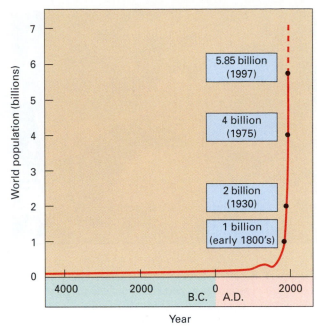

Figure 1–6 The human population has increased rapidly since the 1700s. For most of human history, there were fewer than one-half billion people on Earth. In mid-year 1997, 5.85 billion people inhabited our planet. *(Data from the Population Reference Bureau)* ● Interactive Question: Approximately how many years did it take for the human population to double from 2 billion to 4 billion? If that doubling rate were to continue, how many people would live on Earth in the year 2500? 3000?

microorganisms drift at heights of 10 kilometers or more. But even at these extremes, the biosphere is a very thin layer at the Earth's surface.

Primitive human-like species evolved in East Africa about 4 million years ago. Human populations were initially small and their impact on global ecosystems was minimal. However, within the past few hundred years the human population has soared (Fig. 1–6). Today, farms cover vast areas that were recently forested or covered by natural prairies. People have paved large expanses of land, drained wetlands, dammed rivers, pumped ground water to the surface, and released pollutants into waterways and the atmosphere (Fig. 1–7). These changes have affected even the most remote regions of the Earth, including the Sahara desert, the Amazon rainforest, the central oceans, and the South Pole. Some scientists are concerned that these changes are altering our environment in ways that threaten human well-being.

1.3 Geologic Time

James Hutton was a gentleman farmer who lived in Scotland in the late 1700s. Although trained as a physi-

cian, he never practiced medicine and, instead, turned to geology. Hutton observed that a certain type of rock, called **sandstone,** is composed of sand grains cemented together (Fig. 1–8). He also noted that rocks slowly decompose into sand, and that streams carry sand into the lowlands. He inferred that sandstone is

Figure 1–7 Approximately 50 percent of the Earth's land surface has been transformed from its original condition by humans. Here, suburban development is spreading over the high plains of Colorado toward the Rocky Mountains.

A   B

Figure 1–8   (A) Sandstone cliffs rise above the Escalante River, Utah. (B) This close-up view shows that sandstone is composed of rounded sand grains cemented together. *(Louise K. Broman/Photo Researchers)*

composed of sand grains that originated by the erosion of ancient cliffs and mountains.

Hutton tried to deduce how much time was required to form a thick bed of sandstone. He studied sand grains slowly breaking away from rock outcrops. He watched sand bouncing down stream beds. Finally he traveled to beaches and river deltas where sand was accumulating. By estimating the time needed for thick layers of sand to accumulate on beaches, Hutton concluded that sandstone must be much older than human history. By studying how other rocks form, he reasoned that most rocks are far older than human history. Taking his reasoning one step further, he deduced that our planet is very old. Hutton, overwhelmed by the magnitude of geological time, wrote:

> On us who saw these phenomena for the first time, the impression will not easily be forgotten. . . . We felt ourselves necessarily carried back to the time . . . when the sandstone before us was only beginning to be deposited, in the shape of sand and mud, from the waters of an ancient ocean. . . . The mind seemed to grow giddy by looking so far into the abyss of time. . . .
>
>   We find no vestige of a beginning, no prospect of an end.

Hutton had no way to measure the magnitude of geologic time. However, modern geologists have learned that certain radioactive materials in rocks act as clocks to record the passage of time. Using these clocks and other clues embedded in the Earth's crust, geologists estimate that the Earth formed 4.6 billion years ago.

The primordial planet was vastly different from our modern world. There was no crust as we know it today, no oceans, and the diffuse atmosphere was composed mainly of the light gases hydrogen and helium. There were no living organisms.

No one knows exactly when the first living organisms evolved, but we know that life existed at least as early as 3.8 billion years ago, 800 million years after the planet formed. For the following 3.3 billion years, life evolved slowly, and although some multicellular organisms developed, most of the biosphere consisted of single-celled organisms. Then organisms rapidly became more complex, abundant, and varied about 543 million years ago. The dinosaurs flourished between 225 million and 65 million years ago.

Homo sapiens and our direct ancestors have been on this planet for only about 0.1 percent of its history. In his book *Basin and Range*, about the geology of western North America, John McPhee offers us a metaphor for the magnitude of geologic time. If the history of the Earth were represented by the old English measure of a yard, the distance from the king's nose to the end of his outstretched hand, all of human history could be erased by a single stroke of a file on his middle fingernail. Figure 1–9 summarizes Earth history in graphical form.

Geologists routinely talk about events that occurred millions or even billions of years ago. If you live on the East Coast of North America, you may be familiar with the Appalachian mountains, which started rising from a coastal plain nearly half a billion years ago. If you live in Seattle, the land you are standing on was an island in the Pacific between 80 and 100 million years ago, when dinosaurs ruled the continents.

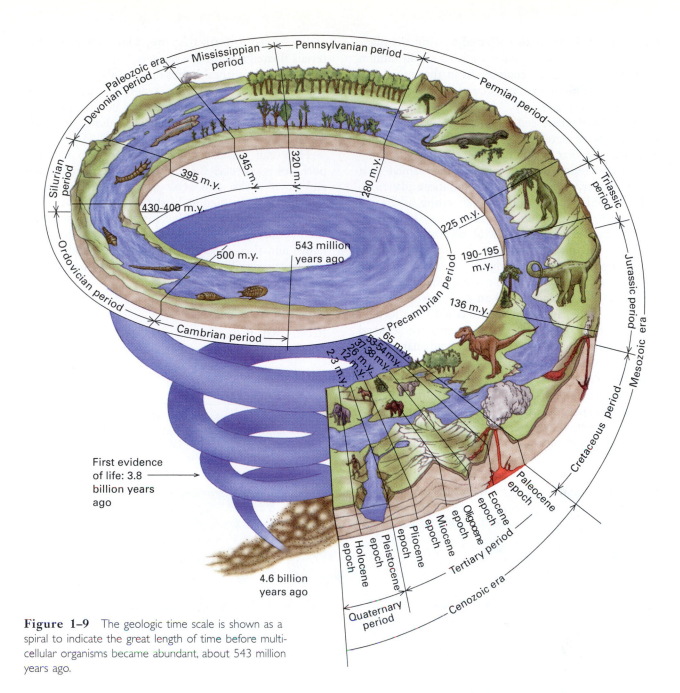

Figure 1–9 The geologic time scale is shown as a spiral to indicate the great length of time before multicellular organisms became abundant, about 543 million years ago.

1.4 The Hot Earth

The Early Earth

No one can go back in time to observe events that occurred millions or billions of years ago. To deduce the Earth's past, scientists study ancient Earth rocks. But on one has found a "genesis rock" that is as old as our planet. Additional information is gathered from observations of nearby planets, comets, meteorites, and asteroids. Astronomers peer into the Sun and also study young stars that might reflect conditions on our

Sun five billion years ago. Finally, scientists use huge computers to simulate the early Solar System.

The best evidence indicates that our Solar System coalesced from a frigid cloud of dust and gas that was rotating slowly in space. Material fell toward the center as a result of gravity to form the Sun. At the same time, rotational forces spun material in the outer cloud into a thin disk. When the turbulence of the initial accretion subsided, small grains stuck together to form fist-sized masses. These planetary seeds then accreted to form small rocky clumps, which grew to form larger planetesimals, 100 to 1000 kilometers in diameter.

Finally, the planetesimals collected to form the planets. Scientists estimate that it took about 40 million years for the Earth to coalesce from the initial fist-sized grains. This process was completed about 4.6 billion years ago (Fig. 1–10).

As the Earth coalesced, the rocky chunks and planetesimals were accelerated by gravity so that they slammed together at high speeds. Particles heat up when they collide, so the early Earth warmed as it formed. Later, asteroids, comets, and more planetesimals crashed into the surface, generating additional heat. At the same time, radioactive decay heated the Earth's interior. As a result of all three of these processes, our planet became so hot that all or most of it melted soon after it formed.

The Earth is layered, with a dense, metallic core, a less dense rocky mantle, and an even less dense surface crust. This layered structure could have formed by either of two processes. Astronomers have detected both metallic and rocky meteorites and many think that both metallic and rocky particles coalesced to form the planets. According to one hypothesis, rock and metal accumulated simultaneously during the initial coalescence, forming a homogeneous (non-layered) planet. Later, after the planet melted, dense molten iron and nickel gravitated toward the center

and collected to form the core, while less dense materials floated toward the surface.

A second hypothesis states that metallic particles initially accumulated to create the metallic core, and later, rocky particles collected around the core to form the rocky mantle. If this hypothesis is accurate, the Earth has always been layered.

How can we determine which of the two hypotheses is correct? By studying modern meteorites and lunar rocks, two geologists recently estimated that the core formed at least 62 million years after the Earth coalesced.[1] This interpretation supports the hypothesis that our planet was initially homogeneous and then separated into the core and mantle at a later date. The crust formed much later, as discussed in Chapter 8.

The Modern Earth

Recall that three processes heated the primordial Earth: gravitational coalescence, bombardment by

[1] Der-Chuen Lee and Alex N. Halliday, "Hafnium-tungsten Chronometry and the Timing of Terrestrial Core Formation," *Science*, Vol. 378, Dec 21–28, 1995, p. 771.

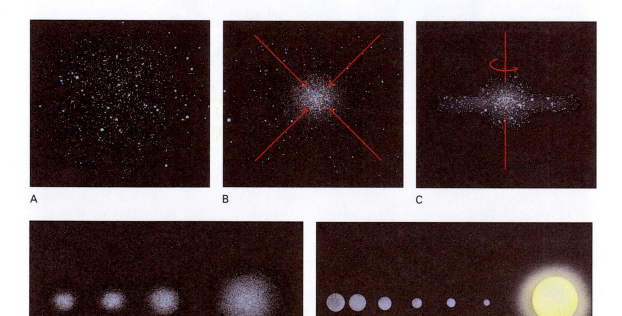

A   B   C

D   E

Figure 1–10 Formation of the Solar System. (A) The Solar System began as a cloud of dust and gas. (B) Gravity caused the dust and gas to coalesce. (C) The shrinking mass rotated and distorted into a disk. (D and E) The central part of the disk condensed to become the Sun; the outer portion became the planets.

meteorites and comets, and radioactive decay in the interior. The heat generated from gravitational coalescence was trapped in the interior and has been slowly escaping ever since. During the first 800 million years of Earth history, extraterrestrial bombardment was significant as our planet's gravity attracted debris from a cluttered early Solar System. As bombardment slowed down, this heat source also slowed. Over billions of years, radioactive materials in the Earth's interior gradually decompose, so this heat source diminishes very slowly. As a result, the Earth is gradually cooling. Today, the temperature of the Earth's core is 6000°C, about as hot as the surface of the Sun.

The Earth's crust and outermost mantle are cool because heat has escaped into space. When rock is cool, it is relatively hard and strong. In contrast, the mantle is so hot that it is weak and plastic and flows slowly, like cold honey.

According to a theory developed in the 1960s, the Earth's outer, cool, rigid shell is broken into several segments called **tectonic plates.** These tectonic plates float on the weak, plastic, mantle rock beneath, and glide across the Earth, riding on their plastic foundation. For example, North America is currently drifting toward China about as fast as your fingernail grows. In a few hundred million years—almost incomprehensibly long on a human time scale, but brief compared with planetary history—Asia and North America may collide, crumpling the edges of the continents and building a giant mountain range. In later chapters we will learn how plate tectonic theory explains earthquakes, volcanic eruptions, the formation of mountain ranges, and many other phenomena (Fig. 1–11).

The outer core is composed of molten metal, but the inner core is under such intense pressure that it is solid. Motion and heat transfer in the outer core may drive movement of the lower mantle and contribute to plate tectonic motion. Geophysical evidence indicates that flowing metal in the core forms the Earth's magnetic field.

1.5 Earth Systems

A **system** is any combination of interrelated, interacting components. For example, the human body is a system composed of stomach, bones, nerves, and many other organs. Each organ is discrete, yet all the organs interact to produce a living human. Blood nurtures the stomach; the stomach helps provide energy to maintain the blood.

In order to understand a system, we must study the exchange of energy and matter both within the system and with its surroundings (Fig. 1–12). For example, a human ingests food and oxygen, two forms of matter that react to produce chemical energy needed by the body. Energy and matter cycle within the body and eventually waste products are excreted.

Systems vary dramatically in size. A single bacterium and the entire Universe are both systems. A large system such as the Universe or the Solar System is so complex that no individual can study all its components in detail. Therefore scientists divide these huge systems into smaller components. Each planet is discrete enough to be a separate system. Most of this book focuses on Earth. Although small amounts of gases escape from our outer atmosphere, and some material is added by meteor bombardment, exchanges of matter with the rest of the Solar System are small compared with the movement of rock, water, and gas within the planet.

Two sources of energy drive Earth systems. The Sun warms the Earth's atmosphere and hydrosphere. The solar heat causes winds to blow, drives ocean waves, and evaporates water. Plants use solar energy to build energy-rich organic tissue. A raging hurricane, a flowing river, and a wood fire are all powered, ultimately, by the Sun. In contrast, the Earth's interior is heated by residual heat from the primordial coalescence and bombardment, and by continuing radioactive decay. This interior heat drives movements of tectonic plates, earthquakes, and volcanic eruptions and builds the Earth's mountain ranges.

In turn, the Earth is composed of smaller systems. Geosphere, hydrosphere, atmosphere, and biosphere

Figure 1–11 Movements of tectonic plates caused this eruption of Ngauruhoe volcano, New Zealand. *(Don Hyndman)*

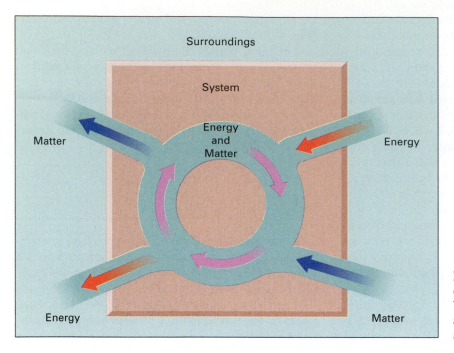

Figure 1–12 A system receives energy and matter from its surroundings. The system uses the energy and matter, alters them, and releases both energy and waste materials.

are all discrete components that can be studied separately. A single continent or ocean basin is a system composed of even smaller systems such as lakes or river valleys.

Earth scientists study individual systems and then try to understand how one system interacts with the whole. For example, a volcanic eruption occurs because molten magma forms in the upper mantle and then rises to the Earth's surface. Lava spewing to the surface may build a mountain and alter nearby drainage systems. Hot lava kills living organisms in its path. Gas and dust rise into the atmosphere, reflecting sunlight and cooling the Earth. Climate models indicate that the eruption of Mt. Pinatubo in the Philippines in 1991 cooled the Earth by a few tenths of a degree for several years. Changes in temperature and evaporation rates altered rainfall patterns. Thus a single event, powered by heat hundreds or even thousands of kilometers beneath the surface, can initiate change throughout many different planetary systems.

1.6 Rates of Change in Earth History

Recall that James Hutton deduced that sandstone forms when rocks slowly decompose to sand, the sand is transported to lowland regions, and the grains cement together. This process occurs step by step—over many years.

Hutton's conclusions led him to formulate a principle now known as **gradualism** or **uniformitarianism.** The principle states that geologic change occurs over long periods of time, by a sequence of almost imperceptible events. Hutton surmised that geologic processes operating today also operated in the past. Thus scientists can explain events that occurred in the past by observing changes occurring today. Sometimes this idea is summarized in the statement, "The present is the key to the past."

However, not all geologic change is gradual. William Whewell, another early geologist, argued that geologic change was sometimes rapid. He wrote that the geologic past may have "consisted of epochs of paroxysmal and catastrophic action, interposed between periods of comparative tranquility." Earthquakes and volcanoes are examples of catastrophic events but, in addition, Whewell argued that occasionally huge catastrophes alter the course of Earth history. He couldn't give an example because they happen so infrequently that none had occurred within human history (Fig. 1–13).

Today, geologists know that both Hutton's uniformitarianism and Whewell's **catastrophism** are correct. Over the great expanses of geologic time, slow, uniform processes alter the Earth. In addition, infrequent catastrophic events radically modify the path of slow change.

Gradual Change in Earth History

Recall that tectonic plates creep slowly across the Earth's surface. Since the first steam engine was built

Figure 1–13 Fifty thousand years ago, a meteorite crashed into the northern Arizona desert, creating Meteor Crater. Catastrophic events have helped shape Earth history. *(Meteor Crater National Monument)*

200 years ago, North America has migrated 8 meters westward, a distance a sprinter can run in 1 second. Thus, the movement of tectonic plates is too slow to be observed except with sensitive instruments. However, over geologic time, this movement alters the shapes of ocean basins, forms lofty mountains and plateaus, generates earthquakes and volcanic eruptions, and affects our planet in many other ways.

Catastrophic Change in Earth History

Chances are small that the river flowing through your city will flood this spring, but if you lived to be 100 years old, you would probably see a catastrophic flood. In fact, many residents of the Midwest saw such a flood in the summer of 1993, California residents experienced floods in the winters of 1994–95, 1996–97, and again in 1997–98, and a huge flood inundated North Dakota and Manitoba, Canada, in 1997.

When geologists study the 4.6 billion years of Earth history, they find abundant evidence of catastrophic events that are highly improbable in a human lifetime or even in human history. For example, clues preserved in rock and sediment indicate that giant meteorites have smashed into our planet, vaporizing portions of the crust and spreading dense dust clouds over the sky. Geologists have suggested that some meteorite impacts have almost instantaneously driven millions of species into extinction. Huge floods have occurred, millions of times more devastating than any floods in historic times. Catastrophic volcanic eruptions have changed conditions for life across the globe.

Threshold and Feedback Effects

Gradualism and catastrophism are simple conceptual models. However, Earth systems do not always behave in such straightforward manners. Instead, many Earth changes occur by complex, interacting processes. Two examples are threshold and feedback effects.

Imagine that some process gradually warmed the atmosphere. Imagine further that the summer temperature of coastal Greenland glaciers was −1.0°C. If gradual change warmed the atmosphere by 0.2°C, the glaciers wouldn't be affected appreciably because ice melts at 0°C and the summer temperature would remain below freezing, −0.8°C. But now imagine that gradual warming brought summer temperature to

On an afternoon field trip, you may find several different types of rocks or watch a river flow by. But you can never see the rocks or river as they existed in the past or as they will exist in the future. Yet a geologist might explain to you how the rocks formed millions or even a few billion years ago and might predict how the river valley will change in the future.

Scientists not only study events that they have never observed and never will observe, but they also study objects that can never be seen, touched, or felt. In this book we describe the center of the Earth 6400 kilometers beneath our feet, even though no one has ever visited it and no one ever will.

Much of science is built on inferences about events and objects outside the realm of direct experience. An inference is a conclusion based on thought and reason. How certain are we that a conclusion of this type is correct?

Scientists develop an understanding of the natural world according to a set of guidelines known as the **scientific method,** which involves three basic steps: (1) observation, (2) forming a hypothesis, and (3) testing the hypothesis and developing a theory.

Observation

All modern science is based on observation. Suppose that you observed an ocean current carrying and depositing sand. If you watched for some time, you would see that the sand accumulates slowly, layer by layer, on the beach. You might then visit Utah or Nevada and see cliffs of layered sandstone hundreds of meters high. Observations of this kind are the starting point of science.

Forming a Hypothesis

A scientist tries to organize observations to recognize patterns. You might note that the sand layers deposited along the coast look just like the layers of sand in the sandstone cliffs. Perhaps you would then infer that the thick layers of sandstone had been deposited in an ancient ocean. You might further conclude that, since the ocean deposits layers of sand slowly, the thick layers of sandstone must have accumulated over a long time.

If you were then to travel, you would observe that thick layers of sandstone are abundant all over the world. Because thick layers of sand accumulate so slowly, you might infer that a long time must have been required for all that sandstone to form. From these observations and inferences you might form the hypothesis that the Earth is old.

A **hypothesis** is a tentative explanation built on strong supporting evidence. Once a scientist or group of scientists proposes a hypothesis, others test it by comparison with observations and experiments. If it explains some of the facts but not all of them, it must be altered, or if it cannot be changed satisfactorily, it must be discarded and a new hypothesis developed.

Testing the Hypothesis and Forming a Theory

If a hypothesis explains new observations as they accumulate and is not substantively contradicted, it becomes elevated to a **theory**. Theories differ widely in form and content, but all obey four fundamental criteria:

−0.2°C. At this point, the system is at a threshold. Another small increment of change would raise the temperature to the melting point of ice. As soon as ice starts to melt, large changes occur rapidly: Glaciers recede and global sea level rises. A **threshold** effect occurs when the environment initially changes slowly (or not at all) in response to a small perturbation, but after the threshold is crossed, an additional small perturbation causes rapid change.

A **feedback** mechanism occurs when a small initial perturbation affects another component of the Earth systems, which amplifies the original effect, which perturbs the system even more, which leads to an even greater effect, and so on. (Picture a falling stack of dominos.) For example, according to one hypothesis, changes in the Earth's orbit caused the advances and retreats of the most recent ice ages. However, scientists have recently calculated that orbital variations, by themselves, may not be significant enough to cause ice ages. Instead, when changes in the Earth's orbit initiated slight cooling, high-latitude forests died and were replaced by tundra. Tundra reflected more sunlight than trees. When more sunlight was reflected, the Earth cooled. Glaciers started to grow. Ice reflected sunlight even more efficiently than tundra, and the cooling accelerated. Soon the entire planet cooled and the glaciers advanced.

Threshold and feedback mechanisms remind us that Earth systems often change in ways that are difficult to predict. These complexities provide the challenge and fascination of Earth Science.

When Earth formed 4.6 billion years ago, it was a barren, hot, rocky sphere. Over eons of time, the surface cooled, oceans and continents formed, an

1. A theory must be based on a series of confirmed observations or experimental results.

2. A theory must explain all relevant observations or experimental results.

3. A theory must not contradict any relevant observations or other scientific principles.

4. A theory must be internally consistent. Thus, it must be built in a logical manner so that the conclusions do not contradict any of the original premises.

For example, the theory of plate tectonics states that the outer layer of the Earth is broken into a number of plates that move horizontally relative to one another. As you will see in Chapters 5 through 8, this theory is supported by many observations and seems to have no major inconsistencies.

Many theories cannot be absolutely proven. For example, even though scientists are just about certain that their image of atom structure is correct, no one has or ever will watch an individual electron travel in its orbit. Therefore, our interpretation of atomic structure is called atomic theory.

However, in some instances, a theory is elevated to a scientific law. A **law** is a statement of how events always occur under given conditions. It is considered to be factual and correct. A law is the most certain of scientific statements. For example, the law of gravity states that all objects are attracted to one another in direct proportion to their masses. We cannot conceive of any contradiction to this principle, and none has been observed. Hence, the principle is called a law.

Sharing Information

The final step in the scientific process is to share observations and conclusions with other scientists and the general public. Typically, a scientist communicates with colleagues to discuss current research by phone, at annual meetings, or by electronic mail. When the scientist feels confident in his or her conclusions, he or she publishes them in a scientific journal. Colleagues review the material before it is published to ensure that the author has followed the scientific method. If the results are of general interest, the scientist may publish them in popular magazines or newspapers. The authors of this text have read many scientific journals and now pass the information on to you, the student.

Focus Question:

Obtain a copy of a news article in a weekly news magazine. Underline the facts with one color pencil and the author's opinions with another. Did the authors follow the rules for the scientific method outlined here to reach their conclusion?

---

atmosphere collected and evolved, life appeared and blossomed. Earth Science is a study of our home and how it has changed. It also allows us to peer into the future. What will Earth be like a hundred, or a thousand, or a million years from now? Will human activity alter the Earth's surface, waters, and atmosphere? Of course these are difficult questions. But curiosity is part of the human spirit, and from the beginning of time people have attempted to understand the origins and future of our world.

SUMMARY

The **geosphere** is composed of a thin surface crust, a large **mantle** that comprises 80 percent of the Earth's volume, and a dense, hot, central **core.** The crust and mantle are composed of rock while the core is composed of iron and nickel. The **hydrosphere** is mostly ocean water. Most of the Earth's fresh water is locked in glaciers; streams, lakes, and rivers account for only 0.01 percent of the planet's water. The **atmosphere** is a mixture of gases, mostly nitrogen and oxygen. The Earth's atmosphere supports life and regulates climate. The **biosphere** is the thin zone inhabited by life.

The Earth is about 4.6 billion years old; life formed at least 3.8 billion years ago; abundant multicellular life evolved about 543 million years ago, and humanoids have been on this planet for a mere 4 million years.

The Earth's heat originated from gravitational co-alescence, bombardment by comets and meteorites, and radioactive decay. This heat causes continents to travel across the globe, ocean basins to open and close, mountain ranges to rise, volcanoes to erupt, and earthquakes to shake the planet.

A **system** is composed of interrelated, interacting components. In order to understand a system, we must study the exchange of energy and matter both within the system and with the outside surroundings.

The principle of **gradualism** or **uniformitarianism** states that geologic change occurs over a long period of time by a sequence of almost imperceptible events. In contrast, **catastrophism** postulates that geologic change occurs mainly during infrequent catastrophic events. In many instances, **threshold** and **feedback** effects modify simple gradualistic or catastrophic processes.

KEY TERMS

geosphere 1	crust 3	ground water 4	system 9	catastrophism 10
hydrosphere 1	mantle 4	sandstone 5	gradualism 10	threshold 12
atmosphere 1	magma 4	tectonic plates 9	uniformitarianism 10	feedback 12
biosphere 1	core 4			

FOR REVIEW

1. List and briefly describe each of the Earth's four realms.

2. List the three major layers of the Earth. Which is/are composed of rock, which is/are metallic? Which is the largest; which is the thinnest?

3. List six types of reservoirs that collectively contain most of the Earth's water.

4. What is ground water? Where in the hydrosphere is it located?

5. What two gases comprise most of the Earth's atmosphere?

6. How thick is the Earth's atmosphere?

7. Briefly discuss the size and extent of the biosphere.

8. How old is the Earth? When did life first evolve? How long have humans and their direct ancestors been on this planet?

9. How did the Earth form?

10. What evidence has convinced many scientists that the Earth was originally homogeneous and later became layered?

11. Define a system and explain why a systems approach is useful in earth science.

12. Compare and contrast uniformitarianism and catastrophism. Give an example of each type of geologic change.

13. Briefly explain threshold and feedback effects.

FOR DISCUSSION

1. Only 0.64 percent of Earth's water is fresh and liquid; the rest is salty sea water or is frozen in glaciers. What are the environmental implications of such a small proportion of fresh water?

2. List five ways that organisms, including humans, change the Earth. What kinds of Earth processes are unaffected by humans and other organisms?

3. The radioactive elements that are responsible for the heating of the Earth decompose very slowly, over a period of billions of years. How would the Earth be different if these elements decomposed much more rapidly, say, over a period of a few million years?

4. Twelve generic types of interactions are possible among the four realms. These are:

(a) Changes in the solid Earth perturb the hydrosphere.

(b) Changes in the hydrosphere perturb the solid Earth.

(c) Changes in the solid Earth perturb the atmosphere.

(d) Changes in the atmosphere perturb the solid Earth.

(e) Changes in the solid Earth perturb the biosphere.

(f) Changes in the biosphere perturb the solid Earth.

(g) Changes in the hydrosphere perturb the atmosphere.

(h) Changes in the atmosphere perturb the hydrosphere.

(i) Changes in the hydrosphere perturb the biosphere.

(j) Changes in the biosphere perturb the hydrosphere.

(k) Changes in the atmosphere perturb the biosphere.

(l) Changes in the biosphere perturb the atmosphere.

Write an example to illustrate each possibility.

5. Give an example of a threshold or feedback effect in science, politics, human relationships, or any field you can think of.

Earth Materials and Time

Sedimentary rocks make up about 5 percent of the Earth's crust. However, because they form on the surface, they spread out in a thin veneer over large portions of the continents. The sandstone in Bryce Canyon National Park has weathered and eroded to form spectacular spires. *(Sandra Nykerk/Dembinski Photo Associates)*

Earth Rocks, Earth History, and Mass Extinctions

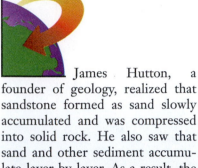

James Hutton, a founder of geology, realized that sandstone formed as sand slowly accumulated and was compressed into solid rock. He also saw that sand and other sediment accumulate layer by layer. As a result, the oldest layers are on the bottom and the youngest on top. Thus, the deepest layers of sedimentary rocks are the oldest and the layers become younger toward the Earth's surface. Modern geologists recognize that this is true only if tectonic forces have not overturned the rocks.

Early geologists also found ancient shells, bones, and other fossils in sandstone and in other kinds of sedimentary rocks. They realized that if rocks formed sequentially, layer by layer, and if fossils were preserved as the rock formed, then fossils in deep layers must be older than the fossils in the upper layers. Therefore, scientists can trace the history of life on Earth by studying fossils imbedded in the rock. This history is now summarized in a geologic time scale, shown in Chapter 4.

Nineteenth-century geologists found discontinuities in the sedimentary rock layers and in the fossils present in the layers. In many locations, they found rocks with abundant fossils. Curiously, just above, the younger rocks contain few fossils. Even more surprising, fossils of many of the most abundant plants and animals in lower layers simply disappeared, never to

appear again in younger rocks. Moving even higher in the sequence, they found abundant fossils again. But most of the younger fossils are very different from those in the older rocks.

Thus, the fossil record suggests that sudden, catastrophic events had decimated life on Earth, and that following these extinctions, new life forms emerged. However, scientists didn't immediately accept the idea that catastrophic events had caused mass extinctions. Instead, they proposed that the rock record might not be complete. Maybe some rocks had been destroyed by erosion, or perhaps sedimentary rocks hadn't formed for a long period of time. According to this reasoning, the part of the record that was lost could have contained evidence for the gradual decline of the species that became extinct, and the gradual emergence of new species. As a result, they concluded that the extinctions of old life forms and emergence of new ones occurred slowly, as a result of gradually changing conditions.

But if pieces of the rock record were missing in some locations, they must exist in other places on Earth. Geologists searched for rocks to fill in the gaps and provide evidence for gradual extinctions, but they didn't find them. Eventually, the evidence for near-instantaneous extinctions became compelling. From studies of this

type, scientists learned that at least five times in Earth history, catastrophic extinctions decimated life on Earth (Fig. 1).

The most dramatic extinction occurred about 245 million years ago, at the end of the Permian period. At that time, 90 percent of all species in the oceans suddenly died out. On land, about two thirds of reptile and amphibian species and 30 percent of insect species vanished. One modern scientist exclaimed, this extinction event was "the closest life has come to complete extermination since its origin."[1]

The death of the dominant life forms left huge ecological voids in the biosphere. Ocean ecosystems changed as new organisms emerged in an ocean relatively free of predators and competition. On land, dinosaurs and many other terrestrial animals and plants emerged and proliferated.

About 160 million years later, another catastrophic extinction wiped out about one fourth of all species, including the dinosaurs. This mass extinction occurred 65 million years ago. Small mammals survived this disaster, facing a new world free of the efficient predators that had hunted them. The number of new mammal species increased

---

[1] Douglas Erwin, *The Great Paleozoic Crisis* (New York: Columbia University Press, 1993), p. 187.

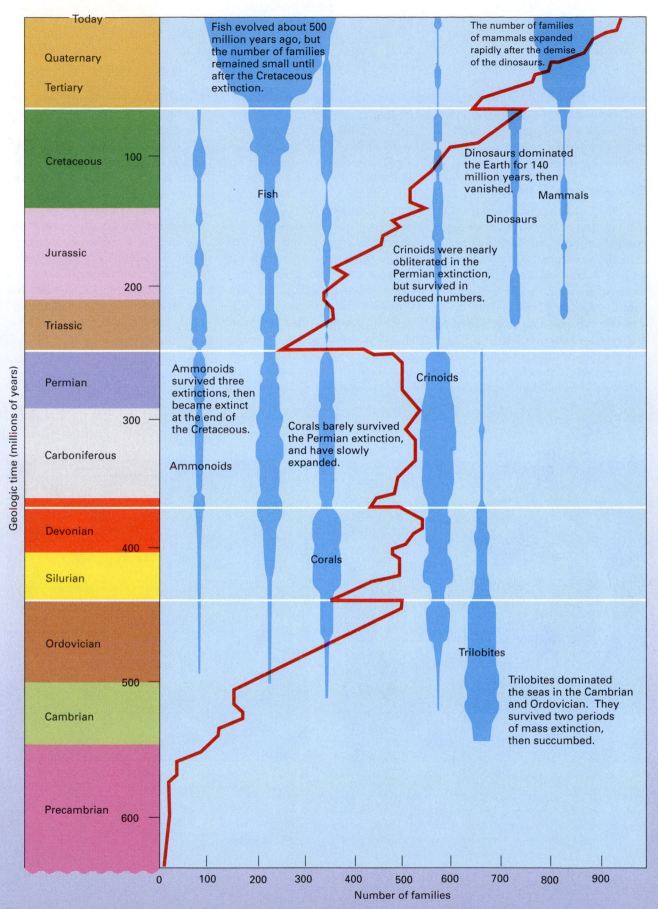

Figure 1 The red line shows that the number of families of organisms has varied throughout the history of the Earth. Sudden leftward shifts of the red line indicate mass extinctions. The number of families within different classes of organisms is indicated by the thickness of the shaded areas.

rapidly after this extinction event, eventually leading to the evolution of humans.

How could conditions on Earth change so rapidly to cause these extinctions? How could species that had survived for tens of millions of years suddenly disappear? Consider the following three hypotheses as possible explanations:

Extraterrestrial Impacts

In 1977, a father-and-son team, Walter and Luis Alvarez, were studying rocks that formed at the time the dinosaurs became extinct. In one layer, they found abundant dinosaur fossils. Just above it, they found very few fossils of any organism and no dinosaur fossils. Between these two rock layers, they found a thin, sooty, clay layer. They took samples of the clay to the laboratory and found that it contained high concentrations of the element iridium. This discovery was surprising because iridium is rare in Earth rocks. Where did it come from? Although iridium is rare in the Earth's crust, it is more abundant in meteorites.

Alvarez and Alvarez suggested that 65 million years ago, a meteorite 10 kilometers in diameter hit the Earth with energy equal to 10,000 times that of today's global nuclear arsenal. The collision vaporized both the meteorite and the Earth's crust at the point of impact, forming a plume of hot dust and gas that ignited fires around the planet (Fig. 2). Soot from the global wildfires and iridium-rich meteorite dust rose into the upper atmosphere, circling the globe. The thick, dark cloud blocked out the Sun and halted photosynthesis for as much as a year. Surface waters froze, and many plants and animals died. This event, called the terminal Cretaceous extinction, killed off the dinosaurs and many other life forms. Then the sooty, iridium-rich dust settled to Earth to form the distinctive clay layer.

Other scientists have found evidence of a huge meteorite crater that formed exactly 65 million years ago in the Caribbean Sea. Most scientists now agree with the meteorite impact hypothesis for the terminal Cretaceous extinction.

Some scientists have suggested that an extraterrestrial impact also caused the Permian extinction, but the evidence is less compelling.[2]

---

[2] Douglas Erwin, *The Great Paleozoic Crisis* (New York: Columbia University Press, 1993).

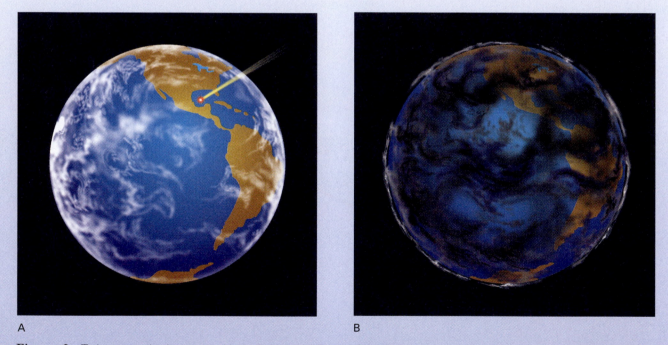

A   B

Figure 2 Today, most scientists agree with the hypothesis that the Cretaceous extinction was caused by a giant meteorite that slammed into the Earth. In this artist's rendition, (A) a large meteorite struck just north of the Yucatan peninsula. (B) The impact pulverized both the meteorite and rocks of the Earth's crust, creating a thick, dark cloud that blocked out the sunlight, causing the Earth's surface to freeze.

Volcanic Eruptions

A volcanic eruption ejects gas and fine volcanic ash into the atmosphere. The volcanic ash reflects sunlight and cools the Earth's surface and atmosphere. One volcanic gas, sulfur dioxide, forms small particles called aerosols that also reflect sunlight and cool the Earth. This cooling occurs rapidly—within a few weeks or months of the event.

Scientists have noted that several mass extinctions coincided with unusually high rates of volcanic activity. For example, massive flood basalts erupted onto the Earth's surface in Siberia at the same time that the Permian extinction occurred. According to one hypothesis, explosive volcanic eruptions blasted massive amounts of volcanic ash and sulfur aerosols into the air. The ash and aerosols spread like a pall throughout the upper atmosphere, blocking out the sun. The temperature plummeted, plants withered, and animals starved or froze to death. If this hypothesis is correct, the Permian extinction (and perhaps others) was initiated by events occurring within the geosphere. Eruptions of subterranean lava altered atmospheric composition sufficiently to decimate the biosphere.

Not all geologists are convinced that the volcanic eruptions were catastrophic enough to have obliterated most life both on the continents and in the seas. As a result, scientists have offered other explanations. One of these is described next.

Atmosphere–Ocean–Continent System Interactions

In the modern oceans, cold polar seawater sinks to the sea floor and flows as a deep ocean current toward the equator. This current transports oxygen to the deep ocean basins and forces the deep water to rise near the equator, where it is warmed. The current thus mixes both heat and gases throughout the oceans and the atmosphere.

One factor that controls ocean currents is the positions of the continents. Recall from Chapter 1 that the continents slowly drift across the globe. The map we see today will be very different 100 million years from now. Several times in Earth history, all the continents have joined together to form one giant supercontinent. One of these supercontinents assembled during Permian time. Computer calculations indicate that with the arrangement of continents at that time, there was little mixing between the surface water and the ocean depths.[3]

Surface marine organisms absorb atmospheric carbon dioxide and use it to produce organic tissue. These organisms then die and settle to the sea floor. This process removes carbon dioxide from the atmosphere, converts the carbon to organic tissue, and transports it to the sea floor. Organisms living near the deep-sea floor then consume the fallen litter and release carbon dioxide into the deep ocean water. Much of this gas dissolves in seawater and is held there by the pressure of overlying water. Modern ocean currents mix deep and shallow seawater, returning the carbon dioxide to the atmosphere. However, because there was little vertical mixing in the Permian oceans, carbon dioxide accumulated in ever-increasing concentration in the deep oceans (Fig. 3A).

This process continued to remove carbon dioxide from the atmosphere and stored it in the deep oceans during late Permian time. But carbon dioxide absorbs heat in the atmosphere and warms the Earth. Conversely, as carbon dioxide was removed from the air, the atmosphere and the oceans cooled. Some scientists have suggested that continental ice sheets formed as a result of the global cooling. Sea level fell, exposing continental shelves and stressing populations of shallow-dwelling marine organisms. The cool air cooled the surface of the sea. When water cools, it becomes denser. Eventually a threshold was reached at which polar surface water became denser than the deep water. At this point the cool, dense surface water sank. The sinking surface waters forced the carbon dioxide–rich deep water to rise to the sea surface, where it released massive amounts of carbon dioxide in a relatively short time (Fig. 3B). According to this hypothesis, the carbon dioxide asphyxiated life both in the seas and on the continents, killing most life on Earth. Later, new organisms evolved when the carbon dioxide level fell.

All three of these hypotheses explaining mass extinctions involve Earth systems interactions, and the first includes an extraterrestrial factor. The third hypothesis, if correct, is a classic example of both Earth systems feedback mechanisms and a threshold effect. The feedback process occurred as tectonic plate movement altered the shapes of the ocean basins. Ocean currents changed when the basins changed. The alteration in currents, in turn, changed atmospheric composition, and unleashed a threshold event that rapidly killed most living organisms on Earth.

[3] A.H. Knoll, R.K. Bambach, D.E. Canfield, and J.P. Grotzinger, "Comparative Earth History and Late Permian Mass Extinction," *Science*, Vol. 273, July 26, 1966, p. 452 ff.

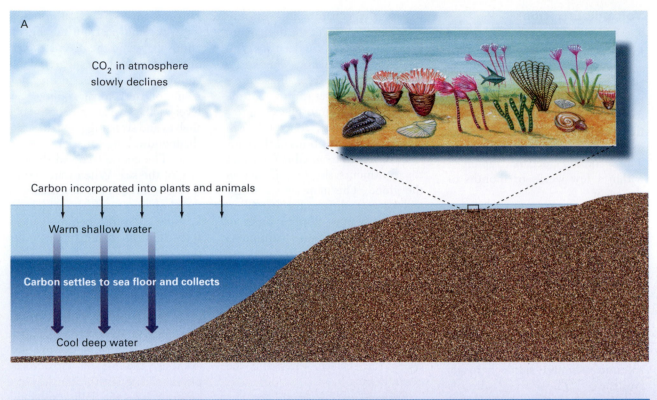

A

CO₂ in atmosphere slowly declines

Carbon incorporated into plants and animals

Warm shallow water

Carbon settles to sea floor and collects

Cool deep water

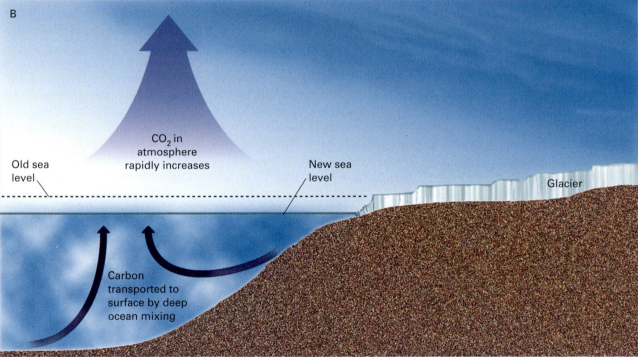

B

CO₂ in atmosphere rapidly increases

Old sea level

New sea level

Glacier

Carbon transported to surface by deep ocean mixing

Figure 3 (A) The assembly of all continents into a supercontinent during Permian time prevented vertical mixing of ocean water. Near-surface aquatic plants absorbed atmospheric carbon dioxide and incorporated the carbon into organic tissue. This carbon collected on the sea floor. As a result, the carbon dioxide concentration in the air declined and the atmosphere cooled. (B) When the atmosphere cooled sufficiently, polar surface water became dense enough to sink, stirring up the deep ocean water and releasing vast amounts of carbon dioxide back toward the surface. This carbon dioxide asphyxiated both marine and terrestrial organisms.

Minerals

Native gold, a metallic mineral, often forms with quartz, a common silicate mineral. *(Breck P. Kent)*

Pick up any rock and look at it carefully. You will probably see small, differently colored specks like those in granite (Fig. 2–1B). Each speck is a mineral. In this photo, the white grains are the mineral feldspar, the black ones are biotite, and the glassy-gray ones are quartz. A **rock** is a mixture of minerals. Most rocks contain two to five abundant minerals plus minor amounts of several others. A few rocks are made of only one mineral.

Earth scientists study minerals and rocks because they are the materials that make up the geosphere— the solid Earth. In addition, many resources that are essential to modern life are produced from rocks and minerals: iron, aluminum, gold, and all other metals, concrete, fertilizers, coal, and nuclear fuels. Both oil and natural gas are recovered from reservoirs in rock. The discoveries of ways to extract and use these Earth materials profoundly altered the course of human history. The Stone Age, the Bronze Age, and the Iron Age are historical periods named for the rock and mineral resources that dominated those times. The Industrial Revolution occurred when humans discovered how to convert the energy stored in coal into useful work. The Automobile Age occurred because humans found vast oil reservoirs in shallow rocks of the Earth's crust. The Computer Age relies on the movement of electrons through crystals of silicon and other elements. ●

2.1 What Is a Mineral?

A **mineral** is a naturally occurring inorganic solid with a definite chemical composition and a crystalline structure. Chemical composition and crystalline structure are the two most important properties of a mineral: They distinguish any mineral from all others. Before discussing them, however, let's briefly consider the other properties of minerals described by this definition.

Natural occurrence A synthetic diamond can be identical to a natural one, but it is not a true mineral because a mineral must form by natural processes. Like diamond, most gems that occur naturally can also be manufactured by industrial processes. Natural gems are valued more highly than manufactured ones. For this reason, jewelers should always tell their customers whether a gem is natural or artificial, and they usually preface the name of a manufactured gem with the term synthetic.

Inorganic Organic substances are made up mostly of carbon that is chemically bonded to hydrogen. Inorganic compounds do not contain carbon-hydrogen bonds. Although organic compounds can be produced in laboratories and by industrial processes, plants and animals create most of the Earth's organic material. The tissue of most plants and animals is organic. When the organisms die, they decompose to form other organic substances. Both coal and oil form by the decay of plants and animals, and are not minerals because of their organic properties. In addition, oil is not a mineral because it is not solid, and has neither a crystalline structure nor a definite chemical composition.

Some material produced by organisms is not organic. For example, limestone, one of the most

A

B

Figure 2–1 (A) The Bugaboo Mountains of British Columbia are made of granite. (B) Each of the differently colored grains in this close-up photo of granite is a different mineral. The white grains are feldspar, the black ones are biotite, and the glassy-gray ones are quartz.

common sedimentary rocks, is usually composed of the shells of dead corals, clams, and similar marine organisms. Shells, in turn, are made of the mineral calcite or a similar mineral called aragonite. Although produced by organisms, the calcite and aragonite are minerals: inorganic solids that form naturally, and have definite chemical compositions and crystalline structures.

Solid All minerals are solids. Thus, ice is a mineral, but neither water nor water vapor are minerals.

2.2 Elements, Atoms, and the Chemical Composition of Minerals

To consider the chemical composition and crystalline structure of minerals, we must understand the nature of chemical **elements**. An element is a fundamental component of matter and cannot be broken into simpler particles by ordinary chemical processes. Most common minerals consist of a small number—usually two to five—of different chemical elements.

A total of 88 elements occur naturally in the Earth's crust. However, eight elements—oxygen, silicon, aluminum, iron, calcium, magnesium, potassium, and sodium—make up more than 98 percent of the crust (Table 2–1).

A complete list of all elements is given in Table 2–2. Each element is represented by a one- or two-letter symbol, such as O for oxygen and Si for silicon. The table shows a total of 112 elements, not 88, because

TABLE 2–1

The Eight Most Abundant Chemical Elements in the Earth's Crust*

Element	Symbol	Weight Percent	Atom Percent	Volume Percent[†]
Oxygen	O	46.60	62.55	93.8
Silicon	Si	27.72	21.22	0.9
Aluminum	Al	8.13	6.47	0.5
Iron	Fe	5.00	1.92	0.4
Calcium	Ca	3.63	1.94	1.0
Sodium	Na	2.83	2.64	1.3
Potassium	K	2.59	1.42	1.8
Magnesium	Mg	2.09	1.84	0.3
Total		98.59	100.00	100.00

From *Principles of Geochemistry* by Brian Mason and Carleton B. Moore. © 1982 by John Wiley & Sons, Inc.

*Abundances are given in percentages by weight, by numbers of atoms, and by volume.

[†]These numbers will vary somewhat as a function of the ionic radii chosen for the calculations.

TABLE 2–2

The Periodic Table

Of the 112 elements that appear here, only 88 occur naturally in the Earth's crust. The other 24 are produced in nuclear reactors or explosions. The eight most abundant elements are shaded in blue.

The names and symbols of elements 104–109 are those recommended by their discoverers. Elements 110–112 have not yet been named.

Focus On

Chemical Bonds

Chemical bonds join atoms together. A **molecule** is the smallest particle of matter that can exist in a free state; it can be a single atom, or a group of atoms bonded together. Four types of chemical bonds are found in minerals: ionic, covalent, and metallic bonds, and van der Waals forces.

Ionic Bonds Cations and anions are attracted by their opposite electronic charges, and thus bond together. This union is called an **ionic bond.** An ionic compound (made up of two or more ions) is neutral be-cause the positive and negative charges balance each other. For example, when sodium and chlorine form an ionic bond, the sodium atom loses one electron to become a cation and chlorine gains one to become an anion. When they combine, the +1 charge balances the −1 charge (Fig. 1).

Covalent Bonds A **covalent bond** develops when two or more atoms share their electrons to produce the effect of filled outer electron shells. For example, carbon needs four electrons to fill its outermost shell. It can

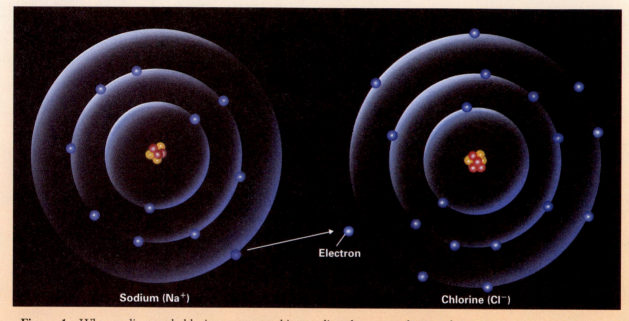

Electron

Sodium (Na⁺)

Chlorine (Cl⁻)

Figure 1 When sodium and chlorine atoms combine, sodium loses one electron, becoming a cation, Na⁺. Chlorine acquires the electron to become an anion, Cl⁻.

24 elements are produced in nuclear reactors but do not occur naturally.

An **atom** is the basic unit of an element. An atom is tiny; the diameter of the average atom is about 10^{-10} meters (1/10,000,000,000 meters). A penny contains about 1.5×10^{22} (15,000,000,000,000,000,000,000) atoms. An atom consists of a small, dense, positively charged center called a **nucleus** surrounded by negatively charged **electrons** (Fig. 2–2).

An electron is a tiny particle that orbits the nucleus, but not in a clearly defined path like that of the Earth around the Sun. Scientists usually portray electron orbits as a cloud of negative charge surrounding the nucleus.

The nucleus is made up of several kinds of particles; the two largest are positively charged **protons** and uncharged **neutrons.** A neutral atom contains equal numbers of protons and electrons. The positive and negative charges balance each other so that a neutral atom has no overall electrical charge.

Electrons concentrate in shells around the nucleus. Each shell can hold a certain number of electrons. An atom is most stable when its outermost shell is completely filled with electrons. But in their neutral states, most atoms do not have a filled outer shell. Some atoms may fill their outer shells by acquiring extra electrons until the shell becomes full. Alternatively, an atom may give up electrons until the outermost shell becomes achieve this by forming four covalent bonds with four adjacent carbon atoms. It "gains" four electrons by sharing one with another carbon atom at each of the four bonds. Diamond consists of a three-dimensional network of carbon atoms bonded into a network of tetrahedra, similar to the silicate framework structure of quartz (Fig. 2). The strength and homogeneity of the bonds throughout the crystal make diamond the hardest of all minerals.

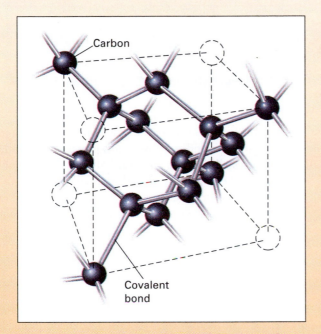

Figure 2 Carbon atoms in diamond form a tetrahedral network similar to that of quartz.

In most minerals, the bonds between atoms are partly covalent and partly ionic. The combined characteristics of the different bond types determine the physical properties of those minerals.

Metallic Bonds   In a **metallic bond,** the outer electrons are loose; that is, they are not associated with particular atoms. The metal atoms sit in a "sea" of outer-level electrons that are free to move from one atom to another. That arrangement allows the nuclei to pack together as closely as possible, resulting in the characteristic high density of metals and metallic minerals, such as pyrite. Because the electrons are free to move through the entire crystal, metallic minerals are excellent conductors of electricity and heat.

van der Waals forces   Weak electrical forces called **van der Waals forces** also bond molecules together. These weak bonds result from an uneven distribution of electrons around individual molecules, so that one portion of a molecule may have a greater density of negative charge while another portion has a partial positive charge. Because van der Waals forces are weak, minerals in which these bonds are important, such as talc and graphite, tend to be soft and cleave easily along planes of van der Waals bonds.

Focus Question:

Why do some minerals, such as native gold, silver, and graphite, conduct electricity, whereas others, such as quartz and feldspar, do not? Discuss relationships among other physical properties of minerals and the types of chemical bonds found in these minerals.

---

empty. In this case, the next lower shell, which is full, becomes the outermost shell. Thus, an atom can become stable with a full outer shell by either acquiring or releasing the proper number of electrons. When an atom loses one or more electrons, its protons outnumber its electrons and it therefore has a positive charge. If an atom gains one or more extra electrons, it is negatively charged. A charged atom is called an **ion.**

A positively charged ion is a **cation.** All of the abundant elements in the Earth's crust except oxygen release electrons to become cations, as shown in Table 2–3. For example, each potassium atom (K) loses one electron to form a cation with a charge of +1. Each silicon atom loses four electrons, forming a cation with

a +4 charge. In contrast, oxygen *gains* two extra electrons to acquire a −2 charge. Atoms with negative charges are called **anions.**

Atoms and ions rarely exist independently. Instead, they unite to form **compounds.** The forces that hold atoms and ions together to form compounds are called **chemical bonds.**

Most minerals are compounds. When ions bond together to form a mineral, they do so in proportions so that the total number of negative charges exactly balances the total number of positive charges. Thus, minerals are always electrically neutral. For example, the mineral quartz contains one (4+) silicon cation for every two (2−) oxygen anions.

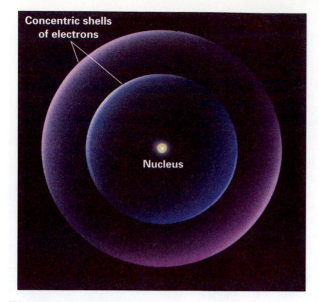

Figure 2–2 Electrons concentrate in layers or shells around the nucleus of an atom.

TABLE 2–3

Element	Chemical Symbol	Common Ion(s)
Oxygen	O	O^{2-}
Silicon	Si	Si^{4+}
Aluminum	Al	Al^{3+}
Iron	Fe	Fe^{2+} and Fe^{3+}
Calcium	Ca	Ca^{2+}
Magnesium	Mg	Mg^{2+}
Potassium	K	K^+
Sodium	Na	Na^+

The Most Common Ions of the Eight Most Abundant Chemical Elements in the Earth's Crust

Recall that a mineral has a definite chemical composition. A substance with a definite chemical composition is made up of chemical elements that are bonded together in definite proportions. Therefore, the composition can be expressed as a chemical formula, which is written by combining the symbols of the individual elements. A few minerals, such as gold and silver, consist of only a single element. Their chemical formulas, respectively, are Au (the symbol for gold) and Ag (the symbol for silver). Most minerals, however, are made up of two to five essential elements. For example, the formula of quartz is SiO_2: It consists of one atom of silicon (Si) for every two of oxygen (O). Quartz from anywhere in the Universe has that exact composition. If it had a different composition, it would be some other mineral. The compositions of some minerals, such as quartz, do not vary by even a fraction of a percent. The compositions of other minerals vary slightly, but the variations are limited.

The 88 elements that occur naturally in the Earth's crust can combine in many ways to form many different minerals. In fact, about 3500 minerals are known! However, the eight abundant elements commonly combine in only a few ways. As a result, only nine **rock-forming minerals** (or mineral "groups") make up most rocks of the Earth's crust. These minerals (or groups) are olivine, pyroxene, amphibole, mica, the clay minerals, quartz, feldspar, calcite, and dolomite.

2.3 Crystals: The Crystalline Nature of Minerals

A **crystal** is any substance whose atoms are arranged in a regular, periodically repeated pattern. All minerals are crystalline. The mineral halite (common table salt) has the composition NaCl: one sodium ion (Na^+) for every chlorine ion (Cl^-). Figure 2–3A is an "exploded" view of the ions in halite. Figure 2–3B is more realistic, showing the ions in contact. In both sketches the sodium and chlorine ions alternate in orderly rows and columns intersecting at right angles. This arrangement is the **crystalline structure** of halite.

Think of a familiar object with an orderly, repetitive pattern, such as a brick wall. The rectangular bricks repeat themselves over and over throughout the wall. As a result, the whole wall also has the shape of a rectangle or some modification of a rectangle. In every crystal, a small group of atoms, like a single brick in a wall, repeats itself over and over. This small group of atoms is called a **unit cell.** The unit cell for halite is shown in Figure 2–3A. If you compare Figures 2–3A and 2–3B, you will notice that the simple halite unit cell repeats throughout the halite crystal.

Most minerals initially form as tiny crystals that grow as layer after layer of atoms are added to their surfaces. A halite crystal might grow, for example, as salty sea water evaporates from a tidal pool. At first, a tiny grain might form, similar to the sketch of halite in Figure 2–3B. This model shows a halite crystal containing 125 atoms; it would be only about one millionth of a millimeter long on each side. As evaporation continued, more and more sodium and chlorine ions would precipitate onto the faces of the growing crystal. When minerals crystallize from

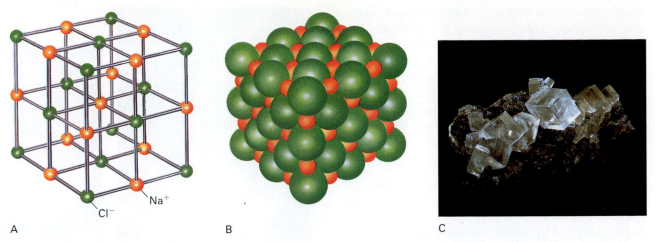
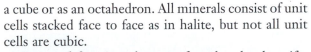

Figure 2–3 (A and B) The orderly arrangement of sodium and chlorine ions in halite. (C) Halite crystals. The crystal model in (A) is exploded so that you can see into it; the ions are actually closely packed as in (B). Note that ions in (A) and (B) form a cube, and the crystals in (C) are also cubes. *(C, American Museum of Natural History)*

magma, atoms also bond in concentric layers to the growing crystal faces.

The shape of a large, well-formed crystal like that of halite in Figure 2–3C is determined by the shape of the unit cell and the manner in which the crystal grows. For example, it is obvious from Figure 2–4A that the stacking of small cubic unit cells can produce a large cubic crystal. Figure 2–4B shows that different stacking of the same cubes can produce an eight-sided crystal, called an octahedron. Halite can crystallize as a cube or as an octahedron. All minerals consist of unit cells stacked face to face as in halite, but not all unit cells are cubic.

A **crystal face** is a planar surface that develops if a crystal grows freely in an uncrowded environment like the halite crystal growing in evaporating sea water. The sample of halite in Figure 2–3C has well-developed crystal faces. In nature, the growth of crystals is often impeded by adjacent minerals that are growing simultaneously or that have previously formed. For this reason, minerals rarely show perfect development of crystal faces.

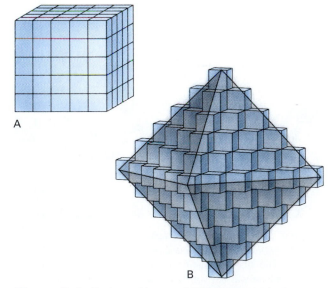

Figure 2–4 Both a cubic crystal (A) and an octahedron (B) can form by different kinds of stacking of identical cubes.

2.4 Physical Properties of Minerals

How does a geologist identify a mineral in the field? Chemical composition and crystal structure distinguish each mineral from all others. For example, halite always consists of sodium and chlorine in a one-to-one ratio, with the atoms arranged in a cubic fashion. But if you pick up a crystal of halite, you cannot see the ions. You could identify a sample of halite by measuring its chemical composition and crystal structure by laboratory procedures, but such analyses are expensive and time consuming. Instead, geologists commonly identify minerals by visual recognition, and confirm the identification with simple tests.

Most minerals have distinctive appearances. Once you become familiar with common minerals, you will recognize them just as you recognize any familiar object. For example, an apple just *looks* like an apple, even though apples come in many colors and shapes. In the same way, quartz looks like quartz to a geologist. The color and shape of quartz may vary from sample to sample, but it still looks like quartz. Some minerals, however, look enough alike that their physical properties must be examined further to make a correct identification. Geologists commonly use properties such as crystal habit, cleavage, and hardness to identify minerals.

Crystal Habit

Crystal habit is the characteristic shape of a mineral, and the manner in which aggregates of crystals grow. If a crystal grows freely, it develops a characteristic shape controlled by the arrangement of its atoms, as in the cubes of halite shown in Figure 2–3C. Figure 2–5 shows three common minerals with different crystal habits. Some minerals occur in more than one habit. For example, Figure 2–6A shows quartz with a prismatic (pencil-shaped) habit, and Figure 2–6B shows massive quartz.

A   B

Figure 2–6   (A) *Prismatic* quartz grows as elongated crystals.   *(Arkansas Geological Commission, J. M. Howard, Photographer)*
(B) *Massive* quartz shows no characteristic shape.   *(Geoffrey Sutton)*

When crystal growth is obstructed by other crystals, a mineral cannot develop its characteristic habit. Figure 2–7 is a photomicrograph (a photo taken through a microscope) of a thin slice of granite in which the crystals fit like pieces of a jigsaw puzzle. This interlocking texture developed because some crystals grew around others as magma solidified.

A

B

C

Figure 2–5   (A) *Equant* garnet crystals have about the same dimensions in all directions. (B) Asbestos is *fibrous.* (C) Kyanite forms *bladed* crystals.   *(Geoffrey Sutton)*

Figure 2–7   A photomicrograph of a thin slice of granite. When crystals grow simultaneously, they commonly interlock and show no characteristic habit. To make this photo, a thin slice of granite was cut with a diamond saw, glued to a microscope slide, and ground to a thickness of 0.02 mm. Most minerals are transparent when such thin slices are viewed through a microscope.

Cleavage

Cleavage is the tendency of some minerals to break along flat surfaces. The surfaces are planes of weak bonds in the crystal. Some minerals, such as mica, have one set of parallel cleavage planes (Fig. 2–8). Others have two, three, or even four different sets, as shown in Figure 2–9. Some minerals, like the micas, have excellent cleavage. You can peel sheet after sheet from a mica crystal as if you were peeling layers from an onion. Others have poor cleavage. Many minerals have no cleavage at all because they have no planes of weak bonds. The number of cleavage planes, the quality of cleavage, and the angles between cleavage planes all help in mineral identification.

A flat surface created by cleavage and a flat surface that is a crystal face can appear identical. However, a cleavage surface is duplicated when a crystal is broken, whereas a crystal face is not. So, if you are unsure which type of flat surface you are looking at, break the sample with a hammer, unless, of course, you want to save it.

Figure 2–8 Mica has a single, perfect cleavage plane. This large crystal is the variety of mica called muscovite. *(Geoffrey Sutton)*

Fracture

Fracture is the manner in which a mineral breaks other than along planes of cleavage. Many minerals fracture into characteristic shapes. Conchoidal fracture creates smooth, curved surfaces (Fig. 2–10). It is characteristic of quartz and olivine. Some minerals break into splintery or fibrous fragments. Most fracture into irregular shapes.

Hardness

Hardness is the resistance of a mineral to scratching. It is easily measured, and is a fundamental property of each mineral because it is controlled by the bond strength between the atoms in the mineral. Geologists commonly gauge hardness by attempting to scratch a mineral with a knife or other object of known hardness. If the blade scratches the mineral, the mineral is softer than the knife. If the knife cannot scratch the mineral, the mineral is harder.

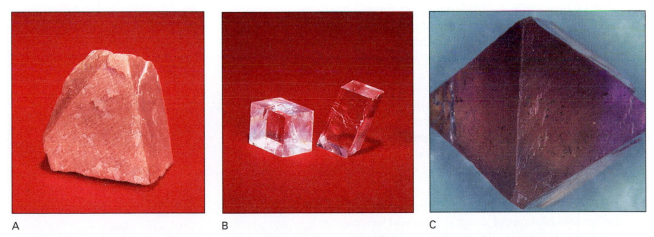

A   B   C

Figure 2–9 Some minerals have more than one cleavage plane. (A) Feldspar has two cleavages intersecting at right angles. (B) Calcite has three sets of cleavage planes. Because the three sets do not intersect at right angles, these cleaved crystals have the appearance of deformed boxes. (C) Fluorite has four cleavage planes. Each cleavage face has a parallel counterpart on the opposite side of the crystal, and thus, perfectly cleaved fluorite forms a double pyramid. *(Arthur R. Hill, Visuals Unlimited)*

Figure 2–10 Quartz typically fractures along smoothly curved surfaces, called conchoidal fractures. This sample is smoky quartz. *(Breck P. Kent)*

To measure hardness more accurately, geologists use a scale based on 10 minerals, numbered 1 through 10 (Table 2–4). Each mineral is harder than those with lower numbers on the scale, so 10 (diamond) is the hardest and 1 (talc) is the softest. The scale is known as the **Mohs hardness scale,** after F. Mohs, the Austrian mineralogist who developed it in the early 19th century.

The Mohs hardness scale shows that a mineral scratched by quartz but not by orthoclase has a hardness between 6 and 7 (Table 2–4). Because the minerals of the Mohs scale are not always handy, it is useful to know the hardness values of common materials. A fingernail has a hardness of slightly more than 2, a pocketknife blade slightly more than 5, window glass about 5.5, and a steel file about 6.5. If you practice with a knife and the minerals of the Mohs scale, you can develop a "feel" for minerals with hardnesses of 5 and under by how easily the blade scratches them.

When testing hardness, it is important to determine whether the mineral has actually been scratched by the object, or whether the object has simply left a trail of its own powder on the surface of the mineral. To check, simply rub away the powder trail and feel the surface of the mineral with your fingernail for the groove of the scratch. Fresh, unweathered mineral surfaces must be used in hardness measurements because weathering often produces a soft rind on minerals.

Specific Gravity

Specific gravity is the weight of a substance relative to that of an equal volume of water. If a mineral weighs 2.5 times as much as an equal volume of water, its specific gravity is 2.5. You can estimate a mineral's specific gravity simply by hefting a sample in your hand. If you practice with known minerals, you can develop a feel for specific gravity. Most common minerals have specific gravities of about 2.7. Metals have much greater specific gravities; for example, gold has the highest specific gravity of all minerals, 19. Lead is 11.3, silver is 10.5, and copper is 8.9.

Color

Color is the most obvious property of a mineral, but is commonly unreliable for identification. Color would be a reliable identification tool if all minerals were pure and had perfect crystal structures. However, both small amounts of chemical impurities and imperfections in crystal structure can dramatically alter color. For example, corundum (Al_2O_3) is normally a cloudy, translucent, brown or blue mineral. Addition of a small amount of chromium can convert corundum to the beautiful, clear, red gem known as ruby. A small quantity of iron or titanium turns corundum into the striking blue gem called sapphire.

Streak

Streak is the color of a fine powder of a mineral. It is observed by rubbing the mineral across a piece of unglazed porcelain known as a streak plate. Many minerals leave a streak of powder with a diagnostic color on the plate. Streak is commonly more reliable than the color of the mineral itself for identification.

Luster

Luster is the manner in which a mineral reflects light. A mineral with a metallic look, irrespective of color, has a metallic luster. Pyrite is a yellowish mineral with a metallic luster (Fig. 2–11). As a result, it looks like gold and is commonly called "fool's gold." The luster of nonmetallic minerals is usually described by self-

TABLE 2–4

Minerals of the Mohs Hardness Scale

Minerals of Mohs Scale	Common Objects
1. Talc	
2. Gypsum	Fingernail
3. Calcite	Copper penny
4. Fluorite	
5. Apatite	Knife blade
	Window glass
6. Orthoclase	Steel file
7. Quartz	
8. Topaz	
9. Corundum	
10. Diamond	

Figure 2–11 Pyrite, or "fool's gold," has a metallic luster. This well-formed pyrite crystal was found on Italy's Isle of Elba. *(Ward's Natural Science Establishment, Inc.)*

explanatory words such as glassy, pearly, earthy, and resinous.

Other Properties

Properties such as reaction to acid, magnetism, radioactivity, fluorescence, and phosphorescence can be characteristic of specific minerals. Calcite and some other carbonate minerals dissolve rapidly in acid, releasing visible bubbles of carbon dioxide gas. Minerals containing radioactive elements such as uranium emit radioactivity that can be detected with a scintillometer. Fluorescent materials emit visible light when they are exposed to ultraviolet light. Phosphorescent minerals continue to emit light after the external stimulus ceases.

2.5 Rock-Forming Minerals

Geologists classify minerals according to their anions (negatively charged ions). Anions can be either simple or complex. A simple anion is a single negatively charged ion such as O^{2-}. Alternatively, two or more atoms can bond firmly together and acquire a negative charge to form a complex anion. Two common examples are the **silicate**, $(SiO_4)^{4-}$, and **carbonate**, $(CO_3)^{2-}$, complex anions. Each mineral group (except the native elements) is named for its anion. For example, the oxides all contain O^{2-}, the silicates contain $(SiO_4)^{4-}$, and the carbonates contain $(CO_3)^{2-}$. Table 2–5 lists important mineral groups.

Although about 3500 minerals are known to exist in the Earth's crust, only a small number—between 50 and 100—are common or valuable. The nine rock-forming minerals make up the bulk of most rocks in the Earth's crust. They are important to geologists simply because they are the most common minerals. They are olivine, pyroxene, amphibole, mica, clay minerals, feldspar, quartz, calcite, and dolomite. The first six minerals in this list are actually "groups" of silicate minerals. Each group contains several varieties with similar chemical compositions, crystalline structures, and appearances. The seventh, quartz, exists in only one type; it has no compositional or structural variation. The last two minerals, calcite and dolomite, are carbonate minerals.

Silicates

The silicate minerals contain the $(SiO_4)^{4-}$ complex anion. Silicates make up about 95 percent of the Earth's crust. They are so abundant for two reasons. First, silicon and oxygen are the two most plentiful elements in the crust. Second, silicon and oxygen combine readily.

To understand the silicate minerals, remember four principles:

1. Every silicon atom surrounds itself with four oxygens. The bonds between each silicon and its four oxygens are very strong.
2. The silicon atom and its four oxygens form a pyramid-shaped structure called the **silicate tetrahedron**

TABLE 2–5

Important Mineral Groups

Group	Member	Formula	Economic Use
Oxides	Hematite	Fe_2O_3	Ore of iron
	Magnetite	Fe_3O_4	Ore of iron
	Corundum	Al_2O_3	Gemstone, abrasive
	Ice	H_2O	Solid form of water
	Chromite	$FeCr_2O_4$	Ore of chromium
Sulfides	Galena	PbS	Ore of lead
	Sphalerite	Zns	Ore of zinc
	Pyrite	FeS_2	Fool's gold
	Chalcopyrite	$CuFeS_2$	Ore of copper
	Bornite	Cu_5FeS_4	Ore of copper
	Cinnabar	HgS	Ore of mercury
Sulfates	Gypsum	$CaSO_4 \cdot 2H_2O$	Plaster
	Anhydrite	$CaSO_4$	Plaster
	Barite	$BaSO_4$	Drilling mud
Native elements	Gold	Au	Electronics, jewelry
	Copper	Cu	Electronics
	Diamond	C	Gemstone, abrasive
	Sulfur	S	Sulfa drugs, chemicals
	Graphite	C	Pencil lead, dry lubricant
	Silver	Ag	Jewelry, photography
	Platinum	Pt	Catalyst
Halides	Halite	NaCl	Common salt
	Fluorite	CaF_2	Used in steel making
	Sylvite	KCl	Fertilizer
Carbonates	Calcite	$CaCO_3$	Portland cement
	Dolomite	$CaMg(CO_3)_2$	Portland cement
	Aragonite	$CaCO_3$	Portland cement
Hydroxides	Limonite	$FeO(OH) \cdot nH_2O$	Ore of iron, pigments
	Bauxite	$Al(OH)_3 \cdot nH_2O$	Ore of aluminum
Phosphates	Apatite	$Ca_5(F,Cl,OH)(PO_4)_3$	Fertilizer
	Turquoise	$CuAl_6(PO_4)_4(OH)_8 \cdot 4H_2O$	Gemstone
Silicates	(See Figures 2–13 and 2–14 for silicate minerals.)		

with silicon in the center and oxygens at the four corners (Fig. 2–12). The silicate tetrahedron has a 4– charge, and forms the $(SiO_4)^{4-}$ complex anion. The silicate tetrahedron is the fundamental building block of all silicate minerals.

3. To make silicate minerals electrically neutral, other cations must combine with the silicate tetrahedron to balance its negative charge. (The lone exception is quartz, in which the positive charges on the silicons exactly balance the negative ones on the oxygens. How this occurs is described below.)

4. Silicate tetrahedra commonly link together by sharing oxygens. For instance, two tetrahedra may share a single oxygen, bonding the tetrahedra together.

Rock-Forming Silicate Minerals   Rock-forming silicates (and most other silicate minerals) fall into five classes, based on five ways in which tetrahedra share oxygens (Fig. 2–13). Each class contains at least one of the rock-forming mineral groups.

1. In independent tetrahedra silicates, adjacent tetrahedra do not share oxygens (Fig. 2–13A). Olivine is an independent tetrahedra mineral that occurs in small quantities in basalt of both continental and oceanic crust (Fig. 2–14A). However, rocks composed of olivine and pyroxene are thought to make up most of the mantle.

2. In the single-chain silicates, each tetrahedron shares oxygens with two adjacent tetrahedra, forming a

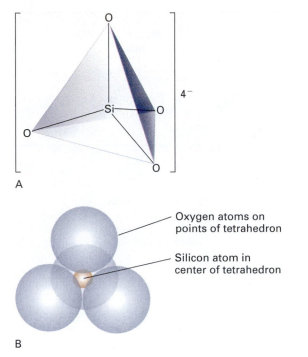

A

B

Figure 2–12 The silicate tetrahedron consists of one silicon atom surrounded by four oxygens. It is the fundamental building block of all silicate minerals. (A) A schematic representation. (B) A proportionally accurate model.

Oxygen atoms on points of tetrahedron

Silicon atom in center of tetrahedron

continuous chain (Fig. 2–13B). The pyroxenes are a group of similar minerals with single-chain structures (Fig. 2–14B). Pyroxenes are a major component of both oceanic crust and the mantle, and are also abundant in some continental rocks.

3. The double-chain silicates consist of two single chains cross-linked by the sharing of additional oxygens (Fig. 2–13C). The amphiboles (Fig. 2–14C) are a group of double-chain silicates with similar properties. They occur commonly in many continental rocks. One variety of amphibole grows as sharply pointed needles, and is a type of asbestos (see *Focus On Asbestos and Cancer*).

Pyroxene and amphibole can resemble each other so closely that they are difficult to tell apart. Both groups have similar chain structures and similar chemical compositions. In addition, both commonly grow as pencil-shaped crystals that are elongate parallel to the chains.

4. In the sheet silicates, each tetrahedron shares oxygens with three others in the same plane, forming a continuous sheet (Fig. 2–13D). All of the atoms within each sheet are strongly bonded, but each sheet is only weakly bonded to those above and

below. Therefore, sheet silicates have excellent cleavage in one plane. The micas are sheet silicates, and typically grow as plate-shaped crystals, with the flat surfaces parallel to the silicate sheets (Fig. 2–14D). Mica is common in continental rocks. Clay is similar to mica in structure, composition, and platy habit (Fig. 2–14E). Individual clay crystals are so small that they can barely be seen with a good optical microscope. As a result, clay has an earthy appearance. Most clay forms when other minerals weather at the Earth's surface. Thus, clay minerals are abundant near the Earth's surface and are an important component of soil and of sedimentary rocks.

5. In the framework silicates, each tetrahedron shares all four of its oxygens with adjacent tetrahedra (Fig. 2–13E). Because tetrahedra share oxygens in all directions, minerals with the framework structure tend to grow blocky crystals with similar dimensions in all directions. Feldspar and quartz have framework structures.

The feldspars (Fig. 2–14F) make up more than 50 percent of the Earth's crust. The different varieties of feldspar are named according to whether potassium or a mixture of sodium and calcium is present in the mineral. Orthoclase is a common feldspar containing potassium. Feldspar containing calcium and/or sodium is called plagioclase. Plagioclase and orthoclase often look alike and can be difficult to tell apart.

Quartz is the only common silicate mineral that contains no cations other than silicon; it is pure SiO_2 (Fig. 2–14G). It has a ratio of one 4+ silicon for every two 2− oxygens, so the positive and negative charges neutralize each other perfectly. Quartz is widespread and abundant in continental rocks but rare in oceanic crust and the mantle.

Carbonates

Carbonate minerals are much less common than silicates in the Earth's crust, but they are important rock-forming minerals because they form limestone and dolostone, two sedimentary rocks that lie in a thin veneer over portions of the continents. Although limestone and dolostone form in shallow seas from the remains of marine organisms (see Chapter 3), these rocks have been uplifted in many regions to form lofty mountains (Fig. 2–15).

The complex carbonate anion $(CO_3)^{2−}$ is the basis for the two common carbonate minerals. Calcite $(CaCO_3)$ forms limestone, and dolomite, $CaMg(CO_3)_2$, makes up the similar rock called dolostone (Figs. 2–16A and B). Limestone is mined as a raw ingredient of cement.

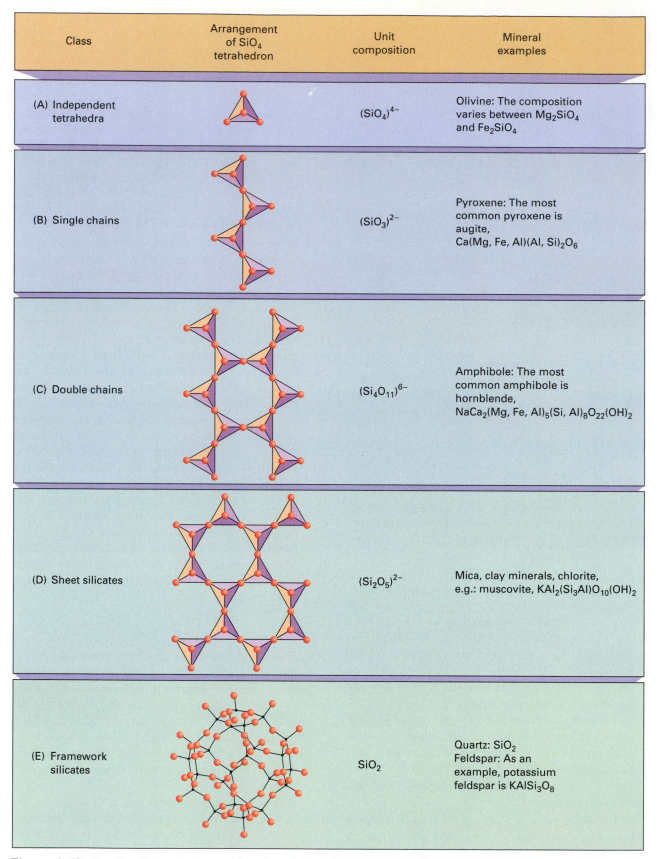

Class	Arrangement of SiO₄ tetrahedron	Unit composition	Mineral examples
(A) Independent tetrahedra		$(SiO_4)^{4-}$	Olivine: The composition varies between Mg_2SiO_4 and Fe_2SiO_4
(B) Single chains		$(SiO_3)^{2-}$	Pyroxene: The most common pyroxene is augite, $Ca(Mg, Fe, Al)(Al, Si)_2O_6$
(C) Double chains		$(Si_4O_{11})^{6-}$	Amphibole: The most common amphibole is hornblende, $NaCa_2(Mg, Fe, Al)_5(Si, Al)_8O_{22}(OH)_2$
(D) Sheet silicates		$(Si_2O_5)^{2-}$	Mica, clay minerals, chlorite, e.g.: muscovite, $KAl_2(Si_3Al)O_{10}(OH)_2$
(E) Framework silicates		SiO_2	Quartz: SiO_2 Feldspar: As an example, potassium feldspar is $KAlSi_3O_8$

Figure 2–13   The five silicate structures are based on sharing of oxygens among silicate tetrahedra. (A) Independent tetrahedra share no oxygens. (B) In single chains, each tetrahedron shares two oxygens with adjacent tetrahedra, forming a chain. (C) A double chain is a pair of single chains that are cross-linked by additional oxygen sharing. (D) In the sheet silicates, each tetrahedron shares three oxygens with adjacent tetrahedra. (E) A three-dimensional silicate framework shares all four oxygens of each tetrahedron.
● Interactive Question: Why does mica have one perfect cleavage plane? Base your answer on the nature of the bonding among the silicate tetrahedra. Consider the chemical bonding in quartz and explain why quartz has no cleavage.

Figure 2–14 The seven rock-forming silicate mineral groups. (A) Olivine. *(Geoffrey Sutton)* (B) Pyroxene. *(Jeffrey Scovill)* (C) Amphibole. *(Jeffrey Scovill)* (D) Black biotite is one common type of mica. White muscovite (Fig. 2–8) is the other. *(Breck P. Kent)* (E) Clay. *(Geoffrey Sutton)* (F) Feldspar, represented here by orthoclase feldspar. *(Breck P. Kent)* (G) Quartz. *(Jeffrey Scovill)*

Figure 2–15 The Three Sisters in British Columbia are made of limestone.

Figure 2–16 Calcite (A) and dolomite (B) are two rock-forming carbonate minerals. *(A, Mark A. Schneider/Dembinski Photo; B, Breck P. Kent)*

Many minerals, such as quartz, are relatively inactive in biological systems. Others, such as apatite (a calcium phosphate mineral mined to produce phosphate fertilizer), provide nutrients necessary for plant growth. A few, such as lead and arsenic minerals and asbestos, contain poisonous elements or compounds or possess other properties that threaten health. In nature, these minerals are usually buried beneath rock and soil, but they can become environmental hazards when they are mined and concentrated.

Asbestos is an industrial name for a group of minerals that crystallize as long, thin fibers. The two most common types are fibrous varieties of the minerals chrysotile and amphibole (Fig. 1). Chrysotile has a crystal structure and composition similar to those of the micas. The fibers of chrysotile form tangled, curly bundles, whereas amphibole asbestos occurs as straight, sharply pointed needles. Chrysotile has been the more commonly used type of asbestos in industrial applications.

Asbestos is commercially valuable because it is flameproof, chemically inert, and extremely strong. For example, a chrysotile fiber is eight times stronger than a steel wire of equivalent diameter. Asbestos has been used to manufacture brake linings, fireproof clothing, insulation, shingles, tile, pipe, and gaskets but is now allowed only in brake pads, shingles, and pipe.

In the early 1900s, asbestos miners and others who worked with asbestos learned that prolonged exposure to the fibers caused asbestosis, an often lethal lung disease. Later, in the 1950s and 1960s, studies showed that asbestos also causes lung cancer and other forms of cancer. One reason that so much time passed before scientists recognized the cancer-causing properties of asbestos

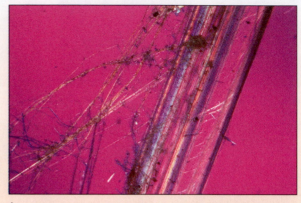

A

B

Figure 1 (A) Chrysotile asbestos occurs as long, curly fibers whereas amphibole asbestos (B) occurs as short, sharply pointed needles. *(© 1995 Kent Wood)*

2.6 Commercially Important Minerals

Many minerals are commercially important even though they are not abundant. Our industrial society depends on metals, such as iron, copper, lead, zinc, gold, and silver. **Ore minerals** are minerals from which metals or other elements can be profitably recovered. A few, such as native gold and native silver, are composed of a single element. However, most metals are chemically bonded to anions. Iron is commonly bonded to oxygen. Copper, lead, and zinc are commonly bonded to sulfur to form sulfide ore minerals (Fig. 2–17).

Although sulfide ore minerals are essential to modern industrial societies, the mining and refining of

Figure 2–17 Galena is the most important ore of lead. *(Ward's Natural Science Establishment, Inc.)*

is that cancer commonly does not develop until decades after the first exposure.

Experiments have shown that lung cancer and related diseases are caused by the fibrous nature of asbestos, not by its chemical composition. For example, amphibole samples of identical composition can occur in both fibrous and nonfibrous forms. In a laboratory study, a group of rodents was exposed to fibrous amphibole and another group to identical amounts of nonfibrous amphibole. The group exposed to the fibrous type developed cancers, but the other group did not.

Other experiments have shown that, although fibrous amphibole asbestos is a potent cause of lung cancer, no correlation exists between chrysotile asbestos and cancer. Apparently, the curly chrysotile fibers dissolve quickly in the lungs and are expelled from the body, whereas the sharp amphibole needles remain in the lungs. Although it is not clear *how* fibrous amphibole asbestos causes cancer, it seems that the shape of the fibers plays an important role and that the sharp needles of amphibole asbestos are carcinogenic, whereas the far more common chrysotile asbestos may be harmless.

In response to growing awareness of the health effects of asbestos, the Environmental Protection Agency (EPA) banned its use in building construction in 1978. In doing so, the EPA failed to distinguish between the chrysotile and amphibole varieties of asbestos. Additionally, the ban did not address the issue of what should be done with the asbestos already installed. In 1986, Congress passed a ruling called the Asbestos Hazard Emergency Response Act, requiring that all schools be inspected for asbestos. Public response has resulted in hasty programs to remove asbestos from schools and other buildings. The EPA estimates that removal of asbestos from schools and public and commercial buildings will cost between $50 billion and $150 billion. But what is the real level of hazard?

Most asbestos in use is the less dangerous (or perhaps even "safe") chrysotile. In addition, most asbestos in buildings is woven into cloth or glued into a tight matrix, and often the surface has been further stabilized by painting. Therefore, the fibers are not free to blow around. The levels of airborne asbestos in most buildings are no higher than those in outdoor air. Many scientists argue that asbestos insulation poses no health danger if left alone, but when the material is removed, it is disturbed and asbestos dust escapes. Not only are workers endangered, but airborne asbestos persists in the building for months after completion of the project.

Thus, when assessing the health effects of asbestos, we should realize that not all varieties of asbestos are carcinogenic, and understand how it is transported and incorporated into living tissue. Amphibole asbestos is unquestionably deadly in a mine where rock is drilled and blasted and dust hangs heavy in the air. However, in a school or commercial building, even the amphibole variety may be harmless until, in the interest of public safety, workers release fibers as they disturb the insulation during removal.

Focus Question:

Evaluate the hazards and discuss the cost-benefit relationships of removing asbestos from an elementary school compared to leaving the asbestos in place.

these minerals can create serious air and water pollution problems. When sulfide ore minerals are mined or refined without adequate pollution control devices, the sulfur escapes into streams, ground water, and the atmosphere, where it forms hydrogen sulfide and sulfuric acid. The atmospheric sulfur compounds then contribute to acid precipitation. In addition, many sulfide ore deposits contain small amounts of other toxic elements such as cadmium and arsenic, which can escape from mine wastes and smelters into the atmosphere and water. Most of the sulfur and toxic metals can be removed by pollution control devices, which are required by law in the United States. These topics are discussed in Chapters 19 through 21.

Industrial minerals are commercially important although they are not considered "ore" because they are mined for purposes other than the extraction of metals. Halite is mined for table salt, and gypsum is mined for plaster and sheetrock. Apatite and other phosphorus minerals are sources of the phosphate fertilizers crucial to modern agriculture. Limestone is the raw material of cement. Native sulfur, used to manufacture sulfuric acid, insecticides, fertilizer, and rubber, is mined from the craters of dormant and active volcanoes where it is deposited from gases emanating from the vents (Fig. 2–18).

A **gem** is a mineral that is prized primarily for its beauty, although some gems like diamonds are also

Figure 2–18 Yellow native sulfur is forming today in the vent of Ollague volcano on the Chile-Bolivia border.

Figure 2–19 Sapphire is one of the most costly precious gems. *(Smithsonian Institution)*

Figure 2–20 Topaz, a colorless crystal, is a popular semi-precious gem. *(American Museum of Natural History)*

used industrially. Depending on its value, a gem can be either precious or semiprecious. Precious gems include diamond, emerald, ruby, and sapphire (Fig. 2–19). Several varieties of quartz, including amethyst, agate, jasper, and tiger's eye, are semiprecious gems. Garnet, olivine, topaz, turquoise, and many other minerals sometimes occur as aesthetically pleasing semiprecious gems (Fig. 2–20).

When you look at a lofty mountain or a steep cliff, you might not immediately think about the tiny mineral grains that form the rocks. Yet minerals are the building blocks of the Earth. In addition, some minerals provide the basic resources for our industrial civilization. As we will see throughout this book, the Earth's surface and the human environment change as minerals react with the air and water, with other minerals, and even with living organisms.

SUMMARY

Minerals are the substances that make up rocks. A mineral is a naturally occurring, inorganic solid with a definite chemical composition and a crystalline structure. Each mineral consists of chemical elements bonded together in definite proportions, so that its chemical composition can be given as a chemical formula. The **crystalline structure** of a mineral is the orderly, periodically repeated arrangement of its atoms. A **unit cell** is a small structural and compositional module that repeats itself throughout a crystal. The shape of a crystal is determined by the shape and

arrangement of its unit cells. Every mineral is distinguished from others by its chemical composition and crystal structure.

Most common minerals are easily recognized and identified visually. Identification is aided by observing a few physical properties, including **crystal habit, cleavage, fracture, hardness, specific gravity, color, streak,** and **luster.**

Although about 3500 minerals are known in the Earth's crust, only the nine **rock-forming mineral groups** are abundant in most rocks. They are **feldspar,** quartz, pyroxene, amphibole, mica, clay, olivine, calcite, and dolomite. The first seven on this list are silicates; their structures and compositions are based on the silicate tetrahedron, in which a silicon atom is surrounded by four oxygens to form a pyramid-shaped structure. Silicate tetrahedra link together by sharing oxygens to form the basic structures of the silicate minerals. The silicates are the most abundant minerals because silicon and oxygen are the two most abundant elements in the Earth's crust and bond together readily to form the silicate tetrahedron. Two carbonate minerals, calcite and dolomite, are also sufficiently abundant to be called rock-forming minerals. Ore minerals, industrial minerals, and gems are important for economic reasons.

KEY TERMS

rock 21
mineral 22
element 23
atom 24
nucleus 24
electron 24
proton 24
neutron 24
ion 25

cation 25
anion 25
compound 25
chemical bond 25
rock-forming mineral 26
crystal 26
crystalline structure 26
unit cell 26

crystal face 27
crystal habit 28
cleavage 29
fracture 29
hardness 29
Mohs hardness scale 30
specific gravity 30
color 30

streak 30
luster 30
silicate 31
carbonate 31
silicate tetrahedron 39
ore mineral 39
industrial mineral 39
gem 39

Rock-Forming Mineral Groups

feldspar	amphibole	olivine
quartz	mica	calcite
pyroxene	clay minerals	dolomite

FOR REVIEW

1. What properties distinguish minerals from other substances?

2. Explain why oil and coal are not minerals.

3. What does the chemical formula for quartz, SiO_2, tell you about its chemical composition? What does $KAlSi_3O_8$ tell you about orthoclase feldspar?

4. What is an atom? An ion? A cation? An anion? What roles do they play in minerals?

5. What is a chemical bond? What role do chemical bonds play in minerals?

6. Every mineral has a "crystalline structure." What does this mean?

7. What are the factors that control the shape of a well-formed crystal?

8. What is a crystal face?

9. What conditions allow minerals to grow well-formed crystals? What conditions prevent their growth?

10. List and explain the physical properties of minerals most useful for identification.

11. Why do some minerals have cleavage and others do not? Why do some minerals have more than one set of cleavage planes?

12. Why is color often an unreliable property for mineral identification?

13. List the rock-forming mineral groups. Why are they called "rock-forming"? Which are silicates? Why are so many of them silicates?

14. Draw a three-dimensional view of a single silicate tetrahedron. Draw the five different arrangements of tetrahedra found in the rock-forming silicate minerals. How many oxygen ions are shared between adjacent tetrahedra in each of the five configurations?

15. Make a table with two columns. In the left column list the basic silicate structures. In the right column list one or more rock-forming minerals with each structure.

1. Diamond and graphite are two minerals with identical chemical compositions, pure carbon (C). Diamond is the hardest of all minerals, and graphite is one of the softest. If their compositions are identical, why do they have such profound differences in physical properties?

2. List the eight most abundant chemical elements in the Earth's crust. Are any unfamiliar to you? List familiar elements that are not among the eight. Why are they familiar?

3. Table 2–1 shows that silicon and oxygen together make up nearly 75 percent by weight of the Earth's crust. But silicate minerals make up more than 95 percent of the crust. Explain the apparent discrepancy.

4. Quartz is SiO_2. Why does no mineral exist with the composition SiO_3?

5. If you were given a crystal of diamond and another of quartz, how would you tell which is diamond?

6. Would you expect minerals found on the Moon, Mars, or Venus to be different from those of the Earth's crust? Explain your answer.

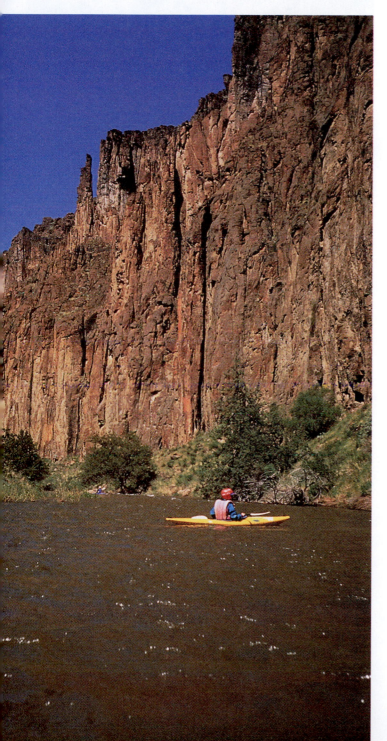

Chapter 3

Rocks

The Earth is almost entirely rock to a depth of 2900 kilometers, where the solid mantle gives way to the liquid outer core. Even casual observation reveals that rocks are not all alike. The great peaks of the Sierra Nevada in California are hard, strong granite. The red cliffs of the Utah desert are soft sandstone. The top of Mount Everest is limestone containing clamshells and the remains of other small marine animals.

The marine fossils of Mount Everest tell us that the limestone formed in the sea. What forces lifted the rock to the highest point of the Himalayas? Where

Steep rhyolite cliffs border the Bruneau River in southern Idaho.

did the vast amounts of sand in the Utah sandstone come from? How did the granite of the Sierra Nevada form? All of these questions ask about the processes that formed the rocks and changed them throughout geologic history. In this chapter we will study rocks: how they form and what they are made of. In later chapters we will use our understanding of rocks to interpret the Earth's history. ●

3.1 Rocks and the Rock Cycle

A rock is a consolidated mixture of one or more minerals. Geologists group rocks into three categories on the basis of how they form: igneous rocks, sedimentary rocks, and metamorphic rocks.

Under certain conditions, hot rocks of the upper mantle and lower crust melt, forming a molten liquid called **magma** (Fig. 3–1). An **igneous rock** forms when magma solidifies.

Rocks of all kinds decompose, or **weather,** at the Earth's surface, mostly as a result of interactions between water, atmospheric gases, and the rocks. Weathering breaks rock down to smaller particles: gravel, sand, and clay. **Sediment** is a collective term for all particles weathered and eroded from rock. Sand on a beach and mud on a lake bottom are examples of sediment. Weathering processes also dissolve some of the rock. Streams, wind, glaciers, and gravity then erode and carry the sediment and dissolved ions downhill and deposit them at lower elevations. A **sedimentary rock** forms when sediment becomes cemented or compacted into solid rock (Fig. 3–2).

A **metamorphic rock** forms when heat, pressure, or hot water alter any preexisting rock. For example, crustal movement can depress the Earth's surface to form a basin that may be hundreds of kilometers in diameter and thousands of meters deep. Sediment accumulates in the depression, and becomes cemented to form sedimentary rock. As more sediment accumulates, it buries the deepest layers under a huge weight. As a result of burial, the temperature and pressure rise, altering both the minerals and the texture of the rock (Fig. 3–3).

No rock is permanent over geologic time; instead, all rocks change slowly from one of the three rock types to another. This continuous process is called the **rock cycle** (Fig. 3–4). In the example of metamorphism, sediment accumulated to form sedimentary rock. As the basin sank, new sediment buried the sedimentary rock to greater depths where rising temperature and pressure converted it to metamorphic rock.

Figure 3–1 Magma rises from Pu'u'O'o vent during an eruption in June 1986. *(U.S. Geological Survey, J.D. Griggs)*

If the temperature rose sufficiently, the metamorphic rock would melt to form magma. The magma would then rise and solidify to become an igneous rock. Millions of years later, movement of the Earth's crust might raise the igneous rock to the surface, where it would weather to form sediment. Rain and streams would wash the sediment into a new basin, starting the cycle over again.

The rock cycle can follow many different paths. For example, weathering may turn a metamorphic rock to sediment, which then becomes cemented to form a sedimentary rock. An igneous rock may be metamorphosed. The rock cycle simply expresses the idea that rock is not permanent, but changes over geologic time.

The rock cycle is an excellent example of the ways in which the atmosphere, biosphere, and hydrosphere interact with rocks of the geosphere. Rain and gases of the atmosphere, aided by acids and other chemicals secreted by plants, act both physically and chemically on solid rock to reduce it to sediment. Additional rain washes the sediment into streams, which carry it to a sedimentary basin such as the Mississippi River delta, on the edge of a continent. The energy that drives these processes comes from the Sun. Solar heat evaporates moisture to form rain, which, in turn, feeds flowing streams.

As the basin sinks under the weight of additional sediment, the deeper rocks become heated and metamorphosed by the Earth's internal heat. The same heat may melt the rocks eventually to produce magma. But then, the magma rises upward and perhaps even erupts onto the Earth's surface from a volcano. In this way, heat is transferred from the Earth's interior to the surface.

More importantly for the Earth's surface environment, however, is the fact that volcanic eruptions release large amounts of carbon dioxide into the atmosphere. As we will learn in Chapter 15, carbon dioxide is a greenhouse gas that absorbs infrared radiation and warms the atmosphere. The greatest known episodes of volcanic eruptions in Earth history released enough carbon dioxide to raise average global atmospheric temperatures by approximately 10°C. These events and processes are described in Chapters 15, 17, and 18.

These interactions among the atmosphere, biosphere, hydrosphere, and geosphere that drive the rock cycle illustrate the point that, although we often describe the Earth in terms of several "systems," those systems continuously exchange both energy and material so that the Earth is a single, integrated whole.

Figure 3–2 Sedimentary layers of sandstone and coal form steep cliffs near Bryce, Utah.

IGNEOUS ROCKS

3.2 Magma: The Source of Igneous Rocks

If you drilled a well deep into the crust, you would find that Earth temperature rises about 30°C for every kilometer of depth. Below the crust, temperature continues to rise, but not as rapidly. In the mantle between depths of 100 and 350 kilometers, the temperature is so high that in certain environments rocks melt to form magma.

The temperature of magma varies from about 600° to 1400°C, depending on its chemical composition and the depth at which it forms. As a comparison, an iron bar turns red-hot at about 600°C and melts at slightly over 1500°C.

When rock melts, the resulting magma expands by about 10 percent. The magma is then less dense than

Figure 3–3 Metamorphic rocks are commonly contorted as a result of movements of the Earth's crust that deform them.

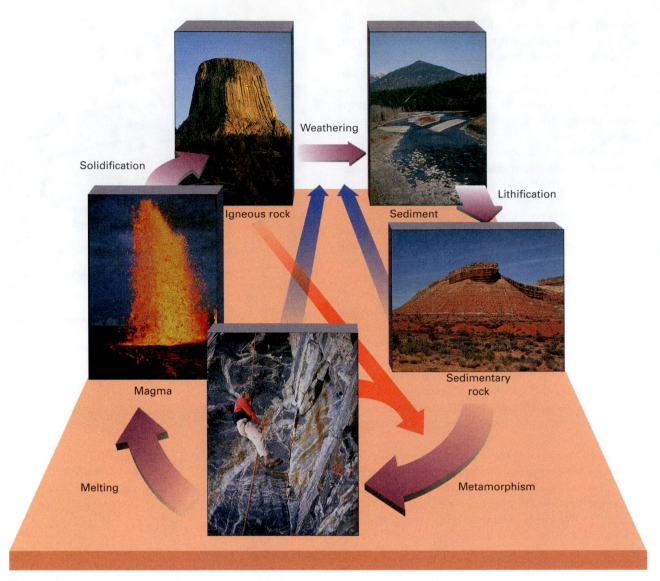

Figure 3–4 The rock cycle shows that rocks change over geologic time. The arrows show paths that rocks can follow as they change.

the rock around it, so it rises as it forms—much as a hot air balloon ascends in the atmosphere. When magma rises, it enters the cooler environment near the Earth's surface where it solidifies to form solid igneous rock.

Because the Earth melted shortly after it formed, the early crust was quite different from our modern crust, and consisted entirely of igneous rocks. Later geological activity modified the original igneous crust to form sedimentary, metamorphic, and younger igneous rocks. However, about 95 percent of the Earth's crust is still igneous rock or metamorphosed igneous rock. Even though today much of this igneous foundation is buried by a relatively thin layer of sedimentary rock, igneous rocks are easy to view because they make up some of the world's most spectacular mountains (Fig. 3–5).

Figure 3–5 The peaks of Sam Ford Fiord, Baffin Island, are composed of granite.

TABLE 3–1

Igneous Rock Textures Based on Grain Size

Grain Size	Name of Texture
No mineral grains (obsidian)	Glassy
Too fine to see with naked eye	Very fine grained
Up to 1 millimeter	Fine-grained
1–5 millimeters	Medium-grained
More than 5 millimeters	Coarse-grained
Relatively large grains in a finer-grained matrix	Porphyry

3.3 Types of Igneous Rocks

Magma can either rise all the way through the crust to erupt onto the Earth's surface, or it can solidify within the crust. An **extrusive igneous rock** forms when magma erupts and solidifies on the Earth's surface. Because extrusive rocks are so commonly associated with volcanoes, they are also called **volcanic rocks,** after Vulcan, the Roman god of fire.

An **intrusive igneous rock** forms when magma solidifies *within* the crust. Intrusive rocks are sometimes called **plutonic rocks** after Pluto, the Roman god of the underworld.

Textures of Igneous Rocks

The **texture** of a rock refers to the size, shape, and arrangement of its mineral grains, or crystals (Table 3–1). Some igneous rocks consist of mineral grains that are too small to be seen with the naked eye; others are made up of thumb-size or even larger crystals.

Extrusive (Volcanic) Rocks   **Lava** is fluid magma that flows from a crack or a volcano onto the Earth's surface. The term also refers to rock that forms when lava cools and becomes solid. After lava erupts onto the relatively cool Earth surface, it solidifies rapidly— perhaps over a few days or years. Crystals form but do not have much time to grow. As a result, many volcanic rocks have fine-grained textures, with crystals too small to be seen with the naked eye. **Basalt** is a common very fine grained volcanic rock (Fig. 3–6). In unusual circumstances, molten lava may solidify within a few hours of erupting. Because the magma hardens so quickly, the atoms have no time to align themselves to form crystals. The result is the volcanic glass called **obsidian** (Fig. 3–7).

If magma rises slowly through the crust before erupting, some crystals may grow while most of the magma remains molten. If this mixture of magma and crystals then erupts onto the surface, it solidifies quickly, forming **porphyry,** a rock with the large crystals, called **phenocrysts,** embedded in a fine-grained matrix (Fig. 3–8).

Intrusive (Plutonic) Rocks   When magma solidifies within the crust, the overlying rock insulates the magma like a thick blanket. The magma then crystallizes slowly and the crystals grow over hundreds of thousands to millions of years. As a result, most plutonic rocks are medium to coarse grained. **Granite,** the most abundant rock in continental crust, is a medium- or coarse-grained plutonic rock. The crystals in granite are clearly visible. Many are a millimeter or so across, although some crystals may be much larger.

Figure 3–6   Basalt is a fine-grained volcanic rock. The holes were gas bubbles that were preserved as the magma solidified in southeastern Idaho.

Figure 3–7   Obsidian is natural volcanic glass. It contains no crystals. *(Geoffrey Sutton)*

Figure 3–8 Porphyry is an igneous rock containing large crystals embedded in a fine-grained matrix. This rock is rhyolite porphyry with large, pink feldspar phenocrysts.

Naming Igneous Rocks

Geologists use both the minerals and texture to name igneous rocks. For example, any medium- or coarse-grained igneous rock consisting mostly of feldspar and quartz is called granite. **Rhyolite** also consists mostly of feldspar and quartz but is very fine grained (Fig. 3–9). The same magma that solidifies slowly within the crust to form granite can also erupt onto the Earth's surface to form rhyolite.

Like granite and rhyolite, most common igneous rocks are classified in pairs, each member of a pair containing the same minerals but having a different texture. The texture depends mainly on whether the rock is volcanic or plutonic. Figure 3–10 shows the minerals and textures of common igneous rocks.

Figure 3–10 shows that granite and rhyolite contain large amounts of *fel*dspar and *si*lica, and so are called **felsic** rocks. Basalt and gabbro are called **mafic** rocks because of their high *ma*gnesium and iron contents (*ferrum* is the Latin word for iron). Rocks with especially high magnesium and iron concentrations are called **ultramafic**. Rocks with compositions between those of granite and basalt are called **intermediate rocks.**

Once you learn to identify the rock-forming minerals, it is easy to name a plutonic rock using Figure 3–10 because the minerals are large enough to see. It is more difficult to name many volcanic rocks because the minerals are too small to identify. A field geologist often uses color to name a volcanic rock. Rhyolite is usually light in color: white, tan, red, and pink are common. Many andesites are gray or green, and basalt is commonly black. The minerals in many volcanic rocks cannot be identified even with a microscope because of their tiny crystal sizes. In this case, definitive identification is based on chemical and X-ray diffraction analyses carried out in the laboratory.

3.4 Common Igneous Rocks

Granite and Rhyolite

Granite contains mostly feldspar and quartz. Small amounts of dark biotite or hornblend often give it a black and white speckled appearance. Granite (and metamorphosed granitic rock) is the most common rock in continental crust. It is found nearly everywhere beneath the relatively thin veneer of sedimentary rocks and soil that covers most of the continents. Geologists often call this rock **basement rock** because it makes

A   B

Figure 3–9 Although granite (A) and rhyolite (B) contain the same minerals, they have very different textures because granite cools slowly and rhyolite cools rapidly.

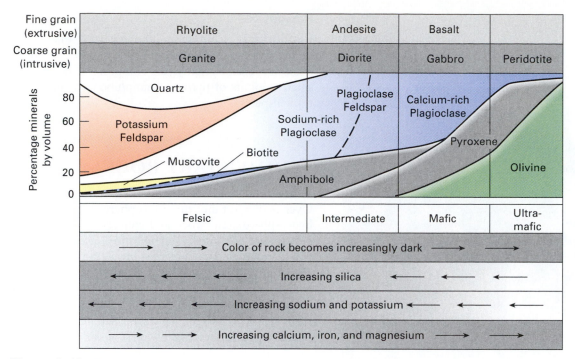

Fine grain (extrusive)	Rhyolite		Andesite	Basalt	
Coarse grain (intrusive)	Granite		Diorite	Gabbro	Peridotite

Figure 3–10 The names of common igneous rocks are based on the minerals and texture of a rock. In this figure, a mineral's abundance in a rock is proportional to the thickness of its colored band beneath the rock name. If a rock has a fine-grained texture, its name is found in the top row of rock names; if it has a coarse-grain texture, its name is in the second row.

up the foundation of a continent.[1] Granite is hard and resistant to weathering; it forms steep, sheer cliffs in many of the world's great mountain ranges. Mountaineers prize granite cliffs for the steepness and strength of the rock (Fig. 3–11).

As granitic magma rises through the Earth's crust, some of it may erupt from a volcano to form rhyolite, while the remainder solidifies beneath the volcano, forming granite. Most obsidian forms from magma with a granitic (rhyolitic) composition.

Basalt and Gabbro

Basalt is a mafic rock that consists of approximately equal amounts of plagioclase feldspar and pyroxene. It makes up most of the oceanic crust, as well as huge basalt plateaus on continents (Fig. 3–12). **Gabbro** is

the plutonic counterpart of basalt; it is mineralogically identical but consists of larger crystals. Gabbro is uncommon at the Earth's surface, although it is abundant in deeper parts of oceanic crust, where basaltic magma crystallizes slowly.

Figure 3–11 The authors on Inugsuin Point Buttress, a granite wall on Baffin Island.

---

[1] The terms basement rock, bedrock, parent rock, and country rock are commonly used by geologists. *Basement rock* is the igneous and metamorphic rock that lies beneath the thin layer of sediment and sedimentary rocks covering much of the Earth's surface, and thus it forms the basement of the crust. *Bedrock* is the solid rock that lies beneath soil or unconsolidated sediments. It can be igneous, metamorphic, or sedimentary. *Parent rock* is any original rock before it is changed by metamorphism or any other geologic process. The rock enclosing or cut by an igneous intrusion or by a mineral deposit is called *country rock*.

Figure 3–12 Lava flows of the Columbia River basalt plateau are well exposed along the Columbia River.

Andesite and Diorite

Andesite is a volcanic rock intermediate in composition between basalt and granite. It is commonly gray or green and consists of plagioclase feldspar and dark minerals (usually biotite, amphibole, or pyroxene). It is named for the Andes Mountains, the volcanic chain on the western edge of South America, where it is abundant. Because it is volcanic, andesite is typically very fine grained.

  Diorite is the plutonic equivalent of andesite. It forms from the same magma as andesite and, consequently, often underlies andesitic mountain chains such as the Andes.

Peridotite

Peridotite is an ultramafic igneous rock that makes up most of the upper mantle but is rare in the Earth's crust. It is coarse grained and composed of olivine and small amounts of pyroxene, amphibole, or mica, but no feldspar.

SEDIMENTARY ROCKS

Over geologic time, the atmosphere, biosphere, and hydrosphere attack rock and convert it to clay, sand, gravel, and ions dissolved in water. Flowing water, wind, gravity, and glaciers then erode the decomposed rock, transport it downslope, and deposit it on the seacoast or in lakes and river valleys. With time, the loose, unconsolidated sediment becomes compacted and cemented, or **lithified,** to form sedimentary rock. Sedimentary rocks make up only about 5 percent of the Earth's crust. However, because they form on the Earth's surface, sedimentary rocks are widely spread in a thin veneer over underlying igneous and metamorphic rocks. As a result, they cover about 75 percent of continents.

3.5 Types of Sedimentary Rocks

Sedimentary rocks are broadly divided into four categories:

1. Clastic sedimentary rocks are composed of fragments of weathered rocks called clasts, which have been transported, deposited, and lithified. Clastic rocks make up more than 85 percent of all sedimentary rocks (Fig. 3–13). This category includes sandstone, siltstone, and shale.

2. Organic sedimentary rocks consist of the lithified remains of plants or animals. Coal is an organic sedimentary rock made up of decomposed and compacted plant remains.

3. Chemical sedimentary rocks form by direct precipitation of minerals from solution. Rock salt, for example, forms when salt precipitates from evaporating seawater or saline lake water.

4. Bioclastic sedimentary rocks are composed of broken shell fragments and similar remains of living organisms. The fragments are clastic, but they are of a biological origin. Most limestone formed from broken shells and is a **bioclastic sedimentary rock.**

Figure 3-13 Sandstone, silt-stone, and shale are clastic rocks that make up more than 85 percent of all sedimentary rocks. Limestone and some "other" sedimentary rocks make up less than 15 percent.

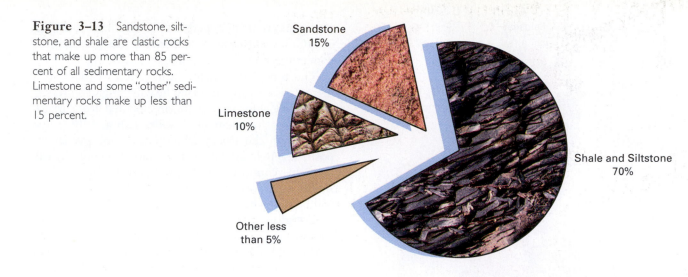

Sandstone
15%

Limestone
10%

Shale and Siltstone
70%

Other less
than 5%

Clastic Sedimentary Rocks

Clastic sediment is called gravel, sand, silt, or clay, in order of decreasing particle size (Table 3–2). As clastic particles ranging in size from coarse silt to boulders tumble downstream, their sharp edges are worn off and they become rounded (Fig. 3–14). Finer silt and clay do not round effectively because they are so small and light that water, and even wind, cushion them as they bounce along.

If you fill a measuring cup with sand or other clastic sediment, you can still add a substantial amount of water that occupies empty space, or **pore space,** among the sand grains (Fig. 3–15A). Commonly, sand and similar sediment have about 20 to 40 percent pore space.

As more sediment accumulates, the weight of overlying layers compresses the buried sediment. Some of the water is forced out, and the pore space shrinks (Fig. 3–15B). This process is called **compaction.** If the

grains have platy shapes, as in clay and silt, compaction alone may lithify the sediment as the platy grains interlock like pieces of a puzzle.

As sediment is buried, water circulates through the pore space. This water commonly contains dissolved ions that precipitate in the pore spaces, cementing the

Figure 3-14 Boulders and cobbles collide as they move downstream. The collisions abrade the sharp edges, producing rounded rocks.

TABLE 3-2

Sizes and Names of Sedimentary Particles and Clastic Rocks

Diameter (mm)	Sediment		Clastic Sedimentary Rock
256– 64– 2–	Boulders Cobbles Pebbles	Gravel (rubble)	Conglomerate (rounded particles or breccia (angular particles)
$\frac{1}{16}$	Sand		Sandstone
$\frac{1}{256}$	Silt	Mud	Siltstone } Claystone }Mudstone or shale }
	Clay		

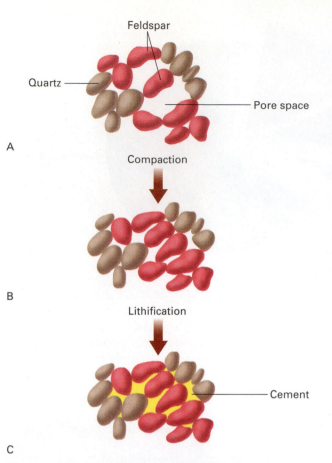

Feldspar

Quartz

Pore space

A

Compaction

B

Lithification

Cement

C

Figure 3–15 (A) Pore space is the open space between sediment grains. (B) Compaction squashes the grains together, reducing the pore space and lithifying the sediment by interlocking the grains. (C) Cement fills the remaining pore space, lithifying the sediment by gluing the grains together.

clastic grains firmly together to form a hard rock (Fig. 3–15C). Calcite, quartz, and iron oxides are the most common cements in sedimentary rocks.

The time required for lithification of loose sediment varies greatly. In some heavily irrigated areas of southern California, calcite precipitated from irrigation water has cemented soils within only a few decades. In the Rocky Mountains, some glacial deposits less than 20,000 years old are cemented by calcite. In contrast, sand and gravel deposited in southwestern Montana about 30 million years ago can still be dug with a hand shovel.

Conglomerate is lithified gravel (Fig. 3–16). Each clast in a conglomerate is usually much larger than the individual mineral grains in the clast. Therefore, the clasts retain most of the characteristics of the parent rock. If enough is known about the geology of an area where conglomerate is found, it may be possible to identify exactly where the clasts originated. For example, a granite clast probably came from nearby granite bedrock.

The next time you walk along a cobbly stream, look carefully at the cobbles. You will probably see sand or silt trapped among the larger clasts. In a similar way, most conglomerates contain fine sediment among the large clasts.

Sandstone consists of lithified sand grains (Fig. 3–17). Of all the common minerals, quartz is the most resistant to weathering. Feldspar and other minerals decompose during weathering and transport. In contrast, about all that happens to quartz grains during weathering and transport is that they become rounded.

Figure 3–16 Conglomerate is lithified gravel. The cobbles were rounded as they bounced downstream, before they were deposited and lithified.

A　　　　　　　　　　　　　　　　　　　　　　　B

Figure 3–17　Sandstone is lithified sand. (A) A sandstone cliff above the Colorado River, Canyonlands, Utah. (B) A close-up photo shows the well-rounded sand grains.

Consequently, most beach sands and most sandstones consist predominantly of rounded quartz grains.

Shale (Fig. 3–18) consists mostly of tiny clay minerals and lesser amounts of quartz. Shale typically splits easily along very fine layering called **fissility**. The fissility of shale results from the parallel orientation of the platy clay particles.

Siltstone is lithified silt. The main component of most siltstones is quartz, although clays are commonly present. Siltstones often show layering but

Figure 3–18　Shale near Drummond, Montana. The larger fragments are about 10 centimeters across.

lack the fine fissility of shales because of their lower clay content.

Figure 3–13 shows that shale and siltstone make up 70 percent of all clastic sedimentary rocks. Their abundance reflects the vast quantity of clay produced by weathering. Shale is usually gray to black due to the presence of partially decayed remains of plants and animals commonly deposited with clay-rich sediment. This organic material in shales is the source of most oil and natural gas. (The formation of oil and gas from this organic material is discussed in Chapter 21.)

Organic Sedimentary Rocks

Organic sedimentary rocks form by lithification of the remains of plants and animals.

Chert is a rock composed of pure silica. It occurs as sedimentary beds and as irregularly shaped lumps called nodules in other sedimentary rocks (Fig. 3–19). Microscopic examination of bedded chert often shows that it is made up of the remains of tiny marine organisms whose skeletons are composed of silica rather than calcium carbonate. In contrast, some nodular chert appears to form by precipitation from silica-rich ground water, most often in limestone. Chert was one of the earliest geologic resources. Flint, a dark gray to black variety, was frequently used for arrowheads, spear points, scrapers, and other tools chipped to hold a fine edge.

Coal　When plants die, their remains usually decompose by reaction with oxygen. However, in warm

Figure 3–19 Red nodules of chert in light-colored limestone.

swamps and in other environments where plant growth is rapid, dead plants accumulate so rapidly that the oxygen is used up long before the decay process is complete. The partially decayed plant remains form peat. As peat is buried and compacted by overlying sediments, it converts to coal, a hard, black, combustible rock. (Coal formation is discussed in more detail in Chapter 21.)

Chemical Sedimentary Rocks

Some common elements in rocks and minerals, such as calcium, sodium, potassium, and magnesium, dissolve during chemical weathering and are carried by

Figure 3–20 An evaporating lake precipitated thick salt deposits on the Salar de Uyuni, Bolivia.

ground water and streams to lakes or to the ocean. Most lakes are drained by streams that carry the salts to the ocean. However, some lakes, such as the Great Salt Lake in Utah, are landlocked. Streams flow into the lake, but no streams exit. As a result, water escapes only by evaporation. When water evaporates, salts remain behind and the lake water becomes steadily more salty. **Evaporites** are rocks that form when evaporation concentrates dissolved ions to the point at which they precipitate from solution (Fig. 3–20). The same process can occur if ocean water is trapped in coastal or inland basins, where it can no longer mix with the open sea.

The most common minerals found in evaporite deposits are gypsum ($CaSO_4 \cdot 2H_2O$) and halite ($NaCl$). Gypsum[2] is used in plaster and wallboard, and halite is common salt. Evaporites form important economic deposits, but they comprise only a small proportion of all sedimentary rocks.

Bioclastic Sedimentary Rocks

Carbonate rocks are primarily made up of carbonate minerals, which contain the carbonate ion $(CO_3)^{2-}$. The most common carbonate minerals are calcite, which is calcium carbonate, $CaCO_3$, and dolomite, which is calcium magnesium carbonate, $CaMg(CO_3)_2$. Calcite-rich carbonate rocks are called **limestone,** whereas rocks rich in the mineral dolomite are also called dolomite. Many geologists use the term **dolostone** for the rock name to distinguish it from the mineral dolomite.

Seawater contains large quantities of dissolved calcium carbonate ($CaCO_3$). Clams, oysters, corals, some types of algae, and a variety of other marine organisms convert dissolved calcium carbonate to shells and other hard body parts. When these organisms die, waves and ocean currents break the shells into small fragments. A rock formed by lithification of such sediment is called bioclastic limestone, indicating that it forms by both biological and clastic processes. Most limestones are bioclastic. The bits and pieces of shells appear as fossils in the rock (Fig. 3–21).

Organisms that form limestone thrive and multiply in warm shallow seas because the sun shines directly on the ocean floor, where most of them live. Therefore, bioclastic limestone typically forms in shallow water along coastlines at low and middle latitudes. It also forms on continents where rising sea level floods land with shallow seas. Limestone makes up many of the world's great mountains, and the summit of Mount Everest is made of limestone containing marine fossils. Leonardo de Vinci puzzled over the presence of

---

[2] The $2H_2O$ in the chemical formula of gypsum means water is incorporated in the mineral structure.

A   B

Figure 3–21 Most limestone is lithified shell fragments and other remains of marine organisms. (A) A limestone mountain in British Columbia, Canada. (B) A close-up of shell fragments in limestone. *(Breck P. Kent)*

fossils on mountain tops and was perhaps the first person to propose that Earth processes actively raise rocks from the sea floor to the tops of mountains, stating, "The silent fossils on snowy windswept summits remind us of the earth's active nature."

Coquina is bioclastic limestone consisting wholly of coarse shell fragments cemented together. **Chalk** is a very fine grained, soft, white bioclastic limestone made of the shells and skeletons of microorganisms that float near the surface of the oceans. When they die, their remains sink to the bottom and accumulate to form chalk. The pale-yellow chalks of Kansas, the off-white chalks of Texas, and the gray chalks of Alabama remind us that all of these areas once lay beneath the sea (Fig. 3–22).

3.6 Sedimentary Structures

Nearly all sedimentary rocks contain **sedimentary structures,** features that developed during or shortly after deposition of the sediment. These structures help us understand how the sediment was transported and deposited. Because sedimentary rocks form at the

▶**Figure 3–22** The Niobrara chalk of western Kansas consists of the remains of tiny marine organisms. *(David Schwimmer)*

Figure 3–23 Sedimentary bedding shows clearly in the walls of the Grand Canyon. *(Donovan Reese/Tony Stone Images)*

Earth's surface, sedimentary structures and other features of sedimentary rocks also contain clues about conditions at the Earth's surface when the rocks formed.

The most obvious and common sedimentary structure is **bedding,** or **stratification**—layering that develops as sediment is deposited (Fig. 3–23). Bedding forms because sediment accumulates layer by layer.

Nearly all sedimentary beds were originally horizontal because most sediment accumulates on nearly level surfaces.

Cross-bedding consists of small beds lying at an angle to the main sedimentary layering (Fig. 3–24A). Cross-bedding forms in many environments where wind or water transports and deposits sediment. For

Figure 3–24 (A) Cross-bedding preserved in lithified ancient sand dunes in Arches National Park, Utah. (B) The development of cross-bedding as the prevailing wind direction changes. • Interactive Question: In what direction was the wind blowing when the dune in (A) formed?

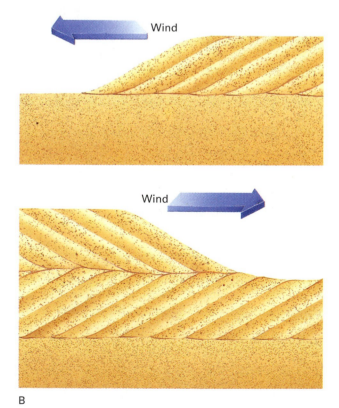

A

B

Figure 3–25   Ripple marks in billion-year-old siltstone in eastern Utah.

Figure 3–26   Mud cracks form when wet mud dries and shrinks.

example, wind heaps sand into parallel ridges called dunes and flowing water forms similar features called sand waves. Figure 3–24B shows that cross-beds are the layering formed by sand grains tumbling down the steep, downstream face of a dune or sand wave. Cross-bedding is common in sands deposited by wind, streams, and ocean currents and by waves on beaches.

Ripple marks are small, nearly parallel ridges and troughs that are also formed by moving water or wind. They are like dunes and sand waves, but smaller. Ripple marks are often preserved in sedimentary rocks (Fig. 3–25).

Mud cracks are polygonal cracks that form when mud shrinks as it dries (Fig. 3–26). They indicate that the mud accumulated in shallow water that periodically dried up. For example, mud cracks are common on intertidal mud flats where sediment is flooded by water at high tide and exposed at low tide. The cracks often fill with sediment carried in by the next high tide and are commonly well preserved in rocks.

Occasionally, very delicate sedimentary structures are preserved in rocks. Geologists have found imprints of raindrops that fell on a muddy surface about 1 bil-lion years ago (Fig. 3–27) and imprints of salt crystals that formed as a puddle of salt water evaporated. Like mud cracks, raindrop and salt imprints show that the mud must have been deposited in shallow water that intermittently dried up.

Fossils are any remains or traces of a plant or animal preserved in rock—any evidence of past life. Fossils include remains of shells, bones, or teeth; whole bodies preserved in amber or ice; and a variety of tracks, burrows, and chemical remains. Fossils are discussed further in Chapter 4.

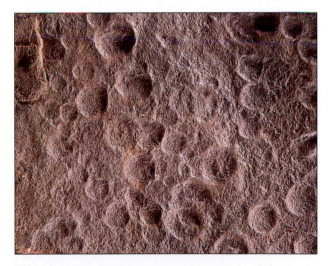

Figure 3–27   Delicate raindrop imprints formed by rain that fell about a billion years ago on a mud flat.   (Breck P. Kent)

METAMORPHIC ROCKS

A potter forms a delicate vase from moist clay. She places the soft piece in a kiln and slowly heats it to 1000°C. As temperature rises, the clay minerals decompose. Atoms from the clay then recombine to form new minerals that make the vase strong and hard. The breakdown of the clay minerals, growth of new minerals, and hardening of the vase all occur without melting of the solid materials.

Metamorphism (from the Greek words for "changing form") is the process by which rising temperature, pressure, or an influx of hot water transforms rocks and minerals. Metamorphism occurs in solid rock, like the transformations in the vase as the potter fires it in her kiln. Small amounts of water and other fluids help metamorphism, but the rock remains solid as it changes. Metamorphism can change any type of parent rock: sedimentary, igneous, or even another metamorphic rock.

3.7 Metamorphism

A mineral that does not decompose or change in other ways, no matter how much time passes, is a stable mineral. Millions of years ago, weathering processes may have formed the clay minerals used by the potter to create her vase. They were stable and had remained unchanged since they formed. A stable mineral can become unstable when environmental conditions change. Three types of environmental change cause metamorphism: rising temperature, rising pressure, and changing chemical composition caused by an influx of hot water.

For example, when the potter put the clay in her kiln and raised the temperature, the clay minerals decomposed because they became unstable at the higher temperature. The atoms from the clay then recombined to form new minerals that were stable at the higher temperature. Like the clay, every mineral is stable only within a certain temperature range. In a similar manner, each mineral is stable only within a certain pressure range.

In addition, a mineral is stable only in a certain chemical environment. If hot water seeping through bedrock carries new chemicals to a rock, those chemicals may react with the original minerals to form new ones that are stable in the altered chemical environment. If hot water dissolves chemical components from a rock, new minerals may form for the same reason.

Metamorphism occurs because each mineral is stable only within a certain range of temperature, pressure, and chemical environment. If temperature or pressure rises above that range, or if chemicals are added or removed from the rock, the rock's original minerals may decompose and their components recombine to form new minerals that are stable under the new conditions.

Metamorphic Grade

The **metamorphic grade** of a rock is the intensity of metamorphism that formed the rock. Temperature is the most important factor in metamorphism, and therefore grade closely reflects the temperature of metamorphism. Since temperature increases with depth in the Earth, a general relationship exists between depth and metamorphic grade (Fig. 3–28). Low-grade metamorphism occurs at shallow depths, less than 10 kilometers beneath the surface, where temperature is no higher than 300° to 400°C. High-grade conditions are found deep within continental crust and in the upper mantle, 40 to 55 kilometers below the Earth's surface. The temperature here is 600° to 800°C, close to the melting point of rock. High-grade conditions can develop at shallower depths, however, in areas adjacent to rising magma or hot intrusive rocks. For example, today metamorphic rocks are forming beneath Yellowstone Park, where hot magma lies close to the Earth's surface.

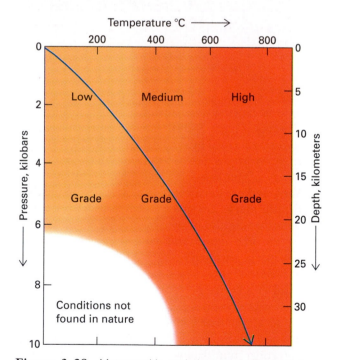

Figure 3–28 Metamorphic grade increases with depth because temperature and pressure rise with depth. The blue arrow traces the path of increasing temperature and pressure with depth in a normal part of the crust. • Interactive Question: In a normal part of the crust, what are the temperature and pressure at a depth of 25 kilometers? What grade of metamorphic rocks would you expect to exist there?

Rocks and Radon

Radon is an invisible, odorless, and tasteless radioactive gas that occurs naturally in bedrock and soil. It seeps from the ground into homes and other buildings where it concentrates and causes an estimated 5000 to 20,000 cancer deaths per year among Americans. The risk of dying from radon-caused lung cancer in the United States is about 0.4 percent over a lifetime, much greater than the risk of dying from cancer caused by asbestos, pesticides, or other air pollutants and nearly as high as the risk of dying in an auto accident or from a fall or a fire at home.

Americans are not all exposed to equal amounts of radon. Some homes contain very low concentrations of the gas; others have high concentrations. The variations in concentration are due to two factors: geology and home ventilation.

Radon is one of a series of elements formed by the radioactive decay of uranium. Thus, it forms wherever uranium occurs. Uranium occurs naturally in all types of rock, but it concentrates in granite and shale. It is also found in soil that formed from granite and shale, and in construction materials made from those rocks, such as concrete and concrete blocks. Radon is itself radioactive, and it decays into other radioactive elements. Because of their radioactivity, radon and its decay products are carcinogenic, and because it is a gas, we breathe radon into our lungs.

As radon forms by slow radioactive decay of uranium in bedrock, soil, or construction materials, it seeps into the basement of a home and circulates throughout the house. Radon concentrations are highest in poorly ventilated homes built on granite or shale, on soil derived from these rocks, and in homes constructed with concrete and concrete blocks containing these types of rocks. The highest home radon concentrations ever measured were found in houses built on the Reading Prong, a uranium-rich granite pluton extending from Reading, Pennsylvania, through northern New Jersey and into New York. The air in one home in this area contained 700 times more radon than the EPA "action level"—the concentration at which the Environmental Protection Agency recommends that corrective measures be taken to reduce the amount of radon in indoor air.

A homeowner should ask two questions in regard to radon hazards: "What is the radon concentration in my house?" and "If it is high, what can be done about it?" Because radon is radioactive, it can be measured with a simple detector available at most hardware stores and from local government agencies for about $25. If the detector indicates excessive radon, three types of solutions can be implemented. The first is to extend a ventilation duct from the basement to the outside of the house. This solution prevents basement air from circulating through the house and also prevents radon from accumulating in the basement. The second solution is to ventilate the house so that indoor air is continually refreshed. However, this method allows hot air to escape and increases heating bills. A third solution is to pump outside air into the house to keep indoor air at a slightly higher pressure than the outside air. This positive pressure prevents gas from seeping from soil or bedrock into the basement.

It is impossible to avoid exposure to radon completely because it is everywhere, in outdoor air as well as in homes and other buildings. But it is relatively easy and inexpensive to minimize exposure and thus avoid a significant cause of cancer.

Focus Question:

Radon has a half-life of 3.8 days, which means it decays naturally at a rate such that half of the radon atoms in a sample transform into other isotopes every 3.8 days. If radon decays so rapidly, why do granites that may be millions or hundreds of millions of years old contain radon?

3.8 Metamorphic Changes

Metamorphism commonly alters both the texture and mineral content of a rock.

Textural Changes

As a rock undergoes metamorphism, some mineral grains grow larger and others shrink. The shapes of the grains may also change. For example, fossils give the fossiliferous limestone shown in Figure 3–21B its texture. Both the fossils and the cement between them are made of small calcite crystals. If the limestone is heated, some of the calcite grains grow larger at the expense of others. In the process, the fossiliferous texture is destroyed.

Metamorphism transforms limestone into a metamorphic rock called **marble** (Fig. 3–29). Like the fossiliferous limestone, the marble is composed of calcite, but the texture is now one of large interlocking grains, and the fossils have vanished.

Metamorphism commonly occurs in regions where the Earth's crust is in motion. The tectonic forces

Figure 3–29 Metamorphism has destroyed the fossiliferous texture of the limestone in Figure 3–21B and replaced it with the large, interlocking calcite grains of marble. *(Adam Hart-Davis/Photo Researchers)*

crush, break, and bend rocks in this environment as the rocks are undergoing metamorphism. This combination of metamorphism and **deformation** creates layering in the rocks.

Micas are common metamorphic minerals that form as many different parent rocks undergo metamorphism. Recall from Chapter 2 that micas are shaped like pie plates. When metamorphism occurs without deformation, the micas grow with random orientations, like pie plates flying through the air (Fig. 3–30A). However, when horizontal force squeezes shale during metamorphism, the rock deforms into folds. At the same time, the clays decompose to form mica. The micas grow with their flat surfaces perpendicular to the direction of squeezing. This parallel alignment of micas (and other minerals) produces the metamorphic layering called **foliation** (Fig. 3–30B). Many metamorphic rocks break easily along the foliation planes. This parallel fracture pattern is called **slaty cleavage** (Fig. 3–31). In most cases, slaty cleavage cuts across the original sedimentary bedding.

The foliation layers range from a fraction of a millimeter to a meter or more thick. Metamorphic foliation can resemble sedimentary bedding, but is different in origin. Foliation results from alignment

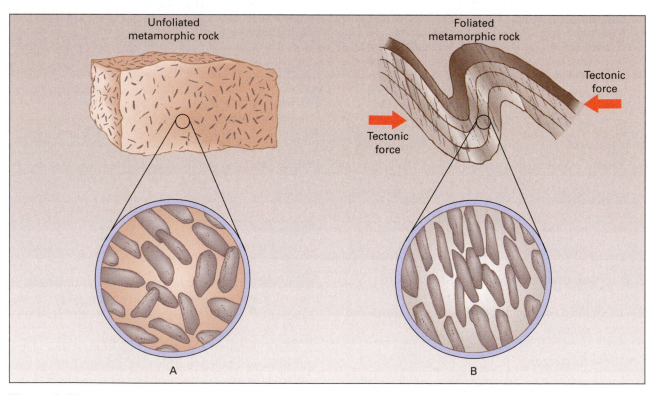

Figure 3–30 (A) When metamorphism occurs without deformation, platy micas grow with random orientations. (B) When deformation accompanies metamorphism, platy micas orient perpendicular to the force squeezing the rocks, forming foliated metamorphic rocks. The light and dark bands in the sketch represent original sedimentary layers that have been folded by the deformation.

Figure 3–31 These metamorphic rocks in Vermont show well-developed slaty cleavage; they are fractured along foliation planes created by parallel alignment of micas.

and segregation of metamorphic minerals during metamorphism, and forms at right angles to the forces acting on the rocks. Sedimentary bedding develops because sediments are deposited in a layer-by-layer process.

Mineralogical Changes

As a general rule, when a parent rock contains only one mineral, metamorphism transforms the rock into one composed of the same mineral, but with a coarser texture. The mineral content does not change because no other chemical components are available during metamorphism. Limestone converting to marble is one example of this generalization. Another is the metamorphism of quartz sandstone to **quartzite,** a rock composed of recrystallized quartz grains.

In contrast, metamorphism of a parent rock containing several minerals usually forms a rock with new

and different minerals *and* a new texture. For example, a typical shale contains large amounts of clay, as well as quartz and feldspar. When heated, some of those minerals decompose, and their atoms recombine to form new minerals such as mica, garnet, and a different kind of feldspar. Figure 3–32B shows a rock called **gneiss** that formed when metamorphism altered both the texture and minerals of shale.

If migrating fluids alter the chemical composition of a rock, new minerals invariably form. These effects are discussed further in Section 3.9.

3.9 Types of Metamorphism and Metamorphic Rocks

Recall that three conditions cause metamorphism: rising temperature, rising pressure, and changing chemical environment caused by an influx of hot water. In addition, deformation resulting from movement of the Earth's crust develops foliation and thus strongly affects the texture of a metamorphic rock. These conditions occur in four geologic environments.

Contact Metamorphism

Contact metamorphism occurs where hot magma intrudes cooler country rock of any type—sedimentary, metamorphic, or igneous. The highest-grade metamorphic rocks form at the contact, closest to the magma. Lower-grade rocks develop farther out. A metamorphic halo around a pluton can range in width from less than a meter to hundreds of meters, depending on the size and temperature of the intrusion and the effects of water or other fluids (Fig. 3–33).

Contact metamorphism commonly occurs without deformation. As a result, the metamorphic minerals

A

B

Figure 3–32 Metamorphism can convert a typical shale (A) to gneiss (B), a rock with different minerals and a different texture from those of shale.

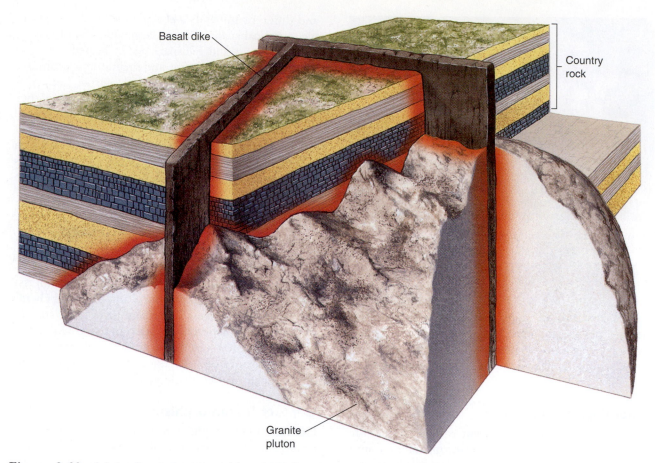

Basalt dike

Country rock

Granite pluton

Figure 3–33 A halo of contact metamorphism, shown in red, surrounds a pluton. The later intrusion of the basalt dike metamorphosed both the pluton and the sedimentary rock, creating a second metamorphic halo.

grow with random orientations—like the pie plates flying through the air—and the rocks develop no metamorphic layering.

Burial Metamorphism

Burial metamorphism results from deep burial of rocks in a sedimentary basin. A large river carries massive amounts of sediment to the ocean every year, where it accumulates on a delta. Over tens or even hundreds of millions of years, the weight of the sediment becomes so great that the entire region sinks, just as a canoe settles when you climb into it. Younger sediment may bury the oldest layers to a depth of more than 10 kilometers in a large basin. Over time, temperature and pressure increase within the deeper layers until burial metamorphism begins.

Burial metamorphism is occurring today in the sediments underlying many large deltas, including the Mississippi River delta, the Amazon Basin on the east coast of South America, and the Niger River delta on the west coast of Africa.

Burial metamorphism occurs without deformation. Consequently, metamorphic minerals grow with random orientations and, like contact metamorphic rocks, burial metamorphic rocks are unfoliated.

Regional Metamorphism

Regional metamorphism occurs where major crustal movements build mountains and deform rocks. It is the most common and widespread type of metamorphism and affects broad regions of the Earth's crust.

In a region where tectonic plates converge, magma rises into the continent and heats large portions of the crust (see Chapter 5). The high temperatures cause new metamorphic minerals to form throughout the region. The rising magma also deforms the hot, plastic country rock as it forces its way upward. At the same time, the movements of the crust squeeze and deform rocks. As a result of all these processes acting together, regionally metamorphosed rocks are strongly foliated and are typically associated with mountains, and with

Figure 3–34 A contact between light-gray granite and dark regional metamorphic rocks is exposed in the peaks of the Bugaboo Mountains of British Columbia.

the igneous rocks that form as the magma solidifies (Fig. 3–34). Regional metamorphism produces zones of foliated metamorphic rocks tens to hundreds of kilometers across.

Common Rocks Formed by Regional Metamorphism　Shale consists of clay minerals, quartz, and feldspar and is the most abundant sedimentary rock. The mineral grains are too small to be seen with the naked eye and can barely be seen with a microscope. Shale undergoes a sequence of changes as metamorphic grade increases.

Figure 3–35 shows the temperatures at which certain metamorphic minerals are stable as shale undergoes metamorphism. Thus, it shows the sequence in which minerals appear, and then decompose, as metamorphic grade increases. As regional metamorphism begins, the clay minerals in shale break down and are replaced by mica and chlorite (Fig. 3–36A). These new, platy minerals grow perpendicular to the direction of squeezing. As a result, the rock develops slaty cleavage and is called **slate** (Fig. 3–36B). With rising temperature and continued deformation, the micas and chlorite grow larger, and wavy or wrinkled surfaces replace the flat, slaty cleavage, giving **phyllite** a silky appearance (Fig. 3–36C).

As temperature continues to rise, the mica and chlorite grow large enough to be seen by the naked eye, and foliation becomes very well developed. Rock of this type is called **schist** (Fig. 3–36D). Schist forms approximately at the transition from low to intermediate metamorphic grades.

At high metamorphic grades, light- and dark-colored minerals often separate into bands that are thicker than the layers of schist, to form a rock called gneiss (pronounced "nice") (Fig. 3–36E). At the highest metamorphic grade, the rock begins to melt, forming small veins of granitic magma. When metamorphism wanes and the rock cools, the magma veins solidify to form **migmatite,** a mixture of igneous and metamorphic rock (Fig. 3–36F).

Quartz sandstone and limestone transform to foliated quartzite and foliated marble, respectively, during regional metamorphism.

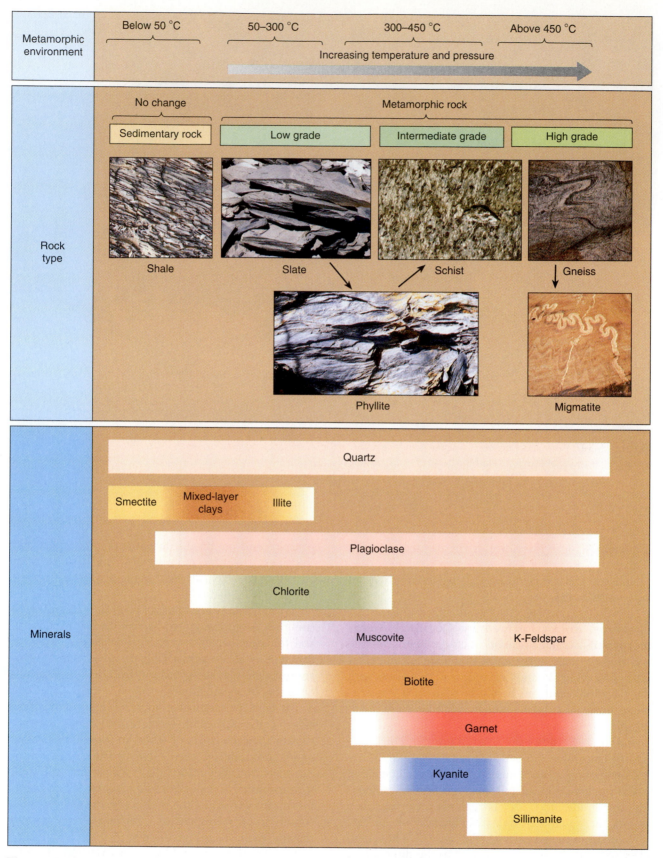

Figure 3–35 Shale undergoes changes both in texture and minerals as metamorphic grade increases. The lower part of the figure shows the stability ranges of common metamorphic minerals.

A

B

C

D

E

F

Figure 3–36 (A) Shale is the most common sedimentary rock. Regional metamorphism progressively converts shale to slate (B) *(Breck P. Kent),* phyllite (C), schist (D), and gneiss (E). Migmatite (F) forms when gneiss begins to melt.

Hydrothermal Metamorphism

Water is a chemically active fluid; it attacks and dissolves rocks and minerals. If the water is hot, it attacks minerals even more rapidly. **Hydrothermal metamorphism** (also called **hydrothermal alteration** and **metasomatism**) occurs when hot water and ions dissolved in the hot water react with a rock to change its chemical composition and minerals. In some hydrothermal environments, water reacts with sulfur minerals to form sulfuric acid, making the solution even more corrosive. Most hydrothermal alteration is caused by circulating **ground water**—the water that

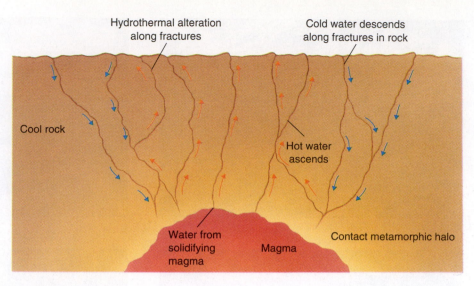

Figure 3–37 Ground water descending through fractured rock is heated by magma and rises through other cracks, causing hydrothermal metamorphism in nearby rock.

Hydrothermal alteration along fractures

Cold water descends along fractures in rock

Cool rock

Hot water ascends

Water from solidifying magma

Magma

Contact metamorphic halo

saturates soil and bedrock. Cold ground water sinks through bedrock fractures to depths of a few kilometers, where it is heated by the deep, hot rocks or in some cases by a hot, shallow pluton. Upon heating, the water expands and rises back toward the surface through other fractures (Fig. 3–37). As it rises, it alters the country rock through which it flows.

Most rocks and magma contain low concentrations of metals such as gold, silver, copper, lead, and zinc. For example, gold makes up 0.0000002 percent of average crustal rock, while copper makes up 0.0058 per-

cent and lead 0.0001 percent. Although the metals are present in very low concentrations, hydrothermal solutions sweep slowly through vast volumes of country rock, dissolving and accumulating the metals as they go. The solutions then deposit the dissolved metals when they encounter changes in temperature, pressure, or chemical environment (Fig. 3–38). In this way, hydrothermal solutions scavenge and concentrate metals from average crustal rocks and then deposit them locally to form ore. Hydrothermal ore deposits are discussed further in Chapter 21.

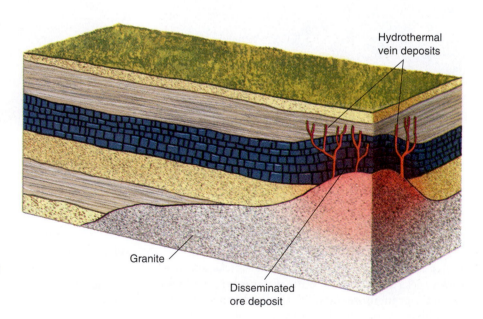

Figure 3–38 Hydrothermal ore deposits form when hot water dissolves metals from country rock and deposits them in fractures and surrounding country rock.

Hydrothermal vein deposits

Granite

Disseminated ore deposit

SUMMARY

Geologists divide rocks into three groups depending upon how they formed. **Igneous rocks** solidify from **magma**. **Sedimentary rocks** form from clay, sand, gravel, and other **sediment** that collects at the Earth's surface. **Metamorphic rocks** form when any rock is altered by temperature, pressure, or an influx of hot water. The **rock cycle** summarizes processes by which rocks continuously recycle in the outer layers of the Earth, forming new rocks from old ones.

Intrusive or **plutonic** igneous rocks are medium- to coarse-grained rocks that solidify within the Earth's crust. **Extrusive** or **volcanic** igneous rocks are fine-grained rocks that solidify from magma erupted onto the Earth's surface. **Granite** and **basalt** are the two most common igneous rocks.

Sediment forms by weathering of rocks and minerals. It includes all solid particles such as rock and mineral fragments, organic remains, and precipitated minerals. It is transported by streams, glaciers, wind, and gravity; deposited in layers; and eventually **lithified** to form sedimentary rock. **Shale, sandstone,** and **limestone** are the most common kinds of sedimentary rock.

When a rock is heated, when pressure increases, or when hot water alters its chemistry, both its minerals and its textures change in a process called **metamorphism**. **Contact metamorphism** affects rocks heated by a nearby igneous intrusion. **Burial metamorphism** alters rocks as they are buried deeply within the Earth's crust. In regions where tectonic plates converge, high temperature and deformation from rising magma and plate movement all combine to cause **regional metamorphism**. **Hydrothermal metamorphism** is caused by hot solutions soaking through rocks and is often associated with emplacement of ore deposits. **Slate, schist, gneiss,** and **marble** are common metamorphic rocks.

KEY TERMS

magma 42
igneous rock 42
weathering 42
sediment 42
sedimentary rock 42
metamorphic rock 42
rock cycle 42
extrusive igneous rock 45
volcanic rock 45
intrusive igneous rock 45
plutonic rock 45
texture 45
lava 45
porphyry 45

phenocryst 45
felsic 46
mafic 46
ultramafic 46
intermediate rock 46
basement rock 46
lithification 48
clastic sedimentary
  rock 48
organic sedimentary rock
  48
chemical sedimentary
  rock 48

bioclastic sedimentary
  rock 48
pore space 49
compaction 49
fissility 51
carbonate rock 52
sedimentary structure 53
bedding (stratification)
  54
cross-bedding 54
ripple marks 54
mud cracks 55
fossils 55

metamorphism 56
metamorphic grade 56
deformation 58
foliation 58
slaty cleavage 58
contact metamorphism
  59
burial metamorphism 60
regional metamorphism
  60
hydrothermal
  metamorphism 63

Common Igneous Rocks

Felsic	Intermediate	Mafic	Ultramafic
obsidian	andesite	basalt	peridotite
granite	diorite	gabbro	
rhyolite			

Common Sedimentary Rocks

Clastic	Bioclastic	Organic	Chemical
conglomerate	limestone	chert	evaporite
sandstone	dolostone	coal	
siltstone	coquina		
shale	chalk		

Common Metamorphic Rocks

Unfoliated	Foliated
marble	slate
quartzite	schist
	gneiss
	migmatite

FOR REVIEW

1. Explain what the rock cycle tells us about Earth processes.
2. What are the three main kinds of rock in the Earth's crust?
3. How do the three main types of rock differ from each other?
4. What is magma, and where does it originate?
5. Describe how igneous rocks are classified and named.
6. What are the two most common kinds of igneous rock?
7. Describe and explain the differences between a plutonic and a volcanic rock.
8. What is sediment? How does it form?
9. How do sedimentary grains become rounded?
10. Where in your own area would you look for rounded sediment?
11. Describe how sediment becomes lithified.
12. What are the differences among shale, sandstone, and limestone?
13. Explain why almost all sedimentary rocks are layered, or bedded.
14. How does cross-bedding form?
15. What is metamorphism? What factors cause metamorphism?
16. What kinds of changes occur in a rock as it is metamorphosed?
17. What is metamorphic foliation? How does it differ from sedimentary bedding?
18. How do contact metamorphism and regional metamorphism differ, and how are they similar?

FOR DISCUSSION

1. Magma usually begins to rise toward the Earth's surface as soon as it forms. It rarely accumulates as large pools in the upper mantle or lower crust, where it originates. Why?
2. In the San Juan Mountains of Colorado, parts of the range are made up of granite, and other parts are volcanic rocks. Explain why these two types of igneous rock are likely to be found together.
3. Suppose you were given a fist-size sample of igneous rock. How would you tell whether it is volcanic or plutonic in origin?
4. How would you tell whether another rock sample is igneous, sedimentary, or metamorphic in origin?
5. Why is shale the most common sedimentary rock?
6. One sedimentary rock is composed of rounded grains, and the grains in another are angular. What can you tell about the history of the two rocks from these observations?
7. What do mud cracks and raindrop imprints in shale tell you about the water depth in which the mud accumulated?
8. How would you distinguish between contact metamorphic rocks and regional metamorphic rocks in the field?
9. What happens to bedding when sedimentary rocks undergo regional metamorphism?
10. How can granite form during metamorphism?
11. Explain how the geosphere, the atmosphere, the hydrosphere, and the biosphere all contribute to the rock cycle.

Geologic Time: A Story in the Rocks

Fossil remains, such as this carniverous dinosaur, *Tarbosaurus bataar,* allow scientists to picture the biosphere 75 million years ago. *(Mahau Kulyk/Photo Researchers)*

Earth scientists commonly study events that occurred in the past. They observe rocks and landforms and ask questions such as "What processes shaped that mountain range?" "How old are the mountains and how high were they earlier in their history?" For example, compare the Appalachians, a low, rounded mountain range, with the Tetons, whose rocky peaks rise precipitously from the valley floor. We might ask, "Do the two ranges seem so different because the Appalachians are older and have been eroding for a longer time? Were the Appalachians once as steep as the Tetons are today? If so, when did their rocky summits rise, and when did erosion wear them away?" ●

4.1 Geologic Time

While most of us think of time in terms of days or years, Earth scientists commonly refer to events that happened millions or billions of years ago. In Chapter 1 you learned that the Earth is approximately 4.6 billion years old. Yet humans and our human-like ancestors have existed for four million years, and recorded history is only a few thousand years old. How do scientists measure the ages of rocks and events that occurred millions or billions of years ago?

Scientists measure geologic time in two different ways. **Relative age** refers only to the *order* in which events occurred. Determination of relative age is based on a simple principle: In order for an event to affect a rock, the rock must exist first. Thus, the rock must be older than the event. This principle seems obvious, yet it is the basis of much geologic work. As you learned in Chapter 3, sediment normally accumulates in horizontal layers. However, the rocks in Figure 4–1 are folded. We deduce that the folding occurred *after* the sediment was deposited. The order in which rocks

and geologic features formed can almost always be interpreted by observation and logic.

Absolute age is age in years. Dinosaurs became extinct 65 million years ago. The Teton Range in Wyoming began rising 6 million years ago. Absolute age tells us both the order in which events occurred and the amount of time that has passed since they occurred.

4.2 Relative Geologic Time

Absolute age measurements have become common only in the second half of the twentieth century. Prior to that time, geologists used field observations to determine relative ages. Even today, with sophisticated laboratory processes available, most field geologists routinely use relative ages. Geologists use a combination of common sense and a few simple principles to determine the order in which rocks formed and changed over time.

Figure 4–1 Limestone was deposited in a shallow sea and then uplifted and folded in the Canadian Rockies, Alberta.

A

B

Figure 4–2 (A) The principle of original horizontality tells us that most sedimentary rocks are deposited with horizontal bedding (San Juan River, Utah). When we see tilted rocks (B), we infer that they were tilted after they were deposited (Connecticut).

The **principle of original horizontality** is based on the observation that sediment usually accumulates in horizontal layers (Fig. 4–2A). If sedimentary rocks lie at an angle, as in Figure 4–2B, or if they are folded as in Figure 4–1, we can infer that tectonic forces tilted or folded them after they formed.

The **principle of superposition** states that sedimentary rocks become younger from bottom to top (as long as tectonic forces have not turned them upside down). This is because younger layers of sediment always accumulate on top of older layers. In Figure 4–3, the sedimentary layers become progressively younger in the order 1, 2, 3, 4, and 5.

The **principle of cross-cutting relationships** is based on the obvious fact that a rock must first exist before anything can happen to it. Figure 4–4 shows a basalt dike cutting across layers of sandstone. Clearly, the dike must be younger than the sandstone. Figure 4–5 shows sedimentary rocks intruded by three granite dikes. Dike 2 cuts dike 1, and dike 3 cuts dike 2, so dike 2 is younger than 1, and dike 3 is the youngest. The sedimentary rocks must be older than all of the dikes.

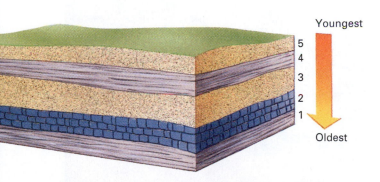

Figure 4–3 In a sequence of sedimentary beds, the oldest bed (labeled 1) is the lowest, and the youngest (labeled 5) is on top. • Interactive Question: Imagine that the horizontal beds in this figure were tilted so they stood vertically. How could you tell which was the oldest and which was the youngest?

Figure 4–4 A basalt dike cutting across sandstone in the walls of Grand Canyon. The dike must be younger than the sandstone.

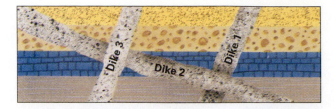

Figure 4–5 Three granite dikes cutting sedimentary rocks. The dikes become younger in the order 1, 2, 3. The sedimentary rocks must be older than all three dikes.

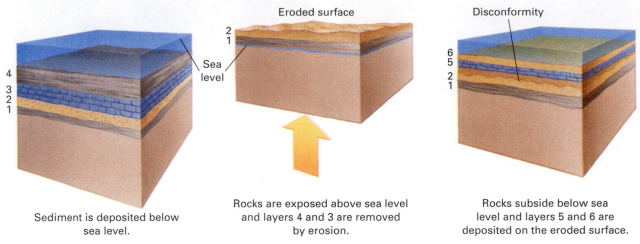

Sediment is deposited below sea level.

Rocks are exposed above sea level and layers 4 and 3 are removed by erosion.

Rocks subside below sea level and layers 5 and 6 are deposited on the eroded surface.

Figure 4–6 A disconformity is created by uplift and erosion, followed by deposition of additional layers of sediment. The disconformity represents a gap in the rock record.

4.3 Unconformities: Gaps in the Time Record

The 2-kilometer-high walls of Grand Canyon are composed of sedimentary rocks lying on older igneous and metamorphic rocks. Their ages range from about 200 million years to nearly 2 billion years. The principle of superposition tells us that the deepest rocks are the oldest and the rocks become progressively younger as we climb up the canyon walls. However, no principle assures us that the rocks formed continuously from 2 billion to 200 million years ago. The rock record may not be complete. Suppose that no sediment were deposited for a period of time, or erosion removed some sedimentary layers before younger layers accumulated. In either case a gap would exist in the rock record. We know that any rock layer is younger than the layer below it, but without more information we do not know how much younger.

▶ **Figure 4–7** A disconformity separates parallel layers of sandstone and overlying conglomerate in Wyoming. Some sandstone layers were eroded away before the conglomerate was deposited.

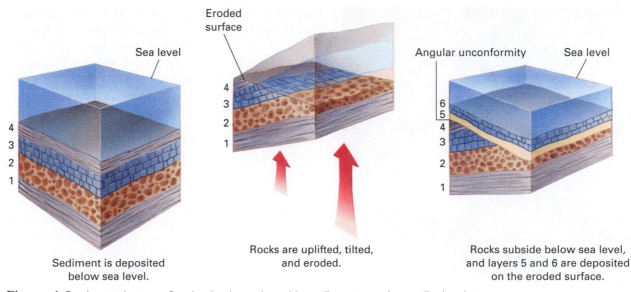

Sea level

Eroded surface

Angular unconformity   Sea level

Sediment is deposited below sea level.

Rocks are uplifted, tilted, and eroded.

Rocks subside below sea level, and layers 5 and 6 are deposited on the eroded surface.

Figure 4–8 An angular unconformity develops when older sedimentary rocks are tilted and eroded before younger sediment accumulates. • **Interactive Question:** Draw a sequence in which older sedimentary rocks are folded and eroded before younger sediment accumulates.

Layers of sedimentary rocks are **conformable** if they were deposited without interruption. An **unconformity** represents an interruption in deposition, usually of long duration. During the interval when no sediment was deposited, some rock layers may have been eroded. Thus, an unconformity represents a long time interval for which no geologic record exists in that place. The lost record may involve hundreds of millions of years.

Several types of unconformities exist. In a **disconformity,** the sedimentary layers above and below the unconformity are parallel (Figs. 4–6 and 4–7). A disconformity may be difficult to recognize unless an obvious soil layer or erosional surface developed. However, geologists identify disconformities by determining the ages of rocks using methods based on fossils and absolute dating, described later in this chapter.

In an **angular unconformity,** tectonic activity tilted older sedimentary rock layers before younger sediment accumulated (Figs. 4–8 and 4–9). A **nonconformity** is an unconformity in which sedimentary rocks lie on igneous or metamorphic rocks. The nonconformity shown in Figure 4–10 represents a time gap of about 1 billion years.

4.4 Fossils and Faunal Succession

Paleontologists study **fossils,** the remains and other traces of prehistoric life, to understand the history of

life and evolution. Fossils also provide information about the ages of sedimentary rocks and their depositional environments. The oldest known fossils are traces of bacteria-like organisms that lived about 3.5 billion years ago. A much younger fossil consists of the frozen and mummified remains of a Bronze Age man recently found in a glacier near the Italian-Austrian border (Fig. 4–11).

Figure 4–9 An angular unconformity near Capitol Reef National Park, Utah.

Figure 4–10 A nonconformity in the cliffs of Grand Canyon, Arizona. You can see sedimentary bedding in the red sandstone above the nonconformity but none in the igneous and metamorphic rocks below.

Interpreting Geologic History from Fossils

Fossils allow geologists to interpret geologic history. Most organisms thrive in specific environments and do not survive in others. For example, the remains of marine animals in limestone of the Canadian Rockies or near the top of Mount Everest tell us that these places once lay submerged beneath the sea. Therefore, we infer that later tectonic processes raised these regions to their present elevations. Or, as another example, corals and other reef-building organisms live in clear, warm, shallow seas. Consequently, fossil coral reefs preserved in sedimentary rocks tell us about the environmental conditions at the time and place the rocks formed.

The theory of **evolution** states that life forms have changed throughout geologic time. Fossils are useful in determining relative ages of rocks because different animals and plants lived at different times in the Earth's history. For example, trilobites lived from 535 million to about 245 million years ago and the first dinosaurs appeared about 225 million years ago.

In a sequence of sedimentary rocks that formed over a long time, different fossils appear and then vanish from bottom to top in the same order in which the organisms evolved and then became extinct through time. Rocks containing dinosaur bones must be younger than those containing trilobite remains. The **principle of faunal succession** states that fossil organisms succeeded one another through time in a definite and recognizable order and that the relative ages of sedimentary rocks can therefore be recognized from their fossils.

4.5 Correlation

Ideally geologists would like to develop a continuous history for each region of the Earth by interpreting rocks that formed in that place throughout geologic time. Unfortunately, there is no single place on Earth where rocks formed and were preserved continuously.

Figure 4–11 The mummified remains of this Bronze Age hunter were recently discovered preserved by glacial ice near the Austrian-Italian border. *(Sygma)*

Erosion has removed some rock layers and at times no rocks formed. Consequently, the rock record in any one place is full of unconformities and historical gaps. To assemble as complete and continuous a record as possible, geologists combine evidence from many localities. To do this, rocks of the same age from different localities must be matched in a process called **correlation.** But how do we correlate rocks over great distances?

If you follow a single continuous sedimentary bed from one place to another, then it is clearly the same layer in both localities. But this approach is impractical over long distances and where rocks are not exposed. Another problem arises when correlation is based on continuity of a sedimentary layer. When rocks are correlated for the purpose of building a geologic time scale, geologists want to show that certain rocks all formed at the same time. Suppose that you are attempting to trace a beach sandstone that formed as sea level rose. The beach would have migrated inland over time and, as a result, the sandstone becomes younger in a landward direction (Fig. 4–12).

Over a distance short enough to be covered in an hour or so of walking, the age difference may be unimportant. But if you traced a similar beach sand laterally over hundreds of kilometers, its age may vary by millions of years. Thus, there are two kinds of correlation: **Time correlation** means age equivalence, but **lithologic correlation** means continuity of a rock unit, such as the sandstone. The two are not always the same because some rock units, such as the sandstone, were deposited at different times in different places. To construct a record of Earth history and a geologic time scale, Earth scientists must find other evidence of the geologic ages of rocks.

Index Fossils   An index fossil indicates the age of rocks containing it. To be useful, an index fossil is produced by an organism that (1) is abundantly preserved in rocks, (2) was geographically widespread, (3) existed as a species or genus for only a relatively short time, and (4) is easily identified in the field. Floating or swimming marine animals that evolved rapidly make the best index fossils. A marine habitat allows rapid and widespread distribution of these organisms. If a species evolved rapidly and soon became extinct, then it existed for only a short time. The shorter the time span that a species existed, the more precisely the index fossil reflects the age of a rock.

In many cases, the presence of a single type of index fossil is sufficient to establish the age of a rock. More commonly, an assemblage of several fossils is used to date and correlate rocks. Figure 4–13 shows an example of how index fossils and fossil assemblages are used in correlation.

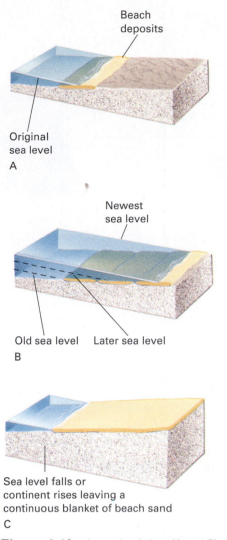

Figure 4–12   As sea level rises (A and B) or falls (C) slowly through time, a beach migrates laterally, forming a single sand layer that is of different ages in different localities.

Key Beds   A key bed is a thin, widespread sedimentary layer that was deposited rapidly and synchronously over a wide area and is easily recognized. Many volcanic eruptions eject large volumes of fine, glassy volcanic ash into the atmosphere. Wind carries the ash over great distances before it settles. Some historic ash clouds have encircled the globe. When the ash settles, the glass rapidly crystallizes to form a pale clay layer that is incorporated into sedimentary rocks. Such volcanic eruptions occur at a precise point in time, so the ash is the same age everywhere. The thin, sooty, iridium-rich clay layer deposited 65 million years ago from debris of a giant meteorite impact is another classic example of a key bed (see essay on page 18).

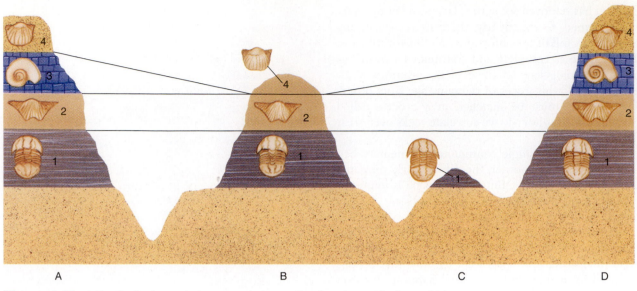

Figure 4–13 Index fossils demonstrate age equivalency of sedimentary rocks from widely separated localities. Sedimentary beds containing the same index fossils are interpreted to be of the same age. The fossils show that at locality B, layer 3 is missing because layer 4 directly lies on top of layer 2. Either layer 3 was deposited and then eroded away before 4 was deposited at locality B, or layer 3 was never deposited here. At locality C all layers above 1 are missing, either because of erosion or because they were never deposited.

4.6 Absolute Geologic Time

How does a geologist measure the absolute age of an event that occurred before calendars and even before humans evolved to keep calendars? Think of how a calendar measures time. The Earth rotates about its axis at a constant rate, once a day. Thus, each time the Sun rises, you know that a day has passed and you check it off on your calendar. If you mark off each day as the Sun rises, you record the passage of time. To know how many days have passed since you started keeping time, you just count the check marks. Absolute age measurement depends on two factors: *a process that occurs at a constant rate* (e.g., the Earth rotates once every 24 hours) and some way to keep *a cumulative record of that process* (e.g., marking the calendar each time the Sun rises). Measurement of time with a calendar, a clock, an hourglass, or any other device depends on these two factors.

Geologists have found a natural process that occurs at a constant rate and accumulates its own record: It is the radioactive decay of elements that are present in many rocks. Thus, many rocks have built-in calendars. We must understand radioactivity to read the calendars.

Recall from Chapter 2 that an atom consists of a small, dense nucleus surrounded by a cloud of electrons. A nucleus consists of positively charged protons and neutral particles called neutrons. All atoms of any given element have the same number of protons in the nucleus. However, the number of neutrons may vary. **Isotopes** are atoms of the same element with different numbers of neutrons. For example, all isotopes of potassium-40 contain 19 protons and 21 neutrons. Potassium-39 has 19 protons but only 20 neutrons.

Many isotopes are stable, meaning that they do not change with time. If you studied a sample of potassium-39 for 10 billion years, all the atoms would remain unchanged. Other isotopes are unstable or **radioactive.** Given time, their nuclei spontaneously decay.[1] Potassium-40 decays naturally to form either of two other isotopes, argon-40 or calcium-40 (Fig. 4–14). A radioactive isotope such as potassium-40 is known as a **parent isotope.** An isotope created by radioactivity, such as argon-40 or calcium-40, is called a **daughter isotope.**

Many common elements, such as potassium, consist of a mixture of radioactive and nonradioactive iso-

---

[1] Radioactive nuclei decay by three processes. In electron capture, the nucleus captures an electron. The electron combines with a proton to create a neutron. The decay of potassium-40 to argon-40 is an example of this process. In beta emission, a neutron ejects an electron from the nucleus, and the neutron becomes a proton. The ejected electron is called a beta particle. The decay of potassium-40 to calcium-40 is an example of this process. In a third type of decay, called alpha emission, a nucleus emits an alpha particle consisting of two protons and two neutrons. Uranium-238 emits 8 alpha particles and 6 beta particles in a sequence of steps to become lead-206.

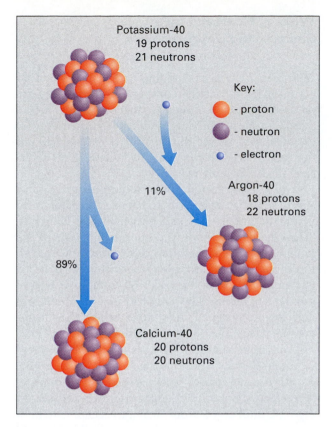

Figure 4–14 Potassium-40 spontaneously decays either to argon-40 or to calcium-40. Eleven percent of the potassium-40 atoms decay to argon-40. In this case, the potassium-40 nucleus captures an electron. The electron combines with a proton to produce a neutron. This process creates an argon-40 nucleus with 18 protons and 22 neutrons. The other 89 percent of the potassium-40 nuclei decay to calcium-40 by expelling an electron from the nucleus. This process converts a neutron to a proton, producing a calcium-40 nucleus containing 20 protons and 20 neutrons.

topes. With time, the radioactive isotopes decay, but the nonradioactive ones do not. Some elements, such as uranium, consist only of radioactive isotopes. The amount of uranium on Earth slowly decreases as it decomposes to other elements, such as lead.

Radioactivity and Half-Life

If you watch a single atom of potassium-40, when will it decompose? This question cannot be answered because any particular potassium-40 atom may or may not decompose at any time. But recall from Chapter 2 that even a small sample of a material contains a huge number of atoms. Each atom has a certain probability of decaying at any time. Averaged over time, half of the atoms in any sample of potassium-40 will decompose in 1.3 billion years. The **half-life** is the time it takes for half of the atoms in a radioactive sample to decompose. The half-life of potassium-40 is 1.3 bil-

lion years. Therefore, if 1 gram of potassium-40 were placed in a container, 0.5 gram would remain after 1.3 billion years, 0.25 gram after 2.6 billion years, and so on. Each radioactive isotope has its own half-life; some half-lives are fractions of a second and others are measured in billions of years.

The Basis of Radiometric Dating

Two aspects of radioactivity are essential to the calendars in rocks. First, the half-life of a radioactive isotope is constant. It is easily measured in the laboratory and is unaffected by geologic or chemical processes. So radioactive decay occurs at a known, constant rate. Second, as a parent isotope decays, its daughter accumulates in the rock. The longer the rock exists, the more daughter isotope accumulates. The accumulation of a daughter isotope is analogous to marking off days on a calendar. Because radioactive isotopes are widely distributed throughout the Earth, many rocks have built-in calendars that allow us to measure their ages. **Radiometric dating** is the process of determining the ages of rocks, minerals, and fossils by measuring the relative amounts of parent and daughter isotopes.

Figure 4–15 shows the relationships between age and relative amounts of parent and daughter isotopes. At the end of one half-life, 50 percent of the parent atoms have decayed to daughter. At the end of two half-lives, the mixture is 25 percent parent and 75 percent daughter. To determine the age of a rock, a geologist measures the proportions of parent and daughter isotopes in a sample and compares the ratio to a similar graph. Consider a hypothetical parent-daughter pair having a half-life of 1 million years. If we determine that a rock contains a mixture of 25 percent parent isotope and 75 percent daughter, Figure 4–15 shows that the age is two half-lives, or 2 million years.

If the half-life of a radioactive isotope is short, that isotope gives accurate ages for young materials. For example, carbon-14 has a half-life of 5730 years. Carbon-14 dating gives accurate ages for materials younger than 50,000 years. It is useless for older materials because by 50,000 years, virtually all the carbon-14 has decayed, and no additional change can be measured.

Isotopes with long half-lives give good ages for old rocks, but not enough daughter accumulates in young rocks to be measured. For example, rubidium-87 has a half-life of 47 billion years. In a geologically short period of time—10 million years or less—so little of its daughter has accumulated that it is impossible to measure accurately. Therefore, rubidium-87 is not useful for rocks younger than about 10 million years. The six radioactive isotopes that are most commonly used for dating are summarized in Table 4–1.

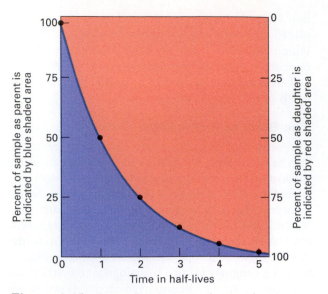

Figure 4–15 As a radioactive parent isotope decays to a daughter, the proportion of parent decreases (blue zone) and the amount of daughter increases (red zone). The half-life is the amount of time required for half of the parent to decay to daughter. At time zero, when the radiometric calendar starts, a sample is 100 percent parent. At the end of one half-life, 50 percent of the parent has converted to daughter. At the end of two half-lives, 25 percent of the sample is parent and 75 percent is daughter. Thus, by measuring the proportions of parent and daughter in a rock, the rock's age in half-lives can be obtained. Because the half-lives of all radioactive isotopes are well known, it is simple to convert age in half-lives to age in years.

What Is Measured by a Radiometric Date?

Biotite is rich in potassium, so the potassium-40/argon-40 (abbreviated K/Ar) parent-daughter pair can be used successfully. Suppose that we collect a fresh sample of biotite-bearing granite, separate out a few grams of biotite, and obtain a K/Ar date of 100 million years. What happened 100 million years ago to start the K/Ar "calendar"?

The granite started out as molten magma, which slowly solidified below the Earth's surface. But at this point, the granite was still buried several kilometers below the surface and was still hot. Later, perhaps over millions of years, the granite cooled slowly as tectonic forces pushed it toward the surface, where we collected our sample. So what does our 100-million-year radiometric date tell us: The age of formation of the original granite magma? The time when the magma became solid? The time of uplift?

We can measure time because potassium-40 decays to argon-40, which accumulates in the biotite crystal. As more time elapses, more argon-40 accumulates. Therefore, the biotite calendar does not start recording time until argon atoms begin to accumulate in biotite. Argon is an inert gas that is trapped in biotite as the argon forms. But if the biotite is heated above a certain temperature, the argon escapes. Biotite retains argon only when it cools below 350°C. Thus, the 100-

TABLE 4–1

The Most Commonly Used Isotopes in Radiometric Age Dating

Isotopes		Half-Life of Parent (Years)	Effective Dating Range (Years)	Minerals and Other Materials That Can Be Dated
Parent	**Daughter**			
Carbon-14	Nitrogen-14	5730 ± 30	0100–70,000	Anything that was once alive: wood, other plant matter, bone, flesh, or shells; also, carbon in carbon dioxide dissolved in ground water, deep layers of the ocean, or glacier ice
Potassium-40	Argon-40 Calcium-40	1.3 billion	50,000–4.6 billion	Muscovite Biotite Hornblende Whole volcanic rock
Uranium-238	Lead-206	4.5 billion	10 million–4.6 billion	Zircon Uraninite and pitchblende
Uranium-235	Lead-207	710 million		
Thorium-232	Lead-208	14 billion		
Rubidium-87	Strontium-87	47 billion	10 million–4.6 billion	Muscovite Biotite Potassium feldspar Whole metamorphic or igneous rock

Carbon-14 Dating

Carbon-14 dating differs from the other parent-daughter pairs shown in Table 4–1 in two ways. First, carbon-14 has a half-life of only 5730 years, in contrast to the millions or billions of years of other isotopes used for radiometric dating. Second, accumulation of the daughter, nitrogen-14, cannot be measured. Nitrogen-14 is the most common isotope of nitrogen, and it is impossible to distinguish nitrogen-14 produced by radioactive decay from other nitrogen-14 in an object being dated.

The abundant stable isotope of carbon is carbon-12. Carbon-14 is continuously created in the atmosphere as cosmic radiation bombards nitrogen-14, converting it to carbon-14. The carbon-14 then decays back to nitrogen-14. Because it is continuously created and because it decays at a constant rate, the ratio of carbon-14 to carbon-12 in the atmosphere remains nearly constant. While an organism is alive, it absorbs carbon from the atmosphere. Therefore, the organism contains the same ratio of carbon-14 to carbon-12 as found in the atmo-

sphere. However, when the organism dies, it stops absorbing new carbon. Therefore, the proportion of carbon-14 in the remains of the organism begins to diminish at death as the carbon-14 decays. Thus, carbon-14 age determinations are made by measuring the ratio of carbon-14 to carbon-12 in organic material. As time passes, the proportion of carbon-14 steadily decreases. By the time an organism has been dead for 50,000 years, so little carbon-14 remains that measurement is very difficult. After about 70,000 years, nearly all of the carbon-14 has decayed to nitrogen-14, and the method is no longer useful.

Focus Question:

Explain why carbon-14 dating would not be useful to date Mesozoic fossils.

million-year age date on the granite is the time that has passed since the granite cooled below 350°C, probably corresponding to cooling that occurred during uplift and erosion.

Two additional conditions must be met for a radiometric date to be accurate: First, when a mineral or rock cools, no original daughter isotope is trapped in the mineral or rock. Second, once the clock starts, no parent or daughter isotopes are added or removed from the rock or mineral. Metamorphism, weathering, and circulating fluids can add or remove parent or daughter isotopes. Sound geologic reasoning and discretion must be applied when choosing materials for radiometric dating and when interpreting radiometric age dates.

4.7 The Geologic Column and Time Scale

As mentioned earlier, no single locality exists on Earth where a complete sequence of rocks formed continuously throughout geologic time. However, geologists have correlated rocks that accumulated continuously through portions of geologic time from many different localities around the world. The information has been combined to create the **geologic column,** which is a nearly complete composite record of geologic time

(Table 4–2). The worldwide geologic column is frequently revised as geologic mapping continues.

Geologists divide all of geologic time into smaller units for convenience. Just as a year is subdivided into months, months into weeks, and weeks into days, large geologic time units are split into smaller intervals. This chronological arrangement is called the **geologic time scale.** The units are named, just as months and days are. The largest time units are **eons,** which are divided into **eras.** Eras are subdivided, in turn, into **periods,** which are further subdivided into **epochs.**

The geologic column and the geologic time scale were originally constructed on the basis of relative age determinations. When geologists developed radiometric dating, they added absolute ages to the column and time scale.

Today, geologists use this time scale to date rocks in the field. Imagine that you are studying a sedimentary sequence. If you find an index fossil or a key bed that has already been radiometrically dated by other scientists, you know the age of the rock and you do not need to send the sample to a laboratory for radiometric dating.

Geologic Time

Look again at the geologic time scale of Table 4–2. Notice that the Phanerozoic Eon, which comprises the

TABLE 4–2

The Geologic Column and Time Scale*

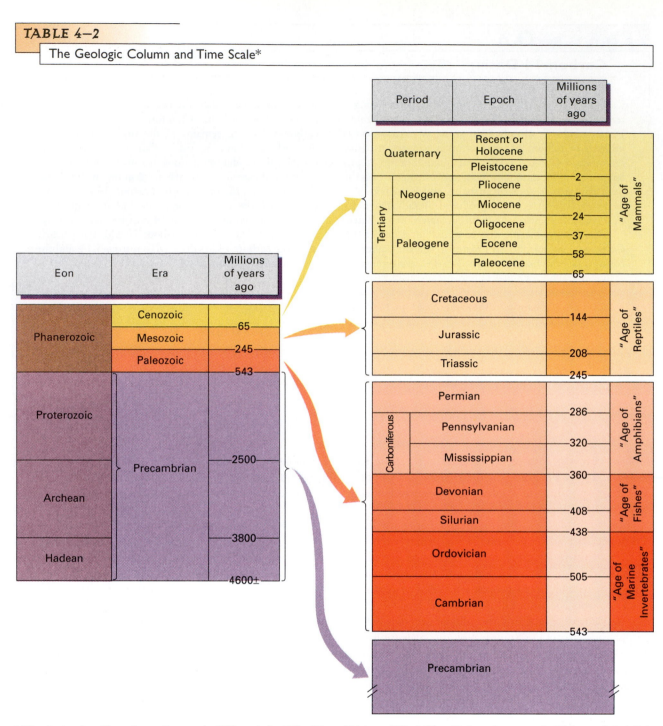

Period		Epoch	Millions of years ago	
Quaternary		Recent or Holocene		"Age of Mammals"
		Pleistocene	2	
Tertiary	Neogene	Pliocene	5	
		Miocene	24	
	Paleogene	Oligocene	37	
		Eocene	58	
		Paleocene	65	
Cretaceous			144	"Age of Reptiles"
Jurassic			208	
Triassic			245	
Permian			286	"Age of Amphibians"
Carboniferous	Pennsylvanian		320	
	Mississippian		360	
Devonian			408	"Age of Fishes"
Silurian			438	
Ordovician			505	"Age of Marine Invertebrates"
Cambrian			543	

Eon	Era	Millions of years ago
Phanerozoic	Cenozoic	
		65
	Mesozoic	
		245
	Paleozoic	
		543
Proterozoic	Precambrian	
		2500
Archean		
		3800
Hadean		
		4600±

Precambrian

* Time is given in millions of years (for example, 1000 stands for 1000 million, which is one billion). The table is *not* drawn to scale. We know relatively little about events that occurred during the early part of the Earth's history. Therefore, the first four billion years are given relatively little space on this chart, while the more recent Phanerozoic Eon, which spans only 543 million years, receives proportionally more space.

most recent 543 million years of geologic time, takes up most of the table and contains all of the named subdivisions. The earlier eons—the Proterozoic, Archean, and Hadean—are often not subdivided at all, even though together they constitute a time interval of 4 billion years, almost 8 times as long as the Phanerozoic. Why is Phanerozoic time subdivided so finely? Or, conversely, what prevents subdivision of the earlier eons?

Earliest Eons of Geologic Time: Precambrian Time

The **Hadean Eon** (Greek for "beneath the Earth") is the earliest time in Earth history and ranges from the

Figure 4–16 In Paleozoic time, the land was inhabited by reptiles and amphibians and covered by ferns, ginkgoes, and conifers. *(Ward's Natural Science Establishment)*

some paleontologists view them as ancestors of organisms alive today. However, other paleontologists believe that the resemblance is only superficial. A few types of Proterozoic shell-bearing organisms have been identified, but shelled organisms did not become abundant until the Paleozoic era.

Commonly, the Hadean, Archean, and Proterozoic Eons are collectively referred to by the informal term **Precambrian,** because they preceded the Cambrian Period, when fossil remains first became very abundant.

Phanerozoic Eon

Sedimentary rocks of the **Phanerozoic Eon** contain abundant fossils (*phaneros* is Greek for "evident"). Four changes occurred at the beginning of Phanerozoic time that greatly improved the fossil record:

1. The number of species with shells and skeletons dramatically increased.

2. The total number of individual organisms preserved as fossils increased greatly.

3. The total number of species preserved as fossils increased greatly.

4. The average sizes of individual organisms increased.

Shels and skeletons are much more easily preserved than plant remains and soft body tissues. Thus, in rocks of earliest Phanerozoic time and younger, the most abundant fossils are the hard, tough shells and skeletons.

Subdivision of the Phanerozoic Eon into three eras is based on the most common types of life during each era. Sedimentary rocks formed during the **Paleozoic era** (Greek for "old life") contain fossils of early life forms, such as invertebrates, fishes, amphibians, reptiles, ferns, and cone-bearing trees (Fig. 4–16). The Paleozoic era ended abruptly about 245 million years ago when a catastrophic mass extinction wiped out 90 percent of all marine species, two thirds of reptile and amphibian species, and 30 percent of insect species. Mass extinctions are discussed in the Earth Systems Interactions essay on pages 16–20.

Sedimentary rocks of the **Mesozoic era** (Greek for "middle life") contain new types of **phytoplankton,** microscopic plants that float at or near the sea surface, and beautiful, swimming cephalopods called **ammonoids.** However, the Mesozoic era is most famous for the dinosaurs that dominated the land (Fig. 4–17). Mammals and flowering plants also evolved during this era. The Mesozoic era ended 65 million years ago with another mass extinction that wiped out the dinosaurs.

planet's origin 4.6 billion years ago to 3.8 billion years ago. Geologic time units are based largely on fossils found in rocks. However, only a few Earth rocks are known that formed during the Hadean Eon and no fossils of Hadean age are known. It may be that erosion or metamorphism destroyed traces of Hadean life, or that Hadean time preceded the evolution of life. In any case, Hadean time is not amenable to subdivision based on fossils.

Most rocks of the **Archean Eon** (Greek for "ancient") are igneous or metamorphic, although a few Archean sedimentary rocks are preserved. Some contain microscopic fossils of single-celled organisms. Although the earliest fossils date to 3.5 billion years ago, other evidence indicates that life began at least 3.8 billion years ago, at the start of the Archean Eon. Fossils are neither numerous nor well-preserved enough to permit much fine tuning of Archean time.

Large and diverse groups of fossils have been found in sedimentary rocks of the **Proterozoic Eon** (Greek for "earlier life"). The most complex not only are multicellular but have different kinds of cells arranged into tissues and organs. Some of these organisms look so much like modern jellyfish, corals, and worms that

Figure 4–17 In this reconstruction, a duckbill dinosaur (*Maiasauras*) nurtures babies in their nest 100 million years ago in Montana. *(Museum of the Rockies)*

During the **Cenozoic era** (Greek for "recent life"), mammals and grasses became abundant (Fig. 4–18). Humans have evolved and lived wholly in the Cenozoic era.

The eras of Phanerozoic time are subdivided into periods, the time unit most commonly used by geologists. Some of the periods are named after special characteristics of the rocks formed during that period. For example, the Cretaceous Period is named from the Latin word for "chalk" *(creta)* after chalk beds of this age in Africa, North America, and Europe. Other periods are named for the geographic localities where rocks of that age were first described. For example, the Jurassic Period is named for the Jura Mountains of France and Switzerland. The Cambrian Period is named for Cambria, the Roman name for Wales, where rocks of this age were first studied.

In addition to the abundance of fossils, another reason that details of Phanerozoic time are better known than those of Precambrian time is that many of the older rocks have been metamorphosed, deformed, and eroded. It is simple probability that the older a rock is, the greater the chance that metamorphism or erosion has destroyed the rock or the features that record Earth's history.

Figure 4–18 During Cenozoic time, grasses proliferated and animals adapted to the new food source. *(National Museum of Natural History, J.H. Matternes mural, with permission)*

SUMMARY

Determinations of **relative time** are based on geologic relationships among rocks and the evolution of life forms through time. The criteria for relative dating are summarized in a few simple principles: the **principle of original horizontality**, the **principle of superposition**, the **principle of cross-cutting relationships**, and the **principle of faunal succession**.

Layers of sedimentary rock are **conformable** if they were deposited without major interruptions. An **unconformity** represents a major interruption of deposition and a significant time gap between formation of successive layers of rock. In a **disconformity**, layers of sedimentary rock on either side of the unconformity are parallel. An **angular unconformity** forms when lower layers of rock are tilted and partially eroded prior to deposition of the upper beds. In a **nonconformity**, sedimentary layers lie on top of an erosion surface developed on igneous or metamorphic rocks.

Fossils are used to date rocks according to the principle of faunal succession. **Correlation** is the demonstration of equivalency of rocks that are geographically separated. **Index fossils** and **key beds** are important tools in **time correlation,** the demonstration that sedimentary rocks from different geographic localities formed at the same time. Worldwide correlation of rocks of all ages has resulted in the **geologic column,** a composite record of rocks formed throughout the history of the Earth.

Absolute time is measured by **radiometric dating,** which relies on the fact that radioactive **parent isotopes** decay to form **daughter isotopes** at a fixed, known rate as expressed by the **half-life** of the isotope. The cumulative effects of the radioactive decay process can be determined because the daughter isotopes accumulate in rocks and minerals.

The major units of the **geologic time scale** are **eons, eras, periods,** and **epochs.** The **Phanerozoic Eon** is finely and accurately subdivided because sedimentary rocks deposited at this time are often well preserved and they contain abundant well-preserved fossils. In contrast, **Precambrian** rocks and time are only coarsely subdivided because fossils are scarce and poorly preserved and the rocks are often altered.

KEY TERMS

relative age 68
absolute age 68
principle of original
   horizontality 69
principle of
   superposition 69
principle of cross-cutting
   relationships 69
conformable 71
unconformity 71
disconformity 71

angular unconformity 71
nonconformity 71
fossil 71
evolution 72
principle of faunal
   succession 72
correlation 73
time correlation 73
lithologic correlation 73
index fossil 73
key bed 73

isotope 73
radioactive 73
parent isotope 73
daughter isotope 74
half-life 75
radiometric dating 75
geologic column 77
geologic time scale 77
eon 77
era 77
period 77

epoch 77
Hadean Eon 78
Archean Eon 79
Proterozoic Eon 79
Precambrian Eon 79
Phanerozoic Eon 79
Paleozoic era 79
Mesozoic era 79
phytoplankton 79
ammonoids 79
Cenozoic era 80

FOR REVIEW

1. Describe the two ways of measuring geologic time. How do they differ?

2. Give an example of how the principle of original horizontality might be used to determine the order of events affecting a sequence of folded sedimentary rocks.

3. How does the principle of superposition allow us to determine the relative ages of a sequence of unfolded sedimentary rocks?

4. Explain the principle of cross-cutting relationships and how it can be used to determine age relationships among sedimentary rocks.

5. Explain a conformable relationship in sedimentary rocks.

6. What geologic events are recorded by an angular unconformity? A disconformity? A nonconformity? What can be inferred about the timing of each set of events?

7. Discuss the principle of faunal succession and the use of index fossils in time correlation.

8. What are the two different types of correlation of rock units? How do they differ?

9. What tools or principles are most commonly used in correlation?

10. Describe the similarities and differences between how a calendar records time and how minerals and rocks containing radioactive isotopes record time.

11. What is radioactivity?

12. What is a stable isotope? An unstable isotope?

13. What is the relationship between parent and daughter isotopes?

14. What is meant by the half-life of a radioactive isotope? How is the half-life used in radiometric age dating?

15. Why are some radioactive isotopes useful for measuring relatively young ages, whereas others are useful for measuring older ages?

16. What geologic event is actually measured by a radiometric age determination of an igneous rock or mineral?

17. Why is the Phanerozoic Eon separated into so many subdivisions in contrast to much longer Precambrian time, which has few subdivisions?

18. List major events that occurred during the Hadean, Archean, and Proterozoic Eons.

19. List four changes that occurred at the beginning of Phanerozoic time that greatly improved the fossil record.

20. Briefly discuss the three major eras of the Phanerozoic Eon.

FOR DISCUSSION

1. Suppose that you landed on the Moon and were able to travel in a vehicle that could carry you over the lunar surface to see a wide variety of rocks, but that you had no laboratory equipment to work with. What principles and tools could you use to determine relative ages of Moon rocks?

2. Imagine that one species lived between 535 and 505 million years ago and another lived between 520 and 245 million years ago. What can you say about the age of a rock that contains fossils of both species?

3. Imagine that someone handed you a sample of sedimentary rock containing abundant fossils. What could you tell about its age if you didn't use radiometric dating and you didn't know where it was collected from? What additional information could you determine if you studied the outcrop that it came from?

4. What geologic events are represented by a potassium-argon age from flakes of biotite in (a) granite, (b) biotite schist, and (c) sandstone?

5. Suppose you were using the potassium-argon method to measure the age of biotite in granite. What would be the effect on the age measurement if the biotite had been slowly leaking small amounts of argon since it crystallized?

6. On rare occasions, layered sedimentary rocks are overturned so that the oldest rocks are on the top and the youngest are buried most deeply. Discuss two ways of determining whether such an event had occurred.

7. Suppose that you measured and described a sequence of middle Paleozoic sedimentary rocks in northern Ohio and later did the same with a sequence of the same age in Wyoming. How would you correlate the two sections?

8. Devise two metaphors for the length of geologic time in addition to the metaphor used in Chapter 1. Locate some of the most important time boundaries in your analogy.

9. Discuss how sedimentary rocks could record major changes in temperature or rainfall in an area.

Internal Processes

U N I T

II

Volcanic eruptions are manifestations of plate tectonic activity and the Earth's internal processes. *(Jacques Jangaux/Tony Stone Images)*

Plate Tectonics and Earth Systems

Recall from our discussion in Chapter 1 that the Earth formed from a diffuse mass of dust and gas. The colliding particles heated as they coalesced to form the Earth. The planet continued to heat up as asteroids, comets, and other debris crashed into its surface. The decay of radioactive isotopes generated additional heat in the early Earth. Then, as bombardment from outer space slowed down and as radioactive isotopes decayed and became less abundant, the Earth began to cool.

Today, although the Earth's surface is cool, the interior remains hot, both from residual heat and from continued decay of the remaining radioactive isotopes. The temperature at the Earth's center is about 6000°C—similar to that of the Sun's surface. This heat drives the Earth's engine, causing earthquakes, volcanic eruptions, mountain building, and continual movements of the continents and ocean basins. These processes, in turn, profoundly affect the Earth's atmosphere, hydrosphere, and biosphere.

How does this great heat engine work? Recall that the mantle is the huge zone in the Earth's interior that extends from the thin crust to a depth of 2900 kilometers, where it adjoins the core. The mantle is hot as a result of its residual heat and radioactive decay. In addition, it is heated from below by the hotter core. The high temperatures render the solid rock plastic so that it flows slowly like road tar or cold honey.

The deepest mantle is hotter than the upper mantle. This deep hot rock expands, becomes buoyant, and rises toward the surface for the same reason that a hot air balloon rises in the atmosphere. The rock flows upward slowly, at a rate

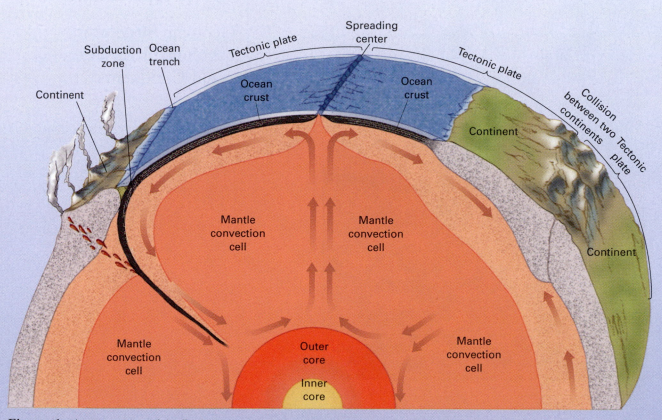

Figure 1 A cross-section of the Earth shows that the mantle convects in huge elliptical cells that rise from the top of the core, then flow across the Earth's surface, and finally sink back to the mantle-core boundary.

of a few centimeters a year, thereby transporting huge amounts of heat toward the Earth's surface (see Fig. 5–12).

As the hot rock approaches the surface, much of its heat radiates into space. Then the rock cools, loses its plasticity, and becomes hard and strong. When it reaches the surface, it has cooled sufficiently that it forms a layer of hard, strong rock about 100 kilometers thick. This layer includes both the continents and the ocean basins, as well as the uppermost part of the mantle.

Driven in part by continued rising of the mantle, this 100-kilometer-thick rock layer then glides horizontally across the Earth's surface. As the moving rock layer continues to cool, it becomes denser until it is more dense than the rock beneath it. Then, it sinks back into the mantle. But it may have migrated several thousand kilometers over the Earth's surface before it turns downward to return to the mantle (Fig. 1).

For many years geologists puzzled over how the heat engine drives the flow of mantle rock and moves the 100-kilometer-thick layer of surface rock. In the mid-1990s, studies of the Earth's interior led to the current model, just described, in which the entire mantle-surface layer system circulates in great cells extending from the core-mantle boundary to the Earth's surface.[1] The term *convection* refers to the upward and downward flow of material in response to heating. The mantle convects in huge elliptical cells that rise from the top of the core, then flow across the Earth's surface, and finally sink back to the mantle-core boundary. A single convection cell of this type may be thousands of kilometers across. The cells transport both rock and heat from the deep mantle to the Earth's surface. The rock then cools as it glides over the surface and finally returns to the lowermost mantle (Fig. 2).

[1] Richard A. Kerr, "Deep-Sinking Slabs Stir the Mantle." *Science*, Vol. 275, January 31, 1997, p. 613.

Richard Monastersky, "Global Graveyard: New Images of Earth's Interior Reveal the Fate of Old Ocean Floor." *Science News*, Vol. 152, July, 1997, p. 46.

Both continents and oceans make up the upper parts of the great cells as they glide horizontally over the Earth's surface. Today, continents are moving across the Earth's surface at rates from less than 1 to 16 centimeters per year. Manhattan Island is now 9 meters farther from London than it was 225 years ago, when the Declaration of Independence was written.

The 100-kilometer-thick portions of the cells that glide over the Earth's surface, carrying continents and ocean basins as their uppermost layers, are called **tectonic plates**, after the Greek term *tektonikos*, meaning "construction." The movements of these plates construct continents, mountain ranges, and ocean basins; tectonic movements also cause earthquakes and volcanic eruptions.

Several cells circulate simultaneously in separate parts of the mantle. Because different mantle cells flow in different directions, the tectonic plates also move in different directions as they glide over the Earth's surface. As a result, in some places on the Earth's surface, two plates are now converging, or

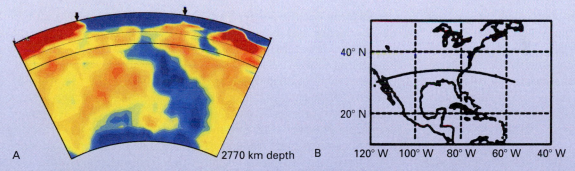

Figure 2 A seismic image of the Earth's mantle beneath North America shows a sinking tectonic plate in blue, descending to the mantle-core boundary at 2770 kilometers depth. The view is looking north, and the cross section runs through the southern United States, as shown by the line in the map at right. The blue color shows a zone of faster-than-average earthquake wave-velocities, which indicate a cool tectonic plate sinking beneath the continent. The red color shows slower-than-average velocities, indicating warmer-than-average zones in the mantle.

The sinking slab is probably the Farallon plate, a tectonic plate that sank beneath western North America over the past 100 million years. No traces of this plate remain at the Earth's surface today; the entire plate has sunk beneath the continent. *Courtesy of Rob van der Hilst (Grand, S.P., Van der Hilst, R.D., and Widiyantoro, S., High resolution global tomography: a snapshot of convection in the Earth, Geological Society of America Today, V. 7, pp. 1–7).*

moving toward one another. In other places, two plates are moving apart from one another. In yet other localities, two plates are sliding horizontally past each other.

If a plate carrying an ocean basin converges with a plate carrying a continent, the ocean basin sinks beneath the continent, beginning its long journey back to the deepest regions of the mantle. On the surface, rocks in the convergence zone buckle and rise upward, creating great mountain chains such as the Andes. The convergence of two plates thrusts rock against rock, and churns the underlying mantle, causing earthquakes and volcanic eruptions.

A new ocean basin opens where two plates separate. The opening is also accompanied by earthquakes and volcanic eruptions. The entire Atlantic Ocean has opened in this way during the last 180 million years. Prior to that time, Europe and Africa were joined to North and South America.

Where plates slide horizontally past one another, rock catches against rock and holds fast, accumulating tremendous amounts of stored elastic energy, like a tightly stretched rubber band. Eventually, too much energy accumulates and the rocks snap and spring back, just as an elastic band snaps back to its original shape when you release one end. This process generates earthquakes. Sometimes, rock buckles and crumples in the region, building small mountains. California's San Andreas fault zone is such a boundary between two plates.

Thus, the Earth's heat engine drives great convecting cells of mantle rock, which in turn drive tectonic plates across the Earth's surface. The movements of the plates create and move continents, ocean basins, and mountain ranges, and generate earthquakes and volcanic eruptions. These processes and events then profoundly affect Earth's surface systems: the atmo-

sphere and global climate, the hydrosphere, and the biosphere.

Although the convecting mantle brings a great quantity of heat to the Earth's surface, it is only $\frac{1}{5000}$ as much as the solar heat that warms the surface and atmosphere. Therefore, the heat brought to the surface by plate tectonic activity contributes little to the total amount that drives the Earth's surface systems—the atmosphere, hydrosphere, and biosphere. The Sun provides almost all of the heat to warm the atmosphere and the oceans, creating weather and climate.

However, tectonic activities strongly affect Earth surface systems in other ways. For example, ocean currents profoundly affect regional climates by carrying warm water from the equator toward the poles and cool water from polar regions toward the equator. These transfers of heat and moisture also influence regional rainfall. Similarly, winds transport heat and moisture over the globe. Both ocean currents and wind are profoundly influenced by the size, shape, and distribution of continents and ocean basins. In addition, mountain ranges deflect winds and affect the distribution of rain and snow. Tectonic movements alter ocean currents and wind by moving continents and ocean basins and by building mountains. In turn, these changes alter regional temperatures and rainfall patterns. Such changes may convert a desert to a rainforest, or a glacier-covered region to a grassland. Streams and drainage patterns must respond to the altered rainfall distribution. Lakes may dry up, or new lakes form. Plants and animals may die or migrate away to be replaced by new species that are adapted to the new climatic conditions.

In some cases, tectonic activity may move a continent from a polar region to an equatorial one. One hundred and twenty million years ago, for example, India lay near the

South Pole. Since that time, it has migrated northward to its present position in the northern hemisphere, passing the equator between 70 and 50 million years ago. During its journey, it passed from a near-polar climatic zone, through an equatorial one, to a tropical environment. The climate and the plant and animal life of India changed profoundly as the continent migrated.

Tectonic events also alter global climate by changing the composition of the Earth's atmosphere. The greatest volcanic episodes in Earth history covered hundreds of thousands of square kilometers of the Earth's surface with basalt lava. These eruptive sequences also emitted millions of tons of carbon dioxide into the atmosphere. Carbon dioxide is a greenhouse gas that causes atmospheric warming. The Earth has experienced several such volcanic episodes. One occurred about 150 million years ago in the Pacific Ocean near South America. Calculations show that the eruptions released enough carbon dioxide to raise Earth's atmospheric temperature between 5° and 10°C. These calculations are supported by evidence preserved in rocks of that age showing that the Earth's climate became at least 10°C warmer during that time.

Despite the fact that we speak of the Earth as composed of independent systems—the geosphere, the atmosphere, the hydrosphere, and the biosphere—it is clear that the Earth is one single integrated system. A change in one part ripples through the entire system, affecting ecosystems and the environment in which we live.

A few billion years in the future, the Earth's mantle will have cooled until it is no longer plastic and capable of flowing. Tectonic activity will stop. The last mountains will erode away; no new ones will form. The Sun will continue to heat the surface until it, too, runs out of fuel about 5 billion years from now.

The Active Earth: Plate Tectonics

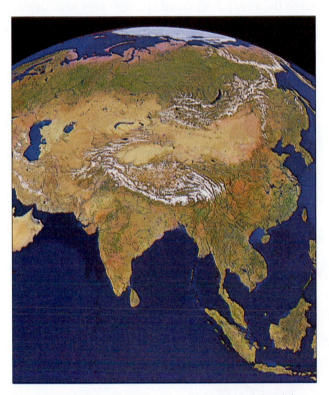

Plate tectonics processes move continents around the globe and create mountain ranges. Fifty-five million years ago, India collided with southern Asia to raise the Himalayas, the Earth's highest mountain chain. *(Tom Van Sant/Geosphere Project, Santa Monica Photo Science Library)*

About one million earthquakes shake the Earth each year; most are so weak that we do not feel them, but the strongest demolish cities and kill thousands of people. Most of us have seen televised volcanic eruptions blasting molten rock and ash into the sky, destroying villages and threatening cities. Over geologic time, mountain ranges rise, continents migrate around the globe, and ocean basins open and close.

In the early 1960s, geologists developed the **plate tectonics theory,** a single, unifying theory that explains earthquakes, volcanic eruptions, mountain building, moving continents, and many other events that affect the Earth and alter our environment. It also allows Earth scientists to identify many natural hazards before they affect humans.

Because plate tectonics theory is so important to modern Earth science, it provides a foundation for many of the following chapters of this book. We describe and explain the basic aspects of the theory in this chapter. This overview is only a beginning. In following chapters we will expand our understanding of plate tectonics and the active Earth. ●

5.1 The Earth's Layers

The energy released by an earthquake travels through the Earth as waves. Geologists have found that both the speed and direction of these waves change abruptly at certain depths as the waves pass through the Earth. Chapter 6 describes how these changes reveal the Earth's layering. Figure 5–1 and Table 5–1 describe the layers. It is necessary to understand the Earth's layers to consider the plate tectonics theory.

The Crust

The **crust** is the outermost and thinnest layer. Because it is relatively cool, the crust consists of hard, strong rock. Crust beneath the oceans differs from that of continents. Oceanic crust is between 4 and 7 kilometers thick and is composed mostly of dark, dense basalt. In contrast, the average thickness of continental crust is about 20 to 40 kilometers, although under mountain ranges it can be as much as 70 kilometers thick. Continents are composed primarily of light-colored, less dense granite.

The Mantle

The **mantle** lies directly below the crust. It is almost 2900 kilometers thick and makes up 80 percent of the Earth's volume. Although the chemical composition may be similar throughout the mantle, Earth temperature and pressure increase with depth. These changes cause the strength of mantle rock to vary with depth, and thus they create layering within the mantle. The upper mantle consists of two layers.

The Lithosphere   The uppermost mantle is relatively cool and consequently is hard, strong rock. In fact, its mechanical behavior is similar to that of the crust. The outer part of the Earth, including both the uppermost mantle and the crust, make up the **lithosphere** (Greek for "rock" layer). The lithosphere nor-

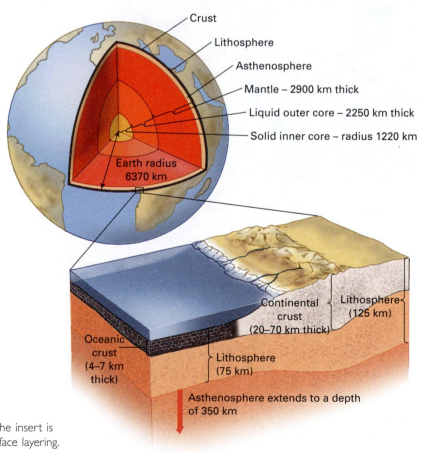

Figure 5–1 The Earth is a layered planet. The insert is drawn on an expanded scale to show near-surface layering.

TABLE 5–1

The Layers of the Earth

	Layer	Composition	Depth	Properties
Crust	Oceanic crust	Basalt	4–7 km	Cool, hard, and strong
	Continental crust	Granite	20–70 km	Cool, hard, and strong
Lithosphere	Lithosphere includes the crust and the uppermost portion of the mantle	Varies; the crust and the mantle have different compositions	75–125 km	Cool, hard, and strong
Mantle	Uppermost portion of the mantle included as part of the lithosphere			
	Asthenosphere	Entire mantle is ultramafic rock. Its mineralogy varies with depth	Extends to 350 km	Hot, weak, and plastic, 1% or 2% melted
	Remainder of upper mantle		Extends from 350 to 660 km	Hot, under great pressure, and mechanically strong
	Lower mantle		Extends from 660 to 2900 km	High pressure forms minerals different from those of the upper mantle
Core	Outer core	Iron and nickel	Extends from 2900 to 5150 km	Liquid
	Inner core	Iron and nickel	Extends from 5150 km to the center of the Earth	Solid

mally varies from about 75 kilometers thick beneath ocean basins to about 125 kilometers under the continents (Fig 5–2).

The Asthenosphere   At a depth varying from about 75 to 125 kilometers the strong, hard rock of the lithosphere gives way to the weak, plastic **asthenosphere** (Greek for "weak" layer). This change in rock properties occurs over a vertical distance of only a few kilometers, and is caused by the Earth's rising temperature with depth. In the asthenosphere, the temperature is hot enough that 1 to 2 percent of the asthenosphere is molten. In addition, because of its high temperature, the solid rock of the asthenosphere is mechanically weak and plastic. Two familiar examples of solid but plastic materials are Silly Putty™ and hot road tar. If you apply force to a plastic solid, it flows slowly. The asthenosphere extends from the base of the lithosphere to a depth of about 350 kilometers.

When building a house, you start with a strong, hard foundation. However, the Earth is not constructed in this way. The strong, hard lithosphere lies on top of the soft, weak asthenosphere. Thus, the lithosphere is not rigidly supported by the rock beneath it. Instead, the lithosphere floats on the soft plastic rock of the asthenosphere. This concept of a floating lithosphere is important to our understanding of plate tectonics and the Earth's internal processes.

Both temperature and pressure increase with depth below the Earth's surface. Increasing temperature causes the transition from the hard, strong rock of the lithosphere to the weak, plastic rock of the asthenosphere. However, at the base of the asthenosphere, increasing pressure overwhelms the effect of rising temperature, and the strength of the mantle increases again. Despite its strength, however, its high temperature makes the entire mantle plastic and capable of flowing slowly, over geologic time. Recent evidence suggests that the lowermost mantle, at the core boundary, is partly molten.[1]

The Core

The **core** is the innermost of the Earth's layers. It is a sphere with a radius of about 3470 kilometers, and

---

[1] Quentin Williams and Edward J. Ganero, "Seismic Evidence for Partial Melt at the Base of the Earth's Mantle." *Science*, Vol. 273, September 13, 1996, p. 1528.

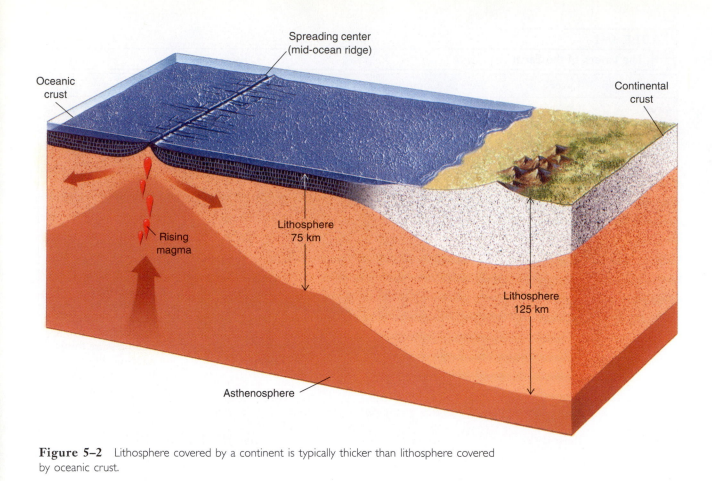

Figure 5–2 Lithosphere covered by a continent is typically thicker than lithosphere covered by oceanic crust.

is composed largely of iron and nickel. The outer core is molten because of the high temperature in that region. Near its center, the core's temperature is about 6000°C, as hot as the Sun's surface. The pressure is more than 1 million times that of the Earth's atmosphere at sea level. The extreme pressure compresses the inner core to a solid despite the fact that it is even hotter than the molten outer core.

5.2 Plates and Plate Tectonics

Like most great unifying scientific ideas, the plate tectonics theory is simple. Briefly, it states that the lithosphere is a shell of hard, strong rock about 100 kilometers thick that floats on the hot, plastic asthenosphere. The lithosphere is broken into seven large (and several smaller) segments called **tectonic plates.** They are also called "lithospheric plates" and simply "plates"—the terms are interchangeable (Fig. 5–3). The lithospheric plates glide slowly over the asthenosphere at rates ranging from less than 1 to about

16 centimeters per year, like sheets of ice drifting across a pond. Continents and ocean basins make up the upper parts of the plates. As a tectonic plate glides over the asthenosphere, the continents and oceans move with it.

A **plate boundary** is a fracture that separates one plate from another. Neighboring plates can move relative to one another in three different ways. At a **divergent boundary,** two plates move apart from each other. At a **convergent boundary,** two plates move toward each other, and at a **transform boundary,** they slide horizontally past each other. Table 5–2 summarizes characteristics and examples of each type of plate boundary.

The great forces generated at a plate boundary build mountain ranges and cause volcanic eruptions and earthquakes. These processes and events are called *tectonic* activity, from the ancient Greek word for "construction." Tectonic activity "constructs" mountains and ocean basins. In contrast to plate boundaries, the interior portion of a plate is usually tectonically quiet because it is far from the zones where two plates interact.

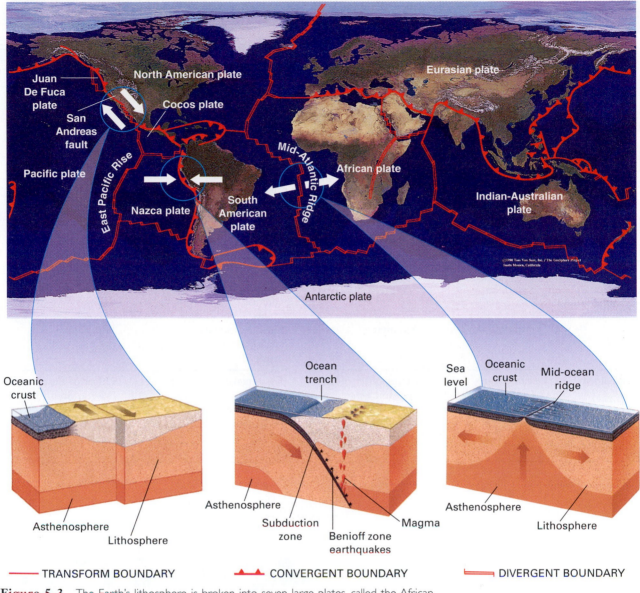

—— TRANSFORM BOUNDARY   ▲▲ CONVERGENT BOUNDARY   ⊏⊐ DIVERGENT BOUNDARY

Figure 5–3 The Earth's lithosphere is broken into seven large plates, called the African, Eurasian, Indian-Australian, Antarctic, Pacific, North American, and South American plates. A few of the smaller plates are also shown. White arrows show that the plates move in different directions. The three different types of plate boundaries are shown below the map: Two plates separate at a divergent boundary. Two plates come together at a convergent boundary. At a transform plate boundary, rocks on opposite sides of the fracture slide horizontally past each other. *(Photo by Tom Van Sant, Geosphere Project)*

Divergent Plate Boundaries

At a divergent plate boundary (also called a **spreading center** and a **rift zone**), two lithospheric plates spread apart from one another as shown in the center portion of Figure 5–4. The underlying asthenosphere then oozes upward to fill the gap between the separating plates. As the asthenosphere rises between the separating plates, some of it melts to form magma.[2]

Most of the magma rises to the Earth's surface, where it cools to form new crust. Most of this activity occurs beneath the seas because most divergent plate boundaries lie in the ocean basins.

---

[2] It seems counterintuitive that the rising, cooling asthenosphere should melt to form magma, but the melting results from decreasing pressure rather than a temperature change. This process is discussed in Chapter 7.

TABLE 5-2

Characteristics and Examples of Plate Boundaries

Type of Boundary	Types of Plates Involved	Topography	Geologic Events	Modern Examples
Divergent	Ocean-ocean	Mid-oceanic ridge	Sea-floor spreading, shallow earthquakes, rising magma, volcanoes	Mid-Atlantic ridge
	Continent-continent	Rift valley	Continents torn apart, earthquakes, rising magma, volcanoes	East African rift
Convergent	Ocean-ocean	Island arcs and ocean trenches	Subduction, deep earthquakes, rising magma, volcanoes, deformation of rocks	Western Aleutians
	Ocean-continent	Mountains and ocean trenches	Subduction, deep earthquakes, rising magma, volcanoes, deformation of rocks	Andes
	Continent-continent	Mountains	Deep earthquakes, deformation of rocks	Himalayas
Transform	Ocean-ocean	Major offset of mid-oceanic ridge axis	Earthquakes	Offset of East Pacific rise in South Pacific
	Continent-continent	Small deformed mountain ranges, deformations along fault	Earthquakes, deformation of rocks	San Andreas fault

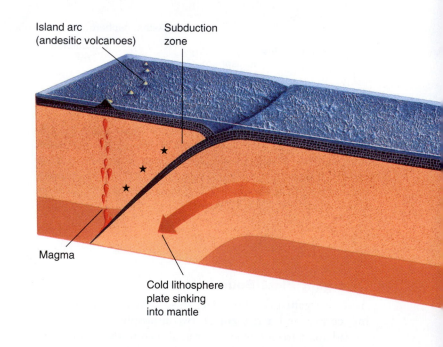

Figure 5-4 Lithospheric plates move away from a spreading center by gliding over the weak, plastic asthenosphere. In the center of the drawing, new lithosphere forms at a spreading center. At the sides of the drawing, old lithosphere sinks into the mantle at subduction zones.

Both the lower lithosphere (the part beneath the crust) and the asthenosphere are parts of the mantle and thus have similar chemical compositions. The main difference between the two layers is one of mechanical strength. The cool lithosphere is strong and hard, but the hot asthenosphere is weak and plastic. As the asthenosphere rises between two separating plates, it cools, gains mechanical strength, and, therefore, transforms into new lithosphere. In this way, new lithosphere continuously forms at a divergent boundary.

At a spreading center, the rising asthenosphere is hot, weak, and plastic. Only the upper 10 to 15 kilometers cools enough to gain the strength and hardness of lithosphere rock. As a result, the lithosphere, including the crust and the upper few kilometers of mantle rock, can be as little as 10 or 15 kilometers thick at a spreading center. But as the lithosphere spreads, it cools from the top downward. As it cools, it thickens because the boundary between the cool, strong lithosphere and the hot, weak asthenosphere migrates downward. Consequently, the thickness of the lithosphere increases as it moves away from the spreading center. Think of ice freezing on a pond. On a cold day, water under the ice freezes and the ice becomes thicker. The lithosphere continues to thicken until it attains a steady-state thickness of about 75 kilometers beneath an ocean basin, and as much as 125 kilometers beneath a continent.

The Mid-Oceanic Ridge: Rifting in the Oceans

The new lithosphere at an oceanic spreading center is hot and therefore of low density. Consequently, it floats to a high level, forming an undersea mountain chain called the **mid-oceanic ridge** (Fig. 5–5). In a few places, such as Iceland, the ridge rises above sea level. But as lithosphere migrates away from the spreading center, it cools and becomes denser; as a result, it sinks. Thus, the average depth of the sea floor away from the mid-oceanic ridge is about 5 kilometers. The mid-oceanic ridge rises 2 to 3 kilometers above the surrounding sea floor and comes within 2 kilometers of the sea surface.

If you could place bright red balls on each side of the ridge axis, and then watch them over millions of years, you would see the balls migrate away from the rift as the plates separated. The balls would also sink to greater depths as the sea floor cools and moves away from the ridge axis (Fig. 5–6).

Oceanic rifts completely encircle the Earth like the seam on a baseball. As a result, the mid-oceanic ridge system is the Earth's longest mountain chain. Basaltic magma that oozes onto the sea floor at the ridge creates 6.5×10^{18} (6,500,000,000,000,000,000) tons of new oceanic crust each year.

Splitting Continents: Rifting in Continental Crust

A divergent plate boundary can rip a continent in half in a process called **continental rifting**. A **rift valley** develops in a continental rift zone because continental crust stretches, fractures, and sinks as it is pulled apart. Continental rifting is now taking place along the East African rift (see Fig. 5–7). If the rifting continues, eastern Africa will separate from the main portion of the continent, and a new ocean basin will open between the separating portions of Africa.

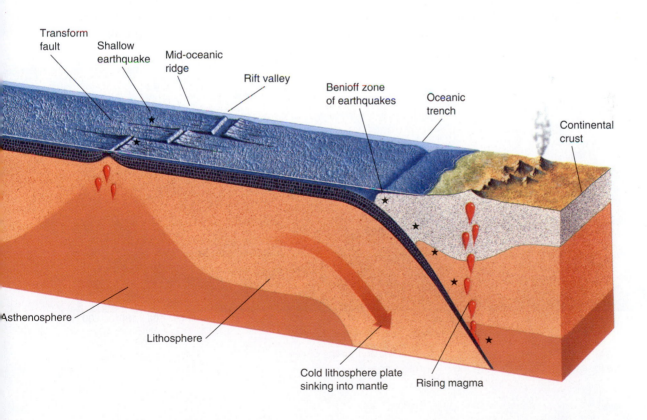

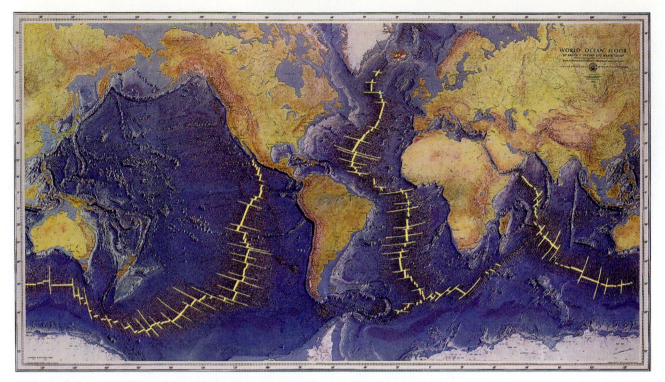

Figure 5–5 An artist's rendition of the sea floor. The mid-oceanic ridge is a submarine mountain chain that circles the globe like the seam on a baseball. *(© Marie Tharp)*

Convergent Plate Boundaries

At a convergent plate boundary, two lithospheric plates move toward each other. As two plates converge, one may sink into the mantle beneath the other. This sinking process is called **subduction,** and is shown in both the right and left sides of Figure 5–4. A **subduction** zone is a long, narrow belt where a lithospheric plate is sinking into the mantle.

Not all lithospheric plates are made of equally dense rock. Where two plates of different densities converge, the denser one sinks in the subduction zone, while the less dense plate remains floating on the asthenosphere.

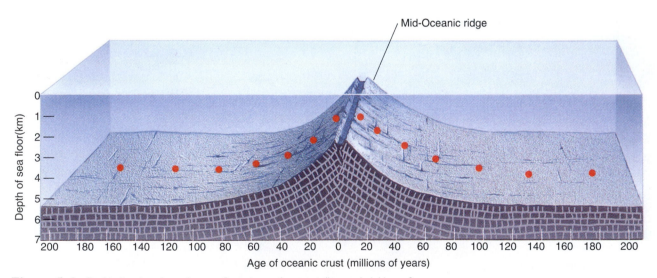

Figure 5–6 Red balls placed on the sea floor trace the spreading and sinking of new oceanic crust as it cools and migrates away from the mid-oceanic ridge.

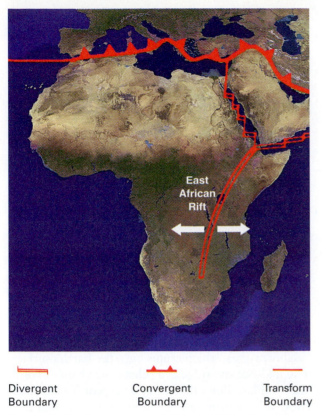

| Divergent Boundary | Convergent Boundary | Transform Boundary |

Figure 5–7 The continent of Africa is splitting apart along the East African rift. *(Tom Van Sant, Geosphere Project)* • Interactive Question: Describe the modern climatic conditions and the plant and animal life in the rift. How will they change if a new ocean basin forms in the rift during the next several million years?

Recall that oceanic crust is generally denser than continental crust. In fact, the entire lithosphere beneath the oceans is denser than continental lithosphere. Consequently, in most cases only oceanic lithosphere can sink into the mantle. A continent does not normally sink into the mantle at a subduction zone for the same reasons that a log doesn't sink in a lake: Both are of lower density than the material beneath them. In certain cases, however, small amounts of continental crust may be forced into the mantle at a subduction zone.

On a worldwide scale, the rate at which old lithosphere sinks into the mantle at subduction zones is equal to the rate at which new lithosphere forms at spreading centers. In this way, the Earth maintains a global balance between the creation of new lithosphere and the destruction of old lithosphere.

The oldest sea-floor rocks on Earth are only about 200 million years old because oceanic crust continuously recycles into the mantle at subduction zones. Rocks as old as 3.96 billion years are found on conti-

nents because subduction consumes little continental crust.

Convergence can occur (1) between a plate carrying oceanic crust and another carrying continental crust, (2) between two plates carrying oceanic crust, and (3) between two plates carrying continental crust.

Convergence of Oceanic Crust with Continental Crust When an oceanic plate converges with a continental plate, the denser oceanic plate sinks into the mantle beneath the edge of the continent. As a result, many subduction zones are located at continental margins. Today, oceanic plates are sinking beneath the western edge of South America; along the coasts of Oregon, Washington, and British Columbia; and at several other continental margins (see Fig. 5–3).

Convergence of Two Plates Carrying Oceanic Crust Recall that newly formed oceanic lithosphere is hot, thin, and of low density, but as it spreads away from the mid-oceanic ridge, it becomes older, cooler, thicker, and denser. Thus, the density of oceanic lithosphere increases with its age. When two oceanic plates converge, the denser one sinks into the mantle. Oceanic subduction zones are common in the southwestern Pacific Ocean.

Convergence of Two Plates Carrying Continents If two converging plates carry continents, neither can sink deeply into the mantle because of their low densities. In this case, the two continents collide and crumple against each other, forming a huge mountain chain. The Himalayas, the Alps, and the Appalachians all formed as results of continental collisions (Fig. 5–8).

Transform Plate Boundaries

A transform plate boundary forms where two plates slide horizontally past one another as they move in opposite directions (Fig. 5–3). California's San Andreas fault is a transform boundary between the North American plate and the Pacific plate. This type of boundary can occur in both oceans and continents.

5.3 The Anatomy of a Tectonic Plate

The nature of a tectonic plate can be summarized as follows:

1. A plate is a segment of the lithosphere; thus, it includes the uppermost mantle and all of the overlying crust.

2. A single plate can carry both oceanic crust and continental crust. The average thickness of lithosphere

Figure 5–8 A collision between India and Asia formed the Himalayas. ● Interactive Question: How does the growth of a great mountain range affect the weather and climate of nearby regions?

covered by oceanic crust is 75 kilometers, whereas that of lithosphere covered by a continent is 125 kilometers. Lithosphere may be as little as 10 to 15 kilometers thick at an oceanic spreading center.

3. A plate is composed of hard, mechanically strong rock.

4. A plate floats on the underlying hot, plastic asthenosphere and glides horizontally over it.

5. A plate behaves like a slab of ice floating on a pond. It may flex slightly, as thin ice does when a skater goes by, allowing minor vertical movements. In general, however, each plate moves as a large, intact sheet of rock.

6. A plate margin is tectonically active. Earthquakes and volcanoes are common at plate boundaries. In contrast, the interior of a lithospheric plate is normally tectonically stable.

7. Tectonic plates move at rates that vary from less than 1 to 16 centimeters per year.

5.4 How Plate Movements Affect Earth Systems

The movements of tectonic plates generate volcanic eruptions and earthquakes. They also build mountain ranges and change the global distributions of continents and oceans. Other less obvious impacts of plate tectonics on Earth systems are also important. For example, changes in the distribution of continents and oceans alter global and regional climate, which, in turn, affect the hydrosphere, atmosphere, and biosphere. The opening essay to this Unit describes many of these Earth systems interactions.

Volcanoes

Volcanic eruptions occur where hot magma rises to the Earth's surface. Volcanic eruptions are common at both divergent and convergent plate boundaries. Three different processes melt rock to form magma. The most obvious is rising temperature. However, hot rocks also melt if pressure decreases or if water is added to them. These processes are described in Chapter 7.

At a divergent boundary, hot asthenosphere oozes upward to fill the gap left between the two separating plates. Pressure decreases as the asthenosphere rises. As a result, portions of the asthenosphere melt to form basaltic magma, which erupts onto the Earth's surface. The mid-oceanic ridge is a submarine chain of volcanoes and lava flows formed at divergent plate boundaries. Volcanoes are also common in continental rifts, such as the East African rift.

At a convergent plate boundary, oceanic lithosphere dives into the asthenosphere. The sinking plate carries water-soaked mud and rock that once lay on the sea floor. As the plate descends into the mantle, it becomes hotter. The heat drives off the water, which rises into the hot asthenosphere beneath the opposite plate. The water is the major factor that causes the asthenosphere rock to melt and form magma in a subduction zone. The magma then rises through the overlying lithosphere. Some solidifies within the crust, and some erupts from volcanoes on the Earth's surface. Volcanoes of this type are common in the Cascade Range of Oregon, Washington, and British Columbia. They are also found in western South America, Japan, the Philippines, and near most other subduction zones (Fig. 5–9).

Earthquakes

Earthquakes are common at all three types of plate boundaries, but uncommon within the interior of a tectonic plate. Quakes concentrate at plate boundaries simply because those boundaries are zones where one plate slips past another. The slippage is rarely smooth and continuous. Instead, the fractures may be locked up for months to hundreds of years. Then, one plate suddenly slips a few centimeters or even a few meters past its neighbor. An earthquake is a vibration in rock caused by these abrupt movements.

Figure 5–9
Mount St. Helens most recently erupted in 1980, and is an active volcano in the Cascade Range of Washington near a convergent plate boundary. *(David Weintraub/Photo Researchers, Inc.)*

Mountain Building

Many of the world's great mountain chains, including the Andes and parts of the mountains of western North America, formed at subduction zones. Several processes combine to build a mountain chain at a subduction zone. The great volume of magma rising into the crust heats and adds material to the crust. Additional crustal thickening may occur where two plates converge for the same reason that a mound of bread dough thickens when you compress it from both sides. Finally, volcanic eruptions build chains of volcanoes. This thick crust then floats upward on the soft, plastic asthenosphere, rising to form a mountain chain.

In some cases, two continents may converge and finally collide after subduction has consumed all of the intervening oceanic crust. Massive mountain chains, including the Himalayas, the Alps, and the Appalachians have formed by such continental collisions.

Great chains of volcanic mountains form at rift zones because the new, hot lithosphere floats to a high level, and large amounts of magma form in these zones. The mid-oceanic ridge and the volcanoes of the East African rift and the Rio Grande rift of the southwestern United States are examples of such mountain chains.

Migrating Continents and Oceans

Continents migrate over the Earth's surface because they are parts of the moving lithospheric plates: They simply ride piggyback on the plates. North America is now moving away from Europe at about 2.5 centimeters per year, as the Mid-Atlantic ridge continues to separate. Thus, during the next 10 years you will ride about 25 centimeters, about one foot, on the back of the North American plate. In the time span of a human life, plate motion is slow. However, in the 65 million years since the extinction of the dinosaurs, North America has migrated about 1620 kilometers, about the distance from New York City to Minneapolis. As the Atlantic Ocean widens, the Pacific is shrinking. Thus, as continents move, ocean basins open and close over geologic time.

Effects on the Hydrosphere, Atmosphere, and Biosphere

In the examples above, we showed that movements of tectonic plates affect the solid Earth in many ways. Plate motion also affects the hydrosphere, atmosphere, and biosphere. For example, we learned that tectonic activity creates mountains, which are generally colder and wetter than nearby lowlands (Fig. 5–10). Streams run faster and cut deeper in the mountains than they do in the valleys; landslides and glaciation are more common and vegetation and wildlife are different. The distribution of continents, oceans, and mountain ranges strongly affects global wind patterns and transfers of warm and cold air masses between low and high latitudes. The positions and movement of continents also affect ocean currents and the ways in which they transfer heat around the globe.

Climate is the long-term average weather pattern, including temperature, rainfall, and humidity, in any region of the Earth. Climatic changes alter the habitats and distribution of plants and animals. Note from Figure 3 in *Focus On: Alfred Wegener and the Origin of an Idea*, that 300 million years ago, North America straddled the equator, and equatorial regions of Africa lay near the South Pole. Therefore, North America was hotter and Africa colder than today. We will learn in Chapter 14 how ocean currents affect climate. The shapes of ocean basins change as tectonic plates migrate, altering the currents and regional climates.

Tectonic events can also directly alter the atmosphere, hydrosphere, and biosphere. A large sequence of volcanic eruptions emits great quantities of carbon dioxide, a greenhouse gas that can cause global warming. One such event during middle Cretaceous time gave off enough carbon dioxide to raise the mean global atmospheric temperature by 5° to 10°C. This global warming event profoundly altered habitats and species distributions throughout the Earth.

Figure 5–10 The snowy peaks of the Andes tower over the dry Bolivian plain.

5.5 Why Plates Move: The Earth as a Heat Engine

Early in the 1960s, geologists formulated the hypothesis that the sea floor continuously spreads outward from the mid-oceanic ridge system.[3] This idea quickly developed into the theory of plate tectonics, in which the entire lithosphere—not just the sea floor—spreads outward from the mid-oceanic ridge. Since the 1960s, Earth scientists have gathered vast quantities of data that support and expand the plate tectonics theory. In the 1980s, laser surveys accurately measured changing distances between continents resulting from movements of lithospheric plates.

After geologists had developed the plate tectonics theory, they began to ask *why* do the great slabs of lithosphere glide over the Earth's surface? Recent research[4] shows that after subduction begins, a tectonic plate sinks all the way to the core-mantle boundary, to a depth of 2900 kilometers. At the same time, equal volumes of hot rock rise from the deep mantle to the surface beneath a spreading center, forming new lithosphere to replace that lost to subduction. The process, called **mantle convection,** continually stirs the entire mantle as old, cold plates sink and hot rock rises toward the Earth's surface. Recent computer models support the hypothesis that the entire mantle convects, from the core-mantle boundary to the lithosphere.[5]

A soup pot on a hot stove illustrates the process of convection. Soup at the bottom of the pot becomes warm and expands. It then rises because it is less dense than the soup at the top. When the hot soup reaches the top of the pot, it flows along the surface until it cools and sinks (Fig. 5–11).

Although the mantle is mostly solid rock, it is so hot that over geologic time it can flow slowly, at rates of a few centimeters per year. Just as the soup on the stove rests on the hot burner, the base of the mantle lies on—and is heated by—the hotter core. Thus, heat from the core, supplemented by additional heat generated by radioactivity within the mantle, drives the entire mantle and lithosphere in huge cells of convecting rock. A tectonic plate is the upper portion of a convecting cell and thus glides over the asthenosphere as a result of the convection (Fig 5–12).

---

[3] This hypothesis, called *sea-floor spreading*, is discussed in Chapter 13.

[4] Summarized by Richard A. Kerr, "Deep Sinking Slabs Stir the Mantle." *Science*, Vol. 275, Jan. 31, 1997, pp. 613–615.

[5] Hans-Peter Bunge and Mark Richards, "The Origin of Large Scale Structure in Mantle Convection: Effects of Plate Motions and Viscosity Stratification." *Geophysical Research Letters*, Vol. 23, Oct. 15, 1996, pp. 2987–2990.

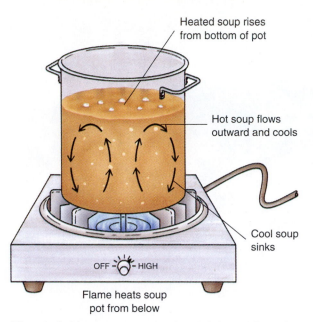

Heated soup rises
from bottom of pot

Hot soup flows
outward and cools

Cool soup
sinks

OFF — HIGH

Flame heats soup
pot from below

Figure 5–11 Soup convects when it is heated from the bottom of the pot.

The upward flow of hot rock transports great quantities of heat from the Earth's interior to its surface, where it radiates into space. If the mantle did not carry heat to the surface in this way, the Earth's interior would be much hotter than it is today.

Two associated processes shown in Figure 5–13 facilitate the movement of tectonic plates. Notice that the base of the lithosphere slopes downward from a spreading center; the grade can be as steep as 8 percent, steeper than most paved highways. Calculations show that even if the slope were less steep, gravity would cause the lithosphere to slide away from a spreading center over the soft, plastic asthenosphere at a rate of a few centimeters per year.

In addition, the lithosphere becomes more dense as it moves away from a spreading center and cools. Eventually, old lithosphere may become denser than the asthenosphere below. Consequently, it can no longer float on the asthenosphere, and sinks into the mantle in a subduction zone, dragging the trailing portion of the plate over the asthenosphere. Both of these processes may contribute to the movement of a lithospheric plate as it glides over the asthenosphere.

To summarize the processes that drive plate tectonics, hot mantle rock flows upward from the core boundary. As the upper portion of the rising mantle nears the surface, it cools to become new lithosphere at a spreading center. The new lithosphere glides over the asthenosphere away from the spreading center. At the same time, the older, leading portion of the same plate sinks deeply into the mantle, replacing the rock that is rising beneath the spreading center. Thus, the mantle convects in huge elliptical cells

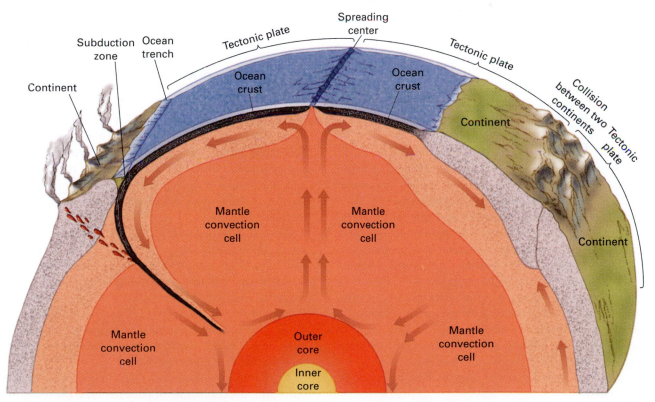

Subduction zone | Ocean trench | Tectonic plate | Spreading center | Tectonic plate | Collision between two Tectonic plates

Continent

Ocean crust | Ocean crust | Continent

Mantle convection cell | Mantle convection cell

Mantle convection cell | Outer core | Inner core | Mantle convection cell

Continent

Figure 5–12 Tectonic plates form the uppermost portion of giant convection cells that circulate through the entire mantle.

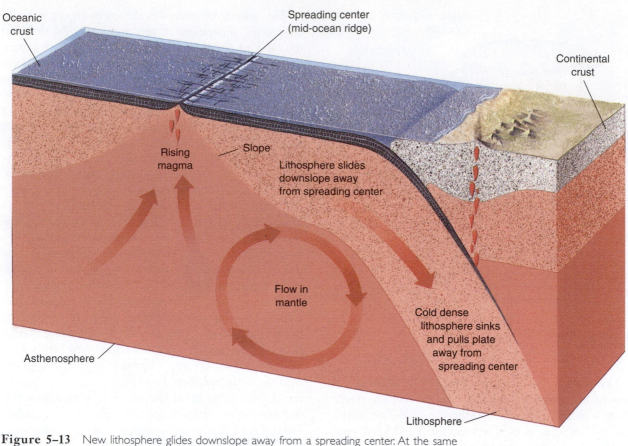

Figure 5–13 New lithosphere glides downslope away from a spreading center. At the same time, the old, cool part of the plate sinks into the mantle at a subduction zone, pulling the rest of the plate along with it. (The steepness of the slope at the base of the lithosphere is exaggerated in this figure.)

that extend from the deepest mantle to the Earth's surface. The convecting system includes the rising mantle, the lithosphere sliding over the asthenosphere, and the lithosphere sinking deeply into the mantle.

Mantle Plumes

In contrast to the huge curtain-shaped mass of mantle that rises beneath a spreading center, a **mantle plume** is a relatively small rising column of hot, plastic mantle rock. Many plumes rise from great depths in the mantle because rock near the core-mantle boundary becomes hotter and more buoyant than surrounding regions of the mantle. Others may form as a result of heating in shallower portions of the mantle. As pressure decreases in a rising plume, magma forms and rises to erupt from volcanoes at a **hot spot** on the Earth's surface. The Hawaiian Island chain is an example of a volcanic center at a hot spot. It erupts in the middle of the Pacific tectonic plate because the plume originates deep in the mantle, away from a plate boundary.

5.6 Supercontinents

Many geologists now suggest that movements of tectonic plates have periodically swept the world's continents together to form a single **supercontinent.** Each supercontinent lasted for a few hundred million years and then broke into fragments. The fragments then separated, each riding away from the others on its own tectonic plate.

Prior to 2 billion years ago, large continents as we know them today may not have existed. Instead, many—perhaps hundreds—of small masses of continental crust and island arcs similar to Japan, New Zealand, and the modern islands of the southwest Pacific Ocean dotted a global ocean basin. Then, between 2 billion and 1.8 billion years ago, tectonic plate movements swept these **microcontinents** together, forming the first supercontinent.

After this supercontinent split up about 1.3 billion years ago, the fragments of continental crust reassembled, forming a second supercontinent about 1 billion years ago. In turn, this continent fractured and the continental fragments reassembled into a third supercontinent about 300 million years ago, 70 million years before the appearance of dinosaurs. This third supercontinent is Alfred Wegener's Pangea, a term meaning "all lands," described in *Focus On: Alfred Wegener and the Origin of an Idea.*

5.7 Isostasy: Vertical Movement of the Lithosphere

If you have ever used a small boat, you may have noticed that the boat settles in the water as you get into it and rises as you step out. The lithosphere behaves in a similar manner. If a large mass is added to the lithosphere, it settles and the underlying astheno-sphere flows laterally away from that region to make space for the settling lithosphere.

But how is mass added or subtracted from the lithosphere? One process that adds and removes mass is the growth and melting of large glaciers. When a glacier grows, the weight of ice forces the lithosphere downward. For example, in the central portion of Greenland, a 3000-meter-thick ice sheet has depressed the continental crust below sea level. Conversely, when a glacier melts, the continent rises—it rebounds. Geologists have discovered Ice Age beaches in Scandinavia tens of meters above modern sea level. The beaches formed when glaciers had depressed the Scandinavian crust. They now lie well above sea level because the land rose as the ice melted. The concept that the lithosphere is in floating equilibrium on the asthenosphere is called **isostasy,** and the vertical movement in response to a changing burden is called **isostatic adjustment** (Fig. 5–14).

(text continued on page 105)

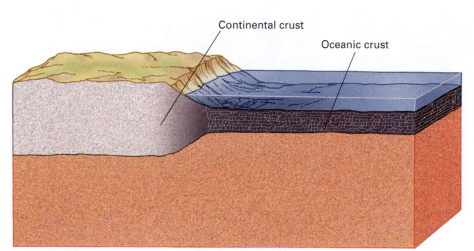

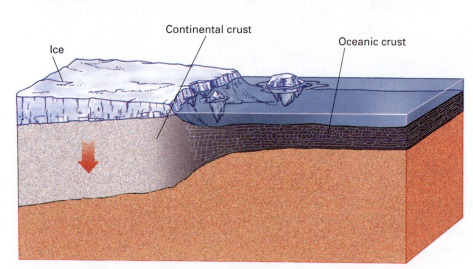

Figure 5–14 The weight of an ice sheet causes continental crust to sink isostatically. ● Interactive Question: How might the growth of a large polar continental sheet affect coastal environments and ecosystems on nearby unglaciated coastal areas, and on an unglaciated continent near the equator?

Alfred Wegener and the Origin of an Idea

In the early twentieth century, a young German scientist named Alfred Wegener noticed that the African and South American coastlines on opposite sides of the Atlantic Ocean seemed to fit as if they were adjacent pieces of a jigsaw puzzle (Fig. 1). He realized that the apparent fit suggested that the continents had once been joined together and had later separated to form the Atlantic Ocean.

Although Wegener was not the first to make this suggestion, he was the first scientist to pursue it with additional research. Studying world maps, Wegener realized that not only did the continents on both sides of the Atlantic fit together, but other continents, when moved properly, also fit like additional pieces of the same jigsaw puzzle (Fig. 2). On his map, all the continents together formed one supercontinent that he called **Pangea** from the Greek root words for "all lands." The northern part of Pangea is commonly called **Laurasia** and the southern part **Gondwanaland.**

Wegener understood that the fit of the continents alone did not prove that a supercontinent had existed. Therefore, he began seeking additional evidence in 1910 and continued work on the project until his death in 1930.

He mapped the locations of fossils of several species of animals and plants that could neither swim well nor fly. Fossils of the same species are now found in Antarctica, Africa, Australia, South America, and India. Why would the same species be found on continents separated by thousands of kilometers of ocean? When Wegener plotted the same fossil localities on his Pangea map, he found that they all lie in the same region of Pangea (Fig. 2). Wegener then suggested that each species had evolved and spread over that part of Pangea rather than mysteriously migrating across thousands of kilometers of open ocean.

Certain types of sedimentary rocks form in specific climatic zones. Glaciers and gravel deposited by glacial ice, for example, form in cold climates and are therefore found at high latitudes and high altitudes. Sandstones that preserve the structures of desert sand dunes form where deserts are common, near latitudes 30° north and south. Coral reefs and coal swamps thrive in near-equatorial tropical climates. Thus, each of these rocks reflects the latitude at which it formed.

Wegener plotted 300-million-year-old glacial deposits on a map showing the modern distribution of continents (Fig. 3A). The white area shows how large the ice mass would have been if the continents had been in their present positions. Notice that the glacier would have crossed the equator, and glacial deposits would have formed in tropical and subtropical zones. Figure 3B shows the same glacial deposits, and other climate-indicating rocks, plotted on Wegener's Pangea map. Here the glaciers cluster neatly about the South Pole. The other rocks are also found in logical locations.

Wegener also noticed several instances in which an uncommon rock type or a distinctive sequence of rocks on one side of the Atlantic Ocean was identical to rocks on the other side. When he plotted the rocks on a Pangea map, those on the east side of the Atlantic were continuous with their counterparts on the west side (Fig. 1). For example, the deformed rocks of the Cape Fold belt of South Africa are similar to rocks found in the Buenos Aires province of Argentina. Plotted on a Pangea map, the two sequences of rocks appear as a single, continuous belt.

Wegener's concept of a single supercontinent that broke apart to form the modern continents is called the theory of **continental drift.** The theory of continental drift was so revolutionary that skeptical scientists demanded an explanation of how continents could move. They wanted an explanation of the mechanism of continental drift. Wegener had concentrated on developing evidence that continents had drifted, not on what caused them to move. Finally, perhaps out of exasperation and as an afterthought to what he considered the important

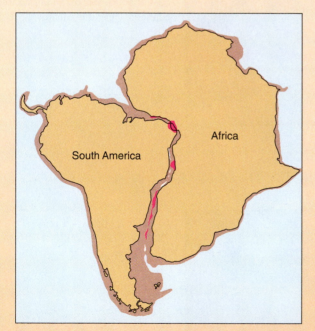

Figure 1 The African and South American coastlines appear to fit together like adjacent pieces of a jigsaw puzzle. The pink areas show locations of distinctive rock types in South America and Africa.

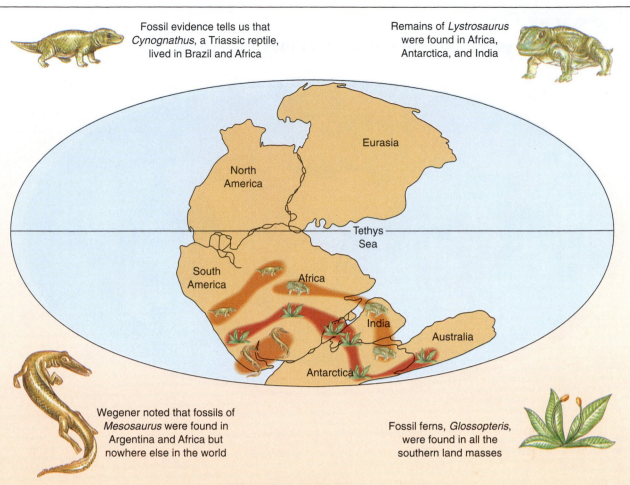

Fossil evidence tells us that *Cynognathus*, a Triassic reptile, lived in Brazil and Africa

Remains of *Lystrosaurus* were found in Africa, Antarctica, and India

Wegener noted that fossils of *Mesosaurus* were found in Argentina and Africa but nowhere else in the world

Fossil ferns, *Glossopteris*, were found in all the southern land masses

Figure 2 Geographic distributions of plant and animal fossils indicate that a single supercontinent, called Pangea, existed about 200 million years ago.

part of his theory, Wegener suggested two alternative possibilities: first, that continents plow their way through oceanic crust, shoving it aside as a ship plows through water; or second, that continental crust slides over oceanic crust. These suggestions turned out to be ill considered.

Physicists immediately proved that both of Wegener's mechanisms were impossible. Oceanic crust is too strong for continents to plow through it. The attempt would be like trying to push a matchstick boat through heavy tar. The boat, or the continent, would break apart. Furthermore, frictional resistance is too great for continents to slide over oceanic crust.

These conclusions were quickly adopted by most scientists as proof that Wegener's theory of continental drift was wrong. Notice, however, that the physicists' calculations proved only that the mechanism proposed by Wegener was incorrect. They did not disprove, or even consider, the huge mass of evidence indicating that

the continents were once joined together. During the 30-year period from about 1930 to 1960, a few geologists supported the continental drift theory, but most ignored it.

Much of the theory of continental drift is similar to plate tectonics theory. Modern evidence indicates that the continents *were* together much as Wegener had portrayed them in his map of Pangea. Today, most geologists recognize the importance of Wegener's contributions.

Focus Question:

Compare the manner in which Wegener developed the theory of continental drift with the processes of the scientific method described in *Focus On: Hypothesis, Theory, and Law* in Chapter 1. Explain why Wegener's theory was first rejected and, more recently, revived.

(Box continued)

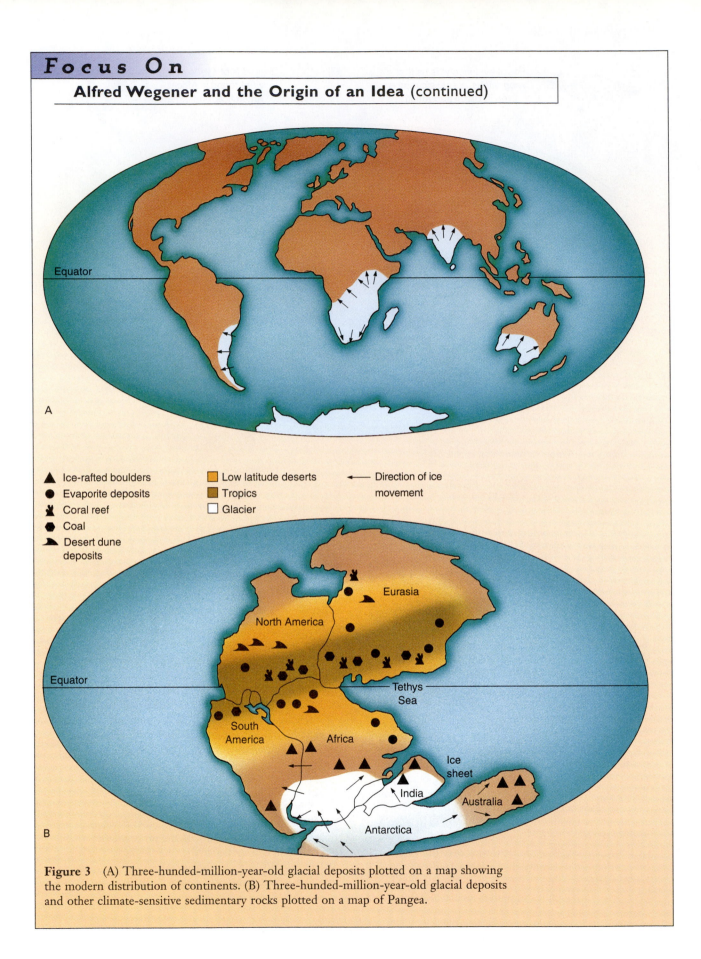

Legend:
▲ Ice-rafted boulders
● Evaporite deposits
♣ Coral reef
⬢ Coal
⬟ Desert dune deposits
▭ Low latitude deserts
▭ Tropics
□ Glacier
← Direction of ice movement

Figure 3 (A) Three-hunded-million-year-old glacial deposits plotted on a map showing the modern distribution of continents. (B) Three-hundred-million-year-old glacial deposits and other climate-sensitive sedimentary rocks plotted on a map of Pangea.

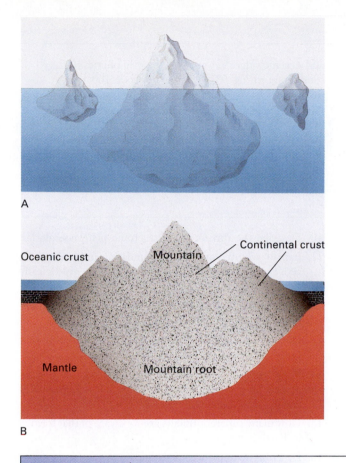

The iceberg pictured in Figure 5–15 illustrates an additional effect of isostasy. A large iceberg has a high peak, but its base extends deeply below the water. The lithosphere behaves in a similar manner. Continents rise high above sea level, and the lithosphere beneath a continent has a "root" that extends as deeply as 125 kilometers into the asthenosphere. In contrast, most ocean crust lies approximately 5 kilometers below sea level, and oceanic lithosphere extends only about 75 kilometers into the asthenosphere. For similar reasons, high mountain ranges have deeper roots than low plains, just as the bottom of a large iceberg is deeper than the base of a small one.

Figure 5–15   (A) Icebergs illustrate some of the effects of isostasy. A large iceberg has a deep root, and also a high peak. (B) In an analogous manner, continental crust extends more deeply into the mantle beneath high mountains than it does under the plains. Oceanic lithosphere is thinner and more dense, and consequently floats at the lower level.

SUMMARY

The **plate tectonics theory** provides a unifying framework for much of modern geology. The Earth is a layered planet. The **crust** is its outermost layer and varies from 4 to 70 kilometers thick. The **mantle** extends from the base of the crust to a depth of 2900 kilometers, where the core begins. The **lithosphere** is the cool, hard, strong outer 75 to 125 kilometers of the Earth; it includes all of the crust and the uppermost mantle. The hot, plastic **asthenosphere** extends to 350 kilometers in depth. The **core** is mostly iron and nickel and consists of a liquid outer layer and a solid inner sphere.

Plate tectonics theory is the concept that the lithosphere floats on the asthenosphere, and is segmented into seven major **tectonic plates,** which move relative to one another by gliding over the asthenosphere. Most of the Earth's major geological activity occurs at **plate boundaries.** Three types of plate boundaries exist: (1) New lithosphere forms and spreads outward at a **divergent boundary,** or **spreading center;** (2) two lithospheric plates move toward each other at a **convergent boundary,** which develops into a **subduction zone** if at least one plate carries oceanic crust; and (3) two plates slide horizontally past each other at a **transform plate boundary.** Volcanoes, earthquakes, and mountain building occur near plate boundaries. Interior parts of lithospheric plates are tectonically stable. Tectonic plates move horizontally at rates that vary from 1 to 16 centimeters per year. Plate movements carry continents across the globe, cause ocean basins to open and close, and affect climate and the distribution of plants and animals.

Mantle convection and movement of lithospheric plates can occur because the mantle is hot, plastic, and capable of flowing. The entire mantle, from the top of the core to the crust, convects in huge cells. Horizontally moving tectonic plates are the uppermost portions of convection cells. Convection occurs because (1) the mantle is hottest near its base, (2) new lithosphere glides downslope away from a spreading center, and (3) the cold leading edge of a plate sinks into the mantle and drags the rest of the plate along. **Supercontinents** may assemble, split apart, and reassemble every 500 million to 700 million years.

The concept that the lithosphere floats on the asthenosphere is called **isostasy.** When weight, such as a glacier, is added to or removed from the Earth's surface, the lithosphere sinks or rises. This vertical movement in response to changing burdens is called **isostatic adjustment.**

plate tectonics
  theory 87
crust 88
mantle 88
lithosphere 88
asthenosphere 89
core 89

tectonic plate 90
plate boundary 90
divergent boundary 90
convergent boundary 90
transform boundary 90
spreading center 91
rift zone 91

mid-oceanic ridge 93
continental rifting 93
rift valley 93
subduction 94
subduction zone 94
mantle convection 98

mantle plume 100
hot spot 100
supercontinent 100
microcontinent 100
isostasy 101
isostatic adjustment 101

FOR REVIEW

1. Draw a cross-sectional view of the Earth. List all the major layers and the thickness of each.

2. Describe the physical properties of each of the Earth's layers.

3. Describe and explain the important differences between the lithosphere and the asthenosphere.

4. What properties of the asthenosphere allow the lithospheric plates to glide over it?

5. Describe some important differences between the crust and the mantle.

6. Describe some important differences between oceanic crust and continental crust.

7. How is it possible for the solid rock of the mantle to flow and convect?

8. Summarize the important aspects of the plate tectonics theory.

9. How many major tectonic plates exist? List them.

10. Describe the three types of tectonic plate boundaries.

11. Explain why tectonic plate boundaries are geologically active and the interior regions of plates are geologically stable.

12. Describe some differences between the lithosphere beneath a continent and that beneath oceanic crust.

13. Describe a reasonable model for a mechanism that causes movement of tectonic plates.

14. Why would a lithospheric plate floating on the asthenosphere suddenly begin to sink into the mantle to create a new subduction zone?

15. How many supercontinents are we aware of that formed in Earth history?

16. Describe the mid-oceanic ridge.

17. Why are the oldest sea-floor rocks only about 200 million years old, whereas some continental rocks are almost 4 billion years old?

FOR DISCUSSION

1. Discuss why a unifying theory, such as the plate tectonics theory, is desirable in any field of science.

2. Central Greenland lies below sea level because the crust is depressed by the ice cap. If the glacier were to melt, would Greenland remain beneath the ocean? Why or why not?

3. At a rate of 5 centimeters per year, how long would it take for a continent to drift the width of your classroom? The distance between your apartment or dormitory and your classroom? The distance from New York to London?

4. Why do most major continental mountain chains form at convergent plate boundaries? What topographic and geologic features characterize divergent and transform plate boundaries in continental crust? Where do these types of boundaries exist in continental crust today?

5. If you were studying photographs of another planet, what features would you look for to determine whether or not the planet is or has been tectonically active?

6. The largest mountain in the Solar System is Olympus Mons, a volcano on Mars. It is 25,000 meters high, nearly three times the elevation of Mount Everest. Speculate on the factors that might permit such a large mountain on Mars.

7. The core's radius is 3470 kilometers, and that of the mantle is 2900 kilometers, yet the mantle contains 80 percent of the Earth's volume. Explain this apparent contradiction.

8. If you built a model of the Earth 1 meter in radius, how thick would the crust, lithosphere, asthenosphere, mantle, and core be?

9. Look at the map in Figure 5–3 and name a tectonic plate that is covered mostly by continental crust. Name one that is mostly ocean. Name two plates that are about half ocean and half continent.

10. Give an example of how tectonic activity might change a portion of the geosphere. Explain how that change could affect the hydrosphere, atmosphere, and biosphere.

Earthquakes and the Earth's Structure

A magnitude 7.2 earthquake destroyed much of the city of Kobe, Japan, on January 18, 1995. Despite earthquake-resistant engineering, this elevated highway toppled onto the street below. *(N. Hosaka/Gamma Liaison)*

Continents glide continuously around the globe, but we cannot feel the motion because it is too slow. Occasionally, however, the Earth trembles noticeably. The ground rises and falls and undulates back and forth, as if it were an ocean wave. Buildings topple, bridges fail, roadways and pipelines snap. An **earthquake** is a sudden motion or trembling of the Earth caused by the abrupt release of energy that is stored in rocks.

Before the plate tectonics theory was developed, geologists recognized that earthquakes occur frequently in some regions and infrequently in others, but they didn't understand why. Modern geologists know that most earthquakes occur along plate boundaries, where huge tectonic plates diverge, converge, or slip past one another. Thus, earthquakes are related to the slow movement of mantle rock that begins 2900 kilometers beneath us at the core-mantle boundary. ●

In this section we will examine how rocks store energy, and why they suddenly release that energy as an earthquake.

Stress is a force exerted against an object.[1] You stress a cable when you use it to tow a neighbor's car. Tectonic forces stress rocks. The movement of lithospheric plates is the most common source of tectonic stress.

If a rock is stressed by tectonic forces, it first undergoes **elastic deformation.** If the stress is removed, the rock springs back to its original size and shape. A rubber band exhibits elastic deformation. The energy used to stretch a rubber band is stored in the elongated rubber. When the stress is removed, the rubber band springs back and releases the stored energy.

Every rock has a limit beyond which it cannot deform elastically. Under certain conditions, when its elastic limit is exceeded, a rock continues to deform like putty. This behavior is called **plastic deformation.** A rock that has deformed plastically retains its new shape when the stress is released (Fig. 6–1). Earthquakes do not occur when rocks deform plastically.

---

[1] More precisely, stress is defined as force per unit area and is measured in units of newtons per square meter (N/m^2).

Under other conditions, an elastically deformed rock may rupture by **brittle fracture** (Fig 6–2). The fracture releases the elastic energy, and the surrounding rock springs back to its original shape. This rapid motion creates vibrations that travel through the Earth and are felt as an earthquake.

When rock fractures and moves, a large crack, called a **fault,** appears. After the earthquake dies away, the fault remains as a weakness in the rock. When more tectonic stress builds, the rock is more likely to move along the fault than to crack again to create a new fault. Thus earthquakes occur repeatedly along established faults. For example, the San Andreas fault in southern California lies along a tectonic plate boundary that has moved many times in the past and will move again in the future (Fig. 6–3).

Although tectonic plates move at rates between 1 and 16 centimeters per year, friction prevents the plates from slipping past one another continuously. As a result, rock near a plate boundary stretches or compresses elastically. When its accumulated elastic energy overcomes the friction that binds plates together, the rock suddenly slips along the fault, generating an earthquake. The rocks may move from a few centimeters to a few meters, depending on the amount of stored energy. Thus, an earthquake is a classic example of a threshold effect. Think of the two tectonic plates moving past each other at the San Andreas fault zone, traveling about as fast as your fingernail grows.

Figure 6–1 This rock deformed plastically when stressed.

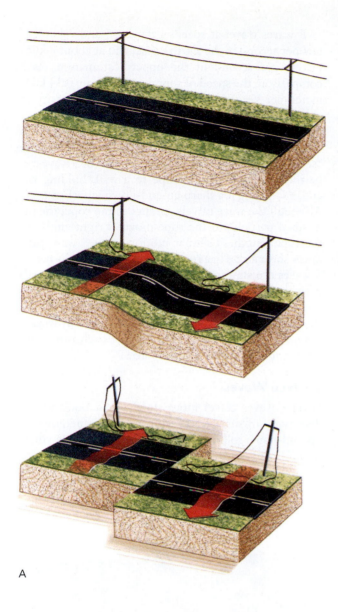

A

B

Figure 6–2 (A) A rock stores elastic energy when it is stressed by a tectonic force. When the rock fractures, it snaps back to its original shape, creating an earthquake. (B) Stressed rock fractured and displaced this roadway during the Loma Prieta earthquake near San Francisco in 1989.

For decades, or even a century or two, the cool, strong rock near the fault zone stretches or compresses but it doesn't fracture or break free from the rock on the other side of the fault. Then, when the strain reaches a critical value, rock snaps loose, sending vibrations through the Earth that may topple buildings and alter the land.

6.2 Earthquake Waves

If you have ever bought a watermelon, you know the challenge of picking out a ripe, juicy one without being able to look inside. One trick is to tap the melon gently with your knuckle. If you hear a sharp, clean sound, it is probably ripe; a dull thud indicates that it may be overripe and mushy. The watermelon illustrates two points that can be applied to the Earth: (1) The energy of your tap travels through the melon, and (2) the nature of the melon's interior affects the quality of the sound.

A wave transmits energy from one place to another. A drumbeat travels through air as a sequence of waves, the Sun's heat travels to Earth as waves, and a tap travels through a watermelon in waves. Waves that travel through rock are called **seismic waves.** Earthquakes and explosions produce seismic waves. **Seismology** is the study of earthquakes and the nature of the Earth's interior based on evidence from seismic waves.

The initial rupture point, where abrupt movement creates an earthquake, typically lies below the surface at a point called the **focus.** The point on the Earth's surface directly above the focus is the **epicenter.** An

Figure 6-3 California's San Andreas fault, the source of many earthquakes, is the boundary between the Pacific plate, on the left in this photo, and the North American plate on the right. *(R. E. Wallace/USGS)*

P waves travel at speeds between 4 and 7 kilometers per second in the Earth's crust and at about 8 kilometers per second in the uppermost mantle. As a comparison, the speed of sound in air is only 0.34 kilometer per second, and the fastest jet fighters fly at about 0.85 kilometer per second.

A second type of body wave, called an **S wave,** is a shear wave. S waves arrive after P waves and are the "secondary" waves to reach an observer. An S wave can be illustrated by tying a rope to a wall, holding the end, and giving it a sharp up-and-down jerk (Fig. 6-6). Although the wave travels parallel to the rope, the individual particles in the rope move at right angles to the rope length. A similar motion in an S wave produces shear stress in a rock and gives the wave its name. S waves are slower than P waves and travel at speeds between 3 and 4 kilometers per second in the crust.

Unlike P waves, S waves move only through solids. Because molecules in liquids and gases are only weakly bound to one another, they slip past each other and thus cannot transmit a shear wave.

Surface Waves

Surface waves travel more slowly than body waves. Two types of surface waves occur simultaneously: an up-and-down rolling motion and a side-to-side vibration. During an earthquake, the Earth's surface rolls

earthquake produces several different types of seismic waves. **Body waves** travel through the Earth's interior and carry some of the energy from the focus to the surface (Fig. 6-4). **Surface waves** then radiate from the epicenter along the Earth's surface. Although the wave mechanisms are different, surface waves undulate across the ground like the waves that ripple across the water after you throw a rock into a calm lake.

Body Waves

Two main types of body waves travel through the Earth's interior. A **P wave** is a compressional elastic wave that causes alternate compression and expansion of the rock (Fig. 6-5). P waves are called "primary" waves because they are so fast that they are the first seismic waves to reach an observer. Consider a long spring such as the popular Slinky™ toy. If you stretch a Slinky™ and strike one end, a compressional wave travels along its length. P waves travel through air, liquid, and solid material. Next time you take a bath, immerse your head until your ears are under water and listen as you tap the sides of the tub with your knuckles. You are hearing P waves.

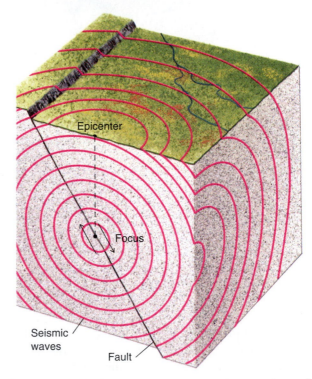

Figure 6-4 Body waves radiate outward from the focus of an earthquake.

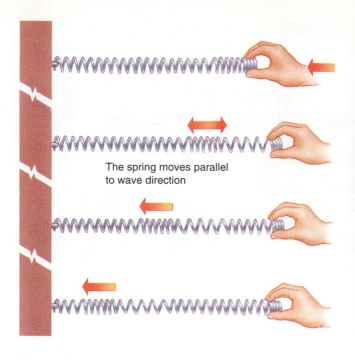

Figure 6–5 Model of a P, or compressional, wave. The spring moves parallel to the direction of wave propagation.

The spring moves parallel to wave direction

The rope moves perpendicular to wave direction

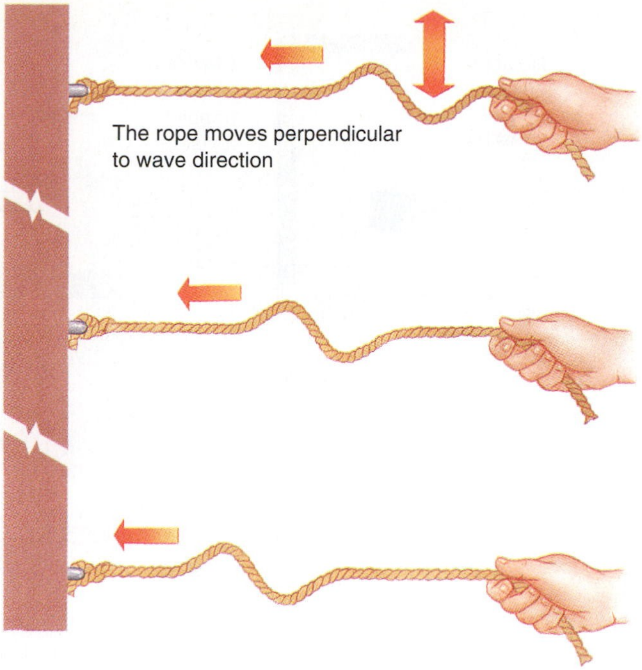

Figure 6–6 Model of an S, or shear, wave. The rope moves perpendicular to the direction of wave propagation.

like ocean waves and writhes from side to side like a snake (Fig. 6–7).

Measurement of Seismic Waves

A **seismograph** is a device that records seismic waves. To understand how a seismograph works, consider the act of writing a letter while riding in an airplane. If the plane hits turbulence, inertia keeps your hand relatively stationary as the plane moves back and forth beneath it, and your handwriting becomes erratic.

Early seismographs worked on the same principle. A heavy weight was suspended from a spring. A pen attached to the weight was aimed at the zero mark on a piece of graph paper (Fig. 6–8). The graph paper was mounted on a rotating drum that was attached firmly to bedrock. During an earthquake, the graph paper jiggled up and down, but inertia kept the weight and its pen stationary. As a result, the paper moved up and down beneath the pen. The rotating drum recorded earthquake motion over time. This record of Earth vibration is called a **seismogram** (Fig. 6–9). Modern seismographs use electronic motion detectors, which transmit the signal to a computer.

Measurement of Earthquake Strength

Over the past century, geologists have devised several scales to express the strength of an earthquake. Before seismographs were commonly available, geologists

evaluated earthquakes on the **Mercalli scale,** which was based on structural damage. On the Mercalli scale, an earthquake that destroyed many buildings was rated as more intense than one that destroyed only a few. This system did not accurately measure the energy released by a quake, however, because structural damage also depends on distance from the focus, the rock or soil beneath the structure, and the quality of construction.

In 1935 Charles Richter devised the **Richter scale** to express earthquake magnitude. Richter magnitude is calculated from the height of the largest earthquake body wave recorded on a specific type of seismograph. The Richter scale is more quantitative than earlier intensity scales, but it is not a precise measure of earth-

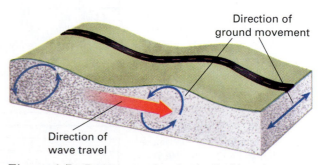

Direction of ground movement

Direction of wave travel

Figure 6–7 During an earthquake the Earth's surface moves up and down and from side to side.

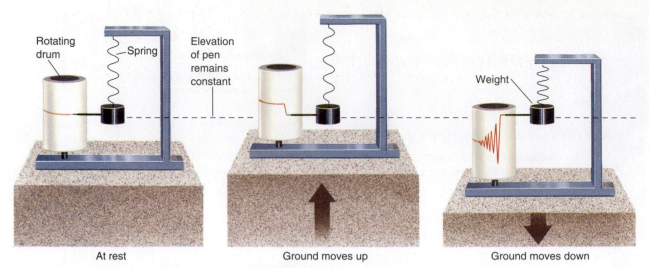

| At rest | Ground moves up | Ground moves down |

Figure 6–8 A seismograph records ground motion during an earthquake. When the ground is stationary, the pen draws a straight line across the rotating drum. When the ground rises abruptly during an earthquake, it carries the drum up with it. But the spring stretches so the weight and pen hardly move. Therefore, the pen marks a line lower on the drum. Conversely, when the ground sinks, the pen marks a line higher on the drum. During an earthquake, the pen traces a jagged line as the drum rises and falls.

quake energy. A sharp, quick jolt would register as a high peak on a Richter seismograph, but a very large earthquake can shake the ground for a long time without generating extremely high peaks. Thus, a great earthquake can release a huge amount of energy that is not reflected in the height of a single peak, and is not adequately expressed by Richter magnitude.

Modern equipment and methods enable seismologists to measure the amount of movement and the surface area of a fault that moved during a quake. The product of these two values allows them to calculate the **moment magnitude.** Most seismologists now use moment magnitude rather than Richter magnitude because it more closely reflects the total amount of

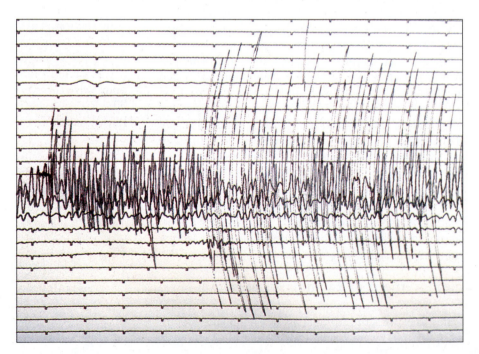

Figure 6–9 This seismogram recorded north-south ground movements during the October 1989 Loma Prieta earthquake. *(Russell D. Curtis/USGS)*

energy released during an earthquake. An earthquake with a moment magnitude of 6.5 has an energy of about 10^{25} (1 followed by 25 zeros) ergs.[2] The atomic bomb dropped on the Japanese city of Hiroshima at the end of World War II released about that much energy.

On both the moment magnitude and Richter scales the energy of the quake increases by a factor of about 30 for each increment on the scale. Thus, a magnitude 6 earthquake releases roughly 30 times more energy than a magnitude 5 earthquake.

The largest possible earthquake is determined by the strength of rocks. A strong rock can store more elastic energy before it fractures than a weak rock. The largest earthquakes ever measured had moment magnitudes of 8.5 to 8.7, about 900 times greater than the energy released by the Hiroshima bomb.

Locating the Source of an Earthquake

If you have ever watched an electrical storm, you may have used a simple technique for estimating the dis-

tance between you and the place where the lightning strikes. After the flash of a lightning bolt, count the seconds that pass before you hear the thunder. Although the electrical discharge produces thunder and lightning simultaneously, light travels much faster than sound. Therefore, light reaches you virtually instantaneously, whereas sound travels much more slowly, at 340 meters per second. If the time interval between the flash and the thunder is 1 second, then the lightning struck 340 meters away and was very close.

The same principle is used to determine the distance from a recording station to both the epicenter and the focus of an earthquake. Recall that P waves travel faster than S waves and that surface waves are slower yet. If a seismograph is located close to an earthquake epicenter, the different waves will arrive in rapid succession for the same reason that the thunder and lightning come close together when a storm is close. On the other hand, if a seismograph is located far from the epicenter, the S waves arrive at correspondingly later times after the P waves arrive, and the surface waves are even farther behind, as shown in Figure 6–10.

Geologists use a **time-travel curve** to calculate the distance between an earthquake epicenter and a seismograph. To make a time-travel curve, a number of seismic stations at different locations record the time

---

[2] An erg is a standard unit of energy in scientific usage. One erg is a small amount of energy; approximately 3×10^{12} ergs are needed to light a 100-watt light bulb for 1 hour. However, 10^{21} is a very large number and 10^{21} ergs represents a considerable amount of energy.

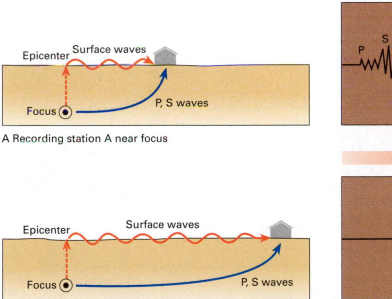

A Recording station A near focus

B Recording station B far from focus

Time of earthquake

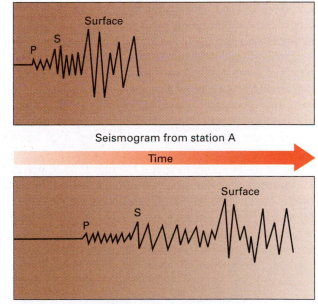

Seismogram from station A

Time

Seismogram from station B

Figure 6–10 The time intervals between arrivals of P, S, and surface waves at a recording station increase with distance from the focus of an earthquake.

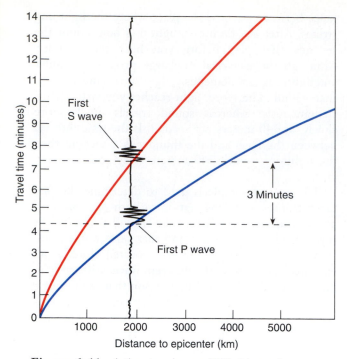

Figure 6–11 A time-travel curve. With this graph you can calculate the distance from a seismic station to the epicenter of an earthquake. In the example shown, a three-minute delay between the first arrivals of P waves and S waves corresponds to an earthquake with an epicenter 1900 kilometers from the seismic station.

of arrival of seismic waves from an earthquake with a known epicenter and occurrence time. Using these data a graph such as Figure 6–11 is drawn. This graph can then be used to measure the distance between a recording station and an earthquake whose epicenter is unknown.

Time-travel curves were first constructed from data obtained from natural earthquakes. However, scientists do not always know precisely when and where an earthquake occurred. In the 1950s and 1960s, geologists studied seismic waves from atomic bomb tests to improve the time-travel curves because they knew both the location and timing of the explosions.

Figure 6–11 shows us that if the first P wave arrives 3 minutes before the first S wave, the recording station is about 1900 kilometers from the epicenter. But this distance does not indicate whether the earthquake originated to the north, south, east, or west. To pinpoint the location of an earthquake, geologists compare data from three or more recording stations. If a seismic station in New York City records an earthquake with an epicenter 6750 kilometers away, geologists know that the epicenter lies somewhere on a circle 6750 kilometers from New York City (Fig. 6–12). The same epicenter is reported to be 2750 kilometers from a seismic station in London and 1700

kilometers from one in Godthab, Greenland. If one circle is drawn for each recording station, the arcs intersect at the epicenter of the quake.

6.3 Earthquakes and Tectonic Plate Boundaries

Although many faults are located within tectonic plates, the largest and most active faults are the boundaries between tectonic plates. Therefore, as Figure 6–13 shows, earthquakes occur most frequently along plate boundaries.

Earthquakes at a Transform Plate Boundary: The San Andreas Fault Zone

The populous region from San Francisco to San Diego straddles the San Andreas fault zone, which is a transform boundary between the Pacific plate and the North American plate (Fig. 6–14). The fault itself is vertical and the rocks on opposite sides move horizontally. A fault of this type is called a **strike-slip fault** (Fig. 6–15). Plate motion stresses rock adjacent to the fault, generating numerous smaller faults, shown by the solid red lines in Figure 6–14. The San Andreas fault and its associated faults form a broad region called the **San Andreas fault zone**.

In the past few centuries, hundreds of thousands of earthquakes have occurred in this zone. Geologists of the United States Geological Survey recorded 10,000

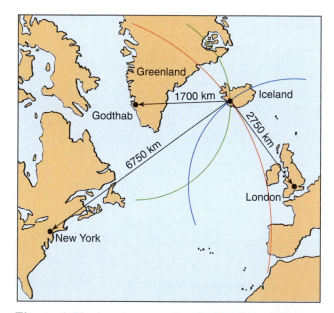

Figure 6–12 Locating an earthquake. The distance from each of three seismic stations to the earthquake is determined from time-travel curves. The three arcs are drawn. They intersect at only one point, which is the epicenter of the earthquake.

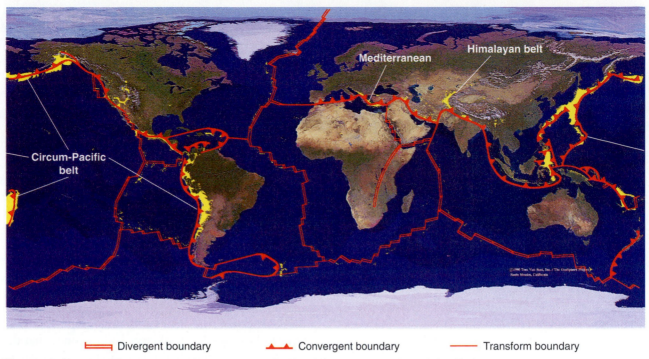

E=1 Divergent boundary ▲▲ Convergent boundary ── Transform boundary

Figure 6–13 The Earth's major earthquake zones coincide with tectonic plate boundaries. Each yellow dot represents a moderate or large earthquake that occurred between 1985 and 1995. *(USGS data overlain on a map by Tom Van Sant, Geosphere Project)* ● **Interactive Question:** What factors, besides earthquake frequency, would you use to locate a tectonic plate boundary?

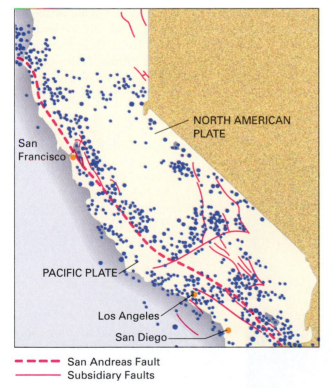

- - - San Andreas Fault
──── Subsidiary Faults

Figure 6–14 Faults and earthquakes in California. The heavy dashed line is the San Andreas fault zone, and lighter red lines are related faults. Blue dots are epicenters of recent earthquakes. *(Redrawn from USGS data)*

earthquakes in 1984 alone, although most were so weak that they could be detected only with seismographs. Severe quakes occur periodically. One shook Los Angeles in 1857, and another destroyed San Francisco in 1906 (Fig 6–16). A large quake in 1989 occurred south of San Francisco, and another rocked Northridge, part of metropolitan Los Angeles, in January 1994. The fact that the San Andreas fault zone is part of a major plate boundary tells us that more earthquakes are inevitable.

The plates move past one another in three different ways along different segments of the San Andreas fault zone:

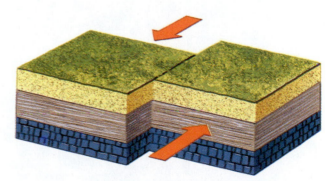

Figure 6–15 A strike-slip fault is vertical, and the rock on opposite sides of the fracture moves horizontally.

Figure 6–16 The 1906 earthquake and fire destroyed most of San Francisco. *(USGS)*

1. Along some portions of the fault zone, rocks slip past one another at a continuous, snail-like pace called **fault creep.** The movement occurs without violent and destructive earthquakes because the rocks move continuously and slowly.

2. In other segments of the fault, the plates pass one another in a series of small hops, causing numerous small, nondamaging earthquakes.

3. Along the remaining portions of the fault, friction prevents slippage of the fault although the plates continue to move past one another. In this case,

rock near the fault deforms and stores elastic energy. Because the plates move past one another at an average of 3.5 centimeters per year, 3.5 meters of elastic deformation accumulate over a period of 100 years. When the accumulated elastic energy exceeds friction, the rock suddenly slips along the fault and snaps back to its original shape, producing a large, destructive earthquake.

CASE STUDY
The Northridge Earthquake of January 1994

In January 1994, a moment magnitude 6.6 earthquake struck Northridge in the San Fernando Valley just north of Los Angeles (Fig. 6–17). Fifty-five people died and property damage was estimated at $8 billion.

As explained earlier, the San Andreas fault is a strike-slip fault that is part of a transform plate boundary. The San Andreas fault lies east of Los Angeles, cutting through outlying suburbs and desert. Because the fault doesn't lie beneath densely populated portions of Los Angeles, geologists had reasoned that the city might be spared the direct impact of a large earthquake. However, in the mid 1980s, geologists discovered buried thrust faults in southern California. A **thrust fault** is a low-angle fault in which rock on one side of the fault slides up and over the rock on the other side. Note that the Santa Monica thrust fault lies west of the main San Andreas fault and directly under Los Angeles (Fig. 6–18). As a result, a major quake on the Santa Monica fault can be more disastrous than one on the main San Andreas fault. The Northridge earthquake occurred when a thrust fault slipped. Immediately after the 1994 quake, geologist James Dolan of the California Institute of Technology ex-

Figure 6–17 On January 17, 1994, a magnitude 6.6 earthquake struck Northridge in the San Fernando Valley just north of Los Angeles, killing 55 people and causing 8 billion dollars in property damage. This building stood on Olympic Boulevard. *(Chromosohm/Sohm)*

Figure 6–18 The focus of the 1994 Northridge quake was on a previously undetected thrust fault west of the San Andreas fault, viewed here from a northerly direction. The same thrust fault also underlies Los Angeles.

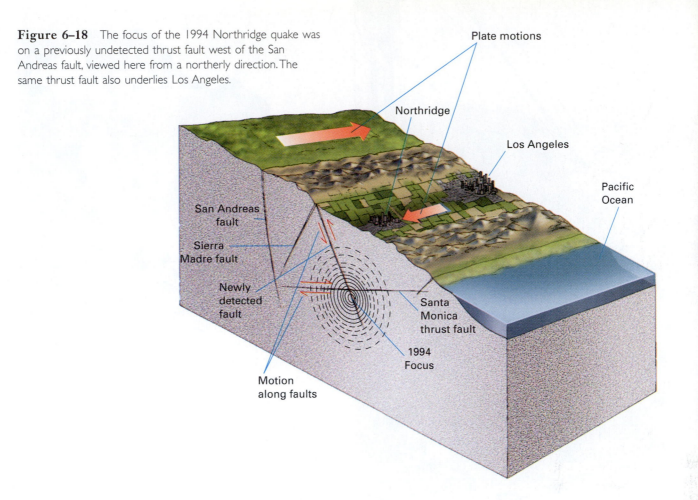

Plate motions

Northridge

Los Angeles

Pacific Ocean

San Andreas fault

Sierra Madre fault

Newly detected fault

Santa Monica thrust fault

1994 Focus

Motion along faults

plained, "There's a whole seismic hazard from buried thrust faults that we didn't even appreciate until six years ago."

The existence of thrust faults indicates that in southern California, both the direction of tectonic forces and the manner in which stress is relieved are more complicated than a simple model expresses. Although many disastrous and expensive earthquakes have shaken southern California in the past few decades, none of them has been "The Big One" that seismologists still fear. ●

Earthquakes at Subduction Zones

In a subduction zone, a relatively cold, rigid lithospheric plate dives beneath another plate and slowly sinks into the mantle. In most places, the subducting plate slips past the plate above it with intermittent jerks, giving rise to numerous earthquakes. The earthquakes concentrate along the upper part of the sinking plate, where it scrapes past the opposing plate (Fig. 6–19). This earthquake zone is called the **Benioff zone** after the geologist who first recognized it. Many

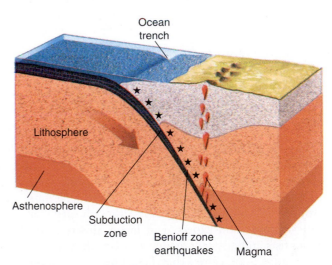

Ocean trench

Lithosphere

Asthenosphere

Subduction zone

Benioff zone earthquakes

Magma

Figure 6–19 A descending lithospheric plate generates magma and earthquakes in a subduction zone. Earthquakes concentrate along the upper portion of the subducting plate, called the Benioff zone. ● **Interactive Question:** Using this illustration show how the position of earthquake epicenters change with the depth of the earthquake focus in a subduction zone.

Figure 6–20 An overhead commuter train came to rest against a heavily damaged rail line following the January 1995 earthquake in Kobe, Japan. *(Mitsuhiko Sato/AP)*

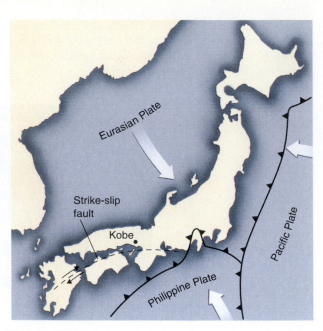

Figure 6–21 Faults and plate motion near Japan, showing the subduction of the Philippine plate beneath the Eurasian plate and the adjacent strike-slip fault that passes close to Kobe. *(Redrawn from USGS data)*

of the world's strongest earthquakes occur in subduction zones.

CASE STUDY
The Kobe Earthquake of 1995

One year after the Northridge quake, a moment magnitude 7 earthquake struck just southwest of Kobe, Japan. Nearly 5000 people died, 25,000 were injured, 300,000 were left homeless, and damage estimates topped $80 billion (Fig. 6-20). Although earthquakes are common and expected in Japan, this one caught scientists and city emergency services off guard.

Figure 6-21 shows that Japan lies in a region of multiple subduction zones and related faults. These faults have generated 23 earthquakes with magnitudes greater than 6.5 since 1900. The 1995 Kobe quake occurred 200 kilometers from the plate boundary, along a secondary strike-slip fault. Recall that the Northridge quake near Los Angeles oc-

curred on a subsidiary thrust fault under the city. The lesson from these two disasters is unsettling. In many earthquake-prone regions, major cities do not lie directly over plate boundaries. But secondary faults crisscross under the cities and, as demonstrated in the past decade, they can be as deadly as the more famous major faults. ●

CASE STUDY
Earthquake Activity in the Pacific Northwest

The small Juan de Fuca plate, which lies off the coasts of Oregon, Washington, and southern British Columbia, is diving beneath North America at a rate of 3 to 4 centimeters per year. Thus, the region should experience subduction zone earthquakes. Yet although small earthquakes occasionally shake the Pacific Northwest, no large ones have occurred in the past 150 to 200 years.

Geologists have offered two hypotheses to explain why earthquakes are relatively uncommon in the Pacific Northwest. Subduction may be occurring slowly and continuously by fault creep. If this is the case, elastic energy would not accumulate in nearby rocks, and strong earthquakes would be unlikely. Alternatively, rocks along the fault may be locked together by friction, accumulating a huge amount of elastic energy that will be released in a giant, destructive quake sometime in the future.

Recently geologists have discovered evidence of great prehistoric earthquakes in the Pacific Northwest. A major coastal earthquake commonly creates violent sea waves, which deposit a layer of sand along the coast. Geologists have found several such sand layers, each burying a layer of peat and mud that accumulated in coastal swamps during the quiet intervals between earthquakes. In addition, they have found submarine landslide deposits lying on the deep sea floor near the coast. These deposits formed when earthquakes triggered submarine landslides that carried sand and mud from the coast to the sea floor. Geologists estimate that 13 major earthquakes, separated by 300 to 900 years, struck the coast during the past 7700 years. There is also evidence for one major historic earthquake. Oral accounts of the native inhabitants chronicle the loss of a small village in British Columbia and a significant amount of ground shaking in northern California. Thus, many geologists anticipate another major, destructive earthquake in the Pacific Northwest during the next 600 years. ●

Earthquakes at Divergent Plate Boundaries

Earthquakes frequently shake the mid-oceanic ridge system as a result of faults that form as the two plates separate. Blocks of oceanic crust drop downward along most mid-oceanic ridges, forming a rift valley in the center of the ridge. Only shallow earthquakes occur along the mid-oceanic ridge because here the asthenosphere rises to levels as shallow as 10 to 15 kilometers below the Earth's surface and is too hot and plastic to fracture.

Earthquakes in Plate Interiors

No major earthquakes have occurred in the central or eastern United States in the past 100 years, and no lithospheric plate boundaries are known in these regions. Therefore, one might infer that earthquake danger there is insignificant. However, the largest historical earthquake sequence in the contiguous 48 states occurred near New Madrid, Missouri. In 1811 and 1812, three shocks with estimated moment magnitudes between 7.3 and 7.8 altered the course of the Mississippi River and rang church bells 1500 kilometers away in Washington, D.C.

Earthquakes in plate interiors are not as well understood as those at plate boundaries, but modern research is revealing some clues. Tectonic forces stretched North America in Precambrian time. As the continent pulled apart, rock fractured to create two huge fault zones that crisscross the continent like a giant X. Although the fault zones failed to develop into a divergent plate boundary, they remain weaknesses in the lithosphere. New Madrid lies at a major intersection of the faults—the center of the X. As the North American plate glides over the asthenosphere, it may pass over irregularities, or "bumps," in that plastic zone, causing slippage and earthquakes along the deep faults. In addition to the quakes recorded in New Madrid in 1811 and 1812, major earthquakes occurred close to the years 1600, 1300, and 900. The zone probably remains active and more earthquakes could occur in the future.

6.4 Earthquake Prediction

Long-Term Prediction

Long-term earthquake prediction recognizes that earthquakes have recurred many times along existing faults and will probably occur in these same regions again. Although some faults exist in plate interiors, the most active faults lie at plate boundaries; hence, earthquakes occur most frequently at plate boundaries (Fig. 6–22).

Long-term prediction tells us *where* earthquakes are likely to occur. However, it gives only a vague idea of *when* the next one will strike. Near New Madrid, earthquakes have occurred separated by 400, 300, and 200 years. The last one struck in 1811. Based on an average 300-year interval, one might expect a quake in the year 2100, but such an analysis could be in error by hundreds of years.

Short-Term Prediction

Short-term predictions are forecasts that an earthquake may occur at a specific place and time. Short-term prediction depends on signals that immediately precede an earthquake.

Foreshocks are small earthquakes that precede a large quake by a few seconds to a few weeks. The cause of foreshocks can be explained by a simple analogy. If you try to break a stick by bending it slowly, you may hear a few small cracking sounds just before the final snap. If foreshocks consistently preceded major earthquakes, they would be a reliable tool for short-term prediction. However, foreshocks preceded only about half of a group of recent major earthquakes selected for study. In addition, some swarms of small shocks thought to be foreshocks were not followed by a large quake.

Another approach to short-term earthquake prediction is to measure changes in the land surface near an active fault zone. Seismologists monitor unusual Earth movements with tiltmeters and laser surveying instruments because distortions of the crust may precede a major earthquake. This method has successfully

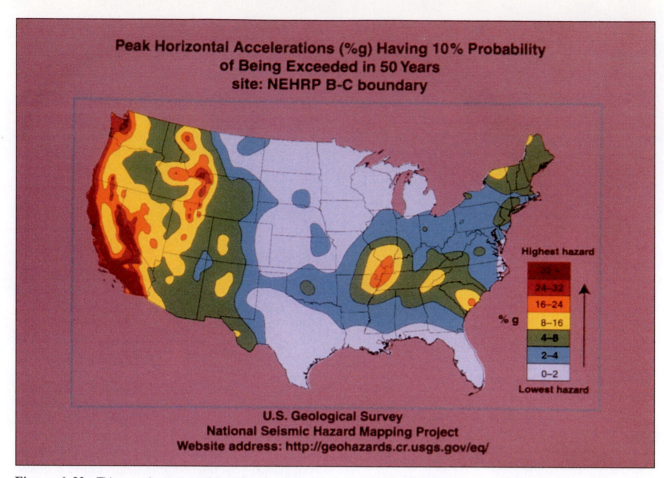

Figure 6–22 This map shows potential earthquake damage in the United States. The predictions are based on records of frequency and magnitude of historical earthquakes. *(USGS)*

predicted some earthquakes, but in other instances predicted quakes did not occur or quakes occurred that had not been predicted.

Other types of signals can be used in short-term prediction. When rock is deformed prior to an earthquake, microscopic cracks may develop as the rock approaches its rupture point. In some cases, the cracks release radon gas previously trapped in rocks and minerals. The cracks may fill with water and cause the water levels in wells to fluctuate. Air-filled cracks do not conduct electricity as well as solid rock, so the electrical conductivity of rock decreases as microscopic cracks form.

In January 1975, Chinese geophysicists recorded swarms of foreshocks and unusual bulges near the city of Haicheng, which had a previous history of earthquakes. When the foreshocks became intense on February 1, authorities evacuated portions of the city. The evacuation was completed on the morning of February 4, and in the early evening of the same day, an earthquake destroyed houses, apartments, and factories but caused few deaths.

After that success, geologists hoped that a new era of earthquake prediction had begun. But a year later, Chinese scientists failed to predict an earthquake in the adjacent city of Tangshan. This major quake was not preceded by foreshocks, so no warning was given, and at least 250,000 people died.[3] Over the past few decades, short-term prediction has not been reliable. Although some geologists continue to search for reliable indicators for short-term earthquake prediction, many other geologists believe that this goal will remain elusive.

6.5 Earthquake Damage and Hazard Mitigation

A large earthquake can displace rock and alter the Earth's surface. As previously mentioned, the New

---

[3] Accurate reports of the death toll are unavailable. Published estimates range from 250,000 to 650,000.

Figure 6–23 The 1985 Mexico City earthquake had a moment magnitude of 8.1 and killed between 8000 and 10,000 people. Earthquake waves amplified within the soil to destroy many buildings in the city. *(David Tennenbaum, AP/Wide World, Inc.)*

Madrid, Missouri, earthquake of 1811 changed the course of the Mississippi River. During the 1964 Alaskan earthquake, some beaches rose 12 meters, leaving harbors high and dry, while other beaches sank 2 meters, causing coastal flooding. Many mountains rise in increments during earthquakes that repeat over and over in the same region over tens of millions of years.

Most earthquake fatalities and injuries occur when falling structures crush people. Structural damage, injury, and death depend on the magnitude of the quake, its proximity to population centers, rock and soil types, topography, and the quality of construction in the region.

How Rock and Soil Influence Earthquake Damage
In many regions, bedrock lies at or near the Earth's surface and buildings are anchored directly to the rock. Bedrock vibrates during an earthquake and buildings may fail if the motion is violent enough. However, most bedrock returns to its original shape when the earthquake is over, so if structures can withstand the shaking, they will survive. Thus, bedrock forms a desirable foundation in earthquake hazard areas.

In many places, structures are built on sand, silt, or clay. Sandy sediment and soil commonly settle during an earthquake. This displacement tilts buildings, breaks pipelines and roadways, and fractures dams. To avert structural failure in such soils, engineers drive steel or concrete pilings through the sand to the bedrock below. These pilings anchor and support the structures even if the ground beneath them settles.

Mexico City provides one example of what can happen to clay-rich soils during an earthquake. The city is built on a high plateau ringed by even higher mountains. When the Spaniards invaded central Mexico, lakes dotted the plateau and the Aztec capital lay on an island at the end of a long causeway in one of the lakes. Over the following centuries, European settlers drained the lake and built the modern city on the water-soaked, clay-rich lake-bed sediment. On September 19, 1985, an earthquake with a moment magnitude of 8.1 struck about 500 kilometers west of the city. Seismic waves shook the wet clay beneath the city and reflected back and forth between the bedrock sides and bottom of the basin, just as waves in a bowl of Jello bounce off the side and bottom of the bowl. The reflections amplified the waves, which destroyed more than 500 buildings and killed between 8000 and 10,000 people (Fig. 6–23). Meanwhile, there was comparatively little damage in Acapulco, which was much closer to the epicenter but is built on bedrock.

If soil is saturated with water, the sudden shock of an earthquake can cause the grains to shift closer together, expelling some of the water. When this occurs, increased stress is transferred to the pore water, and the pore pressure may rise sufficiently to suspend the grains in the water. In this case, the soil loses its shear strength and behaves as a fluid. This process is called **liquefaction.** When soils liquify on a hillside, the slurry flows downslope, carrying structures along with it. During the 1964 earthquake near Anchorage, Alaska, a clay-rich bluff 2.8 kilometers long, 300 meters wide, and 22 meters high liquefied. The slurry carried houses into the ocean and buried some so deeply that bodies were never recovered.

Construction Design and Earthquake Damage

A moment magnitude 6.4 earthquake struck central India in 1993, killing 30,000 people. In contrast, the 1994 moment magnitude 6.6 quake in Northridge (near Los Angeles) killed only 55. The tremendous mortality in India occurred because buildings were not engineered to withstand earthquakes.

Some common framing materials used in buildings, such as wood and steel, bend and sway during an earthquake but resist failure. However, brick, stone, concrete, adobe (dried mud), and other masonry products are brittle and likely to fail during an earthquake. Although masonry can be reinforced with steel, in many regions of the world people cannot afford such structural reinforcement.

Given enough money, engineers can build earthquake-resistant structures. In one approach, a build-ing is set on giant rollers. When the ground shakes, the building rolls back and forth and is less likely to fail than if it were anchored solidly (Fig 6–24).

Fire   Earthquakes commonly rupture buried gas pipes and electrical wires, leading to fire, explosions, and electrocutions (Fig. 6–25). Water pipes may also break, hampering firefighters. Most of the damage from the 1906 San Francisco earthquake resulted from fires.

Again, design features can reduce fire risk. A straight pipe is likely to rupture if the ground stretches and compresses, whereas a zig-zag pipe may flex like an accordion and remain intact.

Landslides   Landslides are common when the Earth trembles. Earthquake-related landslides are discussed in more detail in Chapter 9.

Tsunamis   When an earthquake occurs beneath the sea, part of the sea floor rises or falls. Water is displaced in response to the rock movement, forming a wave. Sea waves produced by an earthquake are often called tidal waves, but they have nothing to do with tides. Therefore, geologists call them by their Japanese name, **tsunami.**

In the open sea, a tsunami is so flat that it is barely detectable. Typically, the crest may be only 1 to 3 meters high, and successive crests may be more than 100 to 150 kilometers apart. However, a tsunami may travel at 750 kilometers per hour. When the wave approaches the shallow water near shore, the base of the

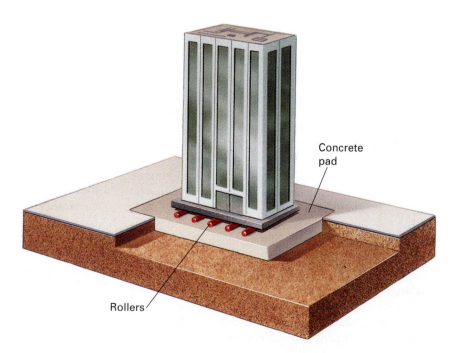

Concrete pad

Rollers

Figure 6–24 Buildings on rollers suffer less damage during an earthquake than those anchored to bedrock or sediment.

Figure 6–25 Ruptured gas and electric lines often cause fires during earthquakes in urban areas. This blaze followed the 1989 San Francisco earthquake. *(Michael Williamson/Sygma)*

wave drags against the bottom and the water stacks up, increasing the height of the wave. The rising wall of water then flows inland (Fig. 6–26). A tsunami can flood the land for as long as 5 to 10 minutes.

Hazard Mitigation

While geologists can't predict when the next quake will strike, engineers can reduce damage and loss of life by building earthquake-resistant structures in high hazard zones. The problems lie in defining the terms "earthquake-resistant" and "high hazard" and then comparing risk assessment and cost.

For example, engineers can construct a highway bridge to withstand a magnitude 8.0 quake with a nearby epicenter. However, the bridge would be very expensive and the risk is small that a major earthquake will occur near *any particular* bridge. So city planners build bridges that will withstand a small nearby quake or a larger quake with a more distant epicenter. If a major quake strikes, bridges near the epicenter will fail while those farther away will survive. Some people will die but construction costs will be affordable (Fig 6–27).

People can't agree on how much protection we should buy. In a city such as Los Angeles, where a large quake is likely in our lifetimes, we must ask, how much is human life worth? In the New Madrid region, where historical frequency has been low but past quakes have been violent, we ask, how much are we willing to gamble that another large quake won't occur within the expected lifespan of a building or bridge? New York and Boston don't lie near any plate boundaries, but quakes up to magnitude 6.5 have occurred in the northeast. So how much should we spend on earthquake-resistant design in these regions? There are no definite answers; we can only do our best to develop approximations that balance geology, chance, and cost.

6.6 Studying the Earth's Interior

Recall that the Earth is composed of a thin crust, a thick mantle, and a core. The three layers are distin-

Figure 6–26 In 1992, a series of tsunamis swamped Nicaragua's Pacific coast, killing approximately 200 people, destroying 700 homes, and leaving 16,000 people homeless. The giant waves originated as the sea floor shifted about 60 kilometers offshore; they rose 10 meters above normal high tide level as they smashed into the coast. *(Paolo Bosio/Gamma Liasion)*

Figure 6–27 This interstate highway overpass collapsed during the 1989 San Francisco earthquake, killing many motorists. It was subsequently rebuilt to a higher standard. *(Paul Scott/Sygma)*

4. P waves are compressional waves and travel through gases, liquids, and solids, whereas S waves are shear waves and travel only through solids.

Discovery of the Crust-Mantle Boundary

Figure 6–28 shows that some waves travel directly from the earthquake focus, through the crust, to a nearby seismograph. Others travel downward into the mantle and then refract back upward to the same seismograph. The route through the mantle is longer than that through the crust. However, seismic waves travel faster in the mantle than they do in the crust. Over a short distance (less than 300 kilometers), waves traveling through the crust arrive at a seismograph before those following the longer route through the mantle. However, for longer distances, the longer route through the mantle is faster because waves travel more quickly in the mantle.

guished by different chemical compositions. In turn, both the mantle and core contain finer layers based on changing physical properties. Scientists have learned a remarkable amount about the Earth's structure even though the deepest well is only a 12-kilometer-deep hole in northern Russia. Scientists deduce the composition and properties of the Earth's interior by studying the behavior of seismic waves. Some of the principles necessary for understanding the behavior of seismic waves are as follows:

1. In a uniform, homogeneous medium, a wave radiates outward in concentric spheres and at constant velocity.

2. The velocity of a seismic wave depends on the nature of the material that it travels through. Thus, seismic waves travel at different velocities in different types of rock. In addition, wave velocity varies with changing rigidity and density of a rock.

3. When a wave passes from one material to another, it refracts (bends) and sometimes reflects (bounces back). Both refraction and reflection are easily seen in light waves. If you place a pencil in a glass half filled with water, the pencil appears bent. Of course the pencil does not bend; the light rays do. Light rays slow down when they pass from air to water, and as the velocity changes, the waves refract. If you look in a mirror, the mirror reflects your image. In a similar manner, boundaries between the Earth's layers refract and reflect seismic waves.

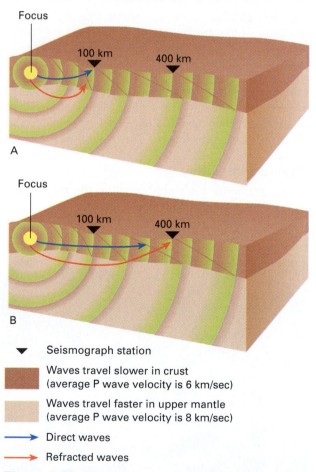

▼ Seismograph station

Waves travel slower in crust (average P wave velocity is 6 km/sec)

Waves travel faster in upper mantle (average P wave velocity is 8 km/sec)

→ Direct waves

→ Refracted waves

Figure 6–28 The travel paths of seismic waves. A The direct waves reach the closer seismic station first. B The refracted waves reach a more distant seismic station before the direct waves. Even though the refracted waves travel a longer distance, they travel at a higher speed through the mantle.

The situation is analogous to the two different routes you may use to travel from your house to a friend's. The shorter route is a city street where traffic moves slowly. The longer route is an interstate highway, but you have to drive several kilometers out of your way to get to the highway and then another few kilometers from the highway to your friend's house. If your friend lives nearby, it is faster to take the city street. But if your friend lives far away, it is faster to take the longer route and make up time on the highway.

In 1909, Andrija Mohorovičić discovered that seismic waves from a distant earthquake traveled more rapidly than those from a nearby earthquake. By analyzing the arrival times of earthquake waves to many different seismographs, Mohorovičić identified the boundary between the crust and the mantle. Today, this boundary is called the **Mohorovičić discontinuity** or the **Moho,** in honor of its discoverer.

The Moho lies at a depth ranging from 4 to 70 kilometers. Oceanic crust is thinner than continental crust, and continental crust is thicker under mountain ranges than it is under plains.

The Structure of the Mantle

The mantle is almost 2900 kilometers thick and comprises about 80 percent of the Earth's volume. Much of our knowledge of the composition and structure of the mantle comes from seismic data. As explained earlier, seismic waves speed up abruptly at the crust-mantle boundary (Fig. 6–29). Seismic waves slow down again when they enter the asthenosphere at a depth between 75 and 125 kilometers. The plasticity and partially melted character of the asthenosphere slow down the seismic waves. At the base of the asthenosphere 350 kilometers below the surface, seismic waves speed up again because increasing pressure overwhelms the temperature effect, and the mantle becomes stronger and less plastic.

At a depth of about 660 kilometers, seismic wave velocities increase again because pressure is great enough that the minerals in the mantle recrystallize to form denser minerals. The zone where the change occurs is called the **660-kilometer discontinuity.** The base of the mantle lies at a depth of 2900 kilometers. Recent research has indicated that the base of the mantle, at the core-mantle boundary, may be so hot that despite the tremendous pressure, rock in this region is liquid.

Discovery of the Core

Using a global array of seismographs, seismologists detect direct P and S waves up to 105° from the focus of an earthquake. Between 105° and 140° is a "shadow zone" where no direct P waves arrive at the Earth's surface. This shadow zone is caused by a discontinuity,

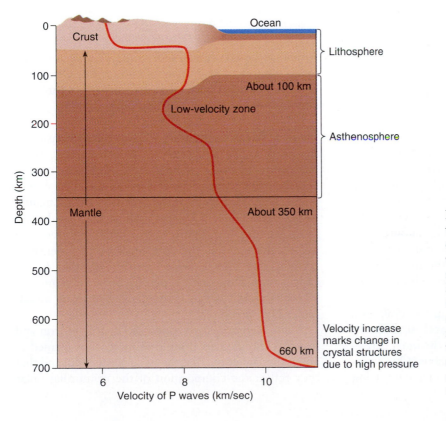

Figure 6–29 Velocities of P waves in the crust and the upper mantle. As a general rule, the velocity of P waves increases with depth. However, in the asthenosphere, the temperature is high enough and the pressure is low enough that rock is plastic. As a result, seismic waves slow down in this region, called the low-velocity zone. Wave velocity increases rapidly at the 660-kilometer discontinuity, the boundary between the upper and the lower mantle, probably because of a change in mineral content due to increasing pressure.

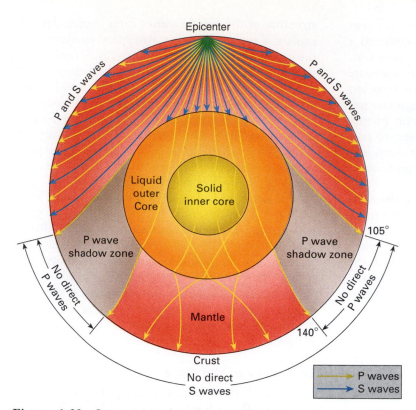

Figure 6–30 *Cross-section of the Earth showing paths of seismic waves. They bend gradually because of increasing pressure with depth. They also bend sharply where they cross major layer boundaries in the Earth's interior. Note that S waves do not travel through the liquid outer core, and therefore direct S waves are only observed within an arc of 105° of the epicenter. P waves are refracted sharply at the core-mantle boundary, so there is a shadow zone of no direct P waves from 105° to 140°.*

which is the mantle-core boundary. When P waves pass from the mantle into the core, they are refracted, or, bent as shown in Figure 6–30. The refraction deflects the P waves away from the shadow zone.

No S waves arrive beyond 105°. Their absence in this region shows that they do not travel through the outer core. Recall that S waves are not transmitted through liquids. The failure of S waves to pass through the outer core indicates that the outer core is liquid.

Refraction patterns of P waves, shown in Figure 6–30, shows that another boundary exists within the core. It is the boundary between the liquid outer core and the solid inner core. Although seismic waves tell us that the outer core is liquid and the inner core is solid, other evidence tells us that the core is composed of iron and nickel.

More detailed studies, conducted in the 1980s and 1990s, show that seismic waves travel at different speeds in different directions through the inner core—thus, the inner core is not homogeneous. One series of measurements suggests that the inner core, lying within its liquid sheath, is rotating at a significantly faster rate than the mantle and crust. Other researchers have proposed that the solid inner core may convect just as the solid mantle does.

Density Measurements   The overall density of the Earth is 5.5 grams per cubic centimeter (g/cm³), but both crust and mantle have average densities less than this value. The density of the crust ranges from 2.5 to 3.0 g/cm³, and the density of the mantle varies from 3.3 to 5.5 g/cm³. Since the mantle and crust account for slightly more than 80 percent of the Earth's volume, the core must be very dense to account for the average density of the Earth. Calculations show that the density of the core must be 10 to 13 g/cm³, which is the density of many metals under high pressure.

Many meteorites are composed mainly of iron and nickel. Cosmologists think that meteorites formed at about the same time that the Solar System did and that they reflect the composition of the primordial Solar

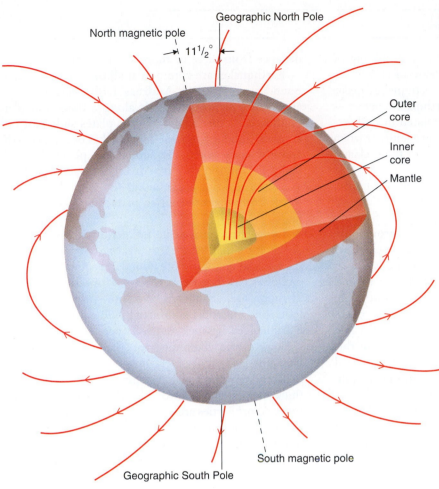

North magnetic pole

Geographic North Pole

$11\frac{1}{2}°$

Outer
core

Inner
core

Mantle

South magnetic pole

Geographic South Pole

Figure 6–31 The magnetic field of the Earth. Note that the magnetic North Pole is offset 11.5° from the geographic pole.

System. Because the Earth coalesced from meteorites and similar objects, scientists believe that iron and nickel must be abundant on Earth. Therefore, they conclude that the metallic core is composed of iron and nickel.

6.7 The Earth's Magnetism

Early navigators learned that no matter where they sailed, a needle-shaped magnet aligned itself in an approximately north-south orientation. From this observation, they learned that the Earth has a magnetic north pole and a magnetic south pole (Fig. 6–31).

The Earth's interior is too hot for a permanent magnet to exist. Instead, the Earth's magnetic field is probably electromagnetic in origin. If you wrap a wire around a nail and connect the ends of the wire to a battery, the nail becomes magnetized and can pick up small iron objects. The battery causes electrons to flow through the wire, and this flow of electrical charges creates the electromagnetic field.

Most likely, the Earth's magnetic field is generated within the outer core. Metals are good conductors of electricity and the metals in the outer core are liquid and very mobile. Two types of motion occur in the liquid outer core. (1) Because the outer core is much hotter at its base than at its surface, the liquid metal convects. (2) The rising and falling metals are then deflected by the Earth's spin. These convecting, spinning liquid conductors are thought to generate the Earth's magnetic field.

Magnetic fields are common in planets, stars, and other objects in space. Our Solar System almost certainly possessed a weak magnetic field when it first formed. The flowing metals of the liquid outer core amplified some of this original magnetic force to initiate the Earth's magnetic field. The Earth's magnetic field is approximately aligned with the Earth's rotational axis because the Earth's spin affects the flow of metal in the outer core.

SUMMARY

An **earthquake** is a sudden motion or trembling of the Earth caused by the abrupt release of slowly accumulated energy in rocks. Most earthquakes occur along tectonic plate boundaries. Earthquakes occur either when the elastic energy accumulated in rock exceeds the friction that holds rock along a fault, or when the elastic energy exceeds the strength of the rock and the rock breaks by **brittle fracture.**

An earthquake starts at the initial point of rupture, called the **focus.** The location on the Earth's surface directly above the focus is the **epicenter. Seismic waves** include **body waves,** which travel through the interior of the Earth, and **surface waves,** which travel on the surface. **P waves** are compressional body waves. **S waves** are body waves that travel slower than P waves. They consist of a shearing motion and travel through solids but not liquids. Surface waves travel more slowly than either type of body wave. Seismic waves are recorded on a **seismograph.** Early in this century, geologists used the **Mercalli scale** to record the extent of earthquake damage. The **Richter scale** expresses earthquake magnitudes. Modern geologists use the **moment magnitude** scale to record the energy released during an earthquake. The distance from a seismic station to an earthquake is calculated by recording the time between the arrival of P and S waves. The epicenter can be located by measuring the distance from three or more seismic stations.

Earthquakes are common at all three types of plate boundaries. The **San Andreas fault zone** is an example of a **strike-slip fault** along a transform plate boundary. At this boundary, two plates slide past one another. Subduction zone earthquakes occur along the **Benioff zone** when the subducting plate slips suddenly. Earthquakes occur at divergent plate boundaries as blocks of lithosphere along the fault drop downward. Earthquakes occur in plate interiors along old faults.

Long-term earthquake prediction is based on the observation that most earthquakes occur on pre-existing faults at tectonic plate boundaries. Short-term prediction is based on occurrences of **foreshocks,** release of radon gas, changes in the land surface, the water table, and electrical conductivity. Earthquake damage is influenced by rock and soil type, construction design, and the likelihood of fires, landslides, and tsunamis.

Recall that the Earth is composed of a thin crust, a thick mantle, and a core. The Earth's internal structure and properties are known by studies of earthquake wave velocities and refraction and reflection of seismic waves as they pass through the Earth. Flowing metal in the outer core generates the Earth's magnetic field.

KEY TERMS

earthquake 107
stress 108
elastic deformation 108
plastic deformation 108
brittle fracture 108
fault 108
seismic wave 109
seismology 109
focus 109

epicenter 109
body wave 110
surface wave 110
P wave 110
S wave 110
seismograph 111
seismogram 111
Mercalli scale 111

Richter scale 111
moment magnitude 112
time-travel curve 113
strike-slip fault 114
San Andreas fault
  zone 114
fault creep 116
thrust fault 116

Benioff zone 117
foreshock 119
liquefaction 122
tsunami 122
Mohorovičić discontinuity
  (Moho) 125
660-kilometer
  discontinuity 125

FOR REVIEW

1. Explain how energy is stored prior to and then released during an earthquake.

2. Describe the behavior of rock during elastic deformation, plastic deformation, and brittle fracture.

3. Give two mechanisms that can release accumulated elastic energy in rocks.

4. Why do most earthquakes occur at the boundaries between tectonic plates? Are there any exceptions?

5. Define focus and epicenter.

6. Discuss the similarities and differences among P waves, S waves, and surface waves.

7. Explain how a seismograph works. Sketch what an imaginary seismogram would look like before and during an earthquake.

8. Describe the similarities and differences between the

Richter and moment magnitude scales. What is actually measured and what information is obtained?

9. Describe how the epicenter of an earthquake is located.

10. Discuss earthquake mechanisms at the three different types of tectonic plate boundaries.

11. Briefly discuss major faults close to Los Angeles.

12. What is the Benioff zone? What type of tectonic boundary does it occur at?

13. Why do only shallow earthquakes occur along the mid-oceanic ridge?

14. Discuss earthquake mechanisms at plate interiors.

15. Discuss the scientific reasoning behind long-term and short-term earthquake prediction.

16. List five different factors that affect earthquake damage. Discuss each briefly.

17. What is the Moho? How was it discovered?

18. Explain how geologists concluded that the core is composed of iron and nickel.

19. Briefly discuss a hypothesis for the origin of the Earth's magnetic field.

FOR DISCUSSION

1. Using the graph in Figure 6–11, determine how far away from an earthquake you would be if the first P wave arrived 5 minutes before the first S wave.

2. The seismograph shown in Figure 6–8 measures up-and-down earth movement. Draw a diagram for a seismograph that would record side-to-side motion. Explain how your seismograph works.

3. Using a map of the United States, locate an earthquake that is 1000 kilometers from Seattle, 1300 kilometers from San Francisco, and 700 kilometers from Denver.

4. Mortality was high in the India earthquake in 1993 because the quake occurred at night when people were sleeping in their homes. However, mortality in the Northridge earthquake was low because it occurred early in the morning rather than during rush hour. Is there a contradiction in these two statements?

5. Explain why the existence of thrust faults west of the San Andreas fault complicates attempts at earthquake prediction near Los Angeles.

6. Imagine that geologists predict a major earthquake in a densely populated region. The prediction may be right or it may be wrong. City planners may heed it and evacuate the city or ignore it. The possibilities lead to four combinations of predictions and responses, which can be set out in a grid as follows:

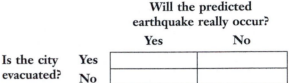

For example, the space in the upper left corner of the grid represents the situation in which the predicted earthquake occurs and the city is evacuated. For each space in the square, outline the consequences of that sequence of events.

7. Discuss the trade-off between money and human lives when considering construction in earthquake-prone zones. Would you be willing to pay twice as much for a house that was earthquake-resistant, as compared with a normal house? Would you be willing to pay more taxes for safer highway bridges? Do you feel that different types of structures (e.g., residential homes and apartments, commercial buildings, nuclear power plants) should be built to different safety standards?

8. Argue for or against placing stringent building codes for earthquake-resistant design in the Seattle area. Would your answer be different for Boston? Houston?

9. From Fig. 6–29, what is the speed of P waves at a depth of 25 kilometers? 200 kilometers? 500 kilometers?

10. Give two reasons why the Earth's magnetic field cannot be formed by a giant bar magnet within the Earth's core.

Volcanoes and Plutons

In Chapters 3 and 5 you learned that magma forms deep within the Earth and then rises toward the surface. In some instances, it erupts onto the Earth's surface to form a volcano and volcanic rocks; in others, it solidifies within the crust to form plutonic rocks.

A volcanic eruption can be one of the most violent of all geologic events. During the past 100 years, eruptions have killed approximately 100,000 people and have caused about $10 billion in damage. Some violent eruptions have buried towns and cities in hot lava or volcanic ash. For example, the 1902 eruption of Mount Pelée on the Caribbean island of Martinique buried the city of Saint Pierre in glowing volcanic ash that killed 29,000 people. On June 25, 1997, the Soufrière Hills volcano on the tiny Caribbean island

Red hot cinders explode into the air as glowing lava flows down the slopes of Arenal Volcano in Costa Rica. *(Schafer & Hill/Tony Stone Images)*

of Montserrat erupted in the most destructive event in a five-year-long sequence of escalating volcanic crises to kill 19 people. An ash plume rose 10 kilometers over the volcano, and volcanic ash buried 4 square kilometers of the island. The 1980 eruption of Mount St. Helens in Washington State and the 1991 Mount Pinatubo eruption in the Philippines were similarly violent events. Other volcanoes erupt gently. Tourists flock to Hawaii and Iceland to photograph fire fountains erupting into the sky, and geologists study advancing lava flows in relative safety (Fig. 7–1).

Volcanic eruptions can trigger other deadly events. The 1883 eruption of Krakatoa in the southwest Pacific Ocean generated tsunamis that killed 36,000 people. In 1985, a small eruption of Nevado del Ruiz in Colombia triggered a mudflow that buried the town of Armero, killing more than 22,000 people.

In contrast to the high visibility of a volcanic eruption, we cannot see plutonic rocks form because they crystallize within the crust, commonly at depths of 5 to 20 kilometers below the surface. However, tectonic forces have raised plutons and erosion has exposed them in many of the world's greatest mountain ranges. California's Sierra Nevada, the Rockies, the Appalachians, the European Alps, and the Himalayas are all composed partly of plutonic rocks.   •

7.1   How Magma Forms

In Chapter 3 we stated that rocks melt in certain environments to form magma. This process is one example of the constantly changing nature of rocks described by the rock cycle. Why do rocks melt, and in what environments does magma form?

Figure 7–1   Two geologists retreat from a slowly advancing lava flow on the island of Hawaii. *(U.S. Geological Survey)*

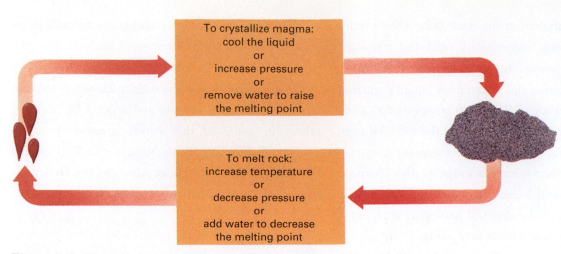

Figure 7–2 The lower box shows that increasing temperature, addition of water, and decreasing pressure all melt rock to form magma. The upper box shows that cooling, increasing pressure, and water loss all solidify magma to form an igneous rock.

Processes that Form Magma

In the asthenosphere (between depths of about 100 to 350 kilometers), the temperature is high enough and the pressure is low enough that 1 or 2 percent of the asthenosphere is molten. Although the majority of the asthenosphere is solid, it is so hot and so close to its melting point that relatively small changes can melt large volumes of rock. In order to understand volcanoes and plutonic rocks, we must first understand three processes that melt the asthenosphere to form magma: increasing temperature, decreasing pressure, and addition of water (Fig. 7–2).

Increasing Temperature   Everyone knows that a solid melts when it becomes hot enough. Butter melts in a frying pan and snow melts under the spring sun. For similar reasons, an increase in temperature will melt a hot rock. Oddly, however, increasing temperature is the least important cause of magma formation in the asthenosphere.

Decreasing Pressure   A mineral is a solid composed of an ordered array of atoms bonded together. When a mineral melts, the atoms become disordered and move freely, taking up more space than the solid mineral. Consequently, magma occupies about 10 percent more volume than the rock that melted to form it (Fig. 7–3).

If a rock is heated to its melting point on the Earth's surface, it melts readily because there is little pressure to keep it from expanding. The temperature in the asthenosphere is more than hot enough to melt rock, but there the high pressure prevents the rock from ex-

panding, and it doesn't melt. However, if the pressure were to decrease, large volumes of the asthenosphere would melt. Melting caused by decreasing pressure is called **pressure-release melting.** In the section "Environments of Magma Formation" we will see how certain tectonic processes decrease pressure on asthenosphere rocks, and melt them.

Addition of Water   A wet rock generally melts at a lower temperature than an otherwise identical dry rock. Consequently, addition of water to rock that is near its melting temperature can melt the rock. Certain tectonic processes, described shortly, add water to the hot rock of the asthenosphere to form magma.

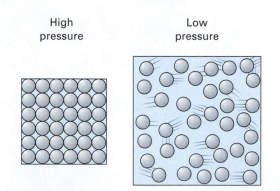

Figure 7–3 When most minerals melt, the volume increases. As a result, high pressure favors the dense, orderly arrangement of a solid mineral and low pressure favors the random, less dense arrangement of molecules in liquid magma.

Environments of Magma Formation

Magma forms abundantly in three tectonic environments: spreading centers, mantle plumes, and subduction zones. Let us consider each environment to see how the three processes just described melt the asthenosphere to create magma.

Magma Production in a Spreading Center   As lithospheric plates separate at a spreading center, hot, plastic asthenosphere oozes upward to fill the gap (Fig. 7–4). As the asthenosphere rises, pressure drops and pressure-release melting forms basaltic magma. (The terms *basaltic* and *granitic* refer to magmas with the chemical compositions of basalt and granite, respectively.) Because the magma is of lower density than the surrounding rock, it rises toward the surface.

Most of the world's spreading centers lie in the ocean basins, where they form the mid-oceanic ridge. The magma created by pressure-release melting erupts from the ridge beneath the sea and solidifies to form new oceanic crust. The oceanic crust then spreads out-ward, riding atop the separating tectonic plates. Nearly all of the Earth's oceanic crust is created in this way at the mid-oceanic ridge. In most places the ridge lies beneath the sea. In a few places, such as Iceland, the ridge is exposed (Fig 7–5). Some spreading centers, like the East African rift, occur in continents, and here, too, basaltic magma erupts onto the Earth's surface.

Magma Production in a Mantle Plume   Recall from Chapter 5 that a mantle plume is a rising column of hot, plastic rock that originates deep within the mantle. The plume rises because it is hotter than the surrounding mantle and, consequently, is less dense and more buoyant. As a plume rises, pressure-release melting forms magma that erupts onto the Earth's surface (Fig. 7–6). A hot spot is a volcanically active place at the Earth's surface directly above a mantle plume. Because mantle plumes form below the asthenosphere, hot spots can occur within a tectonic plate. For example, the Yellowstone hot spot, responsible for the volcanoes and hot springs in Yellowstone

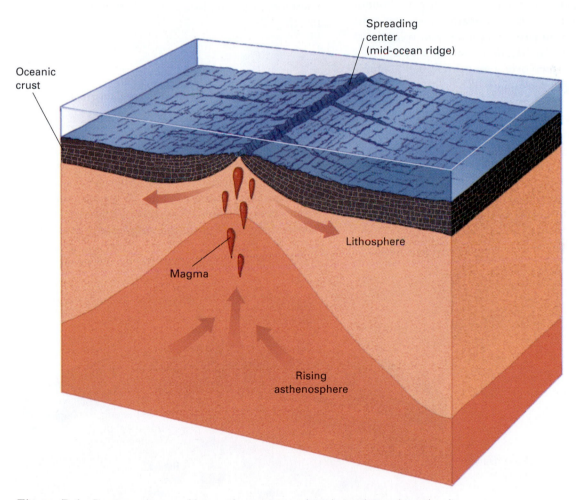

Figure 7–4 Pressure-release melting produces magma where hot asthenosphere rises beneath a spreading center.

Figure 7–5 Iceland is one of the few places on Earth where volcanoes of the mid-oceanic ridge system rise above the sea. This photo shows an eruption of Iceland's Mount Surtsey. *(Ragnar Larusson/Science Source)*

National Park, lies far from the nearest plate boundary. If a mantle plume rises beneath the sea, volcanic eruptions build submarine volcanoes and volcanic islands. The Hawaiian Islands are a chain of hot spot volcanoes that formed over a long-lived mantle plume beneath the Pacific Ocean.

Magma Production in a Subduction Zone   At a subduction zone, a lithospheric plate sinks into the mantle (Fig. 7–7). As you learned in Chapter 5, a subducting plate is covered by water-saturated oceanic crust. As the wet rock dives into the hot mantle, rising temperature drives off the water. The resultant steam ascends into the hot asthenosphere directly above the sinking plate.

As the subducting plate descends, it drags plastic asthenosphere rock down with it, as shown by the elliptical arrows in Figure 7–7. Rock from deeper in the asthenosphere then flows upward to replace the sinking rock. Pressure decreases as this hot rock rises.

Friction generates heat in a subduction zone as one plate scrapes past the opposite plate. Figure 7–7 shows that addition of water, pressure release, and frictional heating combine to melt portions of the asthenosphere, where the subducting plate passes into the asthenosphere. Addition of water is probably the most important factor in magma production in a subduction zone, and frictional heating is the least important.

As a result of these processes, both volcanoes and plutonic rocks are common features of a subduction zone. The volcanoes of the Pacific Northwest, the granite cliffs of Yosemite, and the Andes Mountains are all examples of volcanic and plutonic rocks formed at subduction zones. The "ring of fire" is a zone of active volcanoes that traces the subduction zones encircling the Pacific Ocean basin. About 75 percent of the Earth's active volcanoes (exclusive of the submarine volcanoes at the mid-oceanic ridge) lie in the ring of fire (Fig. 7–8).

7.2 Partial Melting: The Origins of Basalt and Granite

Recall that basaltic magma forms by melting of the asthenosphere. But the asthenosphere is peridotite. Basalt and peridotite are quite different in composition: Peridotite contains about 40 percent silica, but basalt contains about 50 percent. Peridotite contains considerably more iron and magnesium than basalt.

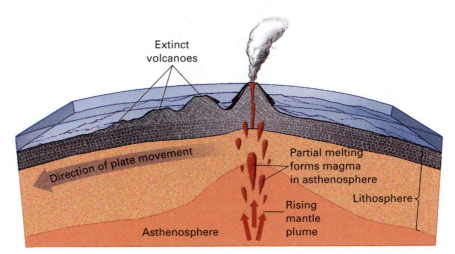

Figure 7–6 Pressure-release melting produces magma in a rising mantle plume. The magma rises to form a volcanic hot spot.

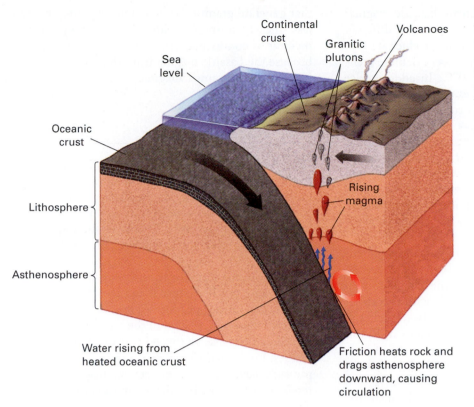

Figure 7–7 Three factors contribute to melting of the asthenosphere and production of magma at a subduction zone: (1) Water rises from oceanic crust on top of the subducting plate; (2) circulation in the asthenosphere decreases pressure on hot rock; (3) friction heats rocks in the subduction zone.

Sea level

Continental crust

Granitic plutons

Volcanoes

Oceanic crust

Lithosphere

Asthenosphere

Rising magma

Water rising from heated oceanic crust

Friction heats rock and drags asthenosphere downward, causing circulation

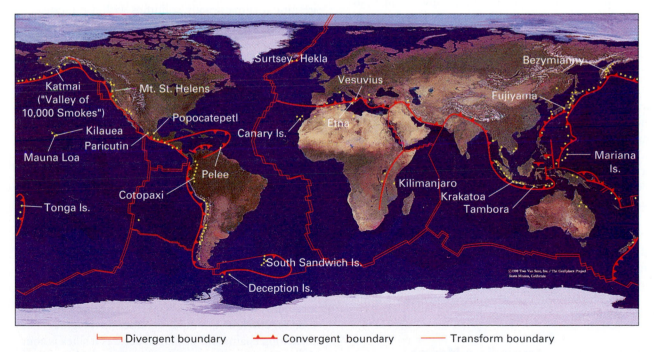

Divergent boundary ▲▲ Convergent boundary — Transform boundary

Figure 7–8 Seventy-five percent of the Earth's active volcanoes (yellow dots) lie in the "ring of fire," a chain of subduction zones at convergent plate boundaries (heavy red lines with teeth) that encircles the Pacific Ocean. *(USGS data superimposed on map by Tom Van Sant/Geosphere Project)*

How does peridotite melt to create basaltic magma? Why does the magma have a composition different from that of the rock that melted to produce it?

Any pure substance, such as ice, has a definite melting point. Ice melts at exactly 0°C. In addition, ice melts to form water, which has exactly the same composition as the ice, pure H_2O. A rock does not behave in this way because it is a *mixture* of several minerals, each of which melts at a different temperature. If you heat peridotite slowly, the minerals with the lowest melting point begin to melt first, while the other minerals remain solid. This phenomenon is called **partial melting.** Of course, if the temperature is high enough, all the minerals will reach their melting points and the whole rock will melt.

In general, minerals with the highest silica contents melt at the lowest temperatures. In parts of the asthenosphere where magma forms, the temperature is only hot enough to melt the minerals with the lowest melting points. As a result, magma is always richer in silica than the rock that melted to produce it. In this way, basaltic magma forms by partial melting of peridotite at a temperature of about 1100°C. When the basaltic magma rises toward the Earth's surface, it leaves silica-depleted peridotite in the asthenosphere.

Figures 7–4 and 7–6 show that basaltic magma forms in this way by pressure-release melting at a spreading center and in a mantle plume. Figure 7–7 shows that basaltic magma also forms by a combination of pressure-release and addition of water in a subduction zone.

Granite and Granitic Magma

Granite contains more silica than basalt and therefore melts at a lower temperature—typically between 700° and 900°C. Thus, basaltic magma is hot enough to melt granitic continental crust. In certain tectonic environments, the asthenosphere melts beneath a continent, forming basaltic magma that rises into continental crust. These environments include a subduction zone, a continental rift zone, and a mantle plume rising beneath a continent.

Because the lower continental crust is hot, even a small amount of basaltic magma melts large quantities of the continent to form granitic magma. Typically, the granitic magma rises a short distance and then solidifies within the crust to form plutonic rocks. Most granitic plutons solidify at depths between 5 kilometers and about 20 kilometers.

Andesite and Intermediate Magma

Igneous rocks of intermediate composition, such as andesite and diorite, form by processes similar to those that generate granitic magma. Their magmas contain less silica than granite, either because they form by melting of continental crust that is lower in silica or because the basaltic magma from the mantle has contaminated the granitic magma.

7.3 Magma Behavior

Why do most granitic magmas crystallize within the Earth to form plutonic rocks, whereas most basaltic magma rises all the way to the surface to erupt from volcanoes? Why do some volcanoes explode violently but others erupt gently? To answer these questions, we must consider the properties and behavior of magma.

Once magma forms, it rises toward the Earth's surface because it is less dense than surrounding rock. As it rises, two changes occur. First, it cools as it enters shallower and cooler levels of the Earth. Second, pressure drops because the weight of overlying rock decreases. Figure 7–2 shows that cooling and decreasing pressure have opposite effects on magma: Cooling tends to solidify it, but decreasing pressure tends to keep it liquid.

So, does magma solidify or remain liquid as it rises toward the Earth's surface? The answer depends on the type of magma. Basaltic magma commonly rises to the surface to erupt from a volcano. In contrast, granitic magma usually solidifies within the crust.

The contrasting behavior of granitic and basaltic magmas is a result of their different compositions. Granitic magma contains about 70 percent silica, whereas the silica content of basaltic magma is only about 50 percent. In addition, granitic magma generally contains up to 10 percent water, whereas basaltic magma contains only 1 to 2 percent water. These differences are summarized in the following table.

Typical Granitic Magma	Typical Basaltic Magma
70% silica	50% silica
Up to 10% water	1–2% water

Effects of Silica on Magma Behavior

In the silicate minerals, silicate tetrahedra link together to form the chains, sheets, and framework structures described in Chapter 2. Silicate tetrahedra link together in a similar manner in magma. They form long chains and similar structures if silica is abundant in the magma, but shorter chains if less silica is present. Because of its higher silica content, granitic magma contains longer chains than does basaltic magma, giving it a higher viscosity (resistance to flow). In granitic magma, the long chains become tangled, making the magma stiff, or viscous. It rises slowly because of its viscosity and has ample time to solidify within the crust before reaching the surface. In contrast, basaltic magma, with its shorter silicate chains, is less viscous and flows easily. Because of its fluidity, it rises rapidly to erupt at the Earth's surface.

Effects of Water on Magma Behavior

A second, and more important, difference between the two is that granitic magma contains more water than basaltic magma. Water lowers the temperature at which magma solidifies. If dry granitic magma solidifies at 700°C, the same magma with 10 percent water may remain liquid at 600°C.

Water tends to escape as steam from hot magma. But deep in the crust where granitic magma forms, high pressure prevents the water from escaping. As magma rises, pressure decreases and water escapes. Because the magma loses water, its solidification temperature *rises*, causing it to crystallize. Water loss causes rising granitic magma to solidify within the crust.

Because basaltic magmas have only 1 to 2 percent water to begin with, water loss is relatively unimportant. As a result, rising basaltic magma usually remains liquid all the way to the Earth's surface, and basalt volcanoes are common.

7.4 Plutons

In most cases, granitic magma solidifies within the Earth's crust to form a large mass of granite called a **pluton** (Fig. 7–9). Many granite plutons measure tens of kilometers in diameter. How can such a large mass of viscous magma rise through solid rock?

If you place oil and water in a jar, screw the lid on, and shake the jar, oil droplets disperse throughout the water. When you set the jar down, the droplets coalesce to form larger bubbles, which rise toward the surface, easily displacing the water as they ascend. Granitic magma rises in a similar way except that the process is slower because it rises through solid rock. Granitic magma forms near the base of continental crust, where surrounding rock is hot and plastic. As the magma rises, it shoulders aside the plastic country rock, which then slowly flows back to fill in behind the rising bubble.

After a pluton forms, tectonic forces may push it upward, and erosion may expose parts of it at the Earth's surface. A **batholith** is a pluton exposed over more than 100 square kilometers of the Earth's surface. A large batholith may be as much as 20 kilometers thick, but an average one is about 10 kilometers thick. A **stock** is similar to a batholith but is exposed over less than 100 square kilometers.

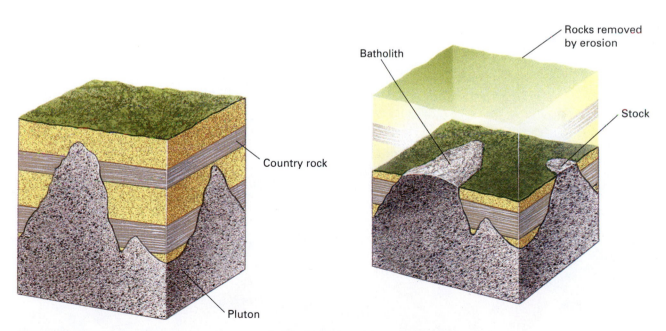

Figure 7–9 A pluton is any intrusive igneous rock. A batholith is a pluton with more than 100 square kilometers exposed at the Earth's surface. A stock is similar to a batholith but has a smaller surface area.

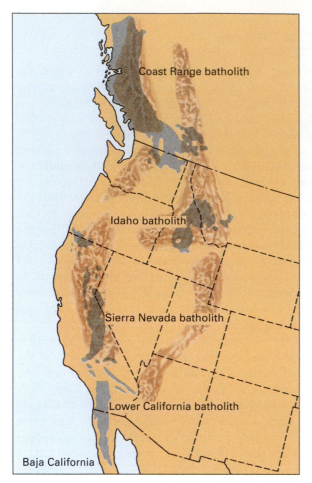

Figure 7–10 The large batholiths in western North America form high mountain ranges.

Figure 7–11 Granite plutons make up most of California's Sierra Nevada. *(Mack Henley/Visuals Unlimited)*

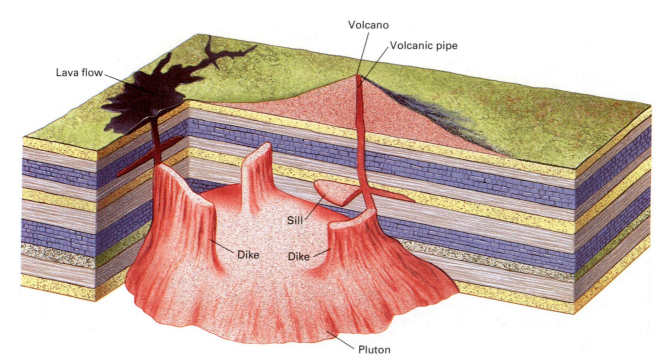

Figure 7–12 A large magma body may crystallize within the crust to form a pluton. Some of the magma may rise to the surface to form volcanoes and lava flows; some intrudes country rock to form dikes and sills.

Figure 7–13 This basalt dike intruded billion-year-old sedimentary rock in Grand Canyon.

A large body of magma pushes country rock aside as it rises. In contrast, a smaller mass of magma may flow into a fracture or between layers in country rock. A **dike** is a tabular, or sheetlike, intrusive rock that forms when magma oozes into a fracture (Fig. 7–12). Dikes cut *across* sedimentary layers or other features in country rock and range from less than a centimeter to more than a kilometer thick (Fig. 7–13). A dike is commonly more resistant to weathering than surrounding rock. As the country rock erodes away, the dike is left standing on the surface (Fig. 7–14).

Magma that oozes between layers of country rock forms a sheetlike rock *parallel* to the layering, called a **sill** (Fig. 7–15). Like dikes, sills vary in thickness from less than a centimeter to more than a kilometer and may extend for tens of kilometers in length and width.

Figure 7–10 shows the major batholiths of western North America. Many mountain ranges, such as California's Sierra Nevada, contain large granite batholiths. A batholith is commonly composed of numerous smaller plutons intruded sequentially over millions of years. For example, the Sierra Nevada batholith contains about 100 individual plutons, most of which were emplaced over a period of 50 million years. The formation of this complex batholith ended about 80 million years ago (Fig. 7–11).

Volcanic Rocks and Volcanoes

The material erupted from volcanoes creates a wide variety of rocks and landforms, including lava plateaus and several types of volcanoes. Many islands, including the Hawaiian Islands, Iceland, and most islands of the southwestern Pacific Ocean, were built entirely by volcanic eruptions.

Lava and Pyroclastic Rocks

Lava is fluid magma that flows onto the Earth's surface; the word also describes the rock that forms when

Figure 7–14 A large dike in central Colorado has been left standing after softer sandstone country rock eroded away. *(© Science Graphics, Inc./Ward's Natural Science Establishment, Inc.)*

Figure 7–15 A basaltic sill has intruded between sedimentary rock layers on Mt. Gould in Glacier National Park, Montana. The light-colored rock above and below the sill was metamorphosed by heat from the magma. *(Breck P. Kent)*

the magma solidifies. Lava with low viscosity may continue to flow as it cools and stiffens, forming smooth, glassy-surfaced, wrinkled, or "ropy" ridges. This type of lava is called **pahoehoe** (Fig. 7–16). If the viscosity of lava is higher, its surface may partially solidify as it flows. The solid crust breaks up as the deeper, molten lava continues to move, forming **aa** lava, with a jagged, rubbly, broken surface. When lava cools, escaping gases such as water and carbon dioxide form bubbles in the lava. If the lava solidifies before the gas escapes, the bubbles are preserved as holes in the rock (Fig. 7–17).

Hot lava shrinks as it cools and solidifies. The shrinkage pulls the rock apart, forming cracks that grow as the rock continues to cool. In Hawaii, geologists have watched fresh lava cool. When a solid crust only 0.5 centimeter thick had formed on the surface of the glowing liquid, five- or six-sided cracks developed. As the lava continued to cool and solidify, the cracks grew downward through the flow. Such cracks, called **columnar joints,** are regularly spaced and intersect to form five- or six-sided columns (Fig. 7–18).

If a volcano erupts explosively, it may eject both liquid magma and solid rock fragments. A rock formed

Figure 7–16 A car buried in pahoehoe lava, Hawaii. *(Kenneth Neuhauser)*

Figure 7–17 Aa lava showing vesicles, gas bubbles frozen into the flow, in Shoshone, Idaho.

A

from this material is called a **pyroclastic rock** (from *pyro*, meaning fire, and *clastic*, meaning particles). The smallest particles, called **volcanic ash,** consist of tiny fragments of glass that formed when liquid magma exploded into the air. **Cinders** vary in size from 2 to 64 millimeters.

Fissure Eruptions and Lava Plateaus

The gentlest type of volcanic eruption occurs when magma is so fluid that it oozes from cracks in the land surface called **fissures** and flows over the land like water. Basaltic magma commonly erupts in this manner because of its low viscosity. Fissures and fissure eruptions vary greatly in scale. In some cases, lava pours from small cracks on the flank of a volcano. Fissure flows of this type are common on Hawaiian and Icelandic volcanoes.

In other cases, however, fissures extend for tens or hundreds of kilometers and pour thousands of cubic kilometers of lava onto the Earth's surface. A fissure eruption of this type creates a **flood basalt,** which covers the landscape like a flood. It is common for many such fissure eruptions to occur in rapid succession and to create a **lava plateau** (or "basalt plateau") covering thousands of square kilometers.

The Columbia River plateau in eastern Washington, northern Oregon, and western Idaho is a lava plateau containing 350,000 cubic kilometers of basalt. The lava is up to 3000 meters thick and covers 200,000 square kilometers (Fig. 7–19). It formed about 15 million years ago as basaltic magma oozed from long fissures in the Earth's surface. The individual flows are between 15 and 100 meters thick. An even larger lava

B

Figure 7–18 (A) Columnar joints at Devil's Postpile National Monument. (B) A view from the top, where the columns have been polished by glaciers.

plateau in Siberia contains one million cubic kilometers of basaltic lava that poured onto the Earth's surface 250 million years ago. Similar large lava plateaus occur in western India, northern Australia, Iceland, Brazil, Argentina, and Antarctica.

A

B

Figure 7–19 (A) The Columbia River basalt plateau covers much of Washington, Oregon, and Idaho. (B) Columbia River basalt in eastern Washington. Each layer is a separate lava flow. *(Larry Davis)* • Interactive Question: What effects might the eruption of this basaltic lava have had on global climate? on the biosphere?

Volcanoes

If lava is too viscous to spread out as a flood, it builds a hill or mountain called a **volcano.** Volcanoes differ widely in shape, structure, and size (Table 7–1). Lava and rock fragments commonly erupt from an opening called a **vent** located in a **crater,** a bowl-like depression at the summit of the volcano (Fig. 7–20). As mentioned previously, lava or pyroclastic material may also erupt from a fissure on the flanks of the volcano.

TABLE 7–1

Characteristics of Different Types of Volcanoes

Type of Volcano	Form of Volcano	Size	Type of Magma	Style of Activity	Examples
Basalt plateau	Flat to gentle slope	100,000–1,000,000 km² in area; 1–3 km thick	Basalt	Gentle eruption from long fissures	Columbia River plateau
Shield volcano	Slightly sloped, 6° to 12°	Up to 9000 m high	Basalt	Gentle, some fire fountains	Hawaii
Cinder cone	Moderate slope	100–400 m high	Basalt or andesite	Ejections of pyroclastic material	Parícutin, Mexico
Composite volcano	Alternate layers of flows and pyroclastics	100–3500 m high	Variety of types of magmas and ash	Often violent	Vesuvius, Mount St. Helens, Aconcagua
Caldera	Cataclysmic explosion leaving a circular depression called a caldera	Less than 40 km in diameter	Granite	Very violent	Yellowstone, San Juan Mountains

Shield Volcanoes   Fluid basaltic magma often builds a gently sloping mountain called a **shield volcano** (Fig. 7–21). The sides of a shield volcano generally slope away from the vent at angles between 6° and 12° from horizontal. Although their slopes are gentle, shield volcanoes can be enormous. The height of Mauna Kea volcano in Hawaii, measured from its true base on the sea floor to its top, rivals the height of Mount Everest.

Although shield volcanoes, such as those of Hawaii and Iceland, erupt regularly, the eruptions are nor-

mally gentle and rarely life threatening. Lava flows occasionally overrun homes and villages, but the flows advance slowly enough to give people time to evacuate.

Cinder Cones   A **cinder cone** is a small volcano composed of pyroclastic fragments. A cinder cone forms when large amounts of gas accumulate in rising magma. When the gas pressure builds up sufficiently, the entire mass erupts explosively, hurling cinders, ash, and molten magma into the air. The particles then fall

Figure 7–21   Mount Skjoldbreidier in Iceland shows the typical low-angle slopes of a shield volcano.   *(Science Graphics, Inc./Ward's Natural Science Establishment, Inc.)*

back around the vent to accumulate as a small mountain of pyroclastic debris. A cinder cone is usually active for only a short time because once the gas escapes, the driving force behind the eruption is gone.

As the name implies, a cinder cone is symmetrical. It also can be steep (about 30°), especially near the vent where ash and cinders pile up (Fig. 7–22). Most are less than 300 meters high, although a large one can be up to 700 meters high. A cinder cone erodes easily and quickly because the pyroclastic fragments are not cemented together.

About 350 kilometers west of Mexico City, numerous extinct cinder cones are scattered over a broad plain. Prior to 1943, a small hole one or two meters in diameter existed on the plain. The hole had been there for as long as anyone could remember, and people grew corn just a few meters away. In February 1943, as two farmers were preparing their field for planting, smoke and sulfurous gases rose from the hole. As night fell, hot, glowing rocks flew skyward, creating spectacular arcing flares like a giant fireworks display. By morning, a 40-meter-high cinder cone had grown where the hole had been. For the next five days, pyroclastic material erupted 1000 meters into the sky and the cone grew to 100 meters in height. After a few months, a fissure opened at the base of the cone, extruding lava that buried the town of San Juan Parangaricutiro. Two years later the cone had grown to a height of 400 meters. After nine years, the eruptions ended, and today the volcano, called El Parícutin, is dormant.

Composite Cones A **composite cone,** sometimes called a **stratovolcano,** forms over a long period of time from alternating lava flows and pyroclastic eruptions. The hard lava covers the loose pyroclastic material and protects it from erosion (Fig. 7–23).

Many of the highest mountains of the Andes and some of the most spectacular mountains of western North America are composite cones. Repeated eruption is a trademark of a composite volcano. Mount St. Helens erupted dozens of times in the 4500 years preceding its most recent eruption in 1980. Mount Rainier, also in Washington, has been dormant in recent times but could become active again.

7.6 Violent Magma: Ash-Flow Tuffs and Calderas

Although granitic magma usually solidifies within the crust, under certain conditions it rises to the Earth's surface, where it erupts violently. The granitic magmas that rise to the surface probably contain only a few percent water, like basaltic magma. They reach the surface because, like basaltic magma, they have little water to lose. "Dry" granitic magma ascends more slowly than basaltic magma because of its higher viscosity. As it rises, decreasing pressure allows the small amount of dissolved water to form steam bubbles in the magma. The bubbles create a frothy mixture of gas and liquid magma that may be as hot as

A

B

Figure 7–22 (A) These cinder cones in southern Bolivia are composed of loose pyroclastic fragments. (B) A large block of lava lies on loose, sand-like volcanic cinders. *(Dembinski Photo Assoc.)*

Figure 7–23 (A) A composite cone consists of alternating layers of lava and loose pyroclastic material. (B) Mt. Rainier rises behind Seattle's skyline. *(Chuck Pefly/Tony Stone Images)*

900°C (Fig. 7–24A). As the mixture rises to within a few kilometers of the Earth's surface, it fractures overlying rocks and explodes skyward through the fractures (Fig. 7–24B).

As an analogy, think of a bottle of beer or soda pop. When the cap is on and the contents are under pressure, carbon dioxide gas is dissolved in the liquid. When you remove the cap, pressure decreases and bubbles rise to the surface. If conditions are favorable, the frothy mixture erupts through the bottleneck.

A large and violent eruption might blast a column of pyroclastic material 10 or 12 kilometers into the sky, and the column might be several kilometers in diameter. A cloud of fine ash may rise even higher—into the upper atmosphere. The force of material streaming out of the magma chamber can hold the column up for hours or even days. Recent eruptions of Mount Pinatubo on the Philippines and the Soufrière Hills volcano on the Caribbean island of Montserrat blasted ash columns high into the atmosphere and held them up for hours.

After an eruption, upper layers of the remaining magma are depleted in dissolved water and the explosive potential is low. However, deeper parts of the

Figure 7–24 (A) When granitic magma rises to within a few kilometers of the Earth's surface, it stretches and fractures overlying rock. Gas separates from the magma and rises to the upper part of the magma body. (B) The gas-rich magma explodes through fractures, rising as a vertical column of hot ash, rock fragments, and gas. (C) When the gas is used up, the column collapses and spreads outward as a high-speed ash flow. (D) Because so much material has erupted from the top of the magma chamber, the roof collapses to form a caldera. ● Interactive Question: How might a large caldera eruption alter local ecosystems? How might it alter global climate?

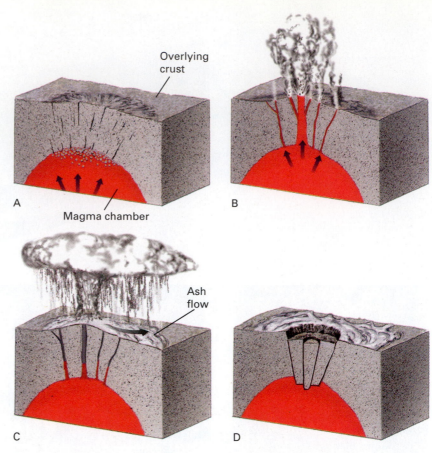

magma continue to release water vapor, which rises and builds pressure again to begin another cycle of eruption. Time intervals between successive eruptions vary from a few thousand to about half a million years.

Ash Flows

When most of the gas has escaped from the upper layers of magma, the eruption ends. The airborne column of ash, rock, and gas that had been sustained by the force of the eruption then falls back to the Earth's surface (Fig. 7–24C). The falling material spreads over the land and flows down stream valleys. Such a flow is called an **ash flow**, or **nuèe ardente**, a French term for "glowing cloud." Small ash flows move at speeds up to 200 kilometers per hour. Large flows have traveled distances exceeding 100 kilometers. The 2000-year-old Taupo flow on New Zealand's South Island leaped over a 700-meter-high ridge as it crossed from one valley into another.

When an ash flow stops, most of the gas escapes into the atmosphere, leaving behind a chaotic mixture of volcanic ash and rock fragments called **ash-flow tuff** (Fig. 7–25). The largest known ash-flow tuff from a single eruption is located in the San Juan Mountains of southwestern Colorado and has a volume greater than 3000 cubic kilometers. Another of comparable size lies in southern Nevada.

Calderas

After the gas-charged magma erupts, nothing remains to hold up the overlying rock, and the roof of the magma chamber collapses (Fig. 7–24D). Because most magma bodies are circular when viewed from above, the collapsing roof forms a circular depression called a **caldera**. A large caldera may be 40 kilometers in diameter and have walls as much as a kilometer high. Some calderas fill up with volcanic debris as the ash column collapses; others maintain the circular depression and steep walls (Fig. 7–26). We usually think of volcanic landforms as mountain peaks, but the topographic depression of a caldera is an exception.

Figure 7–25 Ash-flow tuff forms when an ash flow comes to a stop. The fragments in the tuff are pieces of rock that were carried along with the volcanic ash and gas. *(Geoffrey Sutton)*

Figure 7–27 shows that calderas, ash-flow tuffs, and related rocks occur over a large part of western North America. Consider two well-known examples.

Yellowstone National Park and the Long Valley Caldera

Yellowstone National Park in Wyoming and Montana is the oldest national park in the United States. Its geology consists of three large overlapping calderas and the ash-flow tuffs that erupted from them. The oldest eruption took place 1.9 million years ago and ejected 2500 cubic kilometers of pyroclastic material. This is enough ash to bury all of Yellowstone Park to a depth of 250 meters. The next major eruption occurred 1.3 million years ago and produced about 400 cubic kilometers of ash. The most recent, 0.6 million years ago, ejected 1000 cubic kilometers of ash and other debris and produced the Yellowstone caldera in the center of the park.

Intervals of 0.6 to 0.7 million years separate the three Yellowstone eruptions; 0.6 million years have passed since the most recent one. The park's geysers and hot springs are formed by hot magma beneath Yellowstone, and numerous small earthquakes indicate that the magma is moving. Geologists would not be surprised if another eruption occurred at any time.

A geologic environment similar to that of Yellowstone is found near Yosemite National Park in eastern California. Here the 170-cubic-kilometer Bishop Tuff erupted from the Long Valley caldera 0.7 million years

Figure 7–26 Crater Lake, in Oregon, fills a small, well-preserved caldera. *(Greg Vaughn/Tom Stack and Associates)*

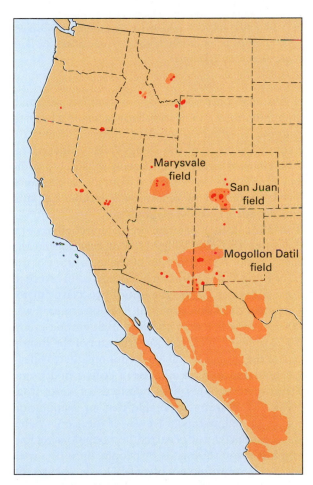

Figure 7–27 Calderas (red dots) and ash-flow tuffs (orange areas) are abundant in western North America.

Figure 7–28 This view of the Sierra Nevada range and the Long Valley caldera was photographed from the space shuttle Columbia. Mono Lake is the lower left of the four large lakes; the caldera occupies the valley just below Mono Lake, and Mammoth Mountain is on the left side of the valley. *(NASA/Science Photo Library)* • Interactive Question: How would another large eruption of this volcanic system affect the human population of California?

ago. Although only one major eruption has occurred, seismic monitoring indicates that magma lies beneath Mammoth Mountain, a popular California ski area, on the southwest edge of the Long Valley caldera (Fig. 7–28). Unusual amounts of carbon dioxide—a common volcanic gas—have escaped from the caldera since 1994. As in the case of Yellowstone, another eruption would not surprise most geologists.

It is difficult to appreciate the potential effects of a Yellowstone or Long Valley type of eruption on the modern world. No event even of the magnitude of the smaller Long Valley eruption has occurred in recorded history. Measured by the amount of pyroclastic material erupted, the 1980 Mount St. Helens eruption was 2500 times smaller than the first Yellowstone event, and 170 times smaller than the eruption of the Bishop Tuff. Yet the Mount St. Helens eruption killed 63 people, caused hundreds of millions of dollars in damage, and disrupted people's lives hundreds of miles from the volcano. It seems inevitable that an eruption on the scale of Yellowstone, or even of the Long Valley caldera, especially if it occurred in a populous region, would kill many thousands of people, entomb towns and cities beneath meters of red-hot ash, change the courses of rivers and streams, and destroy transportation systems and agricultural land. It would probably

raise a dust cloud in the upper atmosphere that would darken the Sun over the entire planet for months or years, cooling the atmosphere, altering global climate, and changing global ecosystems. It is obvious that plants and animals have survived many such events in geologic history, but that is little consolation for the individuals who eventually must experience another such volcanic crisis.

The potential for such a disaster makes a volcanic eruption of this type one of the greatest of geologic hazards. It also makes risk assessment and prediction of volcanic eruptions an important part of modern science.

7.7 Risk Assessment: Predicting Volcanic Eruptions

Approximately 1300 active volcanoes are recognized globally, and 5564 eruptions have occurred in the past 10,000 years. These figures do not include the numerous submarine volcanoes of the mid-oceanic ridge system. Many volcanoes have erupted recently, and we are certain that others will erupt soon. How can geologists predict an eruption and reduce the risk of a volcanic disaster?

Regional Prediction Volcanoes concentrate near subduction zones, spreading centers, and hot spots over mantle plumes but are rare in other places. Therefore, the first step in assessing the volcanic hazard of an area is to understand its tectonic environment. Western Washington and Oregon are near a subduction zone and in a region likely to experience future volcanic activity. Kansas and Nebraska are not.

Furthermore, the potential violence of a volcanic eruption is related to the geologic environment of the volcano. If an active volcano lies on continental crust, the eruptions may be violent because granitic magma may form. In contrast, if the region lies on oceanic crust, the eruptions may be gentle because basaltic volcanism is more likely. Violent eruptions are likely in western Washington and Oregon, but less so on Hawaii or Iceland.

Risk assessment is based both on frequency of past eruptions and on potential violence. However, regional prediction only estimates probabilities and cannot be used to determine when a particular volcano will erupt or the intensity of a particular eruption.

Short-Term Prediction In contrast to regional predictions, short-term predictions attempt to forecast the specific time and place of an impending eruption. They are based on instruments that monitor an active

Figure 7–29 United States Geological Survey geologists accurately predicted the May 1980 eruption of Mount St. Helens. *(USGS)*

volcano to detect signals that the volcano is about to erupt. The signals include changes in the shape of the mountain and surrounding land, earthquake swarms indicating movement of magma beneath the mountain, increased emissions of ash or gas, increasing temperatures of nearby hot springs, and any other signs that magma is approaching the surface.

In 1978, two United States Geological Survey (USGS) geologists, Dwight Crandall and Don Mullineaux, noted that Mount St. Helens had erupted more frequently and violently during the past 4500 years than any other volcano in the contiguous 48 states. They predicted that the volcano would erupt again before the end of the century.

In March 1980, about two months before the great May eruption, puffs of steam and volcanic ash rose from the crater of Mount St. Helens, and swarms of earthquakes occurred beneath the mountain. This activity convinced other USGS geologists that Crandall and Mullineaux's prediction was correct. In response, they installed networks of seismographs, tiltmeters, and surveying instruments on and around the mountain.

In the spring of 1980, the geologists warned government agencies and the public that Mount St. Helens showed signs of an impending eruption. The U.S. Forest Service and local law enforcement officers quickly evacuated the area surrounding the mountain, averting a much larger tragedy (Fig. 7–29). Using sim-

ilar kinds of information, geologists predicted the 1991 Mount Pinatubo eruption in the Philippines, saving many lives.

Although the June 25, 1997, eruption of the Soufrière Hills volcano on the Caribbean island of Montserrat killed 19 people and destroyed many homes and farms, predictions of the eruption by the Montserrat Volcano Observatory saved many additional lives (Fig. 7–30). The island was long known to harbor an active volcano, and the observatory was established to monitor the volcano and to predict

A

B

Figure 7–30 (A) The June 25th, 1997, eruption of the Soufriere Hills volcano on the Carribean island of Montserrat looms over nearby villages. (B) Pyroclastic flows buried and devastated several communities, killing an estimated 19 people. Warnings from the Montserrat Volcano Observatory led to evacuations that saved many other lives. *(Kevin West/Gamma Liaison)* ● Interactive Question: Look up the description of this eruption on the website of the Montserrat Volcano Observatory, and discuss the economic and health effects of this eruption on the inhabitants of Montserrat.

eruptions. In January 1992, observatory seismographs recorded swarms of earthquakes in the southern part of the island. The quakes continued intermittently through the summer of 1995, when the volcano began to emit smoke, steam, and ash. Then, on August 21, 1995, an eruption blanketed the town of Plymouth in a thick ash cloud and darkened the sky for about 15 minutes. Government authorities ordered the first evacuation of southern Montserrat. From September through November, people returned to their homes as observatory scientists noted continued earthquake activity, ash and steam eruptions, and the swelling of a dome in the volcano's crater. Surveyors detected increasing deformation of the land around the volcano. On December 1, southern Montserrat was evacuated for a second time.

In the beginning of January 1996, some people again returned to their homes despite continued volcanic activity. Then, on April 6, the first in a series of pyroclastic flows erupted. Some flowed all the way to the sea. The southern part of the island was evacuated for a third time. Through the summer, small eruptions continued, swelling of the land increased, and earthquakes continually shook the island. Some people returned again to their homes and farms, but many, warned both by the obvious continued activity and by observatory scientists, stayed away.

The observatory recorded continued swelling of the dome, swarms of earthquakes, and many large and small eruptions, including pyroclastic flows, through the rest of 1996 and the first half of 1997. On June 25, 1997, major pyroclastic flows reached to within 50 meters of the airport. Surges and flows devastated several communities, killing 19 people.

Although some people returned to homes and farms during lulls in the long eruptive sequence, there is little doubt that the warnings issued by the observatory and evacuations ordered by the government saved many lives. The volcanic activity continues as this book goes to press.

7.8 Volcanic Eruptions and Global Climate Change

The eruptions of Mount Pinatubo in 1991 produced the greatest ash and sulfur clouds in the latter half of the twentieth century. Satellite measurements show that the total solar radiation reaching the Earth's surface declined by 2 to 4 percent after the Pinatubo eruptions. The following two years, 1992 and 1993, were a few tenths of a degree Celsius cooler than the temperatures of the previous decade. Temperatures rose again in 1994, after the ash and sulfur settled out.

A plot of global temperatures before and after eight recent major volcanic eruptions shows a correlation between global cooling and volcanic eruptions (Fig. 7–31). The correlation substantiates meteorological models showing that high-altitude dust reflects sunlight and cools the atmosphere.

Historic eruptions have been minuscule compared with some in the more distant past. Recall from the opening essay to Unit I that about 245 million years ago, at the end of the Permian period, 90 percent of all marine species and two thirds of reptile and amphibian species died suddenly in the most catastrophic mass extinction in Earth history. This extinction event

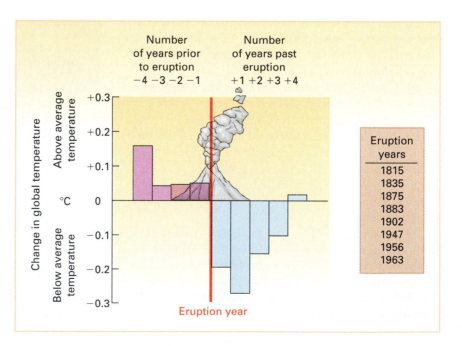

Figure 7–31 A plot of temperatures in the Northern Hemisphere shows that atmospheric cooling follows major volcanic eruptions. *(Michael Rampino,* Annual Review of Earth and Planetary Science, *16, 1988, pp. 73–99)*

coincided with a massive volcanic eruption in Siberia that disgorged a million cubic kilometers of flood basalt onto the Earth's surface to form a basalt plateau of the type described in Section 7.5.[1] The eruption must have released massive amounts of ash and sulfur compounds into the upper atmosphere, leading to cooler global climates. Many geologists think that the Earth cooled enough to cause or at least contribute to the mass extinction. However, as discussed earlier, other scientists have proposed alternative mechanisms for the extinction.

Large reservoirs of carbon are stored in the Earth's mantle, where basaltic magma forms. Modern basalt eruptions emit about 0.44 percent by weight as much carbon dioxide as they do basalt. Thus, huge amounts of carbon dioxide enter the atmosphere during the eruption of a large basalt plateau. Carbon dioxide is a greenhouse gas that causes warming of the Earth's atmosphere.

About 120 million years ago, a massive upwelling of mantle rock created the Mid-Cretaceous superplume, which erupted to form a vast lava plateau beneath the Pacific Ocean, the submarine equivalent of a basalt plateau. These eruptions also released large quantities of carbon dioxide. Mid-Cretaceous fossils indicate that the Earth's climate was several degrees warmer than at present. Dinosaurs flourished in lush swamps, and huge coal deposits formed. Climate models show that this volcanic episode probably warmed Earth's climate during mid-Cretaceous time.

Note that one mechanism links volcanic eruptions with global cooling and, paradoxically, the other associates eruptions with warming. A dark cloud cools the air immediately. Increased atmospheric carbon dioxide warms the air over longer time periods. Some volcanic eruptions may emit more ash and less carbon dioxide, whereas others may emit a higher proportion of gases and fewer solids. In either case, volcanic activity can profoundly alter climate.

We have learned that most volcanic activity is driven by movements of mantle rock deep within the Earth. However, recent research indicates that changes in climate and sea level may affect volcanic activity. In one study, scientists noted a statistical correlation between sea level change and explosive volcanic activity in the Mediterranean.[2] The authors postulated that changes in sea level affect the pressure on magma that is slowly rising through the crust. If sea level rises, pressure increases on magma beneath an island volcano and the magma is less likely to erupt. If sea level falls, pressure decreases and eruptions become more likely. In Chapter 14 we explain how sea level rises and falls in response to climate changes. Therefore, if this hypothesis is correct, climate changes and fluctuations in sea level affect the frequency of volcanic eruptions. Volcanic eruptions, in turn, affect climate. This circle of cause and effect reminds us how complex and interactive Earth systems really are.

7.9 CASE STUDY
Mount Vesuvius

In addition to altering atmospheric composition and global climate throughout geologic time, volcanic eruptions have caused loss of life and destruction of property throughout human history. Table 7–2 sum-

---

[1] A. R. Basu et al., "High −He^3 Plume Origin and Temporal-Spatial Evolution of the Siberian Flood Basalts." *Science*, Vol. 269, p. 5225, August 11, 1995.

[2] W. J. McGuire et al., "Correlation Between Role of Sea-Level Change and Frequency of Explosive Volcanism in the Mediterranean." *Nature*, Vol. 389, Oct. 2, 1997, p. 473.

TABLE 7–2

Some Notable Volcanic Disasters Since the Year A.D. 1500 Involving 5000 or More Fatalities

| Volcano | Country | Year | Primary Cause of Death and Number of Deaths | | | | |
			Pyroclastic Flow	Debris Flow	Lava Flow	Posteruption Starvation	Tsunami
Kelut	Indonesia	1586		10,000			
Vesuvius	Italy	1631			18,000		
Etna	Italy	1669			10,000		
Lakagigar	Iceland	1783				9,340	
Unzen	Japan	1792					15,190
Tambora	Indonesia	1815	12,000			80,000	
Krakatoa	Indonesia	1883					36,420
Pelée	Martinique	1902	29,000				
Santa Maria	Guatemala	1902	6,000				
Kelut	Indonesia	1919		5,110			
Nevado del Ruiz	Colombia	1985	>22,000				

Figure 7–32 Archaeologists uncover molds of Pompeiians killed in the A.D. 79 eruption of Mount Vesuvius. *(UPI/Bettmann)*

marizes the major known volcanic disasters since A.D. 1500.

In A.D. 79, Mount Vesuvius erupted and destroyed the Roman cities of Pompeii and Herculaneum and several neighboring villages near what is now Naples, Italy. Prior to that eruption, the volcano had been inactive for about 700 years, so long that farmers had cultivated vineyards on the sides of the mountain all the way to the summit. During the eruption, an ash flow streamed down the flanks of the volcano, burying the cities and towns under 5 to 8 meters of hot ash. When archaeologists located and excavated Pompeii 17 centuries later, they found molds of inhabitants trapped by the ash flow as they attempted to flee or find shelter (Fig. 7–32). Some of the molds appear to preserve facial expressions of terror. After the A.D. 79 eruption, Mount Vesuvius returned to relative quiescence but became active again in 1631. It was frequently active from 1631 to 1944, and in this century it erupted in 1906, 1929, and 1944.

Mount Vesuvius is a stratovolcano that formed over the past 25,000 years by many eruptions that varied from gently flowing lava to the explosion that buried Pompeii. A recent study of seismic velocities beneath the volcano showed that seismic waves suddenly slow from 6 to 2.7 kilometers per second at a depth of 10 kilometers. The sudden velocity decrease suggests that molten magma still exists at that depth.[3] Since stratovolcanoes erupt frequently and remain active for long periods, and because magma underlies the volcano, geologists consider Mount Vesuvius a high risk for future eruptions. ●

---

[3] A. Zollo et al., "Seismic Evidence for a Low-Velocity Zone in the Upper Crust Beneath Mount Vesuvius," *Science*, Vol. 274, Oct. 25, 1996, pp. 592–594.

SUMMARY

Rocks of the asthenosphere partially melt to produce basaltic magma as a result of three processes: rising temperature, **pressure-release melting,** and addition of water. These processes occur beneath spreading centers, in mantle plumes, and in subduction zones to form both volcanoes and plutons. Basaltic magma forms by partial melting of mantle peridotite. Granitic magma forms by a two-step process in which basaltic magma forms by any of the above processes, then rises into and melts granitic rocks of the lower continental crust.

Basaltic magma usually erupts in a relatively gentle manner onto the Earth's surface from a **volcano.** In contrast, granitic magma typically solidifies within the Earth's crust. When granitic magma does erupt onto the surface, it often does so violently. These contrasts in behavior of the two types of magma are caused by differences in silica and water content.

Any intrusive mass of igneous rock is a **pluton.** A **batholith** is a pluton with more than 100 square kilometers of exposure at the Earth's surface. A **dike** and a **sill** are both sheet-like plutons. Dikes cut across layering in country rock, and sills run parallel to layering.

Magma that rises to the surface may flow onto the Earth's surface as lava or may erupt explosively as **pyroclastic** material. Fluid lava forms **lava plateaus** and **shield volcanoes.** A pyroclastic eruption may form a **cinder cone.** Alternating eruptions of fluid lava and pyroclastic material from the same vent create a **composite cone.** When granitic magma rises to the Earth's surface, it may erupt explosively, forming **ash-flow tuffs** and **calderas.**

Volcanic eruptions are common near a subduction zone, near a spreading center, and at a hot spot over a mantle plume but are rare in other tectonic environments. Eruptions on a continent are often violent, whereas those in oceanic crust are gentle. Such observations form the basis of regional predictions of volcanic hazards. Short-term predictions are made on the basis of earthquakes caused by magma movements, swelling of a volcano, increased emissions of gas and ash from a vent, and other signs that magma is approaching the surface. Large volcanic eruptive episodes are associated with global climate change.

KEY TERMS

pressure-release melting 132

partial melting 136

pluton 137

batholith 137

stock 137

dike 139

sill 139

lava 139

pahoehoe 140

aa 140

columnar joint 140

pyroclastic rock 141

volcanic ash 141

cinders 141

fissure 141

flood basalt 141

lava plateau 141

volcano 142

vent 142

crater 142

shield volcano 143

cinder cone 143

composite cone 144

stratovolcano 144

ash flow 146

nuèe ardente 146

ash-flow tuff 146

caldera 146

FOR REVIEW

1. Describe several different ways in which volcanoes and volcanic eruptions can threaten human life and destroy property.

2. Describe three processes that generate magma in the asthenosphere.

3. Describe magma formation in a spreading center, a hot spot, and a subduction zone.

4. Describe the origin of granitic magma.

5. How much silica does average granitic magma contain? How much does basaltic magma contain?

6. How much water does average granitic magma contain? How much does basaltic magma contain?

7. Why does magma rise soon after it forms?

8. Explain why basaltic magma and granitic magma behave differently as they rise toward the Earth's surface.

9. Many rocks, and even entire mountain ranges, at the Earth's surface are composed of granite. Does this observation imply that granite forms at the surface?

10. Explain the difference between a dike and a sill.

11. How do a shield volcano, a cinder cone, and a composite cone differ from one another? How are they similar?

12. How does a composite cone form?

13. Explain why and how granitic magma forms ash-flow tuffs and calderas.

14. How does a caldera form?

15. Explain why additional eruptions in Yellowstone Park seem likely. Describe what such an eruption might be like.

FOR DISCUSSION

1. How and why does pressure affect the melting point of rock and, conversely, the solidification temperature of magma? How does the explanation differ for basaltic and granitic magma?

2. Why does water play an important role in magma generation in subduction zones, but not in the other two major environments of magma generation?

3. How could you distinguish between a sill exposed by erosion and a lava flow?

4. Imagine that you detect a volcanic eruption on a distant planet but have no other data. What conclusions could you draw from this single bit of information? What types of information would you search for to expand your knowledge of the geology of the planet?

5. Explain why some volcanoes have steep, precipitous faces, but many do not.

6. Compare and contrast the danger of living 5 kilometers from Yellowstone National Park with the danger of living an equal distance from Mount St. Helens. Would your answer differ for people who live 50 kilometers or those who live 500 kilometers from the two regions?

7. Use long-term prediction methods to evaluate the volcanic hazards in the vicinity of your college or university.

8. Discuss some possible consequences of a large caldera eruption in modern times. What is the probability that such an event will occur?

9. We explained how volcanic eruptions can influence climate and how climate can influence volcanic eruptions. Discuss the roles of the biosphere, the hydrosphere, the atmosphere, and the geosphere in these interactions.

Geologic Structures, Mountain Ranges, and Continents

Tectonic forces folded these limestone layers in the Canadian Rockies.

Mountains form some of the most majestic landscapes on Earth. People find peace in the high mountain air and quiet valleys. But mountains also display nature's power. Millions of years ago, tremendous tectonic forces folded the rocks and raised the mountains skyward. Today, storms swirl among the peaks, creating rain and snow that slowly erode the rocks away. Glaciers and streams scour deep valleys into the bedrock. In the first portion of this chapter, we will study how rocks behave when tectonic forces stress them. In the second portion, we will learn how tectonic processes raised mountain ranges and formed the Earth's continents. ●

GEOLOGIC STRUCTURES

8.1 Stress and Rock Deformation

Recall from Chapter 6 that stress is a force exerted against an object. Tectonic forces exert different types of stress on rocks in different geologic environments. **Confining stress,** or **confining pressure,** develops when rock or sediment is buried (Fig. 8–1A). Confining pressure merely compresses rocks but does not distort them because the compressive force acts equally in all directions, like water pressure on a fish.

In contrast, **directed stress** acts most strongly in one direction. Tectonic processes create three types of directed stress. **Compressive stress** squeezes rocks together and shortens the distance parallel to the squeezing direction (Fig. 8–1B). Compression is common in convergent plate boundaries, where two plates converge and the rock crumples, just as car fenders crumple during a collision. **Extensional stress** (often called **tensional stress**) pulls rock apart and is the opposite of tectonic compression (Fig. 8–1C). Rocks at a divergent plate boundary stretch and pull apart because they are subject to extensional stress. **Shear stress** acts in parallel but opposite directions (Fig. 8–1D). Shearing deforms rock by causing one part of a rock mass to slide past the other part, as in a transform fault or a transform plate boundary.

When a rock is stressed, the rock may deform elastically or plastically, or it may simply break by brittle fracture. Several factors control how rocks respond to stress.

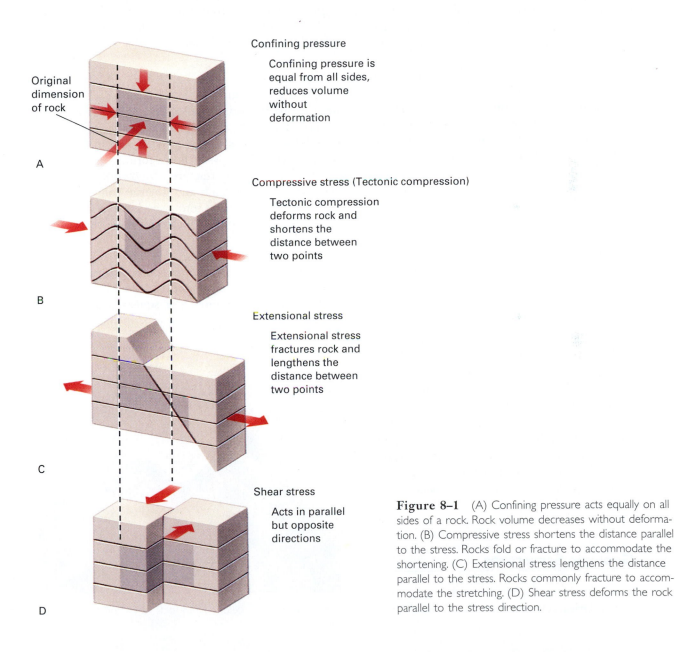

Original dimension of rock

Confining pressure

Confining pressure is equal from all sides, reduces volume without deformation

A

Compressive stress (Tectonic compression)

Tectonic compression deforms rock and shortens the distance between two points

B

Extensional stress

Extensional stress fractures rock and lengthens the distance between two points

C

Shear stress

Acts in parallel but opposite directions

D

Figure 8–1 (A) Confining pressure acts equally on all sides of a rock. Rock volume decreases without deformation. (B) Compressive stress shortens the distance parallel to the stress. Rocks fold or fracture to accommodate the shortening. (C) Extensional stress lengthens the distance parallel to the stress. Rocks commonly fracture to accommodate the stretching. (D) Shear stress deforms the rock parallel to the stress direction.

Figure 8–2 This rock in the Nahanni River, Northwest Territories, Canada, folded plastically and then fractured.

1. **The nature of the material.** Think of a quartz crystal, a gold nugget, and a rubber ball. If you strike quartz with a hammer, it shatters. That is, it fails by brittle fracture. In contrast, if you strike the gold nugget, it deforms in a plastic manner—it flattens and stays flat. If you hit the rubber ball, it deforms elastically and rebounds immediately, sending the hammer flying back at you. Initially, all rocks react to stress by deforming elastically by a slight amount. Near the Earth's surface, where temperature and pressure are low, different types of rocks behave differently with continuing stress. Granite and quartzite tend to behave in a brittle manner. Other rocks, such as shale, limestone, and marble, tend to deform plastically.

2. **Temperature.** The higher the temperature, the greater the tendency of a rock to behave in a plastic manner. It is difficult to bend an iron bar at room temperature, but if the bar is heated in a forge, it becomes plastic and bends easily.

3. **Pressure.** High pressure favors plastic behavior. During burial, both temperature and pressure increase. Both factors promote plastic deformation, so deeply buried rocks have a greater tendency to bend and flow than shallow rocks.

4. **Time.** Stress applied over a long time, rather than suddenly, also favors plastic behavior. Marble park benches in New York City have sagged plastically under their own weight within 100 years. In contrast, rapidly applied stress to a marble bench, such as the blow of a hammer, causes brittle fracture.

8.2 Geologic Structures

Enormous tectonic stresses fracture and deform rocks near the boundaries between tectonic plates. A **geo-** logic structure is any feature produced by rock deformation. Tectonic forces create three types of structures: folds, faults, and joints.

Folds

A **fold** is a bend in rock. Some folded rocks display little or no fracturing, indicating that the rocks deformed in a plastic manner. In other cases, folding occurs by a combination of plastic deformation and brittle fracture (Fig. 8–2). Folds formed in this manner exhibit many tiny fractures. If you hold a sheet of clay between your hands and squeeze your hands together, the clay deforms into a sequence of folds (Fig. 8–3). This demonstration illustrates three characteristics of folds:

1. Folding usually results from compressive stress. For example, tightly folded rocks in the Himalayas indicate that the region was squeezed as tectonic plates converged.

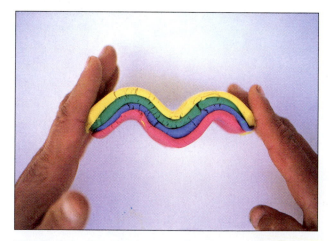

Figure 8–3 Clay deforms into folds when compressed.

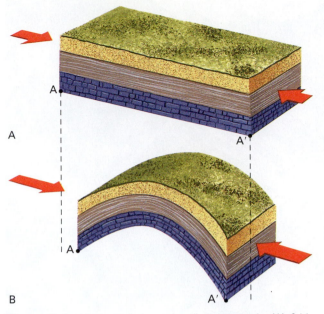

Figure 8–4 Horizontally layered sedimentary rocks (A) fold in response to compressive stress (B). Notice that points A and A′ are closer after folding.

2. Folding always shortens the horizontal distances in rock. Notice in Figure 8–4 that the distance between two points, A and A′, is shorter in the folded rock than it was before folding.

3. A fold usually occurs in a group of many similar folds.

Figure 8–5 summarizes the characteristics of five common types of folds. The figure shows that a fold arching upward is called an **anticline** and one arching downward is a **syncline**. The sides of a fold are called the **limbs**. Even though an anticline is structurally a high point in a fold, anticlines do not always form topographic ridges. Conversely, synclines do not always form valleys. Landforms are created by combinations of tectonic and surface processes. In Figure 8–6, a syncline lies beneath the peak and an anticline forms a saddle between two peaks.

A circular or elliptical anticlinal structure is called a **dome**. Domes resemble inverted bowls. Sedimentary layering dips away from the top of a dome in all directions (Fig. 8–7). A similarly shaped syncline is called a **basin**. Domes and basins can be small structures only a few kilometers in diameter or less. A large basin or dome can be 500 kilometers in diameter. These huge structures form by the sinking or rising of continental crust in response to vertical movements of the underlying mantle. The Black Hills of South Dakota are a large structural dome. The Michigan basin covers much of the state of Michigan, and the Williston basin covers much of eastern Montana, northeastern Wyoming, the western Dakotas, and southern Alberta and Saskatchewan.

Faults

A **fault** is a fracture along which rock on one side has moved relative to rock on the other side (Fig. 8–8). **Slip** is the distance that rocks on opposite sides of a fault have moved. Movement along a fault may be gradual, or the rock may move suddenly, generating an earthquake. Some faults are a single fracture in rock; others consist of numerous closely spaced fractures within a fault zone (Fig. 8–9). Rock may slide hundreds of meters or many kilometers along a large fault zone.

Rock moves repeatedly along many faults and fault zones for two reasons: (1) Tectonic forces commonly persist in the same place over long periods of time (for example, at a tectonic plate boundary). (2) Once a fault forms, it is easier for rock to move again along the same fracture than for a new fracture to develop nearby.

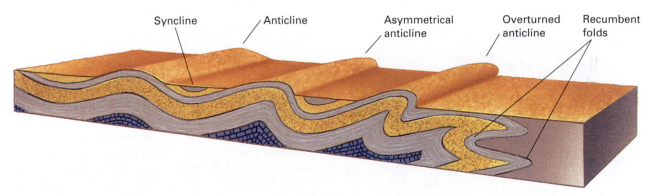

Syncline   Anticline   Asymmetrical anticline   Overturned anticline   Recumbent folds

Figure 8–5 Cross-sectional view of five different kinds of folds. Folds can be symmetrical, as shown on the left, or asymmetrical, as shown in the center. If a fold has tilted beyond the perpendicular, it is overturned.

Figure 8–6 On this peak in the Canadian Rockies, the syncline lies beneath the summit and the anticline forms the low point on the ridge. • Interactive Question: Give a plausible explanation for these relationships.

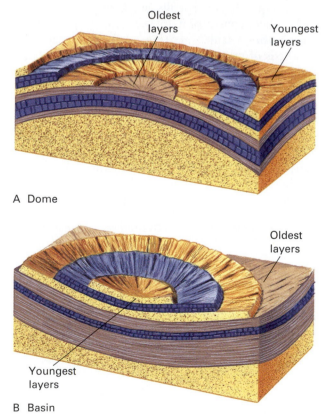

Oldest layers

Youngest layers

A Dome

Oldest layers

Youngest layers

B Basin

Figure 8–7 (A) Sedimentary layering dips away from a dome in all directions. (B) Layering dips toward the center of a basin.

Hydrothermal solutions often precipitate in faults to form rich ore veins. Miners then dig shafts and tunnels along veins to get the ore. Many faults are not vertical but dip into the Earth at an angle. Therefore, many veins have an upper and a lower side. Miners referred to the side that hung over their heads as the **hanging wall** and the side they walked on as the **foot-wall** (Fig. 8–10). These names are commonly used to describe both ore veins and faults.

A **normal fault** forms where extensional stress stretches the Earth's crust, pulling it apart. As the crust stretches and fractures, the hanging wall moves down relative to the footwall. Notice that the horizontal distance between points on opposite sides of the fault, such as A and A′ in Figure 8–10, is greater after normal faulting occurs.

Figure 8–11 shows a wedge-shaped block of rock called a **graben** dropped downward between a pair of normal faults. The word *graben* comes from the German word for "a long depression, ditch, or valley." (Think of a large block of rock settling downward, to form a valley.) If tectonic forces stretch the crust over a large area, many normal faults may develop, allowing numerous grabens to settle downward between the faults. The blocks of rock between the down-dropped grabens then appear to have moved upward relative to the grabens; they are called **horsts.**

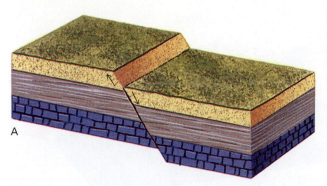

A

B

Figure 8–8 (A) A fault is a fracture along which rock on one side has moved relative to rock on the other side. (B) A small fault has dropped the right side of these volcanic ash layers downward about 60 centimeters relative to the left side. *(Ward's Natural Science Establishment, Inc.)*

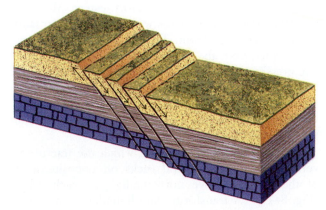

Figure 8–9 Faults with a large slip commonly move along numerous closely spaced fractures, forming a fault zone.

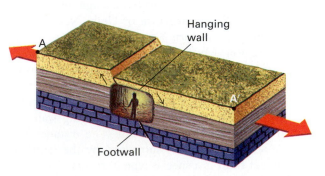

Hanging wall

Footwall

Figure 8–10 A normal fault accommodates extension of the crust. The upper side of a fault is called the hanging wall, and the lower side is called the footwall.

Figure 8–11 (A) Horsts and grabens commonly form where tectonic forces stretch the crust over a broad area. (B) The Basin and Range Province in Nevada is composed of parallel ridges, which are horsts, and valleys, which are grabens. The valleys have no external drainage and are filling with sediment. In spring, water from melting snow drains into the valleys, creating lakes. The water evaporates by mid-summer, leaving dry lake beds that appear as light tan areas in this photograph. *(TSADO/ NASA/Tom Stack and Assoc.)*

B

Normal faults, grabens, and horsts are common where the crust is rifting at a spreading center, such as the mid-oceanic ridge and the East African rift zone, and where tectonic forces stretch a single plate, as in the Basin and Range province of Utah, Nevada, and adjacent parts of western North America.

In a region where compressive tectonic forces squeeze the crust, geologic structures must accommodate crustal shortening. A fold accomplishes shortening. A **reverse fault** also accommodates shortening (Fig. 8–12). In a reverse fault, the hanging wall has moved up relative to the footwall. The distance between points A and A′ is shortened by the faulting.

A **thrust fault** is a special type of reverse fault that is nearly horizontal (Fig. 8–13). In some thrust faults,

the rocks of the hanging wall have moved many kilometers over the footwall. For example, all of the rocks of Glacier National Park in northwestern Montana slid 50 to 100 kilometers eastward along a thrust fault to their present location. This thrust is one of many that formed from about 180 to 45 million years ago as compressive tectonic forces built the mountains of western North America. Most of those thrusts moved large slabs of rock, some even larger than that of Glacier Park, from west to east in a zone extending from Alaska to Mexico.

A **strike-slip fault** is one in which the fracture is vertical, or nearly so, and rocks on opposite sides of the fracture move horizontally past each other (Fig. 8–14). A transform plate boundary is a strike-

Reverse
fault

A

B

Figure 8–12 (A) A reverse fault accommodates crustal shortening when the crust is compressed. (B) A small reverse fault in Zion National Park, Utah.

A

B

Figure 8–13 (A) A thrust fault is a low-angle reverse fault. (B) A small thrust fault near Flagstaff, Arizona. *(Ward's Natural Science Establishment, Inc.)*

depth because rocks become more plastic and less prone to fracturing at deeper levels in the crust.

Joints and faults are important in engineering, mining, and quarrying because they are planes of weakness in otherwise strong rock. Dams constructed in jointed rock often leak, not because the dams themselves have holes, but because water seeps into the

slip fault. As explained previously, the famous San Andreas fault zone is a zone of strike-slip faults that form the boundary between the Pacific plate and the North American plate.

Joints

A **joint** is a fracture in rock and is therefore similar to a fault, except that in a joint, rocks on either side of the fracture have not moved. We discussed columnar joints in basalt in Chapter 7. Tectonic forces also create joints (Fig. 8–15). Most rocks near the Earth's surface are jointed, but joints become less abundant with

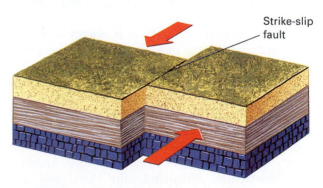

Strike-slip
fault

Figure 8–14 A strike-slip fault is nearly vertical, but movement along the fault is horizontal.

Figure 8–15 Joints such as those in sandstone along the Escalante River in Utah are fractures along which the rock has not slipped.

joints and flows around the dam through the fractures. You can commonly see seepage caused by such leaks in canyon walls downstream from a dam.

Geologic Structures and Plate Boundaries

Each of the three different types of plate boundaries produces different tectonic stresses and therefore different kinds of structures. Extensional stress at a divergent boundary (mid-oceanic ridges and continental rift boundaries) produces normal faults and grabens, but little folding of rocks.

Movement along a transform boundary may fold, fault, and uplift nearby rocks. Forces of this type have formed the San Gabriel Mountains along the San Andreas fault zone, as well as mountain ranges north of the Himalayas.

Compressive stress commonly produces large regions of folds, reverse faults, and thrust faults near a convergent plate boundary. Folds and thrust faults are common in the mountains of western North America, the Appalachian Mountains, the Alps, and the Himalayas, all of which formed at convergent boundaries (Fig. 8–16).

Although plate convergence commonly creates horizontal compression, in some instances crustal exten-

sion and normal faulting also occur. The Himalayas are an intensely folded mountain chain that formed from a collision between India and southern Asia. Yet, normal faults are common in parts of the Himalayas. We will describe how this occurs in Section 8.6.

MOUNTAIN RANGES AND CONTINENTS

8.3 Mountains and Mountain Ranges

Mountain-Building Processes

Mountains grow along each of the three types of tectonic plate boundaries. As you learned in Chapter 5, the world's largest mountain chain, the mid-oceanic ridge, formed at divergent plate boundaries beneath the ocean. Mountain ranges also rise at divergent plate boundaries on land. Mount Kilimanjaro and Mount Kenya, two volcanic peaks near the equator, lie along the East African rift. Other ranges, such as the San Gabriel Mountains of California, form at transform plate boundaries. However, the great continental mountain chains, including the Andes, Appalachians, Alps, Himalayas, and Rockies, all rose near convergent plate boundaries. Folding and faulting of rocks, earthquakes, volcanic eruptions, intrusion of plutons, and metamorphism all occur at a convergent plate boundary. The term **orogeny** refers to the process of mountain building and includes all of these activities.

Because plate boundaries are linear, mountains most commonly occur as long, linear, or slightly curved ranges and chains. For example, the Andes extend in a narrow band along the west coast of South America, and the Appalachians form a gently curving uplift along the east coast of North America.

Most continental mountain ranges rise because the crust thickens as a consequence of plate convergence. Several processes thicken the crust where the plates converge:

1. In a subduction zone, the descending slab generates magma, which cools within the crust to form plutons or erupts onto the surface to form volcanic peaks. Both the plutons and volcanic rocks thicken the continental crust by adding new material to it.

2. Magmatic activity heats the lithosphere above a subduction zone, causing it to become less dense and therefore thicker.

3. In a region where two continents collide, one continent may be forced beneath the other. This process, called **underthrusting**, can double the thickness of continental crust in the collision zone (Fig 8–17).

Figure 8–16 Wildly folded sedimentary rocks form the Neptse-Lhotse wall seen from an elevation of 7600 meters on Mount Everest. *(Galen Rowell/Mountain Light)*

4. Compressive forces fold and crumple rock, squeezing the continent and increasing its thickness. These compressive forces are important in both subduction zones and continent-continent collisions.

Thus, addition of magma, heating, underthrusting, and folding all combine to thicken continental crust and lithosphere.

As the lithosphere thickens, it rises isostatically to form a mountain chain. When a mountain chain grows higher and heavier, eventually the underlying rocks cannot support the weight of the mountains. The crust and underlying lithosphere then spread outward beneath the mountains. As an analogy, consider pouring cold honey onto a table top. At first, the honey piles up into a high, steep mound, but soon it begins to flow

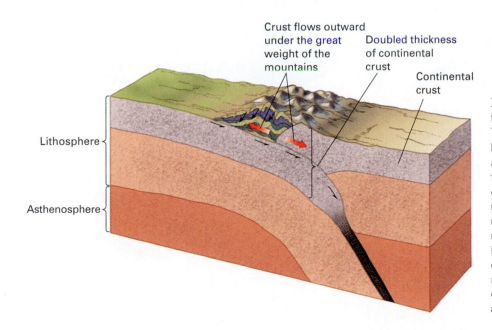

Figure 8–17 Several factors affect the height of a mountain range. Today the Himalayas are being uplifted by continued underthrusting. At the same time, erosion removes the tops of the peaks, and the sides of the range slip downward along normal faults. As these processes remove weight from the range, the mountains rise isostatically. • Interactive Question: Would you expect the height of a mountain range to rise and then decline at constant rates? Defend your answer.

outward under its own weight, lowering the top of the mound.

At the same time streams, glaciers, and landslides erode the peaks as they rise, carrying the sediment into adjacent valleys. Initially, when the mountains erode, they become lighter and rise isostatically, just as a canoe rises when you step out of it. Eventually, erosion wins over isostatic rebound. The Appalachians are an old range where erosion is now wearing away the remains of peaks that may once have been the size of the Himalayas.

With this background, let us look at mountain building in three types of convergent plate boundaries.

8.4 Island Arcs: Subduction in an Ocean Basin

An **island arc** is a volcanic mountain chain that forms where two plates carrying oceanic crust converge. The convergence causes one of the plates to sink into the mantle, creating a subduction zone and an oceanic trench. Magma forms in the subduction zone, and rises to build submarine volcanoes. These volcanoes may eventually grow above sea level, creating an arc-shaped volcanic island chain next to the trench.

A layer of sediment a half kilometer or more thick commonly covers the basaltic crust of the deep sea floor. As the two plates converge, some of the sediment is scraped from the subducting slab and jammed against the inner wall (the wall toward the island arc) of the trench. Occasionally, slices of rock from the oceanic crust, and even pieces of the upper mantle, are scraped off and mixed in with the sea-floor sediment. The process is like a bulldozer scraping soil from bedrock and occasionally knocking off a chunk of bedrock along with the soil. This scraping and compression folds and fractures sediment and rock. The rocks added to the island arc in this way are called a **subduction complex** (Fig. 8–18).

Growth of the subduction complex occurs by addition of the newest slices at the bottom of the complex. This underthrusting bends the subduction complex, forming a sedimentary basin called a **forearc basin** between the subduction complex and the island arc. This process is similar to holding a flexible notebook horizontally between your hands. If you move your hands closer together, the middle of the notebook bends downward to form a topographic depression analogous to a forearc basin. In addition, underthrusting thickens the crust, leading to isostatic uplift of the subduction complex. The forearc basin fills with sediment eroded from the volcanic islands.

Island arcs are abundant in the Pacific Ocean, where convergence of oceanic plates is common. The western Aleutian Islands and most of the island chains of the southwestern Pacific are island arcs (Fig 8–19).

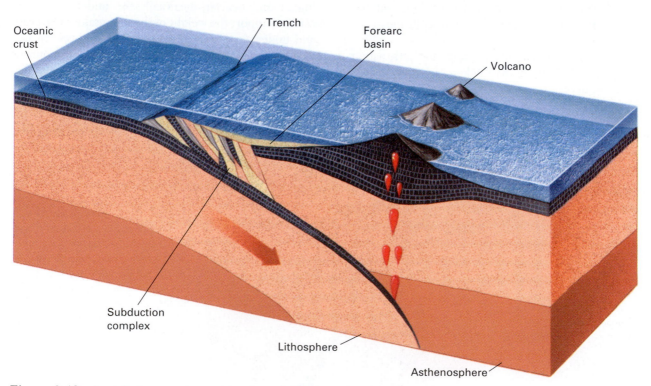

Figure 8–18 A subduction complex contains slices of oceanic crust and upper mantle scraped from the top of a subducting plate.

Figure 8–19 An island arc parallels a subduction zone that curves in a broad crescent from New Guinea to Burma. *(Tom Van Sant)*

8.5 The Andes: Subduction at a Continental Margin

The Andes are the world's second highest mountain chain, with 49 peaks above 6000 meters, nearly 20,000 feet (Fig. 8–20). The highest peak is Aconcagua, at 6962 meters. The Andes rise almost immediately from the Pacific coast of South America, starting nearly at sea level. Igneous rocks make up most of the Andes, although the chain also contains folded sedimentary rocks, especially in the eastern foothills.

In early Jurassic time, the lithospheric plate that included South America started moving westward. To accommodate the westward motion, oceanic lithosphere began to sink into the mantle beneath the west coast of South America. Thus, a subduction zone formed by early Cretaceous time, 140 million years ago (Fig. 8–21A).

By 130 million years ago, the sinking plate was generating vast amounts of basaltic magma (Fig. 8–21B). Some of this magma rose to the surface to erupt from volcanoes. Most of it, however, melted portions of the lower crust to form andesitic and granitic magma. As a result, both volcanoes and plutons formed along the entire length of western South America. As the oceanic plate sank beneath the continent, slices of sea-floor mud and rock were scraped from the subducting plate, forming a subduction complex similar to that of an island arc.

The rising magma heated and thickened the crust beneath the Andes, causing it to rise isostatically and form great peaks. When the peaks became sufficiently high and heavy, the weak, soft rock oozed outward under its own weight. This spreading formed a great belt of thrust faults and folds along the east side of the Andes (Fig. 8–21C).

The Andes, then, are a mountain chain consisting predominantly of igneous rocks formed by subduction at a continental margin. The chain also contains extensive sedimentary rocks on both sides of the mountains; those rocks formed from the sediment eroded from the rising peaks. The Andes are a good general example of subduction at a continental margin, and this type of plate margin is called an **Andean margin.**

Figure 8–20 The Cordillera Apolobamba in Bolivia rises over 6000 meters.

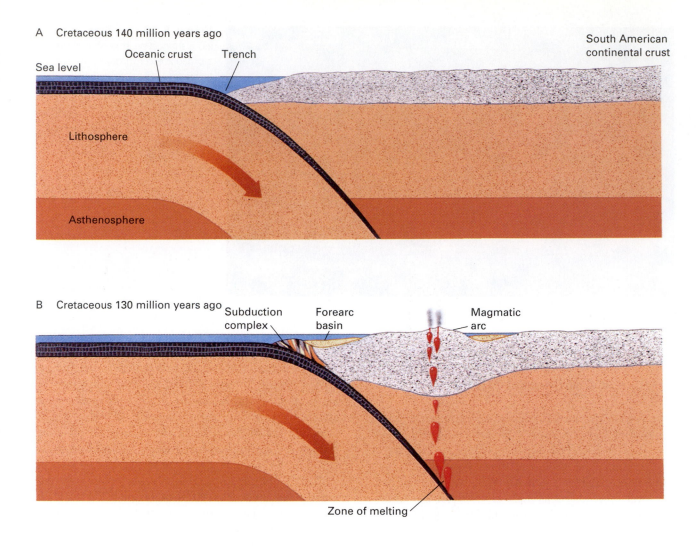

A **Cretaceous 140 million years ago**

Sea level
Oceanic crust   Trench

South American
continental crust

Lithosphere

Asthenosphere

B **Cretaceous 130 million years ago**

Subduction
complex

Forearc
basin

Magmatic
arc

Zone of melting

C **Late Cretaceous 90 million years ago**

New magmatic arc
Old magmatic arc rocks

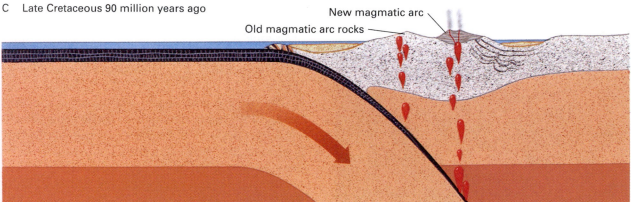

Figure 8–21 Development of the Andes, seen in cross-section looking northward. (A) As the South American lithospheric plate moved westward in early Cretaceous time, about 140 million years ago, a subduction zone and a trench formed at the west coast of the continent. (B) By 130 million years ago, igneous activity began and a subduction complex and forearc basin formed. (C) By 90 million years ago, the trench and region of igneous activity had both migrated eastward. Old volcanoes became dormant and new ones formed to the east. ●
Interactive Question: From the plate map in Chapter 5, would you expect the Andes to be impacted by continent-continent collision, as the Himalayas were? If your answer is yes, give a plausible time frame. If your answer is no, why not?

Figure 8–22 The Himalayas were formed by convergence of a plate carrying oceanic crust with a plate carrying continental crust, followed by a continent-continent collision. Ama Dablam, Nepal. *(Spencer Swanger/Tom Stack and Assoc.)*

8.6 The Himalayas: A Collision Between Continents

The world's highest mountain chain, the Himalayas, separates China from India and includes the world's highest peaks (Fig. 8–22). If you were to stand on the southern edge of the Tibetan Plateau and look southward, you would see the high peaks of the Himalayas. Beyond this great mountain chain lie the rainforests and hot, dry plains of India. If you had been able to stand in the same place 200 million years ago and look southward, you would have seen only ocean. At that time, India was located south of the equator, separated from Tibet by thousands of kilometers of open ocean. The Himalayas had not yet begun to rise (Fig. 8–23).

Figure 8–23 (A) Two hundred million years ago, India was part of a large continent located near the South Pole. (B) At that time India, southern Asia, and the intervening ocean basin were parts of the same lithospheric plate. (In this figure, the amount of oceanic crust between Indian and Asian continental crust is abbreviated to fit the diagram.)

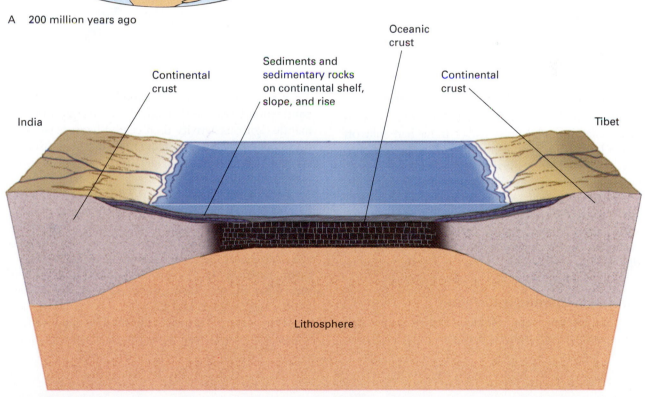

A   200 million years ago

B

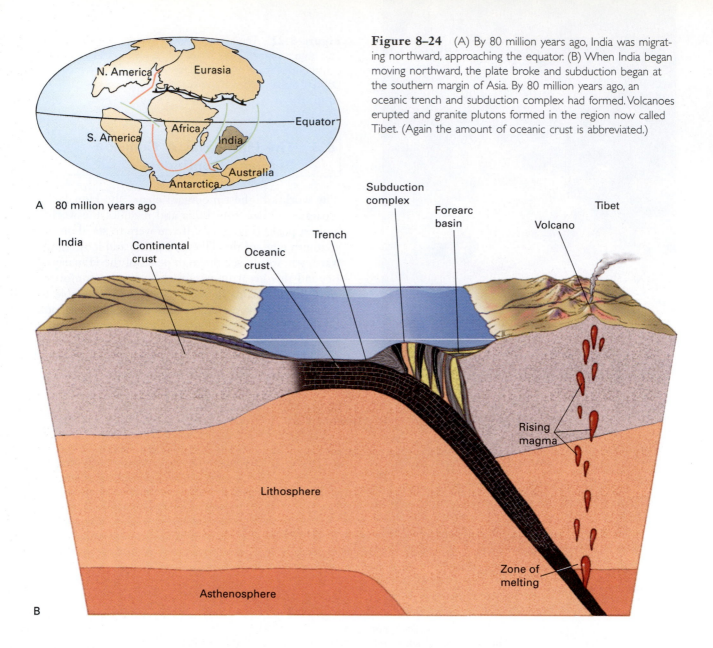

Figure 8–24 (A) By 80 million years ago, India was migrating northward, approaching the equator. (B) When India began moving northward, the plate broke and subduction began at the southern margin of Asia. By 80 million years ago, an oceanic trench and subduction complex had formed. Volcanoes erupted and granite plutons formed in the region now called Tibet. (Again the amount of oceanic crust is abbreviated.)

A   80 million years ago

B

Formation of an Andean-Type Margin

By 80 million years ago, a triangular piece of lithosphere that included present-day India had split off from a large mass of continental crust near the South Pole. It began drifting northward toward Asia at a high speed—geologically speaking—perhaps as fast as 20 centimeters per year. As this Indian plate started to move, its leading edge, consisting of oceanic crust, sank beneath Asia's southern margin. This subduction formed an Andean-type continental margin along the edge of Tibet; volcanoes erupted and granite plutons rose into the continent (Fig. 8–24).

Continent-Continent Collision

By 40 million years ago, subduction had consumed all of the oceanic lithosphere between India and Asia.

Then the two continents collided. Because both are continental crust, neither could sink deeply into the mantle. Igneous activity then ceased because subduction stopped. The collision did not stop the northward movement of India, but it did slow it down to about 5 centimeters per year.

As India continued to move northward, it began to underthrust beneath Tibet, doubling the thickness of continental crust in the region. Thick piles of sediment that had accumulated on India's northern continental shelf were scraped from harder basement rock as India slid beneath Tibet. These sediments were pushed into great folds and thrust faults (Figs. 8–25 B and C).

At the same time, India crushed Tibet and wedged China out of the way along huge strike-slip faults.

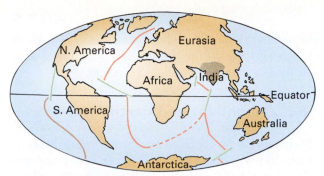

A   40 million years ago

Figure 8–25   (A) By 40 million years ago, India had moved 4000 to 5000 kilometers northward and collided with Asia. (B) When India collided with Tibet, the leading edge of India was underthrust beneath southern Tibet. (C) Continued plate movement has doubled the thickness of continental crust, creating the high Tibetan Plateau and the Himalayas.

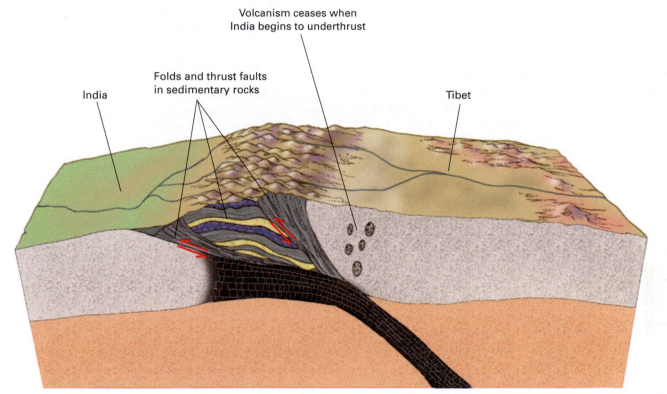

Volcanism ceases when
India begins to underthrust

Folds and thrust faults
in sedimentary rocks

India

Tibet

B   40 million years ago

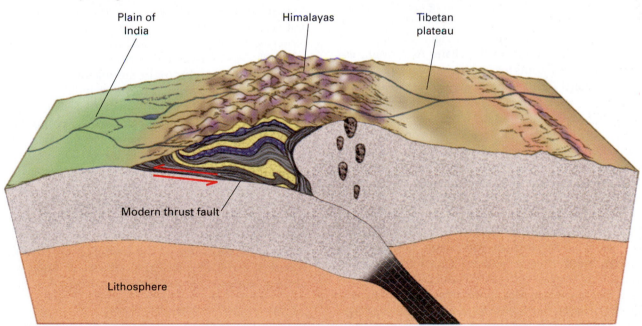

Plain of
India

Himalayas

Tibetan
plateau

Modern thrust fault

Lithosphere

C   Today

India has pushed southern Tibet 1500 to 2000 kilometers northward since the beginning of the collision. These compressional forces have created major mountain ranges and basins north of the Himalayas.

The Himalayas Today

Today, the Himalayas contain igneous, sedimentary, and metamorphic rocks. Many of the sedimentary rocks contain fossils of shallow-dwelling marine organisms that lived in the shallow sea of the Indian continental shelf. Plutonic and volcanic Himalayan rocks formed when the range was an Andean margin. Rocks of all types were metamorphosed by the tremendous stresses and heat generated during subduction and continent-continent collision. The underthrusting of India beneath Tibet and the squashing of Tibet have doubled the thickness of continental crust and lithosphere under the Himalayas and the Tibetan Plateau to the north. Consequently, the region floats isostatically at high elevation (Fig. 8–25C). Even the valleys lie at elevations of 3000 to 4000 meters, and the Tibetan Plateau has an average elevation of 4000 to 5000 meters. One reason that the Himalayas contain all of the Earth's highest peaks is simply that the entire plateau lies at such a high elevation. From the valley floor to the summit, Mount Everest is actually smaller than Alaska's Denali (Mount McKinley), North America's highest peak. Mount Everest rises about 3300 meters from base to summit, whereas Denali rises about 4200 meters. The difference in elevation of the two peaks lies in the fact that the base of Mount Everest is at about 5500 meters, but Denali's base is at 2000 meters.

As the Himalayas rose, they became too heavy to support their own weight. The crust beneath the range then spread outward. This stretching created normal faults in the high mountains. As the crust spread away from the highest parts of the range, it compressed rocks near the margins of the chain. As a result, extensional normal faults formed in the mountains, while compressional thrust faults developed in the foothills. But, at the same time, tectonic forces resulting from continent-continent collision continued to push the mountains upward. These processes continue today and no one knows when India will stop its northward movement or how high the mountains will become. However, when the uplift ends, erosion will eventually lower the lofty peaks to rolling hills.

The Himalayan chain is only one example of a mountain chain built by a collision between two continents. The Appalachian Mountains formed when eastern North America collided with Europe, Africa, and South America between 470 and 250 million years ago. The European Alps formed during repeated collisions between northern Africa and southern Europe beginning about 30 million years ago. The Urals, which separate Europe from Asia, formed by a similar process about 250 million years ago.

8.7 Mountains and Earth Systems

Mountain ranges are lifted by tectonic forces that originate deep within the Earth. However, once mountains rise, they interact with the hydrosphere, atmosphere, and biosphere.

Air rises as it flows across a mountain range. As we will learn in Chapter 16, moisture frequently condenses from rising air to produce rain or snow. Rain forms streams that race down steep hillsides, eroding gullies and canyons, while snow may accumulate to form glaciers that scour soil and bedrock. Thus mountains promote precipitation, which erodes mountains.

The rising of the Himalayas coincided with global cooling. We have described how a rising mountain chain can affect local climate as a result of altered wind patterns, changing precipitation, and the growth of glaciers. In Chapter 18, we will discuss how the rising of the Himalayas may have affected global climate.

Mountains affect and are affected by the biosphere. Cool climates prevail at high elevations, even at the equator. Mean annual temperature changes as much over 1000 meters of elevation as it does over a distance of 1000 kilometers of latitude. As a result, plant and animal communities change rapidly with elevation. For example, if you hike from the top of Grand Canyon to the bottom, you pass through the same ecosystems that you would encounter driving from Grand Canyon to north-central Mexico.

In many mountainous regions, human populations have increased dramatically over the past few decades. As farmland has become scarce, people have cut forests and cultivated steep slopes. Unfortunately, when the protective forest is removed, soil erosion increases dramatically. Mud washes into streams; landslides carry soil downslope (Fig 8–26). In this manner, human activity increases erosion rates. For example, in mountainous regions of Ethiopia, soil loss of 40 tons per hectare[1] has caused annual decrease in crop production of 1 to 2 percent per year. If current rates continue, the land will be barren in 50 to 100 years.

About half of the world's humans rely on water that falls in mountains. When forests are cut and replaced by farmland, rainwater and snowmelt are not retained efficiently. Thus stream flow increases dramatically in the wet season and declines abnormally during dry times. This alteration in the hydrological cycle increases the risk of alternating periods of flood and drought, threatening downstream communities.

---

[1] A hectare equals approximately 2.47 acres.

Figure 8–26 As the human population in mountainous regions increases, people have cut forests on steep hillsides to plant crops. This terraced hillside in Nepal collapsed during heavy monsoon rainfall.

8.8 The Origin of Continents

Recall from earlier discussions that the Earth was hot and active shortly after its formation about 4.6 billion years ago. Magma rose to the surface and then cooled to form the earliest crust. From the evidence of a few traces of old oceanic crust combined with computer models, geologists surmise that the first crust was similar to the rock that now makes up the upper mantle.

When Did Continents Evolve?

The 3.96-billion-year-old Acasta gneiss in Canada's Northwest Territories is the Earth's oldest known rock. It is metamorphosed granitic rock, similar to modern continental crust, and implies that at least some granitic crust had formed by that time.

Geologists have found grains of a mineral called zircon in a sandstone in western Australia. Although the sandstone is younger, the zircon gives radiometric dates of 4.2 billion years. Zircon commonly forms in granite. Geologists infer that the very old zircon initially formed in granite, which later weathered and released the zircon grains as sand. Eventually, the zircon became part of the younger sedimentary rock. The presence of these zircon grains suggests that granitic rocks existed 4.2 billion years ago. Geologists have also found granitic rocks nearly as old as the Acasta gneiss and the Australian zircon grains in Greenland and Labrador.

While these dates tell us that some continental crust existed by 4.2 billion years ago, they don't provide answers to the questions, Did continents exist before 4.2 billion years ago? and How much of the Earth was covered by continental crust early in the planet's history? These questions are difficult to answer because the geologic record is incomplete.

Partial Melting as a Clue to Understanding the Early History of Continents and Oceanic Crust

Recall from Chapter 3 that the mantle, oceanic crust, and continental crust have different compositions. These differences are reviewed below:

Mantle (Peridotite)	Oceanic Crust (Basalt)	Continental Crust (Granite)
Ultramafic	*Mafic*	*Felsic*
High magnesium and iron content; low silica	Less magnesium and iron and more silica compared with mantle rock	Low magnesium and iron; high silica

If the earliest crust had a composition similar to that of the mantle, we must explain how oceanic and continental crust evolved from mantle rocks. When the mantle melts, minerals with the lowest melting point begin to melt first, while other minerals remain solid. Thus the melt has a different composition from the rocks. In Chapter 7 we explained that partial melting always produces magma that contains a higher proportion of silica than the rock from which the melt formed.

Scientists believe that the earliest crust formed from the mantle with little compositional change. Later, partial melting of this primordial crust formed a basaltic crust that was richer in silica. Then, partial melting of the basalt probably formed intermediate rocks such as andesite that underwent another partial melting to form the silica-rich granitic continents. The process of partial melting may explain how the silica-rich continents evolved in a step-wise fashion from the silica-poor mantle. But what tectonic processes caused the sequence of melting steps?

In modern times, magma forms in three geologic environments: spreading centers, subduction zones, and mantle plumes. Similar magma-forming environments may have existed in Archean time, but geologists are uncertain which were most important. Some observations imply that Archean tectonics were similar to modern horizontal plate movements, and most magma formed at spreading centers and subduction zones. Other evidence indicates that horizontal plate motion was minor, and vertical mantle plumes dominated Archean tectonics.

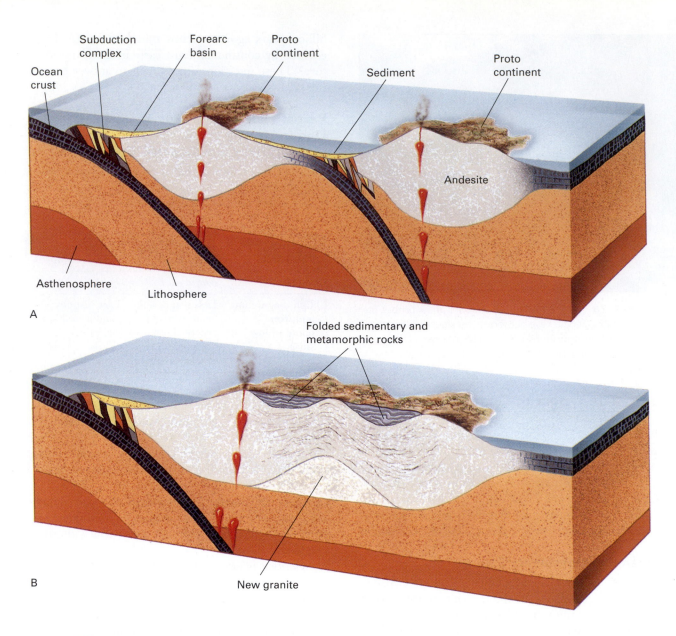

Figure 8–27 (A) According to one model, early continents formed in subduction zones. (B) As plates continued to move, small continents coalesced. Sediments that had eroded from the original islands were folded, and uplifted. Some were subjected to so much heat and pressure that they became metamorphic rocks. New granite formed from partial melting of the crust at the subduction zone.

Horizontal Tectonics

According to one hypothesis, heat-driven convection currents in a hot, active mantle initiated horizontal plate movement in the early crust. The dense primordial crust dove into the mantle in subduction zones, where partial melting created basaltic magma. As a result, the crust gradually became basaltic.

At a later date, the earliest continental crust formed by partial melting of basaltic crust in a new genera-

tion of subduction zones. Recall that island arcs form today in the same manner. Therefore, the first continents probably consisted of small granitic or andesitic blobs—like island arcs—surrounded by basaltic crust (Fig. 8–27A). Gradually, continued plate movement led to further subduction and isolated islands coalesced to form microcontinents. In the process, sediments accumulated and were metamorphosed during the collisions (Fig. 8–27B).

Vertical or Plume Tectonics

Several researchers have proposed, instead, that mantle plumes dominated Archean tectonics. Upwellings of mantle rock, perhaps rising from the mantle-core boundary, led to partial melting of portions of the upper mantle. This magma then solidified to form basaltic crust. Continued partial melting eventually formed granitic continental crust (Fig. 8–28).

As evidence for both models accumulates, some geologists suggest that both mechanisms were important. According to this hypothesis, mantle plumes formed thick basaltic oceanic plateaus, which then oozed outward to initiate horizontal motion. This motion caused subduction and another melting episode that generated continental crust by partial melting of the basalt plateaus.

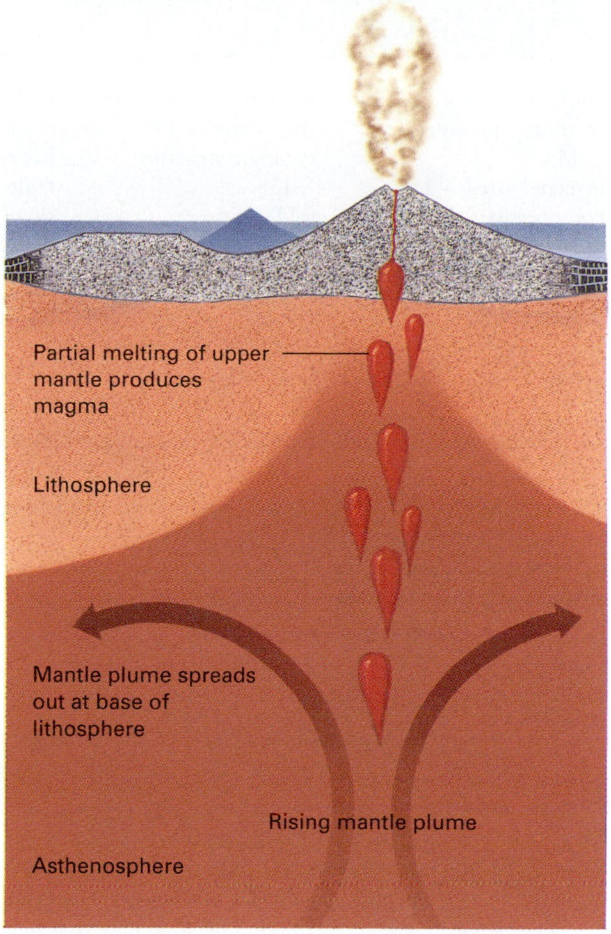

Partial melting of upper mantle produces magma

Lithosphere

Mantle plume spreads out at base of lithosphere

Rising mantle plume

Asthenosphere

Figure 8–28 Another hypothesis contends that early continental crust formed over rising mantle plumes, in a process called "plume tectonics."

SUMMARY

Rocks can be stressed by **confining stress** or by **directed stress.** When directed stress is applied to rocks, the rocks can deform in an elastic or a plastic manner, or they may rupture by **brittle fracture.** The nature of the material, temperature, pressure, and rate at which stress is applied all affect rock behavior under stress.

A **geologic structure** is any feature produced by deformation of rocks. **Folds** reflect plastic rock behavior and usually form when rocks are compressed. A **fault** is a fracture along which rock on one side has moved relative to rock on the other side. **Normal faults** are usually caused by **extensional stress, reverse** and **thrust faults** are caused by **compressional stress,** and **strike-slip faults** form by **shear stress** where blocks of crust slip horizontally past each other along vertical fractures. A **joint** is a fracture in rock where the rock has not been displaced.

Mountains form when the crust thickens and rises isostatically. They become lower when crustal rocks flow outward or are worn away by erosion. If two converging plates carry oceanic crust, a volcanic **island arc** forms. If one plate carries oceanic crust and the other carries continental crust, an **Andean margin** develops. Andean margins are dominated by granitic plutons and andesitic volcanoes. They also contain rocks of a **subduction complex** and sedimentary rocks deposited in a **forearc basin.**

When two plates carrying continental crust converge, an Andean margin develops first as oceanic crust between the two continental masses is subducted. Later, when the two continents collide, one continent is underthrust beneath the other. The geology of mountain ranges formed by continent-continent collisions such as the Himalayan chain is dominated by vast regions of folded and thrust-faulted sedimentary and metamorphic rocks and by earlier-formed plutonic and volcanic rocks.

The Earth's earliest crust was thin, ultramafic oceanic crust. The first continental crust formed by partial melting in subduction zones or over mantle plumes.

FOR REVIEW

1. What is tectonic stress? Explain the main types of stress.

2. Explain the different ways in which rocks can respond to tectonic stress. What factors control the response of rocks to stress?

3. What is a geologic structure? What are the three main types of structures? What type(s) of rock behavior does each type of structure reflect?

4. At what type of tectonic plate boundary would you expect to find normal faults?

5. Explain why folds accommodate crustal shortening.

6. Draw a cross-sectional sketch of an anticline-syncline pair and label the limbs.

7. Draw a cross-sectional sketch of a normal fault. Label the hanging wall and the footwall. Use your sketch to explain how a normal fault accommodates crustal extension. Sketch a reverse fault and show how it accommodates crustal shortening.

8. Explain the similarities and differences between a fault and a joint.

9. In what tectonic environment would you expect to find a strike-slip fault, a normal fault, and a thrust fault?

10. What mountain chain has formed at a divergent plate boundary? What are the main differences between this chain and those developed at convergent boundaries? Explain the differences.

11. Explain why erosion initially causes a mountain range to rise and then eventually causes the peak heights to decrease.

12. Describe the similarities and differences between an island arc and the Andes. Why do the differences exist?

13. Describe the similarities and differences between the Andes and the Himalayan chain. Why do the differences exist?

14. Draw a cross-sectional sketch of an Andean-type plate margin to a depth of several hundred kilometers.

15. Draw a sequence of cross-sectional sketches showing the evolution of a Himalayan-type plate margin. Why does this type of boundary start out as an Andean-type margin?

16. What are the oldest Earth materials found to date? How old are they? What information do they provide us? (What information can we infer from the data?)

17. Briefly outline one model for the formation of the continents.

FOR DISCUSSION

1. Discuss the relationships among types of lithospheric plate boundaries, predominant tectonic stress at each type of plate boundary, and the main types of geologic structures you might expect to find in each environment.

2. Why are thrust faults, reverse faults, and folds commonly found together?

3. Why do most major continental mountain chains form at convergent plate boundaries? What topographic and geologic features characterize divergent and transform plate boundaries in continental crust? Where do these types of boundaries exist in continental crust today?

4. Explain why extensional forces act on mountains rising in a tectonically compressional environment.

5. Explain why many mountains contain sedimentary rocks even though subduction leads to magma formation and the formation of igneous rocks.

6. Give a plausible explanation for the formation of the Ural Mountains, which lie in an inland portion of Asia.

7. Compare and explain the similarities and differences between the Andes and the Himalayan chain. How would the Himalayas, at their stage of development about 60 million years ago, have compared with the modern Andes?

8. Where would you be most likely to find large quantities of igneous rocks in the Himalayan chain—in the northern parts of the chain near Tibet, or southward, near India? Discuss why.

9. Geologist have measured that some Himalayan peaks are now rising as fast as 1 centimeter per year. Compare this rate with the average rate of tectonic uplift. Offer a plausible explanation for any differences.

10. Where would you be most likely to find very old rocks— the sea floor, at the base of a growing mountain range, or within the central portion of a continent?

11. Explain how human development in mountains might endanger mountain residents and, at the same time, impact downstream settlements.

12. Explain how mountains affect climate and how climate impacts mountains. Discuss the relative time frames of these interactions.

Surface Processes

Streams weather and erode rock and transport sediment toward the ocean, cutting valleys and sculpting the landscape. Alpine Creek Cascade in Olympic National Park. *(J. Lotter/Tom Stack and Assoc.)*

Evolution of the Oceans, the Atmosphere, and Life

The primordial Earth, born from colliding planetesimals and heated by radioactive decay, was molten or near molten. The sky, without an atmosphere, was black. There were no oceans, no life. Today, we consider the Earth in terms of four spheres: the geosphere, hydrosphere, atmosphere, and biosphere. Each sphere is as different from the others as a rock is different from a flowing stream, a breath of air, or a butterfly. But to understand the Earth at the very beginning, we need another perspective. Rock and metal, which comprise the geosphere, are nonvolatile, that is, they don't boil and become gases readily. In contrast, air, water, and living organisms are all composed of light, volatile compounds that boil or vaporize at relatively low temperatures. Air is a gas. Water readily becomes a gas by evaporating or boiling. At flame temperatures, the complex molecules in most living organisms break apart and the components evaporate as gases.

For the moment, let's abandon our view of the Earth's four spheres and think of only two: volatile substances and nonvolatile ones. Most scientists agree that the surface of the primordial Earth contained few volatiles. How, then, did enough volatile compounds collect to form a thick atmosphere, vast oceans, and a global cover of living organisms?

For many years, geologists hypothesized that abundant volatiles including water and carbon dioxide were trapped within the Earth's interior. This reasoning was based on three observations and inferences. First, our models show that volatiles were evenly dispersed in the original cloud that coalesced to form the planets. It seemed likely that some of those volatiles would have become trapped within the Earth as it formed. Second, astronomers have detected volatiles on modern comets, meteoroids, and asteroids. If volatiles were trapped within the small objects that passed through our neighborhood in space, it seemed logical to conclude that they also accumulated in the Earth's interior as the original cloud of dust and gas coalesced. Finally, modern volcanic eruptions eject gases and water vapor into the air. Geologists inferred that these gases originated in the mantle and were remnants of the original volatiles trapped during the Earth's formation. Scientists concluded that volatiles escaped from volcanic eruptions early in the Earth's history and that these compounds formed the atmosphere, the oceans, and living organisms.

Today, many scientists question this hypothesis. To understand their questions, let's return to the cloud of dust and gas that coalesced to form the planets. Recall that our region of space heated up as particles accelerated within the collapsing cloud. At the same time, hydrogen fusion began within the Sun and solar energy heated the inner Solar System. The newly born Sun also emitted a stream of ions and electrons called the solar wind, that swept across the inner planets, blowing their volatile compounds into outer regions of the Solar System. As a result, most of the Earth's volatile compounds boiled off and were swept into the cold outer regions of the Solar System.

According to a widely accepted hypothesis, shortly after our planet formed and lost its volatiles, a Mars-sized object smashed into the surface. This cataclysmic impact blasted through the crust and deep into the mantle, ejecting huge quantities of pulverized rock into orbit. The fragments eventually coalesced to form the Moon. The impact also ejected most of the Earth's remaining volatiles into space with enough velocity that they escaped the Earth's gravity. According to this hypothesis, the primordial Earth was barren and rocky, with almost no volatiles either on the surface or in the deep mantle—thus, it had neither water nor an atmosphere. The hot mantle churned, low-density silicate rocks rose to form the crust, and volcanic eruptions repaved the surface with lava, but these events added few volatiles to the Earth's surface. According to one estimate, outgassing of the deep mantle accounted for no more than 10 percent of the Earth's hydrosphere, atmosphere, and biosphere.[1]

If this scenario is correct, why do modern volcanoes emit volatiles? According to one hypothesis, most of the gases given off by modern volcanoes are recycled from the surface. Water, carbon (as carbonate rocks), and other light compounds are transported into the asthenosphere by subducting slabs. These volatiles return to the surface during volcanic eruptions.

---

[1] Paul Thomas, Christopher Chyba, and Christopher McKay, *Comets and the Origin of Life* (New York: Springer-Verlag, 1997).

TABLE 1

Chemical Composition of the Volatile Fraction of Halley's Comet*

Compound	Percent	Compound	Percent
H_2O	79.0	N_2H_4	0.8
H_2CO_2	4.5	HCN	1.0
H_2CO	4.0	N_2	0.5
CO_2	3.5	$H_4C_5N_4$	0.5
CO	1.5	NH_3	1.0
CH_4	2.0	S_2	0.2
C_2H_2	1.5	H_2S	0.2
C_3H_2	0.2	CS_2	0.2

*Morrison/Wolff/Fraknoi, *Abell's Exploration of the Universe*, 7/e, Table 19.2, p. 339.

Therefore, modern volcanic eruptions, like their primordial ancestors, don't outgas appreciable quantities of volatiles from the deep mantle.

Now let's return to the volatiles that streamed away from the hot inner Solar System. As they flew away from the Sun, they entered a cooler region beyond Mars. Most of the volatiles were captured by the outer planets—Jupiter, Saturn, Uranus, and Neptune—but some continued their journey toward the outer fringe of the Solar System. Here, beyond the orbits of the known planets, volatiles from the inner Solar System combined with residual dust and gas to form comets. A comet's nucleus has been compared to a "dirty snowball" because it is composed mainly of ice and rock. Other compounds not common in snowballs exist in comets as well. These include frozen carbon dioxide, ammonia, and simple organic molecules (Table 1).

Volatiles are also abundant in certain types of meteoroids in the region between Mars and Jupiter. Because these meteoroids coalesced in a cool region of the Solar System, they retained the primordial abundances of light, volatile elements. However, they were too small to attract the additional volatiles that streamed away from the inner Solar System.

Astronomers calculate that the early Solar System was crowded with comets, meteoroids, and asteroids—space debris left over from planetary formation (Fig. 1). Many contained volatile compounds. Many crashed into Earth, nearby planets, and moons. When a large object falls through the Earth's atmosphere, it creates a fiery glow, brighter than the full Moon. Any object that creates such a fireball is called a **bolide.** Although nearly all of the early bolide impact craters have been obliterated from Earth by erosion and tectonic processes, abundant craters on the Moon, Mercury, and other neighbors in space chronicle this period of intense bombardment. While bolide accretion added only one thousandth of a percent to the Earth's total mass, it imported 90 percent of its volatiles.

Upon entry and impact, the frozen volatiles in the bolides vaporized, releasing water vapor, carbon dioxide, ammonia, simple organic molecules, and other compounds shown in Table 1. The ammonia molecules broke apart upon contact with the hot lava to release nitrogen and hydrogen. In the conditions present on the hot primordial Earth, all of these molecules were gases and formed a primitive atmosphere. As the planet cooled and the atmospheric pressure increased, water vapor condensed to liquid, forming the first oceans. The atmosphere was quite different from the modern atmosphere, which consists mainly of nitrogen and oxygen. In fact, the primordial atmosphere would have been poisonous to modern life. The evolution of the modern atmosphere from this primordial precursor is discussed in the opening essay for Unit V.

You will see in the following chapters of Unit III that water and air alter and sculpt the Earth's surface. Rain combines with atmospheric gases to form weak acids that attack rock, decomposing minerals and breaking solid rocks into the tiny fragments that form soil. Falling rain and flowing streams then erode the soil, washing it into larger rivers that carry sand and silt to the seas. Glaciers—frozen water—gouge deep valleys into bedrock and deposit gravel on the land surface. Over geologic time, these sediments become lithified to form sedimentary rocks. Thus, not only is water the foundation of life on Earth, but it is the most important agent of the Earth's surface processes: the processes that weather rocks, sculpt landforms, and produce sedimentary rocks.

The light molecules transported to Earth in bolides also provided the raw material for life. Life is composed of a complex molecular framework composed mostly of carbon and hydrogen, with smaller amounts of oxygen, nitrogen, and other elements. All of these elements are abundant in bolides, especially in comets, and all were rare on Earth as it formed.

People have asked, "Did carbon return to Earth as simple molecules or as complex ones?" In many laboratory simulations, chemists have mixed the simple volatiles present on extraterrestrial bodies, and added energy. In all of these experiments the molecules reacted to synthesize building blocks of life such as amino acids and DNA

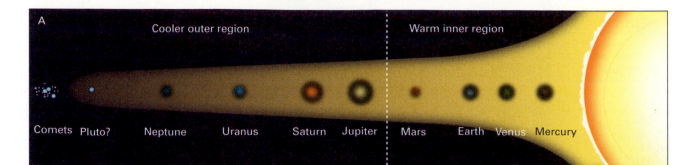

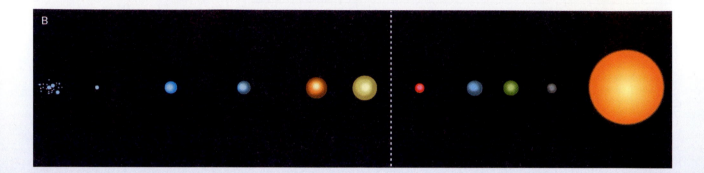

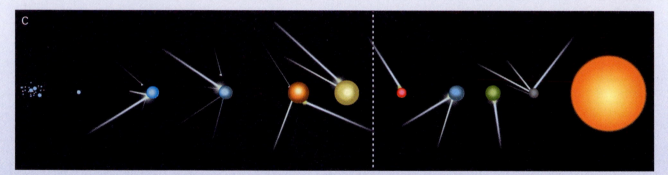

Figure 1 The cometary impact hypothesis for the origin of the Earth's atmosphere and oceans. (A) As the Solar System formed, volatiles boiled away from the inner planets. The solar wind then blew them into the cool region beyond Mars. The outer planets captured some of the volatiles, and some condensed to form comets in the frigid zone beyond Neptune. (B) As a result, the inner four planets were left with no atmospheres or oceans. The outer planets grew to become giants. (Pluto is an anomalous planet and its origin is discussed in Chapter 23.) (C) Comets and meteorites crashed into the Earth and other planets, returning some of the volatiles to their surfaces. The volatiles accumulated to form the atmosphere, oceans, and the foundation of life on Earth. (Sizes and distances in this drawing are not to scale.)

bases. Thus, molecular precursors to life form spontaneously from a variety of mixtures. Spontaneous synthesis occurs in space, and as a result, many comets contain complex organic molecules. These compounds formed when the simple volatiles were bombarded by solar radiation or by internal radioactive decay.

Some scientists speculate that comets carried enough of these complex organic molecules to Earth to provide the basis for the evolution of life. A few scientists have even proposed that life itself evolved on comets, survived the fiery entry, and blossomed in its new home. One argument to support this speculation is based on time and probability. Life appeared on Earth as little as 200 million years after the planet cooled enough to create a favorable environment. But comets were flying through the cold outer Solar System at least half a billion years earlier. The extra time may have allowed extraterrestrial chemical reactions to produce the first primitive living organisms.

Many scientists disagree with the hypothesis that life hitchhiked to Earth on comets. They argue that even if life had evolved on a comet headed for Earth, the organisms would not have survived entry. Many scientists also disagree with the hypothesis that abundant complex molecules arrived on comets. They contend that most complex organics would have been destroyed during entry, so the total influx of these substances must have been low; furthermore, the early Earth's reactive environment would have destroyed the molecules about as fast as they arrived. According to this argument, the arrival rate of complex organic molecules from space was too low to account for their abundance on Earth. Since conditions were favorable for synthesis on Earth,

Figure 2 Halley's Comet. *(Akira Fujii)*

these scientists find it unnecessary to invoke an extraterrestrial origin for the complex molecules. Thus many scientists contend that, while comets transported the light elements needed to produce life, complex molecules and living organisms originated and evolved here on our home planet.

Despite the ambiguities, at least some and probably most of the compounds needed to produce the hydrosphere, the atmosphere, and the biosphere crashed to Earth from outer regions of the Solar System. Later in Earth history, bolide impacts blasted rock and dust into the sky, causing mass extinctions and killing large portions of life on Earth. Thus bolide impacts may have provided the raw materials for the origin of life and later caused mass extinctions.

What about the future? Scientists estimate that there are now about two thousand asteroids in orbits that cross the Earth's path. Each is large enough to cause a

mass extinction—perhaps even to destroy civilization or the human species—if it crashed into Earth. Every few years a large comet passes close enough to be visible to the naked eye or with a small telescope, and tiny comets and meteorites enter our atmosphere routinely (Fig. 2). In October 1989, an asteroid with a diameter of 100 to 400 meters passed within 1.1 million kilometers of Earth. In January 1991, a smaller one passed within 170,000 kilometers, half the distance from the Earth to the Moon. In July 1994, Comet Shoemaker-Levy 9 smashed into Jupiter. These encounters remind us that another catastrophic collision on Earth is possible, even likely, sometime in the future—although "the future" may be millions of years away. Some astronomers have suggested that we establish a search-and-destroy system to detect incoming asteroids and blow them apart or deflect them with nuclear weapons.

The vegetated Nile delta, about 150 kilometers wide, contrasts sharply with the surrounding Egyptian desert. Streams transport most of the sediment that is weathered and eroded from continents to coasts, and deposit it on deltas such as this. *(Landsat Satellite/Tony Stone Worldwide)*

Early in the history of the Solar System, swarms of meteorites, comets, and asteroids crashed into all of the planets and their moons. Today, there are no ancient craters on the Earth's surface, but the Moon is pockmarked with craters of all ages. Why have the craters vanished from the Earth, and why has the Moon retained its craters?

The Earth's tectonic activity continually changes the surface of our planet. Continents migrate across the globe, ocean basins open and close, great mountain chains rise, earthquakes and volcanic eruptions renew the Earth's surface over geologic time.

In addition, Earth is sufficiently massive to have retained its atmosphere and water, which combine to weather and erode surface rocks. The combination of tectonic activity and erosion has eliminated all traces of early craters from the Earth's surface. In contrast, the smaller Moon has lost most of its heat, so tectonic activity is nonexistent. In addition, the Moon's gravitational force is too weak to have retained an atmosphere or water to erode its surface. As a result, the Moon's ancient craters have been preserved for billions of years.

Thus on the Earth, plate tectonics processes shape continents and ocean basins. But the Earth's **surface processes**—weathering and erosion—modify the surface of the land. ●

9.1 Weathering and Erosion

Return to your childhood haunts and you will see that the rock outcrops in woodlands or parks have not changed. Over geologic time, however, air and water attack rocks and break them down at the Earth's surface. The processes that decompose rocks and convert them to loose gravel, sand, clay, and soil are called **weathering** (Fig. 9–1).

Weathering involves little or no movement of the decomposed rocks and minerals. The weathered material accumulates where it forms on bedrock. However, loose soil and other weathered material offer little resistance to rain or wind, and are easily eroded away. Thus, **erosion** processes soon pick up and carry off weathered rocks and minerals. Rain, running water, wind, glaciers, and gravity are the most

Figure 9–1 Weathering processes are breaking this boulder into smaller fragments. Erosion has not removed the fragments, so the original shape of the boulder is still visible.

common agents of erosion. These agents may then transport the weathered material great distances and finally deposit it as layers of sediment at the Earth's surface.

Weathering, erosion, transport, and deposition typically occur in an orderly sequence. For example, water freezes in a crack in granite, weathering the rock and loosening a grain of quartz. A hard rain erodes the grain and washes it into a stream. The stream then transports the quartz to the seashore and deposits it as a grain of sand on a beach (Fig. 9–2).

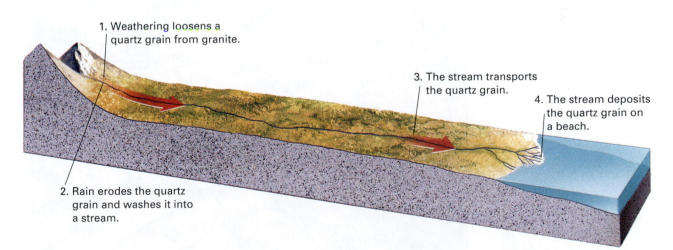

1. Weathering loosens a quartz grain from granite.

2. Rain erodes the quartz grain and washes it into a stream.

3. The stream transports the quartz grain.

4. The stream deposits the quartz grain on a beach.

Figure 9–2 Weathering loosens a quartz grain from granite. Rain erodes the grain, washing it into a stream. The stream transports it and finally deposits the grain on a beach.

Weathering occurs by both mechanical and chemical processes. **Mechanical weathering** (also called *physical weathering*) reduces solid rock to small fragments but does not alter the chemical composition of rocks and minerals. Think of grinding a rock in a crusher; the fragments are no different from the parent rock, except that they are smaller. In contrast, **chemical weathering** occurs when air and water chemically react with rock to alter its composition and mineral content. Chemical weathering is similar to rusting in that the final products differ both physically and chemically from the starting material.

9.2 Mechanical Weathering

Five different processes cause mechanical weathering: pressure-release fracturing, frost wedging, abrasion, organic activity, and thermal expansion and contraction. Two additional processes—salt cracking and hydrolysis-expansion—result from combinations of mechanical and chemical processes.

Pressure-Release Fracturing

Many igneous and metamorphic rocks form deep below the Earth's surface. Imagine, for example, that a granitic pluton solidifies from magma at a depth of 15 kilometers. At that depth, the pressure from the weight of overlying rock is about 5000 times that at the Earth's surface. Over millennia, tectonic forces may raise the pluton to form a mountain range. As the pluton rises, the overlying rock erodes away and the pressure on the buried rock decreases. When the pressure diminishes, the rock expands, but because the rock is now cool and brittle, it fractures as it expands. This process is called **pressure-release fracturing.** Many igneous and metamorphic rocks that formed at depth, but now lie at the Earth's surface, have fractured in this manner (Fig. 9–3).

Frost Wedging

Water expands when it freezes. If water accumulates in a crack and then freezes, its expansion pushes the rock apart in a process called **frost wedging.** In a temperate climate, water may freeze at night and thaw during the day. Ice cements the rock together temporarily, but when it melts, the rock fragments may tumble from a steep cliff. Experienced mountaineers try to travel in the early morning before ice melts. Large piles of loose angular rocks, called **talus,** lie beneath many cliffs (Fig. 9–4). These rocks fell from the cliffs mainly as a result of frost wedging.

Abrasion

Rocks, grains of sand, and silt collide with one another when currents carry them along a stream or beach. During these collisions, their sharp edges and corners wear away and the particles become rounded. The mechanical wearing of rocks by friction and impact is called **abrasion** (Fig. 9–5). Note that water itself is not abrasive; the collisions among rock, sand, and silt cause the weathering.

Figure 9–3 Pressure-release fracturing contributed to the fracturing of this granite in California's Sierra Nevada.

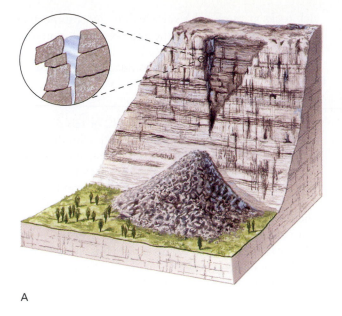

A

B

Figure 9–4 (A) Frost wedging dislodges rocks from cliffs and creates talus slopes. (B) Frost wedging produced this talus cone in the Valley of the Ten Peaks, Canadian Rockies.

Wind hurls sand and other small particles against rocks, sandblasting unusual shapes (Fig. 9–6). Glaciers also abrade rock as they drag rocks, sand, and silt across bedrock.

Organic Activity

If soil collects in a crack in bedrock, a seed may fall there and sprout. The roots work their way into the crack, expand, and may eventually widen the crack (Fig. 9–7). City dwellers often see the results of this process where tree roots raise and crack concrete sidewalks.

Thermal Expansion and Contraction

Rocks at the Earth's surface are exposed to daily and yearly cycles of heating and cooling. They expand when they are heated and contract when they cool.

Figure 9–5 Abrasion rounded these rocks in a stream bed in Yellowstone National Park, Wyoming.

Figure 9–6 Wind abrasion selectively weathered and eroded the base of this rock in Lago Poopo, Bolivia, because windblown sand moves mostly near the ground surface.

Figure 9–7 As this tree grew from a crack in bedrock, its growing roots widened the crack.

When temperature changes rapidly, the surface of a rock heats or cools faster than its interior, and as a result, the surface expands or contracts faster than the interior. The resulting forces may fracture the rock.

In mountains or deserts at mid-latitudes, temperature may fluctuate from −5°C to +25°C during a spring day. Is this 30° difference sufficient to fracture rocks? The answer is uncertain. In one laboratory experiment, scientists heated and cooled granite repeatedly by more than 100°C and no fractures formed. These results imply that normal temperature changes might not be an important cause of mechanical weathering. However, the rocks used in the experiment were small and the experiment was carried out over a brief period of time. Perhaps thermal expansion and contraction are more significant in large outcrops; or daily heating-cooling cycles repeated over hundreds of thousands of years may promote fracturing.

In contrast to small daily temperature changes, fire heats rock by hundreds of degrees. If you line a campfire with granite stones, the rocks commonly break as you cook your dinner. In a similar manner, forest fires or brush fires occur commonly in many ecosystems and are an important agent of mechanical weathering.

9.3 Chemical Weathering

Rock is durable over a single human lifetime. Over geologic time, however, air and water chemically at-tack rocks at the Earth's surface. The most important processes of chemical weathering are dissolution, hydrolysis, and oxidation. Water, acids and bases, and oxygen cause these processes to decompose rocks.

Dissolution

If you put a crystal of halite (rock salt, or table salt) in water, it dissolves to form a solution. Halite dissolves so rapidly and completely in water that it is rare in moist environments. Water also dissolves most other minerals, but much more slowly than it dissolves salt.

Many rocks and minerals dissolve more rapidly when water becomes acidic or basic. An acidic solution contains a high concentration of hydrogen ions (H^+), whereas a basic solution contains a high concentration of hydroxyl ions ($OH)^-$. To understand how acidic or basic water dissolves a mineral, think of an atom on the surface of a crystal. It is held in place because it is attracted to the other atoms in the crystal by the electrical forces called chemical bonds. At the same time, electrical attractions to the outside environment pull the atom away from the crystal. The result is like a tug-of-war. If the bonds between the atom and the crystal are stronger than the attraction of the atom to its outside environment, then the crystal remains intact. If outside attractions are stronger, they pull the atom away from the crystal and the mineral dissolves (Fig. 9–8). Acids and bases dissolve most minerals more effectively than pure water because they provide more electrically charged hydrogen and hydroxyl ions to pull atoms out of crystals. For example, limestone is made of the mineral calcite ($CaCO_3$). Calcite barely dissolves in pure water but is quite soluble in acid. If you place a drop of strong acid on limestone, bubbles of carbon dioxide gas instantly form as the calcite dissolves.

Water found in nature is never pure. Atmospheric carbon dioxide dissolves in raindrops and reacts to form a weak acid called carbonic acid. As a result, even the purest rainwater falling in the Arctic or on remote mountains is slightly acidic. Air pollution can make rain even more acidic.

Thus, water—especially if it is acidic or basic—dissolves ions from soil and bedrock and carries the dissolved material away. Ground water also dissolves rock to produce spectacular caverns in limestone (Fig. 9–9). (Cave formation is discussed further in Chapter 10.)

Hydrolysis

In **hydrolysis,** water reacts with a mineral to form a new mineral with the water as part of its crystal structure. Most common minerals weather by hydrolysis. For example, feldspar, the most abundant mineral in the Earth's crust, weathers by hydrolysis to form clay

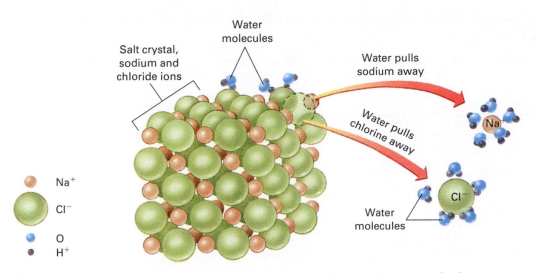

Na⁺
Cl⁻
O
H⁺

Figure 9–8 Halite dissolves in water because the attractions between the water molecules and the sodium and chloride ions are greater than the strength of the chemical bonds in the crystal.

(see *Focus On: Representative Reactions in Chemical Weathering*).

Quartz is the only rock-forming silicate mineral that does not weather to form clay. Quartz is resistant to chemical weathering because it dissolves extremely slowly, and only to a small extent. When granite weathers, the feldspar and other minerals react to form clay but the unaltered quartz grains fall free from the rock. Some granites have been so deeply weathered by hydrolysis that quartz grains can be pried out with a fingernail at depths of several meters (Fig. 9–10). The rock looks like granite but has the consistency of sand.

Because quartz is so tough and resistant to both mechanical and chemical weathering, it is the primary component of sand. Much of this sand is transported to streams and ultimately to the sea coast, where it concentrates on beaches and eventually forms sandstone.

Oxidation

Many elements react with atmospheric oxygen, O_2. Iron rusts when it reacts with water and oxygen. Rusting is an example of a more general process called **oxidation.**[1] Iron is abundant in many minerals; if the iron in such a mineral oxidizes, the mineral decomposes.

Many valuable metals such as iron, copper, lead, and zinc occur as sulfide minerals in ore deposits. When they oxidize during weathering, the sulfur reacts to form sulfuric acid, a strong acid. The sulfuric acid washes into streams and ground water where it may harm aquatic organisms. Many natural ore deposits generate sulfuric acid when they weather, and the reaction may be accelerated when ore is dug up and exposed at a mine site. This problem is discussed further in Chapter 21.

Figure 9–9 Caverns form when ground water dissolves limestone. *(Hubbard Scientific Co.)*

---

[1]Oxidation is properly defined as the loss of electrons from a compound or element during a chemical reaction. In the weathering of common minerals this usually occurs when the mineral reacts with molecular oxygen.

Focus On

Representative Reactions in Chemical Weathering

Dissolution of Calcite

Calcite, the mineral that comprises limestone and marble, weathers in natural environments in a three-step process. In the first two steps, water reacts with carbon dioxide in the air to produce carbonic acid, a weak acid:

$$CO_2 + H_2O \longrightarrow H_2CO_3 \longrightarrow H^+ + HCO_3^-$$

Carbon dioxide   Water   Carbonic acid   Hydrogen ion   Bicarbonate ion

In the third step, calcite dissolves in the carbonic acid solution.

$$CaCO_3 + H^+ \longrightarrow Ca^{2+} + HCO_3^-$$

Calcite   Hydrogen ion   Calcium ion   Bicarbonate ion

Hydrolysis

An example of a reaction in which feldspar hydrolyzes to clay is as follows:

$$2KAlSi_3O_8 + 2H^+ + H_2O \longrightarrow Al_2Si_2O_5(OH)_4 + 2K^+ + 4SiO_2$$

Orthoclase feldspar   Hydrogen ion   Water   Clay mineral   Potassium ion   Silica

Chemical and Mechanical Weathering Operating Together

Chemical and mechanical weathering work together, often on the same rock at the same time. After mechanical processes fracture a rock, water and air seep into the cracks to initiate chemical weathering.

Salt Cracking   In environments where ground water is salty, salt water seeps through pores and cracks in bedrock. When the water evaporates, the dissolved salts crystallize. The growing crystals exert tremendous forces, enough to loosen mineral grains and widen cracks in a process called **salt cracking.** Thus, salt chemically precipitates in rock, and then the expanding salt crystals mechanically loosen mineral grains and break the rock apart.

Many sea cliffs show pits and depressions caused by salt cracking because spray from the breaking waves

Figure 9–10 Coarse grains of quartz and feldspar accumulate directly over weathered granite. The lens cap in the middle illustrates scale.

brings the salt to the rock. Salt cracking is also common in deserts, where surface and ground water often contains dissolved salts (Fig. 9–11).

Exfoliation Granite commonly fractures by **exfoliation,** a process in which large plates or shells split away like the layers of an onion (Fig. 9–12). The plates may be only 10 or 20 centimeters thick near the surface, but they thicken with depth. Because exfoliation fractures are usually absent below a depth of 50 to 100 meters, they seem to be a result of exposure of the granite at the Earth's surface.

Exfoliation is frequently explained as a form of pressure-release fracturing. However, many geologists suggest that hydrolysis may be the main cause of exfoliation. During hydrolysis, feldspars and other silicate minerals react to form clay. As a result of the addition of water, clays have a greater volume than that

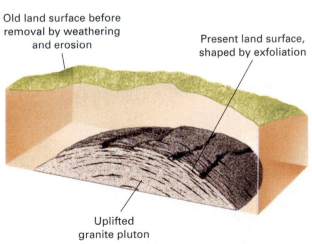

Old land surface before removal by weathering and erosion

Present land surface, shaped by exfoliation

Uplifted granite pluton

A

B

Figure 9–12 (A) Exfoliation occurs when concentric rock layers fracture and become detached from a granite outcrop. (B) Exfoliation has fractured this granite in Pinkham Notch, New Hampshire.

Figure 9–13 A very thin, dark-colored layer of soil overlies light-colored limestone to support this forest in Canada's Northwest Territories. Most plant life is supported by a soil layer only a few centimeters to a few meters thick.

of the original minerals. Thus, a chemical reaction (hydrolysis) forms clay, and the mechanical expansion that occurs as the clay forms may cause exfoliation.

9.4 Soil

Bedrock breaks into smaller fragments as it weathers, and much of it decomposes to clay and sand. Therefore, on most land surfaces, a thin layer of loose rock fragments, clay, and sand overlies bedrock (Fig. 9–13). This material is called **regolith.** Soil scientists define **soil** as upper layers of regolith that support plants, although many Earth scientists and engineers use "soil" and "regolith" interchangeably.

Components of Soil

Soil is a mixture of mineral grains, organic material, water, and gas. The mineral grains include clay, silt, sand, and rock fragments. Clay is so fine-grained and closely packed that water, and even air, do not flow through a clay-rich soil readily, so plants growing in clay soils suffer from lack of oxygen. In contrast, water and air flow easily through sandy soil. The most fertile soil is **loam,** a mixture of sand, clay, silt, and generous amounts of organic matter. Minerals and organic matter contain nutrients necessary for plant growth (Table 9–1).

If you walk through a forest or prairie, you can find bits of leaves, stems, and flowers on the soil surface. This material is called **litter.** When litter decomposes sufficiently that you can no longer determine the origin of individual pieces, it becomes **humus.** Humus is an essential component of most fertile soils. Humus soaks up so much moisture that humus-rich soil swells after a rain and shrinks during dry spells. This alternate shrinking and swelling loosens the soil, allowing roots to grow into it easily. A rich layer of humus also insulates the soil from excessive heat and cold and reduces water loss by evaporation. Humus also retains soil nutrients and makes them available to plants.

TABLE 9–1

Nutrients Essential for Plant Growth

Macronutrients from air and water	
Carbon	The basic building blocks of all
Oxygen	organic tissue
Hydrogen	
Macronutrients from soil	
Nitrogen	Most critical element for growth of plant proteins
Phosphorus	Important for metabolism and development of cell membranes
Potassium	Maintains plant cell permeability
Calcium	Important for plant cell walls
Magnesium	Important for production of chlorophyll
Sulfur	Required for synthesis of plant vitamins and proteins
Micronutrients from soil	
Iron	
Copper	
Manganese	
Boron	
Zinc	
Chlorine	
Molybdenum	

O Horizon. Mostly organic matter.

A Horizon (topsoil). High concentration of organic matter.

B Horizon (subsoil). Clay and cations leached from A horizon accumulated here.

C Horizon (weathered bedrock)

Bedrock

A

B

O

A

B

C

Figure 9–14 (A) A typical soil consists of several horizons, distinguished by color, texture, and chemistry. (B) In this soil, the darkest uppermost layer is the O horizon; the A horizon is less dark; the whiter layer is the B horizon. The B horizon grades downward into the weathered bedrock of the C horizon. The scale is in feet. *(U.S. Department of Agriculture)*

In intensive agriculture, farmers commonly plow the soil and leave it exposed for weeks or months. Humus oxidizes in air and decomposes. At the same time, rain dissolves soil nutrients and carries them away. Farmers replace the lost nutrients with chemical fertilizers but rarely replenish the humus. As a result, much of the soil's ability to absorb and regulate water and nutrients is lost. When rainwater flows over the surface, it transports soil particles, excess fertilizer, and pesticide residues, polluting streams and ground water.

Soil Profiles

A typical mature soil consists of several layers called **soil horizons.** The uppermost layer is called the **O horizon,** named for its **O**rganic component. This layer consists mostly of litter and humus with a small proportion of minerals (Fig. 9–14). The next layer down, called the **A horizon,** is a mixture of humus, sand, silt, and clay. The combined O and A horizons are called **topsoil.** A kilogram of average fertile topsoil contains about 30 percent by weight organic matter, including approximately 2 trillion bacteria, 400 million fungi, 50 million algae, 30 million protozoa, and thousands of larger organisms such as insects, worms, nematodes, and mites.

The third layer, the **B horizon** or subsoil, is a transitional zone between topsoil and weathered parent rock below. Roots and other organic material grow in the B horizon, but the total amount of organic matter is low. The lowest layer, called the **C horizon,** consists of partially weathered rock that grades into unweathered parent rock. This zone contains little organic matter.

When rain falls on soil, it sinks into the O and A horizons, weathering minerals, forming clay, and carrying dissolved ions to lower levels. This downward movement of water and dissolved ions is called **leaching.** The A horizon is sandy because water also carries the clay downward but leaves the sand behind.

Water, dissolved ions, and clay from the A horizon accumulate in the B horizon. This layer retains moisture because of its high clay content. Although moisture retention may be beneficial, if too much clay accumulates the B horizon creates a dense, waterlogged soil.

Soil-Forming Factors

Why are some soils rich and others poor, some sandy and others loamy? Six factors control soil characteristics: parent rock, climate, rates of plant growth and decay, slope aspect and steepness, time, and transport.

Parent Rock   The texture and composition of soil depend partly on its parent rock. For example, when granite decomposes, the feldspar converts to clay and the rock releases quartz as sand grains. If the clay leaches into the B horizon, a sandy soil forms. In contrast, because basalt contains no quartz, soil formed from basalt is likely to be rich in clay and contain only small amounts of sand. Nutrient abundance also depends in part on the parent rock. For example, a pure quartz sandstone contains no nutrients, and soil formed on it must get its nutrients from outside sources.

Climate   Rainfall and temperature affect soil formation. Rain seeps downward through soil, but several other factors pull the water back upward. Roots suck soil water toward the surface, and water near the surface evaporates. In addition, water is electrically attracted to soil particles. This attraction initiates a process called **capillary action,** which also draws water upward toward the soil surface.

During a rainstorm, water seeps through the A horizon, dissolving soluble ions such as calcium, magnesium, potassium, and sodium. In arid and semiarid regions, rain storms typically are of short duration and little rain falls. Consequently, when the rain stops, capillary action and plant roots then draw most of the water back up toward the surface, where it evaporates or is taken up by plants. As the water escapes, many of its dissolved ions precipitate in the B horizon, encrusting the soil with salts. A soil of this type is a **pedocal** (Fig. 9–15A). This process often deposits enough calcium carbonate to form a hard cement called **caliche** in the soil. The Greek word *pedon* means soil; "pedocal" is a composite of *pedon* and the first three letters of calcite, *cal.* In the Imperial Valley in

California, irrigation water contains high concentrations of calcium carbonate. A thick continuous layer of caliche forms in the soil as the water evaporates. To continue growing crops, farmers must then rip this layer apart with heavy machinery.

Because nutrients concentrate when water evaporates, many pedocals are fertile if irrigation water is available. However, salts often concentrate so much that they become toxic to plants (Fig. 9–16). As mentioned previously, all streams contain small concentrations of dissolved salts. If arid or semiarid soils are intensively irrigated, salts can accumulate until plants cannot grow. This process is called **salinization.** Some historians argue that salinization destroyed croplands and thereby contributed to the decline of many ancient civilizations, such as the Babylonian Empire.

In a wet climate, rain storms are of longer duration and more rain falls. As a result, water seeps downward through the soil to the water table, leaching soluble ions from both the A and B horizons. The less soluble elements, such as aluminum, iron, and some silicon, remain behind, accumulating in the B horizon to form a soil type called a **pedalfer** (Fig. 9–15B). The subsoil in a pedalfer is commonly rich in clay, which is mostly aluminum and silicon, and has the reddish color of iron oxide; hence the prefix *ped* is followed by the chemical symbols *Al* for aluminum and *Fe* for iron.

In regions of high temperature and very high rainfall, such as a tropical rainforest, so much water seeps through the soil that it leaches away nearly all the soluble cations. Only very insoluble aluminum and iron minerals remain (Fig. 9–15C). Soil of this type is called a **laterite.** Laterites are often colored rust-red by iron oxide (Fig. 9–17). A highly aluminous laterite, called **bauxite,** is the world's main type of aluminum ore.

The second important component of climate, average annual temperature, affects soil formation in two ways. First, chemical reactions proceed more rapidly in warm temperatures than in cold ones. Second, plant growth and decay are temperature dependent, as discussed next.

Rates of Plant Growth and Decay   In the tropics, plants grow and decay rapidly and the nutrients released by decay are quickly absorbed by growing plants. Heavy rainfall leaches nutrients from the soil, creating a laterite. As a result, little humus accumulates and few nutrients are stored in the soil. Thus, even though the tropical rainforests support great populations of plants and animals, many of these forests are anchored on poor, laterite soils (Fig. 9–18A). The plants depend on a rapid cycle of growth, death, and decay. The Arctic, on the other hand, is so cold that plant growth and decay are slow. Therefore, litter and

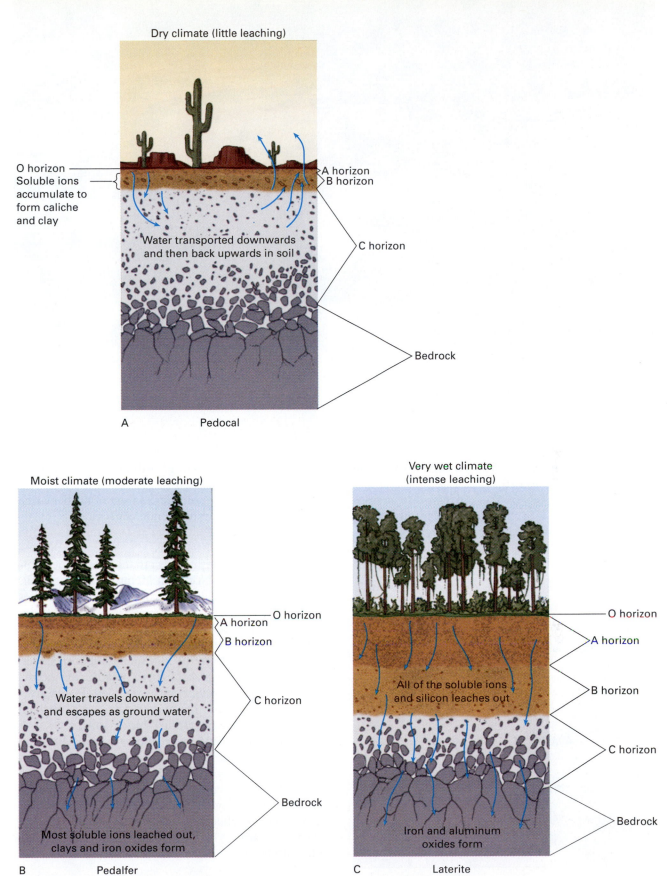

Figure 9–15 (A) Pedocals, (B) pedalfers, and (C) laterites are different types of soils that form in different climates.

Figure 9–16 Salts have poisoned this Wyoming soil. Saline water seeps into the depression and evaporates to deposit white salt crystals on the ground and on the fence posts.

humus form slowly and Arctic soils also contain little organic matter (Fig. 9–18B).

The most fertile soils are those of temperate prairies and forests. There, large amounts of plant litter drop to the ground in the autumn, but decay is slow dur-

Figure 9–17 This aluminum-rich Georgia laterite formed when water leached away the more soluble ions.

ing the winter, and plant growth during the growing season is not fast enough to extract all the nutrients from the soil. As a result, thick layers of nutrient-rich humus accumulate in temperate climates.

Slope Aspect and Steepness **Aspect** is the orientation of a slope with respect to the Sun. In the semiarid regions in the Northern Hemisphere, thick soils and dense forests cover the cool, shady north slopes of hills, but thin soils and grass dominate hot, dry southern exposures (Fig. 9–19). The reason for this difference is that in the Northern Hemisphere more water evaporates from the hot, sunny southern slopes. Therefore, fewer plants grow, weathering occurs slowly, and soil development is retarded. Plants grow more abundantly on the moister northern slopes and more rapid weathering forms thicker soils.

In general, hillsides have thin soils and valleys are covered by thicker soil, because soil erodes from hills and accumulates in valleys. When hilly regions were first settled and farmed, people naturally planted their crops in the valley bottoms, where the soil was rich and water was abundant. Recently, as population has expanded, farmers have moved to the thinner, less stable hillside soils.

Time Chemical weathering is slow in most places, and time is therefore an important factor in deter-

A   B

Figure 9–18 (A) Lateritic tropical soils, such as this one in Vanuatu in the Southwest Pacific Ocean, often support lush growth but contain few nutrients and little humus. (B) Arctic soil of Baffin Island, Canada, supports sparse vegetation and contains little organic matter.

mining the extent of weathering. Recall that most minerals weather to clay. In geologically young soils, weathering may be incomplete and the soils may contain many partly weathered mineral fragments. As a result, young soils are often sandy or gravelly. As soils mature, weathering continues and the clay content increases.

Soil Transport

By studying recent lava flows, scientists have determined how quickly plants return to an area after it has been covered by hard, solid rock. In many cases, plants appear when a lava flow is only a few years old, even before weathering has formed soil. Closer scrutiny shows that the plants have rooted in tiny amounts of soil that were transported from nearby areas by wind or water.

In many of the world's richest agricultural areas, most of the soil was transported from elsewhere. Streams deposit sediment, wind deposits dust, and soil slides downslope from mountainsides into valleys. These foreign materials mix with residual soil, changing its composition and texture. The soils of river flood plains and deltas and the rich windblown soils of both China and the Great Plains are examples of transported soils.

Figure 9–19 In semi-arid climatic regions, thick forests cover the cool, shady north slopes of hills (right), but grass and sparse trees dominate hot, dry southern exposures (left).

Ecologists have studied water and nutrient cycling in the Hubbard Brook Experimental Forest in New Hampshire. That particular ecosystem consists of a series of small valleys, each drained by a single creek. The bedrock beneath the valleys is impermeable, and water cannot escape by seeping underground. Researchers built concrete dams across several of the creeks and anchored the dams on bedrock. They then measured the amount of water flowing from each valley and the amounts of sediment and dissolved nutrients carried by each creek.

Figure 1 shows that in a healthy forest only 25 percent of the rainwater soaks into the soil or escapes as runoff; 75 percent returns back to the atmosphere through transpiration and evaporation. Thus, a forest recycles much of its water back into the atmosphere.

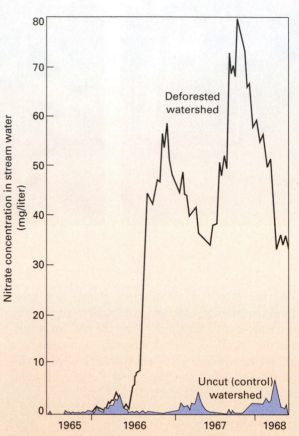

Figure 2 The Hubbard Brook experiment showed that a completely devegetated forest valley lost its soil nutrients much faster than an adjacent undisturbed watershed.

Figure 1 A natural forest returns about 75 percent of rainfall to the atmosphere.

To study the effects of vegetation on this ecosystem, scientists cut all the trees and shrubs in one Hubbard Brook valley and sprayed the soil with herbicides to prevent regrowth. With no plants to absorb water, most of the rainwater flowed downslope quickly. Stream flow increased by 40 percent over that in an adjacent undisturbed valley. In addition, the amount of dissolved nutrients and sediment carried by the stream draining the devegetated valley increased.

The scientists at Hubbard Brook learned that the soils of the deforested valley lost nutrients six to eight times faster than the adjacent undisturbed watershed did (Figure 2). The great increase in water runoff carried large amounts of nutrients from the soil that otherwise would have been conserved by the vegetation. Scientists have recorded similar results for forests, grasslands, and wetlands. These experiments demonstrate the interrelationships between soil and vegetation. Plants maintain healthy soil which, in turn, sustains them.

Figure 9–20 Rain and flowing water erode soil as rapidly as it forms on these slopes in Arches National Monument.

9.5 Erosion

Natural Soil Erosion

Weathering decomposes bedrock, and plants add organic material to the regolith to create soil at the Earth's surface. However, soil does not accumulate and thicken throughout geologic time. If it did, the Earth would be covered by a mantle of soil hundreds or thousands of meters thick, and rocks would be unknown at the Earth's surface. Instead, flowing water, wind, and glaciers erode soil as it forms (Fig. 9–20). In addition, some weathered material simply slides downhill under the influence of gravity. In fact, all forms of erosion combine to remove soil about as fast as it forms. For this reason, soil is usually only a few meters thick or less in most parts of the world.

Once soil erodes, the clay, sand, and gravel begin a long journey as they are carried downhill by the same agents that eroded them: streams, glaciers, wind, and gravity. On their journey they may come to rest in a stream bed, a sand dune, or a lake bed, but those environments are only temporary stops. Eventually, they erode again and are carried downhill until finally, they

are deposited where the land meets the sea. Some of the sediment accumulates on deltas, and coastal marine currents redistribute some of it along the shore. Eventually younger sediment may bury older layers until they become lithified to form sedimentary rocks.

Natural erosion and transport of sediment by streams, glaciers, and wind are the subjects of Chapters 10, 11, and 12. In the last three sections of this chapter, we will discuss landslides: erosion by gravity.

9.6 Erosion by Mass Wasting

Mass wasting is the downslope movement of Earth material, primarily under the influence of gravity. The word **landslide** is a general term for mass wasting and for the landforms created by mass wasting.

Although gravity acts constantly on all slopes, the internal strength of the rock and soil usually hold the slope in place. Eventually, however, natural processes or human activity may destabilize a slope to cause mass wasting. For example, a stream can erode the base of a hillside, undercutting it until it collapses. Rain, melting snow, or a leaking irrigation ditch can add weight and lubricate soil, causing it to slide downslope. Mass wasting occurs naturally in all hilly or mountainous terrain. Steep slopes are especially vulnerable, and landslide scars are common in the mountains.

In recent years, the human population has increased dramatically. As the most desirable land has become overpopulated, large numbers of people have moved to mountains once considered too harsh for homes and farms. In wealthier nations, people have moved into the hills to escape congested cities. As a result, permanent settlements have grown in previously uninhabited steep terrain. Many of these slopes are naturally unstable. Construction and agriculture have destabilized others.

Every year, small landslides destroy homes and farmland. Occasionally, an enormous landslide buries a town or city, killing thousands of people. Landslides cause billions of dollars in damage every year, about equal to the damage caused by earthquakes in 20 years. In many instances, losses occur because people do not recognize dangers that are obvious to a geologist.

Consider three recent landslides that have affected humans:

1. A movie star builds a mansion on the edge of a picturesque California cliff. After a few years, the cliff collapses and the house slides into the valley (Fig. 9–21A).
2. A ditch carrying irrigation water across a hillside in Montana leaks water into the ground. After

Focus On

Soil Erosion and Agriculture

In nature, soil erodes approximately as rapidly as it forms. However, improper farming, livestock grazing, and logging can accelerate erosion. Plowing removes plant cover that protects soil. Logging often removes forest cover, and the machinery breaks up the protective litter layer. Similarly, intensive grazing can strip away protective plants. Rain, wind, and gravity then erode the exposed soil easily and rapidly. Meanwhile, soil continues to form by weathering at its usual slow, natural pace. Thus, increased rates of erosion caused by agriculture lead to net soil loss.

When farmers use proper conservation measures, soil can be preserved indefinitely or even improved. Some regions of Europe and China have supported continuous agriculture for centuries without soil damage. However, in recent years, marginal lands on hillsides, in tropical rainforests, and along the edges of deserts have been brought under cultivation. These regions are particularly vulnerable to soil deterioration, and today, soil is being lost at an alarming rate throughout the world (Figure 1).

The Worldwatch Institute estimates that about 0.4 billion tons of topsoil are created globally by weathering each year. However, soil is eroding more rapidly than it is forming on about 35 percent of the world's croplands. About 25 billion tons are lost every year. The soil lost annually would fill a train of freight cars long enough to encircle our planet 150 times.

In the United States, approximately one third of the topsoil that existed when the first European settlers arrived has been lost. However, as a result of increasing awareness of soil erosion and improved farming and conservation practices, soil erosion rates declined in some parts of the United States between 1982 and 1992 (the last year for which reliable data are available). At present, erosion is continuing in the United States at an average rate of about 10.5 tons per hectare* per year. However, in some regions the rate is considerably higher, and yearly losses of more than 25 tons per hectare are common. To put this number in perspective, a loss of 25 tons per hectare per year would lead to complete loss of the topsoil in about 150 years.

Fertile cropland is also lost when it is developed for roads, cities, and suburban housing. In the United States, 17,000 square kilometers of cropland, an area larger than the state of Connecticut, were urbanized between 1982 and 1992. Today, 86 percent of the fruit and 87 percent of the vegetables grown in the United States are raised in rapidly urbanizing areas, mostly in California and Florida. California's Central Valley produces 8 percent of the total dollar value of all United States agriculture, but California lost 3 percent of its total cropped area to urban development between 1984 and 1992. China lost 38,700 square kilometers of farmland to urbanization between 1987 and 1992, representing an annual production capacity loss of 15 million tons of grain, enough to feed 45 million people. Although these cropland losses are unrelated to erosion, they must be considered in monitoring changes in the Earth's capacity to feed its growing population.

*One hectare equals 2.47 acres.

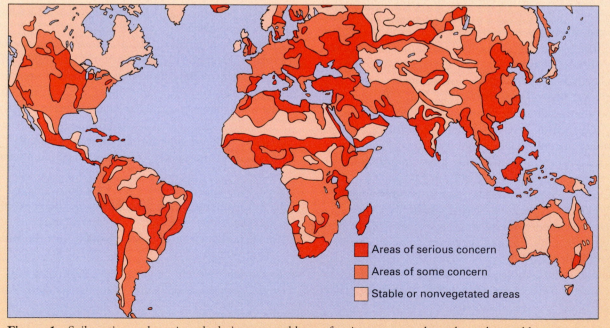

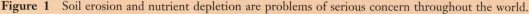

Areas of serious concern

Areas of some concern

Stable or nonvegetated areas

Figure 1   Soil erosion and nutrient depletion are problems of serious concern throughout the world.

A

B

C

Figure 9–21 Landslides cause billions of dollars in damage every year. (A) A few days after this photo was taken, the corner of the house hanging over the gully fell in. *(J.T. Gill, USGS)* (B) A landslide, triggered by a leaking irrigation ditch, threatens a house in Darby, Montana. (C) A landslide destroyed several expensive buildings in Hong Kong. *(Hong Kong Government Information Services)*

years of seepage, the muddy soil slides downslope and piles against a house at the bottom of the hill (Fig. 9–21B).

3. Excavations for roads and high-rise buildings undercut the base of a steep hillside in Hong Kong. Suddenly, the slope slides, destroying everything in its path (Fig. 9–21C).

Why does mass wasting occur? Is it possible to avoid or predict such catastrophes to reduce property damage and loss of life?

Factors That Control Mass Wasting

Imagine that you are a geological consultant on a construction project. The developers want to build a road at the base of a hill, and they wonder whether landslides will threaten the road. What factors should you consider?

Steepness of the Slope   Obviously, the steepness of a slope is a factor in mass wasting. If frost wedging dislodges a rock from a steep cliff, the rock tumbles to the valley below. However, a similar rock is less likely to roll down a gentle hillside.

Type of Rock and Orientation of Rock Layers If sedimentary rock layers dip in the same direction as a slope, the upper layers may slide over the lower ones. Imagine a hill underlain by shale, sandstone, and limestone oriented so that their bedding lies parallel to the slope, as shown in Figure 9–22A. If the base of the hill is undercut (Fig. 9–22B), the upper layers of sandstone and limestone may slide over the weak shale. In contrast, if the rock layers dip at an angle to the hillside, the slope may be stable even if it is undercut (Figs. 9–22C and 9–22D).

Several processes can reduce the stability of a slope. A stream or ocean waves can erode its base. Road building and excavation can also destabilize it. Therefore, a geologist or engineer must consider not only a slope's stability before construction, but how the project might alter its stability.

The Nature of Unconsolidated Materials   The **angle of repose** is the maximum slope or steepness at which loose material remains stable. If the slope becomes steeper than the angle of repose, the material slides. The angle of repose varies for different types of material. Rocks commonly tumble from a cliff to collect at the base as angular blocks of talus. The angular blocks interlock and jam together. As a result, talus typically has a steep angle of repose, up to 45°. In contrast, rounded sand grains do not interlock and therefore have a lower angle of repose, about 30° to 35° (Fig. 9–23).

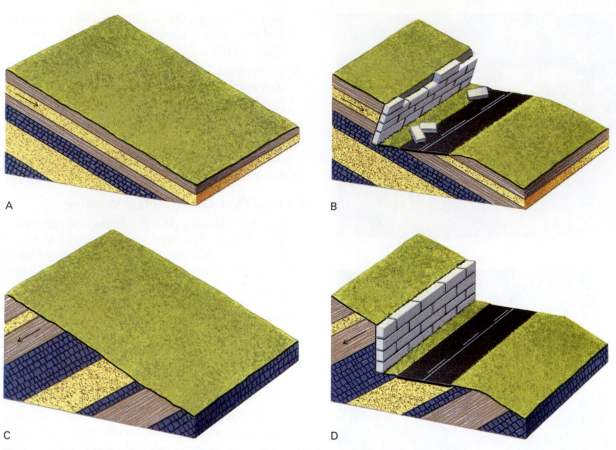

A

B

C

D

Figure 9–22 (A) Sedimentary rock layers dip parallel to this slope. (B) If a road cut undermines the slope, the dipping rock provides a good sliding surface, and the slope may fail. (C) Sedimentary rock layers dip at an angle to this slope. (D) The slope may remain stable even if it is undermined.

Water and Vegetation   To understand how water affects slope stability, think of a sand castle. Even a novice sand-castle builder knows that the sand must be moistened to build steep walls and towers (Fig. 9–24). But too much water causes the walls to collapse. Small amounts of water bind sand grains together because the electrical charges of water molecules attract the grains. However, excess water lubricates the sand[2] and adds weight to a slope. When some soils become water saturated, they flow downslope, just as the sand

_____

[2]Water fills pores in loose material, and is under pressure of overlying material. Excess water raises that pressure, and it is actually the increased pore pressure that lowers slope stability.

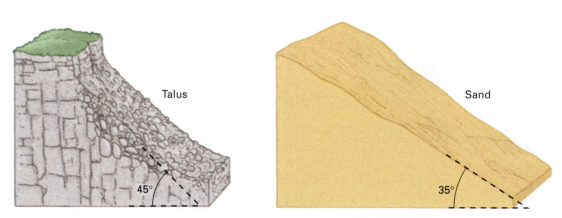

Talus

45°

Sand

35°

Figure 9–23 The angle of repose is the maximum slope that can be maintained by a specific material.

Figure 9–24 The angle of repose depends on both the type of material and its water content. Dry sand forms low mounds, but if you moisten the sand, you can build steep, delicate towers.

castle collapses. In addition, if water collects on impermeable clay or shale, it may provide a weak, slippery layer so that overlying rock or soil can move easily.

Roots hold soil together and plants absorb water; therefore, a highly vegetated slope is more stable than a similar bare one. Many forested slopes that were stable for centuries slid when the trees were removed during logging, agriculture, or construction.

Mass wasting is common in deserts and regions with intermittent rainfall. For example, southern California has dry summers and occasional heavy winter rain. Vegetation is sparse because of summer drought and wildfires. When winter rains fall, bare hillsides often become saturated and slide. Mass wasting occurs for similar reasons during infrequent but intense storms in deserts.

Earthquakes and Volcanoes   An earthquake may cause mass wasting by shaking an unstable slope, causing it to slide. A volcanic eruption may melt snow and ice near the top of a volcano; the water then soaks into the slope to release a landslide. The case studies on pages 204 and 206 describe two disastrous landslides caused by volcanic eruptions and earthquakes.

9.7   Types of Mass Wasting

Mass wasting can occur slowly or rapidly. In some cases, rocks fall freely down the face of a steep mountain. In other instances, rock or soil creeps downslope so slowly that the movement may be unnoticed by a casual observer.

Mass wasting falls into three categories: flow, slide, and fall (Fig. 9–25). To understand these categories, think again of a sand castle. Sand that is saturated with water flows down the face of the structure. During **flow,** loose, unconsolidated soil or sediment moves as a fluid. Some slopes flow slowly—at a speed of 1 centimeter per year or less. On the other hand, mud with a high water content can flow almost as rapidly as water.

If you undermine the base of a sand castle, part of the wall may fracture and slip downward. Movement of coherent blocks of material along fractures is called **slide.** Slide is usually faster than flow, but it still may take several seconds for the block to slide down the face of the castle.

If you take a huge handful of sand out of the bottom of the castle, the whole tower topples. This rapid, free-falling motion is called **fall.** Fall is the most rapid type of mass wasting. In extreme cases, like the face of a steep cliff, rock can fall at a speed dictated solely by the force of gravity and air resistance.

Table 9–2 outlines the characteristics of flow, slide, and fall. Details of these three types of mass wasting are explained in the following sections.

Flow

As the name implies, **creep** is the slow, downhill flow of rock or soil under the influence of gravity. A creeping slope typically moves at a rate of about 1 centimeter per year, although wet soil can creep more rapidly. During creep, the shallow soil or rock layers

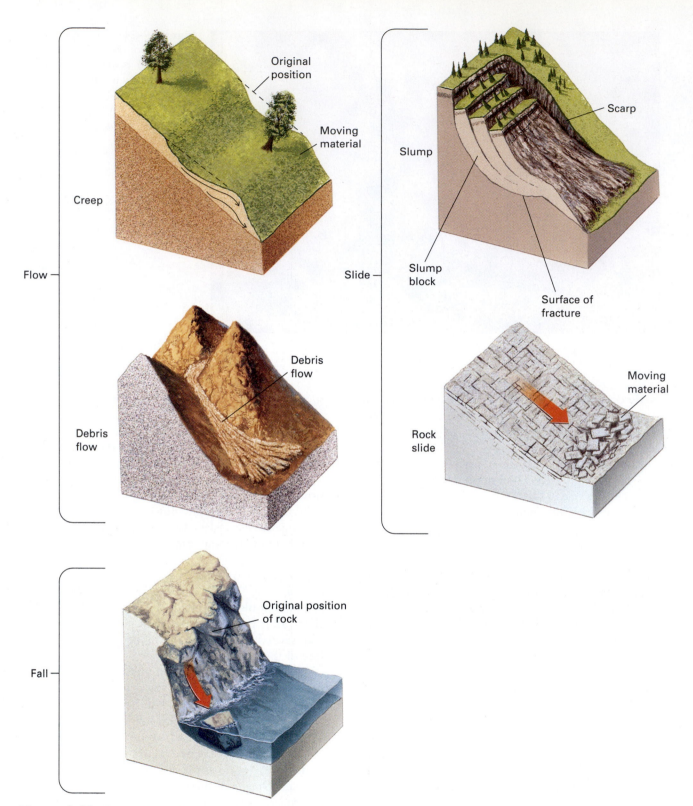

Figure 9–25 Flow, slide, and fall are the three categories of mass wasting.

TABLE 9–2

Categories of Mass Wasting

Type of Movement	Description	Subcategory	Description	Comments
Flow	Individual particles move downslope independently of one another, not as a consolidated mass. Typically occurs in loose, unconsolidated regolith	Creep	Slow, visually imperceptible movement	Trees on creep slopes develop pistol-butt shape
		Debris flow	More than half the particles larger than sand size; rate of movement varies from less than 1 m/year to 100 km/hr or more.	Common in arid regions with intermittent heavy rainfall, or can be triggered by volcanic eruption
		Earthflow and mudflow	Movement of fine-grained particles with large amounts of water	
Slide	Material moves as discrete blocks; can occur in regolith or bedrock	Slump	Downward slipping of a block of Earth material, usually with a backward rotation on a concave surface	Trees on slump blocks remain rooted
		Rockslide	Usually rapid movement of a newly detached segment of bedrock	
Fall	Materials fall freely in air; typically occurs in bedrock	—	—	Occurs only on steep cliffs

move more rapidly than deeper material (Fig. 9–26). As a result, anything with roots or a foundation tilts downhill (Fig. 9–27).

Trees have a natural tendency to grow straight upward. As a result, when soil creep tilts a growing tree, the tree develops a J-shaped curve in its trunk called pistol butt (Fig. 9–28). If you ever contemplate buying hillside land for a home site, examine the trees. If they have pistol-butt bases, the slope is probably creeping, and creeping soil may tear a building apart.

In temperate regions, creep can also result from freeze–thaw cycles during the spring and fall. Recall that water expands when it freezes. When damp soil freezes, expansion pushes it outward at a right angle to the slope. However, when the Sun melts the frost, the particles sink vertically, as shown in Figure 9–29. This movement creates a net downslope displacement. The displacement in a single cycle is small, but the soil may freeze and thaw once a day for a few months, leading to a total movement of a centimeter or more every year.

Figure 9–26 Creep has bent layering in sedimentary rocks in a downslope direction. *(Ward's Natural Science Establishment, Inc.)*

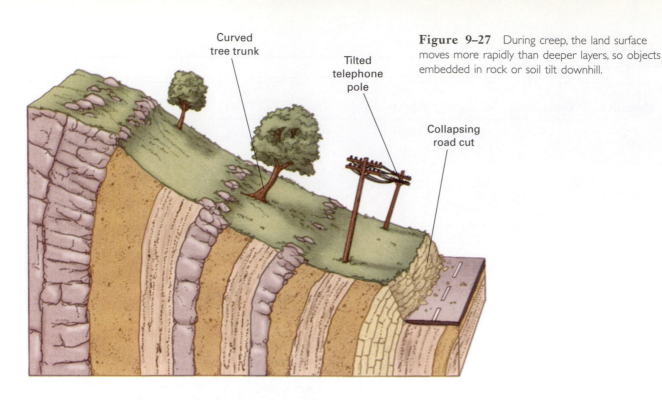

Curved tree trunk

Tilted telephone pole

Collapsing road cut

Figure 9–27 During creep, the land surface moves more rapidly than deeper layers, so objects embedded in rock or soil tilt downhill.

Figure 9–28 If a hillside creeps as a tree grows, the tree develops pistol butt.

Other factors that cause creep include expansion and shrinking of clay-rich soils during alternating wet and dry seasons, and activities of burrowing animals. Both of these processes move soil downslope in a manner similar to that of freeze–thaw cycles.

If heavy rain falls on unvegetated soil, the water can saturate the soil to form a slurry of mud and rocks, often called a **mudflow.** A slurry is a mixture of water and solid particles that flows as a liquid. Wet concrete is a familiar example of a slurry. It flows easily and is routinely poured or pumped from a truck.

The advancing front of a mudflow often forms tongue-shaped lobes (Fig. 9–30). A slow-moving mudflow travels at a rate of about 1 meter per year, but others can move as fast as a car speeding along an interstate highway. A mudflow can pick up boulders and automobiles and smash houses, filling them with mud or even dislodging them from their foundations.

Slide

In some cases, a large block of rock or soil, sometimes an entire mountainside, breaks away and slides downslope as a coherent mass or as a few intact blocks. Two types of slides occur: slump and rockslide.

A **slump** occurs when blocks of material slide downhill over a gently curved fracture in rock or regolith (Fig. 9–31). Trees remain rooted in the moving blocks. However, because the blocks rotate on the concave fracture, trees on the slumping blocks are tilted

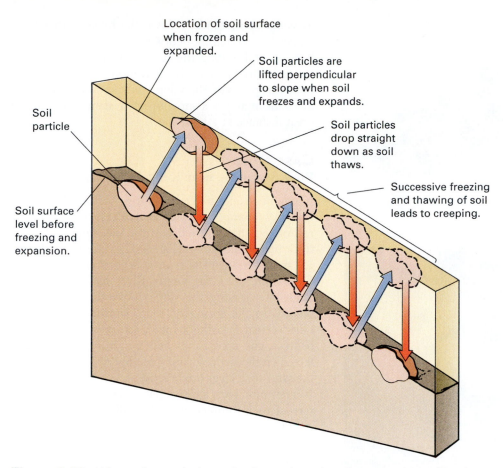

Location of soil surface when frozen and expanded.

Soil particles are lifted perpendicular to slope when soil freezes and expands.

Soil particles drop straight down as soil thaws.

Successive freezing and thawing of soil leads to creeping.

Soil particle

Soil surface level before freezing and expansion.

Figure 9–29 When soil expands due to freezing or absorption of water by clays, soil particles move outward, perpendicular to the slope. But when the soil shrinks again, particles sink vertically downward. The net result is a small downhill movement with each expansion-contraction cycle.

Figure 9–30 The 1980 eruption of Mount St. Helens melted large quantities of glacial ice near the summit of the peak. The meltwater mixed with soil to create lobe-shaped mudflows. *(M. Freidman, USGS)*

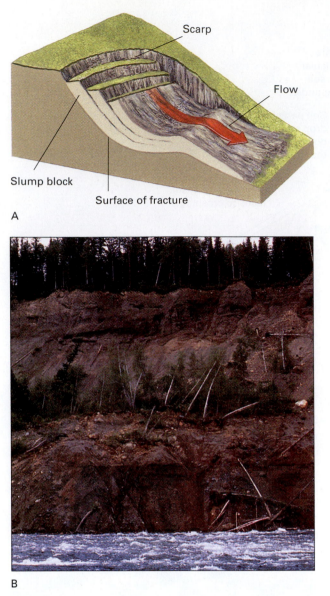

Figure 9–31 (A) In slump, blocks of soil or rock remain intact as they move downslope. (B) Trees tilt back into the hillside on this slump along the Quesnell River, British Columbia.

mass of rubble tumbles down the hillside. In a large avalanche, the falling debris traps and compresses air beneath and among the tumbling blocks. The compressed air reduces friction and allows some avalanches to attain speeds of 500 kilometers per hour. The same mechanism allows a snow or ice avalanche to cover a great distance at a high speed.

CASE STUDY
Rock Avalanche near Kelly, Wyoming

A mountainside above the Gros Ventre River near Kelly, Wyoming, was composed of a layer of sandstone resting on shale, which in turn was supported by a thick bed of limestone (Fig. 9–32). The rocks dipped 15° to 20° toward the river and parallel to the slope. Over time, the Gros Ventre River had undercut the sandstone, leaving the slope above the river unsupported. In the spring of 1925, snowmelt and heavy rains seeped into the ground, saturating the soil and bedrock and increasing their weight. The water collected on the shale, forming a slippery surface. Finally, the sandstone layer broke loose and slid over the shale. In a few moments, approximately 38 million cubic meters of rock tumbled into the valley. The sandstone crumbled into blocks that formed a 70-meter-high natural dam across the Gros Ventre River. Two years later, the lake overflowed the dam, washing it out and creating a flood downstream that killed several people. •

Fall

If a rock dislodges from a steep cliff, it falls rapidly under the influence of gravity. Several processes commonly detach rocks from cliffs. Recall from our discussion of weathering that frost wedging can dislodge rocks from cliffs and cause rockfall. Rockfall also occurs when ocean waves or a stream undercuts a cliff (Fig. 9–33).

backward. Thus, you can distinguish slump from creep because slump tilts trees uphill, whereas creep tilts them downhill. At the lower end of a large slump, the blocks often break apart and pile up to form a jumbled, hummocky topography.

It is useful to identify slump because it often recurs in the same place or on nearby slopes. Therefore, a slope that shows evidence of past slump is not a good place to build a house.

During a **rockslide**, (or rock avalanche), bedrock slides downslope over a fracture plane. Characteristically, the rock breaks up as it moves and a turbulent

9.8 Mass Wasting Triggered by Earthquakes and Volcanoes

In many cases, an earthquake or volcanic eruption causes comparatively little damage, but it triggers a devastating landslide.

CASE STUDY
The Madison River Slide, Montana

In August 1959, a moderate-size earthquake jolted the area just west of Yellowstone National Park. This

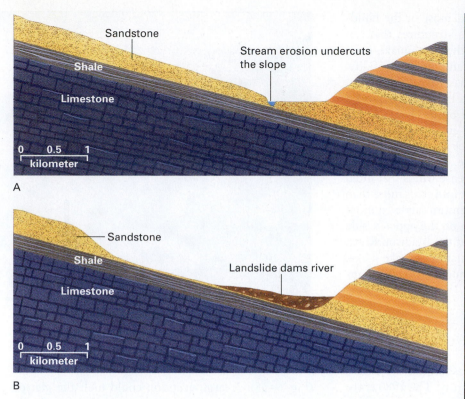

Figure 9–32 A profile of the Gros Ventre hillside (A) before and (B) after the slide. (C) About 38 million cubic meters of rock and soil broke loose and slid downhill during the Gros Ventre slide.

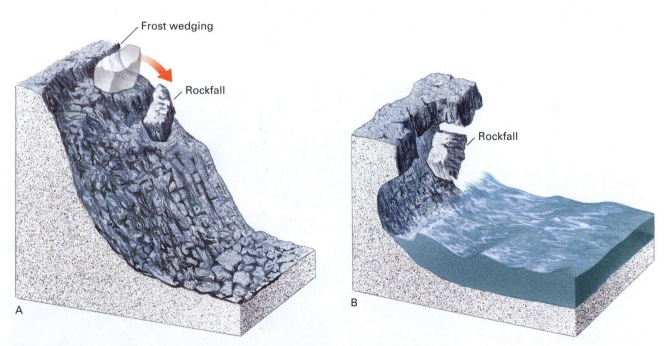

Figure 9–33 (A) Rockfall commonly occurs in spring or fall when freezing water dislodges rocks from cliffs. (B) Undercutting of cliffs by waves, streams, or construction can also cause rockfall.

region is sparsely populated, and most of the buildings in the area are wood-frame structures that can withstand quakes. As a result, the earthquake itself caused little property damage and no loss of life. However, the quake triggered a massive rockslide from the top of Red Mountain, which lay directly above a U.S. Forest Service campground on the banks of the Madison River. About 30 million cubic meters of rock broke loose and slid into the valley below, burying the campground and killing 26 people. Compressed air escaping from the slide created intense winds that lifted a car off the ground and carried it into trees more than 10 meters away. The slide's momentum carried it more than 100 meters up the mountain on the opposite side of the valley. The debris dammed the Madison River, forming a lake that was later named Quake Lake. Figure 9–34 shows the debris and some of the damage caused by this slide. ●

CASE STUDY
Mass Wasting in Washington

Several volcanoes in western Washington State have been active in recent geologic history. The 1980 eruption of Mount St. Helens blew away the entire north side of the mountain (Chapter 7). The heat of the eruption melted glaciers and snowfields near the summit, and the water mixed with volcanic ash and soil to create mammoth mudflows.

Is catastrophic mass wasting possible or likely elsewhere in Washington State? Unfortunately, the answer is yes. Mount Baker and Mount Rainier are active, glacier-covered volcanoes that lie near Seattle

Figure 9–34 This landslide near Yellowstone Park buried a campground, killing 26 people. *(Donald Hyndman)*

(Fig. 9–35). A large eruption could melt the glaciers on either mountain to create flows similar to those that devastated the valleys below Mount St. Helens in 1980. Mount Baker lies 20 kilometers upriver from the town of Glacier, Washington, and less than 50 kilometers upriver from the city of Bellingham, which has a population of more than 50,000. Mount Rainier is situated upstream from numerous small towns. We can imagine an eruption on either peak initiating a mudflow that buries nearby towns.

Figure 9–35 A wisp of steam rises near the summit of Mount Baker, at the bottom of the photograph.

We are not predicting an eruption of Mount Baker or Mount Rainier; we are simply stating that both are geologically plausible. What should be done in response to the hazard? It is impractical to move an entire city. Therefore, the only alternative is to monitor the mountain continuously and hope that it will not erupt or, if it does, that it will provide enough warning that urban areas can be evacuated in time. ●

9.9 Predicting and Avoiding Landslides

Landslides commonly occur in the same area as earlier landslides because the geologic conditions that cause mass wasting tend to be constant over a large area and for long periods of time. Thus, if a hillside has slumped, nearby hills may also be vulnerable to mass wasting. In addition, landslides and mudflows commonly follow the paths of previous slides and flows. If an old mudflow lies in a stream valley, future flows may follow the same valley.

Many towns were founded decades or centuries ago, before geologic disasters were understood. Often the choice of a town site was not dictated by geologic considerations but by factors related to agriculture, commerce, or industry, such as proximity to rivers and ocean harbors and the quality of the farmland. Once a city is established, it is virtually impossible to move it. Furthermore, geologists' warnings that a disaster might occur are often ignored.

Awareness and avoidance are the most effective defenses against mass wasting. Geologists evaluate landslide probability by combining data on soil and bedrock stability, slope angle, climate, and history of slope failure in the area. They include evaluations of the probability of a triggering event, such as a volcanic eruption or earthquake. Building codes then regulate or prohibit construction in unstable areas. For example, according to the *United States Uniform Building Code*, a building cannot be constructed on a sandy slope steeper than 27°, even though the angle of repose of sand is 30° to 35°. Thus the law leaves a safety margin of 3° to 8°. Architects can obtain permission to build on more precipitous slopes if they anchor the foundation to stable rock.

SUMMARY

Weathering is the decomposition and disintegration of rocks and minerals at the Earth's surface. **Erosion** is the removal of weathered rock or soil by flowing water, wind, glaciers, or gravity. After being eroded from the immediate environment, rock or soil may be transported large distances and eventually deposited.

Mechanical weathering can occur by **pressure-release fracturing, frost wedging,** abrasion, organic activity, and thermal expansion and contraction.

Chemical weathering occurs when chemical reactions decompose minerals. A few minerals dissolve readily in water. Acids and bases often markedly enhance the solubility of minerals. Rainwater is slightly acidic due to reactions between water and atmospheric carbon dioxide. During **hydrolysis** water reacts with a mineral to form new minerals. Hydrolysis of feldspar and other common minerals, except quartz, is a form of chemical weathering. **Oxidation** is the reaction with oxygen to decompose minerals.

Chemical and mechanical weathering often operate together. For example, solution seeping into cracks may cause rocks to expand by growth of salts or hydrolysis. Hydrolysis combines with pressure-release fracturing to form **exfoliated** granite.

Soil is the layer of weathered material overlying bedrock. Sand, silt, clay, and **humus** are commonly found in soil. Water **leaches** soluble ions downward through the soil. Clays are also transported downward by water. The uppermost layer of soil, called the **O horizon,** consists mainly of litter and humus. The amount of organic matter decreases downward. Leaching removes dissolved ions and clay from the **A horizon** and deposits them in the **B horizon.**

Six major factors control soil characteristics: parent rock, climate, rates of plant growth and decay, slope aspect and steepness, time, and transport.

In dry climates, **pedocals** form. In pedocals, leached ions precipitate in the B horizon, where they accumulate and may form **caliche.** In moist climates, **pedalfer** soils develop. In these regions, soluble ions are removed from the soil, leaving high concentrations of less soluble aluminum and iron. **Laterite** soils form in very moist climates, where all of the more soluble ions are removed.

Mass wasting is the downhill movement of rock and soil under the influence of gravity. The stability of a slope and the severity of mass wasting depend on (1) steepness of the slope, (2) orientation and type of rock layers, (3) nature of unconsolidated materials, (4) water and vegetation, and (5) earthquakes or volcanic eruptions.

Mass wasting falls into three categories: flow, slide, and fall. During **flow,** a mixture of rock, soil, and water moves as a viscous fluid. **Creep** is a slow type of flow that occurs at a rate of about 1 centimeter per year. A **mudflow** is a fluid mass of sediment and water. **Slide** is the movement of a coherent mass of material. **Slump** is a type of slide in which the moving mass travels on a concave surface. In a **rockslide,** a newly detached segment of bedrock slides along a tilted bedding plane or fracture. **Fall** occurs when particles fall or tumble down a steep cliff.

Earthquakes and volcanic eruptions trigger devastating mass wasting. Damage to human habitation can be averted by proper planning and engineering.

KEY TERMS

surface processes 181
weathering 181
erosion 181
mechanical weathering (physical weathering) 182
chemical weathering 182
pressure-release fracturing 182
frost wedging 182
talus 182
abrasion 182
hydrolysis 184

oxidation 185
salt cracking 186
exfoliation 187
regolith 188
soil 188
loam 188
litter 188
humus 188
soil horizons 189
O horizon 189
A horizon 189
topsoil 189
B horizon 189

C horizon 189
leaching 190
capillary action 190
pedocal 190
caliche 190
salinization 190
pedalfer 190
laterite 190
bauxite 190
aspect 192
mass wasting 195
landslide 195
angle of repose 197

flow 199
slide 199
fall 199
creep 199
mudflow 202
slump 202
rockslide 204

FOR REVIEW

1. Explain the differences among the terms weathering, erosion, transport, and deposition.

2. Explain the differences between mechanical weathering and chemical weathering.

3. List five processes that cause mechanical weathering.

4. What is a talus slope? What conditions favor the formation of talus slopes?

5. List three processes that cause chemical weathering.

6. What is hydrolysis? What happens when granitic rocks undergo hydrolysis? What minerals react? What are the reaction products?

7. What are the components of healthy soil? What is the function of each component?

8. Characterize the four major horizons of a mature soil.

9. List six factors that control soil characteristics and briefly discuss each one.

10. Imagine that soil forms on granite in two regions, one wet and the other dry. Will the soil in the two regions be the same or different? Explain.

11. Explain how soils formed from granite will change with time.

12. What are laterite soils? How are they formed? Why are they unsuitable for agriculture?

13. List and describe each of the factors that control slope stability.

14. What is the angle of repose? Why is the angle of repose different for different types of materials?

15. How does vegetation affect slope stability?

16. Why is mass wasting common in deserts and semiarid lands?

17. How do volcanic eruptions cause landslides?

18. How do earthquakes cause landslides?

19. Discuss the differences among flow, slide, and fall. Give examples of each.

20. Compare and contrast creep and slump. What does a pistol-butt tree trunk tell you about slope stability?

21. Explain how trees are tilted but not killed by slump.

22. How do landslides reach and destroy towns and villages many kilometers from the steep slopes where the slides originate?

FOR DISCUSSION

1. What process is responsible for each of the following observations or phenomena—is the process a mechanical or chemical change? (a) A board sawn in half. (b) A board is burned. (c) A cave is formed when water seeps through a limestone formation. (d) Calcite is formed when mineral-rich water is released from a hot underground spring. (e) Meter-thick sheets of granite peel off a newly exposed pluton. (f) Rockfall is more common in the mountains of a temperate region in the spring than in mid-summer.

2. Most substances contract when they freeze, but water expands. How would weathering be affected if water contracted instead of expanded when it froze?

3. Discuss the similarities and differences between salt cracking and frost wedging.

4. What types of weathering would predominate on the following fictitious planets? Defend your conclusions. (a) Planet X has a dense atmosphere composed of nitrogen, oxygen, and water vapor with no carbon dioxide. Temperatures range from a low of 10°C in the winter to 75°C in the summer. Windstorms are common. No living organisms have evolved. (b) The atmosphere of Planet Y consists mainly of nitrogen and oxygen with smaller concentrations of carbon dioxide and water vapor. Temperatures range from a low of −60°C in the polar regions in the winter to +35°C in the tropics. Windstorms are common. A lush blanket of vegetation covers most of the land surfaces.

5. The Arctic regions are cold most of the year and summers are short there. Thus decomposition of organic matter is slow. In contrast, decay is much more rapid in the temperate regions. How does this difference affect the fertility of the soils?

6. The Moon is considerably less massive than the Earth, and therefore its gravitational force is less. It has no atmosphere and therefore no rainfall. The interior of the Moon is cool, and thus it is geologically inactive. Would you expect mass wasting to be a common or an uncommon event in mountainous areas of the Moon? Defend your answer.

7. Explain how wildfires affect slope stability and mass wasting.

8. What types of mass wasting (if any) would be likely to occur in each of the following environments? (a) A very gradual (2 percent) slope in a heavily vegetated tropical rainforest. (b) A steep hillside composed of alternating layers of conglomerate, shale, and sandstone, in a region that experiences distinct dry and rainy seasons. The dip of the rock layers is parallel to the slope. (c) A hillside similar to that of (b) in which the rock layers are oriented perpendicular to the slope. (d) A steep hillside composed of clay in a rainy environment in an active earthquake zone.

9. Identify a hillside in your city or town that might be unstable. Using as much data as you can collect, discuss the magnitude of the potential danger. Would the landslide be likely to affect human habitation?

10. Explain how the mass wasting triggered by earthquakes and volcanoes can have more serious effects than the earthquake or volcano itself. Is this always the case?

11. Develop a strategy for minimizing loss of life from mass wasting if Mount Baker should show signs of an impending eruption similar to the signs shown by Mount St. Helens in the spring of 1980.

Fresh Water: Streams, Lakes, Ground Water, and Wetlands

Rivers erode rock and sculpt landscapes. The author kayaks on the Bighorn River in Alberta, Canada.

Water circulates constantly through the hydrosphere—the watery part of our planet. Water evaporates from the sea and from land, falls as rain or snow, flows over the land and underground, and eventually returns to the sea. Thus, evaporation transports water from the hydrosphere to the atmosphere, and condensation returns water to the hydrosphere. As a result of this continual exchange, atmospheric changes affect the hydrosphere. For example, changes in wind patterns or air temperature alter evaporation and precipitation, and therefore affect the amount of water supplied to streams and lakes. Changes in the hydrosphere also affect the atmosphere. For example, ocean currents carry heat from one region to another, warming climates in some places and cooling them in others.

When water flows from mountains toward the sea, it erodes rock and soil and carries sediment to the seacoast. Along its route, the water supports terrestrial ecosystems. Not only do organisms need water to live, most are largely composed of water. For example, a tree is about 60 percent water by weight. Humans and most other animals are about 65 percent water.

Fresh water serves humans and shapes our environment in many ways. At home we drink water, cook with it, and wash our bodies, dishes, and clothes in it. Crops need water from rain or irrigation. Many industries use water for washing, cooling, or as a chemical agent. We dam rivers to produce energy. Rivers and lakes also provide transportation and recreation. Wetlands provide habitat for millions of waterfowl and other aquatic organisms. In addition, flowing water shapes the Earth's surface into the familiar landforms that surround us. ●

10.1 The Water Cycle

The constant circulation of water among the sea, land, the biosphere, and the atmosphere is called the **hydrologic cycle,** or the water cycle (Fig. 10–1). About 1.3 billion cubic kilometers of water exist at the Earth's

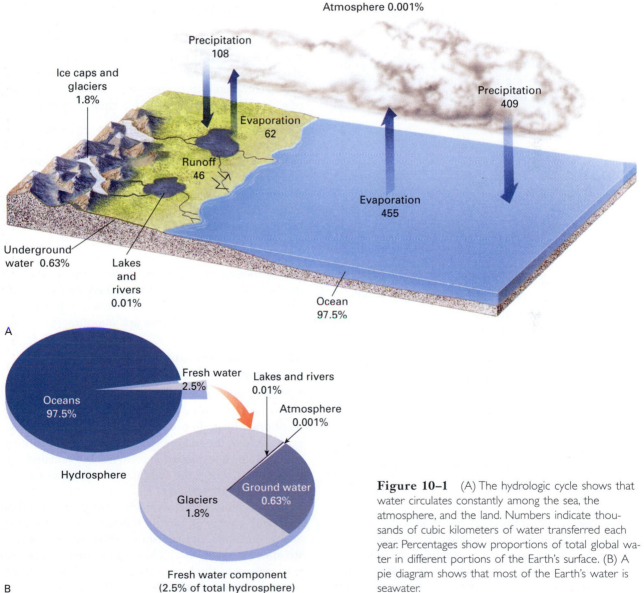

Figure 10–1 (A) The hydrologic cycle shows that water circulates constantly among the sea, the atmosphere, and the land. Numbers indicate thousands of cubic kilometers of water transferred each year. Percentages show proportions of total global water in different portions of the Earth's surface. (B) A pie diagram shows that most of the Earth's water is seawater.

surface. Of this huge quantity, 97.5 percent is salty seawater, and another 1.8 percent is frozen into the great ice caps of Antarctica and Greenland. Thus, although the hydrosphere contains a great amount of water, only a tiny fraction is fresh and available in streams, ground water, lakes, and wetlands.

Water that falls on land can follow three different paths.

1. Surface water flowing to the sea in streams and rivers is called **runoff.** This water may stop temporarily in a lake or wetland, but eventually it flows to the oceans.

2. Some water seeps into the ground to become part of a vast, subterranean reservoir known as **ground water.** Although surface water is more conspicuous, 60 times more water is stored as ground water than in all streams, lakes, and wetlands combined. Ground water seeps through bedrock and soil toward the sea, although it flows much more slowly than surface water.

3. The remainder of water that falls onto land evaporates back into the atmosphere. Water also evaporates directly from plants as they breathe in a process called **transpiration.**

10.2 Streams

Earth scientists use the term **stream** for all water flowing in a channel, regardless of the stream's size. The term *river* is commonly used for any large stream fed by smaller ones called **tributaries.** Most streams run year round, even during times of drought, because they are fed by ground water that seeps into the stream bed. When rainfall is heavy or when snow melts rapidly, a **flood** may occur. During a flood, a stream overflows its banks and spreads over low-lying adjacent land called a **flood plain.**

Stream Flow and Velocity

Three factors control stream velocity.

Gradient is the steepness of a stream. Obviously, if all other factors are equal, water flows more rapidly down a steep slope than a gradual one. A tumbling mountain stream may drop 40 meters or more per kilometer.

Discharge is the amount of water flowing down a stream. It is expressed as the volume of water flowing past a point per unit time, usually in cubic meters per second (m^3/sec). The velocity of a stream increases when its discharge increases. Thus, a stream flows faster during flood, even though its gradient is unchanged.

When Huckleberry Finn floated down the Mississippi on a raft, he probably drifted downstream at about 7 to 9 kilometers per hour (2 to 2.5 meters per second). However, a modern rafter on one of the small mountain tributaries of the Mississippi may drift downstream at only 5.4 kilometers per hour (1.5 meters per second). Thus, although it is counterintuitive, a large, lazy-appearing river may flow more rapidly than a small, steep mountain stream (Fig. 10–2).

The largest river in the world is the Amazon, with an average discharge of 150,000 m^3/sec. In contrast, the Mississippi, the largest river in North America, has an average discharge of about 17,500 m^3/sec, approximately one ninth that of the Amazon.

A stream's discharge can change dramatically from month to month or even during a single day. For example, the Selway River, a mountain stream in Idaho, has a discharge of 100 to 130 m^3/sec during early summer, when mountain snow is melting rapidly. During the dry season in late summer, the discharge drops to about 10 to 15 m^3/sec (Fig. 10–3). A desert stream may dry up completely during summer, but become the site of a flash flood during a sudden thunderstorm.

Channel characteristics refer to the shape and roughness of a stream channel. The floor of the channel is called the **bed,** and the sides of the channel are the **banks.** Friction between flowing water and the stream channel slows current velocity. Consequently, water flows more slowly near the banks than near the center of a stream. If you paddle a canoe down a straight stream channel, you move faster when you stay away from the banks. The amount of friction depends on the roughness and shape of the channel. Boulders on the banks or in the stream bed increase friction and slow a stream down, whereas the water flows more rapidly if the bed and banks are smooth.

Stream Erosion and Sediment Transport

Streams shape the Earth's surface by eroding soil and bedrock. The flowing water carries the eroded sediment grain by grain toward the sea. A stream may deposit some of the sediment on its flood plain, forming a level valley bottom, while it carries the remainder to the seacoast, where the sediment accumulates to form deltas and sandy beaches.

Stream erosion and sediment transport depend on a stream's energy. A rapidly flowing stream has more energy to erode and transport sediment than a slow stream of the same size. The **competence** of a stream is a measure of the largest particle it can carry. A fast-flowing stream can transport cobbles and even boul-

A   B

Figure 10–2 The large, lazy-appearing Colorado River flows more rapidly than Toby Creek in British Columbia because the large river has the greater discharge.

ders in addition to small particles. A slow stream carries only silt and clay.

The **capacity** of a stream is the total amount of sediment it can carry past a point in a given amount of time. Capacity is proportional to both current speed and discharge. Thus, a large, fast stream has a greater capacity than a small, slow one. Because the ability of a stream to erode and carry sediment is proportional to both velocity and discharge, most erosion and sediment transport occur during the few days each year when the stream is in flood. Relatively little erosion and sediment transport occur during the remainder of

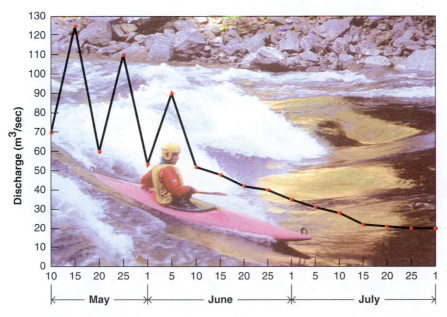

Figure 10–3 The 1988 hydrograph for the Selway River in Idaho shows that the discharge varied from 125 m^3/second in the spring to 15 m^3/second in the summer. The sharp peaks reflect high discharge during periods of rapid snowmelt. *(U.S. Forest Service)* ● Interactive Question: Would you expect the hydrograph to look the same every year? If so, why? If not, how would it differ?

the year. To see this effect for yourself, look at any stream during low water. It will most likely be clear, indicating little erosion or sediment transport. Look at the same stream later, when it is flooding. It will probably be muddy and dark, indicating that the stream is eroding its bed and banks, and transporting the sediment.

After a stream erodes soil or bedrock, it transports the sediment downstream in three different ways (Fig. 10–4). Ions dissolved in water are called **dissolved load.** A stream's ability to carry dissolved ions depends mostly on its discharge and its chemistry, not its velocity. Thus, even the still waters of a lake or ocean contain dissolved substances; that is why the sea and some lakes are salty. Although dissolved ions are invisible, they comprise more than half of the total sediment load carried by some rivers. More commonly, dissolved ions make up less than 20 percent of the total sediment load of streams.

If you place loamy soil in a jar of water and shake it up, the sand grains settle quickly. But the smaller silt and clay particles remain suspended in the water as **suspended load,** giving it a cloudy appearance. Clay and silt are small enough that even the slight turbulence of a slow stream keeps them in suspension. A rapidly flowing stream can carry sand in suspension.

During a flood, when stream energy is highest, the rushing water can roll boulders and cobbles along the bottom as **bed load.** Sand also moves in this way, but if the stream velocity is sufficient, sand grains bounce over the stream bed in a process called **saltation.** Saltation occurs because a turbulent stream flows with many small chaotic currents. When one of these currents scours the stream bed, it lifts sand and carries it a short distance before dropping it back to the bed. The falling grains strike other grains and bounce them up into the current. The overall effect is one of millions of sand grains hopping and bouncing downstream over the stream bed.

The world's two muddiest rivers—the Yellow River in China and the Ganges River in India—each carry more than 1.5 *billion* tons of sediment to the ocean every year. The sediment load of the Mississippi River is about 450 million tons per year (Fig. 10–5). Most streams carry the greatest proportion of sediment in suspension, less in solution, and the smallest proportion as bed load.

Downcutting and Base Level

A stream erodes downward into its bed and laterally against its banks. Downward erosion is called **downcutting** (Fig. 10–6). The **base level** of a stream is the

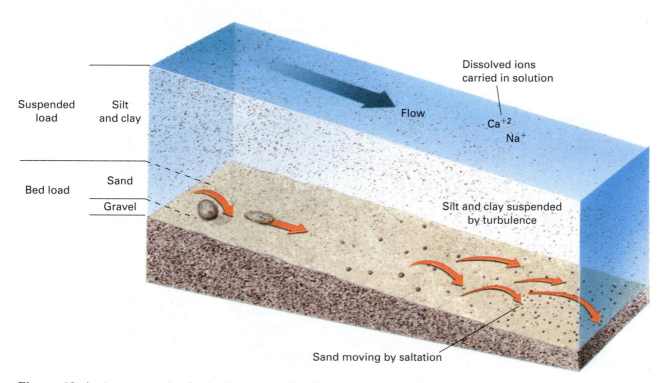

Figure 10–4 A stream carries dissolved ions in solution, silt and clay in suspension, and larger particles as bed load.

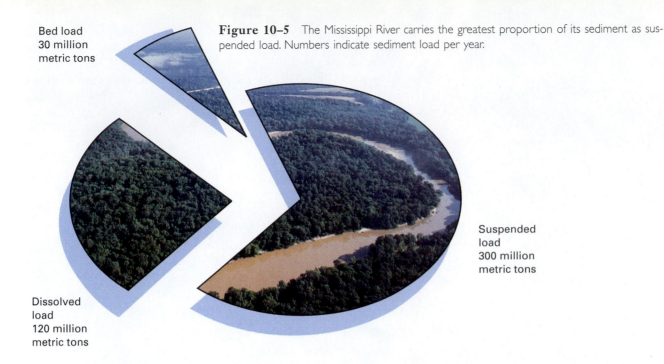

**Bed load
30 million
metric tons**

Figure 10–5 The Mississippi River carries the greatest proportion of its sediment as suspended load. Numbers indicate sediment load per year.

**Suspended
load
300 million
metric tons**

**Dissolved
load
120 million
metric tons**

deepest level to which it can erode its bed. Most streams can't erode below sea level, which is called the ultimate base level.[1] This concept is straightforward. Water can only flow downhill. If a stream were to cut its way down to sea level, it would stop flowing and hence would no longer erode its bed.

In addition to ultimate base level, a stream may have a number of local, or temporary, base levels. For example, a stream stops flowing where it enters a lake. It then stops eroding its channel because it has reached a temporary base level (Fig. 10–7). A layer of rock that resists erosion may also establish a temporary base level because it flattens the stream gradient. Thus, the stream slows down and erosion decreases. The top of a waterfall is a temporary base level commonly established by resistant rock. Niagara Falls is formed by a resistant layer of dolomite overlying softer shale. Although the dolomite resists erosion at the top of the falls, the turbulent falling water erodes the underlying shale. When its foundation is undermined, the dolomite cap collapses and the falls migrate upstream. As a result, Niagara Falls has retreated 11 kilometers upstream since its formation about 9000 years ago (Fig. 10–8). Thus, the falls retreat a little more than one meter per year. The erosion rate is rapid because the water generates considerable energy as it tumbles over the falls.

---

[1]We say "for most streams" because a few empty into valleys that lie below sea level.

Figure 10–6 Deer Creek, a tributary of Grand Canyon, has downcut its channel into solid sandstone.

Figure 10–7 A steep mountain stream in the Canadian Rockies reaches a temporary base level and stops flowing where it enters a lake.

Figure 10–8 Niagara Falls has eroded 11 kilometers upstream in the last 9000 years and continues to erode today. *(Hubertus Kanus/Photo Researchers, Inc.)*

A stream like that in Figure 10–9A erodes rapidly in the steep places where its energy is high, and deposits sediment in the low-gradient stretches where it flows more slowly. Over time, erosion and deposition smooth out the irregularities in the gradient. The resulting **graded stream** has a smooth, concave profile (Fig. 10–9B). Once a stream becomes graded, there is no net erosion or deposition and the stream profile no longer changes. An idealized graded stream such as this does not actually exist in nature, but many streams come close.

Sinuosity of a Stream Channel

A steep mountain stream usually downcuts rapidly compared with the rate of lateral erosion. As a result, it cuts a relatively straight channel with a steep-sided, V-shaped valley (Fig. 10–10). The stream maintains its relatively straight path because it flows with enough energy to erode and transport any material that slumps into its channel.

In contrast, a low-gradient stream is less able to erode downward into its bed. Much of the stream energy is directed against the banks, causing **lateral erosion.** Lateral erosion undercuts the valley sides and

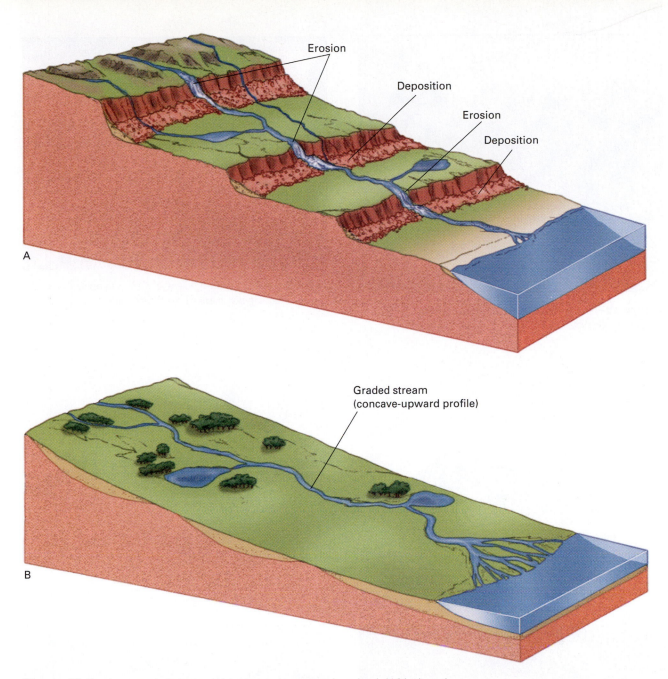

Erosion

Deposition

Erosion

Deposition

A

Graded stream
(concave-upward profile)

B

Figure 10–9 An ungraded stream (A) has many temporary base levels. With time, the stream smooths out the irregularities to develop a graded profile (B).

widens a stream valley. Most low-gradient streams flow in a series of bends, called **meanders** (Fig. 10–11). A meandering stream wanders back and forth across its entire flood plain, forming a wide valley with a flat bottom.

As a stream flows into a meander bend, the inertia of the moving water tends to keep the water moving in a straight path. Consequently, most of the current flows to the outside bank of the meander. As a result, both the velocity and channel depth are greatest near the outside of the bend, and the stream erodes its outside bank. At the same time, sediment accumulates in the slower water on the inside of the meander to form a **point bar.** Because a meandering stream erodes the outside banks of its meanders, and deposits sediment on the insides of the bends, the meanders migrate slowly down the flood plain as the stream constantly modifies its channel. Occasionally, an **oxbow lake** forms where the stream cuts across the neck of a meander and isolates an old meander loop (Fig. 10–12).

Figure 10–10 A steep mountain stream eroded a V-shaped valley into soft shale in the Canadian Rockies.

Figure 10–11 A low-gradient stream commonly flows in a series of looping bends called meanders. This one is in Baffin Island, Canada.

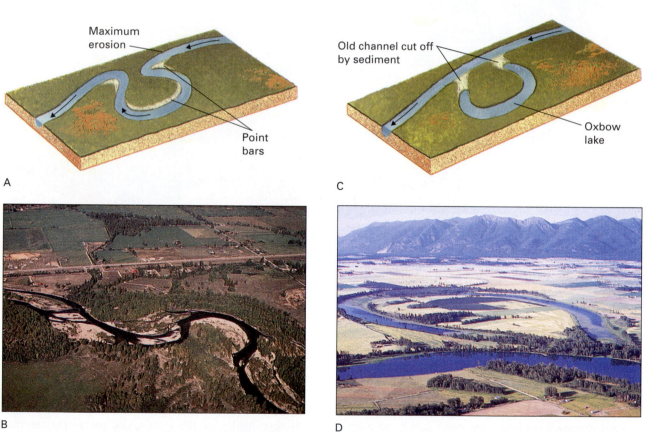

Figure 10–12 (A) A stream erodes the outsides of meanders and deposits sand and gravel on the inside bends to form point bars. (B) Meanders and point bars form the channel of the Bitterroot River, Montana. (C) Over time, a stream may erode through the neck of a meander to form an oxbow lake. (D) This oxbow lake formed in the Flathead River, Montana.

Most streams flow in a single channel. In contrast, a **braided stream** flows in many shallow, interconnecting channels (Fig. 10–13). A braided stream forms where more sediment is supplied to a stream than it can carry. The excess sediment accumulates in the channel, filling it and forcing the stream to overflow its banks and erode new channels. As a result, a braided stream flows simultaneously in several channels and shifts back and forth across its floodplain.

Braided streams are common in both deserts and glacial environments because both produce abundant sediment. A desert yields large amounts of sediment because it has little or no vegetation to prevent erosion. Glaciers grind bedrock into fine sediment, which is carried by streams flowing from the melting ice.

Alluvial Fans and Deltas

If a steep mountain stream flows onto a flat plain, its gradient and velocity decrease abruptly. As a result, it deposits most of its sediment in a fan-shaped mound called an **alluvial fan.** Alluvial fans are common in many arid and semiarid mountainous regions (Fig. 10–14).

Figure 10–14 This alluvial fan in Death Valley formed where a steep mountain stream deposited most of its sediment as it entered the flat valley. A road runs across the lower part of the fan.

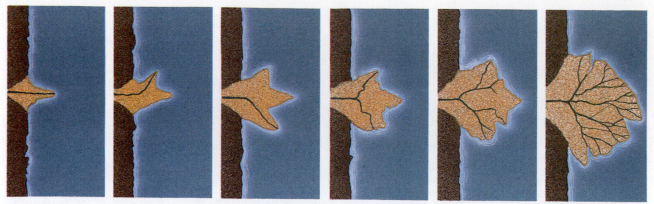

Young delta straight channel

As the delta grows the channel is diverted to one side and then the other

Distributaries form

Mature delta

Figure 10–15 A delta forms and grows with time where a stream deposits its sediment as it flows into a lake or the sea.

A stream also slows abruptly where it enters the still water of a lake or ocean. The sediment settles out to form a nearly flat landform called a **delta.** Part of the delta lies above water level, and the remainder lies slightly below water level. Deltas are commonly fan shaped, resembling the Greek letter Δ.

Both deltas and alluvial fans change rapidly. The stream abandons sediment-choked channels, while new channels develop, as in a braided stream. As a result,

a stream feeding a delta or fan splits into many channels called **distributaries.** A large delta may spread out in this manner until it covers thousands of square kilometers (Fig. 10–15). Most fans, however, are much smaller, covering a fraction of a square kilometer to a few square kilometers.

Figure 10–16 shows that the Mississippi River has flowed through seven different delta channels during the past 6000 years. But in recent years engineers have

Figure 10–16 The Mississippi River has flowed into the sea by seven different channels during the past 6000 years. As a result, the modern delta is composed of seven smaller overlapping deltas that formed at different times. The oldest delta is numbered 1, and the current delta is 7.

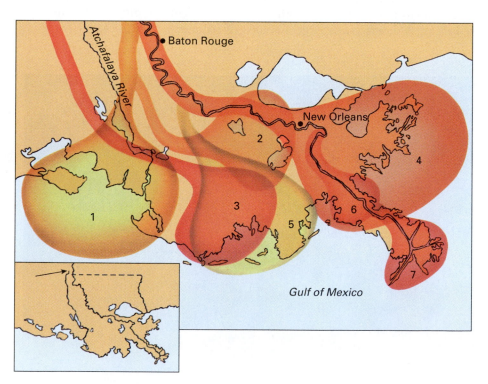

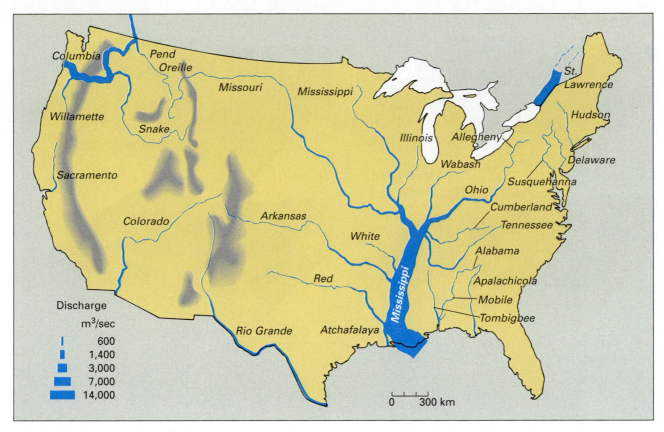

Figure 10–17 Most of the surface water in the United States flows to the sea from approximately a dozen major streams.

built great systems of levees to stabilize the channels. If the Mississippi River were left alone, it would probably abandon the lower 500 kilometers of its present path and cut into the channel of the Atchafalaya River to the west. However, this part of the delta is heavily industrialized, and it is impractical to allow the river to change its course, to flood towns in some areas and leave shipping lanes and wharves high and dry in others.

Drainage Basins

Only a dozen or so major rivers flow into the sea along the coastlines of the United States (Fig. 10–17). Each is fed by a number of tributaries, which are in turn fed by smaller tributaries. Mountain ranges or plateaus separate adjacent river systems. The region drained by a single river is called a **drainage basin.** For example, the Rocky Mountains separate the Colorado and Columbia drainage basins to the west from the Mississippi and Rio Grande basins to the east. Together, those four river systems drain more than three fourths of the United States.

10.3 Stream Erosion and Mountains: How Landscapes Evolve

According to a model popular in the first half of this century, streams erode the Earth's surface and create landforms in an orderly sequence (Fig. 10–18). At first, they cut steep, V-shaped valleys into mountains. Over time, the streams erode the mountains away and widen the valleys into broad flood plains. Eventually, the entire landscape flattens, forming a large, featureless plain. However, if this were the only mechanism affecting the Earth's surface during its 4.6-billion-year history, all landforms would have eroded to a flat plain. Why, then, do mountains, valleys, and high plateaus still exist?

The model tells only half the story. Streams *do* continuously erode the landscape, flattening mountains and widening flood plains. But at the same time, tectonic activity may uplift the land and interrupt the simple, idealized sequence. In this way, the Earth's hydrosphere and atmosphere work together with tectonic processes to create landforms.

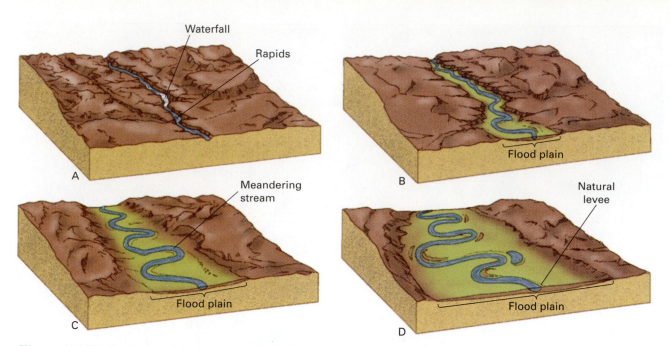

Figure 10–18 If tectonic activity does not uplift the land, over time streams erode the mountains away and widen the valleys into broad flood plains.

Consider the Himalayas. Today, they are a rapidly rising mountain range as a result of tectonic processes and isostatic uplift described in Chapter 8. As the mountains rise, landslides, glaciers, and running water erode the peaks. The water flows into streams, which cut deep, V-shaped valleys through the range, and transport vast quantities of sediment to the sea. Modern streams carry an annual average of 1000 tons of sediment per square kilometer from the Himalayas, creating the vast deltas and submarine alluvial fans of the Indus and Ganges Rivers (Fig 10–19). This erosion rate, averaged over the entire Himalayan Range, amounts to a lowering of the land surface by 0.5 millimeters each year. But the average rate of tectonic and isostatic uplift of the Himalayas is about 5 millimeters per year, 10 times faster than the erosion rate.

Sometime in the future, tectonic uplift will cease. But erosion will continue to remove rock and sediment from the mountains. At first, this erosion won't lower the mountains by much because the range will continue to rise isostatically. Eventually, erosion will dominate over the isostatic uplift. Then, streams, glaciers, and landslides will wear the Himalayas down to low, rounded mountains that will probably resemble the modern Appalachians.

Recall that the Appalachians rose as the result of an ancient collision between continents, much as the Himalayas have risen during a more recent collision.

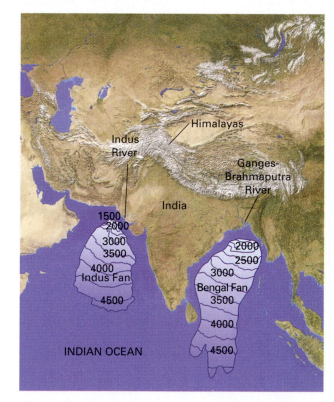

Figure 10–19 The Indus and Ganges-Bramaphutra Rivers have transported millions of cubic kilometers of sediment eroded from the Himalayas to the Indian Ocean. *(Tom Van Sant, Geosphere Project)*

The early Appalachians may have been as lofty as the modern Himalayas, and the modern Appalachians may serve as a model for the future of the Himalayas.

10.4 Floods

When more water flows down a stream than the channel can hold, water overflows onto the flood plain, creating a flood. Floods in the United States cause an average of 85 human deaths and more than $1 billion in property damage each year. In the spring of 1997, the Red River rose 13 meters above flood stage at Fargo, North Dakota. The flood inundated large parts of North Dakota, Minnesota, and Manitoba, causing more than $400 million in damage. The Ohio River basin floods during the winter of 1997 caused more than $425 million in damage and resulted in 24 deaths. Two weeks of torrential rain flooded portions of California in January 1995, killing at least nine people and causing $1.3 billion in damage. Floods plagued California again during the winters of 1996–97 and 1997–98.

While flooding is a natural event, flood frequency and severity are often augmented by logging, farming, and urbanization. Recall from Chapter 9 that when all the vegetation was cut or killed in a New Hampshire forest, stream flow increased by 40 percent. In a similar manner, when forests are cut and replaced by farms, stream flow may increase. Similar conditions develop when prairies are plowed and farmed. Nearly all rainwater runs off heavily paved urban areas directly into nearby streams.

Floods are costly in part because people choose to live in flood plains. Many riverbank cities originally grew as ports, to take advantage of the easy transportation afforded by rivers. In addition, flood plains provide rich soils for farms, flat land for roads and buildings, and access to abundant water for industry and agriculture.

Although we normally think of floods as destructive events, river and flood plain ecosystems depend on floods. For example, cottonwood tree seeds only germinate after a flood, and waterfowl depend on flood plain wetlands. Many species of fish gradually lose out to stronger competitors during normal flows, but have adapted better to floods so their populations increase as a result of flooding. Thus, species diversity is maintained by alternate periods of flooding and normal flow.

In some cases, flooding even benefits humans directly. Frequent small floods dredge bigger channels, which reduce the severity of large floods. Flooding streams carry large sediment loads and deposit them on flood plains to form fertile soil. So, paradoxically, the same floods that cause death and disaster create the rich soils that make flood plains so attractive for farming.

Flood Frequency

Many streams flood regularly, some every year. In any stream, small floods are more common than large ones, and the size of floods can vary greatly from year to year. A 10-year flood is the largest flood that occurs in a given stream on an average of once every 10 years. A 100-year flood is the largest that occurs on an average of once every 100 years. For example, a stream may rise 2 meters above its banks during a 10-year flood, but 7 meters during a 100-year flood. Thus, a 100-year flood is higher and larger, but less frequent, than a 10-year flood.

Flood Control and the 1993 Mississippi River Floods

During the late spring and summer of 1993, heavy rain soaked the upper Midwest. Thirteen centimeters fell in already saturated central Iowa in a single day. In mid-July, 2.5 centimeters of rain fell in 6 minutes in Papillion, Nebraska. As a result of the intense rainfall over such a large area, the Mississippi River and its tributaries flooded. In Fargo, North Dakota, the Red River, fed by a day-long downpour, rose 1.2 meters in 6 hours, flooding the town and backing up sewage into homes and the Dakota Hospital. In St. Louis, Missouri, the Mississippi crested 14 meters above normal and 1 meter above the highest previously recorded flood level. At its peak, the flood inundated nearly 44,000 square kilometers in a dozen states. Damage to homes and businesses on the flood plain reached $12 billion. Forty-five people died (Fig. 10–20).

Figure 10–20 Residents of Hartsburg, Missouri, boating on Main Street during the 1993 flood. *(Stephen Levin)*

Figure 10–21 Levees are effective in containing rivers during small floods. However, large floods often breach the levees and the water spills onto the surrounding flood plain. *(Frank Oberle/Tony Stone Images)*

During the 1993 Mississippi River flood, control projects saved entire towns and prevented millions of dollars in damage. However, in some cases described in the following section, control measures increased damage. Geologists, engineers, and city planners are studying these conflicting results to plan for the next flood, which may be years or decades away.

Artificial Levees and Channels   An **artificial levee** is a wall built along the banks of a stream to prevent rising water from spilling out of the stream channel onto the flood plain. In the past 70 years, the U.S Army Corps of Engineers has spent billions of dollars building 11,000 kilometers of levees along the banks of the Mississippi and its tributaries (Fig. 10–21).

As the Mississippi River crested in July 1993, flood waters surged through low areas of Davenport, Iowa, built on the flood plain. However, the business district of nearby Rock Island, Illinois, remained mostly dry. In 1971, Rock Island had built levees to protect low-lying areas of the town, whereas Davenport had not built levees. Hannibal, Missouri, had just completed

levee construction to protect the town when the flood struck. The $8 million project in Hannibal saved the Mark Twain home and museum and protected the town and surrounding land from flooding.

Unfortunately, levees can create conditions that cause much greater floods in the future, and they can cause higher floods along nearby reaches of a river. In the absence of levees, when a stream floods, it deposits mud and sand on the flood plain (10–22A). When artificial levees are built, the stream cannot overflow during small floods, so it deposits the sediment in its channel, raising the level of the stream bed (10–22B). After several small floods, the entire stream may rise *above* its flood plain, contained only by the levees (Fig. 10–22C). This configuration creates the potential for a truly disastrous flood because if the levee should be breached during a large flood, the entire stream then flows out of its channel and onto the flood plain. As a result of levee building and channel sedimentation, portions of the Yellow River in China now lie 10 meters above its flood plain, and the Mississippi River lies above adjacent parts of New Orleans where it flows

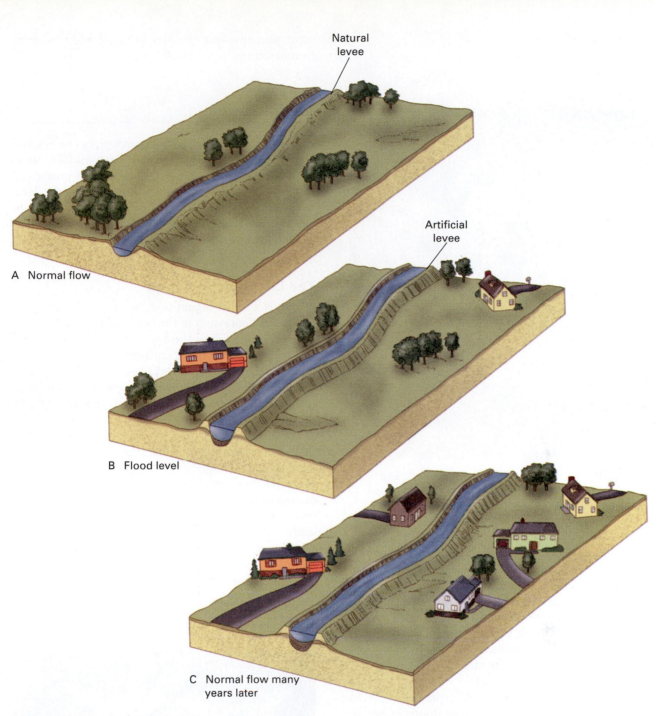

Figure 10–22 (A) In the natural state, a flooding stream carries sediment from the stream channel onto the flood plain. (B) Artificial levees cause sediment to accumulate in a stream channel. (C) Eventually, the channel rises above the flood plain, creating the potential for a disastrous flood.

through the city. Thus, levees may solve flooding problems in the short term, but in a longer time frame they may cause even larger and more destructive floods.

Levees may also cause higher flood levels upstream. A flooding river spreads horizontally over its flood plain, temporarily forming a wide path of flowing wa-

ter. But where levees constrict the river into its narrow channel, they form a partial dam, causing the waters to rise to even higher flood levels upstream (Fig. 10–23).

Engineers have tried to solve the problem of channel sedimentation by dredging **artificial channels** across meanders. When a stream is straightened, its

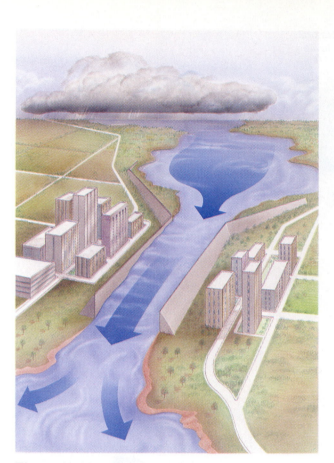

Figure 10–23 Levees force a flooding river into a restricted channel, forming a partial dam that raises the flood level upstream from the restriction.

gradient and velocity increase and it scours more sediment from its channel. This solution, however, also has its drawbacks. A straightened stream is shorter than a meandering one, and consequently the total volume of its channel is reduced. Therefore, the channel cannot contain as much excess water, and flooding is likely to increase both upstream and downstream from the straightened reach of the stream.

Flood Plain Management As explained above, in many cases, attempts at controlling floods either do not work or they shift the problem to a different time or place. An alternative approach to flood control is to abandon some flood control projects and let the river spill out onto its flood plain. Of course the question is, "What land should be allowed to flood?" Every farmer and homeowner on the river wants to maintain the levees that protect his or her land. Currently, federal and state governments are establishing wildlife reserves in some flood plains. Since no development is allowed in these reserves, they will flood during the next high water. However, a complete river management plan involves complex political and economic considerations.

10.5 Lakes

Lakes and lake shores are attractive places to live and play. Clean, sparkling water, abundant wildlife, beautiful scenery, aquatic recreation, and fresh breezes all come to mind when we think of going to the lake. Despite their great value, lakes are fragile and ephemeral. Modern, post–ice age humans live in a special time in Earth history when the Earth's surface is dotted with beautiful lakes.

The Life Cycle of a Lake

A **lake** is a large, inland body of standing water that occupies a depression in the land surface (Fig. 10–24). Streams flowing into the lake carry sediment, which fills the depression in a relatively short time, geologically speaking. Soon the lake becomes a swamp, and with time the swamp fills with more sediment and vegetation to become a meadow or forest with a stream flowing through it.

If most lakes fill quickly with sediment, why are they so abundant today? Most lakes exist in places that were covered by glaciers during the latest ice age. About 18,000 years ago, great continental ice sheets extended well south of the Canadian border, and mountain glaciers scoured their alpine valleys as far south as New Mexico and Arizona. Similar ice sheets and alpine glaciers existed in higher latitudes of the Southern Hemisphere. We are just now emerging from that glacial episode.

Figure 10–24 Lago Nube is a high mountain lake in Bolivia's Cordillera Apolobamba.

Figure 10–25 Gravel deposited by a glacier forms a dam to create this mountain lake in the Sierra Nevada.

The glaciers created lakes in several different ways. Flowing ice eroded numerous depressions in the land surface, which then filled with water. The Finger Lakes of upper New York State and the Great Lakes are examples of large lakes occupying glacially scoured depressions.

The glaciers also deposited huge amounts of sediment as they melted and retreated. Some of these great piles of glacial debris formed dams across stream valleys. When the glaciers melted, streams flowed down the valleys but were blocked by the dams. Many modern lakes occupy glacially dammed valleys (Fig. 10–25).

In addition, the melting glaciers left huge blocks of ice buried in the glacial sediment. As the ice blocks melted, they left depressions that filled with water. Many thousands of small lakes and ponds, called **kettles,** formed in this way. Kettles are common in the northern United States and the southern Canadian prairie (Fig. 10–26).

Most of these glacial lakes formed within the past 10,000 to 20,000 years, and sediment is rapidly filling them. Many smaller lakes have already become swamps. In the next few hundred to few thousand years, many of the remaining lakes will fill with mud. The largest, such as the Great Lakes, may continue to exist for tens of thousands of years. But the life spans of lakes such as these are limited, and it will take another glacial episode to replace them.

Lakes also form by nonglacial means. A volcanic eruption can create a crater that fills with water to form a lake, such as Crater Lake, Oregon. Oxbow lakes form in abandoned river channels. Other lakes, such as Lake Okeechobee of the Florida Everglades, form in flat lands with shallow ground water. These types of lakes, too, fill with sediment and, as a result, have limited lives.

A few lakes, however, form in ways that extend their lives far beyond that of a normal lake. For example, Russia's Lake Baikal is a large, deep lake lying in a depression created by an active fault. Although rivers pour sediment into the lake, movement of the fault repeatedly deepens the basin. As a result, the lake has existed for more than a million years, so long that indigenous species of seals, other animals, and fish have evolved in its ecosystem.

Nutrient Balance in Lakes

When plants and animals die in a lake, their bodies settle slowly to the bottom, carrying essential nutrients with them. If the water is deep enough, the nutrients accumulate below the zone where there is enough sunlight for plant growth. Thus, in a deep lake, sunlight is available near the surface, but nutrients are abundant only on the bottom. Plankton (small, free-floating organisms) grow poorly on the surface due to the lack of nutrients, and bottom-rooted plants cannot grow due to lack of sunlight. Thus, the lakes contain low concentrations of nitrates, phosphates, and other critical nutrients that sustain aquatic food webs. The purity of the water gives the lakes a deep blue color that we associate with a clean, healthy lake. Such a lake is called **oligotrophic,** meaning "poorly nour-

Figure 10–26 Kettle lakes in Montana's Flathead Valley formed when blocks of glacial ice melted.

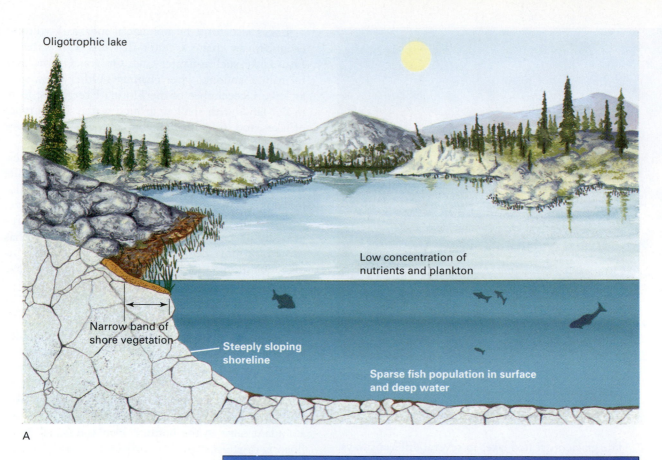

Oligotrophic lake

Low concentration of
nutrients and plankton

Narrow band of
shore vegetation

Steeply sloping
shoreline

Sparse fish population in surface
and deep water

A

Figure 10–27 (A) An oligotrophic, or
nutrient-poor, lake contains pure water and
few organisms. (B) Crater Lake, Oregon, is
an oligotrophic lake with a low nutrient
content, few plants or fish, and clear, blue
water. *(Rich Buzzelli/Tom Stack and Associates)*

B

ished" (Fig. 10–27). Oligotrophic lakes have low productivities, meaning that they sustain relatively few living organisms, although a lake of this type typically
contains a few huge trout or similar game fish.

As a lake fills with sediment, however, it becomes
shallower and sunlight reaches more and more of the
lake bottom. The sunlight allows bottom-rooted
plants to grow. As they die and rot, their litter adds

nutrients to the lake water. Plankton increase in numbers, as do fish and other organisms. The lake becomes
so productive that its surface may become covered with
a green scum of plankton or a dense mat of rooted
plants. The litter contributes to the sediment filling
the lake, and eventually the lake becomes a swamp. A
lake of this kind with a high nutrient supply is called
a **eutrophic** lake (Fig. 10–28). Eutrophication occurs

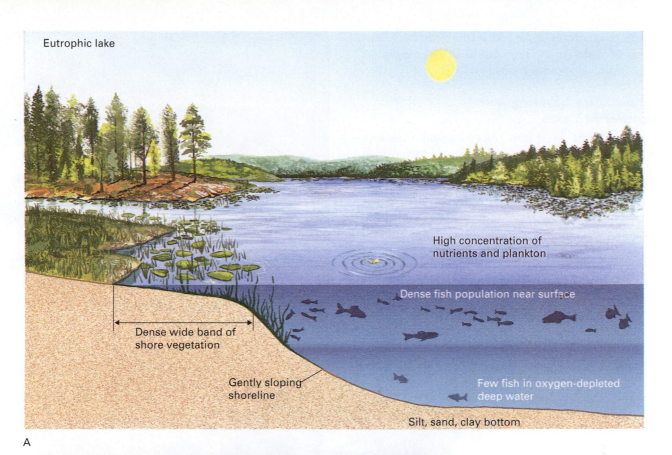

Eutrophic lake

High concentration of nutrients and plankton

Dense fish population near surface

Dense wide band of shore vegetation

Gently sloping shoreline

Few fish in oxygen-depleted deep water

Silt, sand, clay bottom

A

B

Figure 10–28 (A) A eutrophic, or nutrient-rich, lake contains many plants and other organisms. (B) A slimy mat of algae and bacteria covers the surface of a eutrophic lake in Western New York. *(W.A. Banaszewski/Visuals Unlimited)* ●

Interactive Question: Which type of lake, oligotrophic or eutrophic would be more desirable for humans? Discuss and defend your answer.

naturally as part of the life cycle of a lake. However, addition of nutrients in the form of sewage and other kinds of pollution has greatly accelerated the eutrophication of many lakes.

Fresh Water and Salty Lakes

Most lakes contain fresh water because the constant flow of streams both into and out of them keeps salt from accumulating. A few lakes are salty; some, such as Utah's Great Salt Lake, are saltier than the oceans. A salty lake forms when streams flow into the lake but no streams flow out. Streams carry salts into the lake, but water leaves the lake only by evaporation and a small amount of seepage into the ground. Evaporation removes pure water, but no salts. Thus, over time the small amounts of dissolved salts carried in by the streams concentrate in the lake water. Salty lakes usually occur in desert and semiarid basins, where dry air and sunshine evaporate water rapidly (Fig. 10–29).

Temperature Layering and Turnover in Lakes

If you have ever dived into a deep lake on a summer day, you probably discovered that the top meter or so of lake water can be much warmer than deeper water. This occurs because sunshine warms the upper layer of water, making it less dense than the cooler, deeper water. The warm, less dense water floats on the cooler, denser water. The boundary between the warm and cool layers is called the **thermocline** (Fig. 10–30).

In temperate climates, colder autumn weather cools the surface water to a temperature below that of deeper water, so that the surface water becomes more dense than the deeper water. Consequently it sinks, mixing

Figure 10–29 Camels graze near a salty desert lake in the Gobi Desert, northern Mongolia.

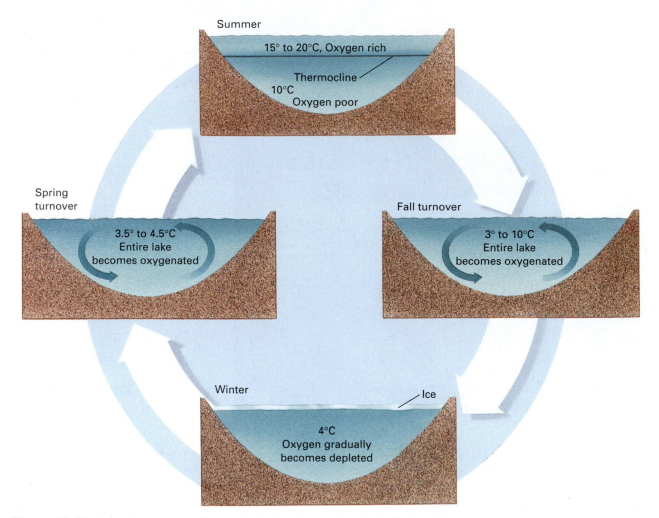

Summer

15° to 20°C, Oxygen rich

Thermocline

10°C

Oxygen poor

Spring turnover

3.5° to 4.5°C Entire lake becomes oxygenated

Fall turnover

3° to 10°C Entire lake becomes oxygenated

Winter

Ice

4°C Oxygen gradually becomes depleted

Figure 10–30 Lakes in temperate climates develop temperature layering in both summer and winter. As a result, bottom waters become depleted in oxygen. In fall and spring, water temperature becomes constant throughout the lake and turnover brings new supplies of oxygen to the deep waters.

the surface and deep waters and equalizing the water temperature throughout the lake. This process is called fall **turnover.** In the winter, ice floats on the surface and temperature layering develops again. In spring, as ice melts on the lake, surface water again becomes more dense than deep lake waters, and spring turnover occurs. As summer comes, the lake again develops thermal layering.

Turnover in temperate lakes illustrates an important Earth systems interaction among the atmosphere, the hydrosphere, and the biosphere. During summer and winter when the lake water is layered, bottom-dwelling organisms may use up most or all of the oxygen in deep waters. At the same time, surface organisms may deplete surface waters of dissolved nutrients. However, surface water is rich in oxygen because it is in contact with the atmosphere, and deep water may be rich in nutrients because it is in contact with bottom sediment. Turnover enriches deep water in oxygen and, at the same time, supplies nutrients to the surface water. The latter effect often becomes evident in the form of an algal bloom—a sudden and obvious increase in the amount of floating green algae on a lake's surface—in spring and fall.

10.6 Ground Water

If you drill a hole into the ground in most places, its bottom fills with water after a short time, usually within a few minutes to a few days. The water appears even if no rain falls and no streams flow nearby. The water that seeps into the hole is part of the vast reservoir of subterranean ground water that saturates the Earth's crust in a zone between a few meters and a few kilometers below the surface.

Ground water is exploited by digging wells and pumping the water to the surface. It provides drinking water for more than half of the population of North America and is a major source of water for irrigation and industry. However, deep wells and high-speed pumps now extract ground water more rapidly than natural processes replace it in many parts of the central and western United States. In addition, industrial, agricultural, and domestic contaminants seep into ground water in many parts of the world. Such pollution is often difficult to detect and expensive to clean up. Ground water use and pollution are discussed in Chapter 20.

Characteristics of Ground Water

Ground water fills small cracks and voids in soil and bedrock. The proportional volume of these open spaces is called **porosity.** Sand and gravel typically have high porosities—40 percent or more. Mud can have a porosity of 90 percent or more. It has such a high porosity because the tiny clay particles are electrically attracted to water, and consequently, clay-rich mud absorbs a very high proportion of water. Most rocks have lower porosities than loose sediment. Sandstone and conglomerate can have 5 to 30 percent porosity (Fig. 10–31). Shale typically has a porosity of less than 10 percent. Igneous and metamorphic rocks have very low porosities unless they are fractured.

Porosity indicates the amount of water that rock or soil can hold. In contrast, **permeability** is the ability of rock or soil to transmit water (or any other fluid). Water can flow rapidly through material with high permeability. Most materials with high porosity also have high permeability. Sand and sandstone have numerous, relatively large, well-connected pores that allow the water to flow through the material. However, if the pores are very small, as in clay and shale, electrical attractions between water and soil particles slow the passage of water. Clay typically has a high porosity, but because its pores are so small it commonly has a very low permeability and transmits water slowly.

The Water Table and Aquifers

When rain falls, much of it soaks into the ground. Water does not descend into the crust indefinitely, however. Below a depth of a few kilometers, the pressure from overlying rock closes the pores, making bedrock both nonporous and impermeable. Water accumulates on this impermeable barrier, filling pores in the rock and soil above it. This completely wet layer of soil and bedrock above the barrier is called the **zone of saturation.** The **water table** is the top of the zone of saturation (Fig. 10–32). The unsaturated zone, or **zone of aeration,** lies above the water table. In this layer, the rock or soil may be moist but is not saturated.

If you dig into the unsaturated zone, the hole does not fill with water. However, if you dig below the water table into the zone of saturation, you have dug a **well,** and the water level in a well is at the level of the water table. During a wet season, rain seeps into the ground to **recharge** the ground water, and the water table rises. During a dry season, the water table falls. Thus, the water level in most wells fluctuates with the seasons.

An **aquifer** is any body of rock or soil that can yield economically significant quantities of water. An aquifer must be both porous and permeable so that water flows into a well to replenish water that is pumped out. Sand and gravel, sandstone, limestone, and highly fractured bedrock of any kind make excellent aquifers. Shale, clay, and unfractured igneous and metamorphic rocks are poor aquifers.

Well-sorted sediment

Poorly sorted sediment

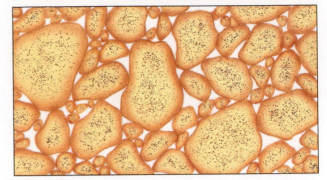

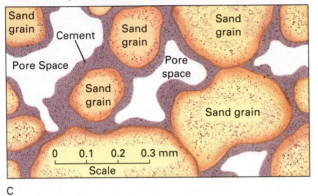

Sedimentary rock with cementing material between grains

Sand grain

Cement

Pore Space

Sand grain

Sand grain

Pore space

Sand grain

Sand grain

| 0 | 0.1 | 0.2 | 0.3 mm |
Scale

A

B

C

Figure 10–31 Different materials have different amounts of open pore space between grains. (A) Well-sorted sediment consists of equal-size grains and has a high porosity, about 30 percent in this case. (B) In poorly sorted sediment, small grains fill the spaces among the large ones, and porosity is lower. In this drawing it is about 15 percent. (C) Cement partly fills pore space in sedimentary rock, lowering the porosity.

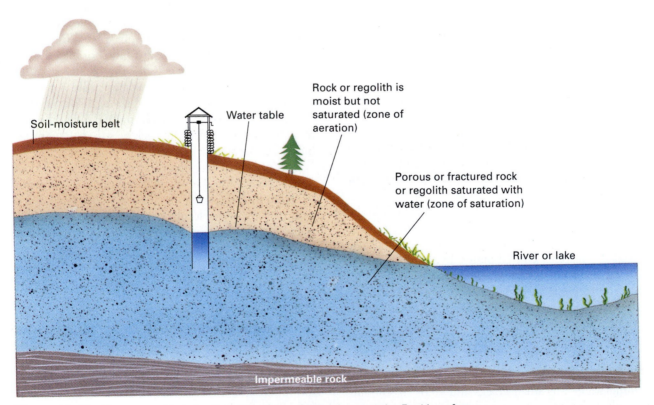

Soil-moisture belt

Water table

Rock or regolith is moist but not saturated (zone of aeration)

Porous or fractured rock or regolith saturated with water (zone of saturation)

River or lake

Impermeable rock

Figure 10–32 The water table is the top of the zone of saturation near the Earth's surface. It intersects the land surface at lakes and streams and is the level of standing water in a well.

Ground Water Movement

Nearly all ground water seeps slowly through bedrock and soil. Ground water flows at about 4 centimeters per day (about 15 meters per year), although flow rates may be much faster or slower depending on permeability. Most aquifers are like sponges through which water seeps, rather than underground pools or streams. However, ground water can flow very rapidly through large fractures in bedrock, and in a few regions, underground rivers flow through caverns.

In general, the water table is higher beneath a hill than it is beneath an adjacent valley. Ground water flows from zones where the water table is highest toward areas where it is lowest. Some ground water flows along the sloping surface of the water table toward the valley. Much of the ground water, however, flows downward beneath the hill. Then the weight of water

Figure 10–34 The Escalante River is an influent or losing stream that flows through the Utah desert.

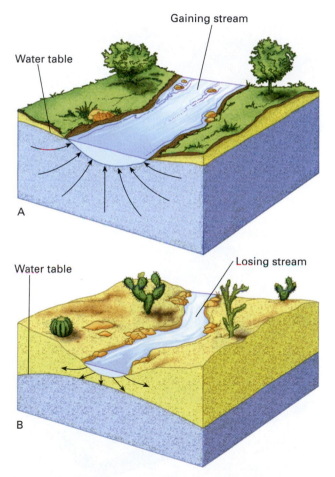

Figure 10–33 The water table follows topography, rising beneath hills and sinking beneath valleys. (A) In a moist climate the water table lies above the stream and ground water seeps into the stream. (B) A desert stream lies above the water table. Water seeps from the stream bed to recharge the ground-water reservoir beneath the desert.

beneath the hill forces the ground water to curve back upward beneath the valley, as shown by the arrows in Figure 10–33A.

Streams flow through most valleys. Because ground water rises beneath the valley floor, it continually feeds the stream. That is why streams continue to flow even when rain has not fallen for weeks or months. A stream that is recharged by ground water is called an **effluent** (or gaining) **stream** (Fig. 10–33A). Ground water also flows into most lakes because lakes occupy low parts of the land.

In a desert, however, the water table commonly lies below a stream bed, and water seeps downward from the stream to the water table (Figs. 10–33B and 10–34). Such a stream is an **influent** (or losing) **stream.** Desert stream channels are dry most of the time. When they do run, the water often flows from nearby mountains where precipitation is greater or when snow melts in the spring, although a desert storm can also fill the channel briefly. Thus, a desert stream feeds the ground-water reservoir, but in temperate climates, ground water feeds the stream.

Springs and Artesian Wells A **spring** occurs where the water table intersects the land surface and water flows or seeps onto the surface (Fig. 10–35A). In some places, a layer of impermeable rock or clay lies above the main water table, creating a locally saturated zone, the top of which is called a **perched water table** (Fig. 10–35B). Hillside springs often flow from a perched water table. Springs also occur where fractured bedrock or cavern systems intersect the land surface (Figs. 10–35 C and D).

Figure 10–36 shows a tilted layer of permeable sandstone sandwiched between two layers of impermeable shale. An inclined aquifer, such as the sandstone layer, bounded top and bottom by impermeable rock is an **artesian aquifer.** Water in the lower part of the aquifer is under pressure from the weight of water above. Therefore, if a well is drilled through the shale and into the sandstone, water rises in the well without being pumped. A well of this kind is called an **artesian well.** If pressure is sufficient, the water spurts out onto the land.

Caverns, Sinkholes, and Karst Topography

Just as streams erode valleys and form flood plains, ground water also creates landforms (Fig. 10–37). Recall that rainwater reacts with atmospheric carbon dioxide to become slightly acidic and capable of dissolving limestone. A **cavern** forms when acidic water seeps into cracks in limestone, dissolving the rock and enlarging the cracks. Mammoth Cave in Kentucky and Carlsbad Caverns in New Mexico are two famous caverns formed in this way.

Although caverns form when limestone dissolves, most caverns also contain features formed by deposition of calcite. When a solution of water, dissolved

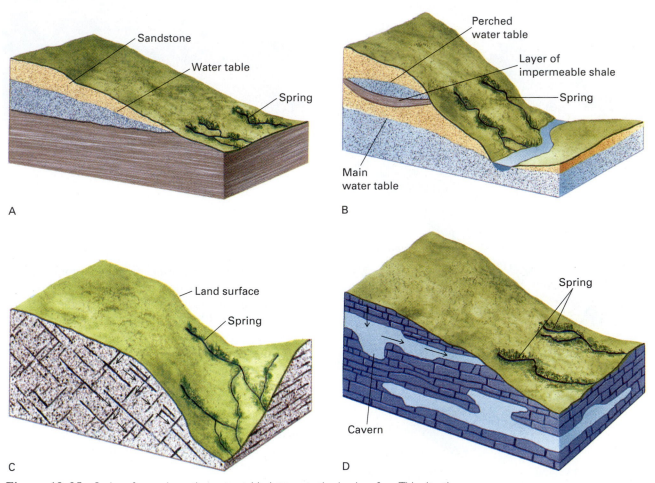

Figure 10–35 Springs form where the water table intersects the land surface. This situation can occur where (A) the land surface intersects a contact between permeable and impermeable rock layers; (B) a layer of impermeable rock or clay lies "perched" above the main water table; (C) water flows from fractures in otherwise impermeable bedrock; and (D) water flows from caverns onto the surface.

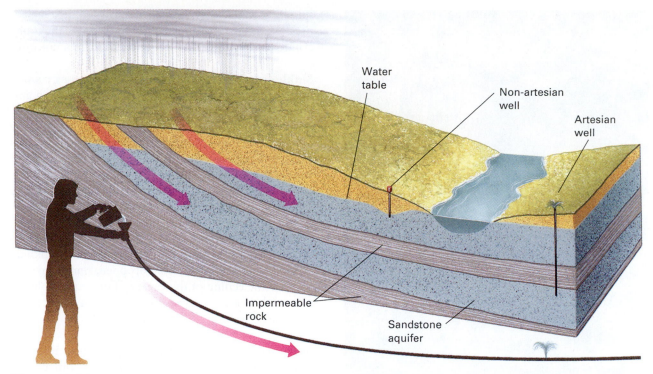

Figure 10–36 An artesian aquifer forms where a tilted layer of permeable rock, such as sandstone, lies sandwiched between layers of impermeable rock, such as shale. Water rises in an artesian well without being pumped. A hose with a hole (inset) shows why an artesian well flows spontaneously.

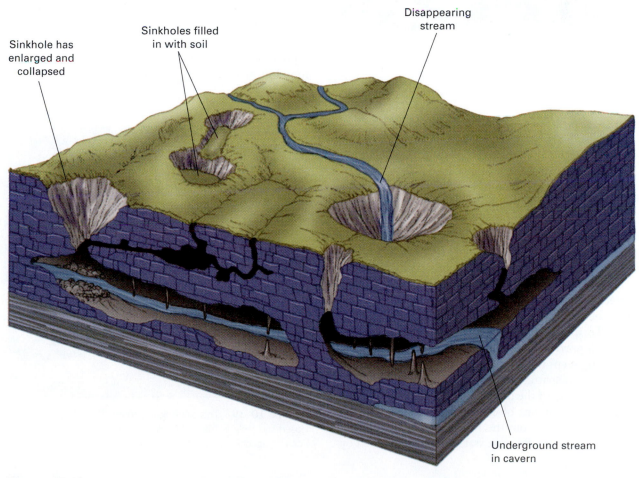

Figure 10–37 Sinkholes and caverns are characteristic of karst topography. Streams commonly disappear into sinkholes and flow through the caverns to emerge elsewhere.

Figure 10–38 Stalactites and stalagmites form as calcite precipitates in a limestone cavern. Luray Caverns, Virginia. *(Breck P. Kent)*

splashing water, they tend to be broader than stalactites. As the two features continue to grow, they may eventually meet and fuse together to form a **column.**

Sinkholes   If the roof of a cavern collapses, a **sinkhole** forms on the Earth's surface. A sinkhole can also form as limestone dissolves from the surface downward. A well-documented sinkhole formed in May 1981 in Winter Park, Florida. During the initial collapse, a three-bedroom house, half a swimming pool, and six Porsches in a dealer's lot all fell into the underground cavern. Within a few days, the sinkhole was 200 meters wide and 50 meters deep, and it had devoured additional buildings and roads (Fig. 10–39).

Although sinkholes form naturally, human activities can accelerate the process. The Winter Park sinkhole formed when the water table dropped, removing support for the ceiling of the cavern. The water table fell as a result of a severe drought augmented by excessive removal of ground water by humans.

Karst Topography   **Karst topography** forms in broad regions underlain by limestone and other readily soluble rocks. Caverns and sinkholes are common features of karst topography (Fig. 10–40). Surface streams often pour into sinkholes and disappear into caverns. In the area around Mammoth Cave in Kentucky, streams are given names such as Sinking Creek, an indication of their fate. The word *karst* is derived from a region in Croatia where this type of topography is well developed. Karst landscapes are found in many parts of the world.

calcite, and carbon dioxide percolates through the ground, it is under pressure from water in the cracks above it. If a drop of this solution seeps into the ceiling of a cavern, the pressure decreases because the drop comes in contact with the air. The high humidity of the cave prevents the water from evaporating rapidly, but the lowered pressure allows some of the carbon dioxide to escape as a gas. When the carbon dioxide escapes, the drop becomes less acidic. This decrease in acidity causes some of the dissolved calcite to precipitate as the water drips from the ceiling. Over time, a beautiful and intricate **stalactite** grows to hang icicle-like from the ceiling of the cave (Fig. 10–38).

Only a portion of the dissolved calcite precipitates as the drop seeps from the ceiling. When the drop falls to the floor, it spatters and releases more carbon dioxide. The acidity of the drop decreases further, and another minute amount of calcite precipitates. Thus, a **stalagmite** builds from the floor upward to complement the stalactite. Because stalagmites are formed by

Figure 10–39 This sinkhole in Winter Park, Florida, collapsed suddenly in May 1981, swallowing several houses and a Porsche agency. *(Wide World Photos/Associated Press)*

Figure 10–40 Rain and ground water dissolved Florida limestone to create this sinkhole. *(A.J. Copley/ Visuals Unlimited)*

Hot Springs, Geysers, and Geothermal Energy

At numerous locations throughout the world, hot water naturally flows to the surface to produce **hot springs.** Ground water can be heated in three different ways:

1. The Earth's temperature increases by about 30°C per kilometer in the upper portion of the crust. Therefore, if ground water descends through cracks to depths of 2 to 3 kilometers, it is heated by 60°C to 90°C. The hot water or steam then rises because it is less dense than cold water. However, it is unusual for fissures to descend so deep into the Earth, and this type of hot spring is uncommon.

2. In regions of recent volcanism, magma or hot igneous rock may remain near the surface and can heat ground water at relatively shallow depths. Hot springs heated in this way are common throughout western North and South America because these regions have been magmatically active in the recent past and remain so today. Shallow magma heats the hot springs and geysers of Yellowstone National Park.

3. Many hot springs have the odor of rotten eggs from small amounts of hydrogen sulfide (H_2S) dissolved in the hot water. The water in these springs is heated by chemical reactions. Sulfide minerals, such as pyrite (FeS_2), react chemically with water to produce hydrogen sulfide and heat. The hydrogen sulfide rises with the heated ground water and gives it the strong odor.

Most hot springs bubble gently to the surface from cracks in bedrock. However, **geysers** violently erupt hot water and steam. Geysers generally form over open cracks and channels in hot underground rock. Before a geyser erupts, ground water seeps into the cracks and is heated by the rock (Fig. 10–41). Gradually, steam bubbles form and start to rise, just as they do in a heated teakettle. If part of the channel is constricted, the bubbles accumulate and form a temporary barrier that allows the steam pressure below to increase. The rising pressure forces some of the bubbles upward past the constriction and short bursts of steam and water spurt from the geyser. This lowers the steam pressure at the constriction, causing the hot water to vaporize, blowing steam and hot water skyward (Fig. 10–42).

The most famous geyser in North America is Old Faithful in Yellowstone National Park, which erupts on the average of once every 65 minutes. Old Faithful is not as regular as people like to believe; the intervals between eruptions vary from about 30 to 95 minutes.

Hot ground water can be used to drive turbines and generate electricity, or it can be used directly to heat homes and other buildings. Energy extracted from the Earth's heat is called **geothermal energy.** In the United States, a total of 70 geothermal plants in California, Hawaii, Utah, and Nevada have a generating capacity of 2500 megawatts, enough to supply over one million people with electricity and equivalent to the power output of $2\frac{1}{2}$ large nuclear reactors.

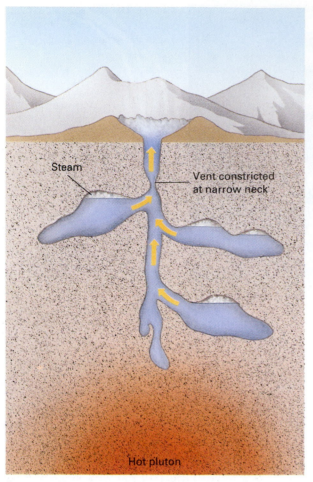

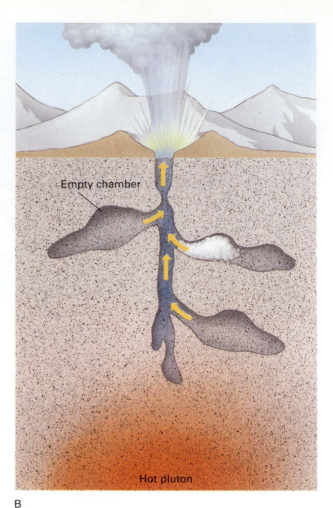

Figure 10–41 Before a geyser erupts, (A) ground water seeps into underground chambers and is heated by hot igneous rock. Foam constricts the geyser's neck, trapping steam and raising pressure. (B) When the pressure exceeds the strength of the blockage, the constriction blows out. Then the hot ground water flashes into vapor and the geyser erupts.

However, this amount of energy is minuscule compared with the potential of geothermal energy.

10.8 Wetlands

Wetlands are known across North America as swamps, bogs, marshes, sloughs, mud flats, and flood plains. They are regions that are water soaked or flooded for part or all of the year. Some wetlands are

Figure 10–42 The geysers in Yellowstone National Park form where ground water is heated by shallow magma. The hot water and steam rise through constricted channels in the rock; Sawmill Geyser. *(Tess Young/Tom Stack and Associates)*

wet only during exceptionally wet years and may be dry for several years at a time.

Wetland ecosystems vary so greatly that the concept of a wetland defies a simple definition. Wetlands share certain properties, however: The ground is wet for at least part of the time; the soils reflect anaerobic (lacking oxygen) conditions; and the vegetation consists of plants such as red maple, cattails, bullrushes, mangroves, and other species adapted to periodic flooding or water saturation. North American wetlands include all stream flood plains, frozen Arctic tundra, warm Louisiana swamps, coastal Florida mangrove swamps, boggy mountain meadows of the Rockies, and the immense swamps of interior Alaska (Fig. 10–43).

Wetlands are among the most biologically productive environments on Earth. Two thirds of the Atlantic fish and shellfish consumed by humans rely on coastal wetlands for at least part of their life cycles. One third of the endangered species of both plants and animals in the United States also depend on wetlands for survival. More than 400 of the 800 species of protected migratory birds and one third of all resident bird species feed, breed, and rest in wetlands.

When European settlers first arrived in North America, 87 million hectares of wetlands existed (exclusive of those in Alaska). Americans have long viewed wetlands as mosquito-infested, malarial swamps occupying land that can be farmed or otherwise developed if drained or filled. In the mid-1800s, the federal government passed legislation known as the Swamp Land Acts, which established an official policy to fill and drain wetlands to convert them to agricultural uses wherever possible.

Over the past century, farmers, ranchers, and developers drained or filled more than half of the original wetlands. California and several upper Midwestern states have lost more than 80 percent of their wetlands (Fig. 10–44A). Wetlands now make up between 6 and 9 percent of the lower 48 states and as much as 60 percent, or about 80 million hectares, of Alaska (Fig. 10–44B). Currently, between about 120,000 and 200,000 hectares of wetlands are destroyed each year.

Government policy and private practice have ignored the beneficial qualities of wetlands. Aquatic organisms consume many pollutants and degrade them to harmless byproducts. Because these organisms abound in wetlands, the ecosystems are natural sewage treatment systems. Wetlands also mitigate flooding by absorbing excess water that might otherwise overrun towns and farms.

Wild ducks and geese and other migratory birds depend on wetlands for breeding, food, and cover. Many wildlife biologists think that the recent population de-

A

B

Figure 10–43 North American wetlands extend from (A) the Everglades to (B) the immense Alaskan swamps. *(A, David Young/Tom Stack and Associates) (B, Corel images)*

cline of these waterfowl may reflect the health of America's wetlands. Between 1975 and 1990, the number of breeding ducks and geese declined from 45 million to 31 million, a drop of 31 percent.

The importance of wetlands in water purification, flood control, and wildlife habitat did not become widely recognized until the 1960s. At present, the focus of federal and state laws has changed from destruction of wetlands to their protection and preservation. But it has proved more difficult to reverse practice than policy, and wetland losses continue today.

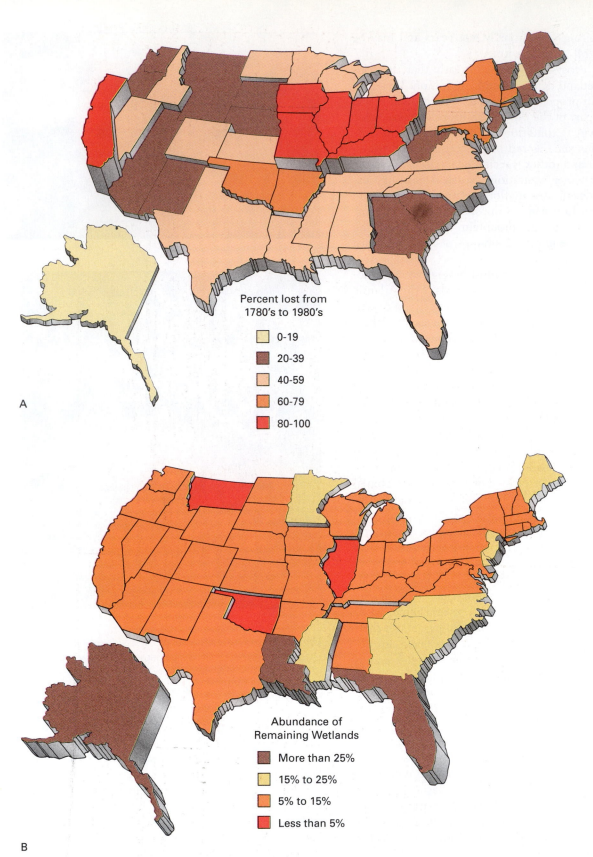

Figure 10–44 (A) The proportion of wetlands lost, by state, in the U.S. in the past 200 years. (B) The abundance of remaining wetlands in the U.S. Percentages indicate the proportional area of each state still covered by wetlands.

SUMMARY

Only about 0.64 percent of the Earth's water is fresh. The rest is salty seawater and glacial ice. Evaporation, **transpiration**, precipitation, and **runoff** continuously recycle water among land, sea, and the atmosphere in the **hydrologic cycle.**

A **stream** is any body of water flowing in a channel. The velocity of a stream is determined by its **gradient** and **discharge.** A **flood** occurs when a stream overflows its banks and flows over its **flood plain.** Floods are the most common and costly of all geologic hazards.

Base level is the lowest elevation to which a stream can erode its bed; it is usually sea level. A lake or resistant rock can form a local, or temporary base level. A **graded stream** has a smooth, concave profile. **Downcutting, lateral erosion,** and mass wasting combine to form a stream valley. Mountain streams downcut rapidly and form V-shaped valleys, whereas lower-gradient streams form wider valleys and **meanders** by lateral erosion and mass wasting. A **drainage basin** is the region drained by a single stream. Streams erode the land surface to flatten rugged topography, but tectonic activity and isostatic rebound uplift the land at the same time that erosion levels it.

Lakes are short-lived landforms because streams fill them with sediment. Many modern lakes were created by recent glaciers; as a result, we live in an unusual time of abundant lakes. The life history of a lake commonly involves a progression from an **oligotrophic** lake to a **eutrophic** lake to a swamp and, finally, to a flat meadow or forest. Temperate lakes normally develop temperature layering during the summer and winter, but experience **turnover** twice a year, which mixes the lake water and supplies oxygen to deep waters and nutrients to surface waters.

Much of the rain that falls on land seeps into soil and bedrock to become **ground water.** Ground water saturates the upper few kilometers of soil and bedrock to a level called the **water table.** Ground water provides drinking water for more than half of the population of the United States and is a major source of irrigation and industrial water. **Porosity** is the proportion of rock or soil that consists of open space. **Permeability** is the ability of soil or bedrock to transmit water. An **aquifer** is a body of rock that can yield economically significant quantities of water. Aquifers are porous and permeable.

Most ground water moves slowly, about 4 centimeters per day. **Springs** occur where the water table intersects the surface of the land. Dipping layers of permeable and impermeable rock can produce an **artesian aquifer.**

Caverns form where ground water dissolves limestone. A **sinkhole** forms when the roof of a limestone cavern collapses. **Karst topography,** with numerous caves, sinkholes, and subterranean streams, is characteristic of limestone regions. **Hot springs** and **geysers** develop when hot ground water rises to the surface. Hot ground water has been tapped to produce **geothermal energy.**

Wetlands are among the most biologically productive environments on Earth. In addition, they are natural water purification systems, and they mitigate flood effects by absorbing flood waters. Despite these facts, government and private efforts have eliminated more than half of the wetlands that existed in the lower 48 states when European settlers first arrived.

KEY TERMS

1. In which physical state (solid, liquid, or vapor) does most of the Earth's free water exist? Which physical state accounts for the least?

2. Describe the movement of water through the hydrologic cycle.

3. Describe the factors that determine the velocity of stream flow and how those factors interact.

4. What is meant by the term "100-year flood"? By the term "10-year flood"?

5. Give two examples of natural features that create temporary base levels. Why are they temporary?

6. Draw a profile of a graded stream and an ungraded stream.

7. Explain how a stream forms and shapes a valley.

8. In what type of terrain would you be likely to find a V-shaped valley? Where would you be likely to find a meandering stream with a broad flood plain?

9. Discuss the pros and cons of flood control policies.

10. Why are most lakes short-lived landforms?

11. What geological conditions create a long-lived lake?

12. Describe the differences between an oligotrophic lake and a eutrophic lake.

13. What is a thermocline? Under what conditions does a thermocline form in a lake?

14. What conditions cause mixing, or turnover, of surface and deep waters in a temperate lake?

15. (a) Draw a cross-section of soil and shallow bedrock, showing the zone of saturation, water table, and zone of aeration. (b) Explain each of the preceding terms.

16. What is an aquifer, and how does water reach it?

17. Explain why bedrock or soil must be both porous and permeable to be an aquifer.

18. Compare the movement of ground water in an aquifer with that of water in a stream.

19. How does an artesian aquifer differ from a normal one? Why does water from an artesian well rise without being pumped?

20. What is karst topography? How can it be recognized? How does it form?

21. Describe three types of heat source for a hot spring.

22. Describe wetlands. What types of environments are included in the wetlands category?

23. What proportion of wetlands that the early European settlers found in the United States no longer exists?

24. Describe how wetlands mitigate flooding.

1. Describe the ways in which (a) a rise and (b) a fall in the average global temperature could affect the Earth's hydrologic cycle.

2. A stream is 50 meters wide at a certain point. A bridge is built across the stream, and the abutments extend into the channel, narrowing it to 40 meters. Discuss the changes that might occur as a result of this constriction.

3. Describe and discuss effects of flood control structures, such as artificial levees, on the damage caused by the 1993 Mississippi River floods.

4. The National Flood Insurance Program (NFIP) is a federally sponsored insurance for people who live on flood plains. In exchange for low-cost insurance policies, the NFIP establishes rigorous building codes for new construction in flood plains. However, older buildings are not subject to these standards. As long as damage from one flood is less than 50 percent of the value of the structure, it is eligible for subsidized insurance against the next flood. Outline potential benefits and drawbacks of this program.

5. Imagine that a 100-year flood has just occurred on a river near your home. You want to open a small business in the area. Your accountant advises that your business and building have an economic life expectancy of 50 years. Would it be safe to build on the flood plain? Why or why not?

6. Most substances become denser when they freeze. Water is anomalous in that ice is less dense than water. Outline the seasonal temperature profile of a lake if ice were denser than water. How would aquatic ecosystems be affected?

7. Imagine that you live on a hill 25 meters above a nearby stream. You drill a well 40 meters deep and do not reach water. Explain.

8. Discuss and explain how a stream can supply water to an underground aquifer in one environment, but an aquifer can supply water to a stream in another environment.

9. Why are caverns, sinkholes, and karst topography most commonly found in limestone terrain?

10. Discuss why wetlands support such a large population of wildlife.

11. Discuss the mechanisms by which wetlands ecosystems purify contaminated water.

12. Discuss how changing public perceptions may have played a role in reversing governmental policy toward wetlands.

Glaciers and Ice Ages

An alpine glacier flows from the Lyell Icefield in the Canadian Rockies.

We often think of glaciers as features of high mountains and the frozen polar regions. Yet anyone who lives in the northern third of the United States is familiar with glacial landscapes. Many low, rounded hills of upper New York State, Wisconsin, and Minnesota are composed of gravel deposited by great ice sheets. In addition, people in this region swim and fish in lakes that were formed by recent glaciers.

Numerous times during Earth history, glaciers have grown to cover large parts of the Earth, and then melted away. Before the most recent major glacial advance, beginning about 100,000 years ago, the world was free of ice except for high mountains and the polar ice caps of Antarctica and Greenland. Then, in a relatively short time—perhaps only a few thousand years—the Earth's climate cooled. As winter snow failed to melt completely in summer, the polar ice caps grew and spread into lower latitudes. At the same time, glaciers formed near mountain summits, even near the equator. They flowed down mountain valleys into nearby lowlands. When the glaciers reached their maximum size 18,000 years ago, they covered one third of the Earth's continents. About 15,000 years ago, Earth's climate warmed again and the glaciers melted rapidly.

Although 18,000 years is a long time compared with a single human lifetime, it is an eyeblink in geologic time. In fact, humans lived through the most recent glaciation. In southwest France and northern Spain, humans developed sophisticated spear heads and carved body ornaments between 40,000 and 30,000 years ago. People first began experimenting with agriculture about 10,000 years ago. ●

11.1 Formation of Glaciers

In most temperate regions, winter snow melts completely in spring and summer. However, in certain cold, wet environments, only a portion of the winter snow melts and the remainder accumulates year after year. During summer, snow crystals become rounded and denser as the snowpack is compressed and alternately warmed during daytime and cooled at night. If snow survives through one summer, it converts to rounded ice grains called **firn**. Mountaineers like firn because the sharp points of their ice axes and crampons sink into it easily and hold firmly. If firn is buried deeper in the snowpack, it converts to glacial ice, which consists of closely packed ice crystals (Fig. 11–1).

A **glacier** is a massive, long-lasting, moving mass of compacted snow and ice. Glaciers form only on land, wherever the amount of snow that falls in winter exceeds the amount that melts in summer.

Mountain glaciers flow downhill. Glaciers on level land flow outward under their own weight.

Glaciers form in two environments. Alpine glaciers form at all latitudes on high, snowy mountains. Continental ice sheets form at all elevations in the cold polar regions.

Alpine Glaciers

Mountains are generally colder and wetter than adjacent lowlands. Near the summits, winter snowfall is deep and summers are short and cool. These conditions create **alpine glaciers** (Fig. 11–2). Alpine glaciers exist on every continent—in the Arctic and Antarctica, in temperate regions, and in the tropics. Glaciers cover the summits of Mount Kenya in Africa and Mount Cayambe in South America, even though both peaks are near the equator.

Some alpine glaciers flow great distances from the peaks into lowland valleys. For example, the Kahiltna Glacier, which flows down the southwest side of Denali (Mount McKinley) in Alaska, is about 65 kilometers long, 12 kilometers across at its widest point, and about 700 meters thick. Although most alpine glaciers are smaller than the Kahiltna, some are larger.

The growth of an alpine glacier depends on both temperature and precipitation. The average annual temperature in the state of Washington is warmer than that in Montana, yet alpine glaciers in Washington are larger and flow to lower elevations than those in Montana. Winter storms buffet Washington from the moisture-laden Pacific. Consequently, Washington's mountains receive such heavy winter snowfall that even though summer melting is rapid, large quantities of snow accumulate every year. In much drier Montana, snowfall is light enough that most of it melts in the summer, and thus Montana's mountains have only a few small glaciers.

Continental Glaciers

Winters are so long and cold and summers so short and cool in polar regions that glaciers cover most of

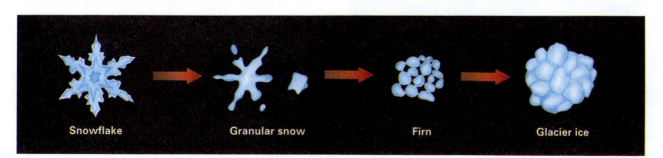

Figure 11–1 Newly fallen snow changes through several stages to form glacier ice.

Figure 11–2 This alpine glacier flows around granite peaks in British Columbia, Canada.

the land regardless of its elevation. An **ice sheet**, or **continental glacier**, covers an area of 50,000 square kilometers or more (Fig. 11–3). The ice spreads outward in all directions under its own weight.

Today, the Earth has only two ice sheets, one in Greenland and the other in Antarctica. These two ice sheets contain 99 percent of the world's ice and about three fourths of the Earth's fresh water. The Greenland sheet is more than 2.7 kilometers thick in places and covers 1.8 million square kilometers. Yet it

Figure 11–3 The Beardmore glacier is a portion of the Antarctic ice sheet. *(Kevin Killelea)*

is small compared with the Antarctic ice sheet, which blankets about 13 million square kilometers, almost 1.5 times the size of the United States. The Antarctic ice sheet covers entire mountain ranges, and the mountains that rise above its surface are islands of rock in a sea of ice.

Whereas the South Pole lies in the interior of the Antarctic continent, the North Pole is situated in the Arctic Ocean. Only a few meters of ice freeze on the relatively warm sea surface, and the ice fractures and drifts with the currents. As a result, no ice sheet exists at the North Pole.

11.2 Glacial Movement

As an experiment, geologists have set two poles in dry ground on opposite sides of a glacier, and a third pole in the ice to form a straight line with the other two. After a few months, the center pole moved downslope, and the three poles formed a triangle. The center pole moved because the glacier flowed downhill.

Rates of glacial movement vary with slope steepness, precipitation, and air temperature. In the coastal ranges of southeast Alaska where annual precipitation is high and average temperature is relatively warm (for glaciers), some glaciers move 15 centimeters to a meter per day. In contrast, in the interior of Alaska where conditions are colder and drier, glaciers move only a few centimeters per day. At these rates, ice can flow the length of an alpine glacier in a few hundred to a few thousand years. In some instances, a glacier may surge at a speed of 10 to 100 meters per day.

Glaciers move by two mechanisms: basal slip and plastic flow. In **basal slip,** the entire glacier slides over bedrock in the same way that a bar of soap slides down a tilted board. Just as wet soap slides more easily than dry soap, water between bedrock and the base of a glacier accelerates basal slip.

Several factors cause water to accumulate near the base of a glacier. The Earth's heat melts ice near bedrock. Friction from glacial movement also generates heat. Water occupies less volume than an equal amount of ice. As a result, pressure from the weight of overlying ice favors melting. Finally, during the summer, water melted from the surface of a glacier may seep downward to its base.

A glacier also moves by **plastic flow,** in which it deforms as a viscous fluid. Plastic flow is demonstrated by two experiments. In one, scientists set a line of poles in the ice (Fig. 11–4). After a few years, the ice moved downslope so that the poles formed a U-shaped array. This experiment shows that the center of the glacier moves faster than the edges. Frictional resistance with the valley walls slows movement along the edges and

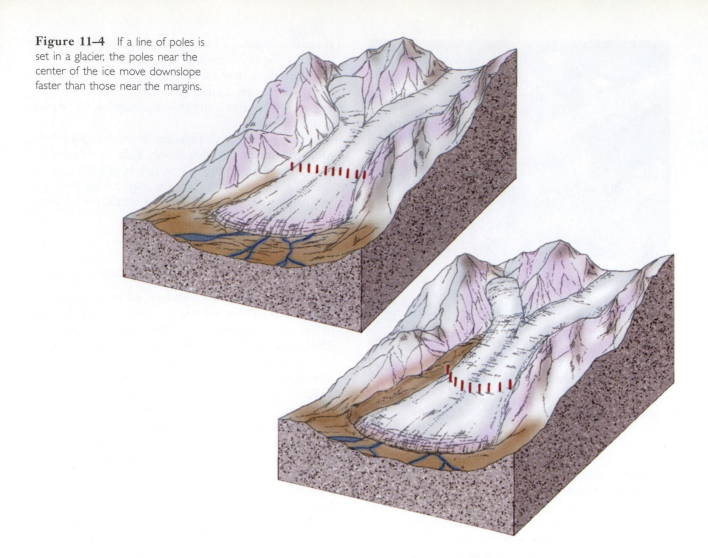

Figure 11–4 If a line of poles is set in a glacier, the poles near the center of the ice move downslope faster than those near the margins.

glacial ice flows plastically, allowing the center to move faster than the sides.

In another experiment, scientists drove a straight but flexible pipe downward into the glacier to study the flow of ice at depth (Fig. 11–5). At a later date, they discovered that the entire pipe moved downslope and also became bent. At the surface of a glacier, the ice is brittle, like an ice cube or the ice found on the surface of a lake. In contrast, at depths greater than about 40 meters, the pressure is sufficient to allow ice to deform plastically. The curvature in the pipe shows that the ice moved plastically and that middle levels of

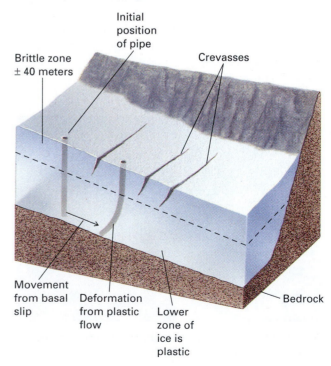

Figure 11–5 In this experiment, a pipe was driven through a glacier until it reached bedrock. The entire pipe moved downslope with the ice, and also became curved. The pipe became curved because friction with bedrock slowed movement of the bottom of the glacier. Middle layers of ice flowed more rapidly because there the ice is plastic. At a depth shallower than 40 meters, the ice does not flow plastically, and the pipe in that zone remained straight.

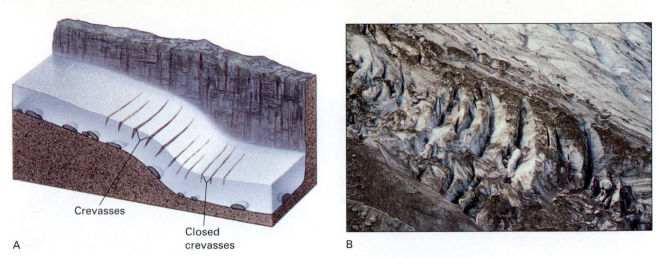

Figure 11–6 (A) Crevasses form in the upper, brittle zone of a glacier where the ice flows over uneven bedrock. (B) Crevasses in the Bugaboo Mountains of British Columbia.

the glacier moved faster than the lower part. The base of the glacier is slowed by friction against bedrock, so it moved more slowly than the plastic portion above it.

The relative rates of basal slip and plastic flow depend on the steepness of the bedrock underlying the glacier and the thickness of the ice. A small alpine glacier on steep terrain moves mostly by basal slip. In contrast, the bedrock beneath portions of the Antarctica and Greenland ice sheets is relatively level, so the ice cannot slide downslope. Thus, these continental glaciers are huge plastic masses of ice (with a thin rigid cap) that ooze outward mainly under the forces created by their own weight.

When a glacier flows over uneven bedrock, the deeper plastic ice bends and flows over bumps, while the brittle upper layer stretches and cracks, forming **crevasses** (Fig. 11–6). Crevasses form only in the brittle upper 40 meters of a glacier, not in the lower plastic zone. Crevasses open and close slowly as a glacier moves. An **ice fall** is a section of a glacier consisting of crevasses and towering ice pinnacles. The pinnacles form where ice blocks break away from the crevasse walls and rotate as the glacier moves. With crampons, ropes, and ice axes, a skilled mountaineer might climb into a crevasse. The walls are a pastel blue, and sunlight filters through the narrow opening above. The ice shifts and cracks, making creaking sounds as the glacier advances. Many mountaineers have been crushed by falling ice while traveling through ice falls.

The Mass Balance of a Glacier

Consider an alpine glacier flowing from the mountains into a valley (Fig. 11–7). At the upper end of the glacier, snowfall is heavy, temperatures are below freezing for much of the year, and avalanches carry large quantities of snow from the surrounding peaks onto the ice. Thus, more snow falls in winter than melts in summer, and snow accumulates from year to year. This higher-elevation part of the glacier is called the **zone of accumulation.** There the glacier's surface is covered by snow year round.

Lower in the valley, the temperature is higher throughout the year, and less snow falls. This lower part of a glacier, where more snow melts in summer than accumulates in winter, is called the **zone of ablation.** When the snow melts, a surface of old, hard glacial ice is left behind. The **snow line** is the boundary between permanent snow and seasonal snow. The snow line shifts up and down the glacier from year to year, depending on weather. Ice exists in the zone of ablation because the glacier flows downward from the accumulation area. Even farther down valley, the rate of glacial flow cannot keep pace with melting, so the glacier ends at its **terminus** (Fig. 11–8).

Glaciers grow and shrink. If annual snowfall increases or average temperature drops, more snow accumulates; then the snow line of an alpine glacier descends to a lower elevation, and the glacier grows thicker. At first the terminus may remain stable, but eventually it advances farther down the valley. The lag time between a change in climate and a glacial advance may range from a few years to several decades depending on the size of the glacier, its rate of motion, and the magnitude of the climate change. If annual snowfall decreases or the climate warms, the accumulation area shrinks and the glacier retreats.

When a glacier retreats, its ice continues to flow downhill, but the terminus melts back faster than the glacier flows downslope. In Glacier Bay, Alaska, glaciers have retreated 60 kilometers in the past 125 years. Near the terminus, newly exposed rock is bare and

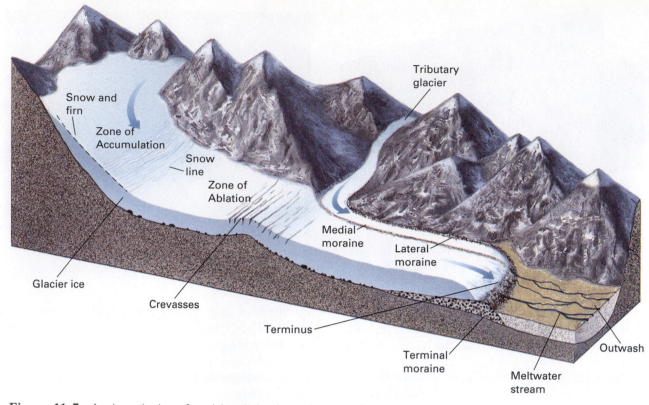

Figure 11–7 A schematic view of an alpine glacier shows the zone of accumulation and zone of ablation, and the erosional and depositional landforms created by the ice.

lifeless. A few kilometers from the glacier, where rock has been exposed for a few decades, scattered lichens grow on otherwise bare rock. Even farther away, seabird droppings have mixed with windblown silt and weathered rock to form thin soil that supports mosses in sheltered cracks. Near the head of the bay, 60 kilometers from the terminus, tidal currents and ocean storms have washed enough sediment over the land to create soil and support stunted trees.

In equatorial and temperate regions, glaciers commonly terminate at an elevation of 3000 meters or higher. However, in a cold, wet climate, a glacier may extend into the sea (Fig. 11–9). Giant chunks of ice break off, forming **icebergs.**

The largest icebergs in the world are those that form from the Antarctic ice shelf. In January 1995, the edge of the 300-meter-thick Larson Ice Shelf cracked and an iceberg almost as big as Rhode Island broke free and floated into the Antarctic Ocean. The tallest icebergs in the world form from tidewater glaciers in Greenland. Some extend 150 meters above sea level; since the visible portion of an iceberg represents only about 10 to 15 percent of its mass, these bergs may be as much as 1500 meters from base to tip.

11.3 Glacial Erosion

Rock at the base and sides of a glacier may have been fractured by tectonic forces and may be loosened by weathering processes, such as frost wedging or pressure-release fracturing. The moving ice then dislodges the loosened rock (Fig. 11–10). Ice is viscous enough to pick up and carry particles of all sizes, from silt-sized grains to house-sized boulders. Thus

Figure 11–8 Crevasses fracture the terminus of a glacier flowing from the Lyell Group in the Canadian Rockies.

Figure 11–9   A kayaker paddles among small icebergs that calved from the Le Conte glacier, Alaska.

Figure 11–10   (A) A glacier plucks rocks from bedrock and then drags them along, abrading both the loose rocks and the bedrock. (B) These crescent-shaped depressions in granite at Le Conte Bay, Alaska, were formed by glacial plucking.

glaciers erode and transport huge quantities of rock and sediment.

Ice itself is not abrasive to bedrock because it is too soft. However, rocks embedded in the ice scrape across bedrock, cutting deep, parallel grooves and scratches called **glacial striations** (Fig. 11–11). When glaciers melt and striated bedrock is exposed, the markings show the direction of ice movement. Glacial striations are used to map the flow directions of glaciers.

Erosional Landforms Created by Alpine Glaciers

Let's take an imaginary journey through a mountain range that was glaciated in the past but is now mostly ice free (Fig. 11–12). We start with a helicopter ride

Figure 11–11 Stones embedded in the base of a glacier gouged these striations in bedrock in British Columbia.

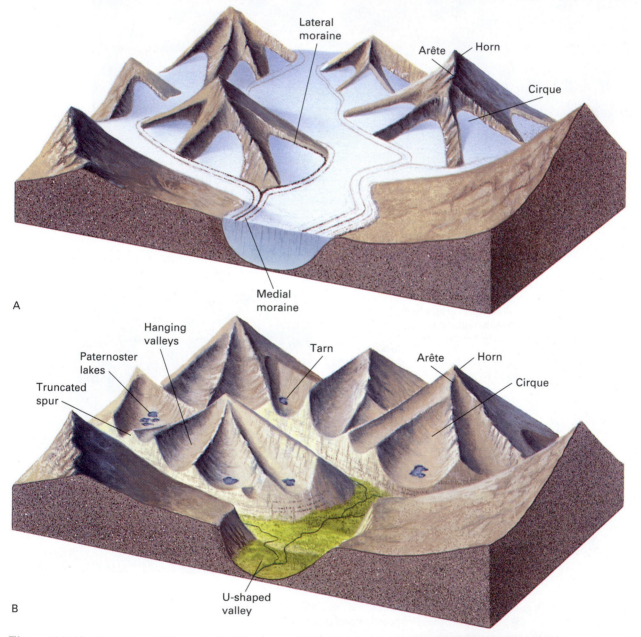

Figure 11–12 Two views of the same glacial landscape. (A) The landscape as it appeared when it was mostly covered by glaciers. (B) The same landscape as it appears now, after the glaciers have melted. • **Interactive Question:** Draw a third diagram showing the landscape in the future. Assume that there is no additional tectonic uplift or glacial advances.

to the summit of a high, rocky peak. Our first view from the helicopter is of sharp, jagged mountains rising steeply above smooth, rounded valleys.

A mountain stream commonly erodes downward into its bed, cutting a steep-sided, V-shaped valley. A glacier, however, is not confined to a narrow stream bed but instead fills its entire valley. As a result, it scours the sides of the valley as well as the bottom, carving a broad, rounded, **U-shaped valley** (Fig. 11–13).

We land on one of the peaks and step out of the helicopter. Beneath us, a steep cliff drops off into a horseshoe-shaped depression in the mountainside called a **cirque.** A small glacier at the head of the cirque reminds us of the larger mass of ice that existed in a colder, wetter time (Fig. 11–14A).

To understand how a glacier creates a cirque, imagine a gently rounded mountain. As snow accumulates

Figure 11–13 A U-shaped valley in the Purcell mountains of British Columbia.

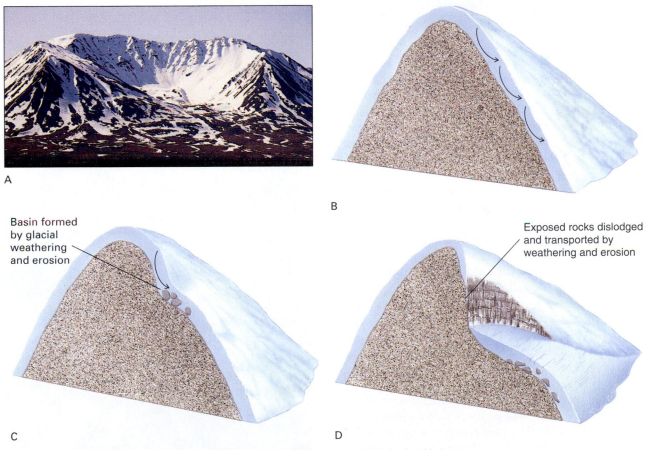

Figure 11–14 (A) A glacier eroded this concave cirque into a mountainside in the Alaska Range. *(McCutcheon/Visuals Unlimited)* (B) To form a cirque, snow accumulates, and a glacier begins to flow from the summit of a peak. (C) Glacial weathering and erosion form a small depression in the mountainside. (D) Continued glacial movement enlarges the depression. When the glacier melts, it leaves a cirque carved in the side of the peak, as in the photograph.

Basin formed by glacial weathering and erosion

Exposed rocks dislodged and transported by weathering and erosion

Figure 11–15 Glaciers eroded bedrock to form this string of paternoster lakes in the Sierra Nevada.

Looking downward from our peak, we may see a waterfall pouring from a small, high valley into a larger, deeper one. A small glacial valley lying high above the floor of the main valley is called a **hanging valley** (Fig. 11–17). The famous waterfalls of Yosemite Valley in California cascade from hanging valleys. A hanging valley forms where a small tributary glacier joined a much larger one. The tributary glacier eroded a shallow valley while the massive main glacier gouged a deeper one. When the glaciers melted, they exposed an abrupt drop where the small valley joins the main valley.

Deep, narrow inlets called **fjords** extend far inland on many high-latitude seacoasts. Most fjords are glacially carved valleys that were later flooded by rising seas as the glaciers melted (Fig. 11–18).

Erosional Landforms Created by a Continental Glacier

A continental glacier erodes the landscape just as an alpine glacier does. However, a continental glacier is

and a glacier forms, the ice flows down the mountainside (Fig. 11–14B). The ice erodes a small depression that grows slowly as the glacier flows (Fig. 11–14C). With time, the cirque walls become steeper and higher. The glacier carries the eroded rock from the cirque to lower parts of the valley (Fig. 11–14D). When the glacier finally melts, it leaves a steep-walled, rounded cirque.

Streams and lakes are common in glaciated mountain valleys. As a cirque forms, the glacier may erode a depression into the bedrock beneath it. When the glacier melts, this depression fills with water, forming a small lake or **tarn** nestled at the base of the cirque. If we hike down the valley below the high cirques, we may encounter a series of lakes called **paternoster lakes,** which are commonly connected by rapids and waterfalls (Fig. 11–15). Paternoster lakes are a sequence of small basins plucked out by a glacier. The name paternoster refers to a string of rosary beads and evokes an image of a string of lakes in a glacial valley. When the glacier recedes, the basins fill with water.

If glaciers erode three or more cirques into different sides of a peak, they may create a steep, pyramid-shaped rock summit called a **horn.** The Matterhorn in the Swiss Alps is a famous horn (Fig. 11–16). Two alpine glaciers flowing along opposite sides of a mountain ridge may erode both sides of the ridge, forming a sharp, narrow **arête** between adjacent valleys.

Figure 11–16 The Matterhorn in Switzerland formed as three alpine glaciers eroded cirques into the peak from three different sides. The ridge between two cirques is called an arête. *(Swiss Tourist Board)*

Figure 11–17 Waterfalls cascade from two hanging valleys in Yosemite National Park. *(Science Graphics/Ward's Natural Science Establishment, Inc.)*

Figure 11–18 A steep-sided fjord bounded by 1000-m-high cliffs in Baffin Island, Canada.

considerably larger and thicker and is not confined to a valley. As a result, it covers vast regions, including entire mountain ranges. The recent ice sheets have scoured deep basins that have filled with water to form the Great Lakes and the Finger Lakes in New York (Fig 11–19).

11.4 Glacial Deposits

In the 1800s, geologists recognized that the large deposits of sand and gravel found in some places had been transported from distant sources. A popular hypothesis at the time explained that this material had drifted in on icebergs during catastrophic floods. The deposits were called "drift" after this inferred mode of transport.

Today we know that continental glaciers covered vast parts of the land only 10,000 to 20,000 years ago, and these glaciers carried and deposited drift. Although the term *drift* is a misnomer, it remains in common use. Now geologists define **drift** as all rock or sediment transported and deposited by a glacier. Glacial drift averages 6 meters thick over the rocky hills and pastures of New England and 30 meters thick over the plains of Illinois.

Drift is divided into two categories. **Till** was deposited directly by glacial ice. **Stratified drift** was first carried by a glacier and then transported and deposited by a stream.

Landforms Composed of Till

Ice is so much more viscous than water that it carries a wide range of particle sizes. When a glacier melts,

it deposits particles of all sizes—from fine clay to huge boulders—in an unsorted, unstratified mass (Fig. 11–20). Within a glacier, each rock or grain of sediment is protected by the ice that surrounds it. Therefore, the pieces do not rub against one another, and glacial transport does not round sediment as a stream does. If you find rounded gravel in till, it

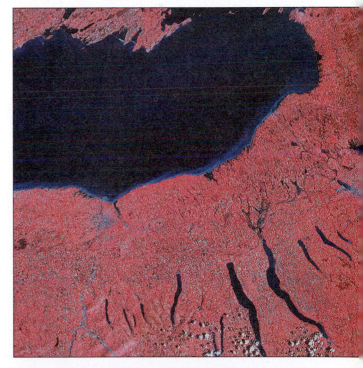

Figure 11–19 The Finger Lakes (bottom right) formed when Pleistocene glaciers scoured deep valleys and the valleys filled with water. Pleistocene glaciers also formed Lake Ontario (top) as discussed in the Case Study on page 261. *(U.S.G.S., EROS Data Center)*

Figure 11–20 Unsorted glacial till. Note that large cobbles are mixed with smaller sediment. The cobbles were rounded by stream action before they were transported and deposited by the glacier.

Figure 11–21 The end moraine of an alpine glacier on Baffin Island, Canada, in mid-summer. Dirty, old ice forms the lower part of the glacier below the snow line, and clean snow lies higher up on the ice in the zone of accumulation. *(Steve Sheriff)*

became rounded by a stream before the glacier picked it up.

Occasionally, large boulders lie on the surface in country that was once glaciated. In many cases the boulders are of a rock type different from the bedrock in the immediate vicinity. Boulders of this type are called **erratics** and were transported to their present locations by a glacier. The origins of erratics can be determined by exploring the terrain in the direction the glacier came from until the parent rock is found. Some erratics were carried 500 or even 1000 kilometers from their point of origin and provide clues to the movement of glaciers.

Moraines   A **moraine** is a mound or a ridge of till. Think of a glacier as a giant conveyor belt. An old-fashioned airport conveyor belt simply carried suitcases to the end of the belt and dumped them in a pile. Similarly, a glacier carries sediment and deposits it at its terminus. If a glacier is neither advancing nor retreating, its terminus may remain in the same place for years. During that time, sediment accumulates at the terminus to form a ridge called an **end moraine** (Fig. 11–21). An end moraine that forms when a gla-

cier is at its greatest advance, before beginning to retreat, is called a **terminal moraine.**

If warmer conditions prevail, the glacier recedes. If the glacier then stabilizes again during its retreat and the terminus remains in the same place for a year or more, a new end moraine, called a **recessional moraine,** forms.

When a glacier recedes steadily, till is deposited in a relatively thin layer over a broad area, forming a **ground moraine.** Ground moraines fill old stream channels and other low spots. Often this leveling process disrupts drainage patterns. Many of the swamps in the northern Great Lakes region lie on ground moraines formed when the most recent continental glaciers receded.

End moraines and ground moraines are characteristic of both alpine and continental glaciers. An end moraine deposited by a large alpine glacier may extend for several kilometers and be so high that even a person in good physical condition would have to climb for an hour to reach the top. Moraines may be dangerous to hike over if their sides are steep and the till is loose. Large boulders are mixed randomly with rocks, cobbles, sand, and clay. A careless hiker can dislodge boulders and send them tumbling to the base.

Figure 11–22 This wooded terminal moraine in New York State marks the southernmost extent of glaciers in that region. *(Science Graphics/Ward's Natural Science, Inc.)*

Figure 11–23 A lateral moraine lies against the valley wall in the Bugaboo Mountains, British Columbia.

The most recent Pleistocene continental glaciers began receding about 18,000 years ago. Their terminal moraines record the southernmost extent of those glaciers. In North America, the terminal moraines lie in a broad, undulating front extending across the northern United States. Enough time has passed since the glaciers retreated that soil and vegetation have stabilized the till and most of the hills are now covered by vegetation (Fig. 11–22).

Figure 11–24 Three separate medial moraines were formed by merging lateral moraines from coalescing glaciers, Baffin Island, Canada.

When an alpine glacier moves downslope, it erodes the valley walls as well as the valley floor. Therefore, the edges of the glacier carry large loads of sediment. Additional debris falls from the valley walls and accumulates on and near the sides of mountain glaciers. Sediment near the glacial margins forms a **lateral moraine** (Fig. 11–23).

If two alpine glaciers converge, their lateral moraines merge into the middle of the larger glacier. This till forms a visible dark stripe on the surface of the ice called a **medial moraine** (Fig. 11–24).

Drumlins   Elongate hills, called **drumlins,** cover parts of the northern United States and are best exposed across the rolling farmland in upstate New York (Fig. 11–25). Drumlins usually occur in clusters. Each one looks like a whale swimming through the ground with its back in the air. An individual drumlin is typically about 1 to 2 kilometers long and about 15 to 50 meters high. Most are made of till, while others

Figure 11–25 Crop patterns emphasize glacially streamlined drumlins in Wisconsin. *(Kevin Horan/Tony Stone Images)*

Figure 11–26 Streams flowing from the terminus of a glacier filled this valley with outwash on Baffin Island.

consist partly of till and partly of bedrock. In either case, when a glacier forms a drumlin, it erodes and deposits sediment to create the elongate, streamlined shape. The glacier generally erodes a steep-sided face as it advances. It then deposits some sediment on the downslope side to form a long, pointed slope. Thus, a geologist can determine the direction of motion of an ancient glacier by studying drumlins.

Landforms Composed of Stratified Drift

Because of the great amount of sediment eroded by a glacier, streams flowing from a glacier are commonly laden with silt, sand, and gravel. The stream deposits this sediment beyond the glacier terminus as **outwash** (Fig. 11–26). Glacial streams carry such a heavy load of sediment that they often become braided, flowing in multiple channels. Outwash deposited in a narrow valley is called a **valley train.** If the sediment spreads out from the confines of the valley into a larger val-

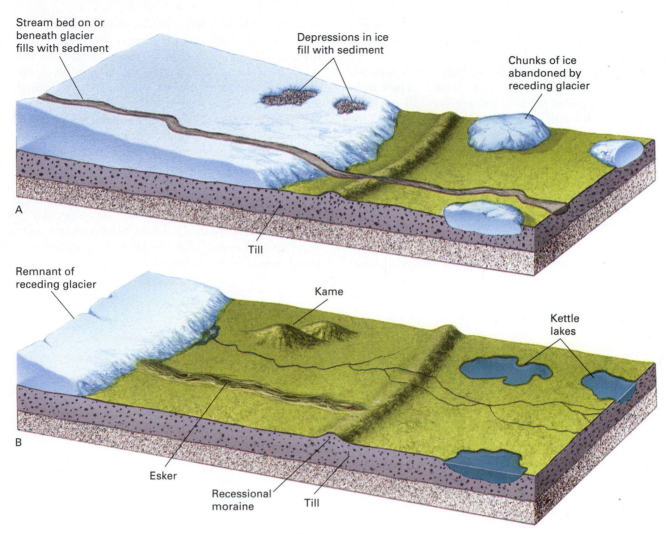

Stream bed on or beneath glacier fills with sediment

Depressions in ice fill with sediment

Chunks of ice abandoned by receding glacier

Till

A

Remnant of receding glacier

Kame

Kettle lakes

Esker

Recessional moraine

Till

B

Figure 11–27 (A) A melting glacier exposes several different glacial landforms that were created beneath the ice. (B) Sediment flowing from the melting ice creates an outwash plain beyond the glacial terminus.

ley or plain, it forms an **outwash plain** (Fig. 11–27). Outwash plains are also characteristic of continental glaciers.

During the summer, when snow and ice melt rapidly, streams form on the surface of a glacier. Many are too wide to jump across. Some of these streams flow off the front or sides of the glacier. Others plunge into crevasses and run beneath the glacier over bedrock or drift. These streams commonly deposit small mounds of sediment, called **kames,** at the margin of a receding glacier or where sediment collects in a crevasse or other depression in the ice. An **esker** is a long, sinuous ridge that forms as the channel deposit of a stream that flowed within or beneath a melting glacier (Fig. 11–28).

Because kames, eskers, and other forms of stratified drift are stream deposits and were not deposited directly by ice, they show sorting and sedimentary bedding,

Figure 11–28 This esker in Manitoba, Canada, is a sinuous ridge of sand and gravel deposited in the bed of a stream that flowed beneath a continental glacier about 20,000 years ago. *(Tim Hauf/Visuals Unlimited)*

which distinguishes them from unsorted and unstratified till. In addition, the individual cobbles or grains are usually rounded.

Large blocks of ice may be left behind in a moraine or an outwash plain as a glacier recedes. When such an ice block melts, it leaves a depression called a **kettle.** Kettles fill with water, forming kettle lakes. A kettle lake is as large as the ice chunks that melted to form the hole. The lakes vary from a few tens of meters to a kilometer or so in diameter, with a typical depth of 10 meters or less.

11.5 The Pleistocene Ice Age

Geologists have found terminal moraines extending across all high-latitude continents. By studying those moraines, as well as lakes, eskers, outwash plains, and other glacial landforms, geologists are certain that massive glaciers once covered large portions of the continents, altering Earth systems. A time when alpine glaciers descend into lowland valleys and continental glaciers spread over land in high latitudes is called an **ice age.** During an ice age, glaciers several kilometers thick spread across the landscape. Beneath the burden of ice, the continents sink deeper into the asthenosphere. The ice weathers rock and erodes soil, altering the landscape.

Geologic evidence shows that the Earth has been warm and relatively ice free for at least 90 percent of the past 1 billion years. However, at least six major ice ages occurred during that time (Fig. 11–29). Each one lasted from 2 to 10 million years.

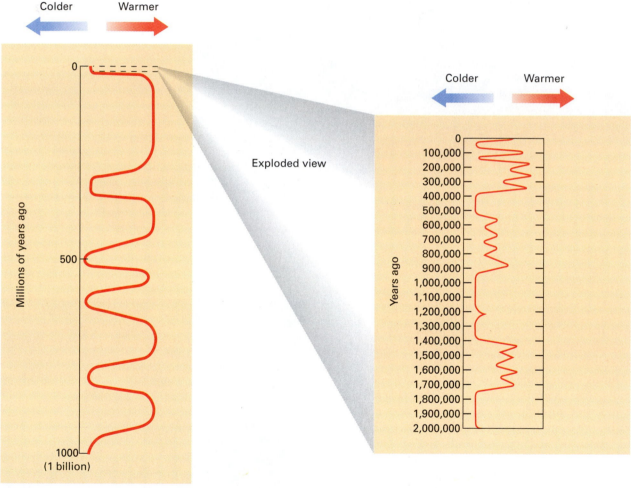

Figure 11–29 Glacial cycles. (A) Scale shows average global temperatures during the past 1 billion years. Cold temperatures caused major ice ages. (B) This expanded scale shows temperatures during the Pleistocene Ice Age. Cold temperatures coincided with Pleistocene glacial advances, and warm intervals coincided with glacial melting. We are probably living within a warm period of the Pleistocene Ice Age, and continental ice sheets may advance once again.
● Interactive Question: Give the approximate dates of the major ice ages within the past one billion years.

The most recent ice age took place mainly during the Pleistocene Epoch and is called the **Pleistocene Ice Age.** It began about 2.6 million years ago (although evidence of an earlier beginning has been found in the Southern Hemisphere). However, the Earth has not been glaciated continuously during the Pleistocene Ice Age; instead, climate has fluctuated and continental glaciers grew and then melted away several times (inset, Fig. 11–29). During the most recent interglacial period, the average temperature was about the same as it is today, or perhaps a little warmer. Then, high-latitude temperature dropped at least 15°C, causing the ice to advance.[1] Most climate models indicate that we are still in the Pleistocene Ice Age and the continental ice sheets may advance again.

Causes of the Pleistocene Ice Age and Glacial Cycles For reasons that are poorly understood, prior to 2.6 million years ago, the Earth's climate had been cooling for tens of millions of years. Then something happened to push the planet over a climate threshold and plunge it into an ice age. David Rea, a geologist from the University of Michigan, recently analyzed sediment cores of the mud on the sea floor in a region of the North Pacific. He found that an increase in volcanic ash 2.6 million years ago coincided closely with a dramatic increase in sediment grains that showed glacial markings. Thus, an increase in volcanic activity occurred at about the same time that glaciers began to spread. According to Dr. Rea, a period of intense volcanic activity injected enough dust into the atmosphere to reflect appreciable quantities of sunlight. This slight additional cooling event may have initiated the Pleistocene Ice Age.

Although volcanic dust may have triggered the *onset* of the Pleistocene glacial epoch, no such events seem to be associated with the repeated growth and melting of glaciers that characterize the Pleistocene Epoch. Instead, scientists have found that slight, periodic variations in the Earth's orbit and orientation coincided with Pleistocene glacial expansion and shrinking.

In the nineteenth century, astronomers detected three periodic variations in the Earth's orbit and spin axis (Fig. 11–30).

1. The Earth's orbit around the Sun is elliptical rather than circular. The shape of the ellipse is called **eccentricity.** The eccentricity varies in a regular cycle lasting about 100,000 years.

2. The Earth's axis is currently tilted at about 23.5° with respect to a line perpendicular to the plane of its orbit around the Sun. The **tilt** oscillates by 2.5° on about a 41,000-year cycle.

3. The Earth's axis, which now points directly toward the North Star, circles like that of a wobbling top. This circling, called **precession,** completes a full cycle every 26,000 years.

These changes affect both the total solar radiation received by the Earth and the distribution of solar energy with respect to latitude and season. Seasonal changes in sunlight reaching higher latitudes can reduce summer temperature. If summers are cool and short, winter snow and ice persist, leading to growth of glaciers.

Early in the twentieth century, a Yugoslavian astronomer, Milutin Milankovitch, calculated that the orbital variations generate alternating cool and warm climates in the mid and higher latitudes. Moreover, the timing of the calculated cooling coincided with that of Pleistocene glacial advances. Therefore, he concluded that orbital variations caused Pleistocene glacial cycles.

Modern calculations indicate that orbital cycles by themselves are not sufficient to cause glaciers to advance and retreat. Instead, orbital cycles disturb other Earth systems, which cause additional cooling. Thus, a relatively small initial disturbance is amplified to cause a major climate change. In one recent study, researchers calculated that orbital variations probably caused high-latitude climate to cool enough to kill vast regions of northern forest. Forests control climate by absorbing solar energy. When the forests died, more solar energy reflected back out to space. This loss of solar energy caused the Earth to cool even more, and the glaciers advanced.[2] The ice reflected more solar energy, leading to additional cooling. Thus an astronomical event altered the biosphere and the altered biosphere triggered additional cooling that led to glaciation. The feedback mechanism continued because glaciation caused even more cooling.

Effects of Pleistocene Continental Glaciers

At its maximum extent about 18,000 years ago, the most recent North American ice sheet covered 10 million square kilometers—most of Alaska, Canada, and parts of the northern United States (Fig. 11–31). At the same time, alpine glaciers flowed from the mountains into the lowland valleys.

---

[1] Mark Chandler, "Glacial cycles; trees retreat and ice advances." *Nature,* Vol. 381, June 6, 1996, pp. 477–478.

[2] R.G. Gallimore and J.E. Kutzbach, "Role of orbitally induced changes in the tundra area in the onset of glaciation." *Nature,* Vol. 381, June 6, 1996, pp. 503–505.

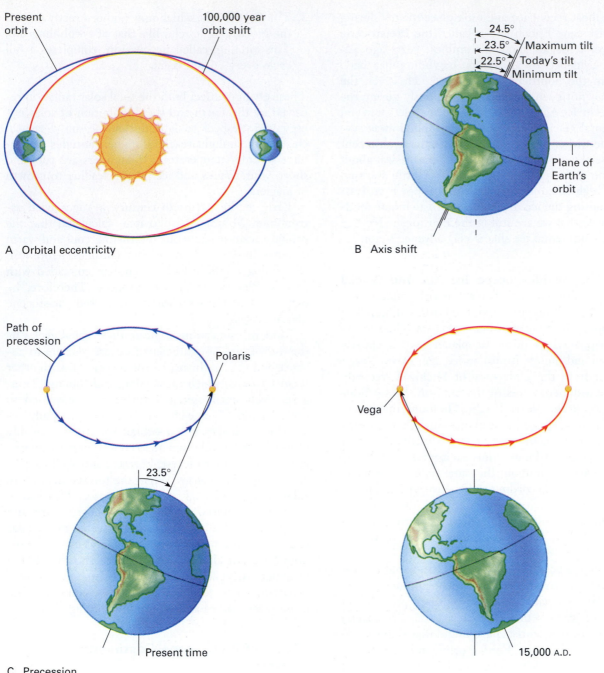

A Orbital eccentricity

B Axis shift

C Precession

Figure 11–30 Earth orbital variations may explain the temperature oscillations and glacial advances and retreats during the Pleistocene Epoch. The Earth's orbit varies in three ways: (A) The elliptical shape of the orbit changes over a cycle of about 100,000 years. (B) The tilt of the Earth's axis of rotation oscillates by about 2° over a cycle of about 41,000 years. (C) The Earth's axis completes a full cycle of precession about every 26,000 years.

The erosional features and deposits created by these glaciers dominate much of the landscape of the northern states. Today, terminal moraines form a broad band of rolling hills from Montana across the Midwest and eastward to the Atlantic Ocean. Long Island, New York, and Cape Cod, Massachusetts, are composed largely of terminal moraines. Kettle lakes or lakes dammed by moraines are abundant in northern Minnesota, Wisconsin, and Michigan. Drumlins dot the landscape in the northern states. Ground moraines, outwash, and loess (windblown glacial silt) cover much of the northern Great Plains. These deposits have

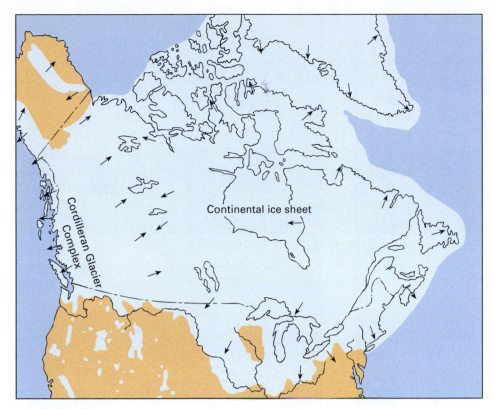

Figure 11–31 Maximum extent of the continental glaciers in North America during the latest glacial advance, approximately 18,000 years ago. The arrows show directions of ice flow.

weathered to form the fertile soil of North America's breadbasket.

Pleistocene glaciers advanced when mid- and high-latitude climates were colder and wetter than today. When the glaciers melted, the rain and meltwater flowed through streams and collected in numerous lakes. Later, as the ice sheets retreated and the climate became drier, many of these streams and lakes dried up.

The basin that is now Death Valley was once filled with water to a depth of 100 meters or more. Most of western Utah was covered by Lake Bonneville. As drier conditions returned, Lake Bonneville shrank to become Great Salt Lake, west of Salt Lake City.

When glaciers grow, they accumulate water that would otherwise be in the oceans, and sea level falls. When glaciers melt, sea level rises again. When the Pleistocene glaciers reached their maximum extent 18,000 years ago, global sea level fell to about 130 meters below its present elevation. As submerged continental shelves became exposed, the global land area increased by 8 percent (although about one third of the land was ice covered).

When the ice sheets melted, most of the water returned to the oceans, raising sea level again. At the same time, portions of continents rebounded isostatically as the weight of the ice was removed. The effect along any specific coast depends upon the relative amounts of sea-level rise and isostatic rebound. Some coastlines were submerged by the rising seas. Others rebounded more than sea level rose. Today, beaches in the Canadian Arctic lie tens to a few hundred meters above the sea. Portions of the shoreline of Hudson's Bay have risen isostatically 300 meters.

CASE STUDY
Pleistocene Glaciers and the Great Lakes

Before the onset of Pleistocene glaciation, several major rivers flowed through what is now the Great Lakes basin and followed the path of the modern St. Lawrence River to the Atlantic Ocean. Then during Pleistocene time, glaciers scoured and deepened the valleys and deposited moraines that dammed the St. Lawrence outlet. As a result, most of the streams changed direction and flowed southward into the Mississippi drainage (Fig. 11–32A).

At the same time, the great accumulation of ice on land had lowered sea level, thereby lowering the base level of the Mississippi River. This effect increased the

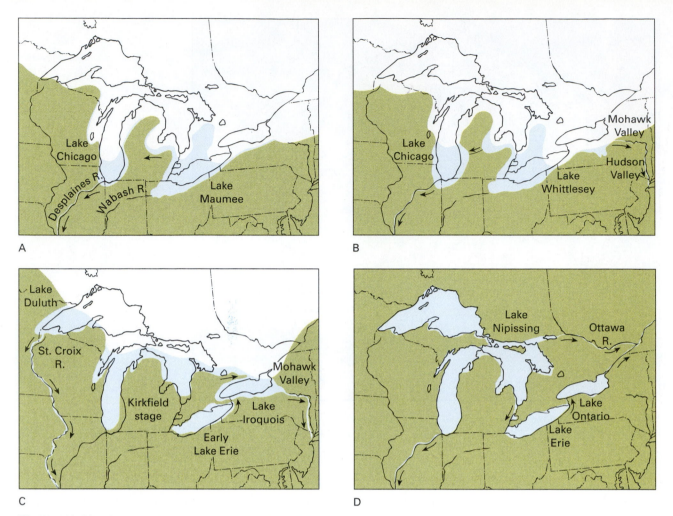

A

B

C

D

Figure 11–32 Continental glaciers scoured the Great Lakes Basin and altered its drainage pattern several times.

gradient of the river, causing it to flow more rapidly. The fast-flowing Mississippi cut rapidly into its bed, forming a V-shaped valley with a narrow flood plain.

About 18,000 years ago, the glaciers began to retreat. The meltwater carried more silt, sand, and gravel into the Mississippi than the current could transport, so the river spread out into many small channels, forming a braided stream and filling the narrow valley with sediment.

As the ice sheet melted, lakes formed behind moraine dams all along the receding ice front. Eventually the rising water level breached a dam near modern Lake Ontario, and much of the water from the Great Lakes region flowed into the Hudson River valley (Fig. 11–32B).

Over the next 5000 years, the ice receded, sea level rose, and the continent rebounded isostatically. These changes altered the drainage patterns several times (Figs. 11–32C and D). About 9000 years ago, the rivers abandoned the Hudson River valley outlet and the Great Lakes drained eastward and southward through the St. Lawrence and Mississippi rivers. A few thousand years later, the Mississippi River link was abandoned and all the water flowed through the St. Lawrence, as it does today.

When its supply of water was reduced, the Mississippi slowed down and carried much less sediment. Gradually, the braided channels coalesced, forming the modern meandering river, with its broad flood plain. ●

Case Study
Glacial Lake Missoula and the Greatest Flood in North America

About 18,000 years ago, a glacier flowed southward from Canada into the United States along the Idaho-Montana border. The ice dammed the Clark Fork

River, forming Glacial Lake Missoula, which was 600 meters deep and contained 2000 cubic kilometers of water—as much as modern Lake Ontario. Ice is a disastrously poor material for a dam because it floats on water. When the lake became deep enough, it floated the ice dam and the water poured across northern Idaho and Washington. But this was no ordinary flood like the 1993 Mississippi River flood—it was orders of magnitude more catastrophic than modern disasters such as the flood that occurred after the Teton Dam collapsed. Geologists estimate that a wall of water 600 meters high raced down valley. The water spread out as the valley opened up, and continued down the Columbia River valley. The raging torrent transported rocks from western Montana and deposited some of them 300 meters above the level of the modern river near the Columbia River Gorge, 450 kilometers from the failed dam. The rushing water eroded deep stream valleys in eastern and central Washington, which now contain tiny streams or no water at all. ●

SUMMARY

If snow survives through one summer, it becomes a relatively hard, dense material called **firn**. A **glacier** is a massive, long-lasting accumulation of compacted snow and ice that forms on land and creeps downslope or outward under the influence of its own weight. **Alpine glaciers** form in mountainous regions; **continental glaciers** cover vast regions. Glaciers move both by **basal slip** and by **plastic flow**. The upper 40 meters of a glacier is too brittle to flow, and large cracks called **crevasses** develop in this layer.

In the **zone of accumulation**, the annual rate of snow accumulation is greater than the rate of melting, whereas in the **zone of ablation**, melting exceeds accumulation. The **snow line** is the boundary between permanent snow and seasonal snow. The end of the glacier is called the **terminus**.

Glaciers erode bedrock, forming **U-shaped valleys**, **cirques**, and other landforms. **Drift** is any rock or sediment transported and deposited by a glacier. The unsorted drift deposited directly by a glacier is **till**. Most glacial terrain is characterized by large mounds of till known as **moraines**. **Terminal moraines, ground moraines, recessional moraines, lateral moraines, medial moraines,** and **drumlins** are all depositional features formed by glaciers. **Stratified drift** consists of sediment first carried by a glacier and then transported, sorted, and deposited by streams. **Valley trains, outwash plains, kames,** and **eskers** are composed of stratified drift. **Kettles** are depressions created by melting of large blocks of ice abandoned by a retreating glacier.

During the past 1 billion years, at least six major ice ages have occurred. The most recent was the **Pleistocene Ice Age.** One hypothesis contends that Pleistocene advances and retreats were caused by climate change induced by variations in the Earth's orbit and the orientation of its rotational axis. Ice sheets isostatically depress continents, which later rebound when the ice melts. Sea level falls when continental ice sheets form and rises again when the ice melts.

Summary of Glacial Features: I	
Erosional Features	**Depositional Features**
Glacial striation	Moraine
U-shaped valley	Drumlin
Cirque	Outwash
Tarns and lake basins	Valley train
Paternoster lake	Kame
Horn	Esker
Arête	
Hanging valley	
Fjord	

Summary of Glacial Features: II		
Primarily Formed by Alpine Glaciers	**Primarily Formed by Continental Glaciers**	**Formed by Both Alpine and Continental Glaciers**
U-shaped valley	Drumlin	Moraine
Cirque		Kame
Tarn		Esker
Paternoster lake		Outwash
Horn		Lake basins
Arête		
Hanging valley		
Valley train		

KEY TERMS

firn 244
glacier 244
alpine glacier 244
ice sheet 245
continental glacier 245
basal slip 245
plastic flow 245
crevasse 247
ice fall 247
zone of accumulation 247
zone of ablation 247
snow line 247

terminus 247
icebergs 248
glacial striation 250
U-shaped valley 251
cirque 251
tarn 252
paternoster lakes 252
horn 252
arête 252
hanging valley 252
fjord 252
drift 253

till 253
stratified drift 253
erratic 254
moraine 254
end moraine 254
terminal moraine 254
recessional moraine 254
ground moraine 254
lateral moraine 255
medial moraine 255
drumlin 255
outwash 256

valley train 256
outwash plain 257
kame 257
esker 257
kettle 258
ice age 258
Pleistocene Ice Age 259
eccentricity 259
tilt 259
precession 259

FOR REVIEW

1. Outline the major steps in the metamorphism of newly fallen snow to glacial ice.

2. Differentiate between alpine glaciers and continental glaciers. Where are alpine glaciers found today? Where are continental glaciers found today?

3. Distinguish between basal slip and plastic flow.

4. Why are crevasses only about 40 meters deep, even though many glaciers are much thicker?

5. Describe the surface of a glacier in the summer and in the winter in (a) the zone of accumulation and (b) the zone of ablation.

6. Describe how glacial erosion can create (a) a cirque, (b) a paternoster lake, and (c) striated bedrock.

7. Describe the formation of arêtes, horns, and hanging valleys.

8. Distinguish among ground, recessional, terminal, lateral, and medial moraines.

9. Why are kames and eskers features of receding glaciers? How do they form?

10. What topographic features were left behind by the continental ice sheets? Where can they be found in North America today?

11. Briefly outline the changes in drainage patterns in the Great Lakes over the past 18,000 years.

FOR DISCUSSION

1. Compare and contrast the movement of glaciers with stream flow.

2. Outline the changes that would occur in a glacier if (a) the average annual temperature rose and the precipitation decreased; (b) the temperature remained constant but the precipitation increased; and (c) the temperature decreased and the precipitation remained constant.

3. Explain why plastic flow is a minor mechanism of movement for thin glaciers but is likely to be more important for a thick glacier.

4. In some low elevation regions of northern Canada, both summer and winter temperatures are cool enough for glaciers to form, but there are no glaciers. Speculate on why continental glaciers are not forming in these regions.

5. If you found a large boulder lying in a field, how would you determine whether or not it was an erratic?

6. A bulldozer can only build a pile of dirt when it is moving forward. Yet a glacier can build a terminal moraine when it is retreating. Explain.

7. Imagine you encountered some gravelly sediment. How would you determine whether it was a stream deposit or a ground moraine?

8. Explain how medial moraines prove that glaciers move.

9. If you were hiking along a wooded hill in Michigan, how would you determine whether or not it was a moraine?

Deserts and Wind

A desert can be warm, cold, sandy, or rocky. This photograph shows sand dunes and rocky cliffs in Coral Pinks State Park, Utah.

Deserts evoke an image of thirsty travelers crawling across lifeless sand dunes. Although this image accurately depicts some deserts, others are rocky and even mountainous, with colorful cliffs or peaks towering over plateaus and narrow canyons. Rain may punctuate the long hot summers, and in winter a thin layer of snow may cover the ground. Although plant life in deserts is not abundant, it is diverse. Cactus, sage, grasses, and other plants may dot the landscape. After a rainstorm, millions of flowers bloom.

A **desert** is any region that receives less than 25 centimeters (10 inches) of rain per year and consequently supports little or no vegetation.[1] Most deserts are surrounded by semi-arid zones that receive 25 to 50 centimeters of annual rainfall, more moisture than a true desert but less than adjacent regions.

---

[1] The definition of *desert* is linked to soil moisture and depends on temperature and amount of sunlight in addition to rainfall. Therefore, the 25-centimeter criterion is approximate.

Deserts cover 25 percent of the Earth's land surface outside of the polar regions and make up a significant part of every continent. If you were to visit the great deserts of the Earth, you might be surprised by their geologic and topographic variety. You would see coastal deserts along the beaches of Chile, shifting dunes in the Sahara, deep, red sandstone canyons in southern Utah, and stark granite mountains in Arizona. The world's deserts are similar only in that they all receive scant rainfall.

Throughout human history, cultures have adapted to the low water and sparse vegetation of desert ecosystems. Traditionally, many desert societies were nomadic, exploiting resources where and when they were available. Others developed irrigation systems to water crops close to rivers and wells. Modern irrigation systems have improved human adaptation to dry environments and enabled 13 percent of the world's population to live in deserts. Two thirds of the world's oil reserves lie beneath the deserts of the Middle East, transforming some of the poorest nations of the world into the richest. In the future, vast arrays of solar cells may convert desert sunlight to electricity. ●

12.1 Why Do Deserts Exist?

Rain and snow are unevenly distributed over the Earth's surface. The wettest place on Earth, Mount Waialeale, Hawaii, receives an average of 1168 centimeters (38 feet) of rain annually. In contrast, ten years or more may pass between rains or snowfalls in the Atacama Desert of Peru and Chile.

Precipitation supplies water that weathers and erodes rock, and creates soil. In addition, both plants and animals need fresh water. Thus, atmospheric processes affect surface processes and ecosystems. Tectonic processes create ocean basins, continents, and mountain ranges. In turn, these features affect precipitation. Thus, tectonic events indirectly create deserts.

Latitude

The Sun shines most directly near the equator, warming air near the Earth's surface. The air absorbs moisture from the equatorial oceans and rises because it is less dense than surrounding air. This warm, wet air cools as it ascends, and the water vapor condenses and falls as rain. For this reason, vast tropical rainforests grow near the equator. The rising equatorial air, which is now drier because of the loss of moisture, flows northward and southward at high altitudes. This air cools, becomes denser, and sinks back toward the Earth's surface at about 30° north and south latitudes (Fig. 12–1). As the air falls, it is compressed and becomes warmer, which enables it to hold more water

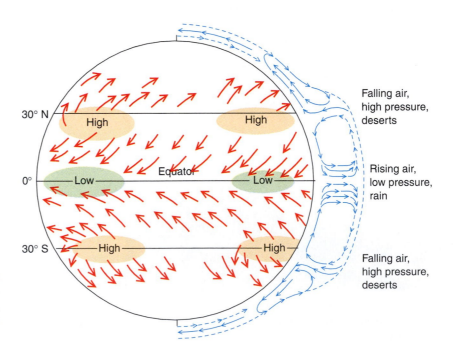

Figure 12–1 Falling air creates deserts at 30° north and south latitudes. The red arrows inside the globe indicate surface winds. The blue arrows on the right show air flow on the surface and at higher elevations.

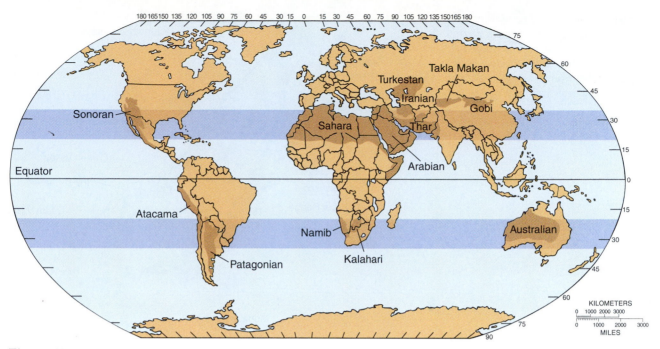

Figure 12–2 The major deserts of the world are concentrated at approximately 30° north and south latitudes.

vapor. As a result, water evaporates from the land surface into the air. Because the sinking air absorbs water, the ground surface is dry and rainfall is infrequent. Thus, many of the world's largest deserts lie at about 30° north and south latitudes (Fig. 12–2).

Mountains: Rain Shadow Deserts

When moisture-laden air flows over a mountain range, it rises. As the air rises, it cools and its ability to hold water decreases. As a result, the water vapor condenses into rain or snow, which falls as precipitation on the windward side and on the crest of the range (Fig. 12–3). This cool air flows down the leeward (or downwind) side and sinks. As in the case of sinking air at 30° latitude, the air is compressed and warmed as it falls and has already lost much of its moisture. This warm, dry air creates an arid zone called a **rain shadow desert** on the leeward side of the range. Figure 12–4 shows the rainfall distribution in California. Note that the leeward valleys are much drier than the mountains to the west.

Figure 12–3 A rain shadow desert forms where warm, moist air from the ocean rises as it flows over mountains. As it rises, it cools and water vapor condenses to form rain. The dry, descending air on the lee side absorbs moisture, forming a desert.

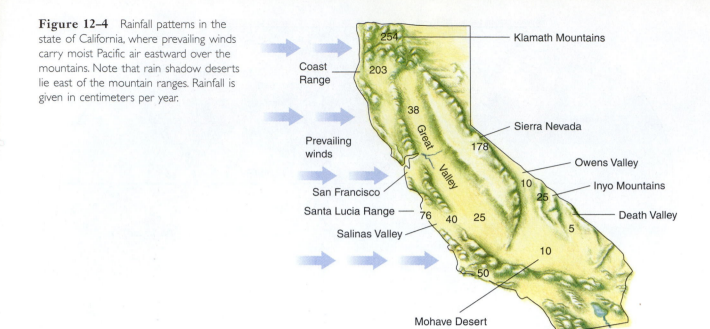

Figure 12–4 Rainfall patterns in the state of California, where prevailing winds carry moist Pacific air eastward over the mountains. Note that rain shadow deserts lie east of the mountain ranges. Rainfall is given in centimeters per year.

Coastal and Interior Deserts

Because most evaporation occurs over the oceans, one might expect that coastal areas would be moist and climates would become drier with increasing distance from the sea. This generalization is often true, but notable exceptions exist.

The Atacama Desert along the west coast of South America is so dry that portions of Peru and Chile often receive no rainfall for a decade or more. Cool ocean currents flow along the west coast of South America. When the cool marine air encounters warm land, the air is heated. The warm, expanding air absorbs moisture from the ground, creating a coastal desert.

The Gobi Desert is a broad, arid region in central Asia. The center of the Gobi lies at about 40°N latitude, and its eastern edge is a little more than 400 kilometers from the Yellow Sea. As a comparison, Pittsburgh, Pennsylvania, lies at about the same latitude and is 400 kilometers from the Atlantic Ocean. If latitude and distance from the ocean were the only factors, both regions would have similar climates. However, the Gobi is a barren desert and western Pennsylvania receives enough rainfall to support forests and rich farmland. The Gobi is bounded by the Himalayas to the south and the Urals to the west, which shadow it from the prevailing winds. In contrast, winds carry abundant moisture from the Gulf of Mexico, the Great Lakes, and the Atlantic Ocean to western Pennsylvania.

Thus, in some regions, deserts extend to the seashore and in other regions the interior of a continent is humid. The climate at any particular place on the Earth results from a combination of many factors. Latitude and proximity to the ocean are important, but the direction of prevailing winds, the direction and temperature of ocean currents, and the positions of mountain ranges also control climate.

12.2 Water and Deserts

Although rain rarely falls in deserts, water plays an important role in these dry environments. Water reaches most deserts from three sources. Streams flow from adjacent mountains or other wetter regions, bringing surface water to some desert areas. Ground water may also flow from a wetter source to an aquifer beneath a desert. Finally, rain and snow fall occasionally on deserts.

Vegetation is sparse in most deserts because of the limited water supply. Thus, much bare soil is exposed, unprotected from erosion. As a result, rain easily erodes desert soils and flowing water is an important factor in the evolution of desert landscapes.

Desert Streams

Large rivers flow through some deserts. For example, the Colorado River crosses the arid southwestern United States, and the Nile River flows through North African deserts (Fig. 12–5). Desert rivers receive most of their water from wetter, mountainous regions bordering the arid lands.

Figure 12–5 The Colorado River, seen from Dead Horse Point near Moab, Utah, flows from the Rocky Mountains through the arid southwestern United States. *(David P. Newell/Visuals Unlimited)*

In a desert, the water table is commonly deep below a stream bed, so that water seeps downward from the stream into the ground. As a result, many desert streams flow for only a short time after a rainstorm or during the spring, when winter snows are melting. A stream bed that is dry for most of the year is called a **wash** (Fig. 12–6).

Desert Lakes

While most lakes in wetter environments lie at the level of the water table and are fed, in part, by ground water, many desert lakes lie above the water table. During the wet season, rain and streams fill a desert lake. Some desert lakes are drained by outflowing streams, while many lose water only by evaporation and seepage. During the dry season, inflowing streams may dry up and evaporation and seepage may be so great that the lake dries up completely. An intermittent desert lake is called a **playa lake,** and the dry lake bed is called a **playa** (Fig. 12–7).

A

B

Figure 12–6 Courthouse Wash, Utah (A) in the spring when rain and melting snow fill the channel with water and (B) in mid-summer, when the creek bed is a dry wash.

Figure 12–7 Mud cracks pattern the floor of a playa in Utah.

wind. However, if a thunderstorm occurs upstream during the night, a flash flood may fill the wash with a wall of water mixed with rocks and boulders, creating disaster for the campers. By mid-morning of the next day, the wash may contain only a tiny trickle, and within 24 hours it may be completely dry again.

In August, 1997, 11 hikers perished near Lake Powell in Utah, when a flash flood filled a narrow, steep-walled canyon with a torrent of water, rocks, and mud. The flood was created by a thunderstorm that occurred 25 miles upstream, and out of sight of the group (Fig. 12–9).

When rainfall is unusually heavy and prolonged, the desert soil itself may become saturated enough to create a mudflow that carries boulders and anything else in its path downslope. Some of the most expensive homes in Phoenix, Arizona, and other desert cities are built on alluvial fans and steep mountainsides, where they have good views but are prone to mudflows during wet years.

Recall from Chapter 10 that streams and ground water contain dissolved salts. When this slightly salty water fills a desert lake and then evaporates, the ions precipitate to deposit the salts on the playa. Over many years, economically valuable mineral deposits, such as those of Death Valley, may accumulate (Fig. 12–8).

Flash Floods

Bedrock or tightly compacted soil covers the surface of many deserts, and little vegetation is present to absorb moisture. As a result, rainwater runs over the surface to collect in gullies and washes. During a rainstorm, a dry stream bed may fill with water so rapidly that a **flash flood** occurs. Occasionally, novices to desert camping pitch their tents in a wash, where they find soft sand to sleep on and shelter from the

Figure 12–8 Borax and other valuable minerals are abundant in the evaporite deposits of Death Valley. Mule teams hauled the ore from the valley in the 1800s. *(U.S. Borax)*

Figure 12–9 In August 1997, 11 hikers perished when a flash flood filled a slot canyon similar to this one in the Utah desert. The steep walls made escape impossible.

Figure 12–10 An alluvial fan forms where a steep mountain stream deposits sediment where it enters a valley. This photograph shows a fan in Death Valley. ● Interactive Question: Would you expect the alluvial fan to grow continuously, month by month, or intermittently? Explain.

Pediments and Bajadas

When a steep, flooding mountain stream empties into a flat valley, the water slows abruptly and deposits most of its sediment at the mountain front, forming an alluvial fan. Although fans form in all climates, they are particularly conspicuous in deserts (Fig. 12–10). A large fan may be several kilometers across and rise a few hundred meters above the surrounding valley floor.

If the mouths of several canyons are spaced only a few kilometers apart, the alluvial fans extending from each canyon may merge. A **bajada** is a broad, gently

Figure 12–11 The bajada in the foreground merges with a gently sloping pediment to form a continuous surface in front of the mountains. This Mongolian basin is filling with sediment from the surrounding mountains because it has no external drainage.

sloping depositional surface formed by merging alluvial fans and extending into the center of a desert valley. Typically, the fans merge incompletely, forming an undulating surface that may follow the mountain front for tens of kilometers. The sediment that forms the bajada may fill the valley to a depth of several thousand meters.

A **pediment** is a broad, gently sloping surface eroded into bedrock. Pediments commonly form along the front of desert mountains. The bedrock surface of a pediment is covered with a thin veneer of gravel that is in the process of being transported from the mountains, across the pediment to the bajada.

Together, a pediment and bajada form a smooth surface from the mountain front to the valley center (Fig. 12–11). The surface steepens slightly near the mountains, so it is concave. To distinguish a pediment from a bajada, you would have to dig or drill a hole. If you were on a pediment, you would strike bedrock after only a few meters, but on a bajada, bedrock may be buried beneath hundreds or even thousands of meters of sediment.

12.3 Two American Deserts

The Colorado Plateau

The Colorado Plateau covers a broad region encompassing portions of Utah, Colorado, Arizona, and New Mexico. During the past one billion years of Earth history, this region has been alternately covered by shallow seas, lakes, and deserts. Sediment accumulated, sedimentary rocks formed, and tectonic forces later uplifted the land to form the Plateau. The Colorado River cut through the bedrock of the Plateau as it rose, to form the 1.6-kilometer-deep Grand Canyon and its tributary canyons. The modern Colorado River receives most of its water from snowmelt and rains in the high Rocky Mountains east and north of the Plateau, and the river then flows through the heart of the great desert to empty into the Gulf of California, in northern Mexico.

A stream forms a canyon by eroding downward into bedrock. If the downcutting stream reaches a resistant rock layer, it may erode laterally, widening the canyon. In some places on the Colorado Plateau, downcutting predominates and streams erode deep, narrow canyons. In other regions, lateral erosion undercuts canyon walls and the rock collapses along vertical joints, to form flat-topped mesas and buttes that rise above a relatively flat plain. The river eventually carries the sediment from the Plateau to the Gulf of California.

The flat tops of mesas and buttes usually form on a bed of sandstone or other sedimentary rock that is relatively resistant to erosion. In many cases, extensive lateral erosion leaves spectacular pinnacles, isolated remnants of once-continuous rock layers that the streams have all but completely eroded away. A **plateau** is a large elevated area of fairly flat land. The term plateau is used for regions as large as the Colorado Plateau as well as for smaller, elevated flat surfaces. A **mesa** is smaller than a plateau and is a flat-topped mountain shaped like a table. A **butte** is also a flat-topped mountain characterized by steep cliff faces and is smaller and more tower-like than a mesa (Fig. 12–12). These landforms are common features of the Colorado Plateau.

Death Valley and the Great Basin

Death Valley lies in the rain shadow of the high Sierras in California. The deepest part of the valley is 82 meters below sea level. It is a classic rain shadow desert, receiving a scant 5 centimeters of rainfall per year. The mountains to the west receive abundant moisture, and during the winter rainy season and spring snow melt, streams flow from the mountains into the Valley.

In the Colorado Plateau, the Colorado River and its tributaries carry sediment away from the desert to the Gulf of California, leaving deep canyons. In contrast, streams flow into Death Valley from the surrounding mountains, but no streams flow out. Because Death Valley has no external drainage, the valley is filling with sediment eroded from the surrounding mountains. The sediment collects to form vast alluvial fans and bajadas (Fig. 12–13). Stream water collects in broad playa lakes that dry up under the hot summer sun.

Like Death Valley, many deserts in the American West have no external drainage. The Great Basin, which includes most of Nevada, the western half of Utah, and parts of California, Oregon, and Idaho, is a large desert region between the Sierra Nevada and the Rocky Mountains. Although streams flow into the region from surrounding mountain ranges and from ranges within the basin, no streams flow out of the Great Basin. Consequently, sediment is not flushed out and has accumulated to become thousands of meters thick.

Bajadas and pediments are common features of the internally drained valleys of the Great Basin. Both fea-

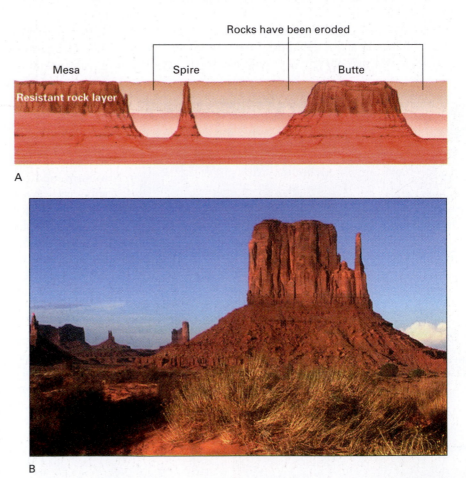

Figure 12–12 (A) Spires and buttes form when streams reach a temporary base level and erode laterally. The streams transport the eroded sediment away from the region. (B) Spires and buttes in Monument Valley, Arizona.

Figure 12–13 Sediment eroded from surrounding mountains is slowly filling Death Valley. *(Frank M. Hanna/Visuals Unlimited)*

tures result from a combination of tectonic, erosional, and depositional processes. The mountains and valleys initially formed by block faulting. The valleys are downdropped blocks of rock called grabens, and the mountain ranges are uplifted blocks called horsts (Fig. 12–14A). Grabens and horsts are discussed in Chapter 8. Desert streams partially eroded the mountains and deposited sediment to form pediments, alluvial fans, and bajadas (Fig. 12–14B).

Faulting continued to deepen the valleys, and at the same time, streams filled them with sediment. As a result, the mountains are slowly drowning in their own sediment (Fig. 12–14C).

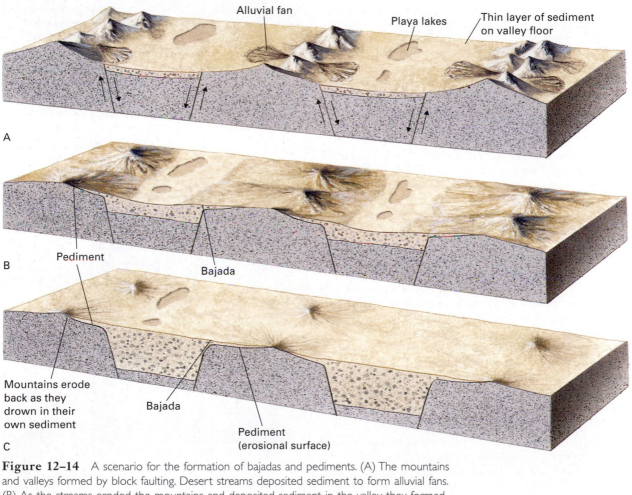

Figure 12–14 A scenario for the formation of bajadas and pediments. (A) The mountains and valleys formed by block faulting. Desert streams deposited sediment to form alluvial fans. (B) As the streams eroded the mountains and deposited sediment in the valley, they formed both the erosional surface called a pediment, and a depositional surface called a bajada. (C) Today, the mountains are drowning in their own sediment. • **Interactive Question:** Explain how the landscape would evolve if a stream flowed through the valley, draining into the ocean.

Six thousand years ago, advanced civilizations thrived along desert rivers such as the Tigris-Euphrates and the Nile. The Babylonians and Egyptians used buckets, wheels, canals, and pumps to irrigate fields near the rivers. Desert nomads herded their flocks in the semi-arid grasslands that lay between the rivers and deserts. Prior to the twentieth century, nomads moved across deserts and semi-arid lands with little regard for national boundaries, traveling with the seasons and abandoning an area after it had been grazed for a short period of time. This constant movement prevented overgrazing. In addition, population levels of the nomadic tribes were stable and low.

The Sahara is the largest desert on the planet. South of the Sahara lies the semi-arid Sahel. During the 1960s, unusually heavy rains caused the Sahel to bloom. People expanded their flocks to take advantage of the additional forage. Rich countries contributed foreign aid. As a result, medical attention and sanitation improved, and the human population grew dramatically. Many people predicted a new era of prosperity for the Sahel, but the favorable rains were an anomaly. In the late 1960s and early 1970s, drought destroyed the range. During this period, governments in North Africa began to enforce national borders more strictly, curtailing nomadism. Civil and international war brought instability to the region, and famine struck. Reports issued in the 1970s and 1980s claimed that the Sahara was expanding southward into the Sahel at a rate of 5 kilometers per year. Scientists argued that overgrazing, farming, and firewood gathering had caused the desert expansion. This growth of the desert caused by human mismanagement has been called **desertification.**

More recent research has shown that overgrazing a semi-arid region causes land degradation but does not cause a desert to expand. The Sahel-Sahara desert boundary is clearly visible on satellite photographs as a boundary separating a region of sparse vegetation from one with almost no plants. Researchers plotted changes in the size of the desert by studying satellite photographs taken between 1980 and 1990. As shown in Figure 1, the desert expands (blue line) when rainfall declines (red line).[1] Thus, decreasing rainfall, not overgrazing, may have been responsible for expansion of the Sahara.

In another project, researchers studied natural and overgrazed range in semi-arid grasslands in southern New Mexico.[2] The natural range consisted mainly of a homogeneous blanket of grass. In regions where cattle

had overgrazed the land, woody creosote and mesquite had replaced the grass. The net primary productivities (the total plant growth) in the natural and overgrazed systems were nearly identical; thus, overgrazing did not reduce plant growth. It did, however, change the types of plants and their spatial distribution. The shrubs that invaded after overgrazing provided poorer forage for cattle and therefore reduced the economic value of the range.

Grasses have shallow roots that absorb falling rain. Excessive grazing destroys the grasses as the cattle eat the grass down to the roots. In addition, the heavy animals pack the soil with their hooves, blocking the natural seepage of air and water. When soil is devoid of vegetation and baked in the sunlight, it becomes so impermeable that water evaporates or runs off before it soaks in. Increased runoff erodes the soil and carries off nutrients. The water, soil, and nutrients collect in shallow depressions and desert washes. In some regions, shrubs accumulate in these moist, nutrient-rich low points. Thus, the original range had a homogeneous distribution of plants, water, and nutrients, whereas the overgrazed range was heterogeneous, alternating thick shrubbery with bare ground. This redistribution of soil and nutrients alters conditions so that the original grassland ecosystem cannot become reestablished quickly or easily. Thus, range deterioration is long lasting, and the limited productivity of the land declines.

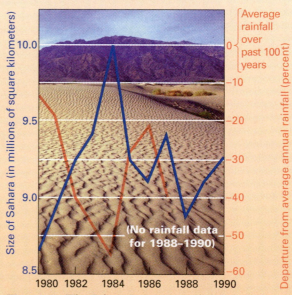

Figure 1 The Sahara Desert expanded and contracted between 1980 and 1990. Note that the Sahara expanded (blue line) when rainfall decreased (red line).

[1] William H. Schlesinger et al., "Biological Feedbacks in Global Desertification," *Science*, Vol. 247, March 1990, p. 1043.

[2] Compton J. Tucker, Harold E. Dregne, and Wilbur W. Newcomb, "Expansion and Contraction of the Sahara Desert from 1980 to 1990," *Science*, Vol. 253, July 19, 1991, p. 299ff.

Wind removes surface sand

Formation of desert pavement complete— no further wind erosion

A

B

Figure 12–15 (A) Wind erodes silt and sand but leaves larger rocks behind to form desert pavement. (B) Desert pavement is a continuous cover of stones left behind when wind blows silt and sand away.

12.4 Wind in Deserts

When wind blows through a forest or across a prairie, the trees or grasses protect the soil from wind erosion. In addition, rain accompanies most windstorms in wet climates; the water dampens the soil and binds particles together. Therefore little wind erosion occurs. In contrast, a desert commonly has little or no vegetation and rainfall, so wind erodes bare, unprotected desert soil.

Wind Erosion

Wind erosion, called **deflation,** is a selective process. Because air is much less dense than water, wind moves only small particles, mainly silt and sand. (Clay particles usually stick together, and consequently wind does not erode clay effectively.) Imagine bare soil contain-

Figure 12–16 Wind abrasion near Grand Canyon, Arizona, selectively eroded the base of this rock because windblown sand moves mostly near the surface.

ing silt, sand, pebbles, and cobbles. When wind blows, it removes only the silt and sand, leaving the pebbles and cobbles as a continuous cover of stones called **desert pavement** (Fig. 12–15). Desert pavement prevents the wind from eroding additional sand and silt, even though this finer sediment may be abundant beneath the layer of stones. As a result of this process, most deserts are rocky and covered with gravel, and sandy deserts are less common.

Transport and Abrasion

Because sand grains are relatively heavy, wind rarely lifts sand more than 1 meter above the ground and carries it only a short distance. In a windstorm, the sand grains bounce over the ground in a process called saltation. (Recall from Chapter 10 that sand also moves by saltation in a stream.) In contrast, wind carries fine silt in suspension. Skiers in the Alps commonly encounter a silty surface on the snow, blown from the Sahara Desert and carried across the Mediterranean Sea.

Windblown sand is abrasive and erodes bedrock. Because wind carries sand close to the surface, wind erosion occurs near ground level. If you see a tall desert pinnacle topped by a delicately perched cap, you know that the top was not carved by wind erosion because it is too high above the ground. However, if the base of a pinnacle is sculpted, wind may be the responsible agent (Fig. 12–16). (Salt cracking at ground level also contributes to the weathering of desert rocks.)

Figure 12–17 Windblown sand formed these dunes near Lago Poopo, Bolivia.

Dunes

A **dune** is a mound or ridge of wind-deposited sand (Fig. 12–17). As explained earlier, wind removes sand from the surface in many deserts, leaving behind a rocky, desert pavement. The wind then deposits the sand in a topographic depression or other place where the wind slows down. Dunes commonly grow to heights of 30 to 100 meters, and some giants exceed 500 meters. In some places they are tens or even hundreds of kilometers long. Approximately 80 percent of the world's desert area is rocky and only 20 percent is covered by dunes. Although some desert dune fields cover only a few square kilometers, the largest is the Rub Al Khali (Empty Quarter) in Arabia, which cov-

ers 560,000 square kilometers, larger than the state of California.

Dunes also form where glaciers have recently melted and along sandy coastlines. A glacier deposits large quantities of bare, unvegetated sediment. A sandy beach is commonly unvegetated because sea salt prevents plant growth. As a result, both of these environments contain the essentials for dune formation: an abundant supply of sand and a windy environment with sparse vegetation.

Dunes form when wind erodes sand from one location and deposits it nearby. A saucer or trough-shaped hollow formed by wind erosion in sand is called a **blowout**. In the 1930s, intense, dry winds eroded

Figure 12–18 Wind eroded sandy soil to create this blowout near Highway 2 in the Sand Hills of western Nebraska. *(Tom Bean)*

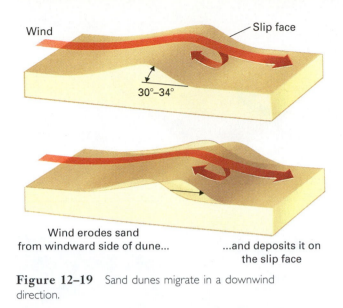

Figure 12–19 Sand dunes migrate in a downwind direction.

Most dunes are asymmetrical. Wind erodes sand from the windward side of a dune, and then the sand slides down the sheltered leeward side. In this way, dunes migrate in the downwind direction (Fig. 12–19). The leeward face of a dune is called the **slip face.** Typically, the slip face is about twice as steep as the windward face. In addition, the sand on a slip face is usually very loose, whereas the sand on the windward face is packed by the wind.

Migrating dunes overrun buildings and highways. For example, near the town of Winnemucca, Nevada, dunes advance across U.S. Highway 95 several times a year. Highway crews must remove as much as 4000 cubic meters of sand to reopen the road. Engineers often attempt to stabilize dunes in inhabited areas. One method is to plant vegetation to reduce deflation and stop dune migration. The main problem with this approach is that desert dunes commonly form in regions that are too dry to support vegetation. Another solution is to build artificial windbreaks to create dunes in places where they do the least harm. For example, a fence traps blowing sand and forms a dune, thereby protecting areas downwind. Fencing is a temporary solution, however, because eventually the dune covers the fence and resumes its migration. In Saudi Arabia, dunes are sometimes stabilized by covering them with tarry wastes from petroleum refining.

Fossil Dunes   When dunes are buried by younger sediment and lithified, the resulting sandstone retains the original sedimentary structures of the dunes. Figure 12–20 shows a rock face in Zion National Park

large areas of the Great Plains and created the Dust Bowl. Deflation formed tens of thousands of blowouts, many of which remain today. Some are small, measuring only 1 meter deep and 2 or 3 meters across, but others are much larger (Fig. 12–18). One of the deepest blowouts in the world is the Qattara Depression in western Egypt. It is more than 100 meters deep and 10 kilometers in diameter. Ultimately, the lower limit for a blowout is the water table. If the bottom of the depression reaches moist soil near the water table, where water binds the sand grains, wind erosion is no longer effective.

Figure 12–20 Cross-bedded sandstone in Zion National Park preserves the sedimentary bedding of ancient sand dunes.

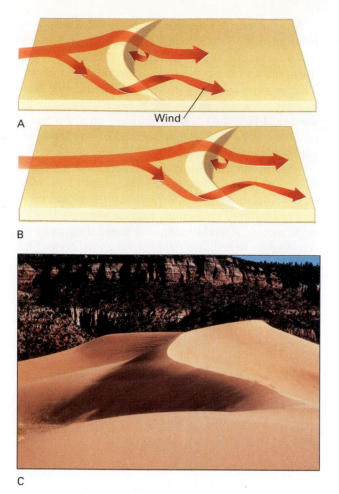

A

B

C

Figure 12–21 (A and B) When sand supply is limited, the tips of a barchan dune travel faster than the center and point downwind. (C) A barchan dune in Coral Pinks, Utah.

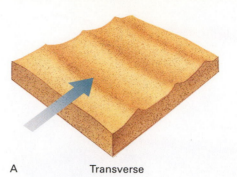

A   Transverse

Figure 12–22 (A) Transverse dunes form perpendicular to the prevailing wind direction in regions with abundant sand. (B) These transverse dunes formed in Death Valley, California. *(Martin G. Miller/Visuals Unlimited)*

in Utah. The steep layering is not evidence of tectonic tilting but is the original steeply dipping layering of the dune slip face. The beds dip in the direction in which the wind was blowing when it deposited the sand. Notice that the planes dip in different directions, indicating changes in wind direction. The layering is an example of **cross-bedding,** described in Chapter 3.

Types of Sand Dunes   Wind speed and sand supply control the shapes and orientation of dunes. **Barchan dunes** form in rocky deserts with little sand. The center of the dune grows higher than the edges (Fig. 12–21A). When the dune migrates, the edges move faster because there is less sand to transport (Fig. 12–21B). The resulting barchan dune is crescent shaped with its tips pointing downwind (Fig. 12–21C). Barchan dunes are not connected to one another, but

instead migrate independently. In a rocky desert, barchan dunes cover only a small portion of the land; the remainder is bedrock or desert pavement.

If sand is plentiful and evenly dispersed, it accumulates in long ridges called **transverse dunes** that

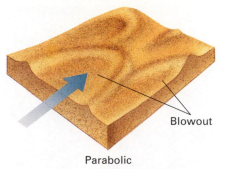

Parabolic

A

B

Figure 12–23 (A) A parabolic dune is crescent-shaped with its tips upwind. It forms where wind blows sand from a blowout, and grass or shrubs anchor the dune tips. (B) Grass and shrubs anchor the tip of this parabolic dune in the southern California desert. *(L. Rhodes)*

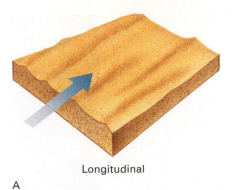

Longitudinal

A

B

Figure 12–24 (A) Longitudinal dunes are long, straight dunes that form where the wind is erratic and sand supply is limited. (B) These longitudinal dunes formed on the Oregon coast. *(Albert Copley/Visuals Unlimited)*

align perpendicular to the prevailing wind (Figs. 12–22 A and B). If desert vegetation is plentiful, the wind may form a blowout in a bare area among the desert plants. As sand is carried out of the blowout, it accumulates in a **parabolic dune,** the tips of which are anchored by plants on each side of the blowout (Figs. 12–23 A and B). A parabolic dune is similar in shape to a barchan dune, except that the tips of the parabolic dune point into the wind. Parabolic dunes are common in moist semi-desert regions and along seacoasts.

If the wind direction is erratic but prevails from the same general quadrant of the compass and the supply of sand is limited, then long, straight **longitudinal dunes** form parallel to the prevailing wind direction (Figs. 12–24 A and B). In portions of the Sahara Desert, longitudinal dunes reach 100 to 200 meters in height and are as much as 100 kilometers long.

Loess

Wind can carry silt for hundreds or even thousands of kilometers and then deposit it as **loess** (pronounced *luss*). Loess is porous, uniform, and typically lacks layering. Often the angular silt particles interlock. As a result, even though the loess is not cemented, it typically forms vertical cliffs and bluffs (Fig. 12–25).

The largest loess deposits in the world, found in central China, cover 800,000 square kilometers and are more than 300 meters thick. The silt was blown from the Gobi and the Takla Makan deserts of central Asia. The particles interlock so effectively that people have dug caves into the loess cliffs to make their homes. However, in 1920 a great earthquake caused the cave system to collapse, burying and killing an estimated 100,000 people.

Figure 12–25 Villagers in Askole, Pakistan, have dug caves in these vertical loess cliffs.

Large loess deposits accumulated in North America during the Pleistocene Ice Age, when continental ice sheets ground bedrock into silt. Streams carried this fine sediment from the melting glaciers and deposited it in vast plains. These zones were cold, windy, and devoid of vegetation, and wind easily picked up and transported the silt, depositing thick layers of loess as far south as Vicksburg, Mississippi.

Loess deposits in the United States range from about 1.5 meters to 30 meters thick (Fig. 12–26). Soils formed on loess are generally fertile and make good farmland. Much of the rich soil of the central plains of the United States and eastern Washington State formed on loess.

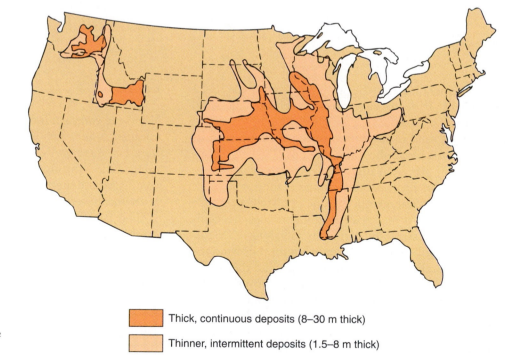

Figure 12–26 Loess deposits cover large areas of the United States.

■ Thick, continuous deposits (8–30 m thick)

■ Thinner, intermittent deposits (1.5–8 m thick)

SUMMARY

Deserts have an annual precipitation of less than 25 centimeters. The world's largest deserts occur near 30° north and south latitudes, where warm, dry, descending air absorbs moisture from the land. Deserts also occur in rain shadows of mountains, continental interiors, and coastal regions adjacent to cold ocean currents.

Desert streams are often dry for much of the year but may develop **flash floods** when rainfall occurs. **Playa lakes** are desert lakes that dry up periodically, leaving abandoned lake beds called **playas.** Alluvial fans are common in desert environments. A **bajada** is a broad depositional surface formed by merging alluvial fans. A **pediment** is a planar erosional surface that may lie at the base of a mountain front in arid and semiarid regions, and merges imperceptibly with a bajada.

The through-flowing Colorado River drains the Colorado Plateau desert. Thus, streams carry sediment away from the region, forming canyons and eroding the plateaus to form **mesas** and **buttes.** Death Valley and the Great Basin have no external drainage, and as a result, the valleys are filling with sediment eroded from the surrounding mountains.

Deflation is erosion by wind. Silt and sand are removed selectively, leaving larger stones on the surface and creating **desert pavement.** Sand grains are relatively large and heavy and are carried only short distances by saltation, seldom rising more than a meter above the ground. Silt can be transported great distances at higher elevations. Wind erosion forms **blowouts.** Windblown particles are abrasive, but because the heaviest grains travel close to the surface, abrasion occurs mainly near ground level.

A **dune** is a mound or ridge of wind-deposited sand. Most dunes are asymmetrical, with gently sloping windward sides and steeper **slip faces** on the lee sides. Dunes migrate. The various types of dunes include **barchan dunes, transverse dunes, longitudinal dunes,** and **parabolic dunes.** Wind-deposited silt is called **loess.**

KEY TERMS

desert 265
rain shadow desert 267
wash 269
playa lake 269
playa 269
flash flood 270

bajada 271
pediment 271
mesa 272
butte 272
plateau 272
deflation 275

desert pavement 275
blowout 276
dune 276
slip face 277
cross-bedding 278
barchan dune 278

transverse dune 278
parabolic dune 279
longitudinal dune 279
loess 279

FOR REVIEW

1. Why are many deserts concentrated along zones at 30° latitude in both the Northern and the Southern Hemispheres?

2. List three conditions that produce deserts.

3. Why do flash floods and debris flows occur in deserts?

4. Why are alluvial fans more prominent in deserts than in humid environments?

5. Compare and contrast floods in deserts with those in more humid environments.

6. Compare and contrast pediments and bajadas.

7. Why is wind erosion more prominent in desert environments than it is in humid regions?

8. Describe the formation of desert pavement.

9. Describe the evolution and shape of a dune.

10. Describe the differences among barchan dunes, transverse dunes, parabolic dunes, and longitudinal dunes. Under what conditions does each type of dune form?

11. Compare and contrast desert plateaus, mesas, and buttes. Describe the formation of each.

12. Compare the effects of stream erosion and deposition in the Colorado Plateau and Death Valley.

1. Coastal regions boast some of the wettest and some of the driest environments on Earth. Briefly outline the climatological conditions that produce coastal rainforests versus coastal deserts.

2. Explain why soil moisture content might be more useful than total rainfall in defining a desert. How could one region have a higher soil moisture content and lower rainfall than another region?

3. Discuss two types of tectonic change that could produce deserts in previously humid environments.

4. Imagine that you lived on a planet in a distant solar system. You had no prior information on the topography or climate of the Earth and were designing an unmanned spacecraft to land on Earth. The spacecraft had arms that could reach out a few meters from the landing site to collect material for chemical analysis. It also had instruments to measure the immediate meteorological conditions and cameras that could focus on anything within a range of 100 meters. The batteries on your radio transmitter had a life expectancy of two weeks. The spacecraft landed and you began to receive data. What information would convince you that the spacecraft had landed in a desert?

5. Deserts are defined as areas with low rainfall, yet water is an active agent of erosion in desert landscapes. Explain this apparent contradiction.

The Oceans

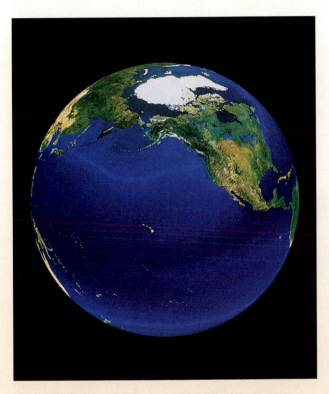

The earth is often called the blue planet because 71 percent of its surface is covered with water. *(Copyright Tom Van Sant/Geosphere Project, Santa Monica/Science Photo Library)*

Flowers Bloomed on Earth While Venus Boiled and Mars Froze

Earth is sometimes called the water planet or the blue planet because of its blue color when seen from space (Fig. 1). Blue predominates because more than two-thirds of our planet is covered by azure seas.

Enough water exists at the Earth's surface that, if the surface were perfectly level, water would form a layer about 2 kilometers thick covering the entire planet. About 97.5 percent of the Earth's water is seawater, and another 1.8 percent is frozen into the great ice caps of Antarctica and Greenland. Only about 0.64 percent is fresh water in streams, underground reservoirs, lakes, and wetlands. Thus, although the hydrosphere contains a great amount of water, only a tiny fraction is fresh.

Exploration of the Solar System has shown that Earth is the only planet or moon in the Solar System with water falling from clouds, running over the land in streams, and collecting in large basins to form oceans. Earth is the only body in the Solar System with liquid water on its surface.

Why is Earth unique? Why are we favored with the abundance of water that makes life possible and creates rain, streams, and oceans that interact with the rocks of the surface? Recall from the opening essay of Unit III that the inner planets were so close to the Sun that the heat and the solar wind swept their volatiles off into space. Thus, Mercury, Venus, Earth, and Mars are mostly solid rock, with metallic cores. In contrast, the planets in the outer reaches of the Solar System orbited in a cooler environment where the solar wind was weaker. As a result, Jupiter, Saturn, Uranus, and Neptune retained thick atmospheres of hydrogen, helium, and other light elements, and even grew as they captured volatiles escaping from the inner planets.

Figure 1 The temperature and atmospheric pressure on Earth create an environment where water exists as ice, liquid, and vapor, and changes readily from one form to another. These conditions are favorable for life and living organisms are abundant on the planet. (*Tom Van Sant/Geosphere Project, Santa Monica/Science Photo Library*)

Recall also that early in the history of the Solar System, many comets, asteroids, and meteoroids crashed into the inner planets and moons, returning the volatiles lost earlier.

The most abundant volatiles in the comets, asteroids, and meteoroids were ice, carbon dioxide, carbon monoxide, ammonia, methane, formic acid, and formaldehyde. These gases were carried in equal proportions to each of the inner planets. Why, then, are the atmospheres of those planets so different today, and why does only Earth have water on its surface?

Mercury is torridly hot because it is so close to the Sun. It is also small, and has a weak gravitational field. Consequently, all of its water and other volatiles have boiled off into space. However, the surface environments of Venus, Earth, and Mars are more different than can be explained simply by their different sizes and distances from the Sun.

Flowers Bloom on Earth

Both carbon dioxide and water vapor are commonly called greenhouse gases because they absorb infrared radiation and warm the atmosphere. Carbon dioxide occurs as an atmospheric gas, it dissolves in seawater, and on Earth it combines with calcium and oxygen to form limestone. Water occurs as a liquid, as water vapor, and as ice.

Earth's orbit lies between those of Venus and Mars. Because of this position, it receives less solar heat than Venus, but more than Mars. After bolides transported water and carbon dioxide to Earth, our planet had oceans and a carbon dioxide–rich atmosphere. Large amounts of carbon dioxide then dissolved in the seas. Some of this dissolved carbon dioxide precipitated to form limestone beds on the sea floor. The partitioning of most of the wa-

ter into the seas, and much of the carbon dioxide into seawater and limestone, removed large amounts of both greenhouse gases from the atmosphere. As a result, Earth's atmospheric temperature stabilized in a range favorable for the existence of liquid water and for the emergence and evolution of life.

Venus Boiled, A Victim of Runaway Greenhouse Warming

Of all the planets in our Solar System, Venus most closely resembles Earth in size, density, and distance from the Sun. Consequently, astronomers once thought that the environment on Venus might be similar to that on Earth, and that both water and life might be found there. However, data obtained from spacecraft reveal that Venus is extremely inhospitable. Venus's atmosphere is 90 times denser than that of Earth (Fig. 2). It consists of more than 97 percent carbon dioxide, with small amounts of nitrogen, helium, neon, sulfur dioxide, and other gases. Sulfuric acid clouds fill the sky. In addition, the Venusian surface is hot enough to melt lead, and hot enough to destroy the complex organic molecules necessary for life.

Why is Venus's surface so different from ours? Early in the history of the Solar System, Venus and Earth probably had similar atmospheres. However, Venus was closer to the Sun, and solar radiation heated Venus's surface to a higher temperature than it did that of Earth. One hypothesis suggests that because of the higher temperature, water vapor never condensed, or if it did, it quickly evaporated again. Because there were no seas for carbon dioxide to dissolve into, most of the carbon dioxide also remained in the atmosphere. Both greenhouse gases

caused Venus's atmosphere to heat up further.

Water has a relatively low molecular weight. Consequently, the solar wind swept it away from Venus, leaving its atmosphere rich in carbon dioxide and sulfur dioxide, searingly hot, strongly acidic, and devoid of water. Both carbon dioxide and sulfur dioxide have higher molecular weights and did not blow away. Thus, Venus lost its water early in its history, and has had a hot and poisonous atmosphere since.

Mars Froze, Victim of a Reverse Greenhouse Effect

Early in its history, Mars must have had a temperate climate somewhat like that of Earth today. Spacecraft images show extinct stream beds and canyons. Near the end of 1997, the Mars Pathfinder lander transmitted close-up images of rounded sand grains and pebbles back to Earth, indicating that streams, wind, or waves carried and rounded sedimentary particles. Estimates of the ages of those features indicate that rivers flowed across Mars's surface about 3.5 billion years ago (Fig. 3). Other images reveal extinct seas, lake beds, and huge alluvial fans that formed when massive floods deposited sediment.

Today, the Martian surface is frigid and dry. How did its climate change so drastically? Mars is farther from the Sun than Earth, and receives less solar warmth. However, bolides transported carbon dioxide and water vapor into the early Martian atmosphere. As a result, a carbon dioxide and water vapor–rich Martian atmosphere may have evolved between 4.0 and 3.5 billion years ago. Both of these greenhouse gases absorbed infrared radiation producing a temperate climate. Rain fell from clouds. Rivers flowed, eroding the land

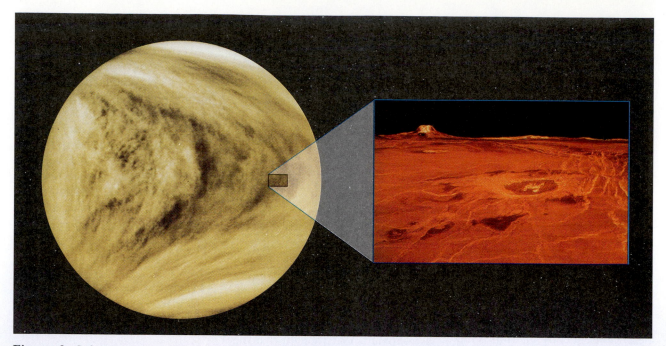

Figure 2 Before astronomers could peer through the Venusian cloud cover or measure its temperature and composition, they speculated that Venus may harbor life. Today we know that the atmosphere is so hot and corrosive that living organisms could not possibly exist. *(JPL/USGS; inset, NASA)*

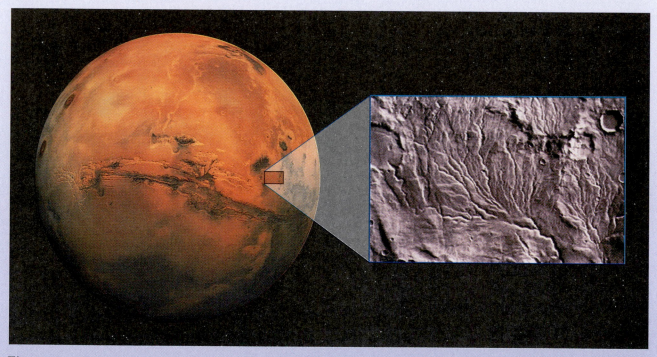

Figure 3 Water once flowed over the surface of Mars, eroding canyons and depositing sediment. However, today the planet is frigid and dry. *(NASA/JPL)*

surface. Perhaps life emerged in this favorable environment.

Astronomers calculate that the Martian atmosphere must have had at least 5 bars of pressure (that is, five times that of the Earth's current atmosphere) to warm the planet above the freezing point of water. Today, the Martian atmosphere contains only 0.007 bars of carbon dioxide and even less water. Sometime in the past Mars lost its atmosphere and froze.

Scientists are divided about how Mars lost its atmosphere while Earth and Venus retained theirs. According to one hypothesis, over a few hundred million years, much of the atmospheric carbon dioxide dissolved into the Martian seas, and the planet began to cool as this greenhouse gas disappeared from the atmosphere. Gradually the cooling pushed Mars across a threshold at which its surface water froze. Additional cooling followed as water vapor (also a greenhouse gas) condensed from the atmosphere. The initial cooling triggered a feedback mechanism that caused the planet to slip into its present deep freeze. This feedback cooling didn't occur on Earth because Earth is closer to the Sun, receives more solar heat, and never slipped below the threshold at which its surface water and atmosphere froze.

In a recent article, David Kass and Yuk Yung suggest an alternative mechanism for the loss of the Martian atmosphere and its water[1]: The magnetic field of a planet deflects the solar wind—the stream of ions and electrons emitted by the Sun—from the planet's surface. Thus, a planet with a strong magnetic field is well protected from the solar wind. Mars has a weaker magnetic field than that of Earth or Venus, and therefore the solar wind more easily strikes the Martian atmosphere. Early in Martian history, the energetic, charged particles of the solar wind broke carbon dioxide and water molecules apart and blew their fragments into space. With less water vapor and carbon dioxide, the atmosphere's infrared absorbing (greenhouse) properties were diminished, and the atmosphere cooled.

Today, Mars's winter ice caps are mostly frozen carbon dioxide, commonly called dry ice. If water is present on Mars, it lies frozen beneath the planet's surface.

Earth's oceans transport heat, cooling the equatorial regions and warming the poles. Water evaporating from the seas alters the composition of the atmosphere and further moderates global climate. Thus, oceans interact with other systems to create the favorable environments that allow humans and other organisms to thrive on Earth. Ocean currents also weather and erode coastlines, and abundant sedimentary rock forms in shallow seas.

On the early Earth, the evolving oceans and atmosphere were caught up in the same Solar System–wide processes that affected its neighboring planets, Mars and Venus. Those processes, however, had very different results on the three planets.

Earth's orbit lies in a nearly perfect location for life to emerge and thrive. Its distance from the Sun resulted in surface temperatures that allowed Earth to retain its liquid water while Venus lost its water and Mars's water froze. Earth's size and the strength of its magnetic field may also be crucial factors in the maintenance of our oceans and atmosphere. We learn how delicate this balance is by looking at our neighboring planets. If a similar planet—like Venus—warms up slightly, a runaway greenhouse effect can evaporate the oceans and cause atmospheric temperature to rise to a level fatal to life. If an Earth-like planet such as Mars cools slightly, atmospheric feedback mechanisms may escalate until the planet's surface freezes.

On Earth, the primordial oceans formed because the planet lay in a fortuitously ideal orbit. But once formed, the oceans became a critical factor in creating and maintaining a favorable environment for life.

---

[1] David Kass and Yuk Yung, "Loss of Atmosphere from Mars Due to Solar Wind–Induced Sputtering." *Science*, Vol. 268, May 5, 1995, p. 697.

Ocean Basins

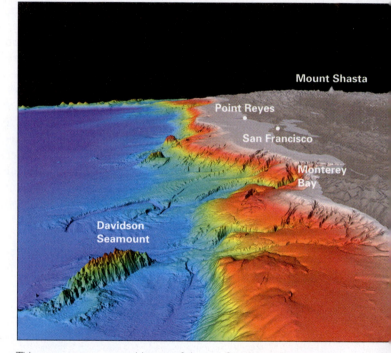

This computer generated image of the sea floor along the California coast is based on sonar mapping, and shows the geologic features in detail. *(William F. Haxby and Lincoln F. Pratson)*

If you were to ask most people to describe the difference between a continent and an ocean, they would almost certainly reply, "Why, obviously a continent is land and an ocean is water!" This observation is true of course, but to a geologist another distinction is more important. He or she would explain that rocks beneath the oceans are different from those of a continent. The accumulation of seawater in the world's ocean basins is a *result* of that difference. ●

13.1 The Earth's Oceans

In the opening essay to Unit III, we described how seawater accumulated on the Earth's surface soon after the planet cooled. Initially, the seas spread uniformly across the surface, forming one global ocean with few or no continents protruding from the watery wilderness. Sometime during the Earth's first half-billion years, partial melting of mantle rock formed the first mini-continents. Over time, tectonic processes replaced the original oceanic crust with new basaltic sea-floor rocks, and enlarged the continents to their present size. Thus, the division of the Earth's surface into ocean basins and continents was not a feature of the primordial Earth, but one created by early tectonic activity. Most Earth scientists think that the ocean basins had filled to their present level by three billion years ago.

Modern oceanic crust is dense basalt and varies from 4 to 7 kilometers thick. Continental crust is made of less dense granite and averages 20 to 40 kilometers thick. In addition, the entire continental lithosphere is both thicker and less dense than oceanic lithosphere. As a result of these differences, the thick, less dense continental lithosphere floats isostatically at high elevations, whereas that of the ocean basins sinks to low elevations. Most of the Earth's water flows downhill to collect in the depressions formed by oceanic lithosphere. Even if no water existed on the Earth's surface, oceanic crust would form deep basins and continental crust would rise to higher elevations.

The Modern Oceans

Oceans cover about 71 percent of the Earth's surface. The sea floor is about 5 kilometers deep in the central parts of the ocean basins, although it is only 2 to 3 kilometers deep above the mid-oceanic ridges, and plunges to 11 kilometers in the Mariana trench (Fig. 13–1).

The ocean basins contain 1.4 billion cubic kilometers of water—18 times more than the volume of all land above sea level. So much water exists at the Earth's surface that if Earth were a perfectly smooth sphere, it would be covered by a global ocean 2000 meters deep.

The size and shape of the Earth's ocean basins change over geologic time. At present, the Atlantic Ocean is growing wider as the sea floor spreads apart at the mid-Atlantic ridge and the Americas move away from Europe and Africa. At the same time, the Pacific is shrinking as oceanic crust sinks into subduction zones around its edges. In short, the Atlantic Ocean basin is now expanding at the expense of the Pacific.

The oceans affect global climate and the biosphere in many ways. The seas reflect and store solar heat differently than do rocks and soil. As a result, oceans are generally warmer in winter and cooler in summer than adjacent land. Most of the water that falls as rain or

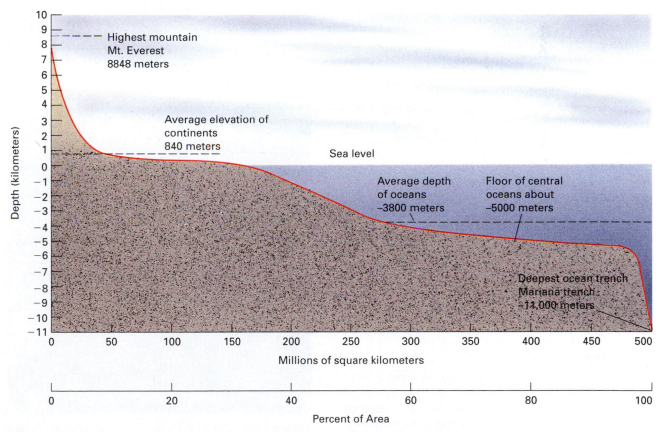

Figure 13–1 A schematic cross-section of the continents and ocean basins. The vertical axis shows elevations relative to sea level. The horizontal axis shows the relative areas of the types of topography. Thus, 30 percent, or about 150 million square kilometers, of the Earth's surface lies above sea level.

Figure 13–2 An oceanographer extracts sediment from a core retrieved from the sea floor. *(Ocean Drilling Program, Texas A&M University)*

snow is water that evaporated from the seas. In addition, ocean currents transport heat from the equator toward the poles, cooling equatorial climates and warming polar environments. Because plate tectonic activities alter the sizes and shapes of ocean basins, they also alter oceanic currents and profoundly affect regional climates over geologic time. In these and other ways, the oceans play a large role in Earth systems interactions.

13.2 Studying the Sea Floor

Seventy-five years ago, scientists had better maps of the Moon than of the sea floor. The Moon is clearly visible in the night sky, and we can view its surface with a telescope. The sea floor, on the other hand, is deep, dark, and inhospitable to humans. Modern oceanographers use a variety of techniques to study the sea floor, including several types of sampling and remote sensing.

Sampling

Several devices collect sediment and rock directly from the ocean floor. A **rock dredge** is an open-mouthed steel net dragged along the sea floor behind a research ship. The dredge breaks rocks from submarine outcrops and hauls them to the surface. Oceanographers sample sea-floor mud by lowering a weighted, hollow steel pipe from a research vessel. The weight drives the pipe into the soft sediment, which is forced into the pipe. The sediment **core** is retrieved from the pipe after it is winched back to the surface. If the core is removed from the pipe carefully, even the most delicate sedimentary layering is preserved.

Sea-floor drilling methods developed for oil exploration also take core samples from oceanic crust. Large drill rigs are mounted on offshore platforms and on research vessels. The drill cuts cylindrical cores from both sediment and rock, which are then brought to the surface for study (Fig. 13–2). Although this type of sampling is expensive, cores can be taken from depths of several kilometers into oceanic crust.

A number of countries, including France, Japan, Russia, and the United States have built small research submarines to carry oceanographers to the sea floor, where they view, photograph, and sample sea-floor rocks, sediment, and deep sea life (Fig. 13–3). More recently, scientists have used deep-diving robots and laser imagers to sample and photograph the sea floor. A robot is cheaper and safer than a submarine, and a laser imager penetrates up to eight times farther through water than a conventional camera.

Remote Sensing

Remote sensing methods do not require direct physical contact with the ocean floor, and for some studies

Figure 13–3 The *Alvin* is a research submarine capable of diving to the sea floor. Scientists on board control robot arms to collect sea-floor rocks and sediment. *(Rod Catanach, Woods Hole Oceanographic Institution)*

Life on the Deep Sea Floor

Oceanographers had long thought that little life could exist on the deep sea floor because no sunlight penetrates to those depths to support photosynthesis. However, scientists using modern diving and sampling techniques have discovered thriving communities and a unique food chain in isolated parts of the deep sea floor.

On the volcanically active mid-oceanic ridge, the hot rocks heat seawater as it circulates through fractures in oceanic crust. The hot water dissolves metals and sulfur from the rocks. Eventually, the hot, metal-and-sulfur-laden water rises back to the sea-floor surface, spouting from fractures as a jet of black water called a black smoker. The black color is caused by precipitation of fine-grained metal sulfide minerals as the solutions cool on contact with seawater (Figure 1).

These scalding, sulfurous waters are as hot as 400°C and would be toxic to life on land. Yet the deep sea floor around a black smoker teems with life (Figure 2). At the vents, bacteria produce energy from hydrogen sulfide in a process called chemosynthesis. Thus, the bacteria release energy from chemicals and are not dependent on photosynthesis. The chemosynthetic bacteria are the foundation of a deep sea food chain—they are either eaten by larger vent organisms or they live symbiotically with them. For example, instead of a digestive tract, the red-tipped tube worm has a special organ that hosts the chemosynthetic bacteria. It provides a home for the bacteria, and in return receives nutrition from the bacteria's wastes. Other vent organisms in this unique food chain include giant clams and mussels, eyeless shrimp, crabs, and fish.

Focus Question:

Discuss similarities and differences between chemosynthesis and photosynthesis.

Figure 1 A black smoker spouts from the East Pacific rise. Seawater is heated as it circulates through hot sea-floor rocks, and dissolves metals and sulfur from the rocks. The ions precipitate as "smoke," consisting of tiny mineral grains, when the hot solution spouts into cold ocean water. *(Dudley Foster, Woods Hole Oceanographic Institution)*

Figure 2 These red tubeworms are part of a thriving plant and animal community living near a black smoker in the Guaymas Basin in the Gulf of California. *(Dudley Foster, Woods Hole Oceanographic Institution)*

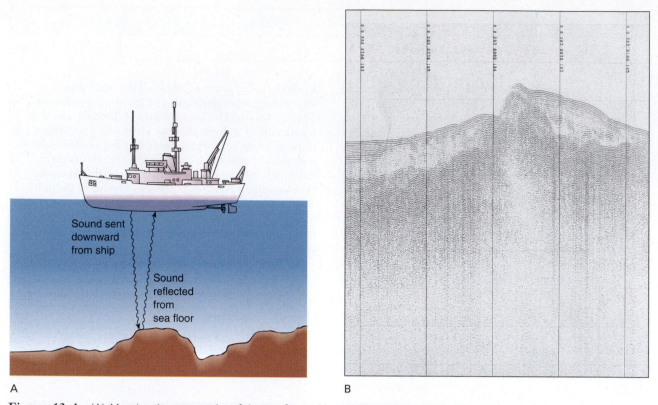

Figure 13–4 (A) Mapping the topography of the sea floor with an echo sounder. A sound signal generated by the echo sounder bounces off the sea floor and back up to the ship, where its travel time is recorded. (B) A seismic profiler records both the sea-floor topography and the layering of sea-floor sediment and rocks. *(Ocean Drilling Program, Texas A&M University)*

this approach is both effective and economical. The **echo sounder** is commonly used to map sea-floor topography. It emits a sound signal from a research ship and then records the signal after it bounces off the sea floor and travels back up to the ship (Fig. 13–4A). The water depth is calculated from the time required for the sound to make the round trip. A topographic map of the sea floor is constructed as the ship steers a carefully navigated course with the echo sounder operating continuously. Modern echo sounders transmit 1000 signals at a time to create more complete and accurate maps.

The **seismic profiler** works in the same way but uses a higher-energy signal that penetrates and reflects from layers in the sediment and rock. This gives a picture of the layering and structure of oceanic crust, as well as the sea-floor topography (Fig. 13–4B and Fig. 13–5).

A **magnetometer** is an instrument that measures a magnetic field. Magnetometers towed behind research ships measure the magnetism of sea-floor rocks.

Satellite-based **microwave radar** instruments have recently been used to measure subtle swells and depressions on the sea surface. These features reflect sea floor topography. For example, the mass of a sea-floor mountain 4000 meters high creates sufficient gravitational attraction to create a gentle 6-meter-high swell on the sea surface directly above it. This technique is used to make modern sea-floor maps (Fig. 13–6).

13.3 Sea-Floor Magnetism

In *Focus On: Alfred Wegener and the Origin of an Idea*, in Chapter 5, we described Wegener's hypothesis that continents had migrated across the globe. Because he was unable to explain *how* continents moved, his theory was ignored by most other scientists. Then, in the 1960s, 30 years after Wegener's death, geologists discovered symmetrical magnetic patterns on the mid-oceanic ridge. Their interpretations of those patterns

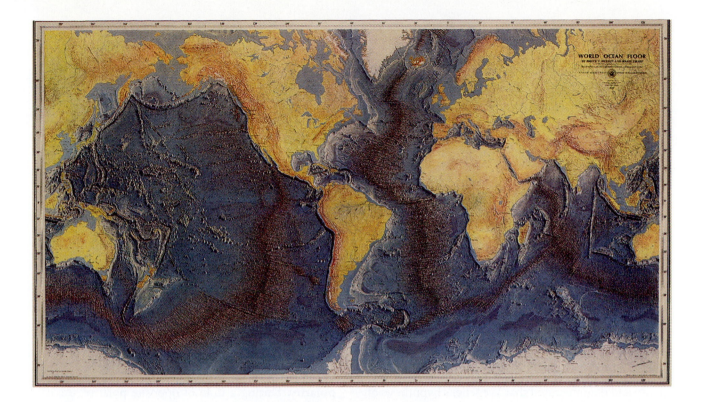

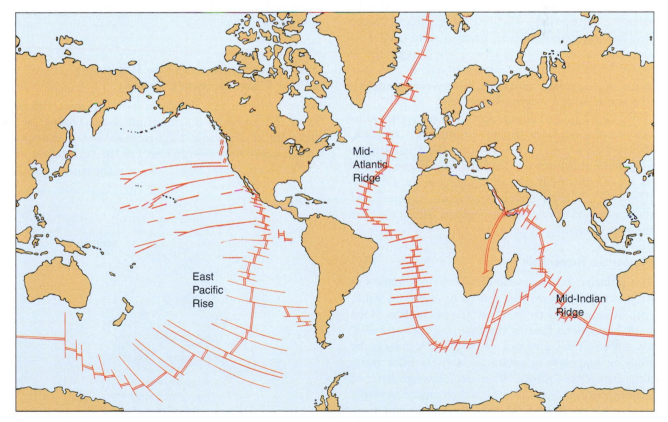

Figure 13–5 An artist rendition of sea-floor topography based on sonar and seismic profiling. *(Marie Tharp)*

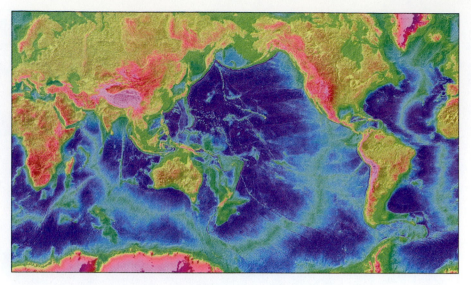

Figure 13–6 The sea floor has as much topographic diversity as the continents. This map was made using satellite-based microwave radar techniques. (W. Smith and D. Sandwell, *Science*, vol. 277, 26 Sept., 1997, pp. 1956–62.)

quickly led to the development of the plate tectonics theory and demonstrated that Wegener's hypothesis of continental drift had been correct.

To understand how magnetic patterns on the sea floor led to the plate tectonics theory, we must consider the relationships between the Earth's magnetic field and magnetism in rocks. Many iron-bearing minerals are permanent magnets. Their magnetism is much weaker than that of a magnet used to stick cartoons on your refrigerator door, but it is strong enough to measure with a magnetometer.

When magma solidifies, certain iron-bearing minerals crystallize and become permanent magnets. When such a mineral cools within the Earth's magnetic field, the mineral's magnetic field aligns parallel to the Earth's field just as a compass needle does (Fig. 13–7). Thus, minerals in an igneous rock record the orientation of the Earth's magnetic field at the time the rock cooled.

Magnetic Reversals

The **polarity** of a magnetic field is the orientation of its positive, or north, end and its negative, or south, end. Because many rocks record the orientation of the Earth's magnetic field at the time the rocks formed, we can construct a record of the Earth's polarity by studying magnetic orientations in rocks from many different ages and places. When geologists constructed such a record, they discovered to their amazement that the Earth's magnetic field has reversed polarity many times throughout geologic history. When a **magnetic reversal** occurs, the north magnetic pole becomes the south magnetic pole, and vice versa. The orientation

of the Earth's field at present is referred to as **normal polarity.** During a time of opposite polarity, the orientation is called **reversed polarity.** The Earth's polarity has reversed about 130 times during the past 65 million years, an average of once every one-half million years.

Sea-Floor Spreading: The Beginning of the Plate Tectonics Theory

As you learned in Chapter 5, most oceanic crust is basalt that formed as magma erupted onto the sea floor from the mid-oceanic ridge system. Basalt contains iron-bearing minerals that record the orientation of the Earth's magnetic field at the time the basalt cooled.

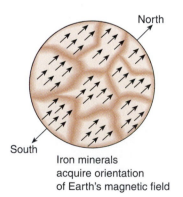

Iron minerals
acquire orientation
of Earth's magnetic field

Figure 13–7 As an igneous rock cools, iron-bearing minerals acquire a permanent magnetic orientation parallel to that of the Earth's field.

Figure 13–8 shows the magnetic polarity of oceanic crust along a portion of the mid-oceanic ridge known as the Reykjanes ridge near Iceland. The black stripes represent rocks with normal polarity, and the intervening stripes represent rocks with reversed polarity. Notice that the stripes form a pattern of alternating normal and reversed polarity, and that the stripes are arranged symmetrically about the axis of the ridge. The central stripe is black, indicating that the rocks of the ridge axis have the same magnetic orientation that the Earth has today.

Shortly after this discovery, geologists suggested that a particular sequence of events created the alternating stripes of normal and reversed polarity in sea-floor rocks:

1. New oceanic crust forms continuously as basaltic magma rises beneath the ridge axis. The new crust then spreads outward from the ridge. This movement is analogous to two broad conveyor belts moving away from one another.

2. As the new crustal basalt cools, it acquires the orientation of the Earth's magnetic field at that time.

3. The Earth's magnetic field reverses orientation periodically.

4. Thus, the magnetic stripes on the sea floor record a succession of reversals in the Earth's magnetic field that occurred as the sea floor spread away from the ridge (Fig. 13–9). To return to our conveyor belt analogy, imagine that a can of white spray paint

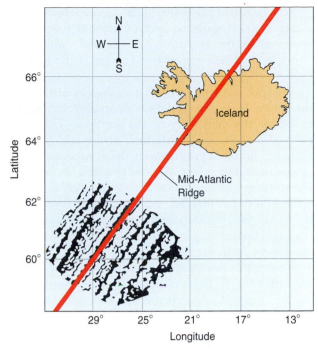

Figure 13–8 The mid-Atlantic ridge, shown in red, runs through Iceland. Magnetic orientations of sea-floor rocks near the ridge are shown in the lower left portion of the map. The black stripes represent sea-floor rocks with normal magnetic polarity, and the intervening stripes represent rocks with reversed polarity. The stripes form a symmetrical pattern of alternating normal and reversed polarity on each side of the ridge. (After Heirtzler *et al.,* 1966, *Deep-Sea Research,* vol. 13.)
● Interactive Question: Explain why the stripes of normal and reversed polarity are not neatly linear as shown in the schematic diagram in Figure 13–9.

Figure 13–9 As new oceanic crust cools at the mid-oceanic ridge, it acquires the magnetic orientation of the Earth's field. Alternating stripes of normal (blue) and reversed (green) polarity record reversals in the Earth's magnetic field that occurred as the crust spread outward from the ridge.

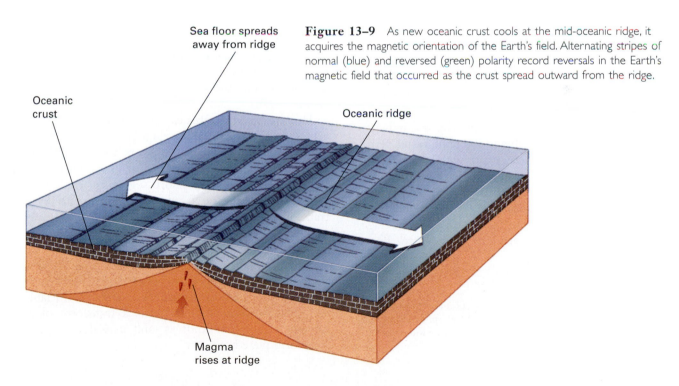

Sea floor spreads away from ridge

Oceanic crust

Oceanic ridge

Magma rises at ridge

were mounted above two black conveyor belts moving apart. If someone sprayed paint at regular intervals as the conveyor belts moved, symmetric white and black stripes would appear on both belts.

At about the same time that geologists discovered the magnetic stripes on the sea floor, they also began to sample the mud lying on the sea floor. They discovered that the mud is thinnest at the mid-Atlantic ridge and becomes progressively thicker at greater distances from the ridge. Mud falls to the sea floor at about the same rate everywhere in the open ocean. It is thinnest at the ridge because the sea floor is youngest there; the mud thickens with increasing distance from the ridge because the sea floor becomes progressively older away from the ridge.

Geologists soon recognized similar magnetic stripes and sediment thickness trends along other portions of the mid-oceanic ridge. As a result, they proposed the hypothesis of **sea-floor spreading** to explain the origin of all oceanic crust.[1] This hypothesis states that new oceanic crust forms at the mid-oceanic ridge system, and continuously spreads outward from the ridge. In a short time, geologists combined Wegener's continental drift hypothesis and the newly developed sea-floor spreading hypothesis to develop the plate tectonics theory. You read in Chapter 5 that this theory explains how and why continents move, mountains rise, earthquakes shake our planet, and volcanoes erupt. It also explains the origin and features of the Earth's largest mountain chain: the mid-oceanic ridge.

13.4 The Mid-Oceanic Ridge

During World War II, naval commanders needed topographic maps of the sea floor to support submarine warfare. Those maps, made with early versions of the echo sounder, were kept secret by the military. When they became available to the public after peace was restored, scientists were surprised to learn that the ocean floor has at least as much topographic diversity and relief as the continents (Fig. 13–5). Broad plains, high peaks, and deep valleys form a varied and fascinating submarine landscape, but the mid-oceanic ridge is the most impressive feature of the deep sea floor.

The mid-oceanic ridge system is a continuous submarine mountain chain that encircles the globe. Its total length exceeds 80,000 kilometers, and it is more than 1500 kilometers wide in places. The ridge rises an average of 2 to 3 kilometers above the surrounding deep sea floor. Although it lies almost exclusively beneath the seas, it is the Earth's largest mountain chain, covering more than 20 percent of the Earth's surface, about two thirds as much as all continents combined. Even the Himalayas, the Earth's largest continental mountain chain, occupy only a small fraction of that area.

A **rift valley** 1 to 2 kilometers deep and several kilometers wide splits many segments of the ridge crest. Oceanographers use small research submarines to dive into the rift valley. They see gaping vertical cracks up to 3 meters wide on the floor of the valley. Recall that the mid-oceanic ridge is a spreading center, where two lithospheric plates are spreading apart from each other. The cracks form as brittle oceanic crust separates at the ridge axis. Basaltic magma then rises through the cracks and flows onto the floor of the rift valley. This basalt becomes new oceanic crust as two lithospheric plates spread outward from the ridge axis.

The new crust (and the underlying lithosphere) at the ridge axis is hot and therefore of relatively low density. Its buoyancy causes it to float high above the surrounding sea floor, elevating the mid-oceanic ridge system 2 to 3 kilometers above the deep sea floor. The new lithosphere cools as it spreads away from the ridge. As a result of cooling, it becomes thicker and denser and sinks to lower elevations, forming the deeper sea floor on both sides of the ridge (Fig. 13–10).

Shallow earthquakes are common at the mid-oceanic ridge because oceanic crust fractures as the two plates separate (Fig. 13–11). Blocks of crust drop downward along the sea-floor cracks, forming the rift valley.

Hundreds of fractures called **transform faults** cut across the rift valley and the ridge (Fig. 13–12). These fractures extend through the entire thickness of the lithosphere. They develop because the mid-oceanic ridge consists of many short segments. Each segment is slightly offset from adjacent segments by a transform fault. Transform faults are original features of the mid-oceanic ridge; they form when lithospheric spreading begins.

Some transform faults displace the ridge by less than a kilometer, but others offset the ridge by hundreds of kilometers. In some cases, a transform fault can grow so large that it forms a transform plate boundary. The San Andreas fault in California is a transform plate boundary.

---

[1] Hess and Dietz proposed the sea-floor spreading hypothesis in 1960, prior to Vine and Matthews' 1963 interpretation of sea-floor magnetic stripes, but the hypothesis received widespread attention only after 1963.

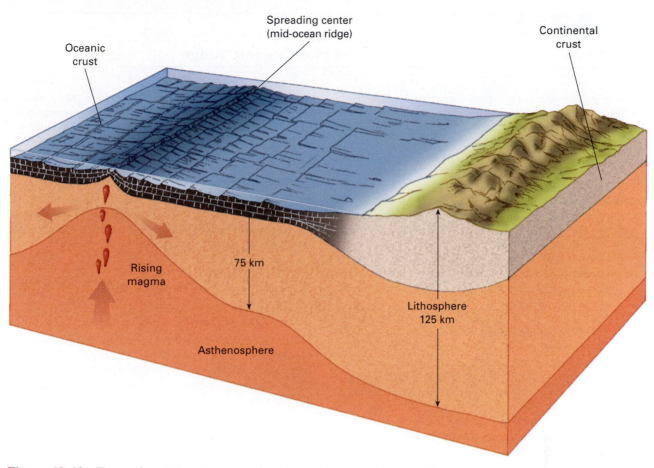

Figure 13–10 The sea floor sinks as it grows older. At the mid-oceanic ridge, new lithosphere is buoyant because it is hot and of low density. It ages, cools, thickens, and becomes denser as it moves away from the ridge and consequently sinks. The central portion of the sea floor lies at a depth of about 5 kilometers. ● **Interactive Question: Explain why the lithosphere thickens as it becomes older and cooler as it migrates from the mid-oceanic ridge.**

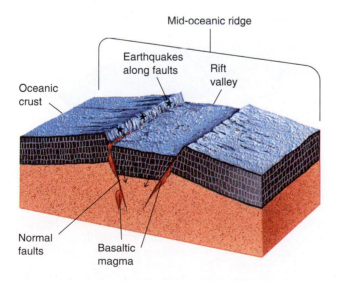

Figure 13–11 A cross-sectional view of the central rift valley in the mid-oceanic ridge. As the plates separate, blocks of rock drop down along the fractures to form the rift valley. The moving blocks cause earthquakes.

Global Sea-Level Changes and the Mid-Oceanic Ridge

A thin layer of marine sedimentary rocks blankets large areas of the Earth's continents. These rocks tell us that those places must have been below sea level when the sediment accumulated.

Tectonic activity can cause a continent to sink, allowing the sea to flood a large area. However, at particular times in the past (most notably during the Cambrian, Carboniferous, and Cretaceous periods), seas flooded low-lying portions of all continents simultaneously. Although our plate tectonics model explains the sinking of individual continents, or parts of continents, it does not explain why all continents should sink at the same time. Therefore, we need to explain how sea level could rise globally by hundreds of meters to flood all continents simultaneously.

The alternating growth and melting of glaciers during the Pleistocene Epoch caused sea level to fluctuate by as much as 200 meters. However, the ages of most marine sedimentary rocks on continents do not coincide with times of glacial melting. Therefore, we must look for a different cause to explain continental flooding.

Recall from Section 13.4 that the new, hot lithosphere at a spreading center is buoyant, causing the mid-oceanic ridge to rise above the surrounding sea floor. This submarine mountain chain displaces a huge volume of seawater. If the mid-oceanic ridge were smaller, it would displace less seawater and sea level would fall. If it were larger, sea level would rise.

The mid-oceanic ridge rises highest at the spreading center, where new lithosphere rock is hottest and has the lowest density. The elevation of the ridge decreases on both sides of the spreading center because the lithosphere cools and shrinks as it moves outward. Now consider a spreading center where spreading is very slow (perhaps 1 to 2 centimeters per year). At such a slow rate, the newly formed lithosphere would cool before it migrated far from the spreading center. As a result, the ridge would be narrow and of low volume, as shown in Figure 1. In contrast, rapid sea-floor spreading of 10 to 20 centimeters per year would create a high-volume ridge because the newly formed, hot lithosphere would be carried a considerable distance away from the spreading center before it cooled and shrank. This high-volume ridge would displace considerably more seawater than a low-volume ridge and would cause a global sea-level rise. If the modern mid-oceanic ridge system were to disappear completely, sea level would fall by about 400 meters.

Sea-floor age data indicate that the rate of sea-floor spreading has varied from about 2 to 16 centimeters per year since Jurassic time, about 200 million years ago. Sea-floor spreading was unusually rapid during Late Cretaceous time, between 110 and 85 million years ago. That rapid spreading should have formed an unusually high-volume mid-oceanic ridge and resulted in flooding of low-lying portions of continents. Geologists have found marine sedimentary rocks of Late Cretaceous age on nearly all continents, indicating that Late Cretaceous time was, in fact, a time of abnormally high global sea level. Unfortunately, because no oceanic crust is older than about 200 million years, the hypothesis cannot be tested for earlier times when extensive marine sedimentary rocks accumulated on continents.

Focus Question:

What processes other than variations in sea-floor spreading rates and growth and melting of glaciers might cause global sea-level fluctuations? What evidence would indicate that other processes caused the fluctuations?

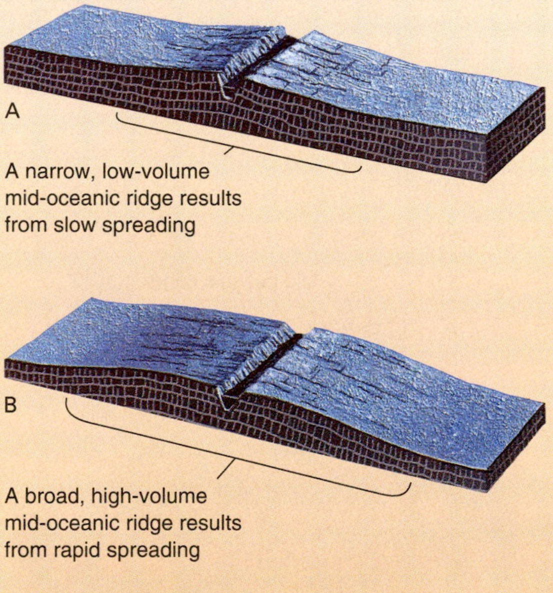

A narrow, low-volume
mid-oceanic ridge results
from slow spreading

A broad, high-volume
mid-oceanic ridge results
from rapid spreading

Figure 1 (A) Slow sea-floor spreading creates a narrow, low-volume mid-oceanic ridge that displaces less seawater and lowers sea level. (B) Rapid sea-floor spreading creates a wide, high-volume ridge that displaces more seawater and raises sea level.

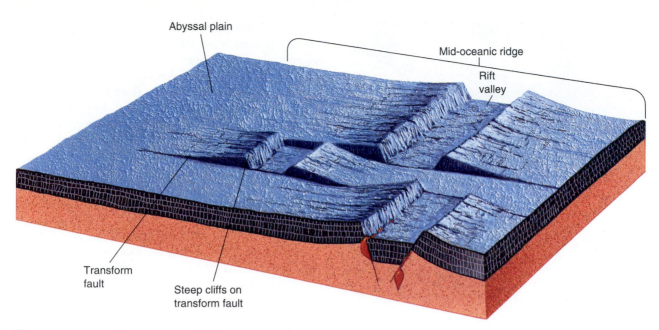

Abyssal plain

Mid-oceanic ridge

Rift valley

Transform fault

Steep cliffs on transform fault

Figure 13–12 Transform faults offset segments of the mid-oceanic ridge. Adjacent segments of the ridge may be separated by steep cliffs 3 kilometers high. Note the flat abyssal plain far from the ridge.

13.5 Sediment and Rocks of the Deep Sea Floor

The Earth is 4.6 billion years old, and rocks as old as 3.96 billion years have been found on continents. Once formed, most continental crust remains near the Earth's surface because of its buoyancy. In contrast, no parts of the sea floor are older than about 200 million years because oceanic crust forms continuously at the mid-oceanic ridge and then recycles into the mantle at subduction zones. Occasionally, old oceanic crust is scraped off the sea floor onto the edge of a continent.

Seismic profiling and sea-floor drilling show that oceanic crust consists of three layers. The uppermost layer consists of sediment, and the lower two are basalt (Fig. 13–13).

Ocean-Floor Sediment

The uppermost layer of oceanic crust, called **layer 1,** consists of two different types of sediment. **Terrigenous sediment** is sand, silt, and clay eroded from the continents and carried to the deep sea floor by gravity and submarine currents. Most of this sediment is found close to the continents. **Pelagic sediment,** on the other hand, collects even on the deep sea floor

Figure 13–13 The three layers of oceanic crust. Layer 1 consists of sediment. Layer 2 is pillow basalt. Layer 3 consists of vertical dikes overlying gabbro. Below layer 3 is the upper mantle.

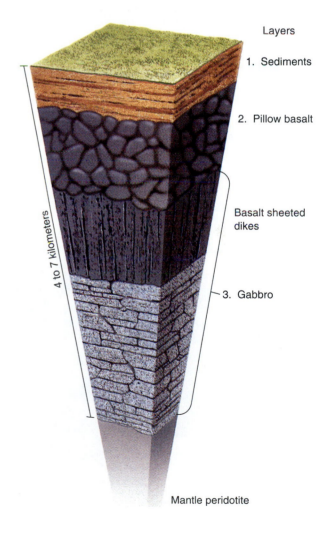

Layers

1. Sediments

2. Pillow basalt

Basalt sheeted dikes

4 to 7 kilometers

3. Gabbro

Mantle peridotite

Figure 13–14 This scanning electron microscope photo shows foraminifera, tiny organisms that float near the surface of the seas. When they die, their remains sink to the sea floor to become part of the pelagic mud layer. *(Texas A&M University Ocean Drilling Program)*

far from continents. It is a gray and red-brown mixture of clay, mostly carried from continents by wind, and the remains of tiny plants and animals that live in the surface waters of the oceans (Fig. 13–14). When these organisms die, their remains slowly settle to the ocean floor.

Pelagic sediment accumulates at a rate of about 2 to 10 millimeters per 1000 years. Near the ridge there is virtually no sediment because the sea floor is so young. The sediment thickness increases with distance from the ridge because the sea floor becomes older as it spreads away from the ridge (Fig. 13–15). Close to shore, pelagic sediment gradually merges with the much thicker layers of terrigenous sediment, which can be 3 kilometers or more thick.

Parts of the ocean floor beyond the mid-oceanic ridge are flat, level, featureless submarine surfaces called the **abyssal plains.** They are the flattest surfaces on Earth. Seismic profiling shows that the basaltic crust is rough and jagged throughout the ocean. On the abyssal plains, however, pelagic sediment buries this rugged profile, forming the smooth abyssal plains. If you were to remove all of the sediment, you would see rugged topography similar to that of the mid-oceanic ridge.

Basaltic Oceanic Crust

Layer 2 lies below layer 1 and is about 1 to 2 kilometers thick. It consists mostly of **pillow basalt,** which forms as hot magma oozes onto the sea floor. Contact with cold seawater causes the molten lava to contract into pillow-shaped spheroids (Fig. 13–16).

Layer 3, 3 to 5 kilometers thick, is the deepest and thickest layer of oceanic crust. It directly overlies the mantle. The upper part consists of vertical basalt dikes, which formed as magma oozing toward the surface froze in the cracks of the rift valley. The lower portion of layer 3 consists of gabbro, the coarse-grained

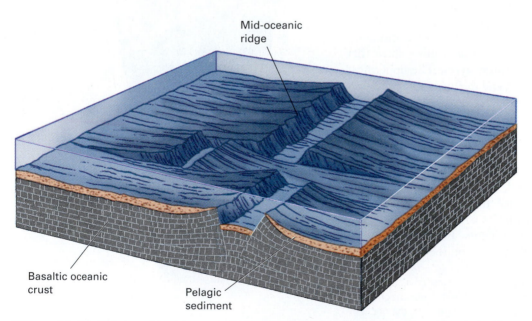

Figure 13–15 Pelagic sediment becomes thicker with increasing distance from the mid-oceanic ridge.

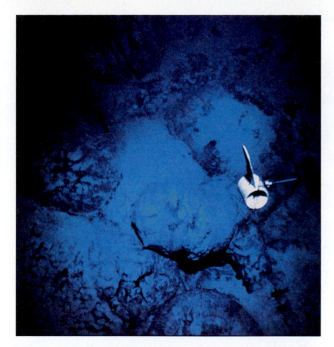

Figure 13–16 Sea-floor pillow basalt is exposed in the Cayman trough. *(Woods Hole Oceanographic Institution)*

equivalent of basalt. The gabbro forms as pools of magma cool slowly because they are insulated by the basalt dikes above them.

The basaltic crust of layers 2 and 3 forms at the mid-oceanic ridge. However, these rocks make up the foundation of all oceanic crust because all oceanic crust forms at the ridge axis and then spreads outward. In some places, chemical reactions with seawater have altered the basalt of layers 2 and 3 to a soft, green rock that contains up to 13 percent water.

13.6 Continental Margins

A **continental margin** is a place where continental crust meets oceanic crust. Two types of continental margins exist. A **passive continental margin** occurs where continental and oceanic crust are firmly joined together. Because it is not a plate boundary, little tectonic activity occurs at a passive margin. Continental margins on both sides of the Atlantic Ocean are passive margins. In contrast, an **active continental margin** occurs at a convergent plate boundary, where oceanic lithosphere sinks beneath the continent in a subduction zone. The west coast of South America is an active continental margin.

Passive Continental Margins

Recall from Chapter 5 that, about 200 million years ago, all of the Earth's continents were joined into the supercontinent called Pangea. Shortly thereafter, Pangea began to rift apart into the continents as we know them today. The Atlantic Ocean opened as the east coast of North America separated from Europe and Africa. As Pangea broke up, the continental crust fractured and thinned near the fractures (Fig. 13–17A). Basaltic magma rose at the new spreading center, forming oceanic crust between North America and Africa (Fig. 13–17B). All tectonic activity then centered at the spreading mid-Atlantic ridge, and no further tectonic activity occurred at the continental margins; hence the term *passive* continental margin (Fig. 13–17C).

The Continental Shelf On all continents, streams and rivers deposit sediment on coastal deltas, like the Mississippi River delta. Then, ocean currents redistribute the sediment along the coast, depositing it both on the thin margin of continental crust and on oceanic crust close to the continent. The sediment forms a shallow, gently sloping submarine surface called a **continental shelf** on the edge of the continent (Fig. 13–18). As sediment accumulates on a continental shelf, the edge of the continent sinks isostatically because of the added weight. This effect keeps the shelf slightly below sea level.

Over millions of years, thick layers of sediment accumulated on the passive east coast of North America, forming a broad continental shelf along the entire coast. The depth of the shelf increases gradually from the shore to about 200 meters at the outer shelf edge. The average inclination of the continental shelf is about 0.1°. A continental shelf on a passive margin can be a large feature. The shelf off the coast of southeastern Canada is about 500 kilometers wide, and parts of the shelves of Siberia and northwestern Europe are even wider.

In some places, a supply of sediment may be lacking, either because no rivers bring sand, silt, or clay to the shelf or because ocean currents bypass that area. In warm regions where sediment does not muddy the water, reef-building organisms thrive. As a result, thick beds of limestone accumulate in tropical and subtropical latitudes where clastic sediment is lacking. Limestone accumulations of this type may be hundreds of meters thick and hundreds of kilometers across and are called **carbonate platforms.** The Florida Keys and the Bahamas are modern-day examples of carbonate platforms on continental shelves (Fig. 13–19).

Some of the world's richest petroleum reserves occur on the continental shelves of the North Sea

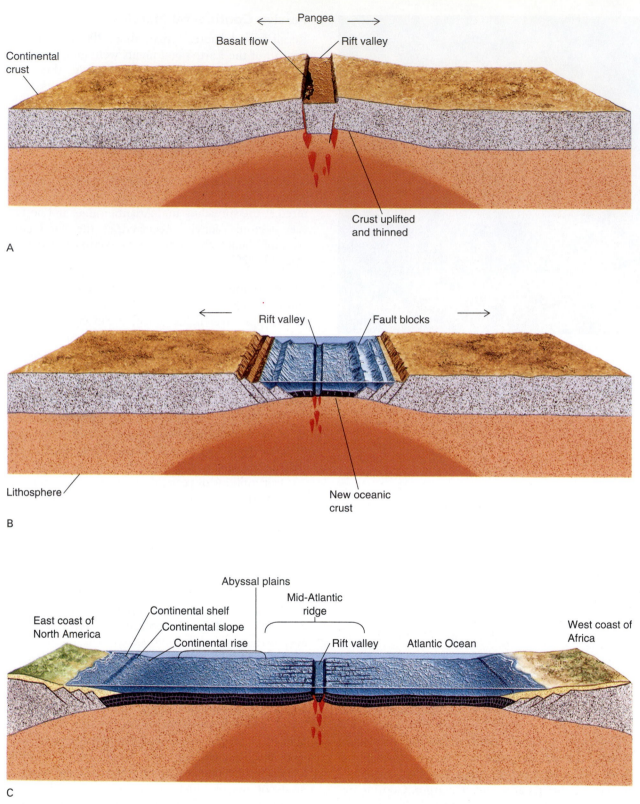

Figure 13–17 (A) Continental crust fractured as Pangea began to rift. (B) Faulting and erosion thinned the crust as it separated. Rising basaltic magma formed new oceanic crust in the rift zone. (C) Sediment eroded from the continents formed broad continental shelves on the passive margins of North America and Africa. ● **Interactive Question: Draw a new diagram; (D) showing a cross-sectional view of North America reversing direction and migrating eastward.**

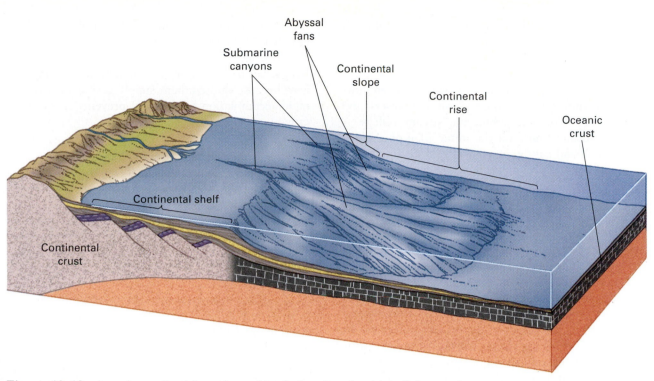

Abyssal fans

Submarine canyons

Continental slope

Continental rise

Oceanic crust

Continental shelf

Continental crust

Figure 13–18 A passive continental margin consists of a broad continental shelf, slope, and rise formed by accumulation of sediment eroded from the continent.

between England and Scandinavia, in the Gulf of Mexico, and in the Beaufort Sea on the northern coast of Alaska and western Canada. In recent years, oil companies have explored and developed these offshore reserves. Deep drilling has revealed that granitic continental crust lies beneath the sedimentary rocks, confirming that the continental shelves are truly parts of the continents despite the fact that they are covered by seawater.

Figure 13–19 Reef-building organisms form large limestone accumulations called carbonate platforms adjacent to many tropical islands. This satellite view shows the great Bahama Bank off Andros Island in the Bahamas. *(NASA/Science Photo Library)*

The Continental Slope and Rise At the outer edge of a shelf, the sea floor suddenly steepens to an average slope of about 4° to 5° as it falls away from 200 meters to about 5 kilometers in depth. This steep region of the sea floor averages about 50 kilometers wide and is called the **continental slope.** It is a surface formed by sediment accumulation, much like the shelf. Its steeper angle is due primarily to thinning of continental crust where it nears the junction with oceanic crust. Seismic profiler exploration shows that the sedimentary layering is commonly disrupted where sediment has slumped and slid down the steep incline.

A continental slope becomes less steep as it gradually merges with the deep ocean floor. This region, called the **continental rise,** consists of an apron of terrigenous sediment that was transported across the continental shelf and deposited on the deep ocean floor at the foot of the slope. The continental rise averages a few hundred kilometers wide. Typically, it joins the deep sea floor at a depth of about 5 kilometers.

In essence, then, the shelf-slope-rise complex is a smoothly sloping submarine surface on the edge of a continent, formed by accumulation of sediment eroded from the continent.

Submarine Canyons and Abyssal Fans　In many places, sea-floor maps show deep valleys called **submarine canyons** eroded into the continental shelf and slope. They look like submarine stream valleys. A canyon typically starts on the outer edge of a continental shelf and continues across the slope to the rise. At its lower end, a submarine canyon commonly leads into an **abyssal fan** (or **submarine fan**), a large, fan-shaped pile of sediment lying on the continental rise.

Most submarine canyons occur where large rivers enter the sea. When they were first discovered, geologists thought the canyons had been eroded by rivers during the Pleistocene Epoch, when accumulation of glacial ice on land lowered sea level by as much as 150 meters. However, this explanation cannot account for the deeper portions of submarine canyons cut into the lower continental slopes at depths of a kilometer or more. Therefore, the deeper parts of the submarine canyons must have formed under water, and a submarine mechanism must be found to explain them.

Geologists subsequently discovered that **turbidity currents** erode the continental shelf and slope to create the submarine canyons. A turbidity current develops when loose, wet sediment tumbles down the slope in a submarine landslide. The movement may be triggered by an earthquake or simply by oversteepening of the slope as sediment accumulates. When the sediment starts to move, it mixes with water. Because the mixture of sediment and water is denser than water alone, it flows down the shelf and slope as a turbulent, chaotic avalanche. A turbidity current can travel at speeds greater than 100 kilometers per hour and for distances up to 700 kilometers.

Sediment-laden water traveling at such speed has tremendous erosive power. Once a turbidity current cuts a small channel into the shelf and slope, subsequent currents follow the same channel, just as a stream uses the same channel year after year. Over time, the currents erode a deep submarine canyon into the shelf and slope. Turbidity currents slow down when they reach the deep sea floor. The sediment accumulates there to form an abyssal fan. Most submarine canyons and fans form near the mouths of large rivers because the rivers supply the great amount of sediment needed to create turbidity currents.

Large abyssal fans form only on passive continental margins. They are uncommon at active margins because in that environment, the sediment is swallowed by the trench. Furthermore, most of the world's largest rivers drain toward passive margins. The largest known fan is the Bengal fan, which covers about 4 million square kilometers beyond the mouth of the

Ganges River in the Indian Ocean east of India. More than half of the sediment eroded from the rapidly rising Himalayas ends up in this fan. Interestingly, the Bengal fan has no associated submarine canyon, perhaps because the sediment supply is so great that the rapid accumulation of sediment prevents erosion of a canyon.

Active Continental Margins

An active continental margin forms in a subduction zone, where an oceanic plate converges with a continent. The oceanic plate sinks into the mantle, forming a long, narrow, steep-sided depression called an **oceanic trench** (Fig. 13–20). A trench can form wherever subduction occurs—where oceanic crust sinks beneath the edge of a continent, or where it sinks beneath another oceanic plate.

Recall that at a passive margin, the continental crust is stretched and thinned and merges gradually with oceanic crust. There is no equivalent thinning at an active margin. Instead, the thick continental crust meets the thin oceanic crust at the subduction zone. As a result, an active margin commonly has a narrow continental shelf. The landward wall (the side toward the continent) of the trench is the continental slope of an active margin. It typically inclines at 4° or 5° in its upper part and steepens to 15° or more near the bottom of the trench. The continental rise is absent because sediment flows into the trench instead of accumulating on the ocean floor.

13.7　Oceanic Trenches and Island Arcs

In many parts of the Pacific Ocean and in some other ocean basins, two oceanic plates converge. One dives beneath the other, forming a subduction zone and an oceanic trench. The deepest place on Earth is in the Mariana trench, north of New Guinea in the southwestern Pacific, where the ocean floor sinks to nearly 11 kilometers below sea level. Depths of 8 to 10 kilometers are common in other trenches.

Huge amounts of magma are generated in the subduction zone. The magma rises and erupts on the sea floor to form submarine volcanoes next to the trench. The volcanoes eventually grow to become a chain of islands, called an **island arc** (Fig. 13–21). The western Aleutian Islands are an example of an island arc. Many others occur at the numerous convergent plate boundaries in the southwestern Pacific (Fig. 13–22).

If subduction stops after an island arc forms, volcanic activity also ends. The island arc may then ride quietly on a tectonic plate until it arrives at another subduction zone at an active continental margin. However, the density of island arc rocks is relatively

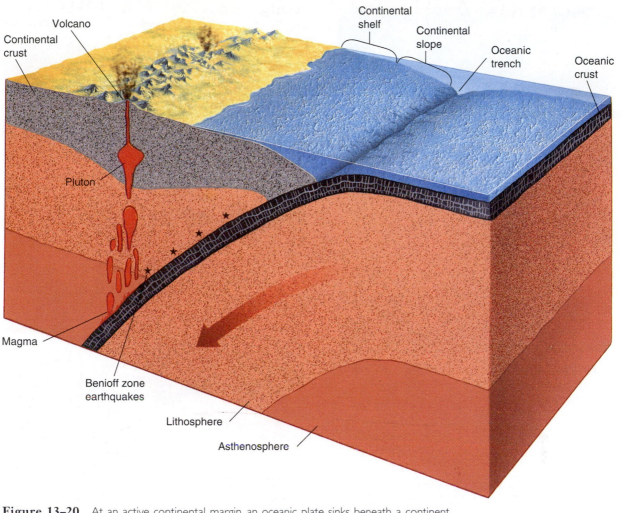

Figure 13–20 At an active continental margin, an oceanic plate sinks beneath a continent, forming an oceanic trench. The continental shelf is narrow, the slope is steep, and the continental rise is nonexistent.

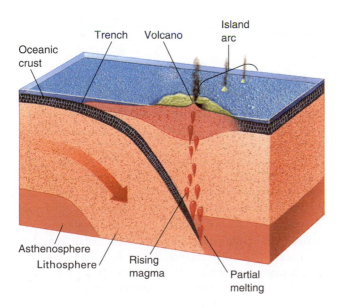

low, making them too buoyant to sink into the mantle. Instead, the island arc collides with the continent (Fig. 13–23). When this happens, the subducting plate commonly fractures on the seaward side of the island arc to form a new subduction zone. In this way, the island arc breaks away from the ocean plate and becomes part of the continent. Much of western California, Oregon, Washington, and western British Columbia were added to North America in this way from 180 million to about 50 million years ago. These late additions to our continent, called **accreted terranes,** are shown in Figure 13–24.

Figure 13–21 An oceanic trench forms at a convergent boundary between two oceanic plates. One of the plates sinks, generating magma that rises to form a chain of volcanic islands called an island arc. ● **Interactive Question:** Explain why the rock that comprises an island arc is generally different from (and less dense than) the surrounding oceanic crust.

Figure 13–22 Mataso is one of many volcanic islands in the Vanuatu island arc that formed along the Northern New Hebrides trench in the South Pacific.

Note the following points:

1. An island arc forms as magma rises from the mantle at an oceanic subduction zone.

2. The island arc eventually migrates toward the edge of a continent and becomes part of it.

3. The continent with the added island arc cannot sink into the mantle at a subduction zone because of its buoyancy.

4. Thus, material is transferred from the mantle to a continent. This aspect of the plate tectonics model suggests that the amount of continental crust has increased throughout geologic time. However, some geologists feel that small amounts of continental crust can also return to the mantle in subduction zones, and that the total amount of continental crust has been approximately constant for the past 2.5 billion years.

13.8 Seamounts, Oceanic Islands, and Atolls

A **seamount** is a submarine mountain that rises 1 kilometer or more above the surrounding sea floor. An **oceanic island** is a seamount that rises above sea level.

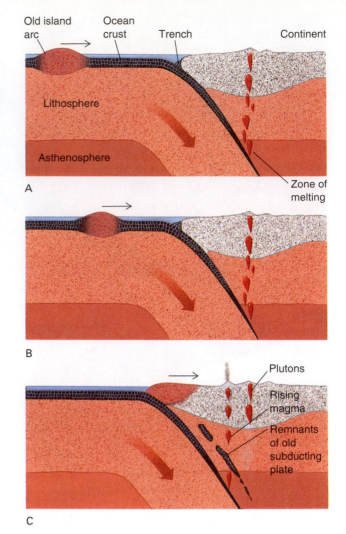

Figure 13–23 (A) An island arc is part of a lithospheric plate that is sinking into a subduction zone beneath a continent. (B) The island arc reaches the subduction zone but cannot sink into the mantle because of its low density. (C) The island arc is jammed onto the continental margin and becomes part of the continent. The subduction zone and trench step back to the seaward side of the island arc.

Both are common in all ocean basins but are particularly abundant in the southwestern Pacific Ocean. Seamounts and oceanic islands sometimes occur as isolated peaks on the sea floor, but they are more commonly found in chains. Dredge samples show that seamounts, oceanic islands, and the ocean floor itself are all made of basalt.

Most seamounts and oceanic islands are volcanoes that formed at a hot spot above a mantle plume, and most form within a tectonic plate rather than at a plate boundary. An isolated seamount or short chain of small seamounts probably formed over a plume that lasted for only a short time. In contrast, a long chain

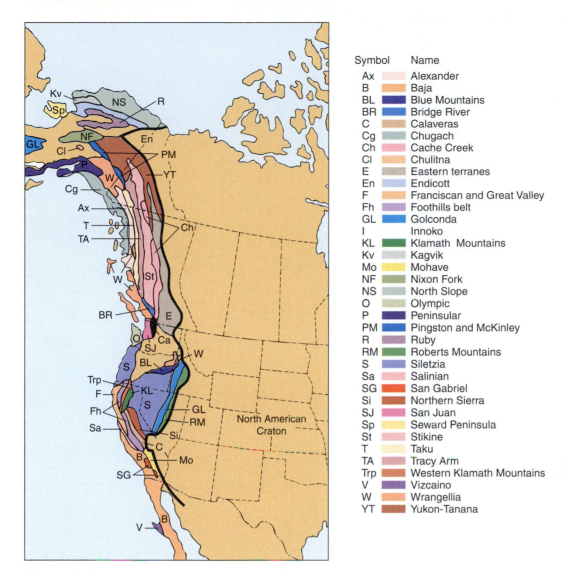

Figure 13–24 The accreted terranes of western North America are microcontinents and island arcs from the Pacific Ocean that were added to the continent.

Symbol	Name
Ax	Alexander
B	Baja
BL	Blue Mountains
BR	Bridge River
C	Calaveras
Cg	Chugach
Ch	Cache Creek
Cl	Chulitna
E	Eastern terranes
En	Endicott
F	Franciscan and Great Valley
Fh	Foothills belt
GL	Golconda
I	Innoko
KL	Klamath Mountains
Kv	Kagvik
Mo	Mohave
NF	Nixon Fork
NS	North Slope
O	Olympic
P	Peninsular
PM	Pingston and McKinley
R	Ruby
RM	Roberts Mountains
S	Siletzia
Sa	Salinian
SG	San Gabriel
Si	Northern Sierra
SJ	San Juan
Sp	Seward Peninsula
St	Stikine
T	Taku
TA	Tracy Arm
Trp	Western Klamath Mountains
V	Vizcaino
W	Wrangellia
YT	Yukon-Tanana

of large islands, such as the Hawaiian Island–Emperor Seamount chain, formed over a long-lasting plume. In this case the lithospheric plate migrated over the plume as the magma continued to rise from a source beneath the lithosphere. Each volcano formed directly over the plume and then became extinct as the moving plate carried it away from the plume. As a result, the seamounts and oceanic islands become progressively younger toward the end of the chain that is volcanically active today (Fig. 13–25).

After a volcanic island forms, it begins to sink. Three factors contribute to the sinking:

1. If the mantle plume stops rising, it stops producing magma. Then the lithosphere beneath the island cools and becomes denser, and the island sinks.

Alternatively, a moving plate may carry the island away from the hot spot. This also results in cooling, contraction, and sinking of the island.

2. The weight of the newly formed volcano causes isostatic sinking.

3. Erosion lowers the top of the volcano.

These three factors gradually transform a volcanic island to a seamount (Fig. 13–26). If the Pacific Ocean plate continues to move at its present rate, the island of Hawaii may sink beneath the sea within 10 to 15 million years. Sea waves may erode a flat top on a sinking island, forming a flat-topped seamount called a **guyot** (Fig. 13–27).

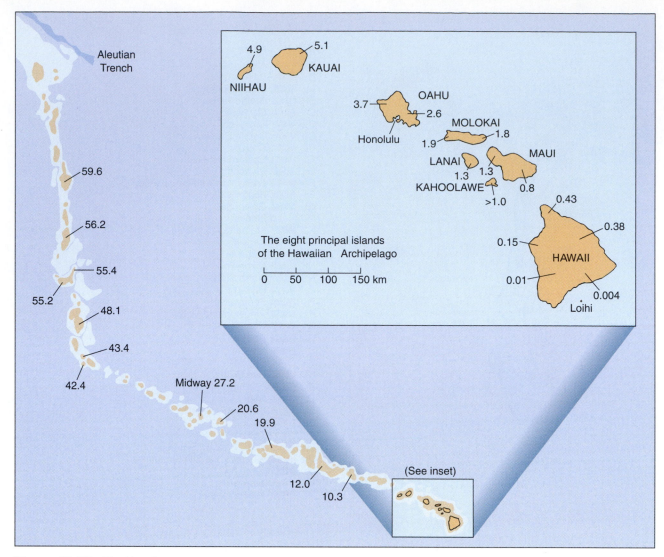

Figure 13–25 The Hawaiian Island—Emperor Seamount chain becomes older in a direction going away from the island of Hawaii. The ages, in millions of years, are for the oldest volcanic rocks of each island or seamount.

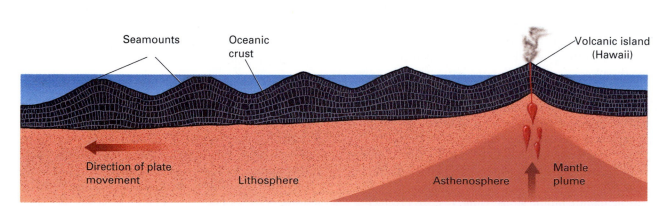

Figure 13–26 The Hawaiian Islands and Emperor Seamounts sink as they move away from the mantle plume.

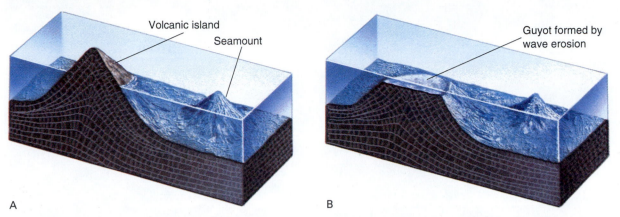

A   B

Figure 13–27 (A) A volcanic island rises above sea level. (B) Waves erode a flat top on a sinking island to form a guyot.

The South Pacific and portions of the Indian Ocean are dotted with numerous islands called atolls. An **atoll** is a circular coral reef that forms a ring of islands around a central lagoon. Atolls vary from 1 to 130 kilometers in diameter and are surrounded by deep water of the open sea. If corals live only in shallow water, how did atolls form in the deep sea? Charles Darwin studied this question during his famous voyage on the *Beagle* from 1831 to 1836. He reasoned that a coral reef must have formed in shallow water on the flanks of a volcanic island. Eventually the island sank, but the

reef continued to grow upward, so that the living portion always remained in shallow water (Fig. 13–28). This proposal was not accepted at first because scientists could not explain how a volcanic island could sink. However, when scientists drilled into a Pacific atoll shortly after World War II and found volcanic rock hundreds of meters beneath the reef, Darwin's hypothesis was revived. It is considered accurate today, in light of our ability to explain why volcanic islands sink (Fig. 13–29).

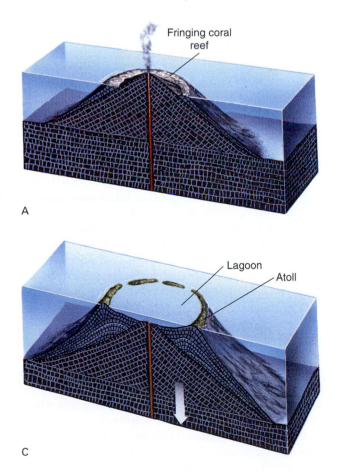

A

C

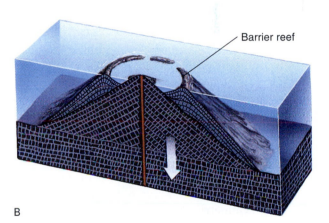

B

Figure 13–28 (A) A fringing reef grows along the shore of a young volcanic island. (B) As the island sinks, the reef continues to grow upward to form a barrier reef that encircles the island. (C) Finally the island sinks below sea level and the reef forms a circular atoll.

Figure 13–29 The Tetiaroa Atoll in French Polynesia formed by the process described in Figure 13–29. Over time, storm waves wash coral sands on top of the reef and vegetation grows on the sand, forming the individual islands in the atoll. *(Tahiti Tourist Board)*

SUMMARY

Continental lithosphere is relatively thick and of relatively low density, whereas oceanic lithosphere is both thinner and of higher density. Consequently, continents float isostatically to high elevations. In contrast, the ocean basins form topographic depressions on the Earth's surface, which fill with water to form oceans. Because of the great depth and remoteness of the ocean floor and oceanic crust, knowledge of them comes mainly from **sampling** and **remote sensing.**

Stripes of normal and reversed **magnetic polarity** that are symmetrically distributed about the mid-oceanic ridge gave rise to the hypothesis of **sea-floor spreading,** which rapidly evolved into the modern plate tectonics theory.

The mid-oceanic ridge is a submarine mountain chain that extends through all of the Earth's major ocean basins. A **rift valley** runs down the center of many parts of the ridge, and the ridge and rift valley are both offset by numerous **transform faults.** The mid-oceanic ridge forms at a spreading center where new oceanic crust is added to the sea floor.

Abyssal plains are flat areas of the deep sea floor where the rugged topography of the basaltic oceanic crust is covered by deep sea sediment. Oceanic crust varies from about 4 to 7 kilometers thick and consists of three layers. The top layer is sediment, which varies from zero to 3 or more kilometers thick. Beneath this lies about 1 to 2 kilometers of **pillow basalt.** The deepest layer of oceanic crust is from 3 to 5 kilometers thick and consists of basalt dikes on top of gabbro. The base of this layer is the boundary between oceanic crust and mantle. The age of sea-floor rocks increases regularly away from the mid-oceanic ridge. No oceanic crust is older than about 200 million years because it recycles into the mantle at subduction zones.

A **passive continental margin** includes a **continental shelf,** a **slope,** and a **rise** formed by accumulation of **terrigenous** sediment. **Submarine canyons** eroded by **turbidity currents** notch continental margins and commonly lead into **abyssal fans,** where the turbidity currents deposit sediments on the continental rise. An **active continental margin,** where oceanic crust sinks in a subduction zone beneath the margin of a continent, usually includes a narrow continental shelf and a continental slope that steepens abruptly into an **oceanic trench.** Trenches are the deepest parts of ocean basins.

Island arcs are common features of some ocean basins, particularly the southwestern Pacific. They are chains of volcanoes formed at subduction zones where two oceanic plates collide. **Seamounts** and **oceanic islands** form in oceanic crust as a result of volcanic activity over mantle plumes. An **atoll** is a circular coral reef growing on a sinking volcanic island.

rock dredge 290
core 290
sea-floor drilling 290
echo sounder 292
seismic profiler 292
magnetometer 292
microwave radar 292
polarity 294
magnetic reversal 294
normal polarity 294

reversed polarity 294
sea-floor spreading 296
rift valley 296
transform fault 296
layer 1 299
terrigenous sediment 299
pelagic sediment 299
abyssal plain 300
layer 2 300
pillow basalt 300

layer 3 300
passive continental margin 301
active continental margin 301
continental shelf 301
carbonate platform 301
continental slope 303
continental rise 303
submarine canyon 304

abyssal fan (submarine fan) 304
turbidity current 304
oceanic trench 304
island arc 304
accreted terrane 305
seamount 306
oceanic island 306
guyot 307
atoll 309

FOR REVIEW

1. Describe the main differences between oceans and continents.

2. Describe a magnetic reversal.

3. Explain how an igneous rock preserves evidence of the orientation of the Earth's magnetic field at the time the rock formed.

4. Describe how the discovery of magnetic patterns on the sea floor confirmed the sea-floor spreading theory.

5. Sketch a cross-section of the mid-oceanic ridge, including the rift valley.

6. Describe the dimensions of the mid-oceanic ridge.

7. Explain why the mid-oceanic ridge is topographically elevated above the surrounding ocean floor. Why does its elevation gradually decrease away from the ridge axis?

8. Explain the origin of the rift valley in the center of the mid-oceanic ridge.

9. Why are the abyssal plains characterized by such low relief?

10. Sketch a cross-section of oceanic crust from a deep sea basin. Label, describe, and indicate the approximate thickness of each layer.

11. Describe the two main types of sea-floor sediment. What is the origin of each type?

12. Compare the ages of oceanic crust with the ages of continental rocks. Why are they so different?

13. Sketch a cross-section of both an active continental margin and a passive continental margin. Label the features of each. Give approximate depths below sea level of each of the features.

14. Explain why a continental shelf is made up of a foundation of granitic crust, whereas the deep ocean floor is composed of basalt.

15. Why does an active continental margin typically have a steeper continental slope than a passive margin?

16. Why does an active margin typically have no continental rise?

17. Explain the relationships among submarine canyons, abyssal fans, and turbidity currents.

18. Why are turbidity currents often associated with earthquakes or with large floods in major rivers?

19. Explain the role played by an island arc in the growth of a continent.

20. Explain the origins of and differences between seamounts and island arcs.

21. Compare the ocean depths adjacent to an island arc and a seamount.

22. Why do oceanic islands sink after they form?

23. Explain how guyots and atolls form.

FOR DISCUSSION

1. The east coast of South America has a wide continental shelf, whereas the west coast has a very narrow shelf. Discuss and explain this contrast.

2. Seismic data indicate that continental crust thins where it joins oceanic crust at a passive continental margin, such as on the east coast of North America. Other than that, we know relatively little about the nature of the junction between the two types of crust. Speculate on the nature of that junction. Consider rock types, geologic structures, ages of rocks, and other features of the junction.

3. Discuss the topography of the Earth in an imaginary scenario in which all conditions are identical to present ones except that there is no water. In contrast, what would be the effect if there were enough water to cover all of the Earth's surface?

4. In Section 13.6 we stated that most of the world's largest rivers drain toward passive continental margins. Explain this observation.

Oceans and Coastlines

Seals nap on a sunny beach near Big Sur, California.

Water was abundant throughout the early Solar System, yet today, Earth is the only planet with both solid continents and liquid seas. No oceans exist on Mercury, Venus, and Mars. No continents exist on Jupiter, Saturn, Uranus, and Neptune. Subterranean oceans may exist on Jupiter's moon, Europa, but there is little liquid water at the surface. Pluto is frozen solid. On Earth, ocean currents carry heat across the globe. Without this exchange, the equator would be unbearably hot and ocean water would evaporate rapidly from equatorial seas. At the same time, the higher latitudes would be frigid and glaciers would spread across the land. Thus, the oceans temper the Earth's climate and help make our planet favorable for life.

The seashore is an attractive place to live or visit. Because the ocean moderates temperature, coastal regions are cooler in summer and warmer in winter than continental interiors. Vacationers and residents sail, swim, surf, and fish along the shore. In addition, the sea provides both food and transportation. For all of these reasons, coastlines have become heavily urbanized and industrialized. About 60 percent of the world's human population lives within 100 kilometers of the coast.

However, coastlines are also among the most geologically active environments on Earth. Rivers deposit great amounts of sediment on coastal deltas. Waves and currents erode the shore and transport sediment. Converging tectonic plates buckle many coastal regions, creating mountain ranges, earthquakes, and volcanic eruptions. Over geologic time, sea level rises and falls, alternately flooding shallow parts of continents, and stranding beaches high above sea level. •

THE OCEANS

14.1 Geography of the Oceans

All of the Earth's oceans are connected, and water flows from one to another, so in one sense the Earth has just one global ocean. However, several distinct ocean basins exist (Fig. 14–1). The largest and deepest ocean is the Pacific. It covers one third of Earth's surface, more than all land combined, and contains more than half of the world's water. The Atlantic Ocean has about half the surface area of the Pacific. The Indian Ocean is slightly smaller than the Atlantic. The Arctic Ocean surrounds the North Pole and extends southward to the shores of North America, Europe, and Asia. Therefore, it is bounded by land, with only a few straits and channels connecting it to the Atlantic and Pacific oceans. The surface of the Arctic Ocean freezes in winter, and parts of it melt for a few months during summer and early fall (Fig. 14–2). The Antarctic Ocean, feared by sailors for its cold and

Figure 14–1 The oceans of the world. The "seven seas" are the North Atlantic, South Atlantic, North Pacific, South Pacific, Indian, Arctic, and Antarctic. However, these designations are related more closely to commerce than to the geology and oceanography of the ocean basins. Geologists and oceanographers recognize four major ocean basins: the Atlantic, Pacific, Indian, and Arctic. *(Tom Van Sant, Inc., The GeoSphere Project, Santa Monica, California)*

Figure 14–2 Ice floes cover large parts of the Arctic Ocean, even in summer.

ferocious winds, has no sharp northern boundary. The northernmost limit of the Antarctic Ocean is the zone where warm currents from the north converge with cold Antarctic water.

14.2 Sea Water

Salinity

The **salinity** of sea water is the total quantity of dissolved salts, expressed as a percentage. In this case, the term "salts" refers to all dissolved ions, not only sodium and chloride.

Dissolved ions make up about 3.5 percent of the weight of ocean water. The six ions listed in Figure 14–3 make up 99 percent of the ocean's dissolved material. However, almost every element found on land is also found dissolved in sea water, albeit mainly in trace amounts. For example, sea water contains about 0.000000004 (4×10^{-9}) percent gold. Although the concentration is small, the oceans are large, and therefore contain a lot of gold. About 4.4 kilograms of gold are dissolved in each cubic kilometer of sea water. Since the oceans contain about 1.3 billion cubic kilometers of water, about 5.7 billion kilograms of gold exist in the oceans. Unfortunately, it would be hopelessly expensive to extract even a small portion of this amount. In addition to trace elements and salts, sea water also contains dissolved gases, especially oxygen and carbon dioxide.

Although the average salinity of the oceans is 3.5 percent, salinity varies geographically. High rainfall near the equator dilutes sea water to 3.45 percent. Conversely, in dry subtropical regions, where evaporation is high and precipitation low, salinity can be as high as 3.6 percent (Fig. 14–4). Salinity varies more dramatically along coastlines. The Baltic Sea is a shallow ocean basin fed by many large fresh water rivers and diluted further by rain and snow. As a result, its salinity is as low as 2.0 percent. In contrast, the Persian Gulf has low rainfall, high evaporation, and few large inflowing rivers, and its salinity exceeds 4.2 percent.

The world's rivers carry more than 2.5 billion tons of dissolved salts to the oceans every year. Underwater volcanoes contribute additional dissolved ions. How-

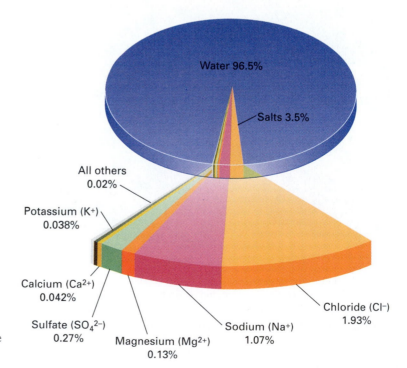

Figure 14–3 Six common ions form most of the salts in sea water.

Water 96.5%

Salts 3.5%

All others 0.02%

Potassium (K+) 0.038%

Calcium (Ca2+) 0.042%

Sulfate (SO4²⁻) 0.27%

Magnesium (Mg2+) 0.13%

Sodium (Na+) 1.07%

Chloride (Cl⁻) 1.93%

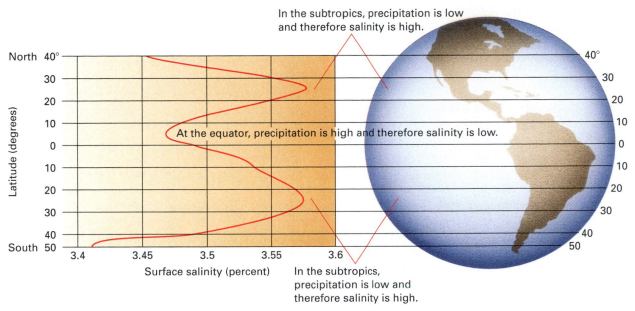

In the subtropics, precipitation is low and therefore salinity is high.

At the equator, precipitation is high and therefore salinity is low.

In the subtropics, precipitation is low and therefore salinity is high.

Figure 14–4 The salinity of the surface of the central oceans changes with latitude.

ever, the salinity of the oceans has been relatively constant throughout much of geologic time because salt has been removed from sea water at the same rate at which it has been added. When a portion of a marine basin becomes cut off from the open oceans, the water evaporates, precipitating thick sedimentary beds of salt. Additionally, large amounts of salt become incorporated into shale and other sedimentary rocks.

Temperature

Recall from Chapter 10 that lakes are comfortably warm for swimming during the summer because warm water floats on the surface and does not readily mix with the deep, cooler water. Oceans also develop a layered temperature profile, but because surface water is stirred by waves and currents, the warm layer in an ocean extends from the surface to a depth as great as 450 meters. Below this layer, temperature drops rapidly with depth in the thermocline. The thermocline extends to a depth of 2 kilometers. Beneath the thermocline, the temperature of ocean water varies from about 1°C to 2.5°C. The cold, dense water in the ocean depths mixes very little with the surface. Thus, there are three distinct temperature zones in the ocean, as shown in Figure 14–5. This layered struc-

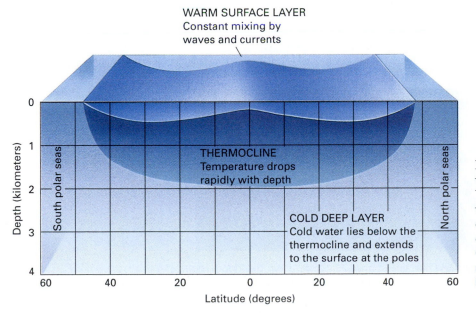

WARM SURFACE LAYER
Constant mixing by waves and currents

THERMOCLINE
Temperature drops rapidly with depth

COLD DEEP LAYER
Cold water lies below the thermocline and extends to the surface at the poles

South polar seas

North polar seas

Figure 14–5 There are three temperature layers in the ocean. The surface 450 meters is warm, temperature cools rapidly with depth in the thermocline, and the ocean depths are cold. ● **Interactive Question:** Explain why the thermocline does not exist at high latitudes.

ture does not exist in the polar latitudes because cold surface water sinks, causing vertical mixing.

14.3 Tides

Even the most casual observer will notice that on any beach the level of the ocean rises and falls on a cyclical basis. If the water level is low at noon, it will reach its maximum height at about 6:13 P.M., and be low again at about 12:26 A.M. These vertical displacements are called **tides.** Most coastlines experience two high tides and two low tides approximately every 24 hours and 53 minutes.

Tides are caused by the gravitational pull of the Moon and Sun. Although the Moon is much smaller than the Sun, it is so much closer to the Earth that its influence predominates. At any time, one region of the Earth (marked A in Fig. 14–6) lies directly under the Moon. Because gravitational force is greater for objects that are closer together, the part of the ocean nearest to the Moon is attracted with the strongest force. The water rises, resulting in a high tide in that region.

But now our simple explanation runs into trouble. As the Earth spins on its axis, a given point on the Earth passes directly under the Moon approximately once every 24 hours and 53 minutes, but the period between successive high tides is only 12 hours and 26 minutes. Why are there ordinarily two high tides in a day? The tide is high not only when a point on Earth is directly under the Moon, but also when it is 180° away. To understand this effect, we must consider the Earth-Moon orbital system. Most people visualize the Moon orbiting around the Earth, but it is more accurate to say that the Earth and the Moon orbit around a common center of gravity. The two celestial partners are locked together like dancers spinning around

in each other's arms. Just as the back of a dancer's dress flies outward as she twirls, the oceans on the opposite side of the Earth from the Moon bulge outward. This bulge is the high tide 180° away from the Moon (point B in Fig. 14–6). Thus, the tides rise and fall twice daily.

High and low tides do not occur at the same time each day, but are delayed by approximately 53 minutes every 24 hours. The Earth makes one complete rotation on its axis in 24 hours, but at the same time, the Moon is orbiting the Earth in the same direction. After a point on the Earth makes one complete rotation in 24 hours, that point must spin for an additional 53 minutes to catch up with the orbiting Moon. This is why the Moon rises approximately 53 minutes later each day. In the same manner, the tides are approximately 53 minutes later each day (Fig. 14–7).

Although the Sun's gravitational pull on the oceans is smaller than the Moon's, it does affect tides. When the Sun and Moon are directly in line with Earth their gravitational fields combine to create a strong tidal bulge. During these times, the variation between high and low tides is large, producing **spring tides** (Fig. 14–8A). When the Moon is 90° out of alignment with the Sun and the Earth, each partially offsets the effect of the other and the differences between the levels of high and low tide are smaller. These relatively small tides are called **neap tides** (Fig. 14–8B).

Tidal variations differ from place to place. For example, the Bay of Fundy is shaped like a giant funnel. The land concentrates the rise and fall of ocean water, and as a result, the tidal variation is as much as 15 meters during a spring tide (Fig. 14–9). On the other hand, tides vary by less than 2 meters along a straight stretch of coast such as that in Santa Barbara, California. In the central oceans, the tides average about 1.0 meters. Mariners consult tide tables that give the time and height of the tides in any area on any day.

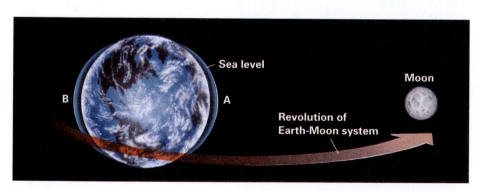

Figure 14–6 The Moon's gravity causes a high tide at point A, directly under the Moon. The motion of the Earth/Moon system causes a high tide at point B, opposite point A.

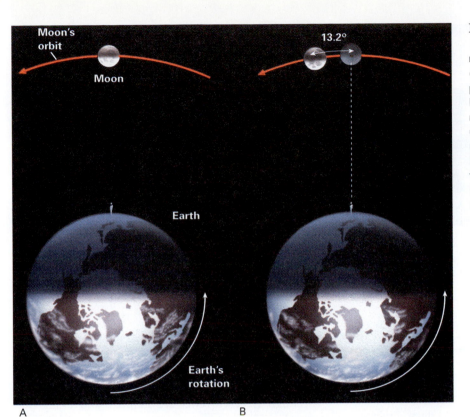

Figure 14–7 The Moon moves 13.2° every day. (A) The Moon is directly above an observer on Earth. (B) One day later, the Earth has completed one complete rotation, but the Moon has traveled 13.2°. The Earth must now travel for another 53 minutes before the observer is directly under it. ● **Interactive Question: Predict how the timing of the tides would be affected if the Moon orbited the Earth at twice its present speed.**

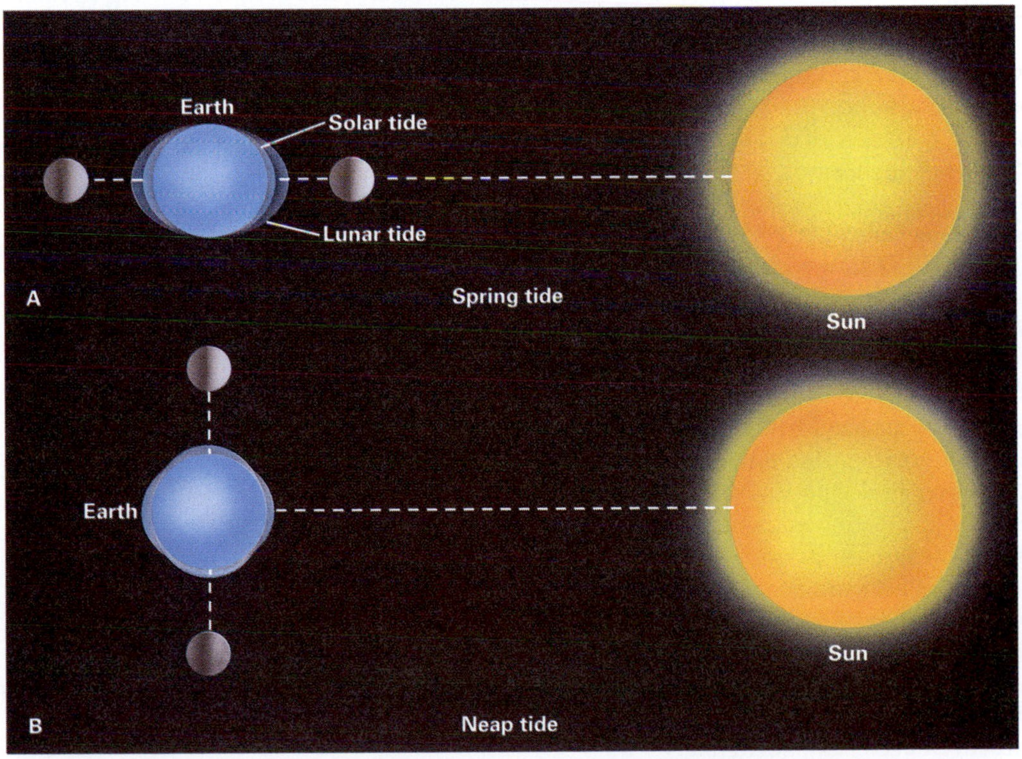

Figure 14–8 (A) Spring tides occur when the Earth, Sun, and Moon are lined up. (B) Neap tides occur when the moon lies at right angles to a line drawn between the Sun and the Earth.

A

B

Figure 14–9 (A) Low tide and (B) high tide vary by as much as 15 meters in the Bay of Fundy, Nova Scotia. *(Geological Survey of Canada, GSC 67898)*

14.4 Waves and Currents

Waves

Most waves develop when wind blows across the water. Waves vary from gentle ripples to destructive giants that can erode coastlines, topple beach houses, and sink ships. In deep water, the size of a wave depends on (1) the wind speed, (2) the length of time that the wind has blown, and (3) the distance that the wind has traveled (sailors call this last factor *fetch*). A 25-kilometer-per-hour wind blowing for 2 to 3 hours across a 15-kilometer-wide bay generates waves about 0.5 meter high. But if a storm blows at 90 kilometers per hour for several days over a fetch of 3500 kilometers, it can generate 30-meter-high waves, as tall as a ship's mast.

The highest part of a wave is called the **crest**; the lowest is the **trough** (Fig. 14–10). The **wavelength** is the distance between successive crests. The **wave height** is the vertical distance from the crest to the trough.

If you tie one end of a rope to a tree and shake the other end, a wave travels from your hand to the tree, but any point on the rope just moves up and down. In a similar manner, a single water molecule in a water wave does not travel in the same direction as the wave. The water molecule moves in circles, as shown in Figure 14–11. That is why a ball on the ocean bobs up and down and sways back and forth as the waves pass, but it doesn't travel along with the waves. In addition, the circles of water movement become smaller with depth. At a depth equal to about one half the wavelength, the disturbance becomes negligible. Thus, if you dive deep enough, you escape wave motion. No one gets seasick in a submarine.

Currents

The water in an ocean wave oscillates in circles, ending up where it started. In contrast, a **current** is a continuous flow of water in a particular direction. Although river currents are familiar and easily observed, ocean currents were not recognized by early mariners.

Surface Currents

In the mid 1700s, Benjamin Franklin lived in London, where he was Deputy Postmaster General for the American colonies. He noticed that mail ships took two weeks longer to sail from England to North America than did merchant ships. Franklin learned that the captains of the merchant ships had discovered a current flowing northward along the east coast of North America and then across the Atlantic to England. When sailing from Europe to North

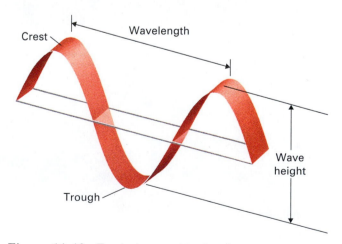

Figure 14–10 Terminology used to describe waves.

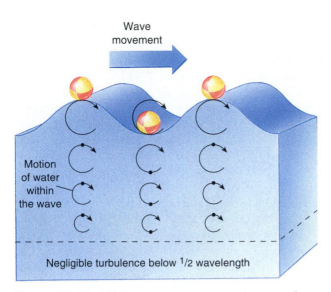

Wave movement

Motion of water within the wave

Negligible turbulence below 1/2 wavelength

Figure 14–11 While a wave moves across the sea surface, the water itself moves only in small circles.

is only partially correct; ocean currents have no well-defined banks, and they carry much more water than even the largest river. The Gulf Stream is 80 kilometers wide and 650 meters deep near the east coast of Florida and moves at approximately 5 kilometers per hour, a moderate walking speed. As it moves northward and eastward, the current widens and slows; east of New York it is more than 500 kilometers wide and travels at less than 8 kilometers per day.

Another important difference between rivers and sea currents is that rivers flow in response to gravity, whereas ocean surface currents are driven primarily by wind. When wind blows across water in a constant direction for a long time, it drags surface water along with it, forming a current. In many regions, the wind blows in the same direction throughout the year, forming currents that vary little from season to season. However, in other places, changing winds cause currents to change direction. For example, in the Indian Ocean prevailing winds shift on a seasonal basis. When the winds shift, the ocean currents follow.

Ocean Currents and the Coriolis Effect

Notice in Figure 14–12 that most open ocean surface currents move in circular paths called **gyres.** Gyres rotate clockwise in the Northern Hemisphere and counterclockwise in the Southern Hemisphere.

To understand gyres, let us start with a few analogies. Imagine riding a skateboard across a smooth parking lot and trying to throw a ball to a friend who is standing a few meters to the side (Fig. 14–13). When

America, the merchant ships saved time by avoiding the current. The captains of the mail ships were unaware of this current and lost time sailing against it on their westward journeys. In 1769, Franklin and his cousin, Timothy Folger, a merchant captain, charted the current and named it the **Gulf Stream.**

As ship traffic increased and navigators searched for the quickest routes around the globe, they discovered other ocean currents (Fig. 14–12). Ocean currents have been described as rivers in the sea. The analogy

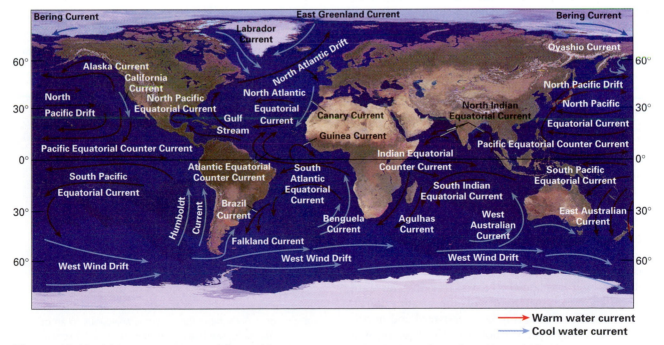

Figure 14–12 Major ocean currents of the world. *(Tom Van Sant, Inc., The GeoSphere Project, Santa Monica, California)*

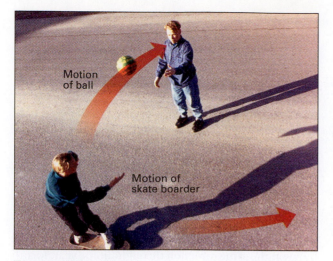

Figure 14–13 When a person on a moving skateboard throws a ball, the ball arcs in the direction of motion.

the ball is in your hand, it is moving with you and the skateboard. After it is thrown, the ball heads toward your friend but at the same time it continues to move in the same direction as the skateboard. As a result, the ball curves, and if your friend is not alert, he or she will not catch it. The effect is dramatic; try it. For an analogy that is closer to the rotating Earth, stand on one side of a spinning playground merry-go-round and have a friend stand in the center (Fig. 14–14). Then throw a ball to the person in the center. Again the ball veers sharply and is difficult to catch. Why? When the merry-go-round makes one complete revolution, the person on the outside travels a distance equal to its circumference. Because the circumference at the center is zero, the person at the center is spinning but is not otherwise moving. Thus, when the ball is thrown, it veers in the direction of the rotation of the merry-go-round.

The spherical Earth is similar to the merry-go-round. The circumference of the Earth is greatest at the equator and decreases to zero at the poles. But all parts of the planet make one complete rotation every day. Since a point on the equator must travel farther than any other point on the Earth in 24 hours, the equatorial region moves faster than regions closer to the poles. At the equator, all objects move eastward with a velocity of about 1600 kilometers per hour; at the poles there is no eastward movement at all and the velocity is 0 kilometers per hour. Now imagine that a rocketship is fired from the equator toward the North Pole. Before it was launched it was traveling eastward at 1600 kilometers per hour with the rotating Earth. As it takes off, it is moving both eastward and northward. But at any distance north of the equator, it is traveling eastward faster than the Earth beneath it.

Thus, the rocket curves toward the east, or the right. In a similar manner, a mass of water or air deflects in an easterly direction as it moves poleward from the equator, as shown in Figure 14–15A.

Conversely, consider an ocean current flowing southward from the Arctic Ocean toward the equator. Since it started near the North Pole, this water moves more slowly than the Earth's surface in tropical regions, and therefore it lags behind as it flows southward. The Earth spins in an easterly direction, and therefore the current veers toward the west, or to the right, as shown in Figure 14–15B.

North-south currents always veer to the right in the Northern Hemisphere. In the Southern Hemisphere currents turn toward the left for the same reason. The deflection of currents caused by the Earth's rotation is called the **Coriolis effect.**

Ocean currents profoundly affect the climates of large regions of the Earth. The Gulf Stream transports one million cubic meters of warm water northward past any point *every second*, warming North America and Europe. For example, Churchill, Manitoba, is a village in the interior of Canada. It is frigid and icebound for much of the year. Polar bears regularly migrate through town. Yet it is at about the same latitude as Glasgow, Scotland, which is warmed by the Gulf Stream and therefore experiences relatively mild winters. In general, the climate in western Europe is warmer than that at similar latitudes in other regions not heated by tropical ocean currents.

Figure 14–14 The Coriolis effect is demonstrated on a playground merry-go-round. The person standing on the rim is analogous to an observer at the equator, and the person standing on a pedestal in the center is analogous to an observer at the pole. The ball arcs in the direction of rotation of the merry-go-round.

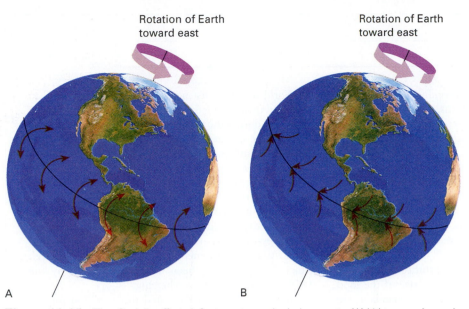

Figure 14–15 The Coriolis effect deflects water and wind currents. (A) Water or air moving poleward from the equator is traveling east faster than the land beneath it and veers to the east (turns right in the Northern Hemisphere and left in the Southern Hemisphere). (B) Water or air moving toward the equator is traveling east slower than the land beneath it and veers to the west (turns right in the Northern Hemisphere and left in the Southern Hemisphere). *(Tom Van Sant, Geosphere Project/Planetary Visions/Science Photo Library)* ● **Interactive Question:** Predict how a wind would be affected if it were moving due East? Due West?

Deep-Sea Currents

Wind does not affect the ocean depths, and oceanographers once thought that deep ocean water was almost motionless. In her book *The Sea Around Us*, Rachael Carson wrote that the ocean depths are "a place where change comes slowly, if at all." However, in 1962 ripples and small dunes were photographed on the floor of the North Atlantic. Since flowing water forms these features, the photographs suggested that water was moving in the ocean depths. More recently, oceanographers have measured the speeds of deep-sea currents directly with flow meters, and photographed moving sand and mud with underwater television cameras.

Wind drives surface currents, but deep-sea currents are driven by differences in water density. Dense water sinks and flows horizontally along the sea floor to form a deep-sea current. Two factors cause water to become dense and sink: cooling temperature and rising salinity. Recall that water is densest when it is cold, close to freezing. Therefore, as tropical surface water moves poleward and cools, it becomes denser and sinks.

In addition, water density increases as salinity increases. Therefore, water sinks when it becomes saltier. Sea water can become saltier if surface water evaporates. Polar seas also become saltier when the surface freezes, because salt does not become incorporated in the ice. Arctic and Antarctic water is dense because the water is both cold and salty.

The Gulf Stream originates in the subtropics. When this water reaches the northern part of the Atlantic Ocean near the tip of Greenland, it cools and sinks. When the sinking water reaches the sea floor, it is deflected to the south to form the North Atlantic Deep Water, which flows along the sea floor all the way to Antarctica (Fig. 14–16). An individual water molecule that sinks near Greenland may travel for 500 to 2000 years before resurfacing half a world away in the south polar sea.

Upwelling

If water sinks in some places, it must rise in others to maintain mass balance. In Figure 14–12, note that the California current flows southward along the coast of California. In the Southern Hemisphere, the Humboldt current moves northward along the west coast of South America. Both currents are deflected westward by the Coriolis effect. As these surface

Life in the Sea

On land, most photosynthesis is conducted by multicellular plants such as mosses, ferns, grasses, and trees. Large animals such as cows, deer, elephants, and bison consume the plants. In contrast, most of the photosynthesis and consumption in the ocean is carried out by small organisms called **plankton.** Many plankton are microscopic; others are up to a few centimeters long. Plankton live mostly within a few meters of the sea surface, where light is available. Surface productivity is limited by nutrients such as nitrates, iron, and phosphates, which are pulled downward toward the sea floor by gravity.

Phytoplankton conduct photosynthesis like land-based plants. Therefore, they are the base of the food chain for aquatic animals. Although phytoplankton are not readily visible, they are so abundant that they supply about 50 percent of the oxygen in our atmosphere. **Zooplankton** are tiny animals that feed on the phytoplankton (Fig. 1). The larger and more familiar marine plants (such as seaweed) and animals (such as fish, sharks, and whales) play a relatively small role in oceanic photosynthesis and consumption.

The shallow water of a continental shelf supports large populations of marine organisms. In addition, many deep-sea fish spawn in shallow water within a kilometer or two of shore. These shallow zones are hospitable to life because they have (1) easy access to the deep sea, (2) lower salinity than the open ocean, (3) a high concentration of nutrients originating from land and sea, (4) shelter, and (5) abundant plant life rooted to the sea floor in addition to the phytoplankton floating on the surface. As a result, about 99 percent of the marine fish caught every year are harvested from the shallow waters of the continental shelves.

World fish harvests increased steadily from 1950 to 1989 and then declined slowly during the 1990s (Fig. 2). However, during the past twenty years, harvests of some species have declined while others have risen. For example, Atlantic cod is a valuable fish that grew abundantly off the northeast coast of North America. The first cod fishermen fished this dangerous but fertile coast 12 years after Columbus had landed in the New World. After three hundred years of intensive fishing, the cod stocks began to decline. When cod populations declined, their prey—herring, mackerel, and squid—thrived. In 1816, Atlantic fishermen began harvesting mackerel, which had previously been used only as bait. By the 1850s, boats set out to net herring.

In 1970, fishermen harvested three million tons of cod. But this catch was biologically unsustainable and declined to one million tons by 1993. In 1995, biologists suspended cod fishing in many areas to allow the fish populations to recover. In 1978, the herring and mackerel stocks began to decline, so boats switched nets to trawl for squid. In the 1980s the squid population began to decline. Thus, fishing pressure has worked through the food chain, disrupting the entire ecosystem.

According to the United Nations Food and Agricultural Organization, world fish harvest is expected to remain fairly constant at around 90 million tons per year until the year 2010. After 2010, ecosystem destruction and predicted overfishing will lead to a sharp decline in populations and harvests.

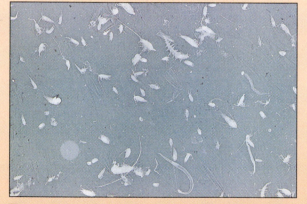

Figure 1 Zooplankton are tiny animals that live near the sea surface and feed on phytoplankton. *(Woods Hole Oceanographic Institution)*

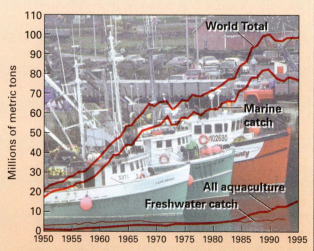

Figure 2 World fish harvest from 1950 to 1995. The marine harvest peaked in 1989 and declined slowly in the following six years. *(Worldwatch Institute)*

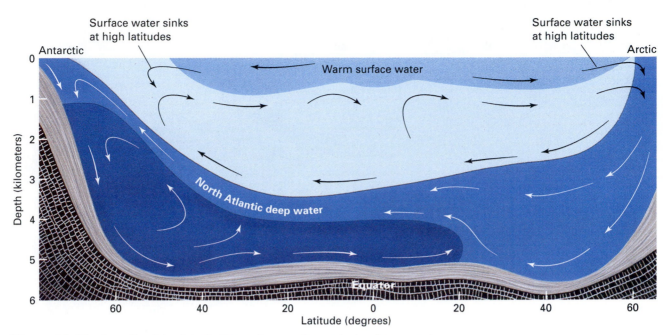

Figure 14–16 A profile of the Atlantic Ocean shows surface and subsurface currents.

currents veer away from shore, water from the ocean depths rises toward the surface. This upward flow of water is called **upwelling.** Upwelling carries cold water from the ocean depths to the surface. In August, water on the east (Atlantic) coast of the United States is warmed by the Gulf Stream and may be a comfortable 21°C. However, on the central California coast, the cool California current combines with the upwelling current to produce water that is only 15°C, and surfers and swimmers must wear wetsuits to stay in the water for long. Upwelling also brings nutrients from the deep ocean to the surface, creating rich fisheries along the coasts of California and Peru.

COASTLINES

Coastlines are among the most geologically active zones on Earth. Subduction zones occur along many continental coasts. Waves and currents weather, erode, transport, and deposit sediment continuously on all coastlines. In addition, the shallow waters are productive ecosystems. Many organisms that live in these regions build hard shells or skeletons of calcium carbonate. Eventually, the remains of these shells and skeletons become lithified to form limestone. Thus, living organisms become part of the rock cycle.

14.5 Weathering and Erosion Along Coastlines

Waves batter most coastlines. If you walk down to the shore, you can watch the turbulent water carry sand grains or even small cobbles along the beach. Even during a calm day, waves frequently steepen as they approach shore and then crash against the beach. When a wave enters shallow water, the bottom of the wave drags against the sea floor. This drag compresses the circular motion of the wave into ellipses. This deformation slows the lower part of the wave, so that the upper part moves more rapidly than the lower. As the front of the wave rises over the base, the wave steepens until it collapses forward, or **breaks** (Fig. 14–17). Chaotic, turbulent waves breaking along a shore are called **surf.**

Most coastal erosion occurs during intense storms because storm waves are much larger and more energetic than normal waves (Fig. 14–18). A 6-meter-high wave strikes shore with 40 times the force of a 1.5-meter-high wave. A giant, 10-meter-high storm wave strikes a 10-meter-wide sea wall with four times the thrust energy of the three main orbiter engines of a space shuttle.

Sea water weathers and erodes coastlines by hydraulic action, abrasion, and solution, three processes that we are familiar with from our discussion of

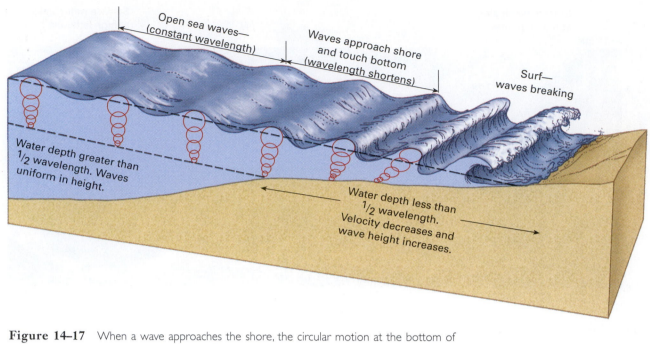

Figure 14–17 When a wave approaches the shore, the circular motion at the bottom of the wave flattens out and becomes elliptical. The bottom of the wave drags against the sea floor. As a result, the wavelength shortens and the wave steepens until it finally breaks, creating surf.

streams. Salt cracking is also significant along coastlines, as will be discussed at the end of this section.

A wave striking a rocky cliff drives water into cracks or crevices in the rock, compressing air in the cracks. The air and water together combine to create hydraulic forces strong enough to dislodge rock frag-

Figure 14–18 Expensive seashore houses are often threatened by coastal erosion that occurs during intense storms; Georgetown in Grand Cayman Island. *(Susan Blanchet/ Dembinsky Photo Assoc.)*

ments or even huge boulders. Storm waves create forces as great as 25 to 30 tons per square meter. Engineers built a breakwater in Wick Bay, Scotland, of car-sized rocks weighing 80 to 100 tons each. The rocks were bound together with steel rods set in concrete, and the sea wall was topped by a steel-and-concrete cap weighing more than 800 tons. A large storm broke the cap and scattered the upper layer of rocks about the beach. The breakwater was rebuilt, reinforced, and strengthened, but a second storm destroyed this wall as well.

On the Oregon coast, the impact of a storm wave tossed a 60-kilogram rock over a 25-meter-high lighthouse. After sailing over the lighthouse, it crashed through the roof of the keeper's cottage, startling the inhabitants.

While images of flying boulders are spectacular, most wave erosion occurs gradually, by abrasion. Water is too soft to abrade rock, but waves carry large quantities of silt, sand, and gravel. Breaking waves roll this sediment back and forth over bedrock, acting like liquid sandpaper eroding the rock. At the same time, smaller cobbles are abraded as they roll back and forth in the surf zone.

Sea water slowly dissolves rock and carries ions in solution. Salt water also soaks into bedrock; when the water evaporates, the growing salt crystals pry the rock apart.

A

B

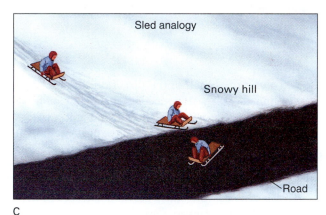

C

Figure 14–19 (A) When a water wave strikes the shore at an angle, one end slows down, causing the wave to bend, or refract. (B) Wave refraction on a lake shore. (C) A sled turns upon striking a paved roadway at an angle because one runner hits the roadway and slows down before the other does.

14.6 Sediment Transport Along Coastlines

Most waves approach the shore at an angle rather than head on. When this happens, one end of the wave encounters shallow water and slows down, while the rest of the wave is still in deeper water and continues to advance at a relatively faster speed. As a result, the wave bends (Figs. 14–19 A and B). This effect is called **refraction.** Consider the analogy of a sled gliding down a snowy hill onto a cleared road. If the sled hits the road at an angle, one runner will reach it before the other. The runner that hits the pavement first slows down, while the other, which is still on the snow, continues to travel rapidly (Fig. 14–19C). As a result, the sled turns abruptly.

As waves approach an irregular coast, they reach the headlands first, breaking against the point and eroding it. The waves then refract around the headland and travel parallel to its sides. Surfers seek such refracted waves because these waves move nearly parallel to the coast and therefore travel for a long distance before breaking. For example, the classic big-wave mecca at Waimea Bay of Oahu, Hawaii, forms along a point of land that juts into the sea. Waves

build when they strike an offshore coral reef, then refract along the point, and charge into the bay (Fig. 14–20).

Refracted waves transport eroded sediment toward the interior of a bay. As the headlands erode and the

Figure 14–20 Surfers in Waimea Bay on Oahu, Hawaii, cherish the long-lasting surf that forms when ocean waves reflect around a headland and break as they enter the bay. (*Warren Bolster/Tony Stone Images*)

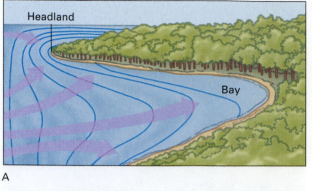

A

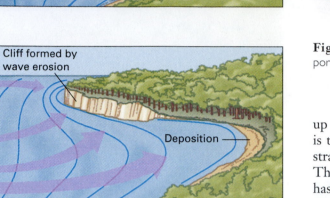

B

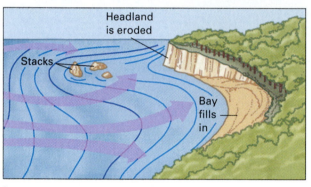

C

Figure 14–21 (A) & (B) When a wave strikes a headland, the shallow water causes that portion of the wave to slow down. Part of the wave breaks against the headland, weathering the rock. (C) A portion of the wave refracts, transporting sediment and depositing it on the beach inside the bay. Eventually, this selective weathering, erosion, and deposition will straighten an irregular coastline.

interiors of bays fill with sediment, an irregular coastline eventually straightens (Fig. 14–21).

When waves strike shore at an angle, they form a **longshore current** that flows parallel to the shore. Longshore currents flow in the surf zone and a little farther out to sea and may travel for tens or even hundreds of kilometers. They transport sand and other sediment along coastlines.

Sediment transport also occurs by **beach drift.** If a wave strikes the beach obliquely, it pushes sediment

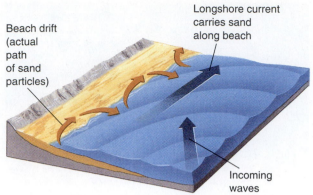

Figure 14–22 Longshore currents and beach drift transport sediment along a coast.

up and along the beach in the direction that the wave is traveling. When water recedes, the sediment flows straight down the beach as shown in Figure 14–22. Thus, at the end of one complete wave cycle, sediment has moved a short distance parallel to the coast. The next wave transports the sediment a little further, until, over time, sediment moves long distances.

As you can see, longshore currents and beach drift work together to transport and deposit sand along a coast. Much of the sand found at Cape Hatteras, North Carolina, originated from the mouth of the Hudson River and from glacial deposits on Long Island and southern New England. Midway along this coast, at Sandy Hook, New Jersey, an average of 2000 tons a day move past any point on the beach. As a result of this process, beaches have been called "rivers of sand."

Tidal Currents

When tides rise and fall along an open coastline, water moves in and out from the shore as a broad sheet. If the flow is channeled by a bay with a narrow entrance or by islands, the moving water funnels into a **tidal current,** which is a flow of ocean water caused by tides. Tidal currents can be intense where large differences exist between high and low tides and narrow constrictions occur in the shoreline. On parts of the west coast of British Columbia, a diesel-powered fishing boat cannot make headway against tidal currents flowing between closely spaced islands. Fishermen must wait until the tide, and hence the tidal currents, reverse direction before proceeding.

14.7 The Water's Edge

Beaches

When most people think about going to the beach, they think of gently sloping expanses of sand. However,

A   B

Figure 14–23 (A) Sandy beaches are common near Santa Barbara, California. (B) Big Sur, to the north, is dominated by rocky beaches.

a **beach** is any strip of shoreline that is washed by waves and tides. Although many beaches are sandy, others are swampy, rocky, or bounded by cliffs (Fig. 14–23).

A beach is divided into two zones, the **foreshore** and the **backshore.** The foreshore, also called the **intertidal zone,** lies between the high- and low-tide lines and is alternately exposed to the air at low tide and covered by water at high tide. The backshore is usually dry but is washed by waves during storms. Many terrestrial plants cannot survive in salt water, so specialized, salt-resistant plants live in the backshore. The backshore can be wide or narrow depending on its topography, the local tidal difference, and the frequency and intensity of storms. In a region where the land rises steeply, the backshore may be a narrow strip. In contrast, if the coast consists of low-lying plains and if coastal storms occur regularly, the backshore may extend several kilometers inland.

Reefs

A **reef** is a wave-resistant ridge or mound built by corals, oysters, algae, or other marine organisms. Because corals need sunlight and warm, clear water to thrive, coral reefs develop in shallow tropical seas where little suspended clay or silt muddies the water (Fig. 14–24). As the corals die, their offspring grow on their remains. Oyster reefs form in temperate estuaries, and can grow in more turbid water.

Globally, coral reefs cover about 600,000 square kilometers, about the area of France, but they spread out in long, thin lines. They are extraordinarily productive ecosystems because the corals provide shelter for many fish and other marine species. Within the past fifty years, 10 percent of the world's coral reefs have been destroyed and an additional 30 percent are in critical condition. Several factors contribute to this destruction:

- Pollutants and silt from cities, urban roadways, farms, and improper logging smother the delicate reef organisms.
- Fertilizer runoff from farms and sewage runoff from cities have added nutrients to coastal waters, feeding coral predators. For example, starfish thrive in nutrient-rich water and starfish eat corals.
- Overfishing or improper fishing can kill reefs. Parrot fish and sea urchins eat algae that smother corals. If fishermen harvest too many parrot fish and sea urchins, the algae grow unchecked and kill the reefs. Also, in many parts of the world, fishermen dynamite coral reefs so their nets don't get tangled. Unfortunately, when the reefs are destroyed, fish populations decline. Therefore, while dynamiting reefs improves short-term gain, the practice diminishes long-term, sustainable harvests.
- In recent years, disease epidemics have destroyed coral reefs throughout the world. Disease is part of any ecosystem and geologists have found evidence that epidemics have affected reefs periodically for

Figure 14–24 (A) Coral reefs form abundantly in clear, shallow, tropical water. This small island on the east coast of Australia is surrounded by reefs that are part of the Great Barrier Reef. *(Manfred Gottschalk/Tom Stack and Assoc.)* (B) Reefs grow in the clear, shallow water near Vanuatu and many other South Pacific Islands.

hundreds of millions of years. However, many oceanographers have suggested that human activity has provoked the recent epidemics. One suggested cause is that sewage provides nutrients for disease organisms that kill coral. Another is that chemical pollutants are altering the species balance in aquatic ecosystems. A third is that sea water temperature has risen in response to global warming and that disease organisms thrive in the warmer sea water.

14.8 Emergent and Submergent Coastlines

Geologists have found drowned river valleys and fossils of land animals on continental shelves beneath the

sea. They have also found fossils of fish and other marine organisms in continental interiors. As a result, we infer that sea level has changed, sometimes dramatically, throughout geologic time. An **emergent coastline** forms when a portion of a continent that was previously under water becomes exposed as dry land. Falling sea level or rising land can cause emergence. As explained in Section 14.9, many emergent coastlines are sandy. In contrast, a **submergent coastline** develops when the sea floods low-lying land and the shoreline moves inland (Fig. 14–25). Submergence occurs when sea level rises or coastal land sinks. A submergent coast is commonly irregular, with many bays and headlands. The coast of Maine, with its numerous inlets and rocky bluffs, is a submergent coastline (Fig. 14–26). Small sandy beaches form in protected coves, but most of the headlands are rocky and steep.

Factors That Cause Coastal Emergence and Submergence

Tectonic processes can cause a coastline to rise or sink. Isostatic adjustment can also depress or elevate a portion of a coastline. About 18,000 years ago, a huge continental glacier covered most of Scandinavia, causing it to sink isostatically. As the lithosphere settled, the displaced asthenosphere flowed southward, causing the Netherlands to rise. When the ice melted, the process reversed as the asthenosphere flowed back from below the Netherlands to Scandinavia. Today, Scandinavia is rebounding and the Netherlands is sinking. The same mechanism is causing much of the United States to sink. During the Pleistocene Ice Age, Canada was depressed by the ice and asthenosphere rock flowed southward. Today, the asthenosphere is flowing back north, much of Canada is rebounding, and much of the United States is sinking.

Sea level can also change globally. A global sea-level change, called **eustatic** change, occurs by three mechanisms: the growth or melting of glaciers, changes in water temperature, and changes in the volume of the mid-oceanic ridge.

During an ice age, vast amounts of water move from the sea to form continental glaciers, and sea level falls resulting in global emergence. Similarly, when glaciers melt, sea level rises globally, causing submergence.

Sea water expands when it is heated and contracts when it is cooled. Although this change is not noticeable in a glass of water, the volume of the oceans is so great that a small temperature change can alter sea level measurably. As a result, global warming causes sea-level rise, and cooling leads to falling sea level.

Temperature changes and glaciation are linked. When global temperature rises, sea water expands and glaciers melt. Thus, even a small global warming can lead to a large sea-level rise. The opposite effect is also

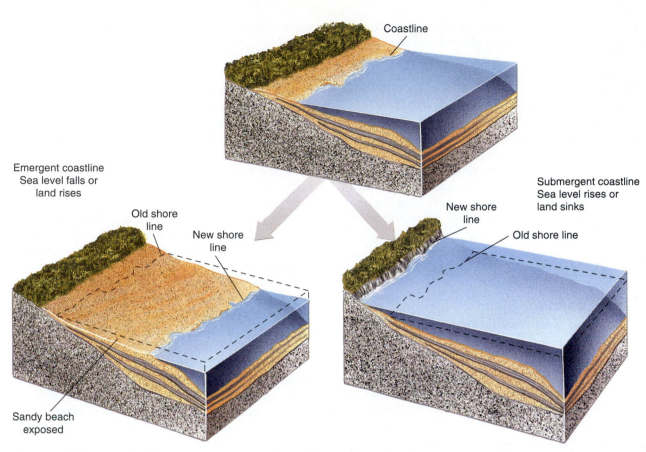

Figure 14–25 If sea level falls or if the land rises, the coastline is emergent. Offshore sand is exposed to form a sandy beach. If coastal land sinks or sea level rises, the coastline is submergent. Areas that were once land are flooded. Irregular shorelines develop and beaches are commonly rocky. ● **Interactive Question: Draw a sequence of diagrams to show how barrier islands could form during coastal emergence, or how small, more circular islands could form during coastal submergence.**

Figure 14–26 The Maine coast is a rocky, irregular, submergent coastline.

true. When temperature falls, sea water contracts and glaciers grow. Both growing glaciers and contracting sea water lead to a fall in sea level.

As explained in Chapter 13, changes in the volume of the mid-oceanic ridge can also affect sea level. The mid-oceanic ridge displaces sea water. When lithospheric plates spread slowly from the mid-oceanic ridge, they create a narrow ridge that displaces relatively little sea water, resulting in low sea level. In contrast, rapidly spreading plates produce a high-volume ridge that displaces more water, causing a global sea-level rise. At times in Earth history, spreading has been relatively rapid, and as a result, global sea level has been high.

14.9 Sediment-Rich and Sediment-Poor Coastlines

If weathering and erosion occur along all coastlines, why are some beaches sediment-rich (sandy) and others sediment-poor (rocky)? The answer is that most coastal sediment is not formed by weathering and erosion at the beach itself. Instead, several processes transport sediment to a seacoast. Major rivers carry large quantities of sand, silt, and clay to the sea and deposit it on deltas that may cover thousands of square kilometers. In some coastal regions, glaciers deposited large quantities of till along coastlines during the Pleistocene Ice Age. In tropical and subtropical latitudes, eroding reefs supply sediment in shallow coastal waters. A sediment-rich (sandy) coastline is one with abundant sediment from any of these sources.

Longshore currents then transport and deposit sediment along a coast. Much of the sand carried by these currents accumulates on underwater offshore bars. Thus, a great deal of sand may be stored offshore from a beach. If such a coastline emerges, this vast supply of sand becomes exposed as dry land. Thus, sandy beaches are abundant on emergent coastlines.

In contrast, sediment-poor (rocky) coastlines occur where sediment from any of these sources is scarce. With no abundant sources of sediment, small concentrations of sand accumulate in protected bays but most of the coast is rocky. On a submergent coastline, rising sea level puts the stored offshore sand even farther out to sea. As a result, submergent coastlines commonly have rocky beaches.

Sediment-Rich (Sandy) Coastlines

A long ridge of sand or gravel extending out from a beach is called a **spit** (Fig. 14–27). As sediment migrates along a coast, the spit may continue to grow. A well-developed spit may rise several meters above high-tide level and may be tens of kilometers long. A spit may block the entrance to a bay, forming a **baymouth bar.** A spit may also extend outward into the sea, creating a trap for other moving sediment.

A **barrier island** is a long, low-lying island that extends parallel to the shoreline. It looks like a beach or spit and is separated from the mainland by a sheltered body of water called a **lagoon** (Fig. 14–28). Barrier islands extend along the east coast of the United States from New York to Florida. They are so nearly continuous that a sailor in a small boat can navigate the entire coast inside the barrier island system and remain protected from the open ocean most of the time. Barrier islands also line the Texas Gulf Coast.

Barrier islands form in several ways. The two essential ingredients are a large supply of sand and waves or currents to transport it. If a coast is shallow for several kilometers outward from shore, breaking storm waves may carry sand toward shore and deposit it just offshore as a barrier island. Alternatively, if a longshore current veers out to sea, it slows down and deposits sand where it reaches deeper water. Waves may then pile up the sand to form a barrier island. Other mechanisms involve sea-level change. Underwater sand bars may be exposed as a coastline emerges. Alternatively, sand dunes or beaches may form barrier islands if a coastline sinks.

The Atlantic coast of the United States is fringed with the longest chain of barrier islands in the world. Many seaside resorts are built on these islands, and developers often ignore the fact that they are transient and changing landforms (Fig. 14–29). If the rate of erosion exceeds that of deposition for a few years in a row, a barrier island can shrink or disappear completely, leading to destruction of beach homes and resorts. In addition, barrier islands are especially vulnerable to hurricanes, which can wash over low-lying islands and move enormous amounts of sediment in a very brief time. In September of 1996, Hurricane Fran flattened much of Topsail Island, a low-lying barrier island in North Carolina. Geologists were not surprised because the homes were not only built on sand, they were built on sand that was virtually guaranteed to move.

Sediment-Poor (Rocky) Coastlines

A sediment-poor (rocky) coastline is one without any of the abundant sediment sources described above. In many areas on land, bedrock is exposed or covered by a thin layer of soil. If this type of sediment-poor terrain is submerged, and if there are no other sources of sand, the coastline is rocky.

A **wave-cut cliff** forms when waves erode the headland into a steep profile. As the cliff erodes, it leaves

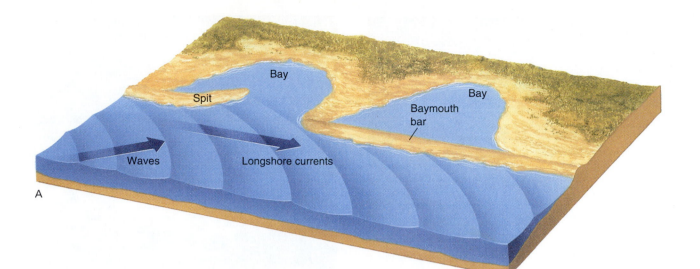

A

B

Figure 14–27 (A) Spits and baymouth bars are common features of sandy emergent coastlines. (B) Aerial view of a spit that formed along a low-lying coast in northern Siberia.

Figure 14–28 This system of barrier islands near Houston, Texas, forms protected waters inside the bay. *(Terranova International/Photo Researchers)*

Figure 14–29 These expensive hotels in Miami were built on geologically transient barrier islands. Much of the rest of the city lies on reclaimed swamps surrounded by lagoons. *(Comstock)*

Figure 14–30 Waves hurl sand and gravel against solid rock to erode cliffs and create a wave-cut platform along the Oregon coast.

Figure 14–31 Massive waves of the Antarctic Ocean eroded cliffs to form these sea stacks near Cape Horn.

a flat or gently sloping **wave-cut platform** (Fig. 14–30). If waves cut a cave into a narrow headland, the cave may eventually erode all the way through the headland, forming a scenic **sea arch.** When an arch collapses or when the inshore part of a headland erodes faster than the tip, a pillar of rock called a **sea stack** forms (Fig. 14–31). As waves continue to batter the rock, eventually the sea stack crumbles.

If the sea floods a long, narrow, steep-sided coastal valley, a sinuous bay called a **fjord** is formed. Fjords are common at high latitudes, where rising sea level has flooded coastal valleys scoured by Pleistocene glaciers. Fjords may be hundreds of meters deep, and often the cliffs drop straight into the sea.

An **estuary** forms where rising sea level or a sinking coastline submerges a broad river valley or other basin. Estuaries are ordinarily shallow and have gently sloping beaches. Streams transport nutrients to the bay, and the shallow water provides habitats for marine organisms. Estuaries also make excellent harbors and therefore are prime sites for industrial activity. As a result, many estuaries have become seriously polluted in recent years.

14.10 Development and Pollution of Coastlines

CASE STUDY
Chesapeake Bay

Chesapeake Bay is an estuary formed by submergence of the Susquehanna River valley (Fig. 14–32). Geologists have recently discovered a series of concentric faults in the bedrock beneath the bay. These faults provide evidence that a meteorite impact 35 million years

ago may have formed a crustal depression. Later, this depression filled with water as sea level rose following the latest glacial retreat.

Chesapeake Bay is approximately 100 kilometers long, averages 10 to 15 kilometers wide, and contains numerous bays and inlets. Despite its great size, it is only 7 to 10 meters deep near its mouth, and its greatest depth is 50 meters. Three major cities—Washington, Baltimore, and Harrisburg—lie along Chesapeake Bay or its tributaries. Between 1950 and 1985, the population in the watershed increased by 50 percent, to 15 million people, and the urban land area tripled. As a result, the bay became polluted.

Water runoff from farms carries silt, fertilizers, and pesticides into Chesapeake Bay. Silt destroys habitats by clogging spaces between rocks and covering plant roots, thereby reducing the flow of oxygen. Fertilizers favor rapid growth of some organisms, thereby disrupting the ecological balance. Pesticides are often toxic to marine organisms. In addition, sewage treatment plants release nutrients into the bay, and numerous factories discharge toxic compounds.

Pollution and overfishing have devastated animal and plant populations in Chesapeake Bay. In 1960, fishermen harvested 2.7 million kilograms of striped

Figure 14–32 Satellite view of Chesapeake Bay. *(Courtesy of Chesapeake Bay Foundation)*

bass from the bay; by 1985, the catch had dwindled by 90 percent to 270,000 kilograms. The bay once supported the richest oyster beds in the world; today the oyster population has dropped by 99 percent.

By the early 1980s, people became concerned with the pollution and ecosystem destruction. In 1984, the Chesapeake Bay Foundation and the Natural Resources Defense Council filed a lawsuit against the Bethlehem Steel Corporation for illegally dumping wastes into the estuary. The environmental groups won the suit: Bethlehem Steel was fined more than a million dollars and was required to reduce the quantity of pollutants emitted. Other industrial polluters began cleanups to avoid similar suits. In the same year, the governments of Maryland, Virginia, Pennsylvania, and the District of Columbia tightened regulations on municipal sewage disposal. State-run agricultural education programs showed farmers that they can maintain high yields with fewer fertilizers and pesticides than they had been using.

All of these efforts have improved the water quality of Chesapeake Bay. Fish populations are recovering slowly. But as long as cities line its coasts, the bay can never return to a pristine state. Developers have built concrete wharves over salt-water marshes. Runoff from roadways, parking lots, and storm drains carries oil and other contaminants into the estuary. Air pollutants drift into the water; runoff from farmers' fields can be reduced but not eliminated. Chemicals that seeped into nearby aquifers decades ago slowly percolate into the estuary's waters. ●

CASE STUDY
Long Island

Long Island extends eastward from New York City and is separated from Connecticut by Long Island Sound. Longshore currents flow westward, eroding sand from glacial deposits at the eastern end of the island and depositing it to form beaches and barrier islands on the south side of the island (Fig. 14–33).

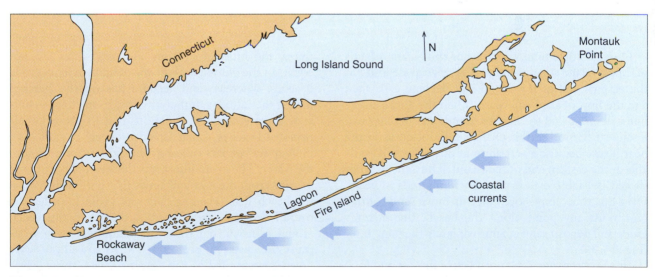

Figure 14–33 Longshore currents carry sand westward along the south shore of Long Island to create a series of barrier islands.

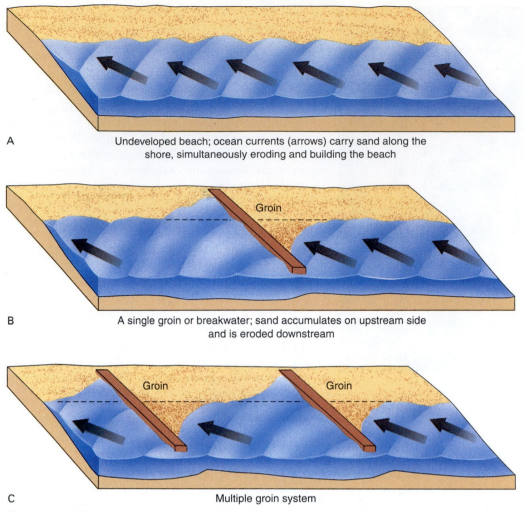

A

Undeveloped beach; ocean currents (arrows) carry sand along the
shore, simultaneously eroding and building the beach

B

Groin

A single groin or breakwater; sand accumulates on upstream side
and is eroded downstream

C

Groin   Groin

Multiple groin system

Figure 14–34 (A) Longshore currents simultaneously erode and deposit sand along an undeveloped beach. (B) A single groin or breakwater traps sand on the upstream side, resulting in
erosion on the downstream side. (C) A multiple groin system propagates the uneven distribution of sand along the entire beach.

Over geologic time, the beaches and barrier islands of Long Island are unstable. The glacial deposits at the eastern end of the island will become exhausted and the flow of sand will cease. Then the entire coastline will erode and the barrier islands and beaches will disappear. However, this change will not occur in the near future because a vast amount of sand is still available at the eastern end of the island. Thus, the beaches should be stable over a period of hundreds of years. Over this time, longshore currents move sand continuously. At any point along the beach, the currents erode and deposit sand at approximately the same rates (Fig. 14–34A).

If we narrow our time perspective further and look at a Long Island beach over a season or during a single storm, it may shrink or expand. Over such short

times, the rates of erosion and deposition are not equal. In the winter, violent waves and currents erode beaches, whereas sand accumulates on the beaches during the calmer summer months. In an effort to prevent these seasonal fluctuations and to protect their personal beaches, Long Island property owners have built stone barriers called **groins** from shore out into the water. The groin intercepts the steady flow of sand moving from the east and keeps that particular part of the beach from eroding. But the groin impedes the overall flow of sand. West of the groin the beach erodes as usual, but the sand is not replenished because the upstream groin traps it. As a result, beaches down-current from the groin erode away (Fig. 14–34B). The landowner living down-current from a groin may then decide to build another groin to pro-

A

B

Figure 14–35 (A) This aerial photograph of a Long Island beach shows sand accumulating on the upstream side of a groin, and erosion on the downstream side. (B) A closeup of the house in (A) shows waves lapping against the foundation.

tect his or her beach (Fig. 14–34C). The situation has a domino effect, with the net result that millions of dollars are spent in ultimately futile attempts to stabilize a system that was naturally stable in its own dynamic manner (Fig. 14–35).

Storms pose another dilemma. Hurricanes may strike Long Island in the late summer and fall, generating storm waves that completely overrun the barrier islands, flattening dunes and eroding beaches. When the storms are over, gentler waves and longshore currents carry sediment back to the beaches and rebuild them. As the sand accumulates again, salt marshes rejuvenate and the dune grasses grow back within a few months.

These short-term fluctuations are incompatible with human ambitions, however. People build houses,

resorts, and hotels on or near the shifting sands. The owner of a home or resort hotel cannot allow the buildings to be flooded or washed away. Therefore, property owners construct large sea walls along the beach. When a storm wave rolls across an undeveloped low-lying beach, it dissipates its energy gradually as it flows over the dunes and transports sand. The beach is like a judo master who defeats an opponent by yielding with the attack, not countering it head on. A sea wall interrupts this gradual absorption of wave energy. The waves crash violently against the barrier and erode sediment at its base until the wall collapses. It may seem surprising that a reinforced concrete sea wall is *more* likely to be permanently destroyed than a beach of grasses and sand dunes, yet this is often the case (Fig. 14–36). ●

Figure 14–36 (A) In a natural beach, the violent winter waves often move sand out to sea. (B) The gentler summer waves push sand toward shore and rebuild the beach. (C) Wave energy concentrates against a sea wall and (D) may eventually destroy it.

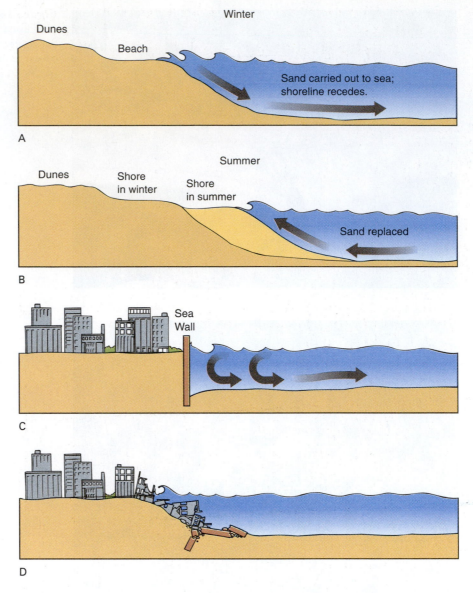

14.11 Global Warming and Sea-Level Rise

Sea level has risen and fallen repeatedly in the geologic past, and coastlines have emerged and submerged throughout Earth history. During the past 40,000 years, sea level has fluctuated by 150 meters, primarily in response to growth and melting of glaciers (Fig. 14–37). The rapid sea-level rise that started about 18,000 years ago began to level off about 7000 years ago. By coincidence, humans began to build cities about 7000 years ago. Thus, civilization has developed during a short time when sea level has been relatively constant.

Shore-based gauging stations and satellite radar studies agree that global sea level is presently rising at about 2 millimeters per year. Thus, if present rates continue, sea level will rise 20 centimeters in 100 years. Such a rise would be significant along very low-lying areas such as the Netherlands and Bangladesh, but most of the world's coastlines wouldn't be appreciably affected.

However, some scientists have suggested that we are on the threshold of a very dramatic rise in sea level; perhaps four meters in the next century (Fig. 14–38). These arguments are based on the predicted behavior of the West Antarctic Ice Sheet.

One hypothesis starts with the assumption that global temperature will rise by a few degrees Celsius in the next century, due to the greenhouse effect. A temperature rise of a few degrees isn't enough to melt large quantities of ice from the surface of the Antarctic

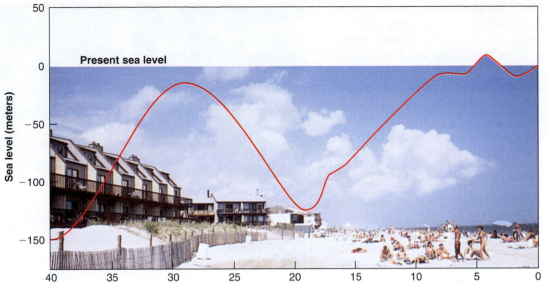

Figure 14–37 Sea-level has fluctuated more than 150 meters during the past 40,000 years. (J.D. Hansom, *Coasts*, Cambridge, U.K.: Cambridge University Press, 1988.)

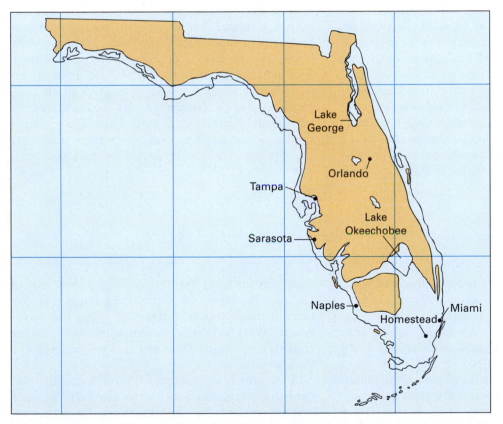

Figure 14–38 The blue shaded area on this map of modern Florida shows the land that would be flooded if sea level rose 4 meters, as some oceanographers predict. Refer back to Figure 14–37 to see that a 4-meter change is small compared to past sea-level fluctuations.

continental glaciers. However, much of the West Antarctic Ice Sheet protrudes into the ocean and rests on underwater bedrock. Glaciers on the Antarctic continent push this ice sheet toward the sea, but floating sea ice provides a restraining barrier and reduces the rate of glacial flow from the land to the sea. If air temperature were to warm by a few degrees, sea surface temperature would warm slightly, and the floating sea ice might melt rapidly during the summer. In turn, if the sea ice melted, large chunks of the West Antarctic Ice Sheet could then flow into the sea, break off, float northward, and melt rapidly in the warm ocean water. Global sea level would rise rapidly, flooding coastal regions worldwide.

The hypothesis outlined above is a classic model of how Earth systems combine to produce a threshold effect. If the model is correct, then a small rise in air temperature will raise the water temperature. This warmer water will melt a small amount of sea ice. When that sea ice melts, it releases a large amount of glacial ice. Thus, sea level rises, altering coastlines and disturbing the biosphere.

Consequences of a sea-level rise vary with location and economics. The wealthy, developed nations could build massive barriers to protect cities and harbors from a small sea-level rise. In regions where global sea-level rise is compounded by local tectonic sinking, dikes are already in place or planned. Portions of Holland lie below sea level, and the land is protected by a massive system of dikes. In London, where the high-tide level has risen by 1 meter in the past century, multimillion-dollar storm gates have been built on the Thames River. However, it is unlikely that people could protect against a dramatic sea-level rise. Coastal cities worldwide would be inundated.

Many poor countries cannot afford coastal protection even for a small sea-level rise. A 1-meter rise in

Figure 14–39 A 1-meter sea-level rise would flood 17 percent of Bangladesh and displace 38 million people. *(Wide World Photos)*

sea level would flood 17 percent of the land area of Bangladesh displacing 38 million inhabitants (Fig. 14–39).

How likely is this scenario? We don't know. First, no one is certain that global warming will continue. Second, air temperature doesn't rise evenly across the globe, but the heat is distributed and concentrated by winds and other factors. Third, ocean temperature doesn't respond evenly to changes in air temperature, but is distributed and concentrated by vertical and horizontal currents. Fourth, our models of glacial flow aren't precise, so scientists aren't sure how the Antarctic glaciers would respond to changes in the sea ice.

SUMMARY

Sea water contains about 3.5 percent dissolved salts. The upper layer of the ocean, about 450 meters thick, is relatively warm. In the thermocline, below the warm surface layer, temperature drops rapidly with depth. Deep ocean water is consistently around 1° to 2.5°C.

Tides are caused by gravitational pull of the Moon and Sun. Two high tides and two low tides occur approximately every day.

Wave size depends on (1) wind speed, (2) the amount of time the wind has blown, and (3) the distance that the wind has traveled. The highest part of a wave is the **crest;** the lowest, the **trough.** The distance between successive crests is called the **wavelength. Wave height** is the vertical distance from the crest to the trough. The water in a wave moves in a circular path.

A **current** is a continuous flow of water in a particular direction. Surface currents are driven by wind and deflected by the **Coriolis effect.** Deep-sea currents are driven by differences in sea water density. Cold, salty water is dense and therefore sinks and flows along the sea floor. When a surface current is deflected offshore, cold deep water rises to replace it in a process called **upwelling.**

When a wave nears the shore, the bottom of the wave slows and the wave **breaks,** creating **surf.** Ocean waves weather and erode rock by hydraulic action, abrasion, solution, and salt cracking.

Refraction is the bending of a wave as it strikes shore at an oblique angle. Refracted waves often form **longshore currents** that transport sediment along a shore. **Tidal currents** also transport sediment in some areas. Irregular coastlines are straightened by erosion and deposition.

A **beach** is a strip of shoreline washed by waves and tides. A **reef** is a wave-resistant ridge or mound built by corals and other organisms. If land rises or sea level falls, the coastline migrates seaward and old beaches are abandoned above sea level, forming an **emergent coastline.** In contrast, a **submergent coastline** forms when land sinks or sea level rises.

Most coastal sediment is transported to the sea by rivers. Glacial drift, reefs, and local erosion also add sediment in certain areas. Coastal emergence may expose large amounts of sand. **Spits, baymouth bars,** and **barrier islands** are common on sandy coastlines. A rocky coast is dominated by **wave-cut cliffs, wave-cut platforms, arches,** and **stacks.** A **fjord** is a submerged glacial valley. An **estuary** is a submerged river bed and floodplain.

Human intervention such as the building of **groins** may upset the natural movement of coastal sediment and alter patterns of erosion and deposition on beaches. Sea level has risen over the past century and may continue to rise into the next.

KEY TERMS

salinity 314
spring tide 316
neap tide 316
crest 318
trough 318
wavelength 318
wave height 318
current 318
Gulf Stream 319
gyres 319

Coriolis effect 320
upwelling 323
breaks 323
surf 323
refraction 325
longshore current 326
beach drift 326
tidal current 326
beach 327
foreshore 327

backshore 327
intertidal zone 327
reef 327
emergent coastline 328
submergent coastline 328
eustatic sea level 328
spit 330
baymouth bar 330
barrier island 330
lagoon 330

wave-cut cliff 330
wave-cut platform 332
sea arch 332
sea stack 332
fjord 332
estuary 332
groin 334

FOR REVIEW

1. Name the major ocean basins and give their locations.

2. Using Figure 14–1, what percentage of the Earth's area (a) lies above an elevation of 840 meters, (b) lies 4 kilometers or more below sea level, (c) lies 5 kilometers or more below sea level?

3. What factors affect the salinity of seawater near shore?

4. Describe the temperature profile of the open oceans.

5. Explain why two high tides occur every day even though the Moon lies directly above any portion of the Earth only once a day.

6. Explain the difference between spring and neap tides.

7. List the three factors that determine the size of a wave.

8. Draw a picture of a wave and label the crest, the trough, the wavelength, and the wave height.

9. Describe the motion of both the surface and the deeper layers of water as a wave passes.

10. Explain the Coriolis effect. What is a gyre?

11. What drives surface currents, deep-sea currents, tidal currents, and longshore currents?

12. Explain how surf forms.

13. What is refraction? How does it affect coastal erosion?

14. How is a tidal current different from a longshore current?

15. Draw a picture of a beach and name the major zones along the interface between land and sea.

16. What is an emergent coastline, and how does it form? What is a submergent coastline, and how does it form?

17. List the major mechanisms of weathering and erosion along coastlines.

18. List three different sources of coastal sediment.

19. Compare and contrast a beach, a barrier island, and a spit.

20. Explain how coastal processes straighten an irregular coastline.

21. Describe some dominant features of sediment-rich and sediment-poor coastlines.

22. What is a groin? How does it affect the beach in its immediate vicinity? How does it affect the entire shoreline?

23. Explain how warming of the atmosphere could lead to a rise in sea level.

FOR DISCUSSION

1. Earthquake waves were discussed in Chapter 6. Compare and contrast earthquake waves with water waves.

2. How can a ship survive 10-meter-high storm waves, whereas a beach house can be smashed by waves of the same size?

3. During World War II, few maps of the underwater profile of shorelines existed. When planning amphibious attacks on the Pacific islands, the Allied commanders needed to know near-shore water depths. Explain how this information could be deduced from aerial photographs of breaking waves and surf.

4. Imagine that an oil spill occurs from a tanker accident. Discuss the effects of mid-ocean currents, deep-sea currents, longshore currents, storm waves, and tides on the dispersal of the oil.

5. Compare and contrast coastal erosion with stream erosion.

6. In Section 14.9 we explained how erosion and deposition tend to smooth out an irregular coastline by eroding headlands and depositing sediment in bays. If coastlines are affected in this manner, why haven't they all been smoothed out in the 4.6-billion-year history of the Earth?

7. The text states that many submergent coastlines are sediment-poor. The east coast of the United States, between Long Island and Georgia, is a predominantly submergent, sediment-rich coastline. Where did the sediment come from?

8. Prepare a three-way debate. Have one side argue that the government should support the construction of groins. Have the second side argue that the government should prohibit the construction of groins. Have the third position defend the argument that groins should be permitted, but not supported.

9. Prepare another debate on whether government funding should be used to repair storm damage to property on barrier islands.

10. Solutions to environmental problems can be divided into two general categories: social solutions and technical solutions. Social solutions involve changes in attitudes and life-styles but do not generally require expensive industrial or technological processes. Technical solutions do not mandate social adjustments, but require advanced and often expensive engineering. Describe a social solution and a technical solution to the problem of coastline changes in Long Island. Which approach do you feel would be more effective?

11. Imagine that sea level rises by 25 centimeters in the next century. How far inland would the shoreline advance if the beach (a) consisted of a vertical cliff, (b) sloped steeply at a 45° angle, (c) sloped gently at a 5° angle, (d) were almost flat and rose only 1°?

12. Explain why sea level does not necessarily rise in a linear relationship to air temperature.

13. Explain how changes in the atmosphere can alter the hydosphere in the form of changing ocean currents. Explain how a feedback mechanism can develop where a change in ocean currents can then change the atmosphere even more, thereby affecting the biosphere.

The Atmosphere

The atmosphere interacts with the biosphere, hydrosphere, and geosphere so that all four spheres are associated components of one complex system. *(Galen Rowell/Peter Arnold, Inc.)*

Origin of Iron Ore and the Evolution of Earth's Atmosphere, Biosphere, and Oceans

Iron, the main ingredient in steel, is the world's most commonly used metal. About 1 billion tons of iron are mined every year, 90 percent from **banded iron formations**, sedimentary layers of iron-rich minerals sandwiched between beds of clay and other silicate minerals. The alternating layers are a few centimeters thick and give the rocks their banded appearance (Fig. 1). A single iron formation of this type may be hundreds of meters thick and cover tens of square kilometers (Fig. 2).

Most of these industrially critical iron deposits accumulated in the seas between 2.6 and 1.9 billion years ago. Few rocks of this type formed before that time, and even fewer formed afterward. What happened during that interval to create these peculiar rocks?

The answer to this question involves two processes: The first is the manner in which oxygen, water, and iron interact at the Earth's surface. The second is the evolution of the Earth's atmosphere from an oxygen-poor composition to an oxygen-rich one.

Oxygen, Water, and Iron

Oxygen dissolves in water. As a result, the oxygen content of the atmosphere is directly linked to the oxygen content of seawater. When the oxygen concentration in the atmosphere rises, more oxygen dissolves in seawater.

In an oxygen-poor environment, iron also dissolves in water. However, when oxygen is abundant, it reacts with dissolved iron and causes the iron to precipitate rapidly.

Thus, if air contains little or no oxygen, there is also little oxygen in seawater, and large amounts of iron dissolve in the seas. If the oxygen concentration of the atmosphere rises, some of the oxygen dissolves in seawater. The oxygen concentration in seawater may rise until it eventually reaches a threshold at which iron precipitates rapidly, forming a layer of iron oxide minerals on the sea floor. With this background, we will consider how the changing atmosphere and evolving biosphere interacted with seawater to form these iron-rich rocks.

The Early Atmosphere

In the opening essay of Unit III, we described how the Earth's early atmosphere accumulated as volatile-rich bolides crashed into Earth, bringing ice, carbon dioxide, and other gases with them. These volatiles, imported from outer regions of the Solar System, made up at least 90 percent of the primitive atmosphere. Oxygen in the form of O_2 (molecular oxygen—the form needed by most modern animals and plants) was present only in tiny amounts. This atmosphere, formed during the first 800 million years of Earth history, would suffocate most modern organisms (Fig. 3A).

Figure 1 In this banded iron formation from Michigan, the red bands are iron oxide minerals and the dark layers are chert (silica). *(Barbara Gerlach/Visuals Unlimited)*

Figure 2 The Mesabi Range in Northern Minnesota is a banded iron formation that is one of the chief iron-producing regions in the world. *(Lillian Bolstad/Peter Arnold, Inc.)*

Over time, water filled the ocean basins and atmospheric composition changed. Some of the carbon dioxide dissolved into the oceans. There, it reacted with other elements in seawater to precipitate as limestone. By 3.8 billion years ago, the atmosphere had lost much of its carbon dioxide. It was rich in nitrogen, but it contained very little molecular oxygen. As a result, seawater also contained little dissolved oxygen, but large quantities of dissolved iron (Fig. 3B).

Life, Oxygen, and Iron Ore

Recall from Chapter 4 that the oldest known fossils date back to about 3.5 billion years ago, although life probably originated earlier. Early organisms developed the ability to conduct photosynthesis, combining carbon dioxide and water in the presence of sunlight to form glucose (sugar) and oxygen. They used glucose for food and released the oxygen into the atmosphere. In this way, photosynthesizers convert solar energy to chemical energy.

As these single-celled organisms evolved and multiplied during the following 3 billion years, they released additional oxygen into the atmosphere (Fig. 3C). At first, there were few organisms and the oxygen content of the early atmosphere and oceans rose only slowly and slightly. However, over geologic time, photosynthesizers became abundant and the concentration of oxygen in both the atmosphere and the oceans increased. About 2.6 billion years ago, enough oxygen had accumulated in seawater to reach a threshold at which it reacted with the dissolved iron. As a result, vast quantities of iron minerals precipitated onto the sea floor, producing a layer of iron-rich minerals (Fig. 3D).

The alternating layers of iron minerals and other minerals may have developed because, for a long time, the oxygen level in the seas hovered near the threshold at which soluble iron converts to the insoluble variety. When the dissolved oxygen concentration increased, the oxygen reacted with the dissolved iron to precipitate iron oxide minerals on the sea floor. But the formation of those minerals extracted oxygen as well as iron from the seawater, and lowered its oxygen concentration below the threshold. Then, dissolved iron accumulated again in the seas while plants slowly replenished the oxy-

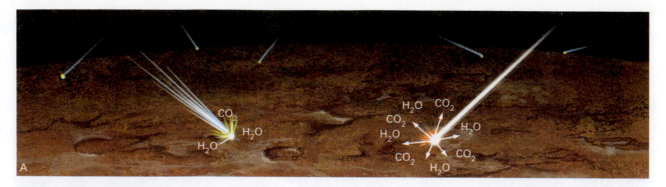

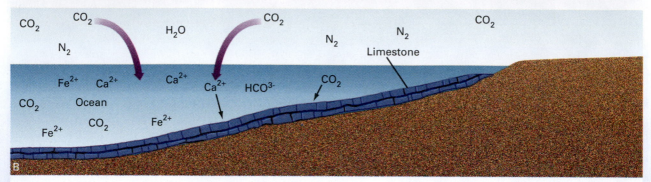

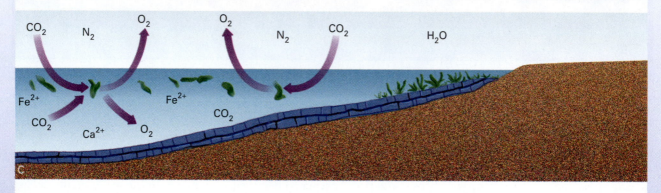

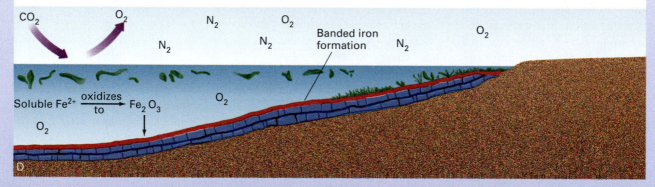

Figure 3 The evolution of the atmosphere. (A) Comets, meteoroids, and asteroids imported the Earth's early atmosphere of water, carbon dioxide, ammonia, simple organic molecules, and other gases. (B) Much of the carbon dioxide dissolved in seawater where it reacted with other compounds to form limestone. As a result, the atmosphere lost much of its carbon dioxide. (C) Later, the oxygen content of the atmosphere gradually increased as Archean and Proterozoic aquatic plants released oxygen. (D) The oxygen concentration in the atmosphere and seawater reached a threshold at which the oxygen combined with dissolved iron to form iron minerals. The iron minerals precipitated to the sea floor to form the first layer of a banded iron formation.

gen. During that time, clay and other minerals washed from the continents and accumulated on the sea floor as they do today, forming the thin layers of silicate minerals that lie between the iron-rich layers. When the oxygen concentration rose above the threshold again, another layer of iron minerals formed.

Banded iron formations contain thousands of alternating layers of iron minerals and silicates. The great thickness of the iron formations, coupled with the fact that they continued to form from 2.6 to 1.9 billion years ago, suggests that these reactions must have kept the levels of dissolved oxygen close to the threshold level for 700 million years. Oxygen levels in both seawater and the atmosphere rose above the threshold only after nearly all of the dissolved iron had been removed from seawater by precipitation of iron minerals.

Thus, the iron-rich rocks that support our industrial society formed by interactions among early photosynthetic organisms, air, and the oceans. Note that the banded iron formations precipitated on the ocean floor. Long after they formed, tectonic processes raised the iron ore deposits onto the land surface, where they are mined today.

Evolution of the Modern Atmosphere

The atmosphere continued to evolve long after precipitation of the banded iron formations ceased about 1.9 billion years ago. Slowly increasing numbers of photosynthetic organisms continued to add molecular oxygen to the atmosphere. But for the following 1.3 billion years, the organisms continued to be biologically simple, and oxygen levels rose at a slow pace.

Then, about 543 million years ago at the beginning of the Cambrian period, multicellular plants and animals suddenly became abundant. According to one hypothesis, abundant multicellular organisms could not have emerged earlier because the oxygen concentration was too low to support them. However, 543 million years ago, the atmospheric oxygen concentration reached a new threshold at which the rapid emergence of new species, and proliferation of great numbers of individuals, became possible. As the numbers and diversity of organisms rose, increasing photosynthetic activity added molecular oxygen to the atmosphere rapidly.

The Earth's atmosphere reached its present oxygen level of 21 percent about 450 million years ago,

93 million years after the sudden bloom of plants and animals. Today, seawater contains very little dissolved iron. The lack of iron limits plant growth in the central oceans and contributes to the fact that these regions are now biological deserts.

Fires burn rapidly if oxygen is abundant; if its concentration were to increase even by a few percent, fires would burn uncontrollably across the planet. If atmospheric oxygen levels were to decrease appreciably, most modern plants and animals would not survive. If the carbon dioxide concentration were to increase by a small amount, atmospheric temperature would rise as a result of greenhouse warming.

The compatibility of Earth's atmosphere with its organisms is not a lucky accident. Organisms did not simply adapt to an existing atmosphere; they partially created it by photosynthesis and respiration. Thus, not only is our delicate oxygen–carbon dioxide balance biologically maintained, but the very presence of oxygen in our atmosphere can be explained only by biological activity. If all life on Earth were to cease, the atmosphere would revert to an oxygen-poor composition and become poisonous to modern plants and animals.

The Earth's Atmosphere

Canada geese flying with cumulus clouds.   *(Robert Sisk/Dembinsky Photo Associates)*

The Earth's atmosphere contains oxygen essential to both plants and animals, and carbon dioxide needed by plants. The atmosphere insulates the Earth's surface, and distributes the Sun's heat around the globe so the surface is neither too hot nor too cold for life. Clouds form from water vapor in the atmosphere, and rain falls from clouds. In addition, the atmosphere filters out much of the Sun's ultraviolet radiation, which can destroy living tissue and cause cancer. The atmosphere carries sound; without air we would live in silence. Without an atmosphere, airplanes and birds could not fly, wind would not transport pollen and seeds, the sky would be black rather than blue, and no reds, purples, and pinks would color the sunset.

When you wake up in the morning and look out the window, you may remark, "It is a nice day," or "What a nasty gray day." Your remarks refer to the **weather,** the state of the atmosphere at a given place and time. Temperature, wind, cloudiness, humidity, and precipitation are all components of weather. The weather changes frequently, from day to day or even hour to hour.

Climate is the characteristic weather of a region, particularly the temperature and precipitation, averaged over several decades. Miami and Los Angeles have warm climates; summers are hot and even the winters are warm. In contrast, New York and Chicago have cool climates with winter snow. Seattle experiences moderate temperatures with foggy, cloudy winters. In this and the following two chapters, we discuss the atmosphere, weather, and climate. In Chapters 18 and 19, we will consider global climate changes, global warming, and the effects of air pollution on our atmosphere. ●

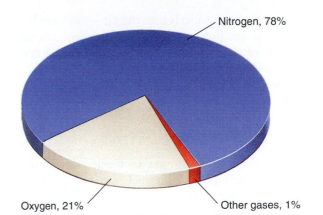

Figure 15–1   The Earth's atmosphere consists mostly of nitrogen and oxygen. Table 15–1 lists the other gases

15.1   The Composition of the Atmosphere

Recall from the opening essay to Unit III that the primordial atmosphere formed mainly from gases transported by comets and meteorites. This original atmosphere then evolved as the gases interacted with water, rock, and living organisms. The modern atmosphere is mostly gas with small quantities of water droplets and dust. The gaseous composition of dry air is roughly 78 percent nitrogen, 21 percent oxygen, and 1 percent other gases (Fig. 15–1 and Table 15–1). Nitrogen, the most abundant gas, does not react readily with other substances. Oxygen, on the other hand, reacts chemically as fires burn, iron rusts, and plants and animals respire.

In addition to the gases listed in Table 15–1, air contains water vapor, water droplets, and dust. The types and quantities of these components vary with both location and altitude. Most natural air contains some water vapor. In a hot, steamy jungle, air may contain 5 percent water vapor by weight, whereas in a desert or cold polar region, only a small fraction of a percent may be present.

If you sit in a house on a sunny day, you may see a sunbeam passing through a window. The visible beam is light reflected from tiny specks of suspended dust. Clay, salt, pollen, bacteria, viruses, bits of cloth, hair, and skin are all components of dust. People travel to the seaside to enjoy the "salt air." Visitors to the Great Smoky Mountains in Tennessee view the bluish, hazy air formed by sunlight reflecting from pollen and other dust particles.

Within the past century, humans have altered the chemical composition of the atmosphere in many different ways. We have increased the carbon dioxide concentration by burning fuels and igniting wildfires.

TABLE 15–1

Gaseous Composition of Natural Dry Air*

Gas	Concentration (%)
Nitrogen, N_2	78.09
Oxygen, O_2	20.94
Inert gases, mostly argon, with much smaller concentrations of neon, helium, krypton, and xenon	0.93
Carbon dioxide, CO_2	0.03
Methane, CH_4, a natural part of the carbon cycle	0.0001
Hydrogen, H_2	0.00005
Oxides of nitrogen, mostly N_2O and NO_2, both produced by solar radiation and by lightning	0.00005
Carbon monoxide, CO, from oxidation of methane and other natural sources	0.00003
Ozone, O_3, produced by solar radiation and by lightning	Trace

*Natural dry air is defined as air without water or industrial pollutants. Carbon dioxide, methane, oxides of nitrogen, carbon dioxide, and ozone are all components of natural air, but they are also industrial pollutants. Therefore, the concentrations of these gases may vary, especially in urban areas. Pollution and its consequences are discussed in Chapter 19.

Factories release chemicals into the air—some are benign, others are poisonous. Smoke and soot change the clarity of the atmosphere. In upcoming chapters we will discuss these changes and consider how they affect our lives.

15.2 Atmospheric Pressure

Gravity pulls atmospheric gas molecules downward, but they are held aloft by thermal (heat) energy (Fig. 15–2). As a result of these two opposing influences, the atmosphere is densest near the Earth's surface and grows less dense with increasing altitude. Anyone who

has ever climbed a high mountain has experienced the effects of thin air. At 3000 meters (about 10,000 feet), even a person in good physical condition notices that breathing is more difficult than at sea level. At 4500 meters (about 15,000 feet), a person's actions slow considerably, and above 6000 meters (about 20,000 feet), climbers move slowly and breathe with difficulty.

The pressure exerted by air is the **barometric pressure,** and a device that measures barometric pressure is a **barometer.** A simple but accurate barometer is constructed from a glass tube that is sealed at one end. The tube is evacuated and the open end is placed in a dish of a liquid such as mercury. The mercury rises in the tube because the weight of the atmosphere pushes

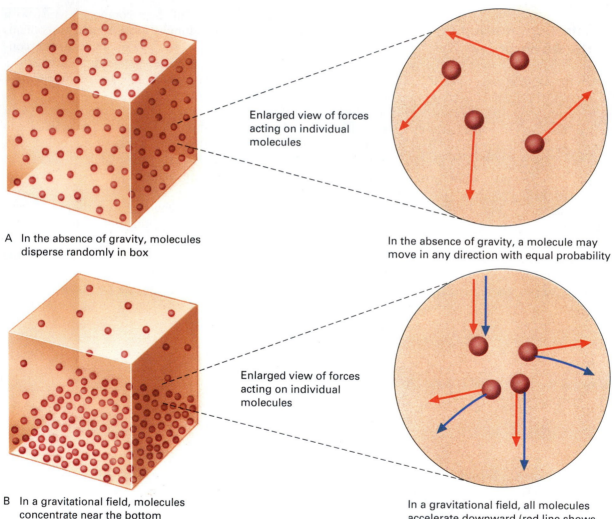

A In the absence of gravity, molecules disperse randomly in box

Enlarged view of forces acting on individual molecules

In the absence of gravity, a molecule may move in any direction with equal probability

Enlarged view of forces acting on individual molecules

B In a gravitational field, molecules concentrate near the bottom

In a gravitational field, all molecules accelerate downward (red line shows movement in the absence of gravity; blue line shows velocity in a gravitational field)

Figure 15–2   The atmosphere is densest near the Earth's surface because gravity pulls air molecules downward. (A) Behavior of a gas in the absence of gravity and (B) in a gravitational field.

downward on the mercury in the dish but there is no air in the tube (Fig. 15–3). At sea level mercury rises approximately 76 centimeters, or 760 millimeters (about 30 inches), into an evacuated tube. Barometric pressure is commonly expressed in three different unit systems. Meteorologists express pressure in inches or millimeters of mercury, referring to the height of a column of mercury in a barometer. They also express pressure in bars and millibars. A bar is approximately equal to sea level atmospheric pressure. A millibar is 1/1000 of a bar.

A mercury barometer is a cumbersome device nearly a meter tall, and mercury vapor is poisonous. A safer and more portable instrument for measuring pressure, called an aneroid barometer, consists of a partially evacuated metal chamber connected to a pointer. When atmospheric pressure increases, it compresses the chamber and the pointer moves in one direction. When pressure decreases, the chamber expands, directing the pointer the other way (Fig. 15–4).

Two different factors cause air pressure to vary. One is changing weather. On a stormy day at sea level, pressure may be 980 millibars (28.94 inches), although it has been known to drop to 900 millibars (26.58 inches) or less during a hurricane. In contrast, during a period of clear, dry weather, a typical high pressure reading may be 1025 millibars (30.27 inches). These changes are discussed in the next chapter.

In addition, pressure decreases with altitude (Fig. 15–5). Mountain climbers and airplane pilots routinely use an **altimeter**, a barometer with a scale calibrated in units of elevation rather than pressure. If you stand on the summit of a mountain 5600 meters (18,400 feet) high, half of the atmosphere lies below you and you must survive on about half as much oxygen as you breathe at sea level. As a reference, the summit of Denali, the highest mountain in North America, is about 6200 meters high. If you ascended in a balloon to 16 kilometers (10 miles) above sea level, you would be above 90 percent of the atmosphere and would need an oxygen mask to survive. At an elevation of 100 kilometers, pressure is only 0.00003 that of sea level, approaching the vacuum of outer space. There is no absolute upper boundary to the atmosphere.

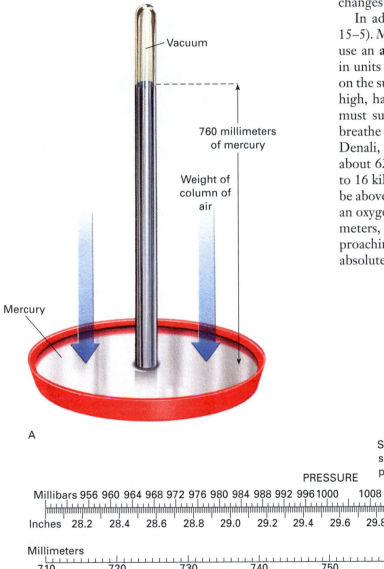

A

Figure 15–3 (A) Atmospheric pressure forces mercury upward in an evacuated glass tube. The height of the mercury in the tube is a measure of air pressure. (B) Three common scales for reporting atmospheric pressure and the conversion among them. ● **Interactive Question:** Express 750 mm in inches and millibars.

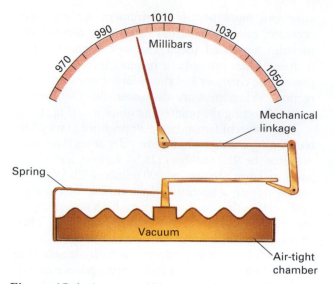

Figure 15–4 In an aneroid barometer, increasing air pressure compresses a chamber and causes a connected pointer to move in one direction. When the pressure decreases, the chamber expands, deflecting the pointer the other way.

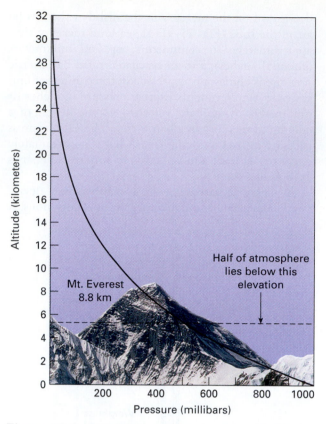

Figure 15–5 Atmospheric pressure decreases with altitude. One-half of the atmosphere lies below an altitude of 5600 meters. *(Ted Kerasote/Photo Researchers)*

15.3 Solar Radiation

Solar energy streams from the Sun in all directions, and Earth receives only 1/2,000,000,000 (one two-billionth) of the total solar output. However, even this tiny fraction warms Earth's surface and makes it habitable.

The space between Earth and the Sun is nearly empty. How does sunlight travel through a vacuum? In the late 1600s and early 1700s, light was poorly understood. Isaac Newton postulated that light consists of streams of particles that he called "packets" of light. Two other physicists, Robert Hooke and Christian Huygens, argued that light travels in waves. Today we know that Hook and Huygens were correct—light behaves as a wave; but Newton was also right—light acts as if it is composed of particles. But how can light be both a wave and a particle at the same time? In a sense this is an unfair question because light is fundamentally different from familiar objects. Light is unique; it behaves as a wave and a particle simultaneously.

Particles of light are called **photons.** In a vacuum, photons travel only at one speed, the speed of light, never faster and never slower. The speed of light is 3×10^8 meters/second. At that rate a photon covers the 150 million kilometers between Sun and Earth in about 8 minutes. Photons are unlike ordinary matter in that they appear when they are emitted and disappear when they are absorbed.

Light also behaves as an electrical and magnetic wave, called **electromagnetic radiation.** Its wavelength is the distance between successive wave crests (Fig. 15–6). The frequency of a wave is the number of complete wave cycles, from crest to crest, that pass by

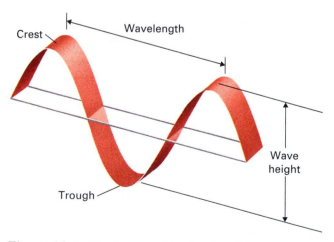

Figure 15–6 The terms used to describe a light wave are identical to those used for water, sound, and other types of waves.

The Upper Fringe of the Atmosphere

How high is the top of the atmosphere? Figure 15–5 shows that atmospheric pressure gradually diminishes with altitude, but never reaches zero, and no upper boundary of the atmosphere is apparent on the graph. One definition of the top of the Earth's atmosphere is that it is where the pressure is about equal to the background pressure in the Solar System.

Most of the gases in space between the planets come from the Sun. The outer fringe of the Sun is the **corona,** which can be observed as a beautiful halo during a full solar eclipse. This region is so hot, about 2,000,000°C, that ordinary matter cannot exist. Electrons are stripped from atoms, and gases become a collection of bare nuclei in a sea of electrons. Many of these nuclei and electrons are so hot and are moving so fast that they escape the Sun's gravity and fly off into space. This stream of charged particles is called the **solar wind.** It blows past

the Earth and other planets and flies outward toward the far reaches of the Solar System. When you think about the solar wind, do not think of a gentle breeze blowing against your face. At the Earth's surface, the Earth's atmosphere contains approximately 10^{19} gas molecules per cubic centimeter, and what you feel when an Earth wind blows is the effect of these molecules striking your cheek. However, the solar wind contains only 5 particles per cubic centimeter and certainly could not be felt against the skin. However, it is forceful enough to move objects in space. For example, it blows volatiles away from a comet to form the characteristic comet's tail. The outer boundary of the Earth's atmosphere is approximately 9600 kilometers (6000 miles) high where atmospheric density approaches 5 particles per cubic centimeter, the density of the solar wind.

any point in a second. (Think of how *frequently* the waves pass by.) Electromagnetic radiation occurs in a wide range of wavelengths and frequencies. The **electromagnetic spectrum** is the continuum of radiation of different wavelengths (Fig. 15–7). At one end of the spectrum, radiation given off by ordinary household current has a long wavelength (5000 kilometers) and low frequency (60 cycles/second). At the other end, cosmic rays from outer space have a short wavelength

(about one trillionth of a centimeter, or 10^{-14} meter), and very high frequency (10^{22} cycles/second). Visible light is a tiny portion (about one millionth of one percent) of the electromagnetic spectrum.

Emission

All objects emit radiant energy at some wavelength (except at a temperature of absolute zero). An iron bar at room temperature emits infrared radiation. This

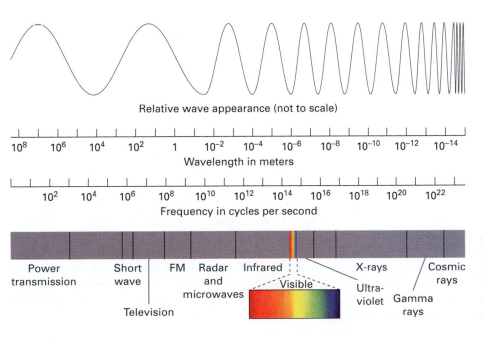

Figure 15–7 The electromagnetic spectrum. The wave shown is not to scale. In reality the wavelength varies by a factor of 10^{22}, and this huge difference cannot be shown.

Figure 15–8 Iron glows red when heated in a forge. If it is heated further, it emits white light. *(Greg Pease/Tony Stone Images)*

radiation has low energy and long wavelengths. It is invisible because the energy is too low to activate the sensors in our eyes. Infrared radiation is sometimes called "radiant heat," or "heat rays." Thus, even at room temperature an iron bar radiates heat.

If you place the bar in a hot flame, it begins to glow with a dull red color when it becomes hot enough. The heat has excited electrons in the iron bar, and the excited electrons then emit visible, red electromagnetic radiation (Fig. 15–8). If you heat the bar further, it gradually changes color until it becomes white. This demonstration shows another property of emitted electromagnetic radiation: The wavelength and color of the radiation are determined by the temperature of the source. With increasing temperature, the energy level of the radiation increases, the wavelength decreases, and the color changes progressively.

Transmission

Radiation can travel through space for trillions upon trillions of kilometers with no change in wavelength or loss of energy. Radiation is also transmitted directly through transparent media. Thus, visible light passes through a window or through clear air. When light is transmitted through a perfectly transparent medium,

no change in wavelength or loss of energy occurs. All materials are transparent to some wavelengths and opaque (not transparent) to others. For example, skin and muscle are opaque to visible light but transparent to X-rays. The walls of your house are opaque to visible light but transparent to radio waves, which is why a radio picks up a signal even though the radio is inside a house.

Absorption

A black cast-iron frying pan is not transparent to visible light. When light strikes its surface, a small amount is reflected, but most is absorbed. When an object absorbs radiation, the photons disappear and convert to another form of energy. If you go outside on a sunny day, your face absorbs energy and feels warm. Similarly rock and soil absorb radiation, and a sunlit cliff can be warm even on a winter afternoon. Thus, radiant energy converts to heat.

The Sun's surface temperature is about 6000°C. Because of its high temperature, the Sun emits relatively high-energy (short-wavelength) radiation, primarily in the ultraviolet and visible portions of the spectrum. When this radiation strikes Earth, it is absorbed by rock and soil. After the radiant energy is absorbed, the rock and soil re-emit it. But the Earth's surface is much cooler than the Sun's. Therefore the Earth emits low-energy infrared heat radiation, which has relatively long wavelength and low frequency. Thus, the Earth absorbs high-energy visible light and emits low-energy, invisible infrared heat radiation.

Reflection

Radiation reflects from many surfaces. We are familiar with the images reflected by a mirror or the surface of a still lake (Fig. 15–9). Some surfaces are better reflectors than others. **Albedo** is the proportional reflectance of a surface. A mirror reflects nearly 100 percent of the light that strikes it, and has an albedo close to 100 percent. Even some dull-looking objects are efficient reflectors. Light is bouncing back to your eye from the white paper of this page, although very little is reflected from the black letters.

Snowfields and glaciers have high albedos and reflect 80 to 90 percent of sunlight. On the other hand, city buildings and dark pavement have albedos of only 10 to 15 percent (Fig. 15–10). The oceans, which cover about two thirds of the Earth's surface, have a low albedo. As a result, they absorb considerable solar energy and thus strongly affect the Earth's radiation balance. If the Earth's albedo were to rise by growth of glaciers or increase in cloud cover, the surface of our planet would cool; alternatively, a decrease in albedo would cause warming.

Figure 15–9 A mountain lake reflects the surrounding landscape of the Beartooth Mountains in southern Montana.

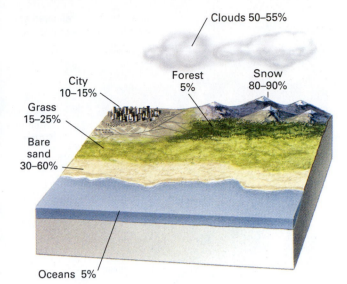

Figure 15–10 The albedos of common Earth surfaces vary greatly.

Scattering

On a clear day the Sun shines directly through windows on the south side of a building, but if you look through a north-facing window, you cannot see the Sun. Even so, light enters through the window, and the sky outside is blue. If sunlight were only transmitted directly, a room with north-facing windows would be dark and the sky outside the window would be black. Atmospheric gases, water droplets, and dust particles scatter sunlight in all directions as shown in Figure 15–11. It is this scattered light that illuminates a room with north-facing windows and turns the sky blue.

The amount of scattering is inversely proportional to the wavelength of light. Short-wavelength blue light, therefore, scatters more than longer-wavelength red light. The Sun emits light of all wavelengths, which combine to make up white light. Consequently, in space the Sun appears white. The sky appears blue from the Earth's surface because the blue component

of sunlight scatters and colors the atmosphere. The Sun appears yellow from Earth because yellow is the color of white light with most of the blue light removed.

15.4 The Radiation Balance

With this background, let us examine the fate of sunlight as it reaches Earth (Fig. 15–12).

The atmosphere and clouds scatter, reflect, and absorb one half of the incoming solar radiation. The other half reaches the Earth's surface. The surface reflects 3 percent of this radiation, and absorbs 47 percent. The absorbed radiation warms rocks, soil, and water.

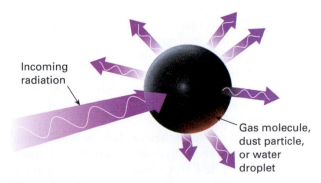

Figure 15–11 Atmospheric gases, water droplets, and dust scatter incoming solar radiation. When radiation scatters, its direction changes but the wavelength remains constant.

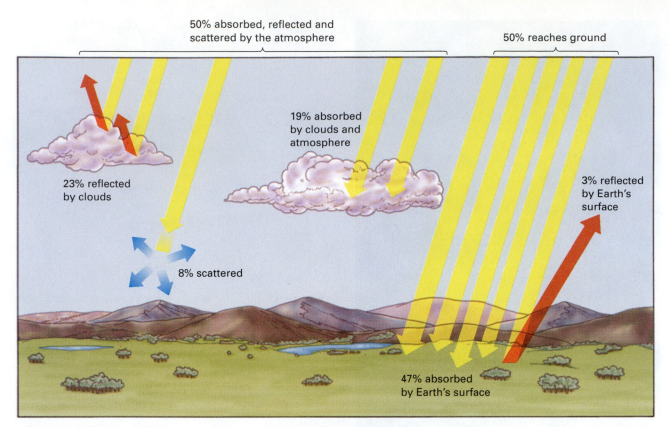

50% absorbed, reflected and scattered by the atmosphere

50% reaches ground

23% reflected by clouds

19% absorbed by clouds and atmosphere

3% reflected by Earth's surface

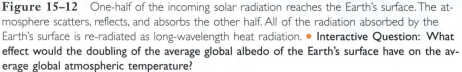
8% scattered

47% absorbed by Earth's surface

Figure 15–12 One-half of the incoming solar radiation reaches the Earth's surface. The atmosphere scatters, reflects, and absorbs the other half. All of the radiation absorbed by the Earth's surface is re-radiated as long-wavelength heat radiation. ● **Interactive Question:** What effect would the doubling of the average global albedo of the Earth's surface have on the average global atmospheric temperature?

If the Earth absorbs radiant energy from the Sun, why doesn't the Earth's surface get hotter and hotter until the oceans boil and the rocks melt? The answer is that rocks, soil, and water re-emit all the energy they absorb. As explained above, most solar energy that reaches the Earth is short-wavelength visible and ultraviolet radiation. The Earth's surface absorbs this radiation and then re-emits the energy mostly as long-wavelength, invisible infrared (heat) radiation. Some of this infrared heat escapes directly into space, but some is absorbed by the atmosphere. The atmosphere traps this heat radiating from Earth and acts as an insulating blanket.

If Earth had no atmosphere, radiant heat loss would be so rapid that the Earth's surface would cool drastically at night. The Earth remains warm at night because the atmosphere absorbs and retains much of the radiation emitted by the ground. If the atmosphere were to absorb even more of the long-wavelength radiant heat from the Earth, the atmosphere and Earth's

surface would become warmer. This warming process is called the **greenhouse effect**[1] (Fig. 15–13).

Some gases in the atmosphere absorb infrared radiation and others do not. Oxygen and nitrogen, which together make up almost 99 percent of dry air at ground level, do not absorb infrared radiation; water, carbon dioxide, methane, and a few other gases do. Thus, they are called greenhouse gases. Water is the most important greenhouse gas because it is so abundant. Carbon dioxide is also important because its abundance in the atmosphere can vary as a result of

---

[1] The comparison between the atmosphere and a greenhouse is only partially correct. Both the glass in a greenhouse and the Earth's atmosphere are transparent to incoming short-wavelength radiation and partially opaque to emitted long-wavelength heat radiation. However, the glass in a greenhouse is also a physical barrier that prevents heat loss through air movement, whereas the atmosphere is not. We use the term greenhouse effect because it has become common, both in atmospheric science and in everyday use.

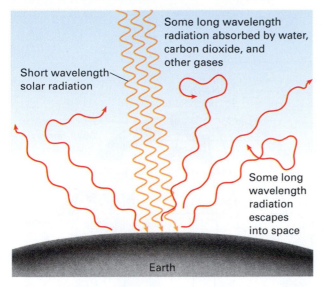

Figure 15–13 The greenhouse effect can be viewed as a three-step process: 1. Rocks, soil, and water absorb short-wavelength solar radiation, and become warmer. 2. The Earth re-radiates the energy as long-wavelength infrared heat rays. 3. Molecules in the atmosphere absorb some of the heat, and the atmosphere becomes warmer. • Interactive Question: What effect would a doubling of the atmospheric concentration of carbon dioxide have on the average global atmospheric temperature?

several natural and industrial processes. Methane has become important because large quantities are released by industry and agriculture. Global climate change is discussed in Chapter 18.

The temperature of the atmosphere changes with altitude (Fig. 15–14). The layer of air closest to the Earth, the layer we live in, is the **troposphere.** Virtually all of the water vapor and clouds exist in this layer, and almost all weather occurs here. The Earth's surface absorbs solar energy, and thus the surface of the planet is warm. But, as explained above, continents and oceans also radiate heat, and some of this energy is absorbed in the troposphere. At higher elevations in the troposphere, the atmosphere is thinner and absorbs less energy; in addition, lower parts of the troposphere have absorbed much of the heat radiating from the Earth's surface. Consequently, temperature decreases at higher levels in the troposphere. Thus, mountaintops are generally colder than valley floors, and pilots flying at high altitudes must heat their cabins.

The top of the troposphere is the **tropopause,** which lies at an altitude of about 17 kilometers at the equator, although it is lower at the poles. At the tropopause the steady decline in temperature with altitude ceases abruptly. Cold air from the upper troposphere is too dense to rise above the tropopause. As a result, little mixing occurs between the troposphere and the layer above it, called the **stratosphere.**

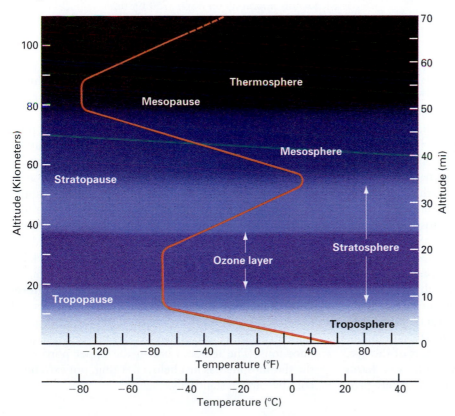

Figure 15–14 Atmospheric temperature varies with altitude. The atmospheric layers are zones in which different factors control the temperature. • Interactive Question: If you were ascending from Earth in a rocket, what layer would you be in at 5 km, 20 km, and 80 km? What would the outside temperature be at each of these elevations?

If someone handed you a perfectly smooth ball with a dot on it and asked you to describe the location of the dot, you would be at a loss to do so because all positions on the surface of a sphere are equal. How, then, can locations on a spherical Earth be described? Even if we ignore irregularities, continents, and oceans, the Earth has points of reference because it rotates on its axis and has a magnetic field that nearly (but not exactly) coincides with the axis of rotation. The North and South Poles lie on the rotational axis and lines of latitude form imaginary horizontal rings around the axis. Mathematicians measure distance on a sphere in degrees. Using this system, the equator is defined as 0° latitude, the North Pole is 90° north latitude, and the South Pole is 90° south latitude (Fig. 1).

No natural east-west reference exists, so a line running through Greenwich, England, was arbitrarily chosen as the 0° line. The planet was then divided by lines of longitude, also measured in degrees. On a globe with the rotational axis vertically oriented, lines of longitude also run vertically. To a navigator they measure east-west angular distance from Greenwich, England.

The system is easy to use. Minneapolis-St. Paul lies at 45° north latitude and 93° west longitude. The latitude tells us that the city lies halfway between the equator (0°) and the North Pole (90°). Since a circle has 360°, the longitude tells us that Minneapolis is about one quarter of the way around the world in a westward direction from Greenwich, England.

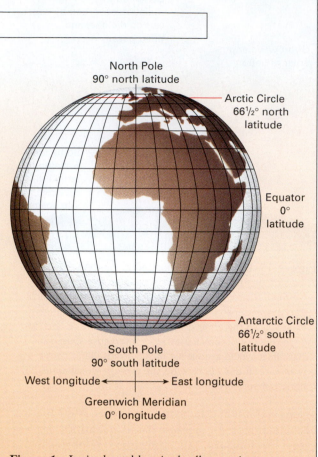

Figure 1 Latitude and longitude allow navigators to identify a location on a spherical Earth.

In the stratosphere, temperature remains constant to 35 kilometers and then increases with altitude until, at about 50 kilometers, it is as warm as that at the Earth's surface. This reversal in the temperature profile occurs because the troposphere and stratosphere are heated by different mechanisms. As already explained, the troposphere is heated primarily from below, by the Earth. The stratosphere, on the other hand, is heated primarily from above, by solar radiation.

Oxygen molecules (O_2) in the stratosphere absorb energetic ultraviolet rays from the Sun. The radiant energy breaks the oxygen molecules apart, releasing free oxygen atoms. The oxygen atoms then recombine to form ozone, (O_3). Ozone absorbs ultraviolet light more efficiently than oxygen does, warming the upper stratosphere. Ultraviolet light is energetic enough to affect organisms. Small quantities give us a suntan, but large doses cause skin cancer and cataracts of the eye, inhibit the growth of many plants, and otherwise harm living tissue. The ozone in the upper atmosphere protects life on Earth by absorbing much of this high-energy radiation before it reaches Earth's surface (Chapter 19).

Ozone concentration declines in the upper portion of the stratosphere, and therefore at about 55 kilometers above the Earth, temperature once more begins to decline rapidly with elevation. This boundary between rising and falling temperature is the **stratopause,** the ceiling of the stratosphere. The second zone of declining temperature is the **mesosphere.** Little radiation is absorbed in the mesosphere, and the thin air is extremely cold. Starting at about 80 kilometers above the Earth, the temperature again remains constant and then rises rapidly in the **thermosphere.** Here the atmosphere absorbs high-energy X-rays and ultraviolet radiation from the Sun. High-energy reactions strip electrons from atoms and molecules to produce ions. The temperature in the upper portion of the thermosphere is just below freezing, not extremely cold by surface standards.

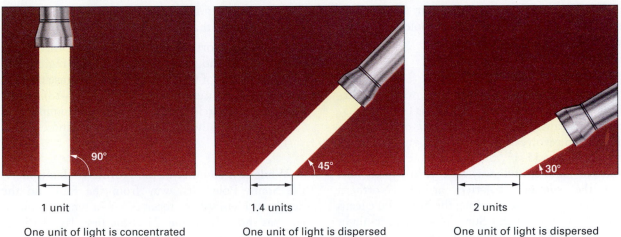

1 unit	1.4 units	2 units
One unit of light is concentrated over one unit of surface	One unit of light is dispersed over 1.4 units of surface	One unit of light is dispersed over 2 units of surface

Figure 15–15 If a light shines from directly overhead, the radiation is concentrated on a small area. However, if the light shines at an angle, or if the surface is tilted, the radiant energy is dispersed over a larger area.

15.6 Temperature Changes with Latitude and Season

Figure 15–14 shows that the temperature at the Earth's surface is about 20°C (68°F). Of course, this is a global average; surface temperature fluctuates with location, time of day, and season. If you travel, you may have experienced bitter-cold arctic air or the oppressive heat of the tropics.

Temperature Changes with Latitude

The region near the equator is warm throughout the year, whereas polar regions are cold and ice-bound even in summer. To understand this temperature dif- ference, consider first what happens if you hold a flash- light above a flat board. If the light is held directly overhead and the beam shines straight down, a small area is brightly lit. If the flashlight is held at an angle to the board, a larger area is illuminated. However, be- cause the same amount of light is spread over a larger area, the intensity is reduced (Fig. 15–15).

Now consider what happens when the Sun shines directly over the equator. The equator, analogous to the flat board under a direct light, receives the most concentrated radiation. The Sun strikes the rest of the globe at an angle and thus, radiation is less concen- trated at higher latitudes (Fig. 15–16). Because the equator receives the most concentrated solar energy,

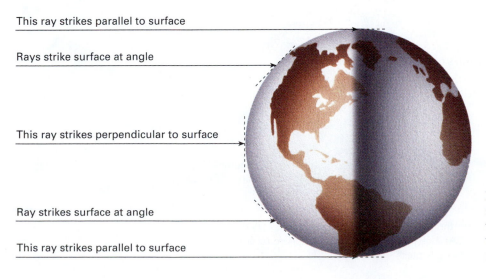

This ray strikes parallel to surface

Rays strike surface at angle

This ray strikes perpendicular to surface

Ray strikes surface at angle

This ray strikes parallel to surface

Figure 15–16 When the Sun shines directly over the equator, the equator receives the most in- tense solar radiation, and the poles receive little.

it is generally warm throughout the year. Average temperature becomes progressively cooler poleward (north and south of the equator).

The Seasons

Earth circles the Sun in a planar orbit, while simultaneously spinning on its axis. This axis is tilted at 23.5° from a line drawn perpendicular to the orbital plane.

The Earth revolves around the Sun once a year. As shown in Figure 15–17, the North Pole tilts toward the Sun in summer and away from it in winter. June 21 is the summer **solstice** in the Northern Hemisphere because at this time, the North Pole leans the full 23.5° toward the Sun. As a result, sunlight strikes the Earth from directly overhead at a latitude 23.5° north of the equator. This latitude is the **Tropic of Cancer.** If you stood on the Tropic of Cancer at noon on June 21, you would cast no shadow. June is warm in the Northern Hemisphere for two reasons:

1. When the Sun is high in the sky, sunlight is more concentrated than in winter.

2. When the North Pole is tilted toward the Sun, it receives 24 hours of daylight. Polar regions are called "lands of the midnight sun" because the Sun never sets in the summertime (Fig. 15–18). Below the Arctic Circle the Sun sets in the summer, but the days are always longer than in winter (Table 15–2).

While it is summer in the Northern Hemisphere, the South Pole tilts away from the Sun and the Southern Hemisphere receives low-intensity sunlight and has short days. June 21 is the first day of winter in the Southern Hemisphere. Six months later, on December 22, the seasons are reversed. The North Pole tilts away from the Sun, giving rise to the winter solstice in the Northern Hemisphere, while it is summer in the Southern Hemisphere. On this day sun-

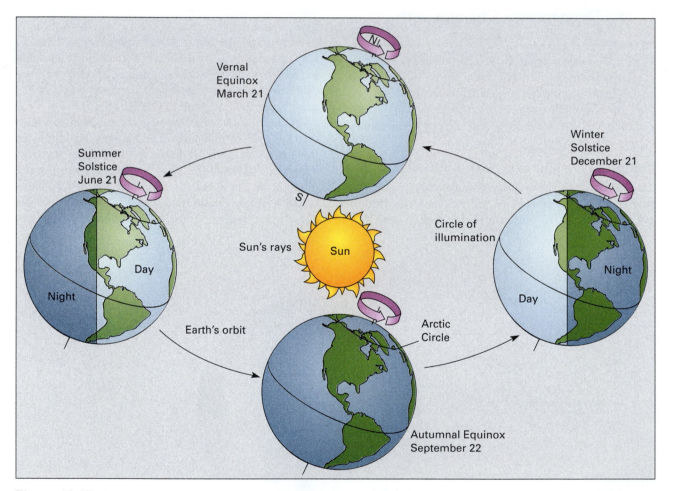

Figure 15–17 Weather changes with the seasons because the Earth's axis is tilted relative to the plane of its orbit around the Sun. As a result, the Northern Hemisphere receives more direct sunlight during summer, but less during winter.

Figure 15–18 The Sun shines at midnight in July at 70° north latitude in the Canadian Arctic.

hours of darkness. For this reason, March 21 and September 21 are called the **equinoxes,** meaning equal nights.

All areas of the globe receive the same total number of hours of sunlight every year. The North and South poles receive direct sunlight in dramatic opposition, 6 months of continuous light and six months of continuous darkness, whereas at the equator, each day and night are close to 12 hours long throughout the year. Although the poles receive the same number of sunlight hours as do the equatorial regions, the sunlight reaches the poles at a much lower angle and therefore delivers much less total energy per unit of surface area.

15.7 Temperature Changes Due to Heat Transport and Storage

Even though all locations at a given latitude receive equal amounts of solar radiation, temperature is not constant with latitude. Figure 15–19 shows temperatures around the Earth in January and in July. Lines called **isotherms** connect areas of the same average temperature. Note that the isotherms loop and dip across lines of latitude. For example, the January 0°C line runs through Seattle, Washington, dips southward across the center of the United States, and then swings northward to northern Norway. Such variations occur with latitude because winds and ocean currents transport heat from one region of the Earth to another, and heat storage varies with location.

Heat Transport

If you place a metal frying pan on the stove, the handle gets hot even though it is not in contact with the burner because the metal conducts heat from the bottom of the pan to the handle. **Conduction** is the transport of heat by atomic or molecular motion. When the frying pan is heated from below, the metal atoms on the bottom of the pan move more rapidly. They then collide with their neighbors and transfer energy

light strikes the Earth directly overhead at the **Tropic of Capricorn,** latitude 23.5° south. At the North Pole the Sun never rises and it is continuously dark, while the South Pole is bathed in continuous daylight.

On March 21 and September 23 the Earth's axis lies at right angles to a line drawn between the Earth and the Sun. As a result, the poles aren't tilted toward or away from the Sun and the Sun shines directly overhead at the equator at noon. If you stood at the equator at noon on either of these two dates, you would cast no shadow. But north or south of the equator, a person casts a shadow even at noon. In the Northern Hemisphere, March 21 is the first day of spring and September 21 is the first day of autumn, whereas the seasons are reversed in the Southern Hemisphere. On the first days of spring and autumn, every portion of the globe receives 12 hours of direct sunlight and 12

TABLE 15–2

Hours of Sunlight Per Day

Latitude	Geographic Reference	Summer Solstice	Winter Solstice	Equinoxes
0°	Equator	12 h	12 h	12 h
30°	New Orleans	13 h 56 min	10 h 04 min	12 h
40°	Denver	14 h 52 min	9 h 08 min	12 h
50°	Vancouver	16 h 18 min	7 h 42 min	12 h
90°	North Pole	24 hr	0 h 00 min	12 h

January

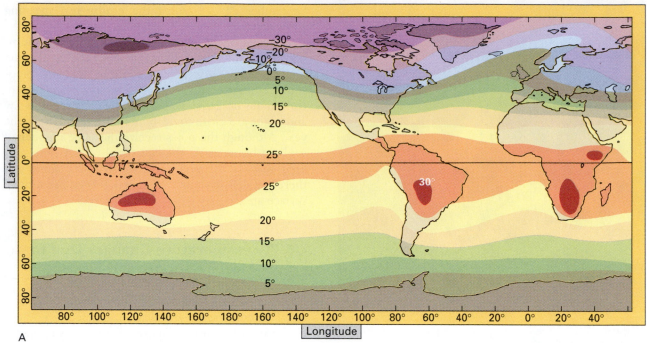

A

July

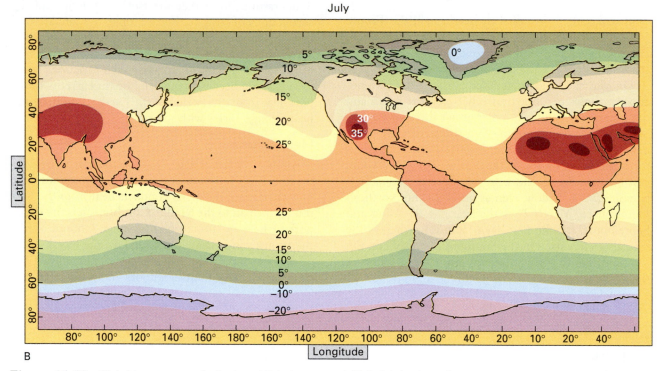

B

Figure 15–19 Global temperature distributions (A) in January, and (B) in July. Isotherm lines connect places with the same average temperatures.

to them. Like a falling row of dominoes, energy is passed from one atom to another throughout the pan until the handle becomes hot. Air, in contrast, is a poor conductor, and energy is not transported from one region of the globe to another by conduction.

To understand how air transports heat, imagine that a heater is placed in one corner of a cold room. The heated air in the corner expands, becoming less dense. This light, hot air rises to the ceiling. It flows along the ceiling, cools, falls, and returns to the stove, where it is reheated (Fig. 15–20). **Convection** is the transport of heat by the movement of currents. Convection occurs readily in liquids and gases. Recall from Chapter 14 that ocean currents transport large quantities of heat northward or southward, thereby altering climate. For example, the Gulf Stream carries tropical water northward and warms the west coast of Europe.

Similar currents occur in the atmosphere. If air in one region is heated above the temperature of surrounding air, this warm air becomes less dense and rises. As the warm air rises, cooler, denser air in another portion of the atmosphere sinks. Air then flows along the surface to complete the cycle. In common speech, this horizontal airflow is called wind. In meteorology, horizontal air flow is called **advection,** whereas *convection* is reserved for vertical air flow. The steady winds that blow across the tropical oceans, a tornado that ravages a city in the Midwest, and a thunderstorm that drops rain and hail on your Sunday picnic are all caused by convective and advective processes in the atmosphere. We will expand our understanding of weather and climate in Chapters 16 and 17.

Changes of State

Given the proper temperature and pressure, most substances can exist in three states: solid, liquid, and gas. However, at the Earth's surface, many substances commonly exist in only one state. In our experience, rock is almost always solid and oxygen is almost always a gas. Water plays an important role in heat transfer and storage because it commonly exists in all three states—as solid ice, as a liquid, and as gaseous water vapor.

If you walk onto the beach after swimming, you feel cool, even on a hot day. Your skin temperature drops because water on your body is evaporating, and evaporation absorbs heat. Similarly, evaporation from any body of water cools the water and the air around it. Conversely, condensation releases heat. The energy released when water condenses into rain during a single hurricane can be as great as the energy release by several atomic bombs.

Latent heat (stored heat) is the energy released or absorbed when a substance changes from one state to another. About 80 calories are required to melt a gram of ice at a constant temperature of 0°C. As a comparison, 100 calories are needed to heat the same amount of water from freezing to boiling (0°C to 100°C). Another 540 to 600 calories are needed to evaporate a gram of water at constant temperature and pressure.[2]

The energy transfers also work in reverse. When one gram of water vapor condenses to liquid, 540 to

---

[2] The heat of vaporization varies with temperature and pressure; at 100°C and one atmosphere pressure the value is 539.5 cal/g.

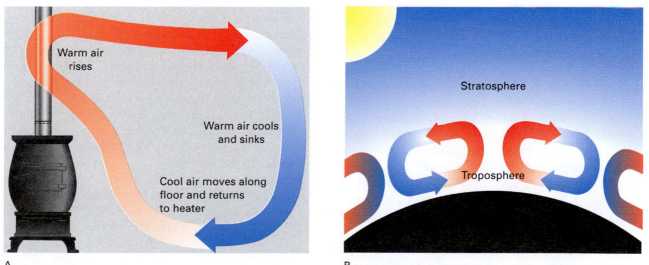

A

Warm air rises

Warm air cools and sinks

Cool air moves along floor and returns to heater

B

Stratosphere

Troposphere

Figure 15–20 (A) Convection currents distribute heat throughout a room. (B) Convection also distributes heat through the atmosphere when the Sun heats the Earth's surface. In this case the ceiling is the boundary between the troposphere and the stratosphere.

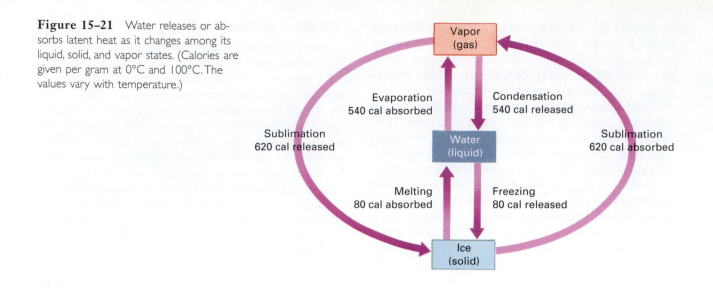

Figure 15–21 Water releases or absorbs latent heat as it changes among its liquid, solid, and vapor states. (Calories are given per gram at 0°C and 100°C. The values vary with temperature.)

Vapor
(gas)

Evaporation
540 cal absorbed

Condensation
540 cal released

Sublimation
620 cal released

Water
(liquid)

Sublimation
620 cal absorbed

Melting
80 cal absorbed

Freezing
80 cal released

Ice
(solid)

600 calories are released. When one gram of water freezes, 80 calories are released (Fig. 15–21).

The energy absorbed and released during freezing, melting, evaporation, and condensation of water is important in the atmospheric energy balance. For example, in the northern latitudes March is usually colder than September, even though equal amounts of sunlight are received in both months. However, in March much of the solar energy is absorbed by melting snow. Snow also has a high albedo and reflects sunlight efficiently. Evaporation cools sea coasts, and the energy of a hurricane comes, in part, from the condensation of massive amounts of water vapor.

Heat Storage

If you place a pan of water and a rock outside on a hot summer day, the rock becomes hotter than the water. Both have received identical quantities of solar radiation. Why is the rock hotter?

1. **Specific heat** is the amount of energy needed to raise the temperature of 1 gram of material by 1°C. Specific heat is different for every substance, and water has an unusually high specific heat. Thus, if water and rock absorb equal amounts of energy, the rock becomes hotter than the water.

2. Rock absorbs heat only at its surface, and the heat travels slowly through the rock. As a result, heat concentrates at the surface. Heat disperses more effectively through water for two reasons. First, solar radiation penetrates several meters below the surface of the water, warming it to this depth. Second, water is a fluid, and transports heat by convection.

3. Evaporation is a cooling process. Water loses heat and cools by evaporation, but rock does not.

Think of the consequences of the temperature difference between rock and water. On a hot summer day you may burn your feet walking across dry sand or rock, but the surface of a lake or ocean is never burning hot. Suppose that both the ocean and the adjacent coastline are at the same temperature in spring. As summer approaches, both land and sea receive equal amounts of solar energy. But the land becomes hotter, just as rock becomes hotter than water. Along the seacoast the cool sea moderates the temperature of the land. The interior of a continent is not cooled in this manner and is generally hotter than the coast (Fig. 15–19B). In winter the opposite effect occurs, and inland areas are generally colder than the coastal regions. Thus, coastal areas are commonly cooler in summer and warmer in winter than continental interiors. This effect is shown by the 0°C temperature line in Figure 15–19A, which dips southward through the continents and rises to higher latitudes over the oceans. The coldest temperatures recorded in the Northern Hemisphere occurred in central Siberia and not at the North Pole, because Siberia is landlocked, whereas the North Pole lies in the middle of the Arctic Ocean.[3] In summer, however, Siberia is considerably warmer than the North Pole. In fact, the average temperature in some places in Siberia ranges from −50°C in winter to +20°C in summer, about the greatest range in the world.

---

[3] Even though the Arctic Ocean is covered with ice, water lies only a few meters below the surface and it still influences climate.

Weather is the state of atmospheric conditions at a specific place and time, and **climate** is the characteristic weather of a region averaged over several decades.

Dry air is roughly 78 percent nitrogen (N_2), 21 percent oxygen (O_2), and 1 percent other gases. Air also contains water vapor, dust, liquid droplets, and pollutants. Atmospheric pressure is the weight of the atmosphere per unit area. Pressure varies with weather and decreases with altitude.

Light is a form of **electromagnetic radiation** and exhibits properties of both waves and particles. When radiation encounters matter, it may be transmitted, reflected, absorbed, or scattered. Earth receives energy in the form of high-energy, short-wavelength solar radiation. Rock and soil re-emit low-energy, long-wavelength radiation. The **greenhouse effect** is the warming of the atmosphere due to the absorption of this long-wavelength radiation by water vapor, carbon dioxide, methane, and other greenhouse gases. About 8 percent of solar radiation is scattered back into space, 19 percent is absorbed in the atmosphere, 26 percent is reflected, and 47 percent is absorbed by soil, rocks, and water. However, the absorbed radiation is eventually re-emitted back into space.

Atmospheric temperature decreases with altitude in the **troposphere.** The temperature rises in the **stratosphere** because ozone absorbs solar radiation. The temperature decreases again in the **mesosphere,** and then in the uppermost layer, the **thermosphere,** temperature increases as high-energy radiation is absorbed.

The general temperature gradient from the equator to the poles results from the decreasing intensity of solar radiation from the equator to the poles. Changes of seasons are caused by the tilt of the Earth's axis relative to the Earth-Sun plane.

Winds and ocean currents transfer heat from one region of the globe to another by **convection** and **advection.** Large quantities of heat are absorbed or emitted when water freezes, melts, vaporizes, or condenses. Oceans affect weather and climate because water has low **albedo** and a high **specific heat.** In addition, currents transport heat both vertically and horizontally. Finally, evaporation cools the sea surface. As a result of all of these factors, coastal areas are generally cooler in summer and warmer in winter than continental interiors at the same latitude.

KEY TERMS

weather 346
climate 347
barometric pressure 348
barometer 348
altimeter 349
photons 350
electromagnetic radiation 350

electromagnetic spectrum 351
albedo 352
greenhouse effect 354
troposphere 355
tropopause 355
stratosphere 355

stratopause 356
mesosphere 356
thermosphere 356
solstice 358
Tropic of Cancer 358
Tropic of Capricorn 359
equinox 359

isotherm 359
conduction 359
convection 361
advection 361
latent heat 361
specific heat 362

FOR REVIEW

1. List the two most abundant gases in the atmosphere. List three other, less abundant gases. List three non-gaseous components of natural air.

2. Draw a graph of the change in pressure with altitude. Explain why the pressure changes as you have shown.

3. What is a barometer and how does it work?

4. What happens to light when it is transmitted, reflected, absorbed, and scattered? Give an example of each of these phenomena.

5. What is the fate of the solar radiation that reaches the Earth?

6. List the layers of the Earth's atmosphere. Discuss the temperature changes within each.

7. Describe the difference between weather and climate. Which is more predictable?

8. Explain how the tilt of the Earth's axis affects climate in the temperate and polar regions.

9. Discuss temperature and lengths of days at the poles, the mid-latitudes, and the equator at the following times of year: June 21, December 21, and the equinoxes.

10. How is a convection current generated? Discuss the effect of wind on world climate.

11. Discuss the effect of the oceans on world climate. Are the coastal regions always warmer than inland areas? Explain.

FOR DISCUSSION

1. Imagine that enough matter vanished from the Earth's core so that the Earth's mass decreased by half. In what ways would the atmosphere change? Would normal pressure at sea level be affected? Would the thickness of the atmosphere change? Explain.

2. An astronaut on a space walk must wear protective clothing as a shield against the Sun's rays, but the same person is likely to relax in a bathing suit in sunlight down on Earth. Explain.

3. Why must climbers wear dark glasses to protect their eyes while they are on high mountains? Why do they get sunburned even when the temperature is below freezing?

4. As the winter ends, the snow generally melts first around trees, twigs, and rocks. Explain why the line of melting radiates outward from these objects, and the snow in open areas melts last.

5. In central Alaska the sky is often red at noon in December. Explain why the sky is red, not blue, at this time.

6. Refer to Figure 15–12. (a) What percent of the solar energy is absorbed by the Earth? (b) Is the Earth growing warmer, or colder, or is the global temperature fairly constant over time? Explain.

7. If the North Pole receives the same number of hours of sunlight per year as do the equatorial regions, why is it so much colder than the equator?

8. Would a large inland lake be likely to affect the climate of the land surrounding it? Deep lakes seldom freeze completely in winter, whereas shallow ones do. Would a deep lake have a greater or a lesser effect on weather than a shallow one?

9. Neither oxygen nor nitrogen is an efficient absorber of visible or infrared radiation. Since these two gases together make up 99 percent of the atmosphere, the atmosphere is largely transparent to visible and infrared radiation. Predict what would happen if oxygen and nitrogen absorbed visible radiation. What would happen if these two gases absorbed infrared radiation?

10. From the information presented in this chapter, explain how the geosphere, hydrosphere, and atmosphere interact to produce climate.

11. How would the Earth's climate be affected if: (a) the Earth had no oceans, (b) the Earth's spin axis were perpendicular to the plane of the Earth's orbit, and (c) the Earth's spin axis were tilted 50 degrees to the plane of the Earth's orbit?

Weather

Weather affects us every day. Extreme weather, such as tornadoes, thunderstorms, hurricanes, and prolonged rain can cause considerable damage and loss of life. *(Keith Kent/ Peter Arnold, Inc.)*

Today's weather may be sunny and warm, a sharp contrast to yesterday, when it was rainy and cold. In New York City, a winter wind from the northwest brings cool air, but when a breeze blows from the southeast, the temperature rises and a storm develops as warm, moist maritime air flows into the city. Moisture, temperature, and wind combine to create the atmospheric conditions called weather.

The most severe weather, such as a winter blizzard or a hurricane, brings heavy precipitation and violent wind. The energy that drives these storms ultimately is derived from the Sun. In this chapter we will learn how the Sun's heat drives a blizzard that sweeps across the land with swirling snow and subzero temperatures, or a hurricane that blackens the sky and flattens houses. ●

16.1 Moisture in Air

Precipitation occurs only when there is moisture in the air. Therefore to understand precipitation, we must first understand how moisture collects in the atmosphere and how it behaves.

Humidity

When water boils on a stove, a steamy mist rises above the pan, and then disappears into the air. The water molecules have not been lost, they have simply become invisible. In the pan, water is liquid, and in the mist above, the water exists as tiny droplets. These droplets then evaporate, and the invisible water vapor mixes with air. Water also evaporates into air from the seas, streams, and lakes, and from soil. Winds then distribute this moisture throughout the atmosphere. Thus, all air contains some water vapor, even over the driest deserts.

Humidity is the amount of water vapor in air. **Absolute humidity** is the mass of water vapor in a given volume of air, expressed in grams per cubic meter (g/m^3).

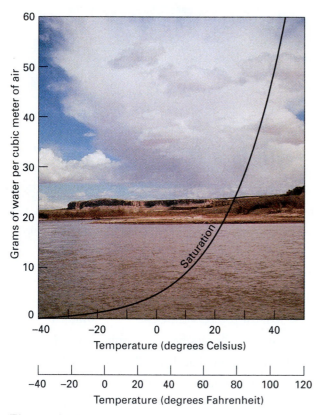

Figure 16–1 Warm air can hold more water vapor than cold air.

Air can hold only a certain amount of water vapor, and warm air can hold more water vapor than cold air. For example, air at 25°C can hold 23 g/m^3 of water vapor, but at 12°C, it can hold only half that quantity, 11.5 g/m^3 (Fig. 16–1). **Relative humidity** is the amount of water vapor in air relative to the maximum it can hold at a given temperature. It is expressed as a percentage:

$$\text{Relative humidity (\%)} = \frac{\text{actual quantity of water per unit of air}}{\text{maximum quantity at the same temperature}} \times 100$$

If air contains half as much water vapor as it can hold, its relative humidity is 50 percent. Suppose that air at 25°C contains 11.5 g/m^3 of water vapor. Since air at that temperature can hold 23 g/m^3, it is carrying half of its maximum, and the relative humidity is 11.5 g/23 g × 100 = 50 percent. Now let us take some of this air and cool it without adding or removing any water vapor. Because cold air holds less water vapor than warm air, the relative humidity increases even though the *amount* of water vapor remains constant. If the air cools to 12°C, and it still contains 11.5 g/m^3, the relative humidity reaches 100 percent because air at that temperature can hold only 11.5 g/m^3.

When relative humidity reaches 100 percent, the air is **saturated**. The temperature at which saturation occurs, 12°C in this example, is the **dew point**. If saturated air cools below the dew point, some of the water vapor may condense into liquid droplets (although, as discussed next, sometimes the relative humidity can rise above 100 percent).

Supersaturation and Supercooling When the relative humidity reaches 100 percent (at the dew point), water vapor condenses quickly onto solid surfaces such as rocks, soil, and airborne particles. Airborne particles such as dust, smoke, and pollen are abundant in the lower atmosphere. Consequently, water vapor may condense easily at the dew point in the lower atmosphere, and there, the relative humidity rarely exceeds 100 percent. However, in the clear, particulate-free air high in the troposphere, condensation occurs so slowly that for all practical purposes it does not happen. As a result, the air commonly cools below its dew point but water remains as vapor. In that case, the relative humidity rises above 100 percent, and the air becomes **supersaturated**.

Similarly, liquid water does not always freeze at its freezing point. Small droplets can remain liquid in a cloud even when the temperature is −40°C. Such water is **supercooled**.

Figure 16–2 Most clouds form as rising air cools. The cooling causes invisible water vapor to condense as visible water droplets or ice crystals, which we see as a cloud.

16.2 Cooling and Condensation

Three atmospheric processes cool air to its dew point and cause condensation: (1) Air cools when it loses heat by radiation. (2) Air cools by contact with a cool surface such as water, ice, rock, soil, or vegetation. (3) Air cools when it rises. Dew, frost, and some types of fog form by radiation and contact cooling. However, clouds, rain, snow, sleet, and hail normally form as a result of the cooling that occurs when air rises (Fig. 16–2).

Radiation Cooling

As described in Chapter 15, the atmosphere, rocks, soil, and water absorb the Sun's heat during the day, and then radiate some of this heat back out toward space at night. As a result of the radiation, air, land, and water become cooler at night, and condensation may occur.

Contact Cooling—Dew and Frost

You can observe condensation on a cool surface with a simple demonstration. Heat water on a stove until it boils and hold a cool drinking glass in the clear air just above the steam. Water droplets will condense on the surface of the glass because the glass cools the hot, moist air to its dew point. The same effect occurs in a house on a cold day. Water droplets or ice crystals appear on windows as warm, moist indoor air cools on the glass (Fig. 16–3).

Figure 16–3 Ice crystals condense on a window on a frosty morning.

In some regions, the air on a typical summer evening is warm and humid. After the Sun sets, plants, houses, windows, and most other objects lose heat by radiation and therefore become cool. During the night, water vapor condenses on the cool objects. This condensation is called **dew.** If the dew point is below freezing, **frost** forms. Thus, frost is not frozen dew, but ice crystals formed directly from vapor.

Cooling of Rising Air

Work and heat are both forms of energy. Work can be converted to heat or heat can be converted to work, but energy is never lost. If you pump up a bicycle tire you are performing work to compress the air. This energy is not lost; much of it converts to heat. Therefore, both the pump and the newly filled tire feel warm. Conversely, if you puncture a tire, the air rushes out. It must perform work to expand, so the air rushing from a punctured tire cools. Variations in temperature caused by compression and expansion of gas are called **adiabatic temperature changes.** Adiabatic means without gain or loss of heat. During adiabatic warming, air warms up because work is done on it, not because heat is added. During adiabatic cooling, air cools because it performs work, not because heat is removed.

As explained in Chapter 15, air pressure decreases with elevation. When surface air rises, it expands, just as air expands when it rushes out of a punctured tire. Rising air performs work to expand, and therefore it cools adiabatically. Dry air cools by 10°C for every 1000 meters it rises (5.5°F/1000 ft). This cooling rate

is called the **dry adiabatic lapse rate.** Thus, if dry air were to rise from sea level to 9000 meters (about the height of Mount Everest), it would cool by 90°C (162°F).

Almost all air contains some water vapor. As moist air rises and cools adiabatically, its temperature eventually decreases to the dew point. At the dew point, moisture begins to condense as droplets, and a cloud forms. But recall that condensing vapor releases latent heat. As the air rises through the cloud, its temperature now is affected by two opposing processes. It cools adiabatically, but at the same time it is heated by the latent heat released by condensation. However, the warming caused by latent heat is less than the amount of adiabatic cooling. The net result is that the rising air continues to cool, but more slowly than at the dry adiabatic lapse rate. The **wet adiabatic lapse rate** is the cooling rate after condensation has begun. It varies from 5°C/1000 m (2.7°F/1000 ft) for air with a high moisture content, to 9°C/1000 m (5°F/1000 ft) for relatively dry air (Fig. 16–4). Thus, once clouds start to form, rising air no longer cools as rapidly as it did lower in the atmosphere. Rising air cools at the dry adiabatic lapse rate until it cools to its dew point and condensation begins. Then, as it continues to rise, it cools at the lesser wet adiabatic lapse rate as condensation continues.

On the other hand, sinking air becomes warmer because of adiabatic compression. Warm air can hold more water vapor than cool air. Consequently, water does not condense from sinking, warming air, and the latent heat of condensation does not affect the rate of

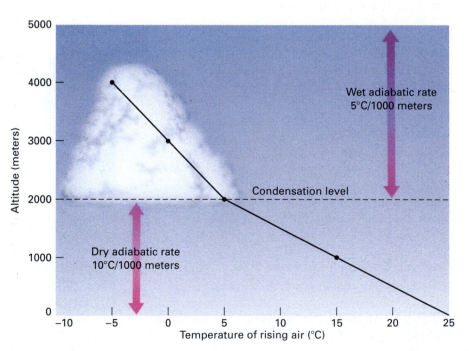

Figure 16–4 A rising air mass initially cools rapidly at the dry adiabatic lapse rate. Then, after condensation begins, it cools more slowly at the wet adiabatic lapse rate.

temperature rise. As a result, sinking air masses always become warmer at the dry adiabatic rate.

16.3 Clouds

A cloud is a visible concentration of water droplets or ice crystals in air. The basic requirements for cloud formation are a supply of humid air and a means of cooling it so that water vapor condenses to form droplets or ice crystals. Most clouds form well above the Earth's surface and so are not cooled by direct contact with the ground. Almost all cloud formation and precipitation occur when air rises.

Why Does Air Rise?

Three mechanisms cause air to rise (Fig. 16–5):

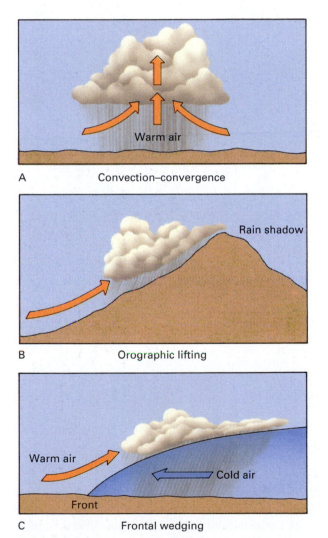

A — Convection–convergence

B — Orographic lifting

C — Frontal wedging

Warm air

Rain shadow

Warm air — Cold air — Front

Figure 16–5 Three mechanisms cause air to rise and cool: (A) convection-convergence, (B) orographic lifting, and (C) frontal wedging.

Convection If one portion of the atmosphere becomes warmer than the surrounding air, the warm air expands, becomes less dense, and rises. Thus a hot air balloon rises because it contains air that is warmer and less dense than the air around it. If the Sun heats one parcel of air near the Earth's surface to a warmer temperature than that of surrounding air, the warm air will rise, just as the hot air balloon rises.

Orographic lifting When air flows over mountains, it is forced to rise.

Frontal wedging A moving mass of cool, dense air may encounter a mass of warm, less dense air. When this occurs, the cool, denser air slides under the warm air mass, forcing the warm air upward to create a weather front. We will discuss weather fronts in more detail later in the chapter.

Rising Air and Clouds

On some days clouds hang low over the land and obscure nearby hills. At other times clouds float high in the sky, well above the mountain peaks. What factors determine the height and shape of a cloud?

Recall that air is generally warmest at the Earth's surface and cools with elevation throughout the troposphere. The rate at which air that is neither rising nor falling cools with elevation is called the **normal lapse rate.** The average normal lapse rate is 6°C/1000 m (3.3°F/1000 ft), and thus is less than the dry adiabatic lapse rate. However, the normal lapse rate is variable. Typically, it is greatest near the Earth's surface, and decreases with altitude. The normal lapse rate also varies with latitude, the time of day, and the seasons. It is important to note that the normal lapse rate is simply the vertical temperature structure of the atmosphere. In contrast, *rising* air cools because of adiabatic cooling.

Figure 16–6 shows two rising warm air masses, one consisting of dry air and the other of moist air. The central part of the figure shows that the normal lapse rate is the same for both air masses: the temperature of the atmosphere decreases rapidly in the first few thousand meters, and then more slowly with increasing elevation. However, the two air masses behave differently because of their different moisture contents.

The dry air mass (A) rises and cools at the dry adiabatic lapse rate of 10°C/1000 m. As a result, in this example, its temperature and density become equal to that of surrounding air at an elevation of 3000 meters. Because the density of the rising air is the same as that of the surrounding air, the rising air is no longer buoyant, and it stops rising. No clouds form because the air has not cooled to its dew point.

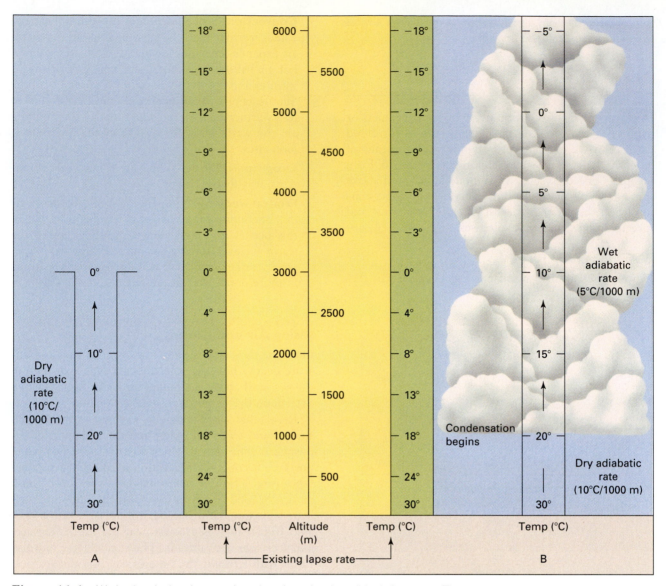

Figure 16–6 (A) As dry air rises, it expands and cools at the dry adiabatic lapse rate. Thus, it soon cools to the temperature of the surrounding air, and it stops rising. (B) As moist air rises, initially it cools at the dry adiabatic lapse rate. It soon cools to its dew point, and clouds form. Then, it cools more slowly at the wet adiabatic lapse rate. As a result, it remains warmer than surrounding air and continues to rise for thousands of meters. It stops rising when all moisture has condensed, and the air again cools at its dry adiabatic rate.

In (B), the rising moist air initially cools at the dry adiabatic rate of 10°C/1000 m, but only until the air cools to its dew point, at an elevation of 1000 meters. At that point, moisture begins to condense, and clouds form. But the condensing moisture releases latent heat of condensation. This additional heat causes the rising air to cool more slowly, at the wet adiabatic rate of 5°C/1000 m. As a result, the rising air remains warmer and more buoyant than surrounding air, and it continues to rise for thousands of meters, creating a towering, billowing cloud with the potential for heavy precipitation.

Types of Clouds

Cirrus clouds are wispy clouds that look like hair blowing in the wind or feathers floating across the sky (from Latin: wisp of hair). Cirrus clouds form at high altitudes, 6000 to 15,000 meters (20,000 to 50,000 feet). The air is so cold at these elevations that cirrus clouds are composed of ice crystals rather than water

Figure 16–7 Cirrus clouds are high, wispy clouds composed of ice crystals. *(D. Wise)*

Figure 16–8 Stratus clouds spread out across the sky in a low, flat layer.

droplets. High winds aloft blow them out into long, gently curved streamers (Fig. 16–7).

Stratus clouds are horizontally layered, sheet-like clouds (from Latin: layer). They form when condensation occurs at the same elevation at which air stops rising, and the clouds spread out into a broad sheet. Stratus clouds form the dark, dull gray, overcast skies that may persist for days and bring steady rain (Fig. 16–8).

Cumulus clouds are fluffy white clouds that typically display flat bottoms and billowy tops (from Latin: heap or pile) (Fig. 16–9). On a hot summer day the top of a cumulus cloud may rise 10 kilometers or more above its base in cauliflower-like masses. The base of the cloud forms at the altitude at which the rising air cools to its dew point and condensation starts. However, in this situation the rising air remains warmer than the surrounding air and therefore continues to rise. As it rises, more vapor condenses, forming the billowing columns.

Other types of clouds are named by combining these three basic terms (Fig. 16–10). **Stratocumulus** clouds are low sheet-like clouds with some vertical structure. The term *nimbo* refers to a cloud that precipitates. Thus, a **cumulonimbus** cloud is a towering rain cloud. If you see one, you should seek shelter, because cumulonimbus clouds commonly produce intense rain, thunder, lightning, and sometimes hail. A **nimbostratus** cloud is a stratus cloud from which rain or snow falls. Other prefixes are also added. *Alti* is derived from the Latin root *altus*, meaning high. An **altostratus** cloud is simply a high stratus cloud.

Types of Precipitation

Rain Why does rain fall from some clouds, whereas other clouds float across a blue sky on a sunny day and produce no rain? The droplets in a cloud are small, about 0.01 millimeter in diameter (about $\frac{1}{7}$ the diameter of a human hair). In still air, such a droplet would require 48 hours to fall from a cloud 1000 meters above the Earth. But these tiny droplets never reach the Earth because they evaporate faster than they fall.

If the air temperature in a cloud is above freezing, the tiny droplets may collide and coalesce. You can observe similar behavior in droplets sliding down a window pane on a rainy day. If two droplets collide, they merge to become one large drop. If the droplets in a cloud grow large enough, they fall as drizzle (0.1 to 0.5 millimeter in diameter) or light rain (0.5 to 2 millimeters in diameter). About one million cloud droplets must combine to form an average-size raindrop.

Figure 16–9 Cumulus clouds are fluffy white clouds with flat bottoms. *(NCAR)*

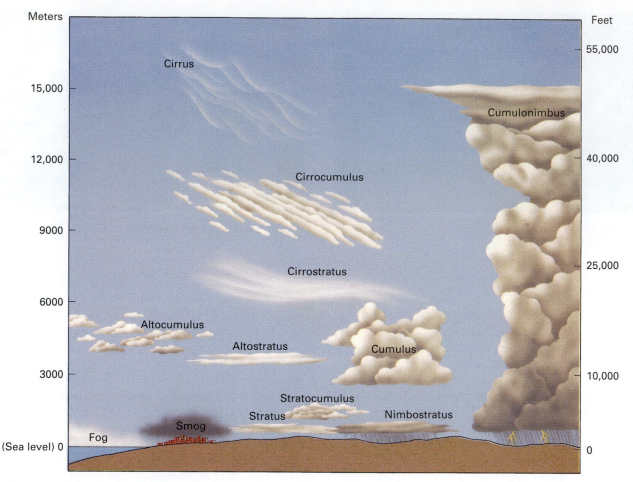

Figure 16–10 Cloud names are based on the shape and altitude of the clouds.

In many clouds, however, water vapor initially forms ice crystals rather than condensing as tiny droplets of supercooled water. Part of the reason for this is that the temperature in clouds is commonly below freezing, but another factor also favors ice formation. Air that is slightly undersaturated with respect to water is slightly supersaturated with respect to ice. For example, if the relative humidity of air is 95 percent with respect to water, it is about 105 percent with respect to ice. Thus, as air cools toward its dew point, all the vapor forms ice crystals rather than supercooled water droplets. The tiny ice crystals then grow larger as more water vapor condenses on them, until they are large enough to fall. The ice then melts to form raindrops as it falls through warmer layers of air.

If you have ever been caught in a thunderstorm, you may remember raindrops large enough to be painful as they struck your face or hands. Recall that a cumulus cloud forms from rising air and that its top may be several kilometers above its base. The tem-

perature in the upper part of the cloud is commonly below freezing. As a result, ice crystals form, and begin to fall. Condensation continues as the crystal falls through the towering cloud, and the crystal grows. If the lower atmosphere is warm enough, the ice melts before it reaches the surface. Raindrops formed in this manner may be 3 to 5 millimeters in diameter, large enough to hurt when they hit.

Snow, Sleet, and Glaze As explained above, when the temperature in a cloud is below freezing, the cloud is composed of ice crystals rather than water droplets. If the temperature near the ground is also below freezing, the crystals remain frozen and fall as snow (Fig. 16–11). In contrast, if raindrops form in a warm cloud and fall through a layer of cold air at lower elevation, the drops freeze and fall as small spheres of ice called **sleet.** Sometimes the freezing zone near the ground is so thin that raindrops do not have time to freeze before they reach the Earth. However, when they land

Cloud Seeding

A cloud contains moisture whether precipitation occurs or not. During droughts, it is frustrating to watch clouds pass overhead, but receive no rain. In the 1940s, meteorologists realized that precipitation could be induced artificially by creating surfaces to enhance the growth of water droplets or ice crystals. Many clouds are so high that air temperature is below freezing and the water droplets are supercooled. In the first cloud-seeding experiments, scientists poured pellets of dry ice from an airplane into a supercooled cloud. (Dry ice is solid carbon dioxide, $-78°C$.) The dry ice froze some of the droplets. The surfaces of the ice crystals formed nuclei for additional deposition from neighboring supercooled droplets. The crystals grew rapidly until they were heavy enough to fall. If the temperature in the lower atmosphere was sufficiently warm, the crystals melted to form raindrops. Because dry ice is expensive,

further experiments were done, which showed that a warm crystal would induce ice formation as long as it had a shape similar to that of an ice crystal. The most effective crystal was silver iodide. Thus, if silver iodide is sprinkled onto a supercooled cloud, ice deposits on the crystal.

Cloud seeding does not produce moisture but generally shifts rainfall from one location to another. One region's gain is then another region's loss. Furthermore, control over the redistribution of rain is not always precise. As a result, conflicts of interest and political problems may arise. For example, ranchers in Colorado objected when ski areas used cloud seeding to increase winter snowfall. They argued that the ski industry was stealing their water. Another problem arises because silver iodide is poisonous and, if highly concentrated, can harm plants and animals.

---

on subfreezing surfaces, they form a coating of ice called **glaze** (Fig. 16–12). Glaze can be heavy enough to break tree limbs and electrical transmission lines. It also coats highways with a dangerous icy veneer. In the winter of 1997–98, a sleet and glaze storm in eastern Canada and the northeastern United States caused

billions of dollars in damage. The ice damaged so many electric lines and power poles that many people were without electricity for a few weeks.

Hail   Occasionally, precipitation takes the form of very large ice globules called **hail**. Hailstones vary from 5 millimeters in diameter to a record 14 centimeters in diameter, weighing 765 grams (more than 1.5 pounds), that fell in Kansas. A 500 gram (1-pound) hailstone crashing to Earth at 160 kilometers (100 miles) per hour can shatter windows, dent car roofs, and kill people and livestock. Even small hailstones can damage crops. Hail falls only from cumulonimbus

Figure 16–11   Snow blankets the ground during the winter in temperate regions. Snow and ice cover the ground year-round in the high mountains and at the Poles.   *(Everett Johnson/Tony Stone Images)*

Figure 16–12   Glaze forms when rain falls on a surface that is colder than the freezing temperature of water.   *(Dave Putnam)*

Figure 16–13 Radiation fog is seen as a morning mist in this field in Idaho. ● Interactive
Question: Why does fog of this type commonly concentrate in low places?

clouds. Because cumulonimbus clouds form in columns with distinct boundaries, hailstorms occur in local, well-defined areas. Thus, one farmer may lose an entire crop while a neighbor is unaffected.

A hailstone consists of concentric shells of ice like the layers of an onion. Two mechanisms have been proposed for their formation. In one, turbulent winds blow falling ice crystals back upward in the cloud. New layers of ice accumulate as additional vapor condenses on the recirculating ice grain. An individual particle may rise and fall several times until it grows so large and heavy that it drops out of the cloud. In the second mechanism, hailstones form in a single pass through the cloud. During their descent, supercooled water freezes onto the ice crystals. The layering develops because different temperatures and amounts of supercooled water exist in different portions of the cloud, and each layer forms in a different part of the cloud.

16.4 Fog

Fog is a cloud that forms at or very close to ground level although most fog forms by processes different from those that create higher-level clouds. **Advection fog** occurs when warm, moist air from the sea blows onto cooler land. The air cools, and water vapor condenses at ground level. San Francisco, Seattle, and Vancouver, B.C., all experience foggy winters as warm, moist air from the Pacific Ocean is cooled first by the cold California current and then by land. The foggiest location in the United States is Cape Disappointment, Washington, where visibility is obscured by fog 29 percent of the time.

Radiation fog occurs when the Earth's surface and air near the surface cool by radiation during the night (Fig. 16–13). Water vapor condenses as fog when the air cools below its dew point. Often the cool, dense foggy air settles into valleys. If you are driving late at night in hilly terrain, beware, because a sudden dip in the roadway may lead you into a thick fog where visibility is low. A ground fog of this type typically "burns off" in the morning. The rising Sun warms the land or water surface which, in turn, warms the low-lying air. As the air becomes warmer, its capacity to hold water vapor increases, and the fog droplets evaporate. Radiation fog is particularly common in areas where the air is polluted because water vapor condenses readily on the tiny particles suspended in the air.

Recall that vaporization of water absorbs heat, and therefore cools surrounding air. **Evaporation fog** occurs when air is cooled by evaporation from a body of water, commonly a lake or river. Evaporation fogs are common in late fall and early winter, when the air has become cool but the water is still warm. The water evaporates, but the vapor cools and condenses to fog almost immediately upon contact with the cold air.

Upslope fog occurs when air cools as it rises along a land surface. Upslope fogs occur both on gradually sloping plains and on steep mountains. For example, the Great Plains rise from sea level at the Mississippi delta to 1500 meters (5000 feet) at the Rocky Mountain front. When humid air moves northwest from the Gulf of Mexico toward the Rockies, it rises and cools adiabatically to form upslope fog. The rapid rise at the mountain front also forms fog.

Warm air rises because it is less dense than surrounding cool air. Rising air exerts less downward force than still air. Consequently, atmospheric pressure is relatively low beneath a rising air mass (Fig. 16–14A). Air

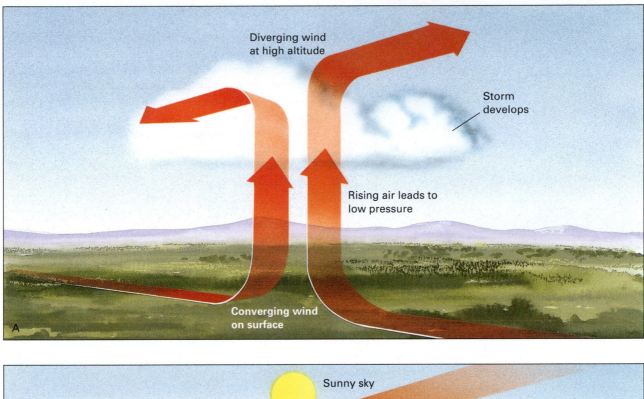

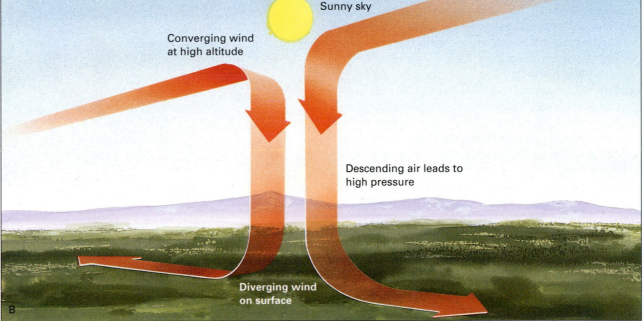

Figure 16–14 (A) Rising air creates low pressure, clouds, and precipitation. Air flows inward toward the low-pressure zone, creating surface winds. (B) Sinking air creates high pressure and clear skies. Air flows outward from the high-pressure zone, and also creates surface winds.

Figure 16-15 Winds vary from gentle zephyrs that cool bathers on a hot summer day to hurricanes and tornadoes that sink ships and destroy homes. *(Alan Klehr/Tony Stone Images)*

rises slowly above a typical low-pressure region, about 1 kilometer per day. In contrast, if air in the upper atmosphere cools, it becomes denser than the air beneath it and sinks. The sinking air exerts a downward force, forming a high-pressure region (Fig. 16-14B).

Air must flow inward over the Earth's surface toward a low-pressure region to replace the rising air mass. But a sinking air mass displaces surface air, pushing it outward from a high-pressure region. Thus, vertical air flow in both high- and low-pressure regions is accompanied by horizontal air flow, called **wind.** Winds near the Earth's surface always flow away from a region of high pressure and toward a low-pressure region. Ultimately, all wind is caused by the pressure differences resulting from unequal heating of the Earth's atmosphere (Fig. 16-15).

Pressure Gradient

Wind blows in response to *differences* in pressure. Imagine that you are sitting in a room and the air is still. Now you open a can of vacuum-packed coffee and hear the hissing as air rushes into the can. Because the pressure in the room is higher than that inside the coffee can, wind blows from the room into the can. But if you blow up a balloon, the air inside the bal-

loon is at higher pressure than the air in the room. When the balloon is punctured, wind blows from the high-pressure zone of the balloon into the lower-pressure zone of the room (Fig. 16-16).

Wind speed is determined by the magnitude of the pressure difference over distance, called the **pressure gradient.** Thus, wind blows rapidly if a large pressure difference exists over a short distance. A steep pressure gradient is analogous to a steep hill. Just as a ball rolls quickly down a steep hill, wind flows rapidly across a steep pressure gradient. To create a pressure-gradient map, air pressure is measured at hundreds of different weather stations. Points of equal pressure are connected by map lines called **isobars.** A steep pressure gradient is shown by closely spaced isobars, whereas a weak pressure gradient is indicated by widely spaced isobars (Fig. 16-17). Pressure gradients change daily, or sometimes hourly, as high- and low-pressure zones move. Therefore, maps are updated frequently.

Coriolis Effect

Recall from Chapter 14 that the Coriolis effect, caused by the Earth's spin, deflects ocean currents. The Coriolis effect similarly deflects winds. In the Northern Hemisphere wind is deflected toward the right, and in the Southern Hemisphere, to the left (Fig.

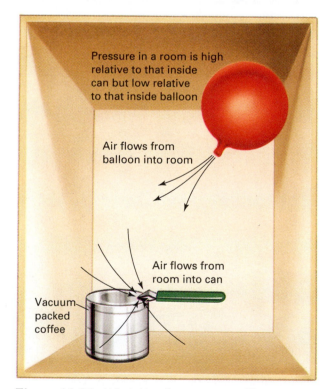

Figure 16-16 Winds blow in response to differences in pressure.

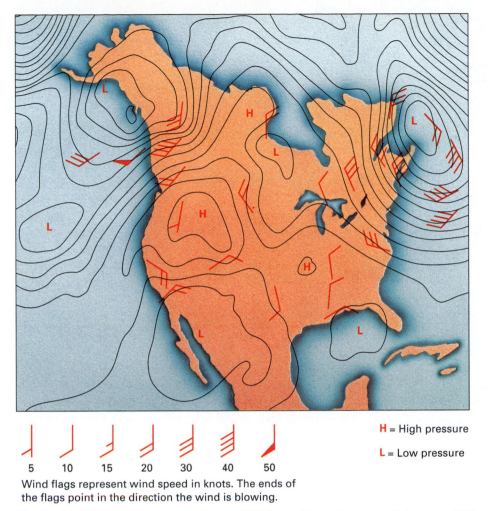

5	10	15	20	30	40	50

Wind flags represent wind speed in knots. The ends of the flags point in the direction the wind is blowing.

H = High pressure

L = Low pressure

Figure 16–17 Pressure map and winds at 5000 feet in North America on February 3, 1992. High-altitude data are shown because the winds are not affected by surface topography and thus the effect of pressure gradient is well illustrated. Note that in the northeast and northwest, steep pressure gradients, shown by closely spaced isobars, cause high winds that spiral counter-clockwise into the low-pressure zones. Widely spaced isobars around high-pressure zones in the central United States cause weaker winds. *(NOAA weather station Missoula, MT)*

16–18). The Coriolis effect alters wind direction, but not its speed.

Friction

Rising and falling air generates wind both along the Earth's surface and at higher elevations. Surface winds are affected by friction with the Earth's surface, whereas high-altitude winds are not. As a result, wind speed normally increases with elevation. This effect was first noted during World War II. On November 24, 1944, U.S. bombers were approaching Tokyo for the first mass bombing of the Japanese capital. Flying between 8000 and 10,000 meters (27,000 to 33,000 feet), the pilots suddenly found themselves roaring

past landmarks 140 kilometers (90 miles) per hour faster than the theoretical top speed of their airplanes! Amid the confusion, most of the bombs missed their targets, and the mission was a military failure. However, this experience introduced meteorologists to **jet streams,** narrow bands of high-altitude wind. The jet stream in the Northern Hemisphere flows from west to east at speeds between 120 and 240 kilometers per hour (75 and 150 mph). As a comparison, surface winds attain such velocities only in hurricanes and tornadoes. Airplane pilots traveling from Los Angeles to New York fly with the jet stream to gain speed and save fuel, whereas pilots moving from east to west try to avoid it.

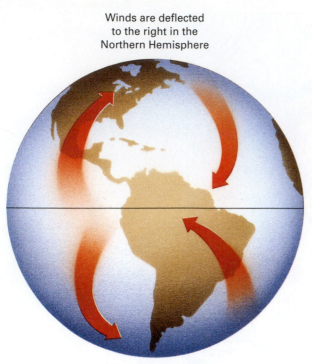

Winds are deflected
to the right in the
Northern Hemisphere

Winds are deflected
to the left in the
Southern Hemisphere

Figure 16–18 The Coriolis effect deflects winds to the right in the Northern Hemisphere, and to the left in the Southern Hemisphere. Only winds blowing due east or west are unaffected.

Cyclones and Anticyclones

Figure 16–19A shows the movement of air in the Northern Hemisphere as it converges toward a low-pressure area. If the Earth did not spin, wind would flow directly across the isobars, as shown by the black arrows. However, the Earth does spin, and the Coriolis effect deflects wind to the right, as shown by the small red arrows. This rightward deflection creates a counterclockwise vortex near the center of the low-pressure region, as shown by the large magenta arrows.

A low-pressure region with its accompanying surface wind is called a **cyclone.** In this usage, "cyclone" means a system of rotating winds, not the violent storms that are sometimes called cyclones, hurricanes, and typhoons. The opposite mechanism forms an **anticyclone** around a high-pressure region. When descending air reaches the surface, it spreads out in all directions. In the Northern Hemisphere, the Coriolis effect deflects the diverging winds to the right, forming a pinwheel pattern with the wind spiraling clockwise (Fig. 16–19B). In the Southern Hemisphere, the Coriolis effect deflects winds leftward, and creates a counterclockwise spiral.

Pressure Changes and Weather

As explained above, wind blows in response to any difference in pressure. However, low pressure generally brings clouds and precipitation with the wind, and sunny days predominate during high pressure. To understand this distinction, recall that when air is heated, it expands and rises, creating low pressure. The rising air cools adiabatically. If the cooling is sufficient, clouds form and rain or snow may fall. Thus, low barometric pressure is an indication of wet weather. Alternatively, when air is cooled, it sinks and the barometric pressure rises. This falling air is compressed and heated adiabatically. Because warm air can hold more water vapor than cold air, the sinking air absorbs moisture and clouds generally do not form over a high-pressure region. Thus, fair, dry weather generally accompanies high pressure.

16.6 Fronts and Frontal Weather

An **air mass** is a large body of air with approximately uniform temperature and humidity at any given altitude. Typically, an air mass is 1500 kilometers or more across and several kilometers thick. Because air acquires both heat and moisture from the Earth's surface, an air mass is classified by its place of origin. Temperature can be either *polar* (cold) or *tropical* (warm). *Maritime* air originates over water and has high moisture content, whereas *continental* air has low moisture content (Fig. 16–20, Table 16–1).

Air masses move and collide. The boundary between a warmer air mass and a cooler one is a **front.** The term was first used during World War I because weather systems were considered analogous to armies that advance and clash along battle lines. When two air masses collide, each may retain its integrity for days before the two begin to mix. During a collision, one of the air masses is forced to rise, which often results in cloudiness and precipitation. Frontal weather patterns are determined by the types of air masses that collide and their relative speeds and directions. The symbols commonly used on weather maps to describe fronts are shown in Figure 16–21.

Warm Fronts and Cold Fronts

Fronts are classified by whether a warm air mass moves toward a stationary (or more slowly moving) cold mass, or vice versa. A **warm front** forms when moving warm air collides with a stationary or slower moving cold air mass. A **cold front** forms when moving cold air collides with stationary or slower moving warm air.

In a warm front, the moving warm air rises over the denser cold air as the two masses collide (Fig. 16–22).

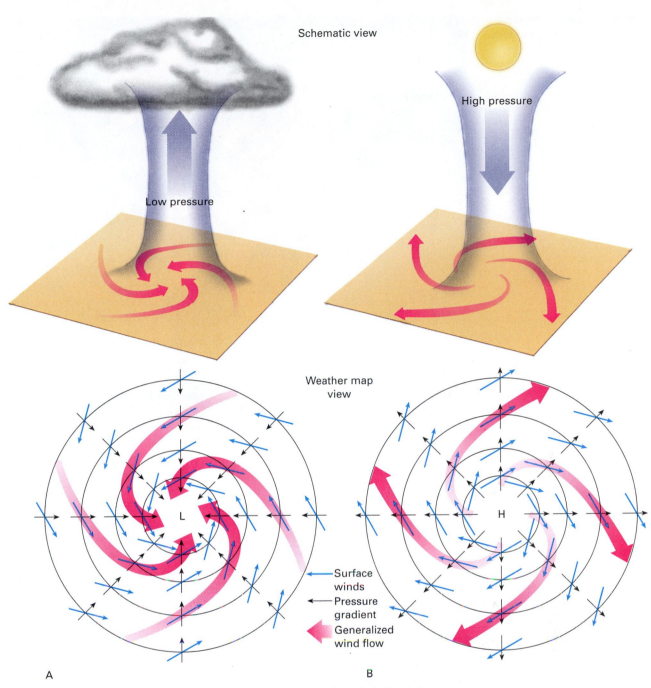

Schematic view

Low pressure

High pressure

Weather map
view

L

H

Surface
winds
Pressure
gradient
Generalized
wind flow

A

B

Figure 16–19   (A) In the Northern Hemisphere, a cyclone consists of winds spiraling counterclockwise into a low pressure region. (B) An anticyclone consists of winds spiraling clockwise out from a high-pressure zone.

The rising warm air cools adiabatically and the cooling generates clouds and precipitation. Precipitation is generally light because the air rises slowly along the gently sloping frontal boundary. Figure 16–22 shows that a characteristic sequence of clouds accompanies a warm front. High, wispy cirrus and cirrostratus clouds

develop near the leading edge of the rising warm air. These high clouds commonly precede a storm. They form as much as 1000 kilometers in front of an advancing band of precipitation falling from thick, low-lying nimbostratus and stratus clouds near the trailing edge of the front. The cloudy weather may last for

Figure 16–20 Air masses are classified by their source regions.

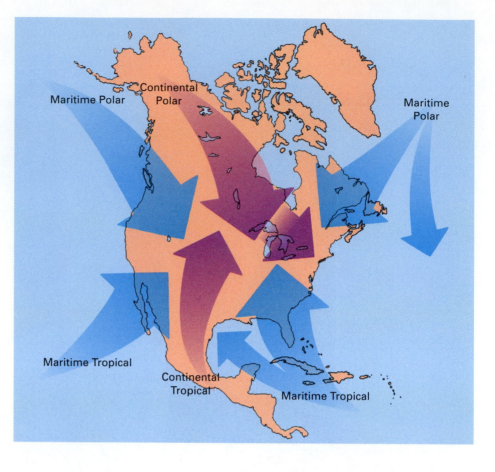

several days because of the gentle slope and broad extent of the frontal boundary.

A cold front forms when faster moving cold air overtakes and displaces warm air. The dense, cold air distorts into a blunt wedge and pushes under the warmer air (Fig. 16–23). Thus, the leading edge of a cold front is much steeper than that of a warm front. The steep contact between the two air masses causes the warm air to rise rapidly, creating a narrow band of violent weather commonly accompanied by cumulus and cumulonimbus clouds. The storm system may be only 25 to 100 kilometers wide, but within this zone downpours, thunderstorms, and violent winds are common.

TABLE 16–1
Classification of Air Masses

Classification according to latitude (temperature):

Polar (P) air masses originate in high latitudes and are cold.
Tropical (T) air masses originate in low latitudes and are warm.

Classification according to moisture content:

Continental (c) air masses originate over land and are dry.
Maritime (m) air masses originate over water and are moist.

Symbol	Name	Characteristics
mP	Maritime polar	Moist and cold
cP	Continental polar	Dry and cold
mT	Maritime tropical	Moist and warm
cT	Continental tropical	Dry and warm

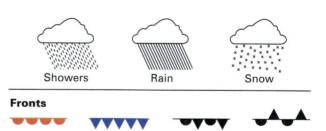

Showers   Rain   Snow

Fronts

Warm   Cold   Occluded   Stationary

Figure 16–21 Symbols commonly used in weather maps. "Warm" and "cold" are relative terms. Air over the central plains of Montana at a temperature of 0°C may be warm relative to polar air above northern Canada but cold relative to a 20°C air mass over the southeastern United States.

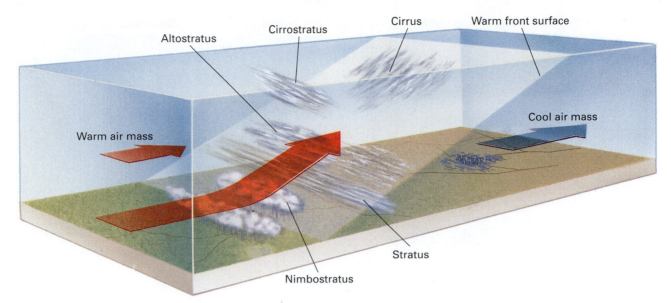

Figure 16–22 In a warm front, moving warm air rises gradually over cold air.

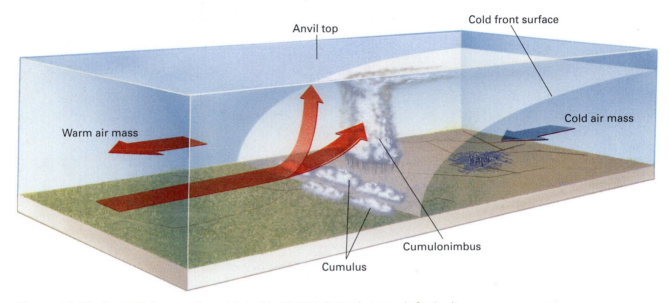

Figure 16–23 In a cold front, moving cold air slides abruptly beneath warm air, forcing it steeply upward.

Occluded Front

An **occluded front** forms when a faster moving cold mass traps a warm air mass against a second mass of cold air. Thus, the warm air mass becomes trapped between two colder air masses (Fig. 16–24). The faster moving cold air mass then slides beneath the warm air, lifting it completely off the ground. Precipitation occurs along both frontal boundaries, combining the narrow band of heavy precipitation of a cold front with the wider band of lighter precipitation of a warm front. The net result is a large zone of inclement weather. A storm of this type is commonly short-lived because the warm air mass is cut off from its supply of moisture evaporating from the Earth's surface.

Stationary Front

A **stationary front** occurs along the boundary between two stationary air masses. Under these conditions, the front can remain over an area for several days. Warm air rises, forming conditions similar to those in a warm front. As a result, rain, drizzle, and fog may occur.

The Life Cycle of a Middle-Latitude Cyclone

Most low-pressure cyclones in the middle latitudes of the Northern Hemisphere develop along a front between polar and tropical air masses. The storm often

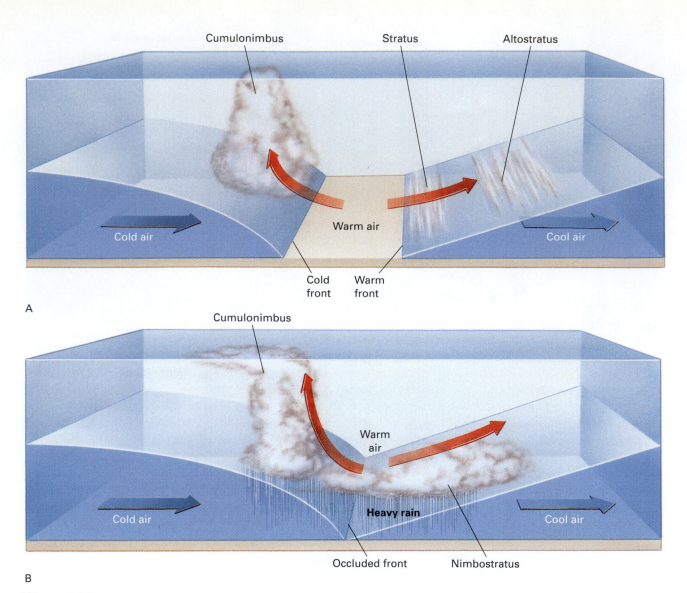

Figure 16–24 An occluded front forms where warm air is trapped and lifted between two cold air masses.

starts with winds blowing in opposite directions along a stationary front between the two air masses (Fig. 16–25A). In this figure, a warm air mass was moving northward, and was deflected to the east by the Coriolis force. At the same time, a cold air mass traveling southward was deflected to the west.

In Figure 16–25B, the cold, polar air continues to push southward, creating a cold front and lifting the warm air off the ground. Then, some small disturbance deforms the straight frontal boundary, forming a wave-like kink in the front. This disturbance may be a topographic feature such as a mountain range, air flow from a local storm, or a local temperature variation. Once the kink forms, the winds on both sides are deflected to strike the front at an angle. Thus, a warm front forms to the east and a cold front forms to the west.

Rising warm air then forms a low-pressure region near the kink (Fig. 16–25C). In the Northern Hemisphere, the Coriolis effect causes the winds to circulate counterclockwise around the kink, as explained in Section 16.5. To the west the cold front advances southward, and to the east the warm front advances northward. At the same time, rain or snow falls from the rising warm air (Fig. 16–25D). Over a period of one to three days, the air rushing into the low-pressure region equalizes pressure differences, and the storm dissipates. Many of the pinwheel-shaped storms seen on weather maps are cyclones of this type. In North America, the jet stream and other prevailing upper-level westerly winds generally move cyclones from west to east along the same paths, called **storm tracks** (Fig. 16–26).

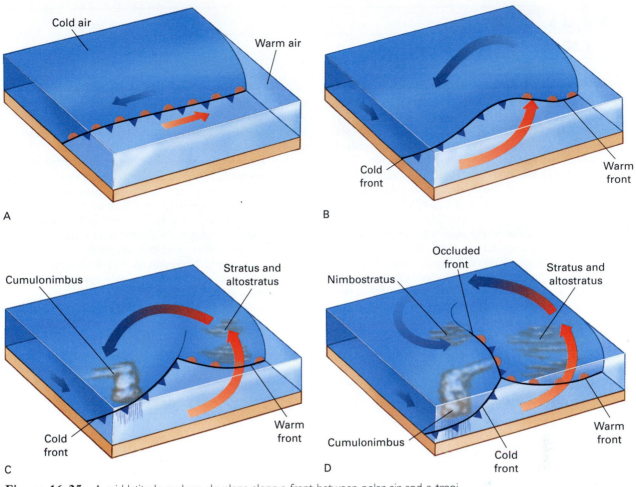

Figure 16–25 A mid-latitude cyclone develops along a front between polar air and a tropical air mass. (A) A front develops. (B) Some small disturbance creates a kink in the front. (C) A low-pressure region and cyclonic circulation develop. (D) An occluded front forms.

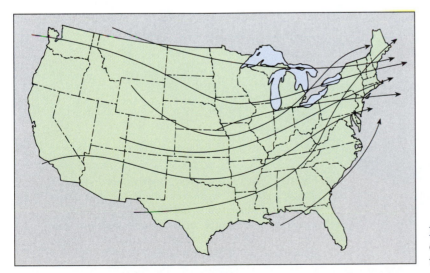

Figure 16–26 Most North American cyclones follow certain paths called storm tracks from west to east.

Focus On

El Niño

Hurricanes are common in the southeastern United States and on the Gulf Coast, but they rarely strike California. Consequently, Californians were taken by surprise in late September of 1997 when Hurricane Nora ravaged Baja California and then, somewhat diminished by landfall, struck San Diego and Los Angeles. The storm brought the first rain to Los Angeles after a record 219 days of drought, then spread eastward to flood parts of Arizona where it caused the evacuation of 1000 people.

Other parts of the world also experienced unusual weather during the autumn of 1997. In Indonesia and Malaysia, fall monsoon rains normally douse fires intentionally set in late summer to clear the rainforest. The rains were delayed for two months in 1997 as the fires raged out of control, filling cities with such dense smoke that visibility at times was no more than a few meters, and even an airliner crash was attributed to the smoke. Severe drought in nearby Australia caused ranchers to slaughter entire herds of cattle for lack of water and feed. At the same time, far fewer hurricanes than

usual threatened Florida and the U.S. Gulf Coast. Floods soaked northern Chile's Atacama Desert, a region that commonly receives no rain at all for a decade at a time, record snowfalls blanketed the Andes, and heavy rains caused floods in Peru and Ecuador.

All of these weather anomalies have been attributed, probably correctly, to El Niño, an ocean current that brings unusually warm water to the west coast of South America. But the current does not flow every year; instead, it occurs about every 3 to 7 years, and its effects last for about a year before conditions return to normal. Although meteorologists paid little attention to the phenomenon until the El Niño year of 1982–83, many now think that El Niño affects weather patterns for nearly three quarters of the Earth.

Meteorologists recorded El Niños in 1982–83, 1986–87, and 1991–92, although they were known to Peruvian fishermen much earlier because they warm coastal waters and diminish fish harvests. The fishermen called the warming events "El Niño" because they commonly occur around Christmas, the birthday of El Niño–

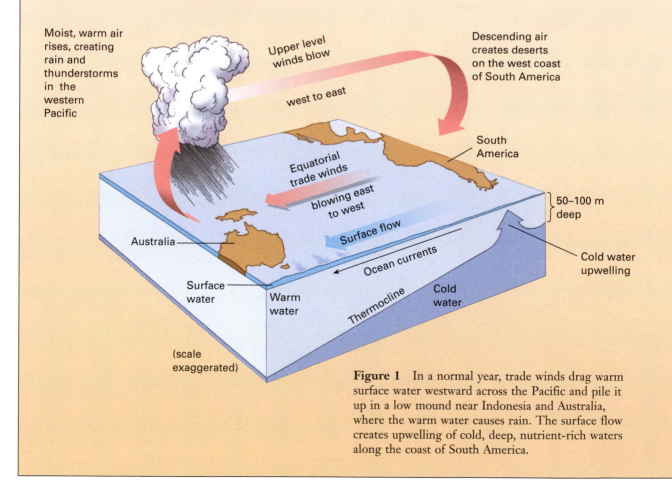

Figure 1 In a normal year, trade winds drag warm surface water westward across the Pacific and pile it up in a low mound near Indonesia and Australia, where the warm water causes rain. The surface flow creates upwelling of cold, deep, nutrient-rich waters along the coast of South America.

the Christ Child. To understand the El Niño effects, first consider interactions between southern Pacific sea currents and weather in a normal, non–El Niño year (Fig. 1).

Normally in fall and winter, strong trade winds blow westward from South America across the Pacific Ocean. The winds drag the warm, tropical surface water away from Peru and Chile, and pile it up in the western Pacific near Indonesia and Australia. In the western Pacific, the warm water forms a low mound thousands of kilometers across. The water is up to 10°C warmer and as much as 60 centimeters higher than the surface of the ocean near Peru and Chile.

As the wind-driven surface water flows away from the South American coast, cold, nutrient-rich water rises from the depths to replace the surface water. The nutrients support a thriving fishing industry along the coasts of Peru and Chile.

Abundant moisture evaporates from the surface of a warm ocean. In a normal year, much of this water condenses to bring rain to Australia, Indonesia, and other lands in the southwestern Pacific, which are adjacent to the mound of warm water. On the eastern side of the Pacific, the cold upwelling ocean currents cool the air above the coasts of Peru and northern Chile. This cool air becomes warmer as it flows over land. The warming lowers the relative humidity and creates the coastal Atacama desert.

In an El Niño year, for reasons poorly understood by meteorologists, the trade winds slacken (Fig. 2). The mound of warm water near Indonesia and Australia then flows downslope—eastward across the Pacific Ocean toward Peru and Chile. The anomalous accumulation of warm water off South America causes unusual rains in normally dry coastal regions, and heavy snowfall in the Andes. At the same time, the cooler water near Indonesia, Australia, and nearby regions causes drought. The mass of warm water that caused the 1997–98 El Niño was about the size of the United States, and caused one of the strongest El Niño weather disturbances in history.

El Niño has global effects that go far beyond regional rainfall patterns. It also modifies global atmospheric

(continued on page 386)

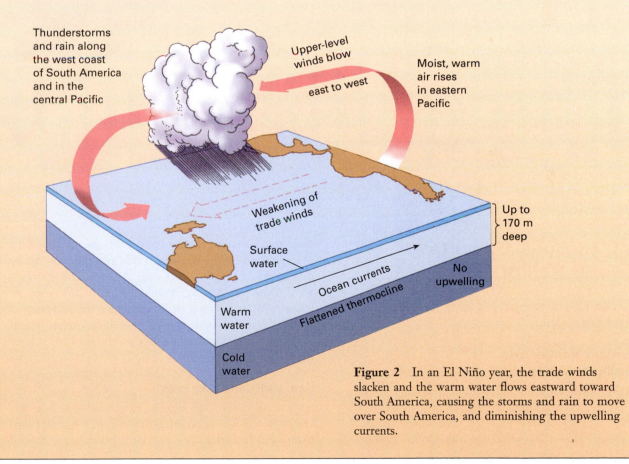

Thunderstorms and rain along the west coast of South America and in the central Pacific

Upper-level winds blow east to west

Moist, warm air rises in eastern Pacific

Weakening of trade winds

Up to 170 m deep

Surface water

Ocean currents

No upwelling

Warm water

Flattened thermocline

Cold water

Figure 2 In an El Niño year, the trade winds slacken and the warm water flows eastward toward South America, causing the storms and rain to move over South America, and diminishing the upwelling currents.

circulation patterns, and thus alters weather on a world-wide scale (Fig. 3). For example, El Niño deflects the jet stream from its normal path as it flows over North America, directing one branch northward over Canada, and the other across southern California and Arizona. Consequently, those regions receive more winter precipitation and storms than usual, while fewer storms and warmer winter temperatures affect the Pacific Northwest, the northern plains, the Ohio River valley, the mid-Atlantic states, and New England. Southern Africa experiences drought, while Ecuador, Peru, Chile, southern Brazil, and Argentina receive more rain than usual.

Altered weather patterns created by El Niño wreak havoc. Globally, 2000 deaths and more than $13 billion in damage are attributed to the 1982–83 El Niño effects. In the United States alone, more than 160 deaths and $2 billion in damage, mostly from storm and flood damage, occurred. In southern Africa, economic losses of $1 billion and uncounted deaths due to disease and starvation have been attributed to the 1982–83 El Niño.

Europe
Not affected by El Niño nearly as much as it is by Atlantic systems

Southern Africa
Drought

Equator

Indian Oc

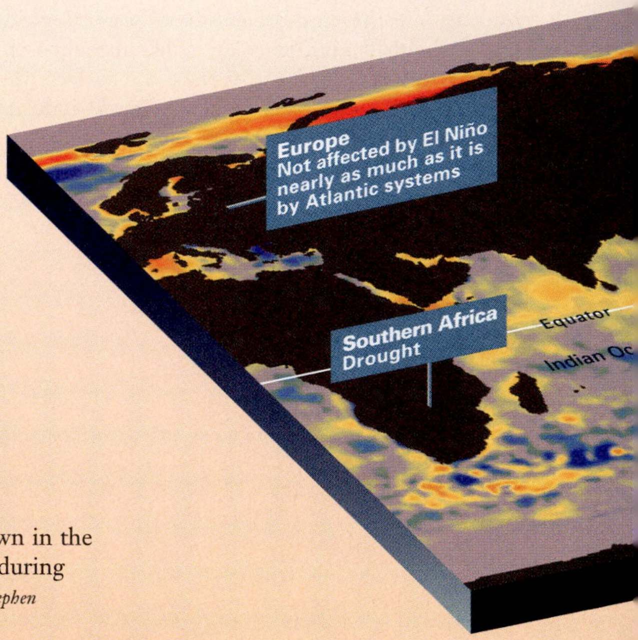

Figure 3 An El Niño event alters worldwide weather patterns, as shown in the boxes. The sea surface temperature deviations from average conditions during the September 1997 El Niño phase are shown in colors. *(Robert Kemp, Stephen Rountree, Richard Gage, U.S. News & World Report. Source data from NOAA)*

16.7 Mountains, Oceans, Lakes, and Weather

Mountain ranges, oceans, and large lakes affect both weather and climate.

Mountain Ranges and Rain-Shadow Deserts

As we described earlier in this chapter, air rises in a process called orographic lifting when it flows over a mountain range. As the air rises, it cools adiabatically, and water vapor may condense into clouds that produce rain or snow. These conditions create abundant precipitation on the windward side and the crest of the range. When the air passes over the crest onto the leeward (downwind) side, it sinks (Fig. 16–27). This air has already lost much of its moisture. In addition, it warms adiabatically as it falls, absorbing moisture and creating a **rain-shadow desert** on the leeward side of the range (Fig. 16–28).

Sea and Land Breezes

Anyone who has lived near an ocean or large lake has encountered winds blowing from water to land and from land to water. Sea and land breezes are caused

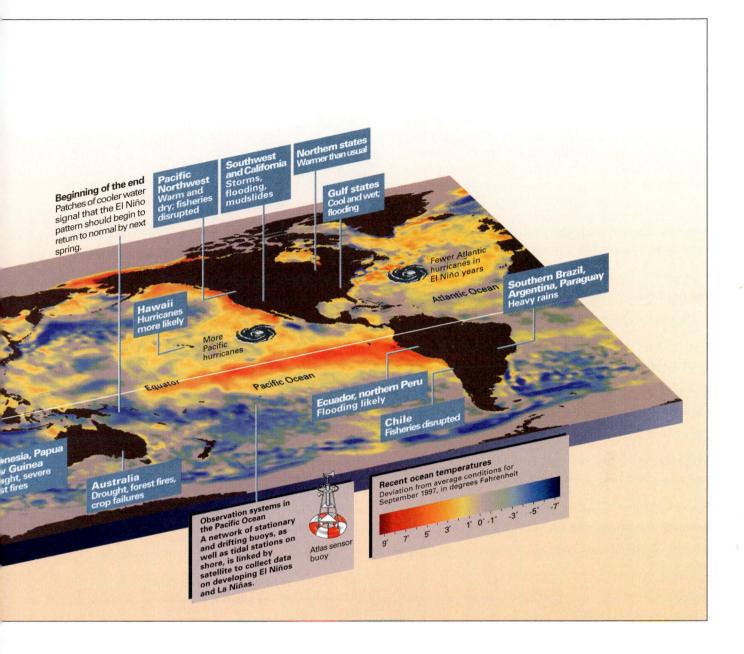

Beginning of the end
Patches of cooler water signal that the El Niño pattern should begin to return to normal by next spring.

Pacific Northwest
Warm and dry; fisheries disrupted

Southwest and California
Storms, flooding, mudslides

Northern states
Warmer than usual

Gulf states
Cool and wet; flooding

Hawaii
Hurricanes more likely

More Pacific hurricanes

Fewer Atlantic hurricanes in El Niño years

Atlantic Ocean

Southern Brazil, Argentina, Paraguay
Heavy rains

Equator

Pacific Ocean

Ecuador, northern Peru
Flooding likely

Chile
Fisheries disrupted

onesia, Papua w Guinea ught, severe st fires

Australia
Drought, forest fires, crop failures

Observation systems in the Pacific Ocean
A network of stationary and drifting buoys, as well as tidal stations on shore, is linked by satellite to collect data on developing El Niños and La Niñas.

Atlas sensor buoy

Recent ocean temperatures
Deviation from average conditions for September 1997, in degrees Fahrenheit

9° 7° 5° 3° 1° 0° -1° -3° -5° -7°

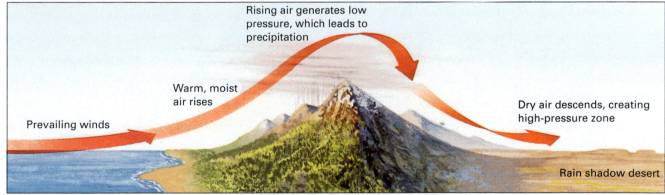

Rising air generates low pressure, which leads to precipitation

Warm, moist air rises

Prevailing winds

Dry air descends, creating high-pressure zone

Rain shadow desert

Figure 16–27   A rain-shadow desert forms where moist air rises over a mountain range and precipitates most of its moisture on the windward side and crest of the range. The dry, descending air on the lee side absorbs moisture, forming a desert.

by uneven heating and cooling of land and water (Fig. 16–29). Recall that land surfaces heat up faster than adjacent bodies of water, and cool more quickly. If land and sea are nearly the same temperature on a summer morning, during the day the land warms and heats the air above it. Hot air then rises over the land, producing a local low-pressure area. Cooler air from the sea flows inland to replace the rising air. Thus, on a hot, sunny day, winds generally blow from the sea onto land. The rising air is good for flying kites or hang gliding, but often brings afternoon thunderstorms.

At night the reverse process occurs. The land cools faster than the sea, and descending air creates a local high pressure area over the land. Then the winds reverse, and breezes blow from the shore out toward the sea.

Monsoons

A **monsoon** is a seasonal wind and weather system caused by uneven heating and cooling of continents and oceans. Just as sea and land breezes reverse direction with day and night, monsoons reverse direction with the seasons. In the summertime the continents become warmer than the sea. Warm air rises over land, creating a large low-pressure area and drawing moisture-laden maritime air inland. When the moist air rises as it flows over the land, clouds form and heavy monsoon rains fall. In winter the process is reversed. The land cools below the sea temperature, and as a result, air descends over land, producing dry continental high pressure. At the same time air rises over the ocean and the prevailing winds blow from

land to sea. More than half of the inhabitants of the Earth depend on monsoons because the predictable heavy summer rains bring water to the fields of Africa and Asia. If the monsoons fail to arrive, crops cannot grow and people starve.

16.8 Thunderstorms

An estimated 16 million thunderstorms occur every year, and at any given moment about 2000 thunderstorms are in progress over different parts of the Earth. A single bolt of lightning can involve several hundred million volts of energy and for a few seconds produces as much power as a nuclear power plant. It heats the surrounding air to 25,000°C or more, much hotter than the surface of the Sun. The heated air expands instantaneously to create a shock wave that we hear as thunder.

Despite their violence, thunderstorms are local systems, often too small to be included on national weather maps. A typical thunderstorm forms and then dissipates in a few hours and covers from about ten to a few hundred square kilometers. It is not unusual to stand on a hilltop in the sunshine and watch rain squalls and lightning a few kilometers away. All thunderstorms develop when warm, moist air rises, forming cumulus clouds that develop into towering cumulonimbus clouds. Different conditions cause these local regions of rising air.

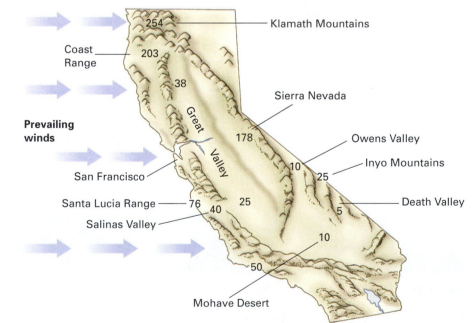

Figure 16–28 Rain-shadow deserts lie east of the California mountain ranges. Rainfall is shown in centimeters per year.

A   B

Figure 16–29 (A) Sea breezes blow inland during the day, and (B) land breezes blow out to sea at night.

1. **Wind convergence** Central Florida is the most active thunderstorm region in the United States. As the subtropical Sun heats the Florida peninsula, rising air draws moist air from both the east and west coasts. Where the two air masses converge, the moist air rises rapidly to create a thunderstorm. Thunderstorms also occur in other environments where moist air masses converge.

2. **Convection** Thunderstorms also form in continental interiors during the spring or summer, when afternoon sunshine heats the ground and generates cells of rising moist air.

3. **Orographic rise** Moist air rises as it flows over hills and mountain ranges, commonly generating mountain thunderstorms.

4. **Frontal thunderstorms** Thunderstorms commonly occur along frontal boundaries, particularly at cold fronts.

A typical thunderstorm occurs in three stages. In the initial stage, moisture in rising air condenses, forming a cumulus cloud (Fig. 16–30A). As the cloud forms, the condensing vapor releases latent heat. In an average thunderstorm, 400,000 tons of water vapor condense within the cloud, and the energy released by the condensation is equivalent to the explosion of 12 atomic bombs the size of the one dropped on Hiroshima during World War II. This heat warms air in the cloud, and fuels the violent convection characteristic of thunderstorms. Large droplets or ice crystals develop within the cloud at this stage, but the rising air keeps them in suspension, and no precipitation falls to the ground.

Eventually, water droplets or hailstones become so heavy that updrafts can no longer support them, and

they fall as rain or hail. During this stage, warm air continues to rise and may attain velocities of over 300 kilometers per hour in the central and upper portions of the cloud. The cloud may double its height in minutes. At the same time, ice falling through the cloud chills the lower regions and this cool air sinks, creating a downdraft. Thus, air currents rise and fall simultaneously within the same cloud. These conditions, known as **wind shear,** are dangerous for aircraft, and pilots avoid large thunderheads (Fig. 16–30B).

As explained earlier, rainfall from a cumulonimbus cloud can be unusually heavy. In one extreme example in 1976, a sequence of thunderstorms dropped 25 centimeters of rain in about 4 hours over Big Thompson Canyon on the eastern edge of the Colorado Rockies. The river flooded the narrow canyon, killing 139 people.

The mature stage of a thunderstorm, with rain or hail and lightning, usually lasts for about 15 to 30 minutes and seldom longer than an hour. The cool downdraft reduces the temperature in the lower regions of the cloud. As the temperature drops, convection weakens and warm, moist air is no longer drawn into the cloud (Fig. 16–30C). Once the water supply is cut off, condensation ceases, and the storm loses it source of latent heat. Within minutes the rapid vertical air motion dies and the storm dissipates. Although a single thundercloud dissipates rapidly, new thunderheads can build in the same region, causing disasters such as that at Big Thompson Canyon.

In 1996, a group of climbers were slowly approaching the summit of Mount Everest. Below them, fluffy white clouds began to obscure the lower peaks. Most of the climbers thought that the clouds were benign and continued on toward the summit. But one of

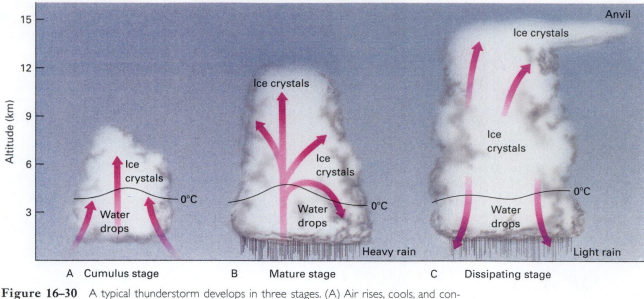

Altitude (km)

A Cumulus stage   B Mature stage   C Dissipating stage

Figure 16–30 A typical thunderstorm develops in three stages. (A) Air rises, cools, and condenses, creating a cumulus cloud. (B) Latent heat of condensation energizes the storm, forming heavy rain and violent wind. (C) The cloud cools, convection weakens, and the storm wanes.
● Interactive Question: Why does hail fall only from cumulonimbus couds?

the climbers was an airplane pilot and realized that the view from 8500 meters was analogous to a view from an airplane and not the perspective that we are normally accustomed to. To his trained eye, the clouds were the tops of rising thunderheads. Realizing that dangerous strong winds and heavy precipitation accompany a thunderstorm, he abandoned his summit attempt and retreated. The others pushed onward. A few hours later the intense storm engulfed the summit ridge and six climbers perished.

Lightning is an intense discharge of electricity that occurs when the buildup of static electricity overwhelms the insulating properties of air (Fig. 16–31). If you walk across a carpet on a dry day, the friction between your feet and the rug shears electrons off the atoms on the rug. The electrons migrate into your body, and concentrate there. If you then touch a metal doorknob, a spark consisting of many electrons jumps from your finger to the metal knob.

In 1752 Benjamin Franklin showed that lightning is an electrical spark. He suggested that charges separate within cumulonimbus clouds and build until a bolt of lightning jumps from the cloud. In the nearly 250 years since Franklin, atmospheric physicists have

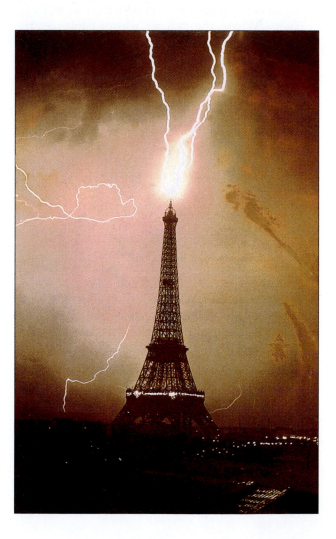

Figure 16–31 A bolt of lightning strikes the Eiffel Tower in Paris, France. Lightning occurs when the potential difference between two groups of electrical charges exceeds the insulating properties of air. *(Jean-Loup Charmet/Science Photo Library)*

been unable to agree upon the exact mechanism of lightning. According to one hypothesis, friction between the intense winds and moving ice crystals in a cumulonimbus cloud generates both positive and negative electrical charges in the cloud, and the two types of charges become physically separated. The positive charges tend to accumulate in the upper portion of the cloud, and the negative charges build up in the lower reaches of the cloud. When enough charge accumulates, the electrical potential exceeds the insulating properties of air, and a spark jumps from the cloud to the ground, from the ground to the cloud, or from one cloud to another.

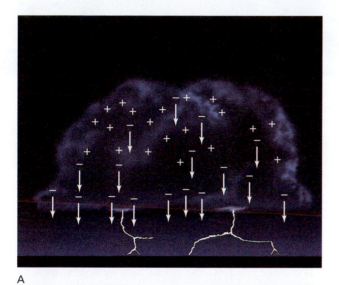

A

B

Figure 16–32 Two hypotheses for the origin of lightning. (A) Friction between intense winds and ice particles generates charge separation. (B) Charged particles are produced from above by cosmic rays and below by interactions with the ground. The particles are then distributed by convection currents.

Another hypothesis suggests that cosmic rays bombarding the cloud from outer space produce ions at the top of the cloud. Other ions form on the ground as winds blow over the Earth's surface. The electrical discharge occurs when the potential difference between the two groups of electrical charges exceeds the insulating properties of air (Fig. 16–32A, B). Perhaps neither hypothesis is entirely correct and some combination of the mechanisms causes lightning.

16.9   Tornadoes and Tropical Cyclones

Tornadoes and tropical cyclones are both intense low-pressure centers. Strong winds follow the steep pressure gradients and spiral inward toward a central column of rising air.

Tornadoes

A **tornado** is a small, short-lived, funnel-shaped storm that protrudes from the base of a cumulonimbus cloud (Fig. 16–33). The base of the funnel can be from 2 meters to 3 kilometers in diameter. Some tornadoes remain suspended in air while others touch the ground. After a tornado touches ground, it may travel for a few meters to a few hundred kilometers across the surface. The funnel travels from 40 to 65 kilometers per hour, and in some cases, as much as 110 kilometers per hour, but the spiraling winds within the funnel are much faster. Few direct measurements have been made of pressure and wind speed inside a tornado. However, we know that a large pressure difference occurs over a very short distance. Meteorologists estimate that winds in tornadoes may reach 500 kilometers per hour or greater. These winds rush into the narrow low-pressure zone and then spiral upward. After a few seconds to a few hours, the tornado lifts off the ground and dissipates.

Tornadoes are the most violent of all storms. One tornado in 1910 lifted a team of horses and then deposited it, unhurt, several hundred meters away. They were lucky. In the past, an average of 120 Americans were killed every year by these storms, and property damage cost millions of dollars. The death toll has decreased in recent years because effective warning systems allow people to seek shelter, but the property damage has continued to increase (Fig. 16–33B). Tornado winds can lift the roof off a house and then flatten the walls. Flying debris kills people and livestock caught in the open. Even so, the total destruction from tornadoes is not as great as that from hurricanes because the path of a tornado is narrow and its duration short.

Although tornadoes can occur anywhere in the world, 75 percent of the world's twisters concentrate

A

B

Figure 16–33 (A) A tornado is a funnel-shaped storm that emanates from the base of a cumulonimbus cloud. This one occurred in 1957 near Union City, Oklahoma. *(NOAA, National Severe Storms Laboratory)* (B) Aftermath of a tornado in residential Raleigh, North Carolina. *(Charles Gupton/Tony Stone Images)*

in the Great Plains, east of the Rocky Mountains. Approximately 700 to 1000 tornadoes occur in the United States each year. They frequently form in the spring or early summer. At that time, continental polar (dry, cold) air from Canada collides with maritime tropical (warm, moist) air from the Gulf of Mexico. As explained previously, these conditions commonly create thunderstorms. Meteorologists cannot explain why most thunderstorms dissipate harmlessly but a few develop tornadoes. However, one fact is apparent: Tornadoes are most likely to occur when large differences in temperature and moisture exist between the two air masses and the boundary between them is sharp.

The probability that any particular place will be struck by a tornado is small. Nevertheless, Codell, Kansas, was struck three years in a row—in 1916, 1917, and 1918—and each time the disaster occurred on May 20! During a 2-day period in 1974, 148 tornadoes occurred in 13 states.

Tropical Cyclones

A **tropical cyclone** (called a *hurricane* in North America and the Caribbean, a *typhoon* in the western Pacific, and a *cyclone* in the Indian Ocean) is less intense than a tornado but much larger and longer-lived (Table 16–2). Tropical cyclones are circular storms that average 600 kilometers in diameter and persist for days

or weeks (Fig. 16–34). Intense low pressure in the center of a hurricane generates wind that varies from 120 to more than 300 kilometers per hour.

The low atmospheric pressure created by a tropical cyclone can raise the sea surface by several meters. Often, as a tropical cyclone strikes shore, strong onshore winds combine with the abnormally high water level created by low pressure to create a **storm surge** that floods coastal areas. In 1969 during Hurricane Camille, sea level rose more than 8 meters above normal on the Gulf Coast as a result of a storm surge.

The deadliest hurricane in the United States struck Galveston, Texas, in September 1900. Eight thousand people died and millions of dollars of damage occurred. One reason the death toll was so high was that the population was caught unaware. A tropical cyclone has a sharp boundary, and even a few hundred kilometers outside that boundary, fluffy white clouds may be floating in a blue sky. Today, in the United States, hurricanes are detected by satellite and people are evacuated from coastal areas well in advance of oncoming storms. Warning cannot eliminate property damage, however. When Hurricane Hugo struck the North Carolina coast in 1989 with 220-kilometer-per-hour winds, waves flooded coastal areas, destroying 1900 homes and killing 40 people. Property damage totaled $3 billion.

In the spring of 1991, more than 100,000 people were killed when a cyclone sent 7-meter waves across

TABLE 16-2

Comparison of Tornadoes and Tropical Cyclones

Feature	Range	
	Tornado	**Tropical Cyclone**
Diameter	2–3 km	400–800 km
Path length (distance traveled across terrain)	A few meters to hundreds of kilometers	A few hundred to a few thousand kilometers
Duration	A few seconds to a few hours	A few days to a week
Wind speed	300–800 km/hr	120–250 km/hr
Speed of motion	0–70 km/hr	20–30 km/hr
Pressure fall	20–200 mb	20–60 mb

the heavily populated, low-lying coast of Bangladesh. Although meteorologists had been tracking the storm, communication was so poor and transportation facilities so inadequate that people were not evacuated in time.

Tropical cyclones form only over warm oceans, never over cold oceans or land. Thus, moist warm air is crucial to development of this type of storm. Recall that a mid-latitude cyclone develops when a small disturbance produces a wave-like kink in a previously linear front. A similar mechanism initiates a tropical cyclone. In late summer, the Sun warms tropical air. The rising hot air creates a belt of low pressure that encircles the globe in those latitudes. In addition, many local low-pressure disturbances move across the tropical oceans at this time of year. If a local disturbance intersects the global tropical low, it creates a bulge in the isobars. Winds are deflected by the bulge

and, directed by the Coriolis effect, begin to spiral inward. Warm, moist air rises from the low. Water vapor condenses from the rising air, and the latent heat warms the air further, which causes even more air to rise. As the low pressure becomes more intense, strong surface winds blow inward to replace the rising air. This surface air also rises, and more condensation and precipitation occur. But the additional condensation releases more heat, which continues to add energy to the storm.

The center of the storm is a region of vertical airflow, called the eye. In the outer, and larger part of the eye, the air that has been rushing inward spirals upward. In the inner eye, air sinks. Thus, the horizontal wind speed in the eye is reduced to near zero (Fig. 16–35). Survivors who have been in the eye of a hurricane report an eerie calm. Rain stops, and the Sun may even shine weakly through scattered clouds.

Figure 16–34 A color-enhanced satellite image of Hurricane Andrew as it approached the Florida Coast in 1992. *(NOAA)*

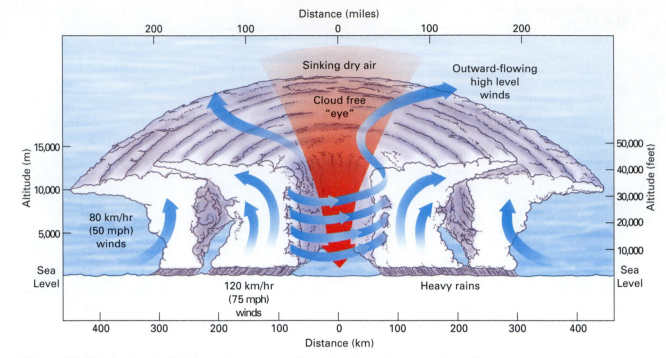

Figure 16–35 Surface air spirals inward toward a hurricane, rises through the towering wall of clouds, and then flows outward above the storm. Falling air near the storm's center creates the eerie calm in the eye of the hurricane.

But this is only a momentary reprieve. A typical eye is only 20 kilometers in diameter, and after it passes the hurricane rages again in full intensity.

Once a hurricane develops, it is powered by the latent heat released by continuing condensation. The entire storm is pushed by prevailing winds, and its path is deflected by the Coriolis effect. It dissipates after it reaches land or passes over colder water because the supply of moist, warm air is cut off. Condensing water vapor in a single tropical cyclone releases as much latent heat energy as that produced by all the electric generators in the United States in six months.

Tropical Cyclones in the United States

Three tropical cyclones, or hurricanes, that struck the United States between 1989 and 1995, Opal (1995), Andrew (1992), and Hugo (1989), caused a total of more than $40 billion in damage. The record-high dollar value of damage from these and other recent hurricanes has led a variety of sources, ranging from *Newsweek* magazine to the United States Senate, to link global warming to increasing hurricane damage.

However, a recent study[1] indicates that hurricane frequency and intensity have actually been *lower* during the past decade than the average for the past century. The increased dollar value of damage occurred because: (1) More Americans lived near the Atlantic and Gulf coasts in the late 1990s than earlier in the century. (2) Inflation increased the dollar value of homes and other structures; and (3) Americans now own more things than ever before and consequently, the dollar value of their possessions is at an all-time high.

The results show that Andrew was not the most damaging hurricane in history. An unnamed tropical cyclone that struck Florida and Alabama in 1926 would have more than doubled Andrew's monetary damage if it had occurred in 1995. Table 16–3 shows the top ten most destructive hurricanes, normalized to 1995 dollars by population increase, inflation, and personal property increases.

This study indicates that, although the non-normalized dollar value of damage done by tropical cyclones during the past decade reached a record high, the high dollar value resulted from cultural changes. In fact, the frequency and intensity of hurricanes during that time was lower than normal. Americans living in Florida and Gulf Coast coastal areas should

---

[1] Roger Pielke and Christopher Landsea, "Normalized Hurricane Damage in the United States: 1925–1995." Proceedings of the Twenty-second Conference on Hurricanes and Tropical Meteorology, American Meteorological Society, May 1997.

TABLE 16–3

Damage Normalized to 1995 Dollars, Caused by the Ten
Most Destructive Hurricanes in United States History

Rank	Hurricane	Year	Category	Damage (billion $)
1.	SE Florida, Alabama	1926	4	$72.303
2.	*Andrew* (SE Florida, Louisiana)	1992	4	$33.094
3.	SW Florida	1944	3	$16.864
4.	New England	1938	3	$16.629
5.	SE Florida, Lake Okeechobee	1928	4	$13.795
6.	*Betsy* (SE Florida, Louisiana)	1965	3	$12.434
7.	*Donna* (Florida, E. United States)	1960	4	$12.048
8.	*Camille* (Mississippi, Louisiana, Virginia)	1969	5	$10.965
9.	*Agnes* (NW Florida, NE United States)	1972	1	$10.705
10.	*Diane* (NE United States)	1955	1	$10.232

Source: National Center for Atmospheric Research

expect even greater financial losses when the frequency and intensity of tropical cyclones returns to normal. Some meteorologists predict that it is only a matter of time before a single $50 billion hurricane strikes Florida or the Gulf Coast.

The numerical "category" of tropical cyclones given in Table 16–3 is based on a rating scheme called the Saffir-Simpson scale, after its developers. This scale, commonly mentioned in weather reports, rates the damage potential of a hurricane or other tropical storm, and typical values of atmospheric pressure, wind speed, and height of storm surge associated with storms of increasing intensity. Table 16–4 shows the ranking and criteria of the Saffir-Simpson scale.

TABLE 16–4

The Saffir-Simpson Hurricane Damage Potential Scale

Type	Category	Damage	Pressure (millibar)	Winds (km/h)	Storm surge (m)
Depression				>56	
Tropical Storm				63–117	
Hurricane	1	minimal	980	119–152	1.2–1.5
Hurricane	2	moderate	965–979	154–179	1.8–2.4
Hurricane	3	extensive	945–964	179–209	2.7–3.7
Hurricane	4	extreme	920–944	211–249	4–5.5
Hurricane	5	catastrophic	<920	>249	>5.5

SUMMARY

Absolute humidity is the mass of water vapor in a given volume of air. **Relative humidity** is the amount of water vapor in air compared to the amount the air could hold at that temperature. Condensation occurs when moist air cools below its **dew point**. When warm air rises, it performs work and therefore cools **adiabatically**.

The characteristics of a cloud depend on the height to which air rises and the elevation at which condensation occurs. The three fundamental types of clouds are **cirrus, stratus,** and **cumulus.** Precipitation occurs when small water droplets or ice crystals coalesce until they become large enough to fall.

When air is heated, it expands and rises, creating low pressure. But the air cools adiabatically as it rises. The cooling may create clouds and rain or snow. Thus, low barometric pressure is an indication that precipitation may soon follow. Alternatively, when cool air falls, creating high pressure, it is compressed and heats up adiabatically. Since warm air can hold more moisture than cold air, clouds generally do not form under high-pressure conditions. Thus, rising barometric pressure generally precedes fair weather.

Uneven heating of the Earth's surface causes pressure differences which, in turn cause wind. Wind speed is determined by the **pressure gradient. Cyclones** and **anticyclones** are low-pressure and high-pressure zones, respectively. Winds spiral into a cyclone, and outward from an anticyclone.

Air cools adiabatically when it rises over a mountain range, often causing precipitation. A **rain-shadow desert** forms where air sinks down the leeward side of a mountain.

When two air masses collide, the warmer air rises along the **front,** forming clouds and often precipitation.

Sea breezes and **monsoons** arise because ocean temperature changes slowly in response to daily and seasonal changes in solar radiation, whereas land temperature changes quickly. A thunderstorm is a small, short-lived storm from a cumulonimbus cloud. Lightning occurs when charged particles separate within the cloud. A **tornado** is a small, short-lived, funnel-shaped storm that protrudes from the bottom of a cumulonimbus cloud and reaches the ground. A tropical cyclone is a larger, longer-lived storm that forms over warm oceans and is powered by the energy released when water vapor condenses to form clouds and rain.

KEY TERMS

humidity 366	wet adiabatic lapse rate 368	sleet 372	air mass 378
absolute humidity 366	convection 369	glaze 373	front 378
relative humidity 366	frontal wedging 369	hail 373	warm front 378
saturation 366	orographic lifting 369	advection fog 374	cold front 378
dew point 366	normal lapse rate 369	radiation fog 374	occluded front 382
supersaturation 366	cirrus cloud 370	evaporation fog 374	stationary front 382
supercooling 366	stratus cloud 371	upslope fog 374	storm tracks 382
dew 368	cumulus cloud 371	wind 376	rain-shadow desert 386
frost 368	stratocumulus cloud 371	pressure gradient 376	monsoon 388
adiabatic temperature changes 368	cumulonimbus cloud 371	isobar 376	wind shear 389
dry adiabatic lapse rate 368	nimbostratus cloud 371	jet stream 377	tornado 391
	altostratus cloud 371	cyclone 378	tropical cyclone 392
		anticyclone 378	storm surge 392

FOR REVIEW

1. Describe the difference between absolute humidity and relative humidity. How can the relative humidity change while the absolute humidity remains constant?

2. List three atmospheric processes that cool air.

3. What is the dew point? How does dew form?

4. List a set of atmospheric conditions that might produce supersaturation or supercooling.

5. What are the differences among dew, frost, and fog?

6. Explain why air cools as it rises.

7. What is an adiabatic temperature change? How does it differ from a nonadiabatic temperature change?

8. Compare and contrast the dry adiabatic lapse rate, the wet adiabatic lapse rate, and the normal lapse rate.

9. Discuss the factors that determine the height at which a cloud forms.

10. Describe cirrus, stratus, and cumulus clouds. Include their shapes and the type of precipitation to be expected from each.

11. How does rain form? How do droplets falling from a stratus cloud differ from those formed in a cumulus cloud?

12. Compare and contrast snow, sleet, and hail.

13. List three mechanisms that cause air to rise.

14. Why does low pressure often lead to rain? Why does high pressure often bring sunny skies?

15. If the wind is blowing southward in the Northern Hemisphere, will the Earth's spin cause it to veer east

or west? If the wind is moving southward in the Southern Hemisphere, which way will it veer? Explain your answer.

16. Compare and contrast sea and land breezes with monsoons.

17. What is a rain-shadow desert? How does it form?

18. How do warm and cold fronts form, and what types of weather are caused by each?

19. Describe the three stages in the life of a thunderstorm.

20. Briefly describe two hypotheses of how lightning forms.

21. Compare and contrast tornadoes with tropical cyclones. How does each form, and how does each affect human settlements?

FOR DISCUSSION

1. Using Figure 16–1, estimate the maximum absolute humidity at 0°C, 10°C, 20°C, and 40°C. Estimate the quantity of water in air, at 50 percent relative humidity, at each of the above temperatures.

2. Explain why frost forms on the inside of a refrigerator (assuming it is an old-fashioned one and not a modern frost-free unit). Would more frost tend to form in (a) summer or winter, and (b) in a dry desert region or a humid one? Explain.

3. Which of the following conditions produces frost? Which produces dew? Explain. (a) A constant temperature throughout the day. (b) A warm summer day followed by a cool night. (c) A cool fall afternoon followed by freezing temperature at night.

4. Draw a chart showing temperatures at cloud level, between the cloud and the ground, and at ground level that will cause condensing water vapor to fall as rain, snow, sleet, and glaze.

5. What is the energy source that powers the wind?

6. Are sea breezes more likely to be strong on an overcast day or on a bright sunny one? Explain.

7. What is an air mass? Describe what would likely happen if a polar air mass collided with a humid subtropical air mass.

8. Study the weather map in today's newspaper, and predict the weather two days from now in your area, Salt Lake City, Chicago, and New York City. Defend your prediction. Check the paper in two days to see if you were right or wrong.

9. Give an example of how the geosphere, the hydrosphere, and the biosphere affect, and are affected by, weather.

Climate

A cool, wet climate creates a temperate rainforest ecosystem near Bella Coola, British Columbia.

Many people watch the weather forecasts on TV, hoping for a warm, sunny day for next Sunday's picnic. However, our expectations are limited by the regional climate. If you live in Montana you would not plan an outdoor barbecue in January. Climate strongly influences many aspects of our lives: our outdoor recreation, the houses we live in, and the clothes we wear. Climate also affects the crops grown in a region and the amount of fuel we burn. Humans tend to migrate toward warm, sunny regions. In the United States, the populations of California and Texas have increased rapidly in the past generation, while North Dakota and Montana have seen little growth.

In Chapter 16, we defined climate as the weather patterns—especially the temperature and precipitation—averaged over several decades. Climates have changed, often dramatically, over geologic time. Deserts have turned to humid swamps that have then been covered by continental ice sheets. This chapter examines the Earth's current climate. Chapter 18 describes climate change. ●

A

B

Figure 17–1 (A) This Costa Rican rainforest lies along the Pacific coast of Central America at about 10° north latitude. (B) The Atacoma Desert, Chile, at 20° south latitude. The contrasting climates and ecosystems result from different ocean currents and winds in the two regions. *(John S. Botkin/Dembinski Photo Associates)*

17.1 Major Factors That Control Earth's Climate

In our study of the atmosphere, we introduced many factors that regulate regional temperature and rainfall. We will review these briefly.

Latitude Ultimately, the Sun controls the Earth's climate. Sunlight warms the Earth and provides the energy to evaporate water and power winds and ocean currents. As we learned in Chapter 15, seasonal sunlight is dependent on latitude. The high latitudes receive less total sunlight than equatorial regions. High latitudes also experience great seasonal differences in sunlight, causing winter and summer seasons. Temperature and precipitation are not constant at a given latitude, however, because many other factors are important:

Wind Wind transports heat and moisture across the globe. We discussed the relationship between wind and weather in Chapter 16; wind and climate are discussed in Section 17.2.

Oceans Oceans affect climate in two ways. Currents transport heat (Section 17.3). However, even in the absence of currents, the ocean stores and releases heat differently than land. As a result, coastlines are generally warmer in winter and cooler in summer than continental interiors.

Oceans also affect precipitation over coastal regions. In regions where warm oceans lie adjacent to cooler land, humid maritime air blows inland, leading to abundant precipitation (Fig. 17–1A). However, not all coastal areas are humid. Large portions of the west coast of South America, from southern Ecuador through northern Chile, contain some of the driest deserts in the world (Fig. 17–1B). These deserts exist because maritime air chilled by a cold Pacific Ocean current warms as it passes over the hot coasts of Ecuador, Peru, and Chile.

Altitude Air is cooler at higher elevations, so mountainous regions and high plateaus are almost always colder than adjacent lowlands. Thus, glaciers exist on Mount Kenya in Africa and Mount Cayambe in South America even though both lie directly on the equator.

Mountains also alter the movement of air and generate local climate zones. In general, air rises on the windward side of a range, causing abundant precipitation. After it has passed over the mountains, it sinks and warms adiabatically, forming a rain-shadow desert on the downwind side of the mountains.

Albedo Temperature not only depends on the amount of sunlight received, but on the albedo, which is the amount reflected. If the albedo of a region is high, much of the solar energy is reflected out into space and the region will be cooler than a similar region with a lower albedo.

Figure 17–2 (A) The Gobi desert and (B) central Pennsylvania lie at the same latitude and the same distance from the ocean. The two regions are influenced by differences in prevailing winds and surrounding mountains. *(Martin Rogers/Tony Stone Images)* ● **Interactive Question:** Locate three other regions with the same latitude and distance from the oceans as the Gobi Desert and central Pennsylvania. Record their climates using Figure 17–8. Explain what factors regulate each climate. (The map of ocean currents in Figure 14–12 may be helpful in answering this question.)

A

B

Climate as a Complex System Many factors affect climate in interactive ways. Thus, each of the factors listed above affects the others. High altitudes are often glacier covered. Glaciers have high albedo. Thus altitude affects albedo. Recall from Chapter 12 that the Gobi Desert and Pittsburgh are geographically similar in latitude and distance from the ocean, but the Gobi is a desert and central Pennsylvania has a moist climate (Fig. 17–2). Prevailing winds, mountain ranges, ocean currents, and many other factors combine to control the climate in any region.

17.2 Global Winds and Climate

In 1735 a British meteorologist, George Hadley, reasoned that global winds are generated solely by the temperature difference between the equator and the poles. Hadley inferred that air is heated at the equator, rises, and then flows poleward at high elevations. At the poles, it cools, sinks, and flows back toward the equator along the Earth's surface (Fig. 17–3). Thus, Hadley reasoned, the atmosphere circulates in two giant convection cells, one in the Northern Hemisphere and one in the Southern. However, when meteorologists mapped atmospheric circulation patterns, they found that the jet stream migrates across the mid-latitudes like a writhing snake (Fig. 17–4). A Norwegian scientist, Carl Rossby, concluded that the Hadley model didn't explain the jet stream because it didn't account for the Earth's rotation and the consequent Coriolis effect.

In the 1950s, global wind systems were modeled experimentally by climatologists at the University of

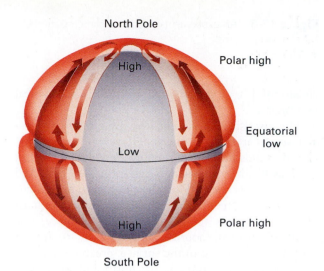

Figure 17–3 The Hadley model for global circulation patterns, proposed in 1735. It has been replaced by the three-cell model.

Chicago. They mounted a circular pan on a variable-speed turntable and placed a heating element around the rim and a cooling coil at the center (Fig. 17–5). The pan represented the Earth, with the rim analogous to the equator and the center analogous to the poles. They filled the pan with water and added dye to trace currents.

When the pan was stationary, water rose at the heated edge, traveled across the surface, sank at the cooled center, and returned to the edge along the bottom, thus forming a Hadley cell. When the pan rotated slowly, this current was deflected by the Coriolis effect, but the single-cell pattern was retained. However, when the scientists increased the rotational speed, the cell broke apart. Midway between the edge and the center—the area representing the Earth's middle latitudes—the current diverged into whirls and eddies that looked very much like the storms observed in the middle latitudes. Today, scientists repeat this experiment mathematically, on giant computers. They then compare the computer data with millions of direct measurements of natural wind patterns.

All climate models start with the observation that the Sun is the ultimate energy source for winds and evaporation. The Sun shines most directly at or near the equator and warms the air near the Earth's surface. The warm air gathers moisture from the equatorial oceans. The warm, moist, rising air forms a vast region of low pressure near the equator, with little horizontal air flow. As the rising air cools adiabatically, the water vapor condenses and falls as rain. Therefore, local squalls and thunderstorms are common, but steady winds are rare. This hot, still region was a serious barrier in the age of sailing ships. Mariners called the equatorial region the **doldrums,** and the old sailing literature is filled with stories recounting the despair and hardship of being becalmed on the vast, windless seas. On land, the frequent rains near the

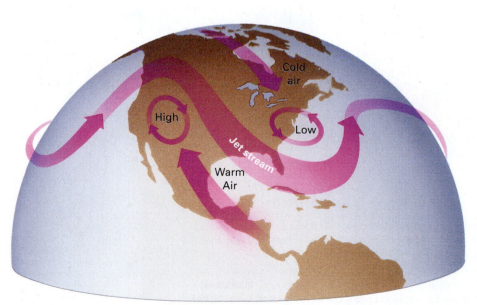

Figure 17–4 The Hadley model cannot explain upper-level winds that snake their way across the Northern Hemisphere.

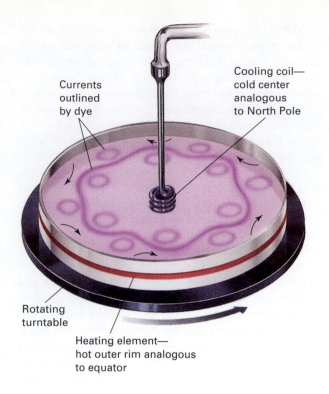

Figure 17–5 An experimental model to demonstrate the Earth's wind patterns.

Currents outlined by dye

Cooling coil— cold center analogous to North Pole

Rotating turntable

Heating element— hot outer rim analogous to equator

equatorial low-pressure zone nurture lush tropical rainforests.

The air rising at the equator splits to flow north and south at high altitudes. However, these high-altitude winds do not continue to flow due north and south as Hadley predicted, because they are deflected by the Coriolis effect. Thus their poleward movement is interrupted. In both the Northern and the Southern Hemispheres, this air veers until it flows due east at about 30° north and south latitudes (Fig. 17–6). It then cools enough to sink to the surface, creating subtropical high-pressure zones at 30° north and south latitudes. The sinking air warms adiabatically, absorbing water and forming clear blue skies. At the center of the high-pressure area the air moves vertically and not horizontally, and therefore few steady surface winds blow. This calm high-pressure belt circling the globe is called the **horse latitudes.** The region was named

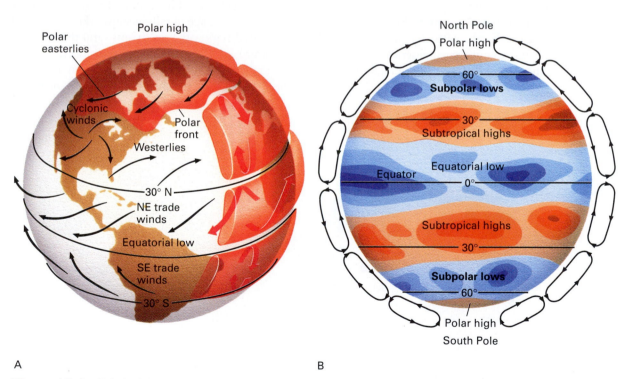

A   B

Figure 17–6 Global wind patterns predicted by the three-cell model. (A) Air rising at the equator moves poleward at high elevations, falls at about 30° north and south latitudes, and returns to the equator, forming trade winds. The orange arrows show both upper level and surface wind patterns. The black arrows show only surface winds. (B) High- and low-pressure belts with surface and upper level wind patterns shown on the edges of the sphere. • Interactive **Question: Predict how the three-cell model would change if the Earth rotated twice as fast as it does now.**

because sailing ships were becalmed and horses transported as cargo often died of thirst and hunger. The warm, dry, descending air in this high-pressure zone forms many of the world's great deserts, including the Sahara in north Africa, the Kalahari in south Africa, and the Australian interior desert.

Descending air at the horse latitudes splits and flows over the Earth's surface in two directions, toward the equator and toward the poles. The surface winds moving toward the equator are deflected by the Coriolis effect, so they blow from the northeast in the Northern Hemisphere and from the southeast in the Southern Hemisphere. Sailors depended on these reliable winds and called them the **trade winds.** The winds moving toward the poles are also deflected by the Coriolis effect forming the **prevailing westerlies.** They flow from the southwest in the Northern Hemisphere and from the northwest in the Southern Hemisphere (Fig. 17–6A).

The poles are cold year round. The cold polar air sinks, creating yet another band of high pressure. The sinking air flows over the surface toward lower latitudes. In the Northern Hemisphere these surface winds are deflected by the Coriolis effect to form the **polar easterlies.** The polar easterlies and prevailing westerlies converge at about 60° latitude. Warm air rises at the convergence, forming a low-pressure boundary zone called the **polar front.**

Recall that the Hadley model for global winds depicts one convection cell in each hemisphere. The more accurate modern model, called the **three-cell model,** depicts three convection cells in each hemisphere. The cells are bordered by alternating bands of high and low pressure (Fig. 17–6B). In the three-cell model, global winds are generated by heat-driven convection currents, and then their direction is altered by the Earth's rotation (Coriolis effect).

Figure 17–6 shows stationary planar boundaries between the cells. In reality these boundaries migrate north and south with the seasons. They are also distorted by surface topography and local air movement. For example, in the Northern Hemisphere, cyclones and anticyclones develop along the polar front as explained in Chapter 16. These storms bring alternating rain and sunshine, conditions that are favorable for agriculture. Thus, the great wheat belts of the United States, Canada, and Russia all lie between 30° and 60° north latitude.

Recall from Chapter 16 that a jet stream is a narrow band of fast-moving high-altitude air. Jet streams form at boundaries between the Earth's climate cells as high-altitude air is deflected by the Coriolis effect. The **subtropical jet stream** flows between the trade winds and the westerlies, and the **polar jet stream** forms along the polar front. When you watch a weather forecast on TV, the meteorologist may show the movement and direction of the polar jet stream as it snakes across North America. Storms commonly occur along this line because the jet stream marks the boundary between cold polar air and the warm, moist westerly flow that originates in the subtropics. The storms develop where the two contrasting air masses converge (Fig. 17–7).

17.3 Ocean Currents and Climate

Both London and central Newfoundland lie near the Atlantic Ocean at 51° north latitude, but Newfoundland has a polar climate whereas London is temperate. Although both regions receive the same amount of solar energy, London is warmed by the Gulf Stream

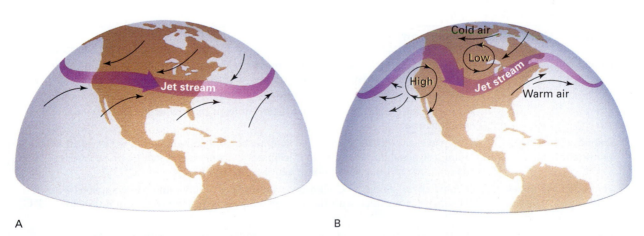

A   B

Figure 17–7 The polar front and the jet stream migrate with the seasons and with local conditions. Storms commonly occur along the jet stream.

and the North Atlantic Drift, related ocean currents that carry tropical water to northern Europe. On the other hand, the Labrador current flows from the North Pole southward to cool Newfoundland.

Winds and ocean currents transport heat, altering regional temperatures. In the absence of these heat transfers, the equatorial regions would be much warmer than they are today and the polar regions would be colder.

Winds and ocean currents also affect rainfall. Recall that a cold ocean current creates the Atacama Desert along the west coast of South America. In contrast, a warm Pacific current is partially responsible for the abundant rain that falls on the northwestern United States and western Canada.

Vertical Mixing of Ocean Water   Recall from Chapter 14 that as the Gulf Stream pushes into the North Atlantic, both the temperature and salinity change so that the water becomes dense enough to sink. Much of this cold, dense water flows at depth all the way to Antarctica. The Gulf Stream water flowing northward is, on average, 8°C warmer than the deep water flowing southward, and this circulation has a flow equal to 100 Amazon Rivers. Therefore, these currents transport immense amounts of heat.

Between 11,000 and 12,000 years ago, the Earth was immersed in a millennium-long cold snap called the Younger Dryas. Evidence from sea-floor sediment indicates that during this period, the North Atlantic deep current slowed down or stopped. When the deep current slowed, the circulation cell was interrupted and the Gulf Stream stopped, slowed, or veered southward. As a result, heat transport to the north Atlantic region slowed. North America and Europe became frigid. At the same time, global temperatures dropped. Thus, a change in ocean circulation caused a severe temperature change in one region. In turn, regional climate change altered winds or other marine currents to cause global change.

17.4   Climate Zones of the Earth

The Earth's major climate zones are classified primarily by temperature and precipitation. But an area with both wet and dry seasons has a different climate from one with moderate rainfall all year long, even though the two areas may have identical total annual precipitation. Therefore, climatic zones are also classified on the basis of seasonal variations in temperature and precipitation. The **Koeppen Climate Classification,** used by climatologists throughout the world, defines five principal groups.

Subclassifications of these five groups are shown in Table 17–1.

Although climate zones are defined by temperature and precipitation, you can often estimate climate types from a photograph of an area. Visual classification is possible because specific plant communities grow in specific climates. For instance, cactus grows in the desert, and trees grow where moisture is more abundant. A **biome** is a community of plants living in a large geographic area characterized by a particular climate. Following is a brief discussion of the major climate zones and their biomes (Fig. 17–8).

The climate in any location is summarized by a climograph that records annual and seasonal temperature and precipitation. Figure 17–9 is a model climograph for Nashville, Tennessee.

A.   Humid Tropical Climates

Tropical Climates with Abundant Rainfall   The large low-pressure zone near the equator causes abundant rainfall, often exceeding 400 centimeters per year,

(text continued on p. 408)

	Koeppen Climate Classification	
A	Humid tropical	Every month is warm with a mean temperature over 18°C (64°F). The temperature difference between day and night is greater than the difference between December and June averages. There is enough moisture to support abundant plant communities.
B	Dry	B climates have a chronic water deficiency; in most months evaporation exceeds precipitation.
C	Humid mid-latitudes with mild winters	C climates have distinct winter and summer seasons and enough moisture to support abundant plant communities. In C climates the winters are mild, with the average temperature in the coldest month above −3°C (27°F).
D	Humid mid-latitudes with severe winters	D climates are similar to C climates, with distinct summer and winter seasons. However, D climates are colder, with the average temperature in the coldest month below −3°C (27°F).
E	Polar	Winters are extremely cold and even the summers are cool, with the average temperature in the warmest month below 10°C (50°F).

TABLE 17–1

World Climates

1st	2nd	3rd	Description	Climate Type	Vegetation
A			**Humid tropical; no winter**		
	f		Wet year round	Af = tropical wet	Rainforest
	m		Wet with short dry season	Am = tropical monsoon	Rainforest
	w		Winter dry season	Aw = tropical wet and dry	Grassland savanna
B			**Dry; evaporation greater than precipitation**		
	S			Steppe	
		h	Semi-arid	BSh = tropical steppe	Steppe grassland
		k	Cool and dry continental interior or rain-shadow location	BSk = mid-latitude steppe	Steppe grassland
	W		Desert		
		h	Warm	BWh = tropical desert	Desert
		k	Cool and dry; continental interior or rain-shadow location	BWk = mid-latitude desert	Desert
C			**Humid mid-latitude; mild winter**		
	f	a	East coast: warm, wet summers with westerly winds and winter cyclonic storms	Cfa = humid subtropical	Forest
	f	b	Cool summer and mild winter; maritime, westerlies, and cyclonic storms	Cfb = marine west coast	Forest to temperate rainforest
	s	a	Dry summer and wet winter with cyclonic storms	Csa = dry summer, wet winter, subtropical	Mediterranean
D			**Humid mid-latitude; severe winter**		
	f	a	Mid-latitude continental warm summer, cold winter	Dfa = humid continental; warm summer, cold winter	Prairie and forest
	f	b	High mid-latitude continental	Dfb = humid continental; cool summer, cold winter	Prairie and forest
	f	c & d	High-latitude continental	Dfc/d = subarctic	Boreal forest taiga
E			**Polar with little warmth, even in summer**		

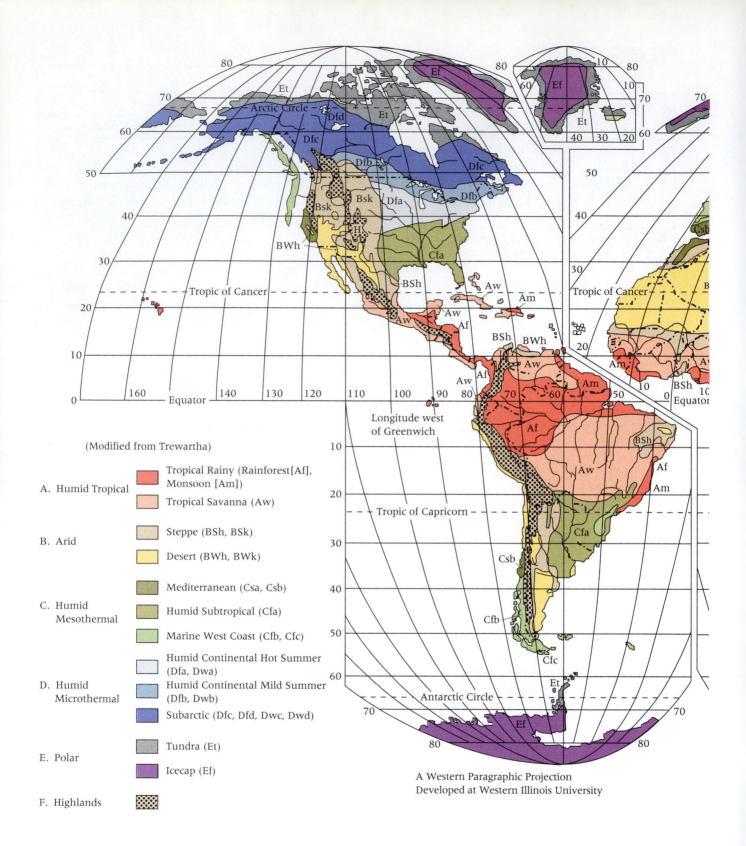

(Modified from Trewartha)

A. Humid Tropical
- ![red] Tropical Rainy (Rainforest[Af], Monsoon [Am])
- ![salmon] Tropical Savanna (Aw)

B. Arid
- ![tan] Steppe (BSh, BSk)
- ![yellow] Desert (BWh, BWk)

C. Humid Mesothermal
- ![olive] Mediterranean (Csa, Csb)
- ![green] Humid Subtropical (Cfa)
- ![light green] Marine West Coast (Cfb, Cfc)

D. Humid Microthermal
- ![light gray] Humid Continental Hot Summer (Dfa, Dwa)
- ![light blue] Humid Continental Mild Summer (Dfb, Dwb)
- ![blue] Subarctic (Dfc, Dfd, Dwc, Dwd)

E. Polar
- ![gray] Tundra (Et)
- ![purple] Icecap (Ef)

F. Highlands
- ![dotted]

Longitude west of Greenwich

A Western Paragraphic Projection
Developed at Western Illinois University

Figure 17–8 Global climate zones. Each climate zone supports a unique biome.

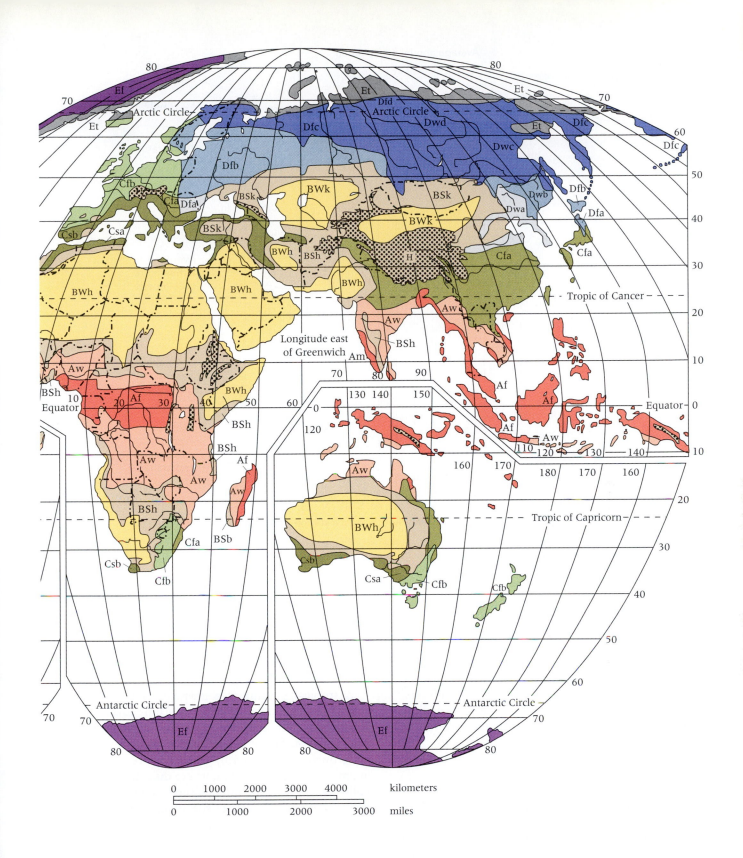

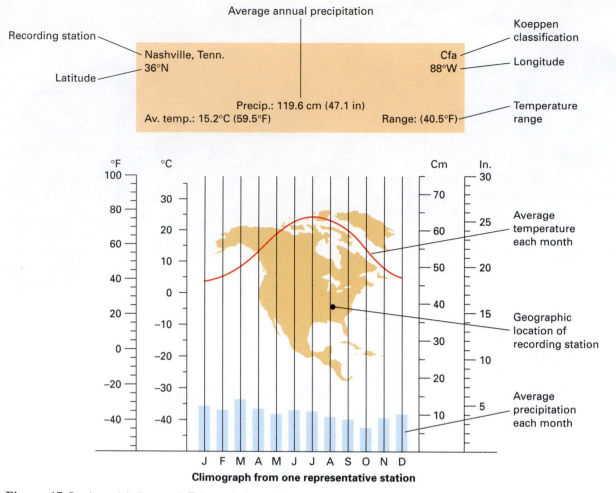

Recording station — Nashville, Tenn.
Latitude — 36°N
Average annual precipitation
Koeppen classification — Cfa
Longitude — 88°W

Precip.: 119.6 cm (47.1 in)
Av. temp.: 15.2°C (59.5°F)   Range: (40.5°F) — **Temperature range**

°F °C   Cm In.

Average temperature each month

Geographic location of recording station

Average precipitation each month

J F M A M J J A S O N D

Climograph from one representative station

Figure 17–9 A model climograph. This graph shows the average monthly temperature (curved line) and average monthly precipitation for Nashville, Tennessee. Other features of the climograph are highlighted by the labels.

that supports **tropical rainforests** (Fig. 17–10). The dominant plants in a tropical rainforest are tall trees with slender trunks. These trees branch only near the top, covering the forest with a dense canopy of leaves. The canopy blocks out most of the light, so as little as 0.1 percent of the sunlight reaches the forest floor, which consequently has relatively few plants. The ground in a tropical forest is soggy, the tree trunks are wet, and water drips everywhere.

Tropical Climates with Distinct Wet and Dry Seasons **Tropical monsoon** (Fig. 17–11) and **tropical savanna** (Fig. 17–12) both have large seasonal variations in rainfall, but a monsoon climate has greater total precipitation and greater monthly variation. Precipitation is great enough in tropical monsoon biomes to support rainforests. The seasonal

precipitation is also ideal for agriculture. Some of the great rice-growing regions in India and southeast Asia lie in tropical monsoon climates.

A tropical savanna is a grassland with scattered small trees and shrubs. Such grasslands extend over large areas, often in the interiors of continents, where rainfall is insufficient to support forests or where forest growth is prevented by recurrent fires. Savannas are most extensive in Africa, where they support a rich collection of grazing animals such as zebras, wildebeest, and gazelles. Grasses sprout and grow with the seasonal rains, and the great African herds migrate with the foliage.

B. Dry Climates

In dry zones where the annual precipitation varies from 25 to 35 centimeters per year, the climate is

Iquitos, Peru   Af
4°S   73°W
Precip.: 262 cm (103.1 in)
Av. temp.: 25°C (77°F)   Range: 2.2°C (4°F)

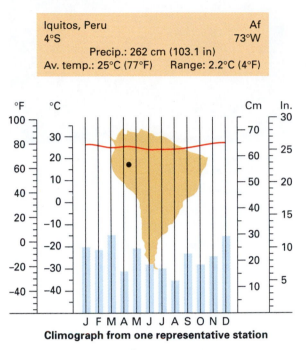

Climograph from one representative station

Figure 17–10 Tropical rainforest.

Tropical rainforest

Latitude: 0° to 15°

Av. temperature difference: 23°C to 28°C
(winter to summer)

Av. annual precip.: 200 cm/yr to 500 cm/yr

General statistics for climate type

Calicut, India   Am
11°N   76°E
Precip.: 301 cm (118.6 in)
Av. temp.: 26.4°C (79.5°F)   Range: 4°C (6.9°F)

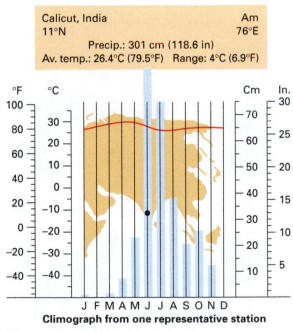

Climograph from one representative station

Figure 17–11 Tropical monsoon.

Tropical monsoon

Latitude: 5° to 30°

Av. temperature difference: 20°C to 30°C
(winter to summer)

Av. annual precip.: 150 cm/yr to 400 cm/yr

General statistics for climate type

Kano, Nigeria   Aw
12°N   8°E
Precip.: 86.5 cm (34 in)
Av. temp.: 26.7°C (80°F)   Range: 9.5°C (17°F)

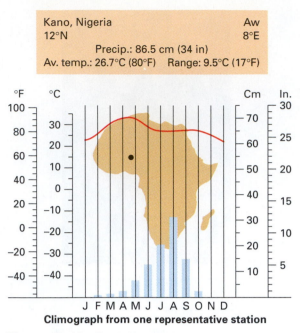

Climograph from one representative station

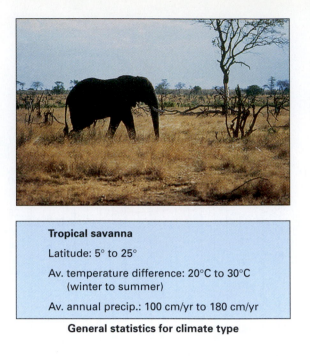

Tropical savanna

Latitude: 5° to 25°

Av. temperature difference: 20°C to 30°C
(winter to summer)

Av. annual precip.: 100 cm/yr to 180 cm/yr

General statistics for climate type

Figure 17–12   Tropical savanna.

semi-arid and grasslands predominate. If the rainfall is less than 25 centimeters per year, deserts form and support only sparse vegetation (Fig. 17–13). The world's largest deserts lie along the 30° latitude high-pressure zones, although rain-shadow and coastal deserts exist in other latitudes.

C. Humid Mid-Latitude Climates with Mild Winters

Humid Subtropics   The southeastern United States has a **humid subtropical climate** (Fig. 17–14). During the summer, conditions can be as hot and hu-

Lima, Peru   BWh
12°S   77°W
Precip.: 4 cm (1.6 in)
Av. temp.: 20°C (68°F)   Range: 9°C (15.5°F)

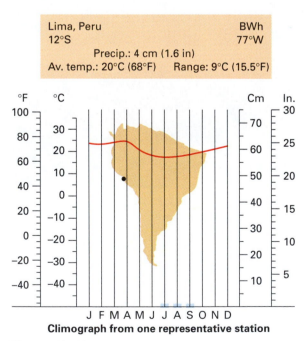

Climograph from one representative station

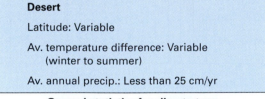

Desert

Latitude: Variable

Av. temperature difference: Variable
(winter to summer)

Av. annual precip.: Less than 25 cm/yr

General statistics for climate type

Figure 17–13   Desert.

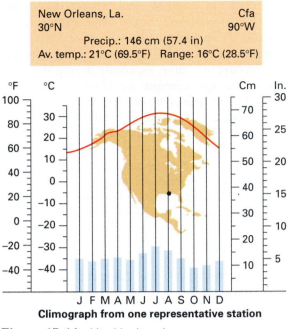

New Orleans, La.   Cfa
30°N   90°W
Precip.: 146 cm (57.4 in)
Av. temp.: 21°C (69.5°F)   Range: 16°C (28.5°F)

Climograph from one representative station

Figure 17–14   Humid subtropics.

Humid subtropics

Latitude: 15° to 40°

Av. temperature difference: 7°C to 32°C
(winter to summer)

Av. annual precip.: 60cm/yr to 250 cm/yr

General statistics for climate type

mid as in the tropics, and rain and thundershowers are common. However, during the winter, arctic air pushes southward, forming cyclonic storms. Although the average monthly temperature seldom falls below 7°C, cold fronts occasionally bring frost and snow. Precipitation is relatively constant year round due to convection-driven thunderstorms in summer and cyclonic storms in winter. This zone supports both trees with needles (conifers) and trees with broad leaves (deciduous) as well as valuable crops such as vegetables, cotton, tobacco, and citrus fruits.

Mediterranean Climate   The **Mediterranean climate** is characterized by dry summers, rainy winters, and moderate temperature (Fig. 17–15). These conditions occur on the west coasts of all continents

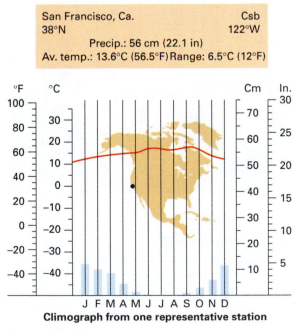

San Francisco, Ca.   Csb
38°N   122°W
Precip.: 56 cm (22.1 in)
Av. temp.: 13.6°C (56.5°F)   Range: 6.5°C (12°F)

Climograph from one representative station

Figure 17–15   Mediterranean climate.

Mediterranean climate

Latitude: 30° to 40°

Av. temperature difference: 5°C to 30°C
(winter to summer)

Av. annual precip.: 35 cm/yr to 75 cm/yr

General statistics for climate type

between latitudes 30° and 40°. In summer the subtropical high migrates to higher latitudes, producing near-desert conditions with clear skies as much as 90 percent of the time. In winter the prevailing westerlies bring warm, moist air from the ocean, leading to fog and rain. Thus, 75 percent or more of the annual rainfall occurs in winter. Although redwoods, the largest trees on Earth, grow in specific environments in central California, the summer heat and drought of Mediterranean climates generally retard the growth of large trees. Instead, shrubs and scattered trees dominate. Fires occur frequently during the dry summers and spread rapidly through the dense shrubbery. During the dry fall of 1991, a particularly devastating fire swept through the hills of Oakland and Berkeley, California, destroying 3354 homes worth $1.5 billion and killing 24 people. If vegetation is destroyed by fire, landslides often occur when the winter rains return. Torrential winter rains frequently bring extensive flooding and landslides to southern California.

Marine West Coast   **Marine west coast climate** zones border the Mediterranean zones and extend poleward to 65° (Fig. 17–16). They are severely influenced by ocean currents that moderate temperature and bring abundant precipitation. Thus, summers are cool and winters warm, and the temperature difference between the seasons is small. Thus, average monthly temperatures vary by only 15.5°C in Port-

land, Oregon. In contrast, Eau Claire, Wisconsin, which is a continental city at the same latitude, experiences a 31.5°C annual temperature range. Seattle and other northwestern coastal cities experience rain and drizzle for days at a time, especially during the winter, when the warm, moist maritime air from the Pacific flows first over cool currents close to shore and then over cool land surfaces. The total rainfall varies from moderate, 50 centimeters per year (20 in/yr), to wet, 250 centimeters per year (100 in/yr). The wettest climates occur where mountains interrupt the maritime air. **Temperate rainforests** grow where rainfall is greater that 100 centimeters per year and is constant throughout the year. Temperate rainforests are common along the northwest coast of North America, from Oregon to Alaska.

D.   Humid Mid-Latitude Climate with Severe Winters

Continental interiors in the mid-latitudes are characterized by hot summers and cold winters (Fig. 17–17). Thus, in the northern Great Plains the temperature can drop to −40°C in winter and soar to 38°C in summer. Even in a given season the temperature may vary greatly as the polar front moves northward or southward. Thus, in winter the northern continental United States may experience arctic cold one day and rain a few days later. If rainfall is sufficient, this climate supports abundant coniferous forests, whereas grasslands

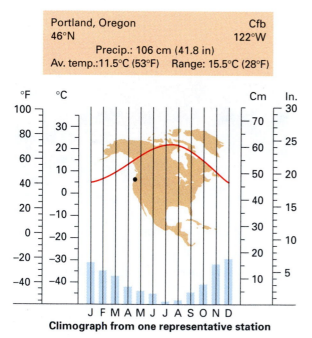

Portland, Oregon   Cfb
46°N   122°W
Precip.: 106 cm (41.8 in)
Av. temp.:11.5°C (53°F)   Range: 15.5°C (28°F)

Climograph from one representative station

Figure 17–16   Marine West Coast.

Marine West Coast

Latitude: 40° to 65°

Av. temperature difference: 0°C to 25°C
(winter to summer)

Av. annual precip.: 50 cm/yr to 250 cm/yr

General statistics for climate type

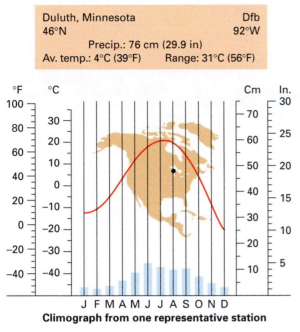

Duluth, Minnesota   Dfb
46°N   92°W
Precip.: 76 cm (29.9 in)
Av. temp.: 4°C (39°F)   Range: 31°C (56°F)

Climograph from one representative station

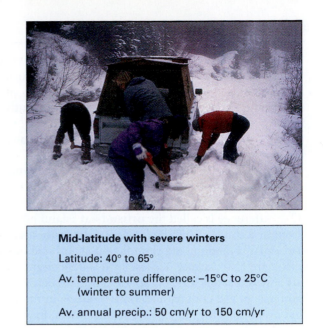

Mid-latitude with severe winters

Latitude: 40° to 65°

Av. temperature difference: −15°C to 25°C
(winter to summer)

Av. annual precip.: 50 cm/yr to 150 cm/yr

General statistics for climate type

Figure 17–17 Mid-latitude with severe winters.

dominate the drier regions. Millions of bison once roamed the continental grasslands of North America, and today wheat and other grains grow from horizon to horizon. The northernmost portion of the D climate zone is the subarctic, which supports the **taiga** biome. Taiga consists of conifers that survive extremely cold winters.

E. Polar Climate

In the Arctic, winters are harsh and long, and the temperature remains above freezing only during a short summer. Trees cannot survive, and low-lying plants such as mosses, grasses, flowers, and a few small bushes cover the land. This biome is called **tundra** (Fig. 17–18).

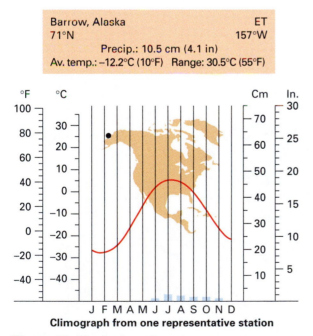

Barrow, Alaska   ET
71°N   157°W
Precip.: 10.5 cm (4.1 in)
Av. temp.: −12.2°C (10°F)   Range: 30.5°C (55°F)

Climograph from one representative station

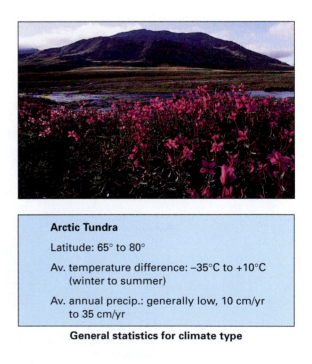

Arctic Tundra

Latitude: 65° to 80°

Av. temperature difference: −35°C to +10°C
(winter to summer)

Av. annual precip.: generally low, 10 cm/yr
to 35 cm/yr

General statistics for climate type

Figure 17–18 Arctic tundra. *(Randy Brandon/Alaska Stock Images)*

17.5 Urban Climates

If you ride a bicycle from the center of a city toward the countryside, you may notice that the air gradually feels cooler and more refreshing as you leave the city streets and enter the green fields or hills of the outlying area. This feeling is not entirely psychological; the climate of cities is measurably different from that of the surrounding rural regions (Table 17–2).

As shown in Figure 17–19, the average winter minimum temperature in the center of Washington, D.C., is more than 3°C warmer than that in outlying areas. This difference is called the **urban heat island effect.** Stone and concrete buildings and asphalt roadways absorb solar radiation, and re-radiate it as infrared heat. Cities are also warmer because little surface water exists and as a result little evaporative cooling occurs. In contrast, in the countryside, water collects in the soil and evaporates for days after a storm. Roots draw water from deeper in the soil, and this water evaporates from leaf surfaces. Urban environments are also warmed by the heat released when fuels are burned. In New York City in winter, the combined heat out-

TABLE 17–2

Average Changes in Climatic Elements Caused by Urbanization

Element	Comparison with Rural Environment
Cloudiness:	
Cover	5–10% more
Fog—winter	100% more
Fog—summer	30% more
Precipitation, total	5–10% more
Relative humidity:	
Winter	2% lower
Summer	8% lower
Radiation:	
Total	15–20% less
Direct sunshine	5–15% less
Temperature:	
Annual mean	0.5–1.0°C higher
Winter minimum (average)	1.0–3.0°C higher
Wind speed:	
Annual mean	20–30% lower
Extreme gusts	10–20% lower
Calms	5–20% higher

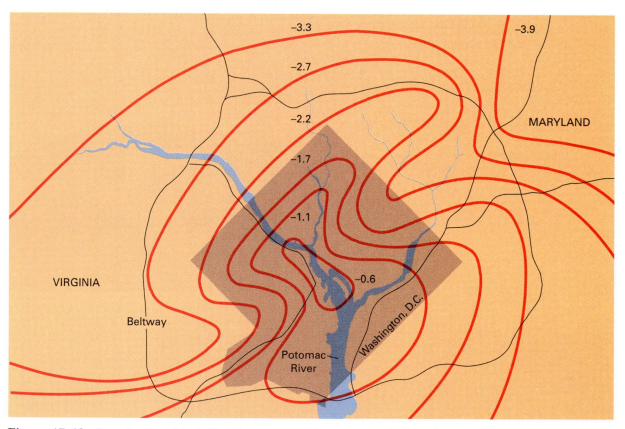

Figure 17–19 The urban heat island effect. The average minimum temperature in and around Washington, D.C., during the winter. *(From C.A. Woollum, Weatherwise 17 [1964]:264)*

put of all the vehicles, buildings, factories, and electrical generators is 2.5 times the solar energy reaching the ground. This heat contributes to the warming of the night air. In addition, the tall buildings block winds that might otherwise disperse the warm air. Finally, air pollutants absorb long-wave radiation (heat rays) emitted from the ground and produce a local greenhouse effect.

As warm air rises over a city, a local low-pressure zone develops, and rainfall is generally greater over the city than in the surrounding areas (Fig. 17–20). Water condenses on dust particles, which are abundant in polluted urban air. Weather systems collide with the city buildings and linger, much as they do on the windward side of mountains. Thus, a front that might pass quickly over rural farmland remains longer over a city and releases more precipitation.

In 1600, less than 1 percent of the global population lived in cities. By 1950, 30 percent of the world's population was urban, and by 1990 that ratio had grown to 50 percent. Therefore, although urban climate change may not affect global climate, it does affect the lives of the many people living in cities.

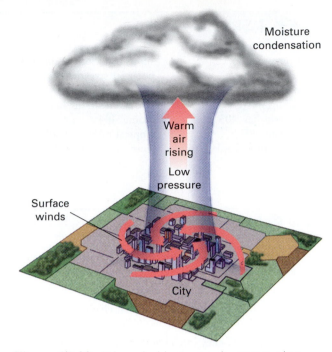

Figure 17–20 Warm air rising over a city creates a low-pressure zone. As a result, precipitation is greater in the city than over the surrounding countryside

SUMMARY

Ultimately, the Sun controls the Earth's climate; seasonal sunlight is dependent on latitude. Temperature and precipitation are not constant at a given latitude, however, because climate is also affected by wind, oceans, altitude, and albedo. Each of these factors interacts with the others.

The **three-cell model** of global wind circulation shows three cells of global air flow bordered by alternating bands of high and low pressure.

Warm air rises near the equator, forming a low-pressure region called the **doldrums.** This air flows north and south at high altitude until it cools and falls at 30° latitude forming a high-pressure region. The falling air splits. A portion flows back toward the equator along the surface, forming the **trade winds** and completing the tropical cell.

The remaining air that falls at 30° north and south latitudes flows poleward along the surface. In the Northern Hemisphere, this air is deflected to form the **prevailing westerlies.** Air rises at a low-pressure region near 60° latitude and returns at high altitude to complete the cell.

A second region of high pressure rises over the poles, where the descending air spreads outward to form the **polar easterlies.** The **polar front** forms where the polar easterlies and the prevailing westerlies converge. The jet stream blows at high altitude along the polar front and along the boundaries between cells at the **horse latitudes**.

Both surface and deep ocean currents affect climate. Vertical mixing can also affect climate, sometimes dramatically.

Climate zones are classified according to annual temperature, annual precipitation, and variability in either of these factors from month to month or season to season. Urban areas are generally warmer and wetter than surrounding countryside.

FOR REVIEW

1. Discuss the factors that control the average temperature in any region.

2. Discuss how oceans affect coastal climate.

3. Explain how changes in albedo can change regional or mean global temperature.

4. Explain the Hadley model and the three-cell model for global circulation. Discuss limitations of each.

5. Describe the trade winds. Why are they so predictable?

6. Why is the doldrums region relatively calm and rainy? Why are the horse latitudes calm and dry?

7. Describe the polar front and the jet stream. How do they affect weather?

8. Explain how horizontal and vertical ocean currents affect temperature.

9. Discuss and describe the five major climate zones and the plant communities associated with each.

10. Explain how urban climate differs from that of the surrounding countryside.

FOR DISCUSSION

1. Give an example of how winds and precipitation are affected by (a) altitude, and (b) ocean currents.

2. Would the exact location of the doldrums low-pressure area be likely to change from month to month? From year to year? Explain.

3. Sailors traveling in the Northern Hemisphere expect to incur predictable winds from the northeast between about 5°N and 30°N latitudes. Should airplane pilots expect northeast trade winds while flying at high altitudes in the same region? Explain.

4. Why doesn't the air that is heated at the equator continue to rise indefinitely?

5. Explain how changes in the temperature in the surface water can lead to rapid vertical mixing in the ocean. Explain why this mechanism can operate as a threshold effect.

6. Predict the climate at the following locations: (a) at 45°N latitude, 200 kilometers from the ocean, on the leeward side of a mountain range; (b) at 45°N latitude, on a coastline influenced by warm currents; (c) on the equator in a continental interior; (d) at 30°N latitude in a continental interior; (e) at 30°N latitude, influenced by warm currents.

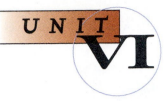

Human Interactions
with Earth Systems

Today, the human population is so great and our technology is so advanced that people dominate large portions of the planet and no ecosystem is immune from human influence. Many cities, such as this one in southeast Asia, are crowded and polluted. *(Paul Chesley/Tony Stone Images)*

417

Human Population and Alteration of Earth Systems

Take a mental journey a few thousand years into the past to a small island in southeast Asia. A young couple, working together, burn a small patch of jungle. They clear the brush, till the soil with sharp sticks, and plant vegetables. On their way home, they wade into the surf and spear a fish. Approaching their grass hut, they meet their children who are hauling water in gourds from a nearby creek. Settling themselves at the hearth, the family wraps their fish in green leaves and cooks it over glowing coals.

During the day's labors, these people have impacted their environment in many ways. They destroyed a portion of the tropical rainforest. Their brush fire released carbon dioxide into the air. Farming altered the soil nitrogen balance. Later, they exploited the Earth's fisheries, while their children were diverting fresh water. The evening fire released more carbon dioxide into the atmosphere.

Before the industrial revolution, humans made only a small impact on Earth because human population was low and technology was primitive (Fig. 1A). If we return to southeast Asia today we see huge cities, such as Singapore with 3 million people and Jakarta with 8 million. Coastlines are paved in concrete, automobile exhaust hangs over the cities, industrial activity releases huge quantities of carbon dioxide into the atmosphere, loggers use chainsaws to cut nearby forests. In the autumn of 1997, as this text was being written, massive fires, ignited during slash and burn agriculture, raged uncontrolled across the region, darkening the sky, destroying forest and fields, and releasing even more carbon dioxide (Fig. 1B).

Today, the human population is so great and our technology is so advanced, that people dominate large portions of the planet and no ecosystem is immune from human influence. Figure 2 describes how humans alter Earth systems. On the top of the chart is the human population itself—5.8 billion in mid-year 1997. Although the increase in the growth rate has slowed recently, population is growing at a

A

B

Figure 1 (A) Primitive agriculture in the western South Pacific. (B) In 1997, massive fires swept across Malaysia, darkening the sky, destroying land, and releasing carbon dioxide into the air. *(Thierry Falise/Gamma Liaison)*

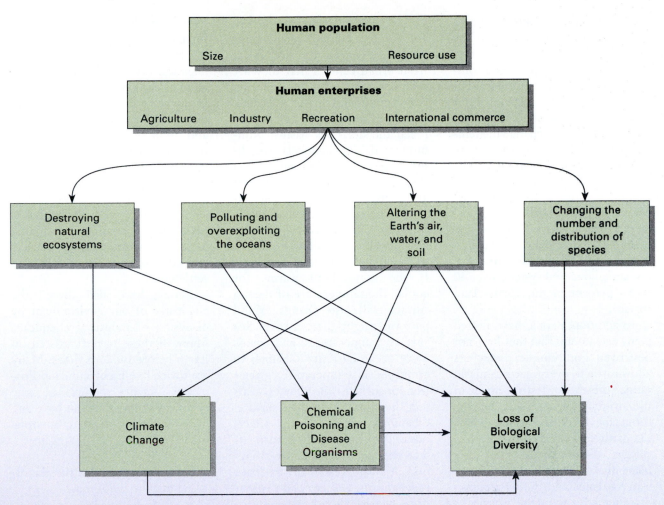

Figure 2 A flow chart showing how human populations impact Earth systems.

rate of 1.4 percent or 80 million people every year. But the number of people on the planet tells only part of the story; we must also consider the resources used by each person. Between 1950 and 1995, the human population increased by a factor of 2.3. During the same time interval, fossil fuel consumption increased by a factor of 4.7, paper consumption increased by a factor of 5, and the number of new cars jumped by a factor of 4.5.[1] If we compare modern consumption with that of preindustrial societies, the difference is even more striking. Even a few hundred years ago, people used horses or oxen to pull plows and cooked their food over small fires. With these technologies, farmers used approximately 12,000 Calories per Person per Day. In 1996, the average person on Earth used 60,000 calories of energy, or five times that of our ancestors. This energy consumption resulted from a combination of fossil fuels, nuclear, hydroelectric, wood, and animal power. (People in wealthy countries used much more.)

As shown in Figure 2, human activities impact Earth systems by four different mechanisms: destroying natural ecosystems; polluting and overexploiting the oceans; altering the Earth's air, water, and soil; and changing the number and distribution of species. Let us consider each in turn.[2]

Destroying Natural Ecosystems

Imagine a stately redwood forest on the northwest coast of California. The giant trees absorb carbon dioxide from the air and convert it to wood and needles. The rich, sponge-like forest soil retains water. The shade from the trees reduces evaporation from the forest floor and preserves soil moisture.

---

[1] Lester Brown et al., *Vital Signs 1997*. WorldWatch Institute (New York: W. W. Norton, 1997).

[2] This figure, the conceptual framework of this essay, and much of the data presented below were adapted from Peter M. Vitousek, Harold A. Mooney, Jane Lubchenco, and Jerry M. Melillo, "Human Domination of Earth Systems." *Science*, Vol. 277, July 25, 1997, p. 494ff.

Roots and trunks transport moisture from the forest floor to the canopy, where water escapes into the air. All of these processes combine to regulate both the local climate and the movement of water through the ecosystem. If people cut the forest, they change the temperature and rainfall in the region, the chemical composition of the air, soil, and water, and the balance of species. If one small patch of forest is cut, global systems aren't affected. But scientists estimate that people have destroyed natural ecosystems and replaced them with human engineered systems over 40 to 50 percent of the Earth's land surface.

In addition, people have altered many ecosystems that they have not destroyed. For example, if ranchers allow cattle to overgraze a semiarid range, the cattle destroy much of the original plant cover, making room for inedible bushes. Thus, the cattle have altered the original species distribution. When the plant distribution changes, the soil water balance and the distribution of other species in the prairies is also affected.

Polluting and Overexploiting the Oceans

About 60 percent of the world's people live within 100 kilometers of the coasts. A large city such as New York, Tokyo, or Calcutta crowds ten million or more people into a single bay or estuary. Thus, in many regions, the water's edge is lined with cities, harbors, factories, and roadways. Pollutants escape—or are dumped—into coastal waters. People have dredged and bulldozed coastal wetlands. Fishing boats ply the productive, protected waters near home. Pollution, habitat destruction, and over-exploitation have decimated coastal fish populations, and possibly killed reefs and fed deadly algal blooms.

The central oceans, in contrast, are so deep, contain such a huge volume of water, and are so far from urban centers that they are relatively unaffected by pollution and habitat destruction. Organic pollutants decompose before they reach the deep sea basins. Non-biodegradable chemicals are diluted by the central oceans so that they do little harm. Yet, scientists are asking whether people are altering the oceans in more subtle ways. In Chapter 18 we will discuss possible links between industrial activity and climate change. If the atmosphere warms (or cools), the temperature of the sea surface will also change. Wind patterns might then change. Sea surface temperatures and winds drive ocean currents. Changes in atmospheric temperature, glaciation, or surface runoff may also affect the salinity of the sea surface. Changing salinity also affects ocean currents, which, in turn, affect climate. Scientists are studying atmosphere–ocean feedback mechanisms, and are asking questions such as: Could a small initial change in atmospheric temperature alter ocean currents? Could these changes trigger global climate change? At present, we don't know the answer to these questions. But evidence for rapid climate change in the past warns us that similar events can occur in the future.

Altering the Earth's Air, Water, and Soil

In a natural ecosystem, plants and animals exchange water, carbon, nitrogen, and other compounds with each other and with the nonliving environment. Animals ingest oxygen and plant sugars and release carbon dioxide and water. Plants use carbon dioxide and water to synthesize sugars. Nutrient cycles are called biogeochemical cycles because they are regulated by bio-logical, geological, and chemical interactions. In modern times, people have impacted global biogeochemical cycles. We have released enough carbon into the air to increase its atmospheric concentration by 14 percent between 1958 and 1996. Farmers add as much fixed nitrogen in the form of fertilizer to terrestrial ecosystems as all natural sources combined. Globally, people now use about half of all the surface fresh water that is reasonably accessible (excluding meltwater from remote glaciers and other sources far from human settlement).

People have also altered the chemistry of our environment by disposing of industrial chemicals. Many of these haven't existed on Earth before modern times. Many are toxic. The Environmental Protection Agency defines a toxic chemical as one that can have any of the following health or environmental effects:

A single exposure of the chemical may be poisonous.

Prolonged exposure may cause cancer or other long-term health effects.

The compound may cause mutations in unborn children.

It may affect adult reproductive systems.

It may affect the nervous system leading to loss of memory or other deleterious effects.

The substance may have adverse effects on aquatic or terrestrial organisms other than humans.

An otherwise benign compound may decompose in the environment to produce a toxic one.

The compound may alter atmospheric chemistry through mechanisms such as ozone depletion.[3]

---

[3] U.S. Environmental Protection Agency, *1994 Toxics Release Inventory*, pp. 43, 44.

In 1994, U.S. manufacturing facilities released about 1 billion kilograms of toxic substances into the air, water, and ground.[4] This figure doesn't account for non-manufacturing releases such as automobile exhaust, pesticide dispersal on lawns and farms, or other pollutants that escape from homes and farms.

No one is certain how alteration of biogeochemical cycles or release of toxic chemicals is affecting human health, the health of other species, or the stability of global ecosystems. We will look at these issues in following chapters.

Changing the Number and Distribution of Species

Over the long history of life, new species have emerged faster than existing species have become extinct. Therefore the number of species is greater now than in the past. However, evolution and extinction do not proceed at linear rates. Instead, periodic mass extinctions have destroyed existing ecosystems. After ecosystems were devastated, new species rapidly evolved into unfilled niches.

Scientists estimate that during the Cenozoic era (from 65 million years ago to the present), an average of ten species in a million became extinct every year. This extinction rate was offset by a more rapid rate of emergence of new species, so the total diversity on the planet has increased. Within the past hundred years, however, species extinction has increased by 50 to 100 times this average rate. In 1996, the World Conservation Union concluded that 34 percent of the world's fish, 25 percent of amphibians, 25 percent of mammals, 20 percent of reptiles, and 11 percent of birds are endangered.

In addition to extinction, people have rearranged the biosphere by transporting plants and animals from one place to another. On many islands, half of the species are invaders from somewhere else, and on continents, people have introduced as much as 20 percent of existing species. In some cases humans import species either intentionally or unintentionally as hitchhikers on international cargo. In other cases, human alteration of ecosystems encourages invasion. A classic example is the digging of a canal that allows marine organisms to swim across a previously impenetrable land barrier.

Biological communities are complex interacting systems that maintain their own habitats. When species are removed, conditions may be less favorable for the remaining species. If the animals in a forest die off, the trees are affected. If the trees are killed, many of the animals die. The biosphere has survived previous catastrophes and will undoubtedly survive current disruptions. But what kind of Earth will be left behind? How will we adapt to changes that we are initiating?

---

[4] Ibid., p. 14.

Climate Change

The average temperature of the Earth's atmosphere has risen by about 0.5°C during the past 100 years. At the same time, mid-latitude continents have grown wetter while the tropics have generally become drier. Continuing climate change could alter ecosystems; drive species into extinction; and impact human settlement, food production, and distribution of disease.

In December of 1997, representatives from 160 nations met in Kyoto, Japan, to discuss global climate change. The major issues were: How seriously are humans altering climate? How will climate change affect

The Earth's climate has changed repeatedly and often dramatically throughout geologic time. Eighteen thousand years ago, continental glaciers stretched across high-latitude portions of the continents and alpine glaciers descended to low elevations even at the equator. Mt. Robson, British Columbia.

humans and global ecosystems? Can the nations of the world cooperate to reduce carbon emissions and defuse the problem?

Climate change is not unique in the twentieth century. The geological record in almost every locality provides evidence that past regional climates were different from those of today. Geologists have discovered sand dunes beneath prairie grasslands near Denver, Colorado, indicating that this semi-arid region was recently desert. Moraines on Long Island, New York, tell us that this temperate region was once glaciated. Fossil ferns in nearby Connecticut indicate that, before the glaciers, the northeastern United States was warm and wet.

Many of these regional climatic fluctuations resulted from global climate change. Thus, 18,000 years ago, the Earth was cooler than it is today and glaciers descended to lower latitudes and altitudes. During Mississippian time, from 360 to 320 million years ago, the Earth was warmer than it is today. Vegetation grew abundantly and some of it collected in huge swamps to form coal. •

18.1 Climate Change in Earth History

Recall from the opening essay to Unit III that the Earth's primordial atmosphere contained high concentrations of carbon dioxide (CO_2) and water vapor (H_2O). Both of these greenhouse gases absorb infrared radiation in the atmosphere. Astronomers have calculated that the Sun was 20 to 30 percent fainter early in Earth history than it is today. Yet Earth oceans did not freeze. Climate models show that the high concentrations of atmospheric carbon dioxide and water vapor retained enough of the Sun's feeble radiation to warm Earth's atmosphere and surface to temperatures that kept the oceans liquid. Luckily for us, the concentration of carbon dioxide and water in the atmosphere declined gradually as the Sun warmed. As a result, the Earth's climate never became so hot that water boiled or so cold that the oceans froze, maintaining a favorable environment for life.

Figure 18–1 is a graph of mean global temperature and precipitation throughout Earth history. Many additional changes occurred but they were of too short a duration to be apparent on this graph. For example, ice core records from Greenland, Antarctica, and alpine glaciers throughout the world show that between 110,000 and 10,000 years ago the mean annual global temperature changed frequently and dramatically. Figure 18–2 gives a close look at a portion of this temperature data. The graph shows that the atmospheric temperature in Greenland changed several times by 5° to 10°C within five to ten years. Many of the cold intervals persisted for 1000 years or more. Additional evidence from sedimentary rocks indicates that rapid climate changes also occurred before the glaciers existed. In a recent article in *Scientific American*, Wallace Broecker writes, "The past 10,000 years are anomalous in the history of our planet. This period, during which civilization developed, was marked by weather more consistent and equable than any similar time span of the past 100 millennia."[1]

The change in global temperature from 1880 to 1997 is shown in Figure 18–3. In this small interval of time, the temperature has risen by 0.5°C ± 0.2°C.

Imagine that the mean annual temperature were 10°C warmer or colder than it is today. If you lived in Florida, the heat might be unbearable or the orange crops might freeze. If you lived in Seattle, winter might be almost as balmy as summer is today or conversely, tremendous piles of snow might accumulate in winter.

Scientists suspect that threshold and feedback mechanisms might be responsible for some of the rapid climate changes in the Earth's past. In Chapter 1 we mentioned that the melting of ice is a threshold phenomenon. If the air above a glacier warms from −1.5°C to −0.5°C, the ice doesn't melt. However, if the air warms another degree to +0.5°C, the ice starts to melt.

Recall that a feedback mechanism occurs when a small initial perturbation affects another component of the Earth's systems, which amplifies the original effect, which perturbs the system even more, which leads to an even greater effect, and so on. For example, recall from Chapter 11 that, according to one hypothesis, changes in the Earth's orbit alone were not sufficient to cause the advances and retreats of the Pleistocene glaciers. However, a slight cooling initiated by changes in the Earth's orbit caused high-latitude forests to die and be replaced by tundra. Tundra has a higher albedo (the capacity to reflect solar radiation) than forest, so it reflects more sunlight and causes additional cooling. Glaciers start to grow. But the albedo of ice is even greater than that of

(text continued on page 426)

---

[1] Wallace Broecker, "Chaotic Climate," *Scientific American*, Nov. 1995, p. 62.

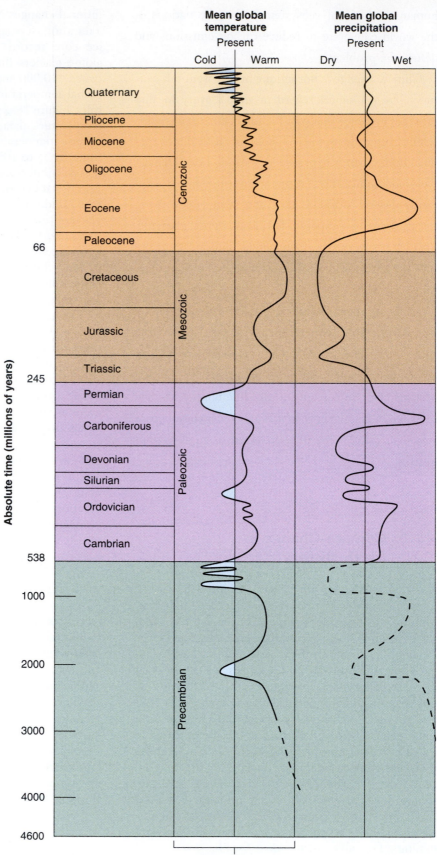

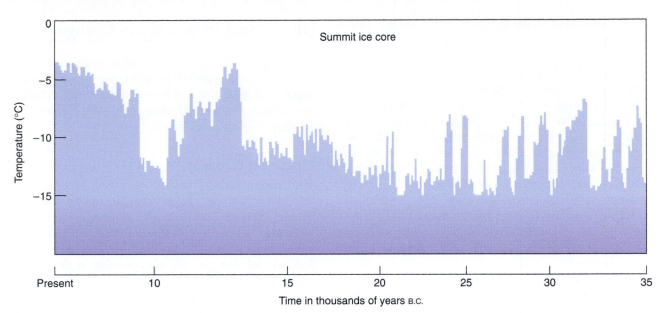

Figure 18–2 Atmospheric temperature fluctuations during the past 35,000 years in Greenland. The data were collected by the Greenland Ice Core Project.

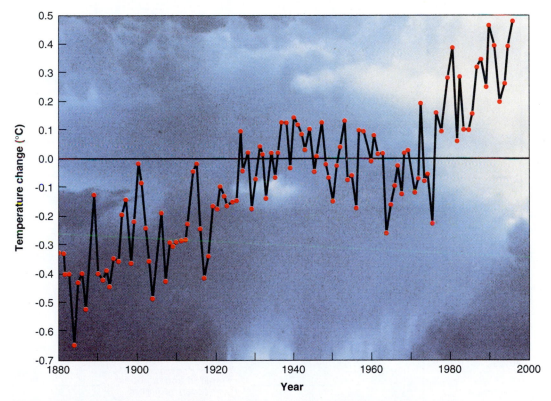

Figure 18–3 Mean global temperature changes from 1880 to 1995. The zero line represents the average from 1951 to 1980, and plus or minus values represent deviations from the average. *(NASA/Goddard Space Center)* ● **Interactive Question: Compare the magnitudes of some of the large temperature changes in Figure 18–2 with the temperature change shown in Figure 18–3. Discuss the comparative magnitudes of human-induced climate change with past examples of natural climate change.**

tundra, so even more sunlight is reflected and the cooling accelerates. Soon the entire planet cools and the glaciers advance.

Another albedo-driven feedback mechanism occurs with sea-level change. Oceans have a lower albedo than land. Imagine that some unrelated mechanism caused global warming. Sea level would then rise as a result of melting glaciers and decreasing seawater density. As sea level rose and flooded low-lying land, the oceans would increase in surface area. Average global albedo would decrease and both the oceans and the atmosphere would become warmer. This would lead to further sea-level rise as more ice melted and warming seawater expanded further. The Earth's albedo would continue to decrease and the Earth would grow even warmer.

Threshold and feedback mechanisms provide a sober warning. There is no guarantee of a linear relationship between cause and effect. In fact, both the geologic record and modern computer models provide abundant evidence that small changes in one component of the Earth's systems may lead to dramatic and rapid global climate changes.

18.2 Measuring Climate Change

For the past 100 years, meteorologists have used instruments to measure temperature, precipitation, wind speed, and humidity. But how do we interpret prehistoric climates? Figure 18–4 reviews several techniques for determining past climate.

Historical Records

Historians search for written records or archeological data that chronicle climate change. In 985 C.E., the Viking explorer Eric the Red sailed to southwest Greenland with a few hundred immigrants. They established two colonies, exported butter and cheese to Iceland and Europe, and the population flourished. Some Vikings sailed farther west, colonized North America, and visited Ellesmere Island, near 80° north latitude. Then within three to four hundred years, the colonies vanished. Sagas tell of heavy sea ice in summer, crop failures, starvation, and death.

During the same time, European glaciers descended into lowland valleys. This period of global

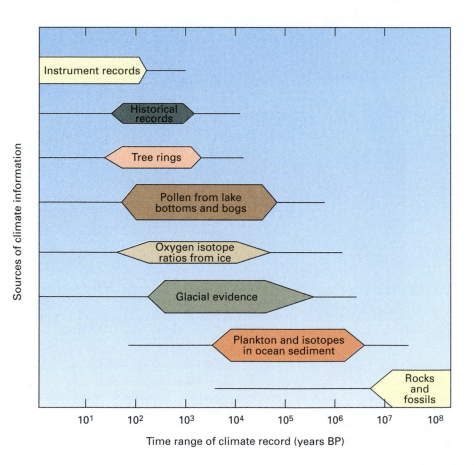

Figure 18–4 Several methods, each with its own useful time range, allow scientists to determine historical and ancient climates. *(Adapted with permission from T. Webb III, J. Kutzbach, and F.A. Street-Perrott in Global Change, T.F. Malone and J.D. Roederer, eds., pp. 182–218. Copyright 1985 by Cambridge University Press, U.K.)*

cooling, called the *Little Ice Age*, is documented by old landscape paintings and writings that depict a glacial advance between the thirteenth and nineteenth centuries. Other historical and archaeological evidence chronicles climate changes at different times and places.

Tree Rings

Growth rings in trees also record climatic variations. Each year, a tree's growth is recorded as a new layer of wood called a tree ring. Trees grow slowly during a cool, dry year and more quickly in a warm, wet year; therefore, tree rings grow wider during favorable years than during unfavorable ones. Paleoclimatologists date ancient logs preserved in ice, permafrost, or glacial till by carbon-14 techniques to determine when the tree died. They then count and measure the rings to reconstruct the history of past climate recorded in the wood. Interpretations of climate change from tree ring data coincide well with historical data. For example, growth rings are narrow in trees that lived during the Little Ice Age.

Plant Pollen

Plant pollen is widely distributed by wind and is coated with a hard waxy cover that resists decomposition. As a result, pollen grains are abundant and well preserved in sediment in lake bottoms and bogs. Figure 18–5 shows an 11,000-year record of tree pollen abundance in a Minnesota bog. Note that 11,000 years ago, spruce was the most abundant tree species. In modern forests, spruce dominates in colder Canadian climates but is less abundant in Minnesota. Therefore, scientists deduce that the climate in Minnesota was colder 11,000 years ago than it is at present. Pollen in younger layers of sediment shows that about 10,500 years ago, pines displaced the spruce, indicating that the temperature became warmer.

Oxygen Isotope Ratios in Glacial Ice

Atmospheric oxygen consists mainly of two isotopes, abundant ^{16}O and rare ^{18}O. Both isotopes are incorporated into water, H_2O. Water molecules containing ^{16}O are lighter and evaporate more easily than those

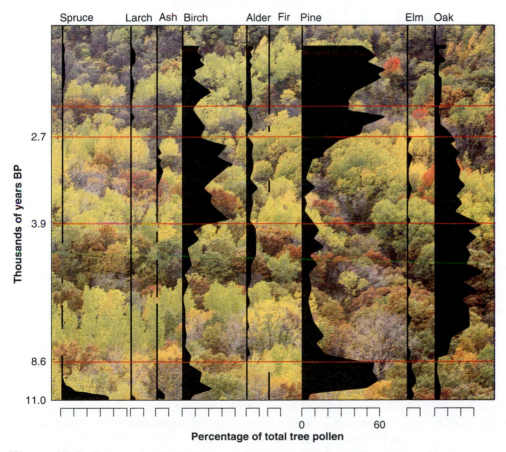

Figure 18–5 Tree species distributions over time interpreted from pollen preserved in a Minnesota bog. *(Adapted with permission from J.H. McAndrews, in* Quaternary Paleontology, *E.J. Cushing and H.E. Wright, Jr., eds., pp. 218–236. Copyright 1967 by University of Minnesota Press.)*

containing ^{18}O. At high temperature, however, evaporating water vapor contains a higher proportion of ^{18}O than it does at lower temperatures. Therefore, the ratio of $^{18}O/^{16}O$ in vapor from warm water is higher than that from cool water. Some of the water vapor condenses as snow, which accumulates in glaciers. Thus the $^{18}O/^{16}O$ ratios in glacial ice reflect water temperature at the time the water evaporated. Because most of the atmospheric water vapor that falls as snow originated from evaporation of ocean water, scientists then use the $^{18}O/^{16}O$ data from glacial ice to estimate mean ocean surface temperatures. In turn, mean ocean surface temperature reflects mean global atmospheric temperature.

Geologists have drilled deep into Greenland and Antarctic glaciers, where the ice is up to 110,000 years old, and have carefully removed ice cores (Fig. 18–6). The age of the ice at any depth is determined by counting annual ice deposition layers or by carbon-14 dating of windblown pollen within the glacier. The oxygen isotope ratios in each layer reflect the air temperature at the time the water evaporated and the snow fell.

Glacial Evidence

Erosional and depositional features created by Pleistocene glaciers are evidence of the growth and retreat of alpine glaciers and ice sheets, which in turn reflect climate. The timing of recent glacial advances can be determined by several methods. One effective technique is carbon-14 dating of logs preserved in glacial till.

Plankton and Isotopes in Ocean Sediment

In a technique that parallels pollen studies, scientists estimate climate by studying fossil abundances in deep sea sediment. The dominant life forms in the ocean are microscopic plants and animals. Many, called plankton, float near the sea surface. Just as pollen ratios change with air temperature, plankton species ratios change with sea surface temperature. Thus, fossil plankton assemblages found in sediment cores reflect sea surface temperature.

Most hard tissues formed by animals and plants, such as shells, exoskeletons, teeth, and bone contain oxygen. Many organisms absorb a high ratio of $^{18}O/^{16}O$ at low temperatures, but the ratio decreases with increasing temperature. For example, foraminifera are tiny marine organisms. During an ice age, their shells contain an average of 2 percent more ^{18}O than similar shells formed during a warm interglacial interval. Thus, just as scientists estimate paleoclimate by measuring oxygen isotope ratios in glacial ice, they

Figure 18–6 Scientists remove an ice core from a glacier in Greenland. Studies of ancient ice provide information about past climate. *(R. Gaillarde/Gamma-Liaison)*

can estimate ancient climate by measuring oxygen isotope ratios in fossil corals, plankton, teeth, and the remains of other organisms. Oxygen is also incorporated into soil minerals so isotope ratios in soil and sea-floor sediment reflect paleoclimate.

The Rock and Fossil Record

Fossils are abundant in many sedimentary rocks of Cambrian age and younger. While fossils do not give accurate temperature and rainfall data, geologists can approximate climate in ancient ecosystems by comparing fossils with modern relatives of the ancient organisms (Fig. 18–7). Fossil reefs were formed by tropical organisms such as corals. Coal deposits and ferns formed in moist tropical environments; cactus indicate that the region was once desert.

Looking backward even farther—into the Proterozoic Era, before life became abundant—it is difficult to measure climate with fossils. Thus, geologists search for clues in rocks. Tillite is a sedimentary rock formed from glacial debris, and thus indicates a cold climate. Lithified dunes were formed in deserts or along coasts.

Sedimentary rocks form in water, so their existence tells us that the temperature was above freezing and below boiling. Carbonate rocks precipitate from carbon dioxide dissolved in sea water. Geochemists know the chemical conditions under which carbon dioxide dissolves and precipitates, so they can calculate a range of atmospheric compositions and temperatures that would have produced carbonate rocks. Some ancient mineral deposits such as banded iron deposits (Unit Opener V) also reflect the chemistry of the ancient atmosphere.

After scientists learned that climates have changed, they began to search further to understand *how* cli-

A

B

Figure 18–7 (A) A fossil fern indicated that a region was wet and warm at the time the fern grew. (B) These fossil sand dune cross beds indicate that this region was dry at the time the dunes formed.

mates change. The remainder of this chapter is devoted to understanding mechanisms that cause climate change.

<table>
<tr><td style="background:#f0a500;padding:4px">18.3</td><td>The Natural Carbon Cycle and Climate</td></tr>
</table>

Carbon circulates among the atmosphere, the hydrosphere, the biosphere, and the solid Earth and is stored in each of these reservoirs (Fig. 18–8). This cycle is especially relevant to climate studies because carbon

dioxide and methane gas absorb infrared radiation and warm the atmosphere.

Carbon in the Atmosphere

As explained previously, oxygen and nitrogen, the most abundant gases in the atmosphere, are transparent to infrared radiation. Carbon exists in the atmosphere mostly as carbon dioxide (CO_2), and in smaller amounts as methane (CH_4). Although only 0.1 percent of the total carbon near the Earth's surface is in

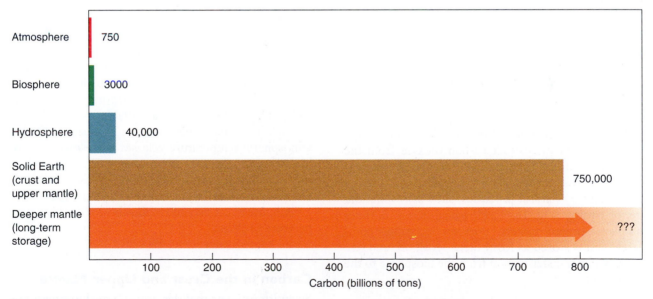

Figure 18–8 Carbon reservoirs in the atmosphere, biosphere, hydrosphere, and solid Earth. The numbers represent billions of tons of carbon. *(Wilfred M. Post et al., "The Global Carbon Cycle." The American Scientist, 78, July-August 1990, p. 310.)* • **Interactive Question: How would the amount of carbon in the atmosphere be affected if 1 percent of the carbon dissolved in sea water were released into the atmosphere?**

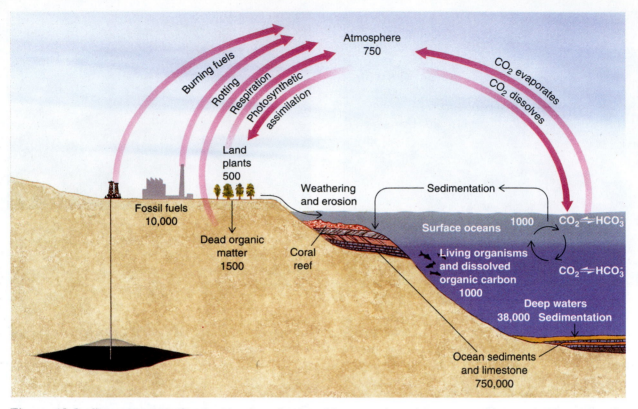

Figure 18–9 The carbon cycle. The numbers show the size of the reservoirs and represent billions of tons of carbon. *(Data taken from U. Siegenthaler and J. L. Sarmiento, "Atmospheric Carbon Dioxide and the Ocean." Nature, 365, September 9, 1993, p. 119.)*

the atmosphere, this reservoir plays an important role in controlling atmospheric temperature because carbon dioxide and methane are greenhouse gases; they absorb infrared radiation and heat the lower atmosphere. If either of these compounds is removed from the atmosphere and collects in water, rock, or living organisms, the atmosphere cools; if they are released into the atmosphere, the air becomes warmer.

Carbon in the Biosphere

Carbon is the fundamental building block for all organic tissue. Plants extract carbon dioxide from the atmosphere and build their body parts predominantly of carbon. Most of the carbon is released back into the atmosphere by natural processes, such as respiration, fire, or rotting (Fig. 18–9). However at certain times and places, organic material does not decompose completely and is stored as fossil fuels—coal, oil, and gas. Thus, plants transfer carbon from the biosphere to the upper crust.

Carbon in the Hydrosphere

Carbon dioxide dissolves in sea water. Most of it then reacts to form bicarbonate, HCO_3^- (commonly found in your kitchen as baking soda or bicarbonate of soda), and carbonate, $(CO_3)^{2-}$.

The amount of carbon dioxide dissolved in the oceans depends in part on the temperature of the atmosphere and the oceans. When sea water warms, it releases dissolved carbon dioxide into the atmosphere, causing greenhouse warming. In turn, greenhouse warming further heats the oceans, causing more carbon dioxide to escape. Warmth evaporates sea water as well, and water vapor also absorbs infrared radiation.

Such a feedback mechanism can escalate. Higher atmospheric temperature releases more carbon dioxide and water vapor from the oceans. These gases heat the atmosphere even more, which releases additional carbon dioxide and water vapor and so on. A runaway greenhouse effect may be responsible for the high temperature on Venus.

Carbon in the Crust and Upper Mantle

Scientists estimate that the atmosphere, biosphere, and hydrosphere, together, contain a little less that 43 trillion tons of carbon. Yet, 17 times that much, 750 trillion tons, exists in the crust and upper mantle. This reservoir is important because if only a small portion of it is released, it can dramatically alter climate.

Carbonate Rocks   Marine organisms absorb calcium and carbonate ions from sea water and convert them into calcium carbonate ($CaCO_3$) in shells and other hard parts. This process removes carbon from sea water and causes more atmospheric carbon dioxide to dissolve into the sea water. Thus, formation of shells removes carbon dioxide from the atmosphere. The shells and skeletons of these organisms gradually collect to form limestone.

When sea level falls or tectonic processes raise portions of the sea floor above sea level, marine limestone weathers in a process that extracts additional carbon dioxide from the atmosphere.[2]

Carbon in Fossil Fuels   Carbon is stored in fossil fuels and carbon dioxide is released when these fuels are burned. Recoverable fossil fuels contain about 4 trillion tons, five times the amount of carbon in the atmosphere today. For this reason, scientists are concerned that burning fossil fuels will raise atmospheric carbon dioxide levels. This topic is discussed in Section 18.4.

Methane in Sea-Floor Sediment   When organic material falls to the sea floor, bacteria decompose it, releasing methane, commonly called natural gas. Between a depth of about 500 meters and 1 kilometer, seawater temperature is low enough and the pressure is favorable to convert methane gas to a frozen solid called methane hydrate. Methane hydrate then gradually collects in sea-floor sediment. After studying both drill samples and seismic wave data, geochemist Keith Kvenvolden of the United States Geological Survey estimates that methane hydrate deposits hold twice as much carbon as all conventional fossil fuels.

At present, commercial extraction of methane hydrates to produce natural gas is impractical. It is expensive to drill in deep water, and a thin layer spread throughout the continental shelves would be prohibitively expensive to exploit. However, scientists are studying links between methane hydrates and climate. Tectonic activity at subduction zones or landslides on continental slopes could release methane from hydrate deposits. Changes in bottom temperatures on continental shelves resulting from warming of sea water could also release methane from the frozen hydrates. In turn, increased atmospheric methane could trigger greenhouse warming. Ice core studies show that global atmospheric methane concentration has changed rapidly in the past, perhaps by sudden releases of oceanic methane hydrates.

About 55 million years ago, near the end of the Paleocene epoch, climate suddenly warmed and many aquatic and terrestrial species became extinct. According to one model, a change in surface circulation caused equatorial waters to remain in low latitudes. High equatorial temperatures evaporated enough water to increase the salinity of the sea surface. When the salinity reached a threshold value where the surface water was denser than the cold deep water, rapid vertical mixing occurred. Warm, salty water sank and cold, less salty water rose. The warm, sinking water melted the methane hydrates and released the methane. Aquatic species were poisoned by the methane in the water and many terrestrial species succumbed to the rapid greenhouse warming.[3]

Carbon in the Deeper Mantle

During subduction, oceanic crust sinks into the mantle (Fig. 18–10). The descending plate may carry carbonate rocks and sediment. As this material sinks to greater depths, the carbonate minerals become hot and release carbon dioxide, which is carried back to the surface by volcanic eruptions. Some carbonate rock may be carried into deeper regions of the mantle during subduction, although geologists are uncertain how much. Large quantities of carbon were trapped within the Earth during its formation. Some of this carbon remains in the deep mantle and may rise from the mantle to the surface during volcanic eruptions. Carbon exchanges between the deep mantle and the surface are an important topic of current research.

18.4 Humans, the Carbon Cycle, and Climate

We have learned that the amount of carbon in the atmosphere is determined by many natural factors, including rates of plant growth, mixing of surface ocean water and deep ocean water, growth rates of marine organisms, weathering, subduction, and volcanic activity. Within the past few hundred years, humans have become an important part of the carbon cycle. Modern industry releases four greenhouse gases—carbon dioxide,

---

[2] The complete reaction is
$$CaCO_3 + CO_2 + H_2O \rightarrow Ca(HCO_3)_2$$
limestone   carbon dioxide   water   calcium bicarbonate (soluble)

[3] Gerald Dickens et al., "A Blast of Gas in the Latest Paleocene; Simulating First-order Effects of Massive Dissociation of Oceanic Methane Hydrate." *Geology*, March 1997, pp. 259–262.

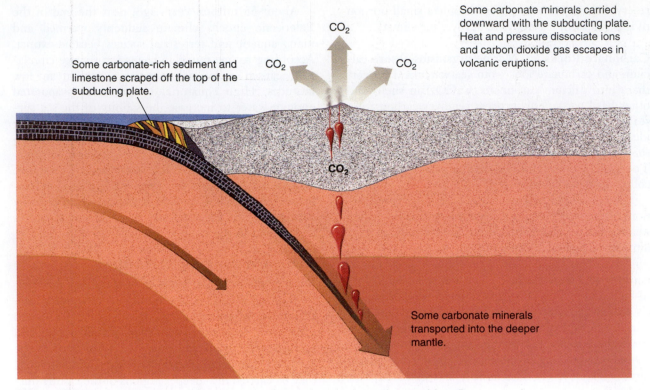

Some carbonate-rich sediment and limestone scraped off the top of the subducting plate.

CO_2

CO_2

CO_2

Some carbonate minerals carried downward with the subducting plate. Heat and pressure dissociate ions and carbon dioxide gas escapes in volcanic eruptions.

CO_2

Some carbonate minerals transported into the deeper mantle.

Figure 18–10 A subducting oceanic plate carries limestone and other carbonate-rich sediment into the mantle. Some of the carbonate minerals are heated to produce carbon dioxide, which escapes during volcanic eruptions. Some of the carbonate minerals may be stored in the mantle.

methane, chlorofluorocarbons (CFCs), and nitrogen oxides (Fig. 18–11).

People release carbon dioxide when they burn fossil fuels. Logging also frees carbon dioxide because stems and leaves are frequently burned and forest litter rots more quickly when it is disturbed by heavy machinery. The recent rise in the concentration of atmospheric carbon dioxide has attracted considerable attention because it is the most abundant industrial greenhouse gas (Fig. 18–12).

Small amounts of methane are released during some industrial processes. Larger amounts are released from the guts of cows, other animals, and termites, and from rotting that occurs in rice paddies. Today, industry and agriculture combined add about 37×10^{12} grams of methane into the atmosphere every year. CFCs, the third most abundant of the greenhouse gases, were, until recently, used as refrigerants and as propellants in aerosol cans (see Chapter 19). N_2O is released from the manufacture and use of nitrogen fertilizer, some industrial chemical syntheses, and from the exhaust of high-flying jet aircraft.

In December 1995, the United Nations Intergovernmental Panel on Climate Change issued a report written by 500 scientists and reviewed by another

500. The report started with the following observations:

1. The atmosphere has warmed by 0.5°C ± 0.2°C during the last century.

2. During the same time period, the atmospheric concentration of carbon dioxide and other greenhouse gases has increased.

3. Laboratory studies show that greenhouse gases absorb infrared radiation.

4. Researchers have measured ancient atmospheric temperatures from oxygen isotope measurements of glacial ice. Air bubbles trapped in the ice provide a simultaneous record of the carbon dioxide content of the atmosphere. In several studies, global temperature and atmospheric carbon dioxide concentration have oscillated in concert.

Starting with these observations, and using computer models to test their hypothesis, the scientists in the U.N. panel concluded that rising atmospheric concentrations of greenhouse gases probably have caused the global temperature rise over the past century. Furthermore, they predicted that the mean global

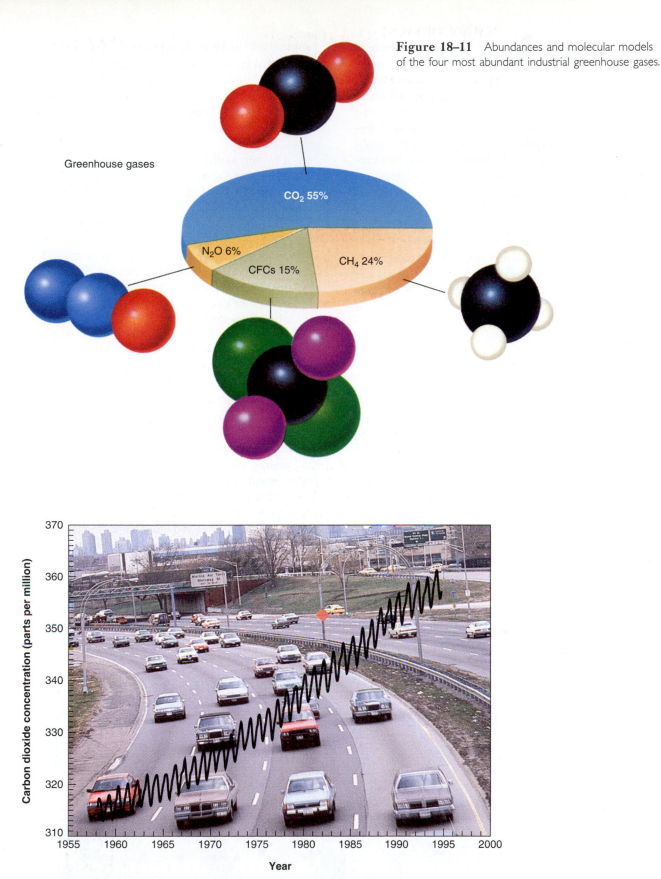

Figure 18–11 Abundances and molecular models of the four most abundant industrial greenhouse gases.

Greenhouse gases

CO₂ 55%

N₂O 6%

CFCs 15%

CH₄ 24%

Figure 18–12 Atmospheric carbon dioxide concentration has risen within the past century. The short-term fluctuations are caused by seasonal changes in carbon dioxide absorption by plants.

temperature will increase by 1 to 3.5°C during the next century.

Not all scientists agree with United Nations report. Some argue that no one can prove that industrial greenhouse gases are responsible for the observed global warming. Recall that climate has changed frequently and drastically throughout Earth history. Therefore a warming global climate may just be a natural event, not a result of greenhouse gas emissions.

Other models of the relationship between greenhouse gas emissions and climate are discussed in Section 18.5.

Consequences of Greenhouse Warming

Climatologists build computer models of the atmosphere, test the models against historical records, and then use them to forecast the future. A recent article in *Scientific American* summarized the consensus of many scientists about the effects of global warming:[4]

Temperature A warmer global climate would mean a longer frost-free period in the high latitudes, which would benefit agriculture. In parts of North America, the growing season is now a week longer than it was a few decades ago. Snowfall in the mid latitudes would diminish, decreasing the winter and spring albedo, which could lead to further warming. On the negative side, warmth would also improve conditions for plant pathogens and parasites, which could decrease crop yields.

Precipitation and Soil Moisture In a warmer world, evaporation and precipitation would increase. However, computer models show that the effects would differ from region to region. Figure 18–13 shows how global precipitation patterns changed

---

[4] Thomas Karl, Neville Nichols, and Jonathan Gregory, "The Coming Climate," *Scientific American*, May 1997, p. 79ff.

from 1900 to 1994. Rainfall increased in the great grain-growing regions in North America and Russia. On the other hand, parts of North Africa, India, and southeast Asia received less rainfall over the same time. These trends are predicted to continue.

In a hotter world, more soil moisture will evaporate. As explained above, in the Northern Hemisphere, scientists predict that global warming will lead to increased rainfall, adding moisture to the soil. At the same time, warming will increase evaporation that removes soil moisture. Most computer models forecast a net loss of soil moisture and depletion of ground water, despite the prediction that rainfall will increase. Dry soils could decrease farm yields significantly, especially because many aquifers, such as the Ogallala in the United States, are being rapidly depleted.

In Africa, India, and southeast Asia, drought has led to famine during recent decades. A continuation of this trend will be disastrous for many of the world's poorest people. As a result of all these factors, most scientists predict that higher mean global temperature would decrease global food production, perhaps dramatically (Fig. 18–14).

Computer models predict that intense rainstorms and flooding will become more common in a warmer, wetter world. In the United States, the number of heavy downpours (defined as 5 centimeters or 2 inches of rain in a single day) increased by 25 percent from 1900 to the 1990s. Worldwide, there were ten times as many catastrophic floods in the decade from 1986 to 1995 as there were in an average decade between 1950 and 1985.

Some species adapt better to climate changes than others. Although population levels would change and some species might become extinct as climate warmed, terrestrial ecosystems have adjusted to far greater perturbations than we are experiencing today.

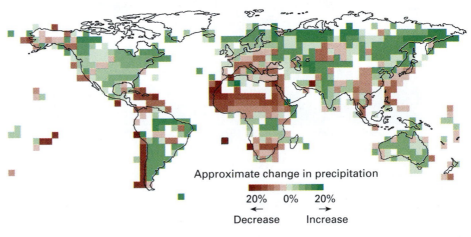

Figure 18–13 Changes in the distribution of global precipitation between 1900 and 1994 reflect a tendency toward more precipitation at high latitudes and less precipitation near the equator. *(Redrawn from: Thomas Karl et al., "The Coming Climate." Scientific American, May, 1997, pp. 78–83.)*

Approximate change in precipitation

20% 0% 20%
← Decrease   Increase →

Figure 18–14 Computer models predict that global warming could cause drought in Africa, India, and southeast Asia, leading to decreased crop production and possibly to increased hunger in an already impoverished region. *(Cedric Galbe/SABA)*

In a warmer world, tropical diseases such as malaria would spread to higher latitudes. If global temperature rises sufficiently, ice would be released from the Antarctic and Greenland ice caps and sea water volume would increase, raising sea level and drowning coastlines.

Can We Avert Greenhouse Warming?

Some people argue that because we cannot prove that greenhouse warming is taking place, and cannot pre-

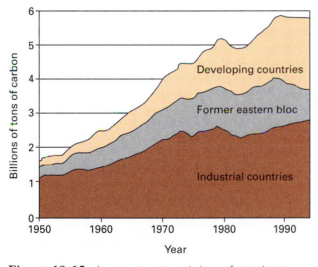

Figure 18–15 In recent years, emissions of greenhouse gases from developing countries have increased faster than emissions from industrial countries. However, the per capita energy consumption from developing countries is still far lower than that of their richer neighbors. *(Redrawn from Christopher Flavin and Odil Tunali,* Climate of Hope, *Worldwatch paper 130, June 1996)* ● **Interactive Question: Discuss the impact on world fuel reserves, air pollution, and global warming if per capita energy consumption in the developing countries rose to that of the developed countries.**

dict the economic effects of such warming, we should study the problem further but delay action. Others argue that the evidence for global warming is convincing and the consequences of rising global temperatures could be severe. Therefore it is better to be safe than sorry, and we should reduce emission of greenhouse gases.

One significant argument at the 1997 Kyoto conference concerned the relative amounts of greenhouse gases produced by the wealthy industrial nations and the less wealthy developing nations (Fig. 18–15). As an example, China has a population of 1.2 billion, a little over four times that of the United States, which has a population of 280 million. Each person in the United States releases about eight times as much carbon dioxide as the average Chinese person (Fig. 18–16). The net result is that people in the United States emit twice as much carbon dioxide as do people in China.

At the Kyoto conference, the United States argued that all countries must decrease emission of greenhouse gases by the same proportion. However, representatives of developing nations, including China,

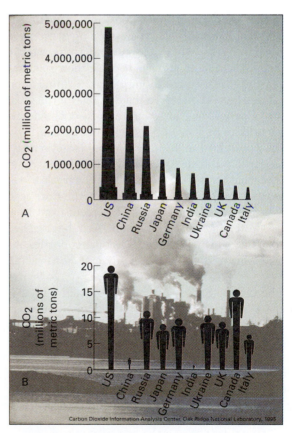

Figure 18–16 (A) Carbon dioxide emissions from the top ten emitting countries in 1994. (B) Per capita carbon dioxide emissions from the top ten emitting countries in 1994.

The most obvious way to reduce atmospheric carbon dioxide concentrations is to burn less fuel. However, some scientists are exploring ways to reduce global warming by removing carbon dioxide from the atmosphere.

Plants remove carbon dioxide from the air and incorporate the carbon into organic tissue. Therefore, if people increase global plant growth, they will reduce the concentration of atmospheric carbon dioxide. Some scientists argue that if another 10 percent of the Earth's land surface were covered with trees, photosynthesis would remove enough carbon from the atmosphere to slow global warming. The major problem with this suggestion is that food shortages threaten many regions. Therefore, people believe they can't afford to convert agricultural land to forest.

If it is impractical to increase photosynthesis on land, what about in the oceans? Microscopic, floating plants called phytoplankton conduct most marine photosynthesis. In 1986, oceanographer John Martin argued that phytoplankton growth is limited because there is very little dissolved iron in the surface waters of the central oceans. When he experimentally added iron to a laboratory sample of phytoplankton, their growth rate increased dramatically.

Oceanographers then spread iron over small segments of the ocean. In one dramatic study, phytoplankton growth rates increased by a factor of thirty after two applications of iron fertilizer. However, it is unclear whether we could reduce the atmospheric carbon dioxide concentration and counteract global warming by spreading iron on the oceans. It would be incredibly expensive to fertilize the oceans. Recent research indicates that nonphotosynthetic bacteria might absorb much of the iron. Perhaps the iron would fall to the sea floor within a short period of time and we would have to repeat the process regularly. Michael MacCracken, director of the U.S. Global Change Research Program, argues that "By attempting to control climate, you could certainly get yourself into an even bigger predicament than you were in to begin with."

---

argued that the United States' position is unfair because it condemns people in developing nations to continued poverty. They reasoned that the world's poor people should be allowed to raise their standard of living first, then worry about global warming later. The United States countered that because of the great number of people in China and the rest of the developing world, even a small per capita increase in fossil fuel use would lead to an unacceptably large increase in total carbon dioxide emissions.

In the final treaty, the United States agreed to reduce greenhouse emissions by 7 percent, the European Union by 8 percent, and Japan by 6 percent from 1990 levels. These targets are scheduled to be reached between 2008 and 2012. The developing countries, including China, did not agree to any emissions reductions.

In the United States, a treaty is not valid until it is ratified by the Senate. Two months after the conference ended, President Clinton proposed a $6.3 billion investment in research and tax incentives to help achieve the United States' commitment. Although the official debate on the treaty probably will not begin before 1999, many senators are already arguing that they cannot vote for emissions reductions without cooperation from the developing countries. They fear that if the United States agrees to emissions reductions, industries will move factories and jobs overseas.

Traditionally, in any nation, wealth and fuel consumption have risen together, and today, the world's richest countries have the highest per capita energy consumption. However, many people argue that if we all use the most efficient technology, we can maintain or increase our wealth without increasing fuel consumption and greenhouse gas emission. Two different types of technologies can be used to achieve this goal.

One approach is to use energy efficiently. For example, automobiles are twice as efficient now as they were 20 years ago. Several manufacturers have built experimental cars that are twice as efficient as the automobiles on the road today. If people used the most efficient technologies in factories, transportation, energy production, and home heating, they could enjoy "the good life" without emitting more carbon dioxide into the air. Moreover, according to this argument, energy efficient technology is profitable in the long run. You may pay more for an efficient home, appliance, or car today, but fuel savings will return your initial investment.

A second approach is to use wind, solar, geothermal, nuclear, and hydroelectric energy to produce energy without burning fossil fuels and emitting carbon dioxide. At the present time, many of these technologies are nearly cost competitive with fossil fuel energy production. Many people argue that we should change our tax structure to reduce personal income taxes and replace them with a tax on carbon emissions. Even a small carbon tax would shift the economics in favor of nonfossil fuel energy sources.

18.5 Climate Change Resulting from Atmosphere-Ocean Interactions

Many of the arguments presented in Section 18.4 assume a linear relationship between the concentration of greenhouse gases in the atmosphere and mean global temperature. In a linear relationship, if you raise the concentration of greenhouse gases a small amount, the mean global temperature will rise by a small amount, and further small increases of greenhouse gas concentrations will cause additional small temperature increases, and so on. However, many scientists argue that the assumption of a linear relationship is incorrect.

For example, according to one argument, a small initial warming would release carbon dioxide and water vapor from the oceans into the atmosphere to cause greenhouse warming. That additional warming would release more carbon dioxide and water vapor from the seas into the atmosphere to cause further warming in an escalating greenhouse effect. James Holdren, a meteorologist from Harvard University, argues that if global economic growth continues, the atmospheric carbon dioxide concentration will reach 700 parts per million (ppm) by the year 2100 (compared to a preindustrial level of 280 ppm) and global temperature will rise by 8°C over the 1950 norm.

Other scientists argue that an increased concentration of greenhouse gases will eventually cause *global cooling*. For example, Jeffrey Kiehl of the National Center for Atmospheric Research argues that greenhouse warming of the oceans will lead to more evaporation and increased cloud cover and that the clouds will reflect enough sunlight to cool the Earth.

Global Cooling Caused by Disruption of Ocean Currents in the North Atlantic

Recall that ice core data show many rapid and dramatic temperature fluctuations within the past 110,000 years. Climatologists have uncovered evidence linking some of these temperature changes to changes in ocean circulation.

In Chapter 14 we learned that warm surface water from the Gulf Stream sinks as it approaches the Arctic because the surface water cools and becomes saltier. In the early 1980s, oceanographers discovered several curious layers of sediment in the North Atlantic that contain boulders derived from limestone and granite bedrock near Hudson's Bay in northeastern Canada. These layers were deposited at the same time that the temperature dropped dramatically over the Greenland ice cap.

Scientists now hypothesize that the Pleistocene ice sheet had picked up the limestone and granite rocks near Hudson's Bay and carried them to the North Atlantic coast of Canada. Then some unknown mechanism caused the ice sheet to release massive armadas of icebergs into the North Atlantic. When the ice melted, the rocks fell to the sea floor. At the same time, the melting icebergs added large quantities of fresh water to the surface of the North Atlantic. The surface water then became less salty and more buoyant and stopped sinking. This change in vertical circulation disrupted the surface currents, and the warm Gulf Stream either stopped flowing or veered far to the south. As a result, the temperature in the Northern Hemisphere fell by 10°C in a short time, perhaps in as little as five to ten years. As the North Atlantic region cooled, winds and other ocean currents shifted and global temperatures dropped.

Changes in vertical ocean currents can be triggered by a threshold mechanism. In the example given above, before large numbers of icebergs floated into the North Atlantic, the ocean surface water was so cool and salty that it was denser than deep water. As icebergs melted and added fresh water, the density of the surface water decreased. For a while, despite this decreased density, the surface water continued to sink. As long as surface water still sank, the vertical and horizontal ocean currents continued to flow and climate remained relatively constant. However, as soon as the surface water reached a threshold at which it floated on the surface, the vertical current stopped. Surface currents were deflected, altering heat and moisture transport across the globe.

Some scientists suspect that similar changes could occur in the future. According to one model, global atmospheric warming would warm the surface water in the North Atlantic and make it more buoyant, just as addition of fresh water does. If the water became warm and buoyant enough, it would stop sinking. This change in vertical circulation would alter the Gulf Stream, as it did in the past. If the Gulf Stream stopped or veered to the south, the Earth would cool rapidly and dramatically.[5] Thus, paradoxically, a small initial warming would cause a rapid cooling.

---

[5] Wallace C. Broecker, "Thermohaline Circulation, the Achilles Heel of our Climate System: Will Man-Made CO_2 Upset the Current Balance?" *Science*, Vol. 278, November 28, 1997, p. 1582.

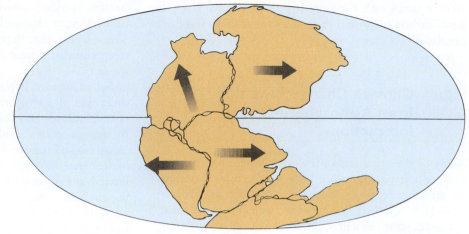

Figure 18–17 Two hundred million years ago, when the Pangea supercontinent was assembled, Africa, South America, India, and Australia were all positioned close to the South Pole.

18.6 Tectonics and Climate Change

Positions of the Continents

A map of Pangea shows that 200 million years ago, Africa, South America, India, and Australia were all clustered near the South Pole (Fig. 18–17). Because climate is colder at high latitudes than near the equator, continental position alters continental climate.

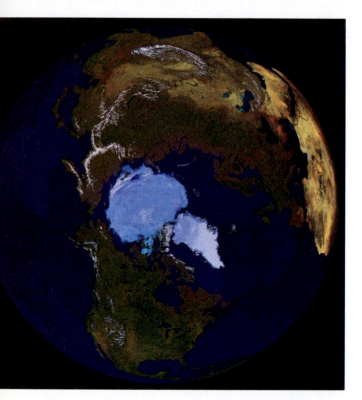

Figure 18–18 The frozen Arctic Ocean is mostly surrounded by land. *(Tom Van Sant/Geosphere Project)*

In addition, continental interiors generally experience colder winters and hotter summers than coastal areas. When all the continents were joined into a supercontinent, the continental interior was huge, and regional climates were different from the climates on many smaller continents with extensive coastlines.

The positions of the continents also influence wind and sea currents, which, in turn, affect climate. For example, today the Arctic Ocean is nearly landlocked, with three straits connecting it with the Atlantic and Pacific oceans (Fig. 18–18). The Bering Strait between Alaska and Siberia is 80 kilometers across, Kennedy Channel between Ellesmere and Greenland is only 40 kilometers across, and a third, wider seaway runs along the east coast of Greenland. Presently, cold currents run southward through Kennedy Channel and the Bering Strait, and the North Atlantic drift carries warm water northward along the coast of Norway. If any of these straits were to widen or close, global heat transfer would be affected. Deep-sea currents also transport heat and are affected by continental positions.

Tectonic plates move from 1 to 16 centimeters per year. A plate that moves 5 centimeters per year travels 50 kilometers in a million years. Thus, continental motion can change global climate within a geologically short period of time by opening or closing a crucial strait. Much longer times are required to modify climate by altering the proximity of a continent to the poles or by creating a supercontinent.

Mountains and Climate

Global cooling during the past 40 million years coincided with the formation of the Himalayas and the North American Cordillera.[6] Mountains interrupt air-

---

[6] William F. Ruddiman and John Kutzbach, "Plateau Uplift and Climatic Change," *Scientific American*, March 1991, p. 66ff.

flow, altering regional winds. Air cools as it rises and passes over high, snow-covered peaks. However, it is unclear whether this regional cooling could account for the global cooling that accompanied this episode of mountain formation.

Large portions of the Himalayas and the North American Cordillera are composed of marine limestone. Recall that when marine limestone weathers, carbon dioxide is removed from the atmosphere. According to one hypothesis, when sea-floor rocks are thrust upwards to form mountains, they become exposed to the air. Rapid weathering then removes enough atmospheric carbon dioxide to cause global cooling.

Volcanoes and Climate

Recall from Chapter 7 that volcanoes emit ash and sulfur compounds that reflect sunlight and cool the atmosphere. For two years after Mt. Pinatubo erupted in 1991, the Earth cooled by a few tenths of a degree Celsius. Temperature rose again in 1994 after the ash and sulfur settled out.

Volcanoes also emit carbon dioxide that warms the atmosphere by absorbing infrared radiation. The net result—warming or cooling—depends on the size of the eruption, its violence, and the proportion of solids and gases released. Some scientists believe that a great eruption in Siberia 250 million years ago cooled the atmosphere enough to cause or contribute to the Permian extinction. The mid-Cretaceous superplume 120 million years ago may have emitted enough carbon dioxide to warm the atmosphere by 7 to 10°C. Dinosaurs flourished in huge swamps, and some of the abundant vegetation collected to form massive coal deposits.

How Tectonics, Sea Level, Volcanoes, and Weathering Interact to Regulate Climate

Tectonics, sea level, volcanoes, and weathering are all part of a tightly interconnected Earth system that affects both global and regional climate. We learned that when tectonic plates spread slowly, the mid-oceanic ridges are so narrow that they displace relatively small amounts of sea water. As a result, sea level falls. When sea level falls, large marine limestone deposits on the continental shelves are exposed as dry land. The limestone weathers. Weathering of limestone removes carbon dioxide from the atmosphere, leading to global cooling. At the same time, when sea-floor spreading is slow, subduction is also slow. Volcanic activity at both the spreading centers and the subduction zones slows down, so relatively small amounts of carbon dioxide are emitted. With small additions of carbon dioxide from volcanic eruptions and removal of atmospheric carbon dioxide by weathering, the atmo-

spheric carbon dioxide concentration decreases and the global temperature cools. In addition, dropping sea level decreases the surface area of the oceans and increases the surface area of the higher-albedo continents. This results in an increase of average global albedo, and, consequently, reinforces the global cooling (Fig. 18–19A). These conditions may have caused the cooling at the end of the Carboniferous period.

In contrast, during periods of rapid sea-floor spreading, a high-volume mid-oceanic ridge raises sea level. Marine limestone beds are submerged, weathering slows down, and weathering removes less carbon dioxide from the atmosphere. Volcanic activity is high during periods of rapid plate movement, so large amounts of carbon dioxide are released into the atmosphere. Rising sea level decreases continental area, and therefore decreases the average global albedo. All of these factors lead to global warming (Fig. 18–19B). These conditions prevailed in the late Cambrian, Carboniferous, and Cretaceous periods.

18.7 Astronomical Causes of Climate Change

Recall from Chapter 11 that variations in the Earth's orbit may have caused the climate fluctuations responsible for the advances and retreats of the Pleistocene Ice Age. Other astronomical factors also cause climate change.

Changes in Solar Radiation

A star the size of our Sun produces energy by hydrogen fusion for about 10 billion years. During this time, its energy output increases slowly. Solar output has increased by 20 to 30 percent during Earth history. This slow evolution has influenced climate over long expanses of geologic time. As discussed in Section 18.4, the atmospheric carbon dioxide concentration was high during the Earth's early history, so temperature was relatively warm even though solar input was low.

Within the past few hundred million years, solar output changed only a fifty-millionth of one percent per century and therefore the variation had no measurable influence on climate change over hundreds to thousands of years.

Over shorter periods of time, solar magnetic storms and other phenomena cause fluctuations in solar output that may affect the Earth's climate. Magnetic storms on the Sun create dark, cool regions on the solar surface called sunspots. Sunspot activity alternates dramatically on an 11-year cycle and on longer cycles spanning hundreds of years. During periods of high sunspot activity, the Sun is slightly cooler than during periods of low activity. Several studies show that

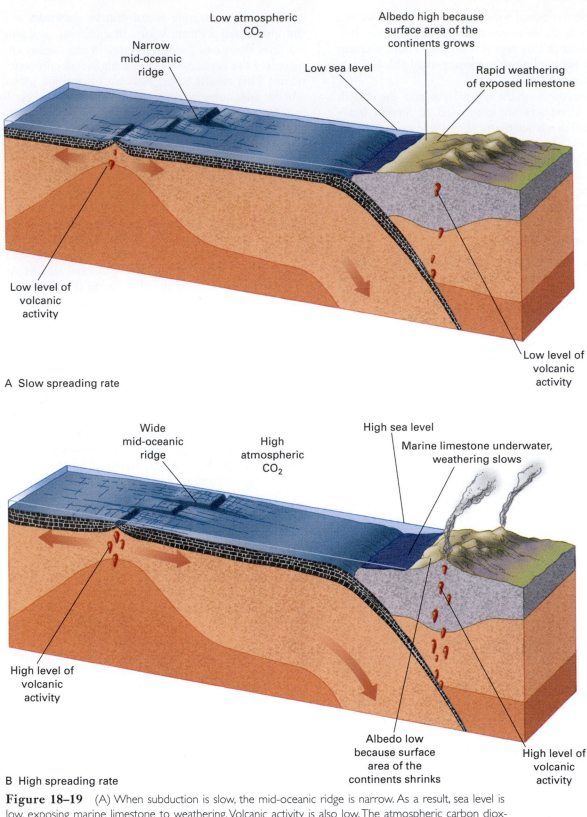

A Slow spreading rate

B High spreading rate

Figure 18–19   (A) When subduction is slow, the mid-oceanic ridge is narrow. As a result, sea level is low, exposing marine limestone to weathering. Volcanic activity is also low. The atmospheric carbon dioxide concentration is low because of rapid weathering and low volcanic activity. Also, global albedo is high because the surface of the continents grows at the expense of shrinking oceans. All of these factors cool the Earth. (B) When subduction is fast, the mid-oceanic ridge is wide. As a result, sea level is high, flooding coastal regions. Volcanic activity is also high. The atmospheric carbon dioxide concentration is high because weathering is slow and volcanic activity is high. Also, global albedo is low because of the low surface area of the continents and greater surface area of the seas. All of these factors warm the Earth.

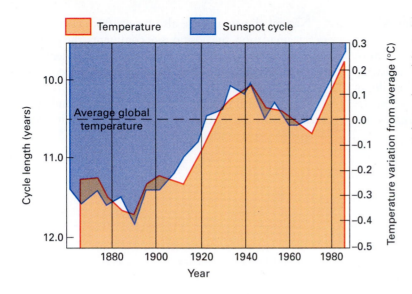

Figure 18–20 Times of maximum sunspot abundance (blue curve) correlate closely to mean global atmospheric temperature (brown curve) between 1860 and 1990. Global temperature is expressed as change from the average annual temperature in the years 1951 to 1980. *(Source: E. Friis-Christensen and K. Lassen, "Length of the Solar Cycle: An Indicator of Solar Activity Closely Associated with Climate." Science 254, Nov. 1, 1991, 698f.)*

changes in global temperatures coincide with changes in sunspot activity (Fig. 18–20). Critics argue that the correlation must be a statistical coincidence because differences in solar output resulting from sunspot cycles are too small to alter the Earth's temperature. Proponents counter that a feedback mechanism must be involved. Perhaps changes in solar radiation affect the stratospheric ozone concentration, which alters heat transfer mechanisms between the stratosphere and the troposphere. The issue remains under vigorous debate.

Bolide Impacts

Recall that evidence strongly suggests that a bolide crashed to Earth about 65 million years ago. The impact blasted enough rock and dust into the sky to block out sunlight and cool the planet. According to one current hypothesis, this cooling led to the extinction of the dinosaurs. Other bolides have caused rapid and catastrophic climate changes throughout Earth history.

Astronomical Changes in the Distant Future

Some astronomers predict that in about 50,000 years, the Solar System will pass through a huge cloud of dust within the Milky Way Galaxy. This dust will fill the space between the Earth and the Sun. Although scientists are not sure how this dust will affect the Earth's climate, Priscilla Frisch of the University of Chicago predicts that it will reshape the Earth's magnetic field, alter atmospheric chemistry, and cool the planet.

In about five billion years, the Sun will pass through a rapid metamorphosis and engulf the Earth. While these events will occur in the far distant future, we are virtually certain that the Earth's climate—which has changed almost continuously since the planet formed—will continue to change until the Earth is engulfed by our fiery Sun.

SUMMARY

Global climate has changed throughout Earth history and is affected by feedback mechanisms and threshold effects.

Past climates are measured by historical records, tree rings, pollen assemblages, oxygen isotope ratios in glacial ice, glacial evidence, plankton assemblages and isotope studies in ocean sediment, and fossils in sedimentary rocks.

Carbon circulates among all four of the Earth's realms. Carbon dioxide gas in the atmosphere is reactive and mobile. Carbon exists in the biosphere as the fundamental building block for organic tissue. Carbon dioxide gas dissolves in sea water to form bicarbonate and carbonate ions. Carbon exists in the crust and mantle in several forms. (1) Marine organisms absorb calcium and carbonate ions and convert them to solid calcium carbonate (shells and skeletons). As a result, large quantities of carbon exist in limestone and marine sediment. (2) Fossil fuels are largely composed of carbon. (3) Large quantities of carbon exist as methane hydrates on the sea floor. (4) Carbon also exists in the deeper mantle. Some of this is primordial carbon trapped during the Earth's formation and some is carried into the mantle on subducting plates.

Humans release carbon dioxide into the atmosphere when they burn fuel. Logging, some industrial processes, and some aspects of agriculture contribute greenhouse gases to the atmosphere. Most scientists agree that human introduction of greenhouse gases has altered climate in the past century.

Natural climate change also results from changes in ocean circulation, the positions of continents, the growth of mountains, and volcanic eruptions. Independent climate-changing factors are linked by Earth systems interactions.

Astronomical causes of climate change include the variations in the Earth's orbit and the tilt of its axis, changes in solar radiation, and bolide impacts.

FOR REVIEW

1. Explain why the early Earth was warm even though the Sun emitted 20 to 30 percent less energy at that time.

2. Explain how feedback and threshold mechanisms operate. Give an example of each.

3. Explain why oxygen isotope measurements provide paleotemperature data both from cores in glacial ice and from coral reefs.

4. Explain how pollen and plankton assemblages are used to measure past climates.

5. List several techniques for measuring climate 100 to 500 million years ago.

6. In what forms is carbon present in the atmosphere, the biosphere, the hydrosphere, and the crust?

7. Draw a rough sketch of the carbon cycle and explain the chemical transformations shown in your cycle.

8. Discuss the scientific arguments for our concern that humans are causing global warming.

9. Discusses the consequences of a warmer Earth. How would temperature changes affect precipitation? How would changes in temperature and precipitation affect agriculture, ecosystems, and human society?

10. Explain two mechanisms whereby an increase in greenhouse gases in the atmosphere could lead to global cooling.

11. Briefly discuss ways to reduce greenhouse gas emissions.

12. Explain how continental positions can affect climate.

13. Explain how weathering can affect atmospheric composition and climate.

14. Explain how volcanic eruptions can cause either a cooling or a warming of the atmosphere.

15. Discuss the role of solar output in determining climate.

FOR DISCUSSION

1. Give an example of a feedback and a threshold effect in a science other than Earth science (such as psychology, political science, or any other science).

2. Based on Figure 18–4, list the useful time ranges for historical records, pollen, and plankton as indicators of paleoclimates. Explain why each of these techniques is not useful farther back in time.

3. Discuss the difficulties and costs involved in reducing carbon dioxide emissions.

4. Explain why a small change in average global temperatures can have environmental, economic, and political effects.

5. Do you feel that the United States should ratify the Kyoto treaty? Defend your answer.

6. In a recent essay (*Natural History*, May 1991, p. 18) Stephen Jay Gould wrote, "During most of the past 600 million years, the Earth has been sufficiently warm so that even the bottom of an orbital cycle produced no ice caps." Explain this statement in your own words.

7. According to David des Marais of Ames Research Center, about 2.2 billion years ago, rapid plate motion led to rapid rise of mountains throughout the world. As the mountains rose, large quantities of organic-rich sediment eroded and were transported into the sea, where they were buried. Predict how this burial might alter atmospheric composition.

Despite laws to designed to curb air pollution in the United States, many industries continue to pollute the air. These plumes come from the stacks of the Louisville Gas and Electric Company in Louisville, Kentucky.   *(Adam Jones/Photo Researchers)*

Air Pollution

One day, in an event now lost in the great expanse of prehistory, a cave dweller brought home a smoldering ember from a forest fire. He or she blew on it and carefully added sticks. A thin wisp of smoke rose into the air and spread across the roof of the cave. Humans had begun to pollute the air.

For hundreds of thousands of years, human contribution to air pollution was insignificant compared with natural sources such as volcanic eruptions, wildfires, and airborne pollen. However, by the Middle Ages, when people concentrated in densely populated northern cities, a pall of wood smoke hung over urban areas. By the start of the Industrial Revolution, coal had replaced wood as a primary fuel for heating and people also burned fossil fuels to power machines. As a result, air pollution became a serious problem.   ●

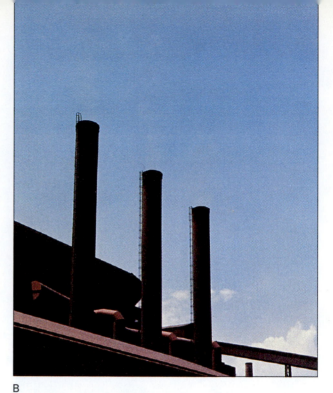

A

B

Figure 19–1 (A) Before the Clean Air Act, most factory smokestacks had no pollution control devices. (B) Pollution controls installed after the Clean Air Act reduced emissions by large proportions. *(John D. Cunningham/Visuals Unlimited)*

19.1 A Brief History of Air Pollution

In 1948, Donora was an industrial town of about 14,000 located 50 kilometers south of Pittsburgh, Pennsylvania. One large factory in town manufactured structural steel and wire and another produced zinc and sulfuric acid. In the hill country of Pennsylvania foggy days are common, especially in the fall. During the last week of October 1948, dense fog settled over the town. But it was no ordinary fog; the moisture contained pollutants from the two factories. After four days, visibility became so poor that people could not see well enough to drive, even at noon with their headlights on. Gradually at first, and then in increasing numbers, residents sought medical attention for nausea, shortness of breath, and constrictions in the throat and chest. On the morning of the fifth day, a retired steel worker died of respiratory problems, and several other deaths followed in rapid succession. Within a week, 20 people had died and about half of the town was seriously ill.

Other incidents similar to that in Donora occurred worldwide. Sixty people died in an air pollution disaster in Belgium in 1930. In the years following World War II, automobile exhaust and emissions from facto-

ries and oil refineries added to the quantity, complexity, and toxicity of air pollution in many cities. In 1952, a dense, polluted fog killed between 3500 and 4000 people in London. These incidents shocked people into realizing that air pollution can be harmful—even deadly.

In response to the growing problem, the United States enacted the Clean Air Act in 1963. This law has been amended and strengthened three times, in 1970, 1977, and 1990. As it stands today, the law addresses air pollution in two categories.

1. **Emission from individual sources** The Clean Air Act limits the quantities of air pollutants that may be emitted by any source. Thus, it directs automobile manufacturers to meet standards for auto emissions, and it regulates smokestack emissions from factories and power plants (Fig. 19–1). Executives of industries that violate the law are subject to a fine of up to $25,000 *per day* and/or imprisonment for up to 1 year.

2. **National Ambient Air Quality Standards** The Clean Air Act directs the Environmental Protection Agency (EPA) to monitor the purity of the air in general. Even if each factory, power plant, and automobile complies with the law, the EPA must set

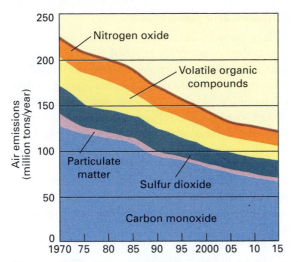

Figure 19–2 Emission of five air pollutants in the United States from 1970 to the present, with projections to the year 2015. The Clean Air Act was first enacted in 1963. (This graph does not show local concentrations in heavily congested areas such as Los Angeles.) *(Source: EPA documents and Chemical and Engineering News, May 12, 1997, page 26.)*

stricter controls if the total pollution level exceeds National Ambient Air Quality Standards.

As a result of the Clean Air Act, total emissions of air pollutants have decreased and air quality across the country has improved (Fig. 19–2). Donora-type incidents have not been repeated. Smog has decreased, and rain has become less acidic. Yet some people believe that we have not gone far enough and that air pollution regulations should be strengthened further.

19.2 Sources of Air Pollution

Sources and types of air pollution are listed in Figure 19–3.

Gases Released When Fuels Are Burned

Since the Industrial Revolution began in about 1750, coal and petroleum have been the major fuels in the developed world. Coal is largely carbon, which, when burned completely, produces carbon dioxide. Petroleum is a mixture of hydrocarbons, compounds composed of carbon and hydrogen. When hydrocarbons burn completely, they produce carbon dioxide and water. Neither is poisonous, but both are greenhouse gases. If fuels were composed purely of compounds of carbon and hydrogen, and if they always burned completely, air pollution from burning of fossil fuels would pose little direct threat to our health. However, fossil fuels contain impurities, and combustion is usually incomplete. As a result, other products form.

Products of incomplete combustion include hydrocarbons such as benzene and methane. Benzene is a carcinogen (a compound that causes cancer), and methane is another greenhouse gas. Incomplete combustion of fossil fuels releases many other pollutants including carbon monoxide, CO, which is colorless and odorless, yet very toxic.

Additional problems arise because coal and petroleum contain impurities that generate other kinds of pollution when they are burned. Small amounts of sulfur are present in coal and, to a lesser extent, in petroleum. When these fuels burn, the sulfur forms

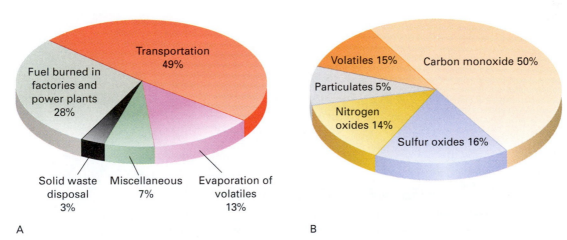

A   B

Figure 19–3 (A) Sources of air pollution in The United States. (B) Types of air pollutants in the United States. (Although carbon dioxide is a greenhouse gas, it is not listed as a pollutant because it is not toxic.) *(Source: EPA)*

oxides, mainly sulfur dioxide, SO_2, and sulfur trioxide, SO_3. High sulfur dioxide concentrations have been associated with major air pollution disasters of the type that occurred in Donora. Today the primary global source of sulfur dioxide pollution is coal-fired electric generators.

Nitrogen, like sulfur, is common in living tissue and therefore is found in all fossil fuels. This nitrogen, together with a small amount of atmospheric nitrogen, reacts when coal or petroleum is burned. The products are oxides of nitrogen, mostly nitrogen oxide, NO, and nitrogen dioxide, NO_2. Nitrogen dioxide is a reddish-brown gas with a strong odor. It therefore contributes to the "browning" and odor of some polluted urban atmospheres. Automobile exhaust is the primary source of nitrogen oxide pollution.

Volatiles

A volatile compound is one that evaporates readily and therefore easily escapes into the atmosphere. Whenever chemicals are manufactured or petroleum is refined, some volatile byproducts escape into the atmosphere. When metals are extracted from ores, gases such as sulfur dioxide are released. When pesticides are sprayed onto fields and orchards, some of the spray is carried off by wind. When you paint your house, the volatile parts of the paint evaporate into the air. As a result of all these processes, tens of thousands of different volatile compounds are present in polluted air: Some are harmless, others are poisonous, and many have not been studied.

Particulates and Aerosols

A particle or particulate is any small piece of solid matter, such as dust or soot. An **aerosol** is any small particle that is larger than a molecule and suspended in air. These three terms are used interchangeably to discuss air pollution. Molecules never settle out of still air, but aerosols do. The settling rate may be so slow, however, that even slight winds keep aerosols aloft (Fig. 19–4).

Many natural processes release aerosols. Wind-blown silt, pollen, volcanic ash, salt spray from the oceans, and smoke and soot from wildfires are all aerosols. Industrial emissions add to these natural sources.

Whenever you build a fire, smoke rises above the flames. Smoke and soot are carcinogenic aerosols consisting mostly of carbon. Coal always contains clay and other noncombustible minerals that accumulated when the coal formed in the muddy bottoms of ancient swamps. When the coal burns, some of these minerals escape from the chimney as **fly ash,** which settles as gritty dust. When metals are mined, the drilling, blasting, and digging raise dust, and this, too, adds to the total load of aerosols. Aerosols also form when evaporated or partially burned gasoline reacts with nitrogen oxides in air and condenses onto dust particles.

Secondary Air Pollutants

Gases and aerosols released during combustion and manufacturing are called **primary air pollutants.**

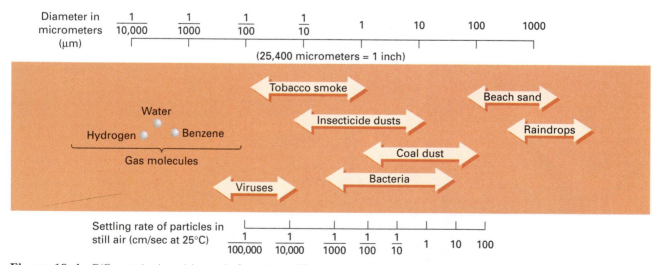

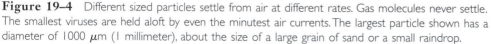

Figure 19–4 Different sized particles settle from air at different rates. Gas molecules never settle. The smallest viruses are held aloft by even the minutest air currents. The largest particle shown has a diameter of 1000 μm (1 millimeter), about the size of a large grain of sand or a small raindrop.

Secondary air pollutants are not released directly into the air by any industrial process, but are generated by reactions within the atmosphere. Two examples are smog and acid precipitation, discussed in sections 19.5 and 19.6.

19.3 The Meteorology of Air Pollution

To assess how a specific air pollutant affects your health, you must be concerned with the amount that you inhale and the effects of that pollutant on your body. The total quantity of sulfur dioxide in the Earth's atmosphere or the concentration in the smokestack of a coal-fired electric generator do not affect you directly because you do not breathe all the world's air, and you don't stick your head in the smokestack. Therefore, you must consider how the atmosphere transports and dilutes pollutants.

Exhaust from an industrial stack or an automobile tailpipe mixes with the surrounding air rapidly if the atmosphere is turbulent, or more slowly if the air is calm. Rapid mixing dilutes the pollutants so that the concentration you breathe is lower. When little mixing occurs, the pollutant concentration remains high near the source.

Under normal conditions, the Sun warms the Earth's surface, which in turn warms the lower atmosphere. The warm air rises to mix with the cooler air above it, creating turbulence and dispersing pollutants near the ground (Fig. 19–5). At night, however, the ground radiates heat and cools. Thus, the ground and the air close to the ground become cooler than the air at higher elevations (Fig. 19–6). This condition, called an **atmospheric inversion** or **temperature inversion,** is stable because the cool air is denser than warm air; therefore, the air near the surface does not rise and mix with the air above it. Pollutants may concentrate in the stagnant layer of cool air next to the ground which is commonly from ten to a few hundred meters thick.

Usually, the morning Sun warms air near the ground and breaks the inversion. However, under some conditions, inversions last for days. For example, a large mass of warm air may move into a region at high altitude and float over the colder air near the ground, keeping the air over a city stagnant. The episode at Donora resulted from such a situation. Inversions are also common along coastlines and large lakes, where the water cools surface air. In the Los

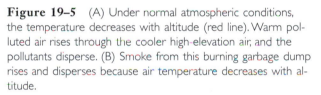

B

Figure 19–5 (A) Under normal atmospheric conditions, the temperature decreases with altitude (red line). Warm polluted air rises through the cooler high-elevation air, and the pollutants disperse. (B) Smoke from this burning garbage dump rises and disperses because air temperature decreases with altitude.

Angeles basin, cool maritime air is often trapped beneath a warm subtropical air mass. The cool air cannot move eastward because of the mountains, and it cannot rise because it is too dense. Pollutants from the city's automobiles and factories then concentrate until the stagnant air becomes unhealthy.

Two types of weather changes can break up an inversion. If the Sun heats the Earth's surface sufficiently, the cool air near the ground warms and rises, dispersing the pollutants. Alternatively, storm winds may dissipate an inversion layer.

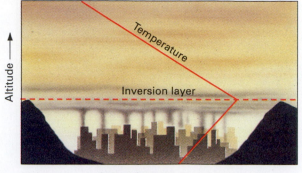

Figure 19–6 (A) During an inversion, cold, dense air is trapped near the ground by overlying warm air. As a result, temperature increases with altitude, then decreases at higher elevations (red line). Polluted air cannot rise above the layer of warm air and remains trapped near the ground. The effect is intensified if the city lies in a valley that confines air movement. (B) Smoke and soot rise from the Comanche Station coal-fired electric generating plant near Pueblo, Colorado. An atmospheric inversion concentrates the pollutants close to ground level, where they are visible as a haze in the background. *(Jim Wark/Peter Arnold)*

19.4 Air Pollution and Human Health

You step outside and notice a layer of brownish haze, the air smells foul, and a nasty cough has lingered for the past week. You wonder if there is a connection between your cough and that brown haze. As explained at the beginning of this chapter, when air pollutants concentrate sufficiently, people sicken and may even die. In addition, many studies show that even moderately polluted air is unhealthy (Fig. 19–7).

One way to test the health effects of a pollutant is through **epidemiology,** the study of disease distribution. In one type of epidemiological study, scientists separate a population into two groups: one that is exposed to the pollutant, and one that is not. A classic example of this type of analysis is the study of cigarette smokers. Because a clear distinction exists between people who smoke and those who do not, it is easy to examine the medical records and death certificates of large numbers of people in these two groups. The knowledge that cigarette smoking is harmful has come from studies of this kind.

In a classic epidemiological study, one quarter of the male employees working in a coal-tar dye factory in the United States between 1912 and 1962 contracted a type of bladder cancer that is rare in the general population. This observation led to the conclusion that coal-tar dyes contain carcinogenic compounds.

Other types of studies compare people who have been exposed to high concentrations of a pollutant with those exposed to lower concentrations. Epidemiologists compare entire populations in polluted areas with those in less polluted environments. Thirteen major chemical plants in the Kanawha Valley in West Virginia release a variety of toxic air pollutants. Between 1968 and 1977 the incidence of respiratory cancer in the region was more than 20 percent above the national average.

A third type of epidemiological study compares diseases in a specific region during periods of high and low air pollution contamination. In August 1986, a strike shut down the Geneva steel plant near Salt Lake City, Utah, for 13 months. When the mill shut down, the atmospheric aerosol concentration decreased and the number of children hospitalized for bronchitis, asthma, pneumonia, and pleurisy declined by 35 per-

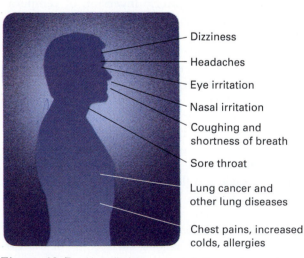

Figure 19–7 Air pollution causes irritation and many types of illnesses.

cent. The study demonstrated a clear epidemiological connection between aerosol air pollution in Salt Lake City and respiratory diseases in children.

If scientists suspect that polluted air is harmful to breath, they try to identify the most harmful pollutants, isolate the suspected toxins, and measure their toxicity in laboratory studies. Studies of this type can be complex.

19.5 Acid Rain

As explained in section 19.2, sulfur and nitrogen oxides are released when coal and petroleum burn, and when metal ores are refined. In moist air, sulfur dioxide reacts to produce sulfuric acid and nitrogen oxides react to form nitric and nitrous acid. All of these secondary air pollutants are strong acids. Atmospheric acids dissolve in water droplets and fall as **acid precipitation**, also called **acid rain** (Fig. 19–8).

Acidity is expressed on the **pH scale.** A solution with a pH of 7 is neutral, neither acidic nor basic. Numbers lower than 7 represent acidic solutions, and numbers higher than 7 represent basic ones. For ex-

ample, soapy water is basic and has a pH of about 10, whereas vinegar is an acid with a pH of 2.4.

Rain reacts with carbon dioxide in the atmosphere to produce a weak acid. As a result, natural rainfall has a pH of about 5.7. However, between 1940 and 1990 the average pH of rain in urban areas in the United States was between 4 and 5. A rainstorm in Baltimore in 1981 had a pH of 2.7, which is almost as acidic as vinegar, and a fog in southern California in 1986 reached a pH of 1.7, which approaches the acidity of hydrochloric acid solutions used to clean toilets.

Consequences of Acid Rain

Sulfur and nitrogen oxides impair lung function, aggravating diseases such as asthma and emphysema. They also affect the heart and liver and have been shown to increase vulnerability to viral infections such as influenza.

Acid rain corrodes metal and rock. Limestone and marble are especially susceptible because they dissolve rapidly in mild acid. Recent air pollution has weathered ancient carvings, doing more damage in the past 20 years than occurred by natural weathering in the previous millennium. In Greece and Italy, government officials have removed statues from their outdoor

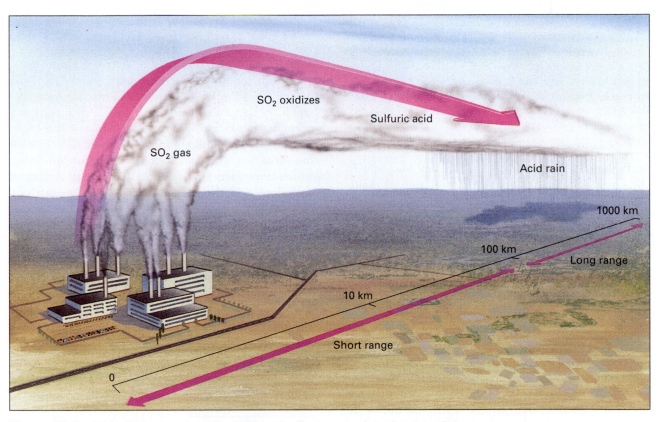

Figure 19–8 Acid rain develops from the addition of sulfur compounds to the atmosphere by industrial smokestacks.

Figure 19–10 Polluted air has caused about $10 billion in damage to German forests. Acid rain killed this forest in Toter Wald. *(Fritz Polking/Peter Arnold)*

Figure 19–9 Acid rain has destroyed this 1817 marble tombstone in England. *(Bruce F. Molnia/Terraphotographics).*
● Interactive Question: In the United States, many large coal-fired electric generating facilities are located in the desert in the Southwest. Argue for or against the statement: Because there is little rain in the desert, sulfur dioxides emitted by these generators are less likely to react to form acid precipitation. Therefore the air pollutants are less likely to cause harm.

settings and sealed them in glass cases in museums. In the United States the cost of deterioration of buildings and materials from acid precipitation is estimated at several billion dollars per year (Fig. 19–9).

Acid rain also affects plants. In the late 1970s, European foresters noticed that the needles and leaves of many trees were yellowing and that trees were dying. In the 1980s, tree mortality increased dramatically. In 1982, about 8 percent of the trees in West Germany were unhealthy. A year later, 34 percent of the trees were affected, and by 1995 more than half of the trees in Germany's western forests were sick or dying (Fig. 19–10). Experiments indicate that acid precipitation, combined with the effects of other air pollutants, killed the trees. Air pollution has caused the loss of about $10 billion worth of timber in Germany alone. But the loss goes beyond the value of the wood. Forests provide recreation for people and

habitat for many animals and plants; they regulate water and protect the soil. Forests also remove carbon dioxide from the atmosphere and release oxygen. Tree death is not as rampant in the United States, but the U.S. Forest Service has reported that pines in the Southeast grew 20 to 30 percent more slowly between 1972 and 1982 than they did between 1961 and 1971, when rain was less acidic.

Acid Rain and Particulates   In the United States, the Clean Air Act has dramatically reduced emission of sulfur and nitrogen oxides. However, some researchers argue that acid rain damage to forests and crops has not diminished as expected.

The Clean Air Act also mandates a reduction of industrial particulates. Many particulates are bases, which neutralize acid rain. As particulate emissions have decreased, their neutralizing effect has been lost. As a result, the reduction of acid emissions has not improved the environment as much as scientists had hoped.[1]

19.6 Smog and Ozone in the Troposphere

Imagine that your great-grandfather had entered the exciting new business of making moving pictures. Old-time photographic film was "slow" and required lots of sunlight, so he would hardly have moved to the in-

---

[1] Lars Hedin and Gene Likens, "Atmospheric Dust and Acid Rain." *Scientific American,* Dec. 1996, p. 88ff.

Figure 19–11 Smog frequently blankets Los Angeles and other large metropolitan areas. *(R. Sager)*

dustrial northeast. Southern California, with its warm, sunny climate and little need for coal, was preferable. Thus, a district of Los Angeles called Hollywood became the center of the movie industry. Its population boomed, and after World War II automobiles became about as numerous as people. Then the quality of the atmosphere began to deteriorate in a strange way. People noted four different kinds of changes:

(1) A brownish haze called **smog** settled over the city (Fig. 19–11); (2) people felt irritation in their eyes and throats; (3) vegetable crops became damaged; and (4) the sidewalls of rubber tires developed cracks.

In the 1950s, air pollution experts worked mostly in the industrialized cities of the East Coast and the Midwest. When they were called to diagnose the problem in southern California, they looked for the sources of air pollution they knew well, especially sulfur dioxide. But they could not find much sulfur dioxide in the air, and the smog was nothing like the pollution they were familiar with. Then, a California chemist, A.J. Haagen-Smit, showed that Los Angeles smog forms when automobile exhaust is exposed to sunlight.

Haagen-Smit piped automobile exhaust into a sealed room equipped with sunlamps (Fig. 19–12).

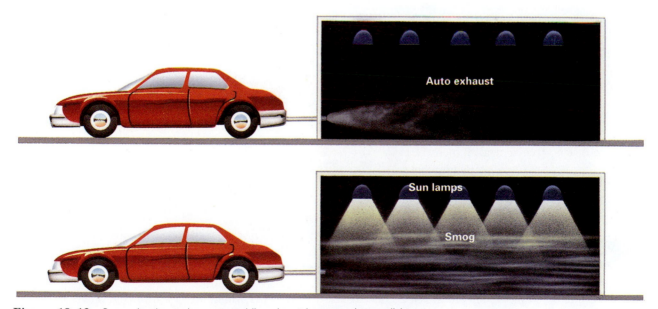

Figure 19–12 Smog develops when automobile exhaust is exposed to sunlight.

The room contained plants and pieces of rubber. It was fitted with mask-like windows that permitted people to breathe the air. With the automobile exhaust in the room and the sunlamps off, the room smelled like automobile exhaust, not like Los Angeles smog. But when Haagen-Smit turned the sunlamps on, smog developed. After a time, the air brought tears to the eyes, the plants showed typical smog damage, and the rubber cracked. Further research showed that incompletely burned gasoline in automobile exhaust reacts with nitrogen oxides and atmospheric oxygen in the presence of sunlight to form ozone, O_3, a gas molecule containing three oxygen atoms. The ozone then reacts further with automobile exhaust to form smog (Fig. 19–13).

We read about the harmful effects of excessive ozone in the air over cities such as Los Angeles. Later in this chapter we will read that ozone in the stratosphere absorbs ultraviolet radiation and protects life on Earth. Is ozone a pollutant to be eliminated, or a beneficial component of the atmosphere that we want to preserve? The answer is that it is both, depending on *where* it is found. Ozone in the troposphere reacts with automobile exhaust to produce smog and therefore it is a pollutant. Ozone in the stratosphere is beneficial and the destruction of the ozone layer creates serious problems.

Consequences of Smog

Ozone irritates the respiratory system, causing loss of lung function and aggravating asthma in susceptible individuals. A recent study showed that nonsmokers exposed to air with high ozone concentrations are twice as likely to develop asthma as nonsmokers who breathe air with low ozone concentrations. Ozone also increases susceptibility to heart disease and is a suspected carcinogen.

In 1995, the total value of crops harvested in the United States was about $80 billion. If ozone pollution were reduced by 25 percent, crop productivity would increase by $2.5 to $3.3 billion, or about 3 to 4 percent. Conversely, if ozone levels increased by 25 percent, total agricultural output would decrease by about $3 billion.

19.7 Air Pollution from Toxic Volatiles

Hundreds of thousands of different chemicals escape into the air during the manufacture of chemicals, waste incineration, oil refining, paint drying, and other activities. Government regulation bans any emissions that cause immediate and acute poisoning. However, long-term effects are open to considerable dispute. Consider the case of dioxin.

Very little dioxin is intentionally manufactured. It is not an ingredient in any herbicide, pesticide, or other industrial formulation. You cannot buy dioxin at your local hardware store or pharmacy. Even a research chemist would find it difficult to obtain dioxin from a chemical supply center. So why is this compound a major environmental problem?

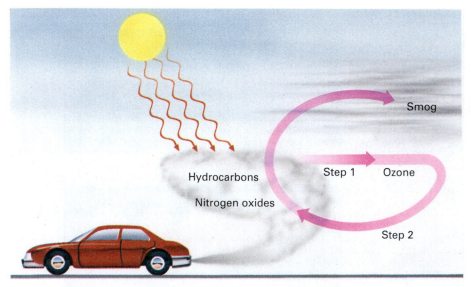

Figure 19–13 Smog forms in a sequential process. *Step 1*: Automobile exhaust reacts with air in the presence of sunlight to form ozone. *Step 2*: Ozone reacts with automobile exhaust to form smog.

Dioxin forms as an unwanted byproduct in the production of certain chemicals and when specific chemicals are burned. The dioxin controversy began during the Vietnam War when massive amounts of the herbicide Agent Orange were sprayed on the rainforests of southeast Asia. Agent Orange is a 50-50 mixture of two common herbicides: 2,4-D and 2,4,5-T. During the manufacture of these two compounds, some chemicals inadvertently combine to form dioxin. If the reacting mixture gets too hot, the mixture produces even more dioxin, just as your morning toast blackens if you leave it in the toaster too long.

Soldiers exposed to high concentrations of dioxin through contact with Agent Orange developed chloracne, a persistent, severe form of acne. Other populations have been exposed to dioxin as the result of herbicide applications and in accidents at chemical plants in Nitro, West Virginia, and in Seveso, Italy. Cancer and birth defects have been linked to these episodes.

In the United States today, garbage incineration is the most common source of dioxin. When a compound containing chlorine, such as the plastic polyvinyl chloride, is burned, some of the chlorine reacts with organic compounds to form dioxin. The dioxin then goes up the smokestack of the incinerator, diffuses into the air, and eventually falls to Earth. Cattle eat grass lightly dusted with dioxin, and store the dioxin in their fat. Humans ingest the compound mostly in meat and dairy products. The EPA estimates that the average U.S. citizen ingests about 100 picograms of dioxin in food every day. One hundred picograms is a tiny amount, 0.0000000001 gram. Dioxin accumulates in human fat, where it concentrates to about 40 ppt (parts per trillion) in the average American. That means that the average American has absorbed forty billionths of a gram of dioxin (0.000000040 gram) for every kilogram of body fat.

In a recent report, the EPA argued that dioxin is the most toxic known chemical and that even these low background levels are close to the levels that cause adverse effects in laboratory animals. These effects include cancer, disruption of regulatory hormones, reproductive and immune system disorders, and birth defects. The report concluded that "Dioxin risks are in a range that we believe is unacceptable." However, it is almost impossible to prove that low levels of a chemical cause a specific disease in humans. As a result, the Chemical Manufacturers Association issued a statement: "On the scale of things to worry about in life, environmental exposure to dioxins is pretty low. There is no direct evidence to show that any of the effects of dioxins occur in humans in everyday levels."

Dioxin is just one of tens of thousands of potentially toxic chemicals released intentionally or accidentally into the atmosphere. Some, such as dioxin or DDT, have been studied intensively. Many have never been studied.

19.8 Air Pollution from Aerosols

In 1988, EPA epidemiologists noted that whenever atmospheric aerosol levels rose above a critical level in Steubenville, Ohio, the number of fatalities—from all causes—rose the following day. Extrapolating this study nationwide, scientists estimated that 60,000 people might die every year from tiny airborne particulates.

In the summer of 1997, after several studies substantiated the Steubenville report, the EPA proposed additional reductions of the ambient aerosol levels in the United States. Opponents argued that it is unfair to target all aerosols because the term covers a wide range of substances from a benign grain of silt to a deadly mist of toxic volatiles. The argument continued that because the chemistry and toxicity of aerosols vary so dramatically, the EPA regulations must be more specific. Researchers are now studying the toxicity of different types of aerosols. The EPA and opponents of the new regulations have agreed to wait until the year 2005 to reassess the proposed regulations.

All aerosols, natural and industrial, reflect sunlight and lead to global cooling. Thus, while greenhouse gases warm the atmosphere, aerosols cool it. Scientists have conducted numerous studies to quantify these effects. The studies show that global warming from greenhouse gases overwhelms cooling from aerosols (Fig. 19–14).

19.9 Depletion of the Ozone Layer

Solar energy breaks oxygen molecules (O_2) apart in the stratosphere, releasing free oxygen atoms (O). The free oxygen atoms combine with oxygen molecules to form ozone (O_3). Ozone absorbs high-energy ultraviolet light. This absorption protects life on Earth because ultraviolet light causes skin cancer, inhibits plant growth, and otherwise harms living tissue.

In 1970, Paul Crutzen from the Max Planck Institute showed that naturally produced nitrous oxide drifts into the upper atmosphere where it breaks ozone molecules apart. Four years later, Sherwood Rowland from the University of California, Irvine, and Mario Molina from MIT showed that organic compounds containing chlorine and fluorine, called

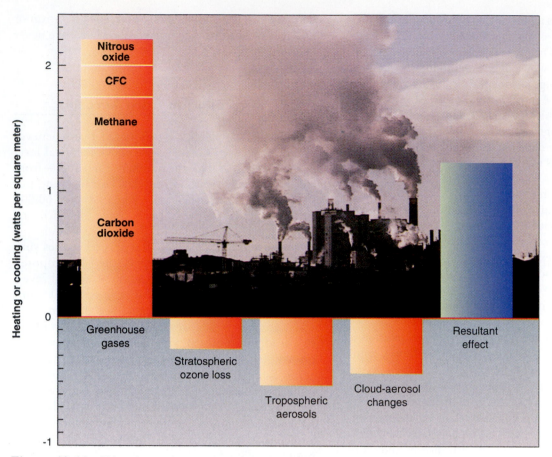

Figure 19–14 Although greenhouse gases warm the atmosphere, stratospheric ozone loss and tropospheric aerosols cool the atmosphere. The combined effect is net atmospheric warming. *(Redrawn from M.P. McCormick, L.W. Thomason, and C.R. Trepte, "Atmospheric Effects of the Mt. Pinatubo Eruption." Nature, 373, February 2, 1995, p. 399ff.)*

chlorofluorocarbons, or **CFCs**, also rise into the upper atmosphere and destroy ozone (Fig. 19–15).[2] Until recently CFCs were used as cooling agents in almost all refrigerators and air conditioners, as propellants in some aerosol cans, as cleaning solvents during the manufacture of weapons, and in plastic foam in coffee cups and some building insulation. Researchers then discovered that organic compounds containing bromine, called halons, also destroy ozone.

In 1985, scientists observed an unusually low ozone concentration over Antarctica, called the **ozone hole**. The ozone concentration over Antarctica continued to

decline between 1985 and 1993, until it was 65 percent below normal over 23 million square kilometers, an area almost the size of North America (Fig. 19–16). Research groups also reported significant increases in ultraviolet radiation from the Sun at ground level in the region. In addition, scientists recorded ozone depletion in the Northern Hemisphere. In March 1995, ozone concentration above the United States was 15 to 20 percent lower than during March 1979, before ozone depletion had started.

Data on global ozone depletion persuaded the industrial nations of the world to limit the use of CFCs and other ozone-destroying compounds. In a series of international agreements signed between 1978 and 1992, many nations of the world agreed to reduce or curtail production of compounds that destroy atmo-

---

[2] These compounds are also referred to as *Freons*, a DuPont trade name.

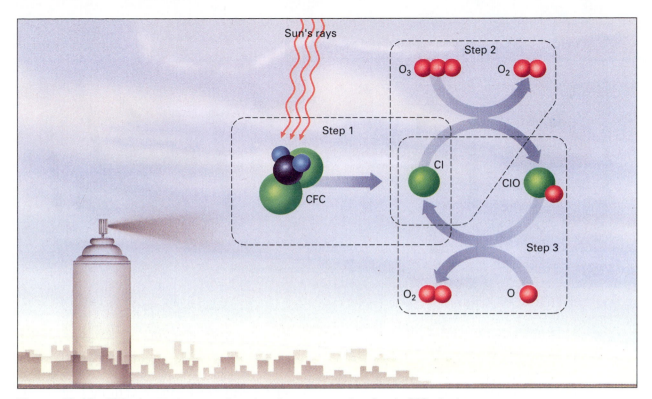

Figure 19–15 CFCs destroy the ozone layer in a three-step reaction. *Step 1:* CFCs rise into the stratosphere. Ultraviolet radiation breaks the CFC molecules apart, releasing chlorine atoms. *Step 2:* Chlorine atoms react with ozone, O_3, to destroy the ozone molecule and release oxygen, O_2. The extra oxygen atom combines with chlorine to produce ClO. *Step 3:* The ClO sheds its oxygen to produce another free chlorine atom. Thus, chlorine is not used up in the reaction, and one chlorine atom reacts over and over again to destroy many ozone molecules.

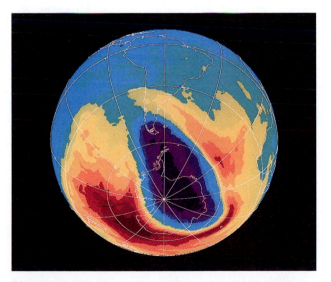

Figure 19–16 The ozone hole grew considerably from 1982 to 1994. The scale is in Dobson units, which measures stratospheric ozone concentration. Purple shows lowest ozone concentration (greatest depletion). *(NASA)*

spheric ozone. Most industrialized countries stopped production of CFCs on January 1, 1996.

The international bans have had positive results. The concentration of ozone-destroying chemicals peaked in the troposphere (lower atmosphere) in 1994 and has been declining ever since. As a result, fewer CFCs have been drifting into the stratosphere. The CFCs that are already in the stratosphere break down slowly but scientists predict that the concentration of ozone-destroying chemicals in the stratosphere should peak between 1997 and 1999, and then should begin to decline. In 1996 the Antarctic ozone hole approached record levels, and the ozone loss over the Northern Hemisphere and the Arctic increased. Despite these trends, calculations show that as CFCs slowly break down, the ozone layer will begin to heal, reaching preindustrial levels by about the year 2050.

In 1995 the three chemists who identified the mechanism of ozone loss—Crutzen, Rowland, and Molina—won the Nobel Prize in chemistry. According to the Royal Swedish Academy of Sciences, "the three

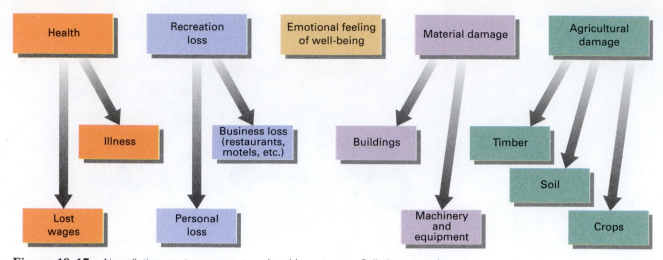

Figure 19–17 Air pollution creates many external and human costs. Pollution control costs money, but reduction of pollution saves money and improves the quality of life.

researchers contributed to our salvation from a global environmental problem that could have had catastrophic consequences."

19.10 Air Pollution: Looking Back—Looking Forward

Chemical companies initially lobbied against international treaties that called for a ban on CFC production. The chemical industry argued that $100 billion worth of equipment, including most refrigerators and air conditioners, relied on CFCs. The argument continued that because there were no viable alternatives, the industrial world could not afford to ban such important chemical compounds. In *Ozone Protection in the United States: Elements of Success*, editor Elizabeth Cook documents how scientific innovation, government policies, economic incentives, and corporate leadership combined to overcome the obstacles and work through the phase-out period with minimal economic disruption.[3]

Almost invariably, when a law is proposed to reduce air or water pollution, leaders of the affected industry complain that compliance will cost money, increase the

price of finished goods, and eventually lead to unemployment, inflation, or both. Compliance with pollution law is expensive, but after 40 years of environmental legislation, the U.S. economy is strong. Inflation and unemployment were generally low in the United States during most of the 1990s.

In 1997, the Environmental Protection Agency summarized the costs and benefits of air pollution control between 1970 and 1990. Over this 20-year period, compliance with air pollution regulations cost industry an estimated $523 billion dollars. However, the economic benefits of the laws saved the American people about $23 trillion. Thus every dollar spent for air pollution control paid back $44 in improved health and reduced damage to materials, forests, and crops (Fig. 19–17).

In view of this economic balance, why does air pollution persist? Consider the problems of ozone and aerosol pollution in the United States. The 1990 amendment to the Clean Air Act mandated tighter emissions restrictions on tropospheric ozone and aerosols. But it takes time to add pollution controls to factories and power plants. If new cars are built to emit fewer pollutants, air quality won't improve significantly until most of the older, more polluting cars wear out and are junked. As a result, the Clean Air Act gave states until 1994 to meet the new standards. However, by 1994, most large urban areas had not complied with the law (Fig. 19–18). In 1997 the EPA proposed even stricter regulations on tropospheric ozone and

---

[3] Elizabeth Cook, ed., *Ozone Protection in the United States: Elements of Success*, Washington, D.C.: World Resources Institute, 1996.

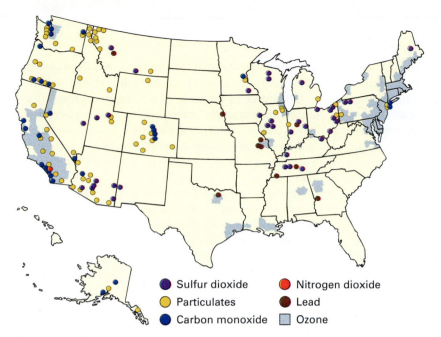

Figure 19–18 Many regions in the United States, particularly urban areas, do not comply with EPA air quality standards. This map shows areas that exceeded EPA standards for eight pollutants during September 1996. *(Scientific American, April, 1997, and EPA)*

aerosols. Many factories refused to comply and most private citizens made no adjustments to their lifestyles.

So what can the EPA do? Shut down power plants and plunge New York City into darkness? Jail otherwise law-abiding citizens in Los Angeles for commuting to work? Of course not. Instead they bring offending industries or states to court, threatening large fines. In recent court cases in 1995 and 1996, lawyers have argued that emissions from large power plants are in compliance with the Clean Air Act but that pollutants have blown in from surrounding areas. Furthermore, they argue that while compliance with

past laws has produced a positive economic return, we have reached a threshold at which additional air pollution control will no longer be economical (Fig. 19–19). The legal, political, and scientific issues are complex, and as this book is being written, ambient air standards remain under dispute.

Figure 19–19 The cost of reducing automobile emissions rises as more pollution is eliminated. A 25 percent reduction in emissions would cost about $2 billion, but a 30 percent reduction would cost $5 billion. *(Source: Alan J. Krupnick and Paul R. Portney, "Controlling Urban Air Pollution: A Benefit-Cost Assessment," Science 252, April 26, 1991: 52219–2).* ● **Interactive Question:** Using the information in this graph, do you feel that there is a limit to the amount of pollution control that the government should require? Where should that limit be: 25% reduction, 30% reduction, 95% reduction? Defend your answer.

SUMMARY

Air pollution prior to and just after Word War II convinced people to pass air pollution control legislation. In the United States, the Clean Air Act regulates emissions from individual sources and ambient air quality. The sources of air pollution include combustion gases released when fuels are burned, toxic volatiles, and aerosols.

Under normal meteorological conditions in the troposphere, air temperature drops steadily with altitude. However, if the upper air is warmer than the air beneath it, the atmosphere near the Earth's surface stagnates and pollutants are trapped under the warmer air. This condition is known as an **atmospheric inversion** or **temperature inversion**.

Laboratory and epidemiological studies show that air pollution can be harmful to human health. Nitrogen and sulfur oxides react in the atmosphere to produce **acid precipitation,** which damages health, weathers materials, and reduces growth of crops and forests. Incompletely burned fuels react with nitrogen oxides in the presence of sunlight and atmospheric oxygen to form ozone. Ozone then reacts further with automobile exhaust to form **smog.** Sunlight provides the energy to convert automobile exhaust to smog.

Dioxin is an example of a compound that is produced inadvertently during chemical manufacture and when certain materials are burned. Some scientists argue that even tiny amounts of dioxin and other toxic volatiles may be harmful to human health, but others disagree.

Scientific studies show that industrial aerosols are harmful to health, but aerosols are so varied that it is difficult to know which ones are most harmful.

Chlorofluorocarbons, compounds containing bromine, and nitrogen oxides in the stratosphere deplete the ozone that filters out harmful UV radiation and protects the Earth. Ozone-destroying chemicals have been regulated by international treaty and their concentration in the troposphere is diminishing.

After 40 years of air pollution control legislation, ambient air pollution levels in the United States have decreased significantly. However, many urban regions still are not in compliance with the law and air pollution remains a problem.

KEY TERMS

aerosol 446
fly ash 446
primary air pollutant 446
secondary air pollutant 447

atmospheric inversion (temperature inversion) 447
epidemiology 448

acid precipitation (acid rain) 449
pH scale 449
smog 451

chlorofluorocarbon (CFC) 454
ozone hole 454

FOR REVIEW

1. Discuss the air pollution disaster in Donora. What pollutants were involved? Where did they come from? How did they become concentrated?

2. Discuss the two major ways that the Clean Air Act addresses air pollution.

3. Briefly list the major sources of air pollution.

4. What is an atmospheric inversion and how does it form?

5. Describe several experiments used to determine the health effects of air pollution.

6. What is acid rain, how does it form, and how does it affect people, crops, and materials?

7. Explain why the effects of acid rain have not diminished dramatically even as acid emissions have diminished.

8. What is smog, how does it form, and how does it differ from automobile exhaust?

9. Describe the experiments that led to our understanding of smog formation.

10. Discuss two pathways through which dioxin escapes into the environment.

11. How much dioxin exists in the body fat of an average American? How harmful is that concentration?

12. Explain how CFCs deplete the ozone layer.

13. Why is ozone in the troposphere harmful to humans while ozone in the stratosphere is beneficial?

14. Discuss the effects of the international ban on ozone-destroying chemicals.

1. State which of the following processes are potential sources of gaseous air pollutants, particulate air pollutants, both, or neither: (a) Gravel is screened to separate sand, small stones, and large stones into different piles. (b) A factory stores drums of liquid chemicals outdoors. Some of the drums are not tightly closed, and others have rusted and are leaking. The exposed liquids evaporate. (c) A waterfall drives a turbine, which makes electricity. (d) Automobile bodies in an assembly plant are sprayed with paint. The automobile bodies then move through an oven that dries the paint. (e) A garbage dump catches fire.

2. Argue for or against this statement: "Because tall chimneys do not collect or destroy pollutants, all they do is protect the nearby areas at the expense of more distant places."

3. Describe the meteorological conditions most conducive to the rapid dispersal of pollutants. Describe those that are least conducive.

4. Imagine that you are performing a study to determine whether a disease is due to natural causes or air pollution. Which of the following experiments would you rely on? Defend your choices. (a) Compare the health records of a population with that of previous years, when population and industrial activity were less. (b) Compare the effects during weekdays, when industrial activity and air pollution emissions are higher, with those on weekends, when it is low. (c) Compare effects during different seasons of the year. (d) Compare effects just before and just after the switch from daylight saving time to see whether there is a sharp 1-hour shift in the data. (e) Compare effects in areas where population and air pollution differ.

5. A report of air pollutant concentrations shows high concentrations of sulfur dioxide, particulate matter, and nitrogen dioxide at the top of a 200-meter stack of a power plant. If you were the health officer or mayor, what action, if any, would you recommend? Would you require any additional information? If so, describe the data you would request.

6. How can sulfur in coal contribute to the acidity of rainwater? What happens in a furnace when the coal is burned? What happens in the outdoor atmosphere?

7. Gasoline vapor plus ultraviolet lamps do not produce the same smog symptoms as do automobile exhaust plus ultraviolet lamps. What is missing from gasoline vapor that helps to produce smog?

8. Imagine that someone planned to build a municipal garbage incinerator in your neighborhood. The facility would burn domestic trash and use the energy to generate electricity. Moreover, stringent air pollution controls would keep toxic emissions to very low levels. Discuss the environmental benefits and drawbacks of this proposed incinerator.

Chapter 20

Water Resources

Clean water is one of our most precious natural resources. Havasu Creek, Grand Canyon.

More than two centuries ago, when fresh water seemed as inexhaustible as the buffalo that roamed the high plains, Benjamin Franklin warned, "When the well's dry, we know the worth of water." Today, as we reach, and in some cases exceed, the limits of our water resources, the well *is* running dry and we are discovering the real value and cost of water.

Water resources can be diminished in two ways: depletion and pollution. We discuss water depletion in Sections 20.1 through 20.4, and water pollution in the last part of this chapter. ●

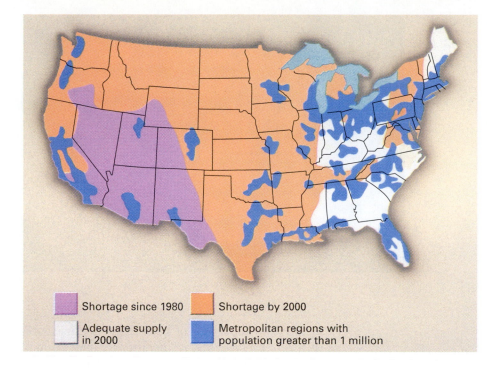

Figure 20–1 Water shortages have affected the American southwest since 1980, and will affect much of the remaining United States, including many metropolitan areas, by the year 2000. *(Compiled from USGS and U.S. Water Resources Council data.)*

Legend:
- Shortage since 1980
- Adequate supply in 2000
- Shortage by 2000
- Metropolitan regions with population greater than 1 million

20.1 Water Supply and Demand in the United States

The problem with water supplies is not that the amount of available water has decreased. The hydrologic cycle continuously replenishes fresh water on land, so that the amount of available fresh water is about the same as it was two centuries ago. The problem is that the demand for water has risen dramatically until it has approached or even exceeded supply in many parts of the world. Two hundred years ago, fewer than 4 million people inhabited the United States; at the beginning of 1998, we numbered approximately 270 million. At the same time, global population has increased by a factor of five. Over the past two centuries, humans have increased their production of food and industrial goods to accommodate the growing population. However, the supply of water has been constant. Thus, as the human population has grown, the amount of available fresh water per person has diminished. In addition, technological advances have made it possible to exploit our water resources at rates that would have been inconceivable to Ben Franklin (Fig. 20–1).

Demands on water resources fall into three categories: domestic, industrial (including cooling water for electric power generation), and agricultural (Fig. 20–2). Combining all three categories, the United States uses 8000 liters per person per day—three times more than the average European country and hundreds

of times more than many developing and less-developed nations.

Water use can be separated into two categories. Any process that uses water, and then *returns it to the Earth locally* is called **withdrawal.** Most of the water used by industry and homes returns to streams or groundwater reservoirs near the place from which it was taken. For example, river water pumped through an electric generating station to cool the exhaust is returned almost immediately to the river. Water used to flush a toilet in a city is pumped to a sewage treatment plant, purified, and discharged into a nearby stream.

In contrast, a process that uses water, and *then returns it to the Earth far from its source* is called **consumption.** Most irrigation water evaporates, disperses with the wind, and returns to the Earth as precipitation

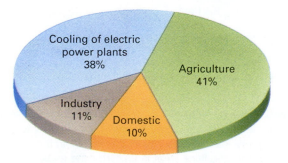

Figure 20–2 Industrial, agricultural, and domestic water use in the United States. *(Compiled from Worldwatch Institute data.)*

Figure 20–3 Most irrigation water evaporates and is carried away by wind.

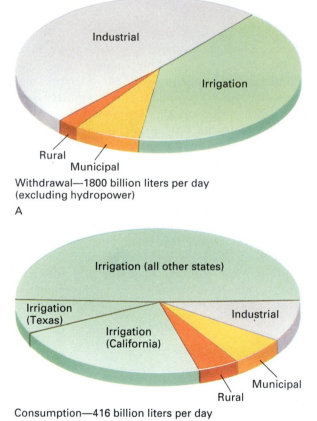

Industrial

Irrigation

Rural

Municipal

Withdrawal—1800 billion liters per day (excluding hydropower)

A

Irrigation (all other states)

Irrigation (Texas)

Irrigation (California)

Industrial

Municipal

Rural

Consumption—416 billion liters per day

B

Figure 20–4 (A) Although industry accounts for about half of all water *withdrawn* in the United States, (B) agriculture accounts for more than three fourths of the water that is *consumed*. Two dry agricultural areas, in California and Texas, consume about one fourth of all water consumed in the United States.

hundreds or thousands of kilometers from its source (Fig. 20–3). Thus, although industry accounts for most of the water withdrawn in the United States, agriculture accounts for most of the water that is consumed and not returned to its place of origin (Fig. 20–4). Globally, about two thirds of the total water that is used is consumed.

Domestic Water Use

Domestic use—for cooking, washing, toilet flushing, and lawn irrigation—accounts for only 10 percent of the water used in the United States (Fig. 20–5). All domestic water uses combined account for an average of 800 liters per day for each person in the United States. In contrast, more than half of the Earth's people get by on less than 100 liters per day. Many of these people must carry their water long distances from public wells or streams to their homes and use water only for drinking and cooking.

Industrial Water Use

Cooling systems of fossil fuel and nuclear electric power generating plants account for 38 percent of all water used in the United States. Although much of this water is returned to the stream from which it was taken, it is considerably warmer, which, as we shall see, affects aquatic ecosystems. All other industrial needs add an additional 11 percent to the national water demand. The quantities of water required to produce some common industrial products are shown in Figure 20–6.

Agricultural Water Use

Agriculture accounts for 41 percent of water use in the United States. Figure 20–7 shows that different agri-

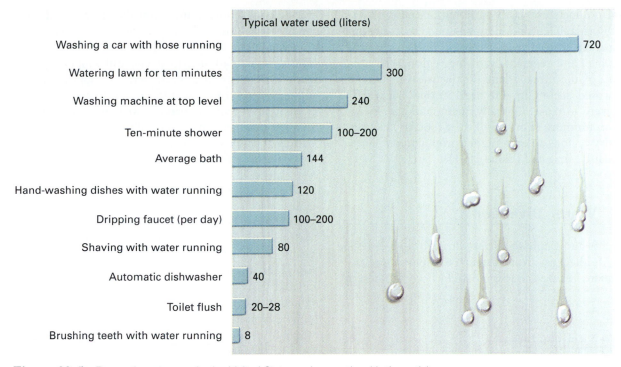

Figure 20–5 Domestic water use in the United States varies greatly with the activity. *(Compiled from American Water Works Association data.)* ● **Interactive Question:** Using the data displayed in this figure, list specific steps that you could take to conserve water this week. Order your list from most effective (greatest water savings) to least effective steps.

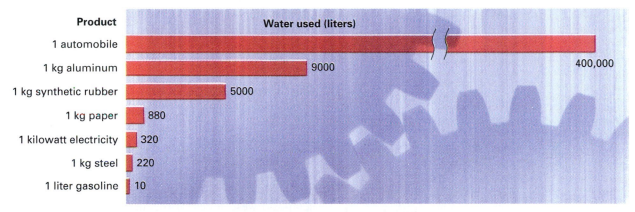

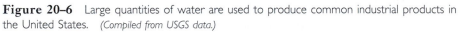

Figure 20–6 Large quantities of water are used to produce common industrial products in the United States. *(Compiled from USGS data.)*

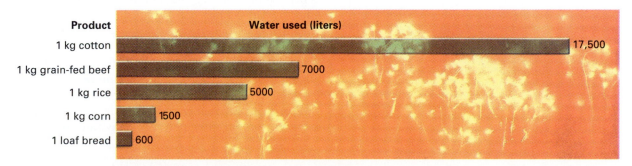

Figure 20–7 Different agricultural products require vastly different amounts of irrigation water in the United States. *(Compiled from USGS data.)*

cultural products use vastly different amounts of water. California's Central Valley was once a desert, but irrigation has transformed it into an immensely productive fruit and vegetable growing region. The productivity of the Great Plains in the western United States would decline by one third to one half if irrigation were to cease. Globally, many nations rely on irrigation for more than half of their food production, and in dry regions an even higher proportion of farmland must be irrigated (Table 20–1).

20.2 Water Diversion

The Tigris and Euphrates rivers flow through deserts of the Middle East from distant mountains. Ancient Babylonians realized that they could farm the desert

TABLE 20–1

Net Irrigated Area, Top 20 Countries and World

Country	Net Irrigated Area* (thousand hectares)	Share of Cropland That Is Irrigated (percent)
China	45,349	47
India	43,039	25
Soviet Union	21,064	9
United States	20,162	11
Pakistan	16,220	78
Indonesia	7,550	36
Iran	5,750	39
Mexico	5,150	21
Thailand	4,230	19
Romania	3,450	33
Spain	3,360	17
Italy	3,100	26
Japan	2,868	62
Bangladesh	2,738	29
Brazil	2,700	3
Afghanistan	2,660	33
Egypt	2,585	100
Iraq	2,550	47
Turkey	2,220	8
Sudan	1,890	15
Other	36,664	7
World	235,299	16

*Area actually irrigated; does not take into account double cropping.
Sources: Intergovernmental Panel on Climate Change, *Policymakers' Summary of the Potential Impacts of Climate Change: Report from Working Group II to IPCC* (Geneva: World Meteorological Organization/U.N. Environment Programme, 1990; Paul E. Waggoner, "U.S. Water Resources Versus an Announced But Uncertain Climate Change," *Science*, March 1, 1991.

if they could move water from the rivers to adjacent land. Archaeologists have uncovered extensive irrigation systems in Babylon dating from about 2000 B.C. The aqueducts of ancient Rome still supply water to some European cities. However, the rapidly improving technology of the past five or six decades has made it possible to move unprecedented amounts of water. Diversion projects have radically changed the distribution of water in many parts of the world.

Water diversion projects extract surface water from a stream or lake and transport it, often long distances, to dry regions where it is needed. In some parts of our country, surface water is scarce but huge reservoirs of ground water lie only a few tens of meters beneath the surface. This ground water can be pumped to the surface for human use. In both cases, water is **diverted** from its natural place and path in the hydrologic cycle and transported to a new place and path to serve human needs.

The United States receives about three times more water from precipitation than it uses. However, precipitation is unevenly distributed geographically over the United States. Some of the driest regions are those that use the greatest amounts of water (Fig. 20–8). In the United States, some of the most productive agricultural regions are located in the deserts of California, Texas, and Arizona where crops are entirely supported by diversion of surface and ground water. Desert cities such as Phoenix, Los Angeles, Albuquerque, and Tucson support large populations that use hundreds of times more water than is locally available and renewable.

Dams and Surface Water Diversion

A diversion system is a pipe, ditch, or canal constructed to transport water. Many use gravity to move the water, while others use pumps to move water uphill. Many diversion systems are augmented by dams that store water. Dams are especially useful in regions of low or seasonal rainfall. For example, in the arid and semi-arid west, much of the annual precipitation occurs in winter and spring. Dams store this water for the summer irrigation season, when crops need the water (Fig. 20–9). In addition, the potential energy of the water in a reservoir can generate electricity. About 5 percent of the energy used in the world today is supplied by dams. Thus, dams are beneficial, but they can create undesirable effects.

Loss of Water   The reservoir formed by a dam provides more surface area for evaporation and more bottom area for seepage into bedrock than did the stream that preceded it. For example, about 270,000 cubic meters of water per year evaporate from Lake

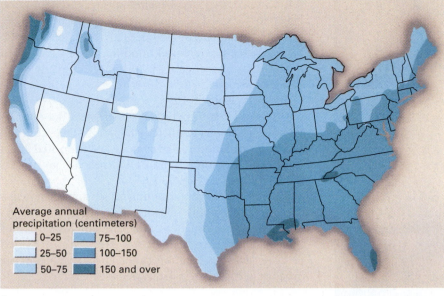

Figure 20–8 (A) Average annual precipitation varies greatly throughout the United States. (B) Average annual water use also varies greatly. But many regions that use the greatest amounts of water receive little rain.

Average annual precipitation (centimeters)

- 0–25
- 25–50
- 50–75
- 75–100
- 100–150
- 150 and over

A

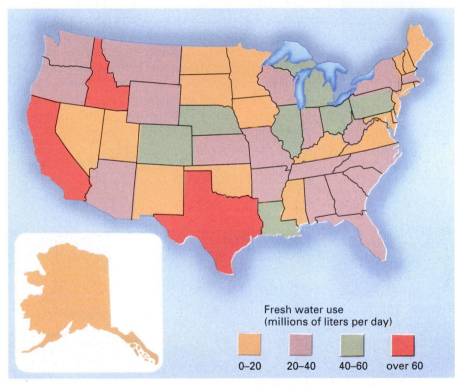

Fresh water use (millions of liters per day)

- 0–20
- 20–40
- 40–60
- over 60

B

Powell, above the Glen Canyon Dam on the Colorado River. Thus, less total water flows to downstream parts of a river. In addition, many canals built to carry water from a reservoir to farmland are simply ditches excavated in soil or bedrock, and they leak profusely.

Salinization   As discussed in Chapter 10, all rivers contain small concentrations of dissolved salts. If desert or semi-desert soils are irrigated for long periods of time, salt accumulates and lowers the soil fertility. In the United States, salinization lowers crop

Figure 20–9 The Glen Canyon Dam forms Lake Powell on the Colorado River, just upstream from Grand Canyon.

yields on 25 to 30 percent of irrigated farmland, more than 5 million hectares (Fig. 20–10). Globally, the problem affects 10 percent of irrigated land, some 25 million hectares, and it is increasing at a rate of 1 to 1.5 million hectares per year.

Silting   A stream deposits its sediment in a reservoir, where the current slows. Rates of sediment accumulation in reservoirs vary with the sediment load of the dammed river. Lake Mead, behind Hoover Dam in Arizona and Nevada, lost 6 percent of its capacity in its first 35 years as a result of sediment accumulation. At a rate such as this, a reservoir lasts for hundreds of years. Although 100 years may be a long time in political or economic terms, it is a brief moment of geological time and human civilization. In a few instances, engineers have made expensive miscalculations of sedimentation rates. The Tarbela Dam in Pakistan took nine years to build, and the reservoir is expected to fill with sediment in 20 years. The reservoir behind the Sanmenxia Dam on the Yellow River in China filled four years after the dam was finished, making both the dam and reservoir useless.

Erosion   A stream erodes the beaches that line its banks during normal streamflow, but it deposits silt and sand to build the beaches back up during a flood. Recently, the beaches along the Colorado River in Grand Canyon were disappearing rapidly because management practices at the Glen Canyon Dam upstream prevented flooding in the Canyon. In a similar way, beaches are vanishing from many other dammed streams.

In the spring of 1996, scientists released enough water from Glen Canyon dam to create a flood in the Grand Canyon. As predicted, the flood waters churned up sediment from the river bed and deposited it on the banks, thereby rebuilding old beaches and creating new ones (Fig. 20–11).

Because a reservoir accumulates silt and sand, it also interrupts the supply of sediment to the sea coast. As a result, coastal beaches erode, but their sand is not replenished by natural processes. This problem has led to expensive rejuvenation projects on some popular beaches in the eastern United States.

Risk of Disaster   A dam can break, creating a disaster in the downstream flood plain. One of the greatest floods since the last ice age roared down Idaho's Teton River canyon when the new Teton dam broke on June 3, 1976. Flood waters swept away 154 of the 155 buildings in the town of Wilford six miles below the dam; the town no longer exists. The flood inundated the lower half of Rexburg, a larger town farther downstream, damaging or destroying 4000 homes and 350 businesses (Fig. 20–12). Damages totaled about $2 billion, and the raging waters stripped the topsoil from tens of thousands of hectares of farmland. Only 11 people died because civil authorities warned and evac-

Figure 20–10 White salts cover a vast expanse of Nevada desert. Salinization has reduced the productivity of millions of hectares of agricultural land globally.

A

B

Figure 20–11 Photos of the same Grand Canyon beach taken before and after the Spring 1996 flood show that the flood more than doubled the size of the beach. *(Courtesy of Mark Manone, Northern Arizona University Sandbar Studies)*

A

B

C

Figure 20–12 (A) The Teton Dam was brand new in the spring of 1976. (B) Rapidly rising water broke the dam on June 3, 1976, sending lake water pouring through the breach. (C) Flood waters from the failed Teton Dam inundated the city of Rexburg, Idaho. *(U.S. Bureau of Reclamation.)*

uated most of the people. But if the flood had occurred in the middle of the night when warnings and evacuations would have been more difficult, thousands might have been killed.

Other dam failures have killed many more people. In 1889, a broken dam flooded Johnstown, Pennsylvania, killing 2200 people; the St. Francis Dam in Saugus, California, collapsed in 1928 and drowned 450 people; in 1963 a rockslide created a huge wave in the lake behind a dam in Vaiont, Italy. The dam withstood the force of the wave, but the wave washed over the top of the dam creating a flood that killed 1800 people.

Recreational and Aesthetic Losses Dams are often built across narrow canyons to minimize engineering and construction costs. But when the canyons are flooded, unique scenery and ecosystems are destroyed. Glen Canyon Dam on the Colorado River created Lake Powell by flooding one of the most spectacular desert canyons in the American west. Today, Glen Canyon is submerged for more than 300 kilometers above the dam and can only be seen in old photos. The flooding of canyons, however, provides lakes that can be used for fishing and other water sports.

Disputes often arise among various groups that use water in dammed reservoirs. To prevent flooding, a reservoir should be nearly empty by spring so that it can store water and fill during spring runoff. Thus, a reservoir should be drawn down slowly during summer and fall. Agricultural users also prefer to have water stored during spring runoff, and supplied for irrigation during summer months. However, recreational users object to a lowering of reservoir levels during summer, when they visit reservoirs most frequently. Managers of hydroelectric dams prefer to run water through their turbines during times of peak demands for electricity, which rarely correspond to flood management, irrigation, or recreational schedules.

Ecological Disruptions   A river is an integral part of the ecosystem that it flows through, and when the river is altered, the ecosystem changes. Dams interrupt water flow during portions of the year, prevent flooding, change the temperature of the downstream water, and often create unnatural daily fluctuations. All of these changes alter relative abundances of aquatic species. In addition, dams may create specific ecological problems in individual rivers.

For example, before Egypt's Aswan Dam was built, the Nile River flooded every spring, depositing nutrient-rich sediment over the flood plain and delta. The delta was in balance between erosion by the sea and addition of new silt from the yearly floods. When the dam and reservoir cut off the silt supply, erosion then dominated, and the Nile delta is now shrinking. In addition, the loss of an annual supply of nutrients eliminated a source of free fertilizer for flood plain farms. Although the energy produced by the dam is used to manufacture commercial fertilizers, many of the poorer farmers on the flood plain cannot afford to purchase what they once received free from the river.

Salmon populations of the Pacific Northwest have decreased from 100 million fish in 1850 to 15 million today. In 1996, the number of coho salmon running up Pacific Northwest streams to breed was only 1 to 3 percent of their historic numbers. The Snake River sockeye and chinook have been declared endangered, and Snake River coho became extinct in 1985. Although overfishing and destruction of breeding habitats by logging, road building, and construction contributed to the decline, wildlife biologists agree that the 18 giant dams on the Columbia River are the primary cause. The dams impede both upstream and downstream migration of the fish to and from their spawning grounds

in the hundreds of small, gravel-bed tributaries of the Columbia. Young fish swimming toward the ocean slow their migration when they enter the lakes above dams, and are vulnerable to predators. In addition, the dam turbines stress young fish passing through them so they become easy prey for predators lurking below the dams.

Ground-Water Diversion

Ground water provides drinking water for more than half of the population of North America and is a major source of water for irrigation and industry. It is a valuable resource because:

1. It is abundant. As mentioned previously, 60 times more fresh water exists underground than in streams and lakes.

2. Because ground water moves so slowly, it is stored below the Earth's surface and remains available during dry periods.

3. In some regions, ground water flows from wet environments to arid ones, making water available in dry areas.

If ground water is pumped to the surface faster than it can flow through the aquifer to the well, a **cone of depression** forms near the well (Fig. 20–13). When the pump is turned off, ground water flows back toward the well in a matter of days or weeks if the aquifer has good permeability, and the cone of depression disappears. On the other hand, the water table drops if water is continuously removed more rapidly than it can flow to the well through the aquifer.

Before the development of advanced drilling and pumping technologies, human impact on ground water was minimal. Today, however, deep wells and high-speed pumps can extract ground water more rapidly than the hydrologic cycle recharges it. Where such excessive pumping is practiced, the water table falls as the ground-water reservoir becomes depleted. In some cases the aquifer is no longer able to supply enough water to support the farms or cities that have overexploited it. This situation is common in the central, western, and southwestern United States (Fig. 20–14).

Most of the high plains in western and midwestern North America receive scant rainfall. Early settlers prospered in rainy years and suffered during drought. In the 1930s, two events combined to change agriculture in this region. One was a great drought that destroyed crops and exposed the unprotected, plowed soil to erosion. Dry winds blew across the land, eroding the parched soil and carrying it for hundreds and even thousands of kilometers. Thousands of families lost their farms, and the region was dubbed the Dust Bowl. The second event was the arrival of inexpensive

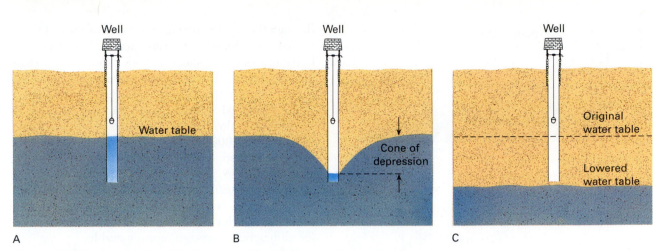

Figure 20–13 (A) A well is drilled into an aquifer. (B) A cone of depression forms because a pump draws water faster than the aquifer can recharge the well. (C) If the pump continues to extract water at the same rate, the water table falls.

technology. Electric lines were built to service rural regions, and relatively cheap pumps and irrigation systems were developed. With the specter of drought fresh in people's memories and the tools available to avert future calamities, the age of modern irrigation began.

The Ogallala aquifer extends almost 900 kilometers from the Rocky Mountains eastward across the prairie, and from Texas to South Dakota (Fig. 20–15). It consists of water-saturated porous sandstone and conglomerate within 350 meters of the surface. The aquifer averages about 65 meters in thickness and

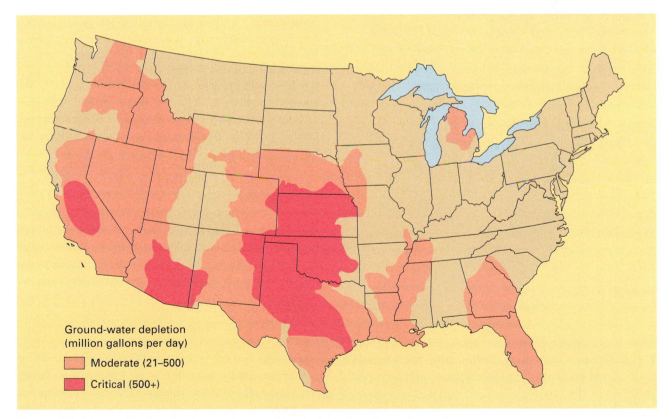

Ground-water depletion
(million gallons per day)

■ Moderate (21–500)

■ Critical (500+)

Figure 20–14 Aquifer depletion is a common problem in the United States. The numbers refer to the amount of water used in excess of the recharge rate.

Figure 20–15 The Ogallala aquifer supplies water to much of the High Plains. (A) A cross-sectional view of the aquifer shows that much of its water originates in the Rocky Mountains and flows slowly as ground water beneath the High Plains. (B) A map showing the extent of the aquifer.

holds a vast amount of water. Between 1930 and 1980, about 170,000 wells were drilled into the Ogallala aquifer, and extensive irrigation systems were installed throughout the region.

Hydrologists estimate that half of the water has already been removed from parts of the Ogallala aquifer, and pumping rates are increasing. Water moves

through the aquifer at an average rate of about 15 meters per year. Most of the water accumulated when the last Pleistocene ice sheet melted, about 15,000 years ago. But because the high plains receive little rain, the aquifer is now mostly recharged by rain and snowmelt in the Rocky Mountains, hundreds of kilometers to the west. At a flow rate of 15 meters per year, ground water takes 60,000 years to travel from the mountains to the eastern edge of the aquifer. At the present time, approximately 9.6 billion liters are returned to the aquifer every year through rainfall and ground-water flow. However, farmers remove 80 billion liters annually. Under such conditions, the deep ground water is, for all practical purposes, nonrenewable.

If the present pattern of water use continues, wells in the Ogallala aquifer will dry up early in the next century. The aquifer irrigates about 5 million hectares of farms and ranches (an area about the size of the states of Massachusetts, Vermont, and Connecticut combined). About 40 percent of the cattle in the United States are fed with corn and sorghum raised in this region, and large quantities of grain and cotton are grown there as well. If the aquifer is depleted and another source of irrigation water is not found, productivity in the central High Plains is expected to decline by 80 percent. Farmers will go bankrupt, and food prices will rise throughout the nation.

Subsidence   Excessive removal of ground water can cause **subsidence,** the sinking or settling of the Earth's surface. When water is withdrawn from an aquifer, rock or soil particles may shift to fill the space left by the lost water. As a result, the volume of the aquifer decreases and the overlying ground subsides.

Subsidence rates can reach 5 to 10 centimeters per year, depending on the rate of pumping and the nature of the aquifer. Subsidence affects large parts of the United States (Fig. 20–16). Some areas in the San Joaquin Valley of California have sunk nearly 10 meters (Fig. 20–17). The land surface has subsided by as much as 3 meters in the Houston-Galveston area of Texas. The problem is particularly severe in cities. For example, Mexico City is built on an old marsh. Over the years, as the weight of buildings and roadways has increased and much of the ground water has been removed, parts of the city have settled as much as 8.5 meters. Many millions of dollars have been spent to maintain the city on its unstable base. Similar problems are affecting Phoenix, Arizona, and other U.S. cities.

Unfortunately, subsidence is irreversible. When rock and soil contract, their porosity is permanently reduced so that ground-water reserves cannot be completely recharged even if water becomes abundant again.

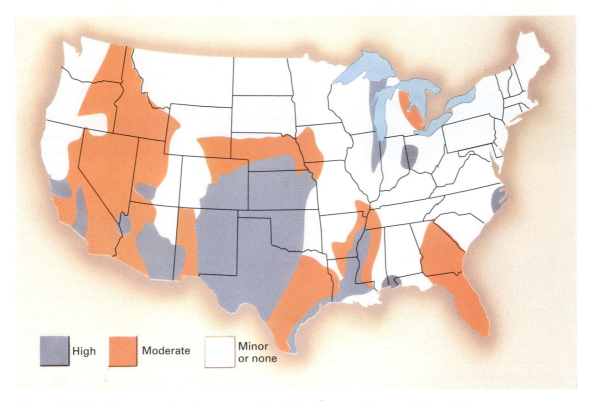

High   Moderate   Minor or none

Figure 20–16 Subsidence affects large parts of the United States. *(Data compiled from USGS and U.S. Water Resources Council.)*

Figure 20–17 The land surface at this point in San Joaquin, California, was at the height of the 1925 marker in the year 1925. Subsidence resulting from ground-water extraction for irrigation lowered the surface by about 10 meters in 52 years. *(U.S. Geological Survey)*

Saltwater Intrusion   Two types of ground water occur in coastal areas: fresh ground water and salty ground water that seeps in from the sea. Fresh water floats on top of salty water because it is less dense. If too much fresh water is removed from the aquifer, salty ground water rises to the level of wells (Fig. 20–18). It is unfit for drinking, irrigation, or industrial use. **Saltwater intrusion** has affected much of south Florida's coastal ground-water reservoirs.

20.3   The Great American Desert

A desert is any region that receives less than 25 centimeters of rain annually, and land with 50 centimeters or less of rain annually is hostile to most kinds of farming. In the years following the Civil War, an American geologist, John Wesley Powell, explored the area between the western mountains (the Sierra Nevada–Cascade crest) and the 100° meridian (the line of longitude running through the Dakotas, Nebraska, Kansas, and Abilene, Texas). He recognized that most of this region is arid or semi-arid, and he called it the Great American Desert.

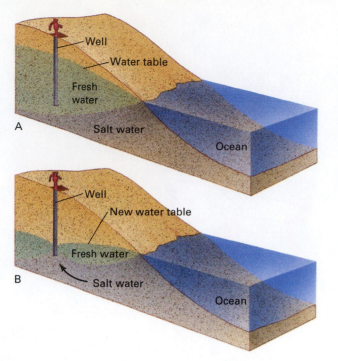

Figure 20–18 Saltwater intrusion can pollute coastal aquifers. (A) Fresh water lies above salt water, and the water in the well is fit to drink. (B) If too much fresh water is removed, the water table falls. The level of salt water rises and contaminates the well.

Despite Powell's warning about inadequate water, Americans flocked to settle the Great American Desert. As mentioned earlier, several parts of this region now use huge amounts of water for cities, industry, and agriculture, although they receive little precipitation. For example, arid and semi-arid California, Nevada, Arizona, Texas, and Idaho use the greatest quantities of water in the United States. Some desert cities, including Phoenix, El Paso, Reno, and Las Vegas, rely on surface-water diversion and/or overpumping of ground water for nearly all of their water (Fig. 20–19).

Billions of dollars have been spent on heroic projects to divert rivers and pump ground water to serve this area of the country. For example, 1200 dams and the two largest irrigation projects in the world have been built in California alone.

CASE STUDY
The Colorado River

The Colorado River runs through the southwestern portion of the Great American Desert. Starting from the snowy mountains of Colorado, Wyoming, Utah, and New Mexico, it flows across the arid Colorado

Figure 20–19 Desert cities, such as Las Vegas, use more water than is locally available in surface and ground-water reservoirs. Consequently they must import much of their water through expensive, publicly funded water diversion systems. *(D.C. Lowe/Tony Stone Images)*

Plateau and southward into Mexico, where it empties into the Gulf of California (Fig. 20–20).

Because the river flows through a desert, farmers, ranchers, cities, and industrial users along the entire length of the river compete for rights to use the water. In the 1920s, the Colorado discharged 18 billion cubic meters of water per year into the Gulf of California. In 1922 the U.S government apportioned 9 billion cubic meters for the Upper Basin (Colorado, Utah, Wyoming, and New Mexico) and the remaining 9 billion cubic meters for Lower Basin users in Arizona, California, and Nevada. No water was set aside for maintenance of the river ecosystem or for Mexican users south of the border. Twenty years later, an international treaty awarded Mexico 1.8 billion cubic meters to be taken equally from Upper and Lower Basin users.

The Colorado and many of its tributaries flow across sedimentary rocks, many of which contain soluble salt deposits. As a result, the Colorado is naturally a salty river. In addition, the United States government built ten large dams on the Colorado, and evaporation from the reservoirs has further concentrated the salty water. In 1961, the water flowing into Mexico contained 27,000 ppm salts, compared with an average salinity of about 100 ppm for large rivers. For comparison, 27,000 ppm is 77 percent as salty as seawater. Mexican farmers used this water for irrigation, and their crops died. In 1973, the United States government built a desalinization plant to reduce the salinity of Mexico's share of the water.

Measurements of the average annual discharge of 18 billion cubic meters were made during a time of abundant rainfall. Drought from 1930 to 1968 reduced

Figure 20–20 (A) The Colorado River drains much of the American Southwest. (B) Water is abundant where the river flows through the Grand Canyon. (C) Use of river water for irrigation often dries up the river completely near the Mexican border. *(Dan Lamont/Matrix.)*

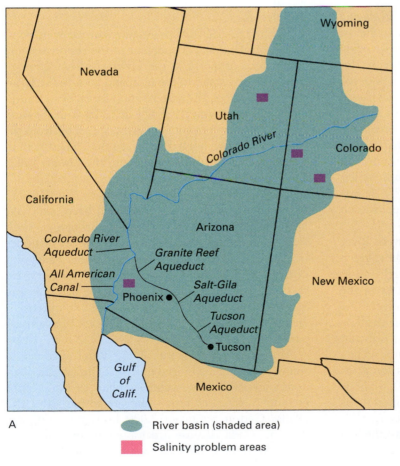

the Colorado's average discharge to 15.6 billion cubic meters, less than the amount of water that had been allocated. As a result, the river completely dried up several times.

Current demands on Colorado River water from farmers and ranchers, cities, recreationists, industry, and hydroelectric dams exceed the total discharge of the river. The population of the American southwest continues to increase, and so do demands for more water in the region. Further compromises will become necessary to share the limited resource. •

CASE STUDY
The Los Angeles Water Project

In the mid-1800s Los Angeles was a tiny, neglected settlement on the California coast. Too far from the California gold fields to attract the miners or their money, it sat in its arid coastal basin with neither a seaport nor a railroad. The average annual precipitation of the Los Angeles area is 65 centimeters, most of which falls during a few winter weeks. The Los Angeles River frequently flooded in winter and diminished to a trickle in summer.

Among the early settlers were Mormons, who had become experts at irrigating dry farmland from their experience in the Utah desert. By the late 1800s, irrigated farms in the Los Angeles basin were producing a wealth of fruits and vegetables. It seemed that almost anything grew bigger and better with the combination of sun, warmth, rich soil, and irrigation. Suddenly, Los Angeles was an attractive, growing town. In 1848 the town had a population of 1600. The population passed 100,000 by 1900 and 200,000 by 1904. Then the city ran out of water. Wells dried up and the river was inadequate to meet demand. To supply human needs, the city government prohibited lawn watering and shut off irrigation wells in the nearby San Fernando Valley. The town was surrounded on three sides by deserts and on the fourth by the Pacific Ocean. There was no nearby source of fresh water.

However, 250 miles to the north, the Owens River flowed out of the Sierra Nevada and watered the green Owens Valley. The city of Los Angeles bought water rights, bit by bit, from farmers and ranchers in the Owens Valley and at the headwaters of the river. The city then spent millions of dollars building the Los Angeles Aqueduct across some of the most difficult, earthquake-prone terrain in North America. When finished, the aqueduct was 357 kilometers long, including 85 kilometers of tunnel (Fig. 20–21). Siphons and pumps carried the water over hills and mountains too treacherous for tunneling. On November 5, 1913, the first water poured from the aqueduct into the San Fernando Valley (Fig. 20–22).

The following 10 to 15 years were unusually rainy in the Los Angeles area and in the Owens Valley. The rain recharged the ground water under the Los Angeles basin, and the Los Angeles River flowed freely again. As a result, little or none of the Owens River water coming through the aqueduct was needed by the city. Instead, most of the water was diverted to farms in the San Fernando Valley. Irrigated farmland in that valley increased from 1200 hectares in 1913, when the aqueduct opened, to more than 30,000 hectares in 1918. In addition, some of the excess water now available to Los Angeles was used to water lawns again. Almost every house had a green lawn, and Santa Monica Boulevard changed from a dusty rut to a palm-lined oasis, creating the impression that Los Angeles wasn't really so dry after all.

In the 1920s, normal dryness and drought returned. Los Angeles and the farms of the San Fernando Valley demanded more water from the Owens River, and the Owens Valley dried up. The mood of the ranchers and farmers in the Owens Valley grew steadily worse, and in 1924 they threatened the aqueduct with sabotage. William Mulholland, superintendent of the Los Angeles Department of Water and Power, remarked that it was too bad that so many of the orchards in the Owens Valley had died of thirst because now there weren't enough trees to hang all the Owens Valley troublemakers.

By the mid-1930s, Los Angeles owned 95 percent of the farmland and 85 percent of the residential and commercial property in the Owens Valley. Although the city leased some of the farmland back to the farmers and ranchers, the water supply became so unpredictable that agriculture slowly dried up with the land. As Los Angeles continued to grow, the water from the Owens River was not sufficient, so the Department of Water and Power drilled wells into the Owens Valley and began pumping its ground-water aquifer dry.

Today, the Owens Valley is a desert (Fig. 20–23). Its remaining citizens pump gas, sell beer, and make up motel beds for tourists driving through on their way to somewhere else. Ironically, the water that once irrigated farms and ranches in the Owens Valley was diverted through the Los Angeles Aqueduct, at great cost to taxpayers, to irrigate farms in the naturally arid San Fernando Valley. The aqueduct converted a farmer's paradise to a desert and a desert to a farmer's paradise.

More than 80 percent of the water diverted to southern California is used to irrigate desert and near-desert cropland. The water supplied to these farms commonly irrigates crops such as cotton, rice, and alfalfa. Cotton requires 17,000 liters per kilogram; rice and alfalfa also use great amounts of water. The U.S. government pays large subsidies to farmers in

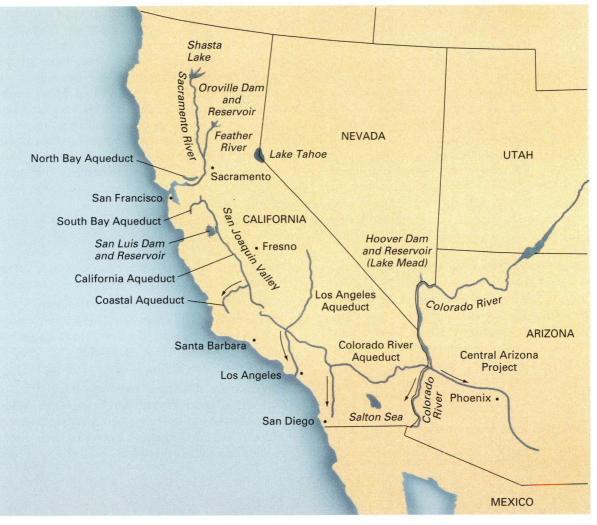

Figure 20–21 The Los Angeles Aqueduct and related water diversion systems carry water to cities and farms in California and Arizona.

Figure 20–22 The Los Angeles aqueduct transports water across the Owens Valley desert to the San Fernando Valley and Los Angeles. *(Charles O'Rear/Westlight.)*

Figure 20–23 Without the Owens River, the Owens Valley is a desert.

naturally wet southeastern and south central states *not* to grow those same crops. The irrigation water used to support cattle ranching and other livestock in California is enough to supply the needs of the entire human population of the state.

In the process of supplying southern California agriculture with water, water diversion projects have dammed and drained streams, lakes, and wetlands; eliminated fish and wildlife habitat, including hundreds of kilometers of salmon spawning beds; and flooded hundreds of kilometers of recreational rivers. The farms and ranches have contaminated waterways with pesticides and fertilizers and have increased the salinity of downstream rivers. However, these losses are balanced by economic gains. Agriculture contributes 2.5 percent to California's total economy.

Controversy over water diversion projects is not limited to the American west. New York City obtains much of its water from reservoirs in upstate New York, and the New York City Board of Water Supply controls those reservoirs and watersheds hundreds of miles from the city. Opposing interests of local land use groups and city water users have provoked heated conflicts in the state of New York. ●

20.4 Water and International Politics

According to Population Action International, 28 nations with populations of 338 million people experienced water shortages in 1990. According to the World Bank, 80 nations with 40 percent of the Earth's population had enough water in 1995 but water availability will limit future economic expansion. By the year 2025, 52 nations with 3.5 billion people may experience chronic water shortages.

Forty percent of the Earth's people use water from rivers that flow through two or more nations. The United States and Mexico share the Colorado River; India and Bangladesh share the Ganges; India and Pakistan share the Indus; Czechoslovakia and Hungary share the Danube; the Jordan River flows through Israel, Jordan, Syria, and Lebanon. In Africa, 57 different lakes and river basins are shared by two or more countries. Frequently, political problems arise when nations must share limited water resources. The problem becomes severe when the nations are unfriendly toward one another for other reasons. Jordan's King Hussein said in 1990 that water was the only issue that could take him to war with Israel. Egyptian president Anwar Sadat stated, after signing the peace accord with Israel, that "the only issue that could take Egypt to war again is water."

Avoidance of conflict over inadequate water resources requires agreements (a) to share the available water and (b) to share the consequences of shortages. However, current international water law offers little help in resolving these issues. For example, the Colorado River originates in the mountains of the southwestern United States but flows into Mexico. So does the water belong to the United States or to Mexico? In general, upstream nations have been unwilling to agree to the principle that common water resources should be cooperatively governed, managed, and shared. Some nations maintain that they have absolute control over the fate of water within their borders and have no responsibility to downstream neighbors.

Any international code of water use must be based on three fundamental principles:

1. Water users in one country must not cause major harm to water users in other countries downstream.

2. Water users in one country must inform neighbors of actions that may affect them before the actions are taken. (For example, if a dam is built, engineers must inform downstream users that they plan to stop or reduce river flow to fill a reservoir.)

3. People must distribute water equitably from a shared river basin. Unfortunately, these principles, especially the last one, are open to such subjective and self-interested interpretations that their intent is easily and often subverted.

20.5 Water Pollution

Pollution is the reduction of the quality of a resource by the introduction of impurities (Fig. 20–24). **Point source pollution** arises from a specific site such as a septic tank, a gasoline spill, or a factory. In contrast, **nonpoint source pollution** is generated over a broad area. Fertilizers and pesticides spread over fields fall into this latter category. Eight categories of pollutants degrade surface and ground waters.

1. Sewage is wastewater from toilets and other household drains. It includes biodegradable organic material such as human and food wastes, soaps, and detergents. Sewage also includes some industrial chemicals because people flush paints, solvents, pesticides, and other chemicals down the toilet. Industrial wastes may end up in the sewer systems because of illegal dumping.

2. Disease organisms, such as typhoid and cholera, are carried into waterways in the sewage of infected people.

Figure 20–24 Pollution is the reduction of the quality of a resource by the introduction of impurities.

3. Plant nutrients, such as phosphates and nitrates, flow into surface and ground waters mainly from nonpoint pollution sources such as farms. Phosphate detergents and phosphates and nitrates from feedlots also fall into this category.

4. Many industrial organic compounds are similar enough to natural materials that decay organisms consume them; thus they are biodegradable. Many others are so foreign to natural food chains that they are not decomposed by environmental chemicals or consumed by decay organisms. Nonbiodegradable pollutants, such as dioxin and DDT, are especially troublesome because they persist in the environment. Many are toxic or carcinogenic. According to the Environmental Protection Agency (EPA), factories in the United States released 33 million kilograms of toxic chemicals into waterways in 1994.

5. Toxic inorganic compounds include mine wastes, road salt, and industrial metals, such as cadmium, arsenic, mercury, and lead.

6. When sediment enters surface waters, the soil particles neither fertilize nor poison the aquatic system. However, the sediment muddies streams and buries aquatic habitats and thus degrades the quality of an ecosystem.

7. Radioactive materials include wastes from the mining of radioactive ores, nuclear power plants, nuclear weapons, and medical and scientific applications.

8. Heat can also pollute water. In a fossil fuel or nuclear electric generator, an energy source (coal, oil, gas, or nuclear fuel) boils water to form steam. The steam runs a generator to produce electricity. The exhaust steam must then be cooled to maintain efficient operation. The cheapest cooling agent is often river or ocean water. A 1000-megawatt power plant heats 10 million liters of water by 35°C every hour. The warm water can kill fish directly, affect their reproductive cycles, and change the aquatic environment to eliminate some native species and favor the introduction of new species.

20.6 How Sewage, Detergents, and Fertilizers Pollute Waterways

Recall from Chapter 10 that an oligotrophic stream or lake contains few nutrients, and clear, nearly pure water (Fig. 20–25A). Sewage, detergents, and agricultural fertilizers are plant nutrients. If humans dump one or more of these nutrients into an oligotrophic waterway, aquatic organisms flourish. These organisms consume most of the dissolved oxygen and the distribution of species changes. Thus an increased supply of nutrients transforms an oligotrophic waterway into a eutrophic waterway (Fig. 20–25B).

As a result, nutrients transform a clear, sparkling stream or lake that supports trout and salmon to an algae-choked waterway with carp and water worms. Note that nutrients do not poison aquatic systems; they nourish them. However, this nourishment alters species distribution and transforms the aquatic ecosystem into one that is less pleasant for humans.

A

B

C

Figure 20–25 (A) An oligotrophic waterway contains few nutrients and clear, pure water. (B) A eutrophic waterway is rich in nutrients, low in dissolved oxygen, and choked with plant life. (C) This stream flowing through La Paz, Bolivia, is an open sewer. Decay organisms have consumed all of the oxygen in the water, and anaerobic bacteria have taken over.

If humans release an excessive amount of sewage, detergent, or fertilizer into a waterway, organisms multiply so rapidly that they consume all the oxygen. As a result, most aquatic life dies. Then, **anaerobic** bacteria (bacteria that live without oxygen) take over, releasing noxious hydrogen sulfide gas as they feast on the organic matter (Fig. 20–25C). Even carp and worms cannot survive in such an environment. As they die and rot, the hydrogen sulfide that bubbles to the surface smells like rotten eggs.

20.7 Toxic Pollutants, Risk Assessment, and Cost-Benefit Analysis

At many mines and industrial sites, toxic chemicals have leaked into streams and ground water. Hydrogeologists (Earth scientists who deal with ground water and related aspects of surface water) measure how fast these compounds spread, and they ask, "Do the chemicals threaten drinking water resources and human health?" "How can they be contained or removed, and how much will it cost?" An evaluation of this type must consider both the concentration of the contaminant and its toxicity. Some contaminants, such as plutonium and dioxin, are poisonous or carcinogenic in extremely low concentrations, whereas others become harmful only at higher levels.

Scientists are uncertain about the concentrations at which many compounds become harmful to human health. The uncertainty results, in part, from delayed effects of some chemicals. For example, if a contaminant increases the risk of a certain type of cancer, the cancer may not develop until 10 to 20 years after exposure. Because scientists can't perform direct toxicity experiments on humans, and because they can't wait decades to observe results, they frequently attempt to assess the carcinogenic properties of a contaminant by

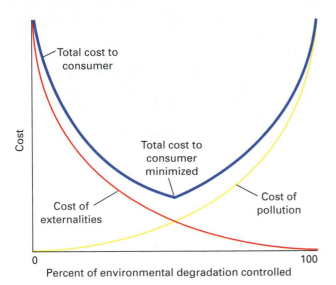

Figure 20–26 Cost of pollution control. Pollution control becomes more expensive as pollution is reduced (yellow curve). Externalities (costs of living in a polluted environment) become less expensive as pollution is reduced (red curve). The blue curve is the sum of the other two curves and represents the total costs to the consumer.

feeding it to laboratory rats. If rats are fed very high doses, sometimes hundreds of thousands of times more concentrated than environmental pollutants, then they may contract cancer in weeks or months rather than in years or decades. Suppose that a rat gets cancer after drinking the equivalent of 100,000 glasses of polluted well water each day for a few weeks. Can we say that the same contaminant will cause cancer in humans who drink 10 glasses a day for 20 years? No one knows. It may or may not be legitimate to extrapolate from high doses to low doses and from one species to another.

Scientists also use epidemiological studies to assess the risk of a pollutant. For example, if the drinking water in a city is contaminated with a pesticide and a high proportion of people in the city develop an otherwise rare disease, then the scientists may infer that the pesticide caused the disease, and that its presence in drinking water constitutes a high level of risk to human health.

Because neither laboratory nor epidemiological studies can prove that low doses of a pollutant are harmful to humans, scientists are faced with a dilemma: "Should we spend money to clean up the pollutant?" Some argue that such expenditure is unnecessary until we can prove that the contaminant is harmful. Others invoke the precautionary principle, "It's better to be safe than sorry." Proponents of the latter approach argue that people commonly act on the basis of incomplete proof. For example, if a mechanic told you that your brakes were likely to fail within the next 1000 miles, you would recognize this as an opinion, not a fact. Yet would you wait for the brakes to fail or replace them now?

Pollution control is expensive. However, pollution is also expensive. If a contaminant causes people to sicken, the cost to society can be measured in terms of medical bills and loss of income resulting from missed work. Many contaminants damage structures, crops, and livestock. People in polluted areas also bear expense because tourism diminishes and land values are reduced when people no longer want to visit or live in a contaminated area. All of these costs are called **externalities.**

Cost-benefit analysis compares the cost of pollution control with the cost of externalities. Figure 20–26 shows that the cost of externalities is very high when pollution levels are high and uncontrolled. As more controls are used to produce a cleaner environment, the cost of externalities decreases, but the cost of the controls rises. At a point in the middle of the blue curve, the total cost to the consumer, combining costs of externalities and pollution controls, reaches a minimum. Some people suggest that we should minimize the total cost even though this approach accepts significant pollution. Others argue that cost-benefit analysis is flawed because it ignores both the quality of life and the value of human life. How, they ask, can you place a dollar value on loss of recreation if people can no longer swim and fish in our waterways? Or even more poignant, how can you measure the economic value of good health or of a life cut short by cancer? If noneconomic costs of pollution are considered, then an increased level of pollution control becomes desirable despite its higher dollar cost.

20.8 Ground-Water Pollution

CASE STUDY
Love Canal

Love Canal in Niagara Falls, New York, was excavated to provide water to an industrial park that was never built. In the 1940s, the Hooker Chemical Company purchased part of the old canal as a site to dispose of toxic wastes created by its manufacturing processes. The engineers at Hooker considered the relatively impermeable ground surrounding the canal to be a reasonably safe disposal area. During the following years, the company disposed of approximately 19,000 tons of chemical wastes by loading them into 55-gallon steel drums and dumping the drums in the canal. In 1953, the company covered one of the sites with dirt and sold the land to the Board of Education of Niagara Falls for $1, after warning the city of the buried toxic wastes. The city then built a school and playground on the site.

In the process of installing underground water and sewer systems to serve the school and growing neighborhood, the relatively impermeable soil surrounding the old canal and waste burial site was breached. Gravel used to backfill the water and sewer trenches then provided a permeable path connecting the toxic waste dump to surrounding parts of the city.

During the following decades, the buried drums rusted through and the chemical wastes seeped into the ground water. In the spring of 1977, heavy rains raised the water table to the surface, and the area around Love Canal became a muddy swamp. But it was no ordinary swamp; the leaking drums had contaminated the ground water with toxic and carcinogenic compounds. The poisonous fluids soaked the playground, seeped into basements of nearby homes, and saturated gardens and lawns. Children who attended the school and adults who lived nearby developed epilepsy, liver malfunctions, skin sores, rectal bleeding, and severe headaches. In the years that followed, an abnormal number of pregnant women suffered miscarriages, and large numbers of babies were born with birth defects. The Love Canal tragedy was not simply a case of careless disposal of toxic wastes, but rather, the result of a series of poor disposal practices, flawed hydrogeology, and unwise decisions by both the Hooker Chemical Company and local officials.

The Love Canal incident is not unique. In December 1979, the U.S. Congress passed the Comprehensive Environmental Response, Compensation, and Liability Act (CERCLA), commonly known as the **Superfund.** This law provides an emergency fund to clean up chemical hazards and imposes fines for maintaining a dump site that pollutes the environment. By October 1993, the EPA had identified 37,537 hazardous waste sites for review, and the EPA anticipated adding 1200 new sites to the list each year. By 1995, the EPA had labeled about 1300 of those sites as the nation's worst hazardous dumps, but only about 200 of those highest priority sites had been cleaned up. It has taken an average of 11 years for the EPA to complete a cleanup project, from start to finish. The average cost of the completed projects is $27 million per site. ●

Aquifer Contamination

Water in a sponge saturates tiny pores and passages. To clean a contaminant from a sponge thoroughly, it would be necessary to clean every pore of the sponge that came into contact with contaminated water. Because this is nearly impossible to achieve completely, no one would wash dishes with a sponge that had been used to clean a toilet.

Many different types of sources contaminate ground water (Fig. 20–27). Because these contaminants permeate an aquifer in a manner analogous to

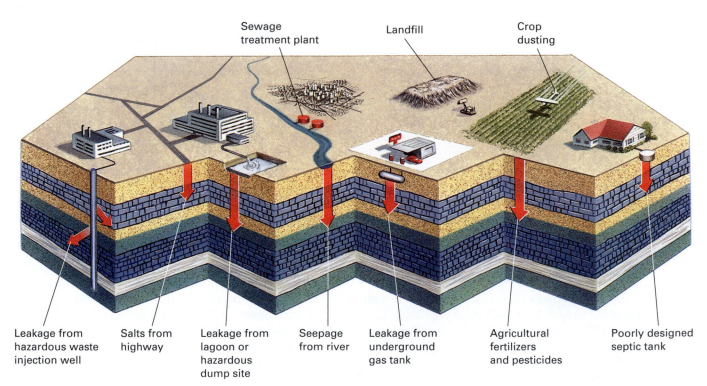

Sewage treatment plant

Landfill

Crop dusting

Leakage from hazardous waste injection well

Salts from highway

Leakage from lagoon or hazardous dump site

Seepage from river

Leakage from underground gas tank

Agricultural fertilizers and pesticides

Poorly designed septic tank

Figure 20–27 Many different sources contaminate ground water.

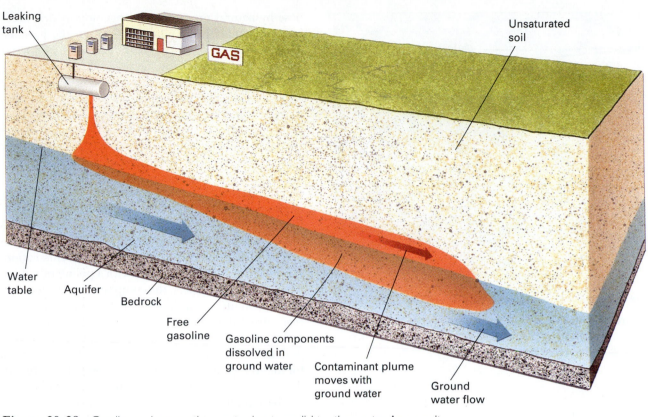

Figure 20–28 Gasoline and many other contaminants are lighter than water. As a result, they float and spread on top of the water table. Soluble components may dissolve and migrate with ground water. • **Interactive Question:** Sketch a similar figure showing dispersion of a trichlorethylene plume in this same environment.

the way in which contaminated water saturates a sponge, the removal of a contaminant from an aquifer is extremely difficult and expensive. Once a pollutant enters an aquifer, the natural flow of ground water disperses it as a growing **plume** of contamination. Because ground water flows slowly, usually at a few centimeters per day, the plume also spreads slowly (Fig. 20–28).

Many contaminants persist in a polluted aquifer for much longer than in a stream or lake. The rapid flow of water through streams and lakes replenishes their water quickly, but ground water flushes much more slowly. In addition, oxygen, which decomposes many contaminants, is less abundant in ground water than in surface water.

As a result, despite 20 years of government concern and action, 45 percent of municipal ground-water supplies in the United States are contaminated with organic chemicals. Wells in 38 states contain pesticide levels high enough to threaten human health. Every major aquifer in New Jersey is contaminated. In Florida, where 92 percent of the population drinks

ground water, more than 1000 wells have been closed because of contamination, and over 90 percent of the remaining wells have detectable levels of industrial or agricultural chemicals.

Treating a Contaminated Aquifer

The treatment, or **remediation,** of a contaminated aquifer commonly occurs in a series of steps.

Elimination of the Source The first step in treating an aquifer is to eliminate the pollution source so additional contaminants do not escape. If an underground tank is leaking, the remaining liquid in the tank can be pumped out and the tank dug from the ground. If a factory is discharging toxic chemicals into an unlined dump, courts may issue an injunction ordering the factory to stop using the dump (Fig. 20–29).

Elimination of the source prevents additional pollutants from entering the ground water, but it does not solve the problem posed by the pollutants that have already escaped. For example, if a buried gasoline tank has leaked slowly for years, many thousands of gallons

Figure 20–29 An oil refinery in New Jersey. New Jersey suffers from some of the worst ground-water pollution in North America as a result of heavy industry.

of gas may have contaminated the underlying aquifer. Once the tank has been dug up and the source eliminated, people must deal with the gasoline in the aquifer.

Monitoring When aquifer contamination is discovered, a hydrogeologist monitors the contaminants to determine how far, in what direction, and how rapidly the plume is moving and whether the contaminant is becoming diluted. The hydrogeologist may take samples from domestic wells. If too few wells surround a pollution source, the hydrogeologist may drill wells to monitor the plume. He or she analyzes the water samples for the contaminant and can thereby monitor the movement of the plume through the aquifer.

Modeling After measuring the rate at which the contaminant plume is spreading, the hydrogeologist develops a computer model to predict future dispersion of the contaminant through the aquifer. The model considers the permeability of the aquifer, directions of ground-water flow, and mixing rates of ground water to predict dilution effects.

Remediation Several processes are currently used to clean up a contaminated aquifer. Contaminated ground water can be contained by building an underground barrier to isolate it from other parts of the aquifer. If the trapped contaminant does not decom-

pose by natural processes, hydrogeologists drill wells into the contaminant plume and pump the polluted ground water to the surface. The contaminated water is then treated to destroy the pollutant.

Bioremediation uses microorganisms to decompose a contaminant. Specialized microorganisms can be fine-tuned by genetic engineers to destroy a particular contaminant without damaging the ecosystem. Once a specialized microorganism is developed, it is relatively inexpensive to breed it in large quantities. The microorganisms are then pumped into the contaminant plume, where they attack the pollutant. When the contaminant is destroyed, the microorganisms run out of food and die, leaving a clean aquifer. Bioremediation can be among the cheapest of all cleanup procedures.

Chemical remediation is similar to bioremediation. If a chemical compound reacts with a pollutant to produce harmless products, the compound can be injected into an aquifer to destroy contaminants. Common reagents used in chemical remediation include oxygen and dilute acids and bases. Oxygen may react with a pollutant directly or provide an environment favorable for microorganisms, which then degrade the pollutant. Thus, contamination can sometimes be reduced simply by pumping air into the ground. Acids or bases neutralize certain contaminants or precipitate dissolved pollutants.

In some extreme cases, reclamation teams dig up the entire contaminated portion of an aquifer. The contaminated soil is treated by incineration or with chemical processes to destroy the pollutant. The treated soil is then used to fill the hole.

20.9 Ground Water and Nuclear Waste Disposal

In a nuclear reactor, radioactive uranium nuclei split into smaller nuclei, many of which are also radioactive. Most of these radioactive waste products are useless and must be disposed of without exposing people to the radioactivity. In the United States, military processing plants, 109 commercial nuclear reactors, and numerous laboratories and hospitals generate approximately 3000 tons of radioactive wastes every year.

Chemical reactions cannot destroy radioactive waste because radioactivity is a nuclear process, and atomic nuclei are unaffected by chemical reactions. Therefore, the only feasible method for disposing of radioactive wastes is to store them in a place safe from geologic hazards and human intervention and to allow them to decay naturally. The U.S. Department of Energy defines a permanent repository as one that will isolate radioactive wastes for 10,000 years.[1] For a repository to keep radioactive waste safely isolated for such a long time, it must meet at least three geologic criteria:

1. It must be safe from disruption by earthquakes and volcanic eruptions.

2. It must be safe from landslides, soil creep, and other forms of mass wasting.

3. It must be free from floods and seeping ground water that might corrode containers and carry wastes into aquifers.

CASE STUDY
The Yucca Mountain Repository

In December 1987, the U.S. Congress chose a site near Yucca Mountain, Nevada, about 175 kilometers from Las Vegas, as the national burial ground for all spent reactor fuel unless sound environmental objections were found. Since that time, scientists have conducted

---

[1] This number is derived from human and political considerations more than scientific ones. The National Academy of Sciences issued a report stating that radioactive wastes will remain harmful for one million years because of the long half-lives of some waste isotopes.

numerous studies of the geology, hydrology, and other aspects of the area.

The Yucca Mountain site is located in the Basin and Range province, a region noted for faulting and volcanism related to ongoing tectonic extension of that part of the western United States. Bedrock at the Yucca Mountain site is welded tuff, a hard volcanic rock. The tuffs erupted from several volcanoes that were active from 16 to 6 million years ago. Later volcanism created the Lathrop Wells cinder cone 24 kilometers from the proposed repository. The last eruption near Lathrop Wells occurred 15,000 to 25,000 years ago. In addition, geologists have mapped 32 faults that have moved during the past 2 million years adjacent to the Yucca Mountain site. The site itself is located within a structural block bounded by parallel faults (Fig. 20–30). On July 29, 1992, a magnitude 5.6 earthquake occurred about 20 kilometers from the Yucca Mountain site, and damaged the Department of Energy (DOE) facilities on the site. Critics of the Yucca Mountain site argue that recent earthquakes and volcanoes prove that the area is geologically active.

The environment is desert dry, and the water table lies 550 meters beneath the surface. The repository will consist of a series of tunnels and caverns dug into the tuff 300 meters beneath the surface and 250 meters above the water table. Thus, it is designed to isolate the waste from ground water. However, geologists

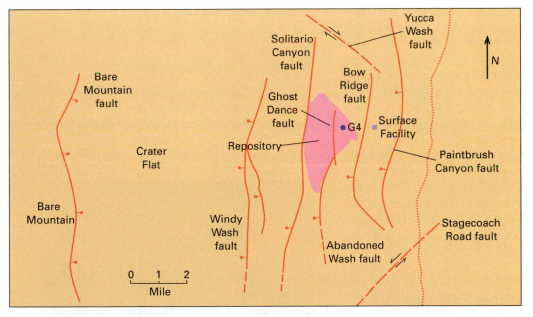

Figure 20–30 A map of the proposed Yucca Mountain repository shows numerous faults where rock has fractured and moved during the past 2 million years. • Interactive Question: In some endeavors there is a high probability that a minor accident will occur. For example, in the transportation industry, trucks are involved in accidents every day. In other endeavors, there is a low probability that a catastrophic disaster will occur. Discuss risk assessment and evaluation in this scenario and relate your discussion to the Yucca Mountain repository.

have suggested that an earthquake could drive deep ground water upward, where it would become contaminated by the radioactive wastes. Because radioactive decay produces heat, the wastes may be hot enough to convert the water to steam. Steam trapped underground could build up enough pressure to rupture containment vessels and cavern walls.

Other scientists are concerned that slow seepage of water from the surface will percolate through the repository site to the water table sometime between 9000 and 80,000 years from now. The lower end of this estimate is within the 10,000-year mandate for isolation. If the climate becomes appreciably wetter, which is possible over thousands of years, the water table may rise. If rocks beneath the site were fractured by an earthquake, then contaminated ground water might disperse more rapidly than predicted. Furthermore, critics point out that construction of the repository will involve blasting and drilling, and these activities could fracture underlying rock, opening conduits for flowing water.

To stop development of the Yucca Mountain site, the state of Nevada refused to issue air quality permits to operate drilling rigs at the repository. In December 1995, U.S. Energy Secretary Hazel O'Leary announced that permanent storage for spent nuclear fuel cannot begin at Yucca Mountain before the year 2015. However, under the current DOE plan, an assessment of the site is to be completed in 1998, and if Congress agrees, emplacement of waste will begin in 2010.

In addition to squabbles among politicians about whether and when to begin storing nuclear wastes at Yucca Mountain, a major problem is what to do with the wastes until governmental agencies can agree on a permanent storage facility. Congress considered this problem in 1997, but failed to act. At present, 33,000 tons of radioactive waste lie in unstable temporary storage facilities across the United States. The amount increases by 3000 tons each year.

Supporters of the Yucca Mountain repository argue that we need nuclear power, and therefore as a society we must accept a certain level of risk. They argue further that the Yucca Mountain site is safer than the temporary storage sites now being used. Opponents argue that the site is geologically unsound, and new alternatives must be found. ●

20.10 Surface Water Pollution

In the early days of the Industrial Revolution, factories and sewage lines dumped untreated wastes into rivers. The first sewage treatment plant in the United States was built in Washington, D.C., in 1889, more than 100 years after the Revolutionary War. Soon other cities followed suit, but few laws regulated industrial waste discharge.

In November 1952, an oily film on the Cuyahoga River in Cleveland, Ohio, caught fire, spreading flame and smoke across the water (Fig. 20–31). The image of a burning river triggered public awareness of water pollution, and people called for action. Nearly 20 years later, in 1970, President Nixon declared that "the 1970s absolutely must be the years when America pays its debt to the past by reclaiming the purity of its air, its waters, and our living environment. It is literally now or never."

The **Clean Water Act,** passed in 1972 over Nixon's veto, stated that

> The objective of this Act is to restore and maintain the chemical, physical, and biological integrity of the Nation's waters. In order to achieve this objective it is hereby declared that . . .
> (1) it is the national goal that the discharge of pollutants into the navigable waters be eliminated by 1985;
> (2) it is the national goal that wherever attainable, an interim goal of water quality which provides for the protection and propagation of fish, shellfish, and wildlife and provides for recreation in and on the water be achieved by July 1, 1983;
> (3) it is the national policy that the discharge of toxic pollutants in toxic amounts be prohibited; . . .

The Clean Water Act set an ambitious agenda for cleaning the nation's rivers, lakes, and wetlands. The results of 25 years of the Clean Water Act are illustrated by two short case studies.

CASE STUDY
Pollution of the Great Lakes

The Great Lakes—Lakes Superior, Michigan, Huron, Erie, and Ontario—lie along the U.S.-Canadian border, filling ancient stream valleys that were scoured

Figure 20–31 Fireboats fight flames on the burning Cuyahoga River in November 1952. *(Cleveland Plain Dealer)*

Figure 20–32 The Great Lakes drainage basin is colored green. All streams within the green area run into the lakes or into streams connecting the lakes. Streams outside the green area flow away from the lakes. Lake Superior is the largest fresh-water lake in the world; it is also the least polluted of the Great Lakes. Because the lakes drain eastward, Lakes Erie and Ontario contain pollutants flowing from the three upstream lakes plus pollutants that are locally generated.

and deepened by glaciers during the Pleistocene Ice Age. Connected by streams that ultimately flow to the Atlantic Ocean through the St. Lawrence River, the lakes cover about 244,000 square kilometers and contain about one fifth of the Earth's fresh surface water (Fig. 20–32).

More than 33 million people live in the Great Lakes watershed, and about 38 million people obtain their drinking water from the lakes. Major industrial cities, including Duluth, Milwaukee, Chicago, Cleveland, Erie, Buffalo, and Toronto line the shores of the lakes, and much of the rural countryside is heavily farmed. Since the mid-1800s, the cities have poured industrial and human wastes into the lakes, and agriculture has added fertilizers and pesticides to the waters. As a result, the lake waters and bottom sediments have become heavily contaminated.

By the 1960s, pollution of the Great Lakes had become a conspicuous problem. Although all the lakes were contaminated by thousands of toxic chemicals and human wastes, Lakes Erie and Ontario and shallow parts of Lakes Huron and Michigan were the most strongly polluted. Lake Erie was the worst of all because it has the most strongly concentrated human population and industrial activity along its shores, and it is the shallowest of the Great Lakes. Eutrophication was common. Waterborne bacteria had become a health hazard to swimmers and to those people drinking inadequately treated lake water; fish were dying by the thousands. Nearly 50 percent of the wildlife species living in or near the lake showed severe birth defects caused by the contamination.

After years of negotiation, Canada and the United States began a joint cleanup effort in the entire Great

TABLE 20–2
Major Contaminants That Persist in the Great Lakes

Chemicals Usually Found*
Mercury (heavy metal)
Dieldrin (pesticide)
PCBs (industrial chemicals)
Chlordane (pesticide)
DDT (pesticide)
Dioxins (chemical contaminants)
Furans (industrial chemicals)

*Based on contaminant monitoring of fish tissues at multiple sites in the Great Lakes and tributary river mouths.

Lakes watershed, beginning in 1972. As a result of these efforts, contamination by human wastes, detergents, industrial pollutants, and agricultural pesticides has decreased steadily in the lakes since the inception of the program. In 1995, the U.S. EPA and Environments Canada issued the *State of the Great Lakes 1995*, reporting that both the health of people living in the region and the water quality of the lakes was showing a slow but steady improvement. DDT levels in women's breast milk and PCB levels in some fish and water bird species have declined since 1967. Populations of bald eagles and other birds that depend on the lakes for food and shelter have increased.

The Great Lakes watershed does not have a clean bill of health yet, however. Forty-three coastal areas are still heavily polluted and designated "areas of concern" by the EPA. Many persistent toxic compounds remain in the lake waters and bottom sediment, and cities, industry, and agriculture continue to add more (Table 20–2). Mercury levels are currently increasing. Because the lake waters are still contaminated, government agencies continue to warn people not to eat fish caught in many parts of the lakes. ●

CASE STUDY
Surface Water Pollution from Mines—The Largest Superfund Site in the United States

Both metal ore and coal are commonly covered and surrounded by soil and rock that has no commercial value. As a result, miners must dig up and discard large amounts of waste rock as they expose and extract the resource. Before pollution controls were enacted, the waste rock from both surface and underground mines was piled up near the mine. The wastes were easily eroded, and the muddy runoff poured into nearby streams, silting aquatic habitats. Toxic heavy metals, such as lead, cadmium, zinc, and arsenic are common in many metal ores. Rain can leach them from the mine wastes and carry them to streams and ground water. In addition, sulfur is abundant in many metal ores as well as in coal. The sulfur reacts with water in the presence of air to produce sulfuric acid (H_2SO_4). If pollution control is inadequate, the sulfuric acid then runs off into streams and ground water below the mine or mill.

Today, responsible mining companies use several methods to stabilize mine and mill wastes. For example, they use crushed limestone to neutralize acid waters; they build well-designed settling ponds to trap silt; and they backfill abandoned mines and settling ponds, cover the fill with topsoil, and restore vegetation to the site. These measures can be costly, but they greatly reduce the quantity of pollutants that escape into streams and ground water.

The largest complex of toxic waste sites in the United States was not produced by a chemical factory or petroleum refinery but by the mines and smelters in Butte and Anaconda in Montana (Fig. 20–33). Mining and refining began in Butte in the early 1800s, before pollution control laws existed. Some of the ore from the mines in Butte contained as little as 0.3 percent copper, with even smaller concentrations of manganese, zinc, arsenic, lead, cadmium, and a variety of other metals. When the ore was extracted, up to 99.7 percent of the rock was waste. Miners simply discarded this material in huge piles near the mine mouths and smelters. Rain and streams eroded the mine spoils into the Clark Fork River, silting the stream bed and destroying aquatic habitats. But that was only the be-

Figure 20–33 Mill wastes surround the now-abandoned smelter at Anaconda, Montana.

ginning of the problem, because the smelting process generated additional wastes.

During the boom days in Butte (the late 1800s and early 1900s), metal smelting was inefficient, so large quantities of arsenic, cadmium, zinc, lead, copper, and other metals were discarded in waste heaps. The Butte ores also contained large amounts of sulfur, which was discarded with the other wastes. Furthermore, Butte's copper ore contains gold, which the early miners extracted by dissolving it in mercury, and the waste mercury was also dumped on the piles. Today in the Butte-Anaconda area there are about 25 square kilometers (about 10 square miles) of mine spoils and mill wastes averaging 15 meters (about 50 feet) high, or about as tall as a five-story building.

Wastes from these piles have washed into aquatic systems and poisoned both the ground water and streams flowing from the Butte-Anaconda area. The sulfur compounds have reacted with water to form sulfuric acid, which has spread into streams and groundwater reservoirs. Metal contamination has killed and stunted plants, the animals that eat those plants, and the microorganisms in the soil (Fig. 20–34).

The sulfur and metals have migrated up to 220 kilometers downstream from Butte and Anaconda along the Clark Fork River (Fig. 20–35). Windblown dust from the spoils has contaminated entire counties. In the 1980s, arsenic levels had risen so high in wells of Milltown—a small town on the Clark Fork River about 200 kilometers downstream from Butte—that government officials ordered the residents to stop drinking well water. An alternative drinking water supply was

Figure 20–35 In a healthy ecosystem, lush vegetation grows on riverbanks. Heavy metals and silt from the Anaconda smelter have poisoned the banks of the upper Clark Fork River in Montana. *(Johnnie Moore)* ● **Interactive Question:** Estimate how long it would take for heavy metals to naturally wash out of a river and riparian ecosystem. How long would it take for an ecosystem to naturally clean itself of (a) sewage, or (b) organic chemicals such as pesticides.

installed at great public cost. No one knows how to clean up such a mess or how much it will cost. The problem is that thousands of square kilometers of land surface are contaminated. The Clark Fork Coalition, a local environmental organization, has estimated that it will cost more than $1 billion to clean up the Butte mine, the Anaconda smelter, and the Clark Fork drainage. Even a partial cleanup, under way now, will cost tens to hundreds of millions of dollars. ●

From these two case studies, we see that analyzing the results of the Clean Water Act today is like trying to decide whether a glass is half empty or half full. Between 1972 and 1994 the proportion of the U.S. population served by sewage treatment plants jumped from 32 to 74 percent. In the same period, aquatic emissions of industrial toxic pollutants declined by 99 percent, and discharge of metals from mining and metal refining declined by 98 percent. However, in 1995, EPA officials estimated that 30 percent of the nation's streams, 42 percent of lakes, and 32 percent of estuaries do not meet the standards of the Clean Water Act. Large quantities of toxic pollutants continue to be released into our surface waters. In addition, fertilizers, pesticides, silt from farms and cities, and sewage degrade surface and ground water. As a result 40 percent of the country's fresh water remained unsafe for human use.

Figure 20–34 Copper leached from the Anaconda mill wastes colors this soil blue-green. Copper, arsenic, cadmium, and other heavy metals have poisoned the soil so that nothing can grow. *(Johnnie Moore)*

Growing populations have created demands for water that stretch, and in some cases exceed, the amount of water that is available, both globally and in the United States. The United States receives about three times as much water, in the form of precipitation, as it uses. But some of the driest regions use the greatest amounts of water. As a result, those regions must import water from wetter regions, or pump it from ground-water reservoirs.

Most of the water used by homes and industry is **withdrawn,** and then returned to streams or ground-water reservoirs near the site of withdrawal. But most of the water used by agriculture is **consumed,** because it evaporates. Water use falls into three categories. Domestic use accounts for 10 percent of U.S. water consumption, industrial use accounts for 49 percent, and agricultural uses require 41 percent.

Water **diversion** projects collect and transport surface and ground water from places where water is available to places where it is needed. In a surface diversion project, a dam is built on a stream or lake outlet to create a reservoir. Canals, aqueducts, or pipes then carry the water to consumers. Dams supply large amounts of water to places where it is needed, and some generate hydroelectric energy. But they can create undesirable effects, including water loss, salinization, silting, erosion, destruction when a dam fails, recreational and aesthetic losses, and ecological disruptions.

Ground-water projects pump water to the Earth's surface for human use, but because ground water flows so slowly, the extraction often creates a **cone of depression** that grows as pumping depletes the aquifer and causes **subsidence.** Depletion of coastal aquifers causes **saltwater intrusion.**

The Great American Desert is a mostly arid and semi-arid region of the United States that reaches from the western mountains to the 100° meridian. Americans have built great cities and extensive farms and ranches in the region, all supplied by irrigation systems. Diminishing water reserves and increasing costs of water diversion projects suggest that their future is uncertain.

Risk assessment and **cost-benefit analysis** are used to measure the hazards and cost of ground- and surface-water contamination. The Comprehensive Environmental Response, Compensation, and Liability Act (CERCLA), commonly known as the **Superfund,** provides funding and guidelines to clean up hazardous waste sites. Pollution can originate from both **point sources** and **nonpoint sources.** A ground-water pollutant normally spreads slowly into an aquifer as a contaminant **plume.** Because many pollutants persist in an aquifer and render the water unfit for use, expensive and difficult **remediation** efforts are commonly undertaken to cleanse a polluted aquifer. Radioactive wastes must be isolated from water resources because they are impossible to destroy, and persist for long times. The **Clean Water Act** regulates discharge of pollutants into surface waters.

KEY TERMS

withdrawal 461	saltwater intrusion 471	externalities 479	bioremediation 482
consumption 461	pollution 476	risk assessment 479	Clean Water Act 484
diversion 464	point source pollution 476	cost-benefit analysis 479	
salinization 465	nonpoint source pollution 476	Superfund 480	
cone of depression 468		plume 481	
subsidence 470	anaerobic 478	remediation 481	

FOR REVIEW

1. The United States receives three times more water in the form of precipitation than it uses. Why do water shortages exist in many parts of the country?

2. Describe the three main categories of water use. What proportion of total U.S. water use falls into each category?

3. Explain the differences among water use, water withdrawal, and water consumption.

4. Why is agriculture responsible for the greatest proportion of water consumption, whereas industry is responsible for the greatest proportion of water withdrawal?

5. What are the two main sources of water exploited by water diversion projects?

6. Describe the beneficial effects of dams and their associated water-delivery systems.

7. Describe the negative effects of dams and their associated water-delivery systems.

8. Why does salinization commonly result from desert irrigation?

9. Describe the factors that make ground water a valuable resource.

10. Describe three problems caused by excessive pumping of ground water.

11. Why does ground-water depletion cause longer-term problems than might result from the draining of a surface reservoir?

12. How and why does land subside when ground water is depleted?

13. If land subsides when an underlying aquifer is depleted, will it rise to its original level when pumping stops and the aquifer is recharged? Explain your answer.

14. Why does saltwater intrusion affect only coastal areas?

15. Describe the geographic region that John Wesley Powell called "the Great American Desert."

16. List the three major goals of the Clean Water Act.

17. List eight categories of water pollutants. What is/are the source(s) of each, and what are the harmful effects?

18. What is eutrophication? How could a nontoxic substance such as cannery waste become a water pollutant?

19. Discuss the progress under the Clean Water Act between 1972 and 1998.

20. Explain why pollutants persist in ground water longer than they do in surface water.

FOR DISCUSSION

1. Consider how (a) a rise or (b) a fall in average global temperature might affect surface- and ground-water reserves in the United States.

2. Discuss the relative merits and disadvantages of surface- and ground-water diversion projects.

3. No major dam has been built in the United States without federal support, and the price charged to irrigators and other users for the water from a major dam has never paid for the cost of building the dam. Discuss why we build dams.

4. Develop scenarios for the future of farming and ranching in regions currently obtaining water from the Ogallala aquifer.

5. Discuss how water conservation in each of the three main categories of water use in the United States can affect the total amount of water available for all types of uses.

6. Discuss environmental problems that arise when cotton and rice are grown in semi-arid California at the same time that farmers in naturally wet southeastern and south central states are paid federal subsidies to not grow the same crops.

7. In the book *Cadillac Desert*, Marc Reisner describes the transformation of the American west from its natural desert and semi-desert environment to a region with great cities and a vast agricultural economy. He argues that the west cannot sustain the cities or the agriculture. Discuss how this transformation occurred, and the validity of his hypothesis that the current system cannot continue indefinitely.

8. Consider reasons why desert and semi-desert cities such as Phoenix, Las Vegas, and Los Angeles have become so heavily populated despite the lack of adequate water resources.

9. Abundant water exists in the Columbia River basin in the northwestern United States and in the Fraser River basin in Canada, whereas the Colorado River basin to the south is water deficient. Argue for or against a proposal to divert water from the Pacific Northwest to the arid southwest.

10. Discuss the role that water plays in the relations between the United States and its neighboring countries.

11. Of the eight categories of pollutants listed at the beginning of Section 20.5, which are most likely to originate from point sources, and which are most likely to originate from nonpoint sources?

12. Imagine that you live in a major industrial city and your drinking water has a slight "off" taste. Bottled water is expensive and the city health officials assure you that the tap water is safe. What facts would you search for in trying to make your own personal assessment of the safety of the water? Discuss the factors that you would weigh in making a personal decision of whether to drink city water or bottled water.

13. The EPA authorized U.S. industry to dispose of 160 million kilograms of toxic waste into deep injection wells. An injection well is approved if it is drilled into a porous rock layer protected from ground water by an impermeable rock layer. Discuss the pros and cons of this method of disposal.

Chapter

21

Geologic Resources

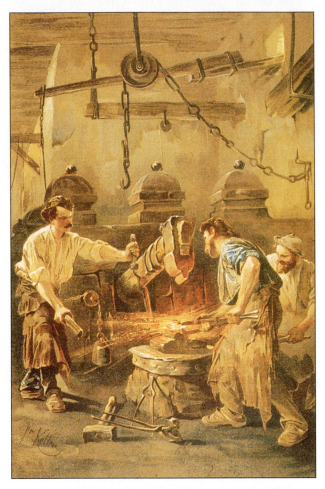

Geological resources have been one of the foundations of human development from the stone age to the space age. Steelworkers shape iron with a forge hammer in a California steel mill around 1880. Colored lithograph. *(Corbis-Bettmann)*

With time, the human use of geologic resources has become increasingly sophisticated. Prehistoric people used flint and obsidian to make weapons and tools. About 7000 B.C., people learned to shape and fire clay to make pottery. Archaeologists have found copper ornaments in Turkey from 6500 B.C.; 1500 years later, Mesopotamian farmers used copper farm implements. Today, the silicon chip that operates your computer, the titanium valves in a space probe, and the gasoline that powers your car are all derived from Earth resources. ●

21.1 Geologic Resources

We use two different types of geologic resources: mineral resources and energy resources. **Mineral resources** include all useful rocks and minerals. Table 21–1 lists important elements that are obtained from mineral resources, and their uses.

About 40 metals are commercially important. Some, such as iron, lead, copper, aluminum, silver, and gold, are familiar. Others, such as vanadium, titanium, and tellurium, are less well known but are vital to industry.

If you pick up any rock and send it to a laboratory for analysis, the report will probably show that the rock contains measurable amounts of iron, gold, silver, aluminum, and other valuable metals. However, the concentrations of these metals are so low in most rocks that the extraction cost would be much greater than the income gained by selling the metals. In certain locations, however, geologic processes have enriched metals many times above their normal concentrations (Table 21–2).

Ore is rock sufficiently enriched in one or more minerals to be mined profitably (Fig. 21–1). Geologists usually use the term *ore* to refer to metallic mineral deposits, and the term is commonly accompanied by the name of the metal—for example, iron ore or silver ore. Table 21–2 shows that the concentration of a metal in ore may exceed its average abundance in ordinary rock by a factor of more than 100,000. **Mineral reserves** are the known supply of ore in the ground. The term can refer to the amount of ore remaining in

TABLE 21–1

Some Important Elements and Their Uses

Mineral	Type	Some Uses
Aluminum (Al)	Metal	Structural materials (airplanes, automobiles), packaging (beverage cans, toothpaste tubes), fireworks
Borax ($Na_2B_4O_7$)	Nonmetal	Diverse manufacturing uses—glass, enamel, artificial gems, soaps, antiseptics
Chromium (Cr)	Metal	Chrome plate, pigments, steel alloys (tools, jet engines, bearings)
Cobalt (Co)	Metal	Pigments, alloys (jet engines, tool bits), medicine, varnishes
Copper (Cu)	Metal	Alloy ingredient in gold jewelry, silverware, brass, and bronze; electrical wiring, pipes, cooking utensils
Gold (Au)	Metal	Jewelry, money, dentistry, alloys, specialty electronics
Gravel	Nonmetal	Concrete (buildings, roads)
Gypsum ($CaSO_4\text{-}2H_2O$)	Nonmetal	Plaster of Paris, wallboard, soil treatments
Iron (Fe)	Metal	Basic ingredient of steel (buildings, machinery)
Lead (Pb)	Metal	Pipes, solder, battery electrodes, pigments
Magnesium (Mg)	Metal	Alloys (aircraft), firecrackers, bombs, flashbulbs
Manganese (Mn)	Metal	Steel, alloys (steamship propellers, gears), batteries, chemicals
Mercury (Hg)	Liquid metal	Thermometers, barometers, dental inlays, electric switches, street lamps, medicine
Molybdenum (Mo)	Metal	Steel alloys, lamp filaments, boiler plates, rifle barrels
Nickel (Ni)	Metal	Money, alloys, metal plating
Phosphorus (P)	Nonmetal	Medicine, fertilizers, detergents
Platinum (Pt)	Metal	Jewelry, delicate instruments, electrical equipment, cancer chemotherapy, industrial catalyst
Potassium (K)*	Nonmetal	Salts used in fertilizers, soaps, glass, photography, medicine, explosives, matches, gunpowder
Common salt (NaCl)	Nonmetal	Food additive
Sand (largely SiO_2)	Nonmetal	Glass, concrete (buildings, roads)
Silicon (Si)	Nonmetal	Electronics, solar cells, ceramics, silicones
Silver (Ag)	Metal	Jewelry, silverware, photography, alloys
Sulfur (S)	Nonmetal	Insecticides, rubber tires, paint, matches, papermaking, photography, rayon, medicine, explosives
Tin (Sn)	Metal	Cans and containers, alloys, solder, utensils
Titanium (Ti)	Metal	Paints; manufacture of aircraft, satellites, and chemical equipment
Tungsten (W)	Metal	High-temperature applications, lightbulb filaments, dentistry
Zinc (Zn)	Metal	Brass, metal coatings, electrodes in batteries, medicine (zinc salts)

*Potassium, which is very reactive chemically, is never found free in nature; it is always combined with other elements.

TABLE 21–2

Comparison of Concentrations of Specific Elements in Earth's Crust with Concentrations Needed to Operate a Commercial Mine

Element	Natural Concentration in Crust (% by Weight)	Concentration Required to Operate a Commercial Mine (% by Weight)	Enrichment Factor
Aluminum	8	24–32	3–4
Iron	5.8	40	6–7
Copper	0.0058	0.46–0.58	80–100
Nickel	0.0072	1.08	150
Zinc	0.0082	2.46	300
Uranium	0.00016	0.19	1,200
Lead	0.00010	0.2	2,000
Gold	0.0000002	0.0008	4,000
Mercury	0.000002	0.2	100,000

a particular mine, or it can be used on a global or national scale.

A **nonmetallic resource** is any useful rock or mineral that is not a metal, such as salt, building stone, sand, and gravel. When we think about "striking it rich" from mining, we usually think of gold. However, more money has been made mining sand and gravel than gold. For example, in the United States in 1994, sand and gravel produced $4.26 billion in revenue, but gold produced $4.1 billion. Sand and gravel are mined from stream and glacial deposits, sand dunes, and beaches.

Portland cement is made by heating a mixture of crushed limestone and clay. Concrete is a mixture of cement, sand, and gravel. Reinforced with steel, it is used to build roads, bridges, and buildings.

Many buildings are faced with stone—usually granite or limestone, although marble, slate, sandstone, and other rocks are also mined from quarries cut into bedrock (Fig. 21–2).

All mineral resources are **nonrenewable:** We use them up at a much faster rate than natural processes create them. Even though mineral resources are limited, many can be recycled and used over and over again.

We use **energy resources** for heat, light, work, and communication. Petroleum, coal, and natural gas are called **fossil fuels** because they formed from the remains of plants and animals. Fossil fuels are not only nonrenewable, but they cannot be recycled. When a lump of coal or a liter of oil is burned, the energy dissipates into the air and is, for all practical purposes, lost. Thus our fossil fuel supply inexorably diminishes. **Nuclear fuels** are radioactive isotopes used to generate electricity in nuclear reactors. Uranium is the most commonly used nuclear fuel. These energy resources, like mineral resources, are nonrenewable. **Alternative energy resources,** such as solar, wind, and geothermal energy, are renewable.

Figure 21–1 In the early 1900s, miners extracted gold, copper, and other metal ores from underground mines such as this one 600 meters below the surface in Butte, Montana. *(Montana Historical Society)*

Figure 21–2 A Chinese quarryman splits a large granite block with a sledge hammer. After he splits the rock, circular saws in the background cut it into thin slabs for floors and walls.

21.2 How Ore Forms

One of the primary objectives of many geologists is to find new ore deposits. Successful exploration requires an understanding of the processes that concentrate metals to form ore. For example, platinum concentrates in certain types of igneous rocks. Therefore, if you were exploring for platinum, you would focus on those rocks rather than on sandstone or limestone.

Magmatic Processes

Magmatic processes form mineral deposits as liquid magma solidifies to form an igneous rock. These processes create metal ores as well as some gems and valuable sulfur deposits.

Some large bodies of igneous rock, particularly those of mafic (basaltic) composition, solidify in layers. Each layer contains different minerals, and is of a different chemical composition from adjacent layers. Some of the layers may contain rich ore deposits.

The layering can develop by at least two processes.

1. Recall from Chapter 3 that cooling magma does not solidify all at once. Instead, higher-temperature minerals crystallize first, and lower-temperature minerals form later as the temperature drops. Most minerals are denser than magma. Consequently, early-formed crystals may sink to the bottom of a magma chamber in a process called **crystal settling** (Fig.

21–3). In some instances, ore minerals crystallize with other early-formed minerals and accumulate in layers near the bottom of a pluton.

2. Some large bodies of mafic magma crystallize from the bottom upward. Thus, early-formed ore minerals become concentrated near the base of the pluton.

The largest ore deposits found in mafic layered plutons are the rich chromium and platinum reserves of South Africa's Bushveld intrusion. The pluton is about 375 by 300 kilometers in area—roughly the size of the state of Maine—and about 7 kilometers thick. The Bushveld deposits contain more than 20 billion tons of chromium and more than 10 billion grams of platinum, the greatest reserves in any known deposit on Earth. The world's largest known nickel deposit occurs in a layered mafic pluton at Sudbury, Ontario, and rich platinum ores are mined from layered plutons in southern Montana and Norilsk, Russia.

Hydrothermal Processes

Hydrothermal processes are probably responsible for the formation of more ore deposits, and a larger total quantity of ore, than all other processes combined. To

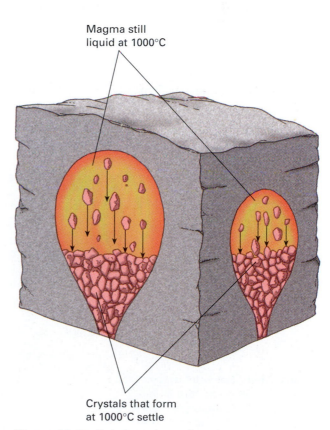

Magma still liquid at 1000°C

Crystals that form at 1000°C settle

Figure 21–3 Early-formed crystals settle and concentrate near the bottom of a magma chamber.

form a **hydrothermal ore deposit,** hot water (hence the roots *hydro* for water and *thermal* for hot) dissolves metals from rock or magma. The metal-bearing solutions then seep through cracks or through permeable rock until they precipitate to form an ore deposit.

Hydrothermal water comes from three sources:

1. Granitic magma contains more dissolved water than solid granite rock. Thus, the magma gives off hydrothermal water as it solidifies. For this reason, hydrothermal ore deposits are commonly associated with granite and similar igneous rocks.

2. Ground water can seep into the crust where it is heated and forms a hydrothermal solution. This is particularly true in volcanic areas where hot rock or magma heat ground water at shallow depths. For this reason, hydrothermal ore deposits are also common in volcanic regions.

3. In the oceans, hot, young basalt near the mid-oceanic ridge heats seawater as it seeps into cracks in the sea floor.

Although water by itself is capable of dissolving minerals, most hydrothermal waters also contain dissolved salts, which greatly increase their ability to dissolve minerals. Therefore, hot, salty, hydrothermal water is a very powerful solvent, capable of dissolving and transporting metals.

Table 21–2 shows that tiny amounts of all metals are found in average rocks of the Earth's crust. For example, gold makes up 0.0000002 percent of the crust, while copper makes up 0.0058 percent and lead 0.0001 percent. Although the metals are present in very low concentrations in country rock, hydrothermal solutions percolate through vast volumes of rock, dissolving and accumulating the metals. The solutions then deposit the metals when they encounter changes in temperature, pressure, or chemical environment (Fig. 21–4). In this way, hydrothermal solutions scavenge metals from large volumes of normal crustal rocks, and then deposit them locally to form ore.

Types of Hydrothermal Ore Deposits   A **hydrothermal vein deposit** forms when dissolved metals precipitate in a fracture in rock. Ore veins range from less than a millimeter to several meters in width. A single vein can yield several million dollars worth of gold or silver. The same hydrothermal solutions may also soak into pores in country rock near the vein to create a large but much less concentrated **disseminated ore deposit.** Because they commonly form from the same solutions, rich ore veins and disseminated deposits are often found together. The history of many mining districts is one in which early miners dug shafts and tunnels to follow the rich veins. After the veins were exhausted, later miners used huge power shovels to extract low-grade ore from disseminated deposits surrounding the veins.

Disseminated copper deposits, with ore veins, are abundant along the entire western margin of North and South America (Fig. 21–5). They are most commonly associated with large granitic plutons. Both the plutons and the copper ore formed as a result of subduction that occurred as tectonic plates carrying North

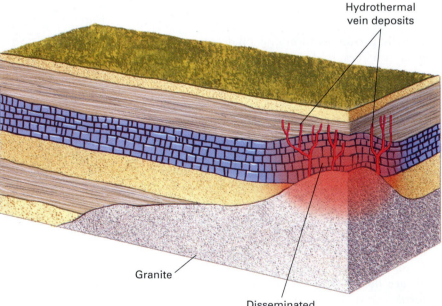

Figure 21–4 Hot water deposits metallic minerals in veins that fill fractures in bedrock. It also deposits low-grade disseminated metal ore in large volumes of rock surrounding the veins.

Hydrothermal vein deposits

Granite

Disseminated ore deposit

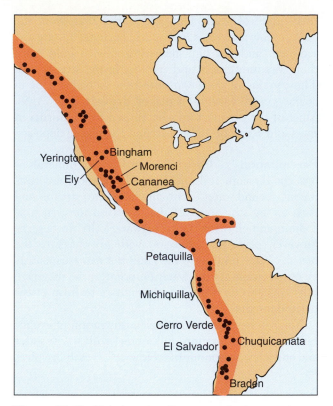

Figure 21–5 Rich deposits of copper and other metals exist throughout the Cordillera of North, Central, and South America. These deposits formed along modern or ancient tectonic plate boundaries. Dots indicate mines.

Ore deposits rich in copper, lead, zinc, gold, silver, and other metals have formed in submarine volcanic environments throughout time. For example, on land an ore deposit containing 2.5 percent zinc is rich enough to mine commercially; but some undersea zinc deposits contain 55 percent zinc. The cost of operating machinery beneath the sea is so great that these deposits cannot be mined profitably. However, in some places tectonic forces have lifted submarine ore deposits to the Earth's surface. The ancient Romans mined copper, lead, and zinc ores of this type in the Apennine Mountains of Italy. This geologic wealth contributed to the political and military ascendancy of Rome.

Manganese Nodules

About 25 to 50 percent of the Pacific Ocean floor is covered with golf ball to bowling ball sized **manganese nodules** (Fig. 21–7). A typical nodule contains 20 to 30 percent manganese, 6 percent iron, about 1 percent each of copper and nickel, and lesser amounts of other metals. (Much of the remaining 60 to 70 percent consists of oxygen and other anions chemically bonded to the metals.) The metals are probably added to seawater by volcanic activity at the mid-oceanic ridge, perhaps by the black smokers. Chemical reactions between seawater and sea-floor sediment precipitate the dissolved metals to form the nodules.

The nodules grow by about 10 layers of atoms per year, which amounts to 3 millimeters per million years. Curiously, they are found only on the surface of, but

and South America migrated westward to converge with oceanic plates. Other metals, including lead, zinc, molybdenum, gold, and silver, are found with the copper ore. Examples of such deposits occur at Butte, Montana; Bingham, Utah; Morenci, Arizona; and Ely, Nevada.

Hydrothermal Ore on the Sea Floor

In volcanically active regions of the sea floor, near the mid-oceanic ridge and submarine volcanoes, seawater circulates through the hot, fractured oceanic crust. The hot seawater dissolves metals from the rocks and then, as it rises through the upper layers of oceanic crust, cools and precipitates the metals to form submarine hydrothermal ore deposits.

The metal-bearing solutions can be seen today as jets of black water, called **black smokers**, spouting from fractures in the mid-oceanic ridge (Fig. 21–6). The black color is caused by precipitation of fine-grained metal sulfide minerals as the solutions cool upon contact with seawater. The precipitating metals accumulate as chimney-like structures near the hot water vent.

Figure 21–6 A black smoker spouts from the East Pacific rise. Seawater becomes hot as it circulates through the hot basalt near a spreading center or submarine volcano. The hot water dissolves sulfur, iron, zinc, copper, and other ions from the basalt. The ions then precipitate as "smoke," consisting of tiny metallic mineral grains, when the hot solution meets cold ocean water. The hot, nutrient-rich water sustains thriving plant and animal communities. *(Dudley Foster, Woods Hole Oceanographic Institution)*

Figure 21-7 Manganese nodules cover many parts of the sea floor, in both the Pacific and Atlantic oceans. These lie at a depth of 5500 meters in the northeastern Atlantic. *(Institute of Oceanographic Sciences/NERC/Science Photo Library)*

small grain of gold. The gold settles first when the current slows down. Over years, currents agitate the sediment and the high-density grains of gold work their way into cracks and crevices in the stream bed. Thus, grains of gold concentrate near bedrock or in coarse gravel, forming a **placer deposit** (Fig. 21–9). Most of the prospectors who rushed to California in the Gold Rush of 1849 searched for placer deposits.

Precipitation   Ground water dissolves minerals as it seeps through soil and bedrock. In most environments, ground water eventually flows into streams and then to the sea. Some of the dissolved ions, such as sodium and chloride, make seawater salty. In deserts, however, playa lakes develop with no outlet to the ocean. Water flows into the lakes but can escape only by evaporation. As the water evaporates, the dissolved salts concentrate until they precipitate to form **evaporite deposits.**

You can perform a simple demonstration of evaporation and precipitation. Fill a bowl with warm water and add a few teaspoons of table salt. The salt dissolves

never within, the sediment on the ocean floor. Because sediment accumulates much faster than the nodules grow, why doesn't it bury the nodules as they form? Photographs show that animals churn up sea-floor sediment. Worms burrow into it, and other animals pile sediment against the nodules to build shelters. Some geologists suggest that these activities constantly lift the nodules onto the surface.

A trillion or more tons of manganese nodules lie on the sea floor. They contain several valuable industrial metals that could be harvested without drilling or blasting. One can imagine that undersea television cameras locate the nodules, and giant vacuums scoop them up and lift them to a ship. But, because the sea floor is a difficult environment in which to operate complex machinery, harvest of manganese nodules is not profitable at the present time.

Sedimentary Processes

Two types of sedimentary processes form ore deposits: sedimentary sorting and precipitation.

Sedimentary Sorting: Placer Deposits   Gold is denser than any other mineral. Therefore, if you swirl a mixture of water, gold dust, and sand in a gold pan, the gold falls to the bottom first (Fig. 21–8). Differential settling also occurs in nature. Many streams carry silt, sand, and gravel with an occasional

Figure 21-8 Jeffery Embrey panning for gold near his cabin in Park City, Montana, in 1898. *(Maud Davis Baker/Montana Historical Society)*

and you see only a clear liquid. Set the bowl aside for a few days until the water evaporates. The salt precipitates and encrusts the sides and bottom of the bowl.

Evaporite deposits formed in desert lakes include table salt, borax, sodium sulfate, and sodium carbonate. These salts are used in the production of paper, soap, and medicines, and for the tanning of leather.

Several times during the past 500 million years, shallow seas covered large regions of North America and all other continents. At times, those seas were so poorly connected to the open oceans that water did not circulate freely between them and the oceans. Consequently, evaporation concentrated the dissolved salts until they precipitated as marine evaporites. Periodically, storms flushed new seawater from the open ocean into the shallow seas, providing a new supply of salt. Thick marine evaporite beds, formed in this way, underlie nearly 30 percent of North America. Table salt, gypsum (used to manufacture plaster and sheetrock), and potassium salts (used in fertilizer) are mined extensively from these deposits.

Most of the world's supply of iron is mined from sedimentary rocks called **banded iron formations.** These iron-rich rocks precipitated from the seas between 2.6 and 1.9 billion years ago, as a result of ris-ing atmospheric oxygen concentrations. The processes that formed these important ores are closely related to the development of life on Earth, and are described in the opening essay to Unit V.

Weathering Processes

In environments with high rainfall, the abundant water dissolves and removes most of the soluble ions from soil and rock near the Earth's surface. This process leaves the relatively insoluble ions in the soil to form **residual deposits.** Both aluminum and iron have very low solubilities in water. **Bauxite,** the principal source of aluminum, forms as a residual deposit, and in some instances iron also concentrates enough to become ore. Most bauxite deposits form in warm, rainy, tropical or subtropical environments where chemical weathering occurs rapidly. Thus bauxite ores are common in Jamaica, Cuba, Guinea, Australia, and parts of the southeastern United States (Fig. 21–10). Some bauxite deposits are found today in regions with dry, cool climates. Most of them, however, formed when the regions had a warm, wet climate, and they reflect climatic change since their origin.

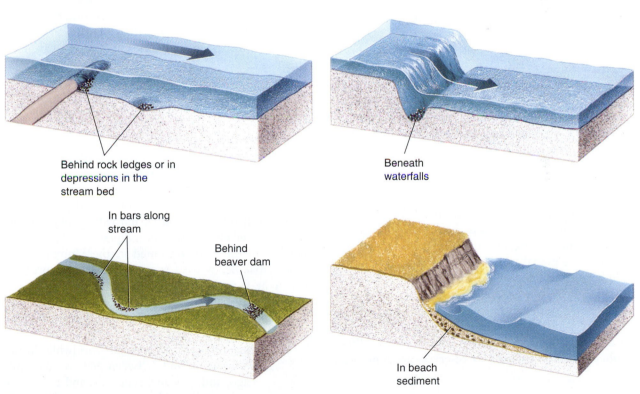

Behind rock ledges or in depressions in the stream bed

Beneath waterfalls

In bars along stream

Behind beaver dam

In beach sediment

Figure 21–9 Placer deposits form where water currents slow down and deposit high-density minerals.

Figure 21–10 Bauxite forms by intense weathering of aluminum-rich rocks. *(H. E. Simpson/USGS)*

The Geopolitics of Metal Resources

The Earth's mineral resources are unevenly distributed, and no single nation is self-sufficient in all minerals. For example, almost two thirds of the world's molybdenum reserves, and more than one third of the lead reserves, are located in the United States. More than half of the aluminum reserves are found in Australia and Guinea. The United States uses 40 percent of all aluminum produced in the world, yet it has no large bauxite deposits. Zambia and the Democratic Republic of Congo supply half of the world's cobalt, although neither nation uses the metal for its own industry.

Five nations—the United States, Russia, South Africa, Canada, and Australia—supply most of the mineral resources used by modern societies. Many other nations have few mineral resources. For example, Japan has almost no metal or fuel reserves; despite its thriving economy and high productivity, it relies entirely on imports for both.

Developed nations consume most of the Earth's mineral resources. Four nations—the United States, Japan, Germany, and Russia—consume about 75 percent of the most intensively used metals, although they account for only 14 percent of world population.

Currently, the United States depends on 25 other countries for more than half of its mineral resources. Some must be imported because we have no reserves of our own. We do have reserves of others, but we consume them more rapidly than we can mine them, or we can buy them more cheaply than we can mine them.

21.3 Mineral Reserves

Mining depletes mineral reserves by decreasing the amount of ore remaining in the ground; but reserves may increase in two ways. First, geologists may discover new mineral deposits, thereby adding to the known amount of ore. Second, subeconomic mineral deposits—those in which the metal is not sufficiently concentrated to be mined at a profit—can become profitable if the price of that metal increases, or if improvements in mining or refining technology reduce extraction costs.

Consider an example of the changing nature of reserves. In 1966 geologists estimated that global reserves of iron were about 5 billion tons.[1] At that time, world consumption of iron was about 280 million tons per year. Assuming that consumption continued at the 1966 rate, the global iron reserves identified in 1966 would have been exhausted in 18 years, and we would have run out of iron ore in 1984. But iron ore is still plentiful and cheap today because new and inexpensive methods of processing lower-grade iron ore were developed. Thus, deposits that were subeconomic in 1966, and therefore not counted as reserves, are now ore.

21.4 Coal

The three major fossil fuels are coal, petroleum, and natural gas. All form from the partially decayed remains of organisms. **Coal** is a combustible rock composed mainly of carbon. Humans began using coal before they used petroleum and natural gas because coal is easily mined and can be burned without refining.

Coal-fired electric generating plants burn about 60 percent of the coal consumed in the United States. The remainder is used to make steel or to produce steam in factories. Although it is easily mined and abundant in many parts of the world, coal emits air pollutants that can be removed only with expensive control devices.

Large quantities of coal formed worldwide during the Carboniferous period, between 360 and 285 million years ago, and later in Cretaceous and Paleocene times, when warm, humid swamps covered broad areas of low-lying land. Coal is probably forming today in some places, such as in the Ganges River delta in India, but the process is much slower than the rate at

---

[1] B. Mason, *Principles of Geochemistry*, 3rd ed. (New York: John Wiley, 1966), Appendix III.

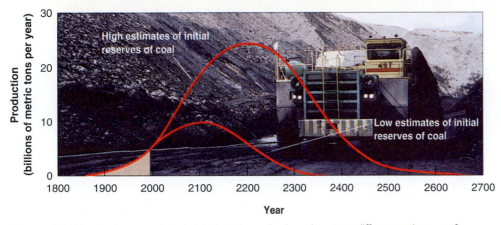

Figure 21–11 Past and predicted global coal supplies based on two different estimates of reserves. Shaded area shows coal already consumed. *(Adapted from M. King Hubbard)*

which we are consuming coal reserves. As shown in Figure 21–11, widespread availability of this fuel is projected at least until the year 2200.

Coal Formation

When plants die in forests and grasslands, organisms consume some of the litter, and chemical reactions with oxygen and water decompose the remainder. As a result, little organic matter accumulates except in the topsoil. In some warm swamps, however, plants grow and die so rapidly that newly fallen vegetation quickly buries older plant remains. The new layers prevent atmospheric oxygen from penetrating into the deeper layers, and decomposition stops before it is complete, leaving brown, partially decayed plant matter called **peat.** Commonly, peat is then buried by mud deposited in the swamp.

Plant matter is composed mainly of carbon, hydrogen, and oxygen and contains large amounts of water. During burial, rising pressure expels the water and chemical reactions release most of the hydrogen and oxygen. As a result, the proportion of carbon increases until coal forms (Fig. 21–12). The grade of coal, and

A Litter falls to floor of stagnant swamp

B Debris accumulates, barrier forms, decay is incomplete

C Sediment accumulates, organic matter is converted to peat

D Peat is lithified to coal

Figure 21–12 Peat and coal form as sediment buries organic litter in a swamp.

TABLE 21–3

Classification of Coal by Grade, Heat Value, and Carbon Content

Type	Color	Water (%)	Other Volatiles and Noncombustible Compounds (%)	Carbon (%)	Heat Value (BTU/lb)
Peat	Brown	75	10	15	3,000–5,000
Lignite	Dark Brown	45	20	35	7,000
Bituminous (soft coal)	Black	5–15	20–30	55–75	12,000
Anthracite (hard coal)	Black	4	1	95	14,000

the heat that can be recovered by burning coal, varies considerably depending on the carbon content (Table 21–3).

21.5 Mines and Mining

Miners extract both coal and metal ore from **underground mines** and **surface mines**. A large underground mine may consist of tens of kilometers of interconnected passages that commonly follow ore veins or coal seams (Fig. 21–13). The lowest levels may be several kilometers deep. In contrast, a surface mine is a hole excavated into the Earth's surface. The largest human-created hole is the open-pit copper mine at Bingham Canyon, Utah. It is 4 kilometers in diameter and 0.8 kilometer deep. The mine produces 230,000 tons of copper a year and smaller amounts of gold, silver, and molybdenum (Fig. 21–14). Most modern coal mining is done by large power shovels that extract coal from huge surface mines (Fig. 21–15).

In the United States, the Surface Mining Control and Reclamation Act requires that mining companies restore mined land so that it can be used for the same purposes for which it was used before mining began. In addition, a tax is levied to reclaim land that was mined and destroyed before the law was enacted. Although modern mining companies comply with the regulations, reclamation of older mines has been slow and incomplete. In the United States, more than 6000 unrestored coal and metal surface mines cover an area of about 90,000 square kilometers, slightly smaller than the state of Virginia. This figure does not include

Figure 21–14 The Bingham Canyon, Utah, open-pit copper mine is the largest human-created hole on Earth. It is 4 kilometers in diameter and 0.8 kilometer deep. *(Agricultural Stabilization and Conservation Service/USDA)*

Figure 21–13 Machinery extracts coal from an underground coal mine.

The 1872 Mining Law

The 1872 Mining Law governs the mining of metal ores on public land in the United States. In the mid-1800s, when pioneers were exploring the western frontier, the intent of the law was to encourage the exploration and development of America's natural resources and to foster economic growth. The law states that any individual or corporation can obtain mineral rights to metal ores on public land. The miner need only prove that he or she can make a profit by mining the deposit and do a small amount of development work on the claim each year. The miner pays no royalties or other fees to the government or public for the ore extracted. In some cases, corporations or citizens can buy land for as little as $5 per acre.

In recent years, conservationists and other public interest groups have attacked the 1872 law, claiming that it is one of the biggest giveaway programs in the United States. For example, in 1994 the federal government sold 2000 acres near Elko, Nevada, to American Barrick Resources, Inc., for $9765. Geologists estimate that the property contains $10 billion worth of gold, although profits will be much less because of mining expenses. In another instance, a mining company bought 160 acres near Keystone, Colorado. They operated a mine on a portion of the land and sold the rest for ski condominiums for $11,000 per acre.

In response, defenders of the 1872 law argue that mineral exploration entails a great deal of risk and expense because the probability of finding new ore is low. In addition, mining companies must invest large sums of money locating and developing an ore deposit before any profit is earned. The potential rewards, as guaranteed by the 1872 law, must be great to make exploration attractive to investors. Defenders also argue that the law has worked well for more than a century and has produced one of the world's most productive and efficient mining industries.

Focus Question:

How does the public benefit from mining of resources on public land?

---

abandoned sand and gravel mines and rock quarries, which probably account for an even larger area.

Although underground mines do not directly disturb the land surface, some abandoned mines collapse, and occasionally buildings have fallen into the holes (Fig. 21–16). Over 800,000 hectares (2 million acres) of land in central Appalachia have settled into underground mine shafts.

Figure 21–16 This house tilted and broke in half as it sank into an abandoned underground coal mine. *(Chuck Meyers/U.S. Department of the Interior)*

Figure 21–15 A huge power shovel dwarfs a person standing inside the Navajo Strip Mine in New Mexico. *(H. E. Malde/USGS)*

Figure 21–17 An oil refinery converts crude oil into useful products such as gasoline.

21.6 Petroleum and Natural Gas

The first commercial oil well was drilled in the United States in 1859, ushering in a new energy age. Crude oil, as it is pumped from the ground, is a gooey, viscous, dark liquid made up of thousands of different chemical compounds. It is then refined to produce propane, gasoline, heating oil, and other fuels (Fig. 21–17). Many petroleum products are used to manufacture plastics, nylon, and other useful materials.

The Origin of Petroleum

Streams carry organic matter from decaying land plants and animals to the sea and to some large lakes, and deposit it with mud in shallow coastal waters. Marine plants and animals die and settle to the sea floor, adding more organic matter to the mud. Over millions of years younger sediment buries this organic-rich mud to depths of a few kilometers, where rising temperature and pressure convert the mud to shale. At the same time, the elevated temperature and pressure convert the organic matter to liquid **petroleum** that is dispersed throughout the rock (Fig. 21–18). The activity of bacteria may enhance the process. Typically, petroleum forms in the temperature range from 50 to 100°C. At temperatures above about 100°C, oil begins to convert to natural gas. Consequently, many oil fields contain a mixture of oil and gas.

The shale or other sedimentary rock in which oil originally forms is called the **source rock**. Oil dispersed in shale cannot be pumped from an oil well because shale is relatively impermeable; that is, liquids do not flow through it rapidly. But under favorable conditions, petroleum migrates slowly to a nearby layer of permeable rock—usually sandstone or limestone—where it can flow readily. Because petroleum is less dense than water or rock, it then rises through

the permeable rock until it is trapped within the rock or escapes onto the Earth's surface.

Many **oil traps** form where a layer of impermeable rock such as shale prevents the petroleum from rising further. Oil or gas then accumulates in a petroleum **reservoir**. Folds and faults create several types of oil traps (Fig. 21–19). In some regions, large, lightbulb-shaped bodies of salt have flowed upward through solid rocks to form salt domes. The rising salt folded the surrounding rock to form an oil trap (Fig. 21–19D). The salt originated as a sedimentary bed of marine evaporite, and it rose because salt is less dense than the surrounding rocks. An oil reservoir is not an underground pool or lake of oil. It consists of oil-saturated permeable rock that is like an oil-soaked sponge.

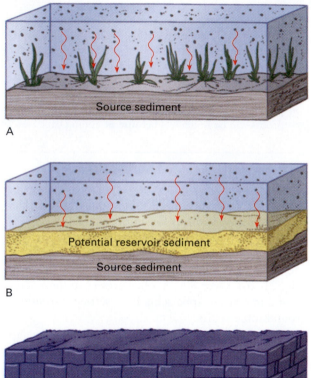

A

B

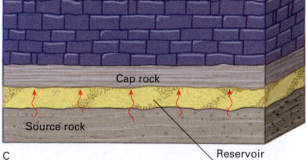

C

Figure 21–18 (A) Organic matter from land and sea settles to the sea floor and mixes with mud. (B) Younger sediment buries this organic-rich mud. Rising temperature and pressure convert the mud to shale, and the organic matter to petroleum. (C) The petroleum is trapped in the reservoir by an impermeable cap rock.

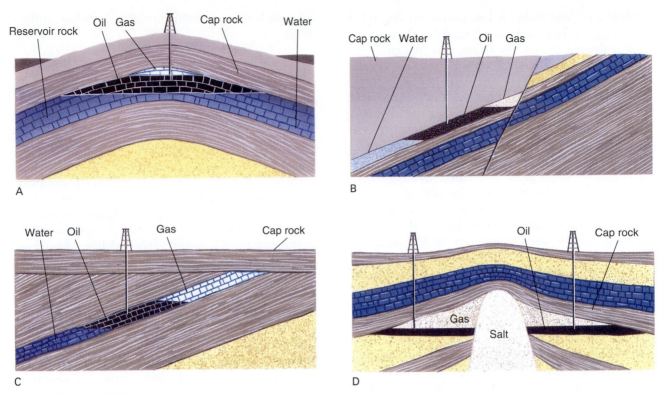

Figure 21–19 Four different types of oil traps. (A) Petroleum rises into permeable limestone capped by impermeable shale in a structural dome. (B) A trap forms where a fault has moved impermeable shale against permeable sandstone. (C) Horizontally bedded shale traps oil in tilted limestone. (D) Sedimentary salt rises and deforms overlying strata to create a trap.

Geologic activity can destroy an oil reservoir as well as create one. A fault may fracture the reservoir rock, or tectonic forces may uplift the reservoir and expose it to erosion. In either case, the petroleum escapes once the trap is destroyed. Sixty percent of all oil wells are found in relatively young rocks that formed during the Cenozoic era. Undoubtedly, much petroleum that had formed in older Mesozoic and Paleozoic rocks escaped long ago and decomposed at the Earth's surface.

Petroleum Extraction, Transport, and Refining

To extract petroleum, an oil company drills a well into a reservoir and pumps the oil to the surface. Fifty years ago, many reservoirs lay near the surface and oil was easily extracted from shallow wells. But these reserves have been exploited, and modern oil wells are typically deeper. For example, in 1949, the average oil well drilled in the United States was 1116 meters deep. In 1994, the average well was 1629 meters deep. After the hole has been bored, the expensive drill rig is removed and replaced by a pumper that slowly extracts the petroleum.

In 1989, the U.S. Geological Survey (USGS) reported that oil reserves in the United States were being rapidly depleted. Petroleum companies found half as much oil as they did in the 1950s for every meter of exploratory drilling. However, improved drilling and recovery techniques have increased our oil reserves.

Most oil wells are drilled vertically into the reservoir. Oil then flows through the reservoir to the well, where pumps raise it to the surface. The amount of oil that reaches the well is limited by the permeability of the reservoir rock and by the viscosity of the oil.

On the average, more than half of the oil in a reservoir is too viscous to be pumped to the surface by conventional techniques. This oil is left behind after an oil field has "gone dry." Recently, oil companies have developed methods of drilling horizontally through reservoirs, allowing access to vast amounts of oil left by earlier wells. Additional oil can be extracted by **secondary and tertiary recovery** techniques. In one simple secondary process, water is pumped into one well, called the injection well. The pressurized water floods the reservoir, driving oil to nearby wells, where both the water and oil are extracted. At the surface, the water is separated from the oil and reused, while the oil is sent to the refinery. One tertiary process forces superheated steam into the injection well. The steam heats the oil and makes it less viscous so that it can flow through the rock to an adjacent well. Because energy is needed to heat the steam, this type of extraction is not always cost effective or energy efficient. Another tertiary process pumps detergent into the reservoir. The detergent dissolves the remaining oil and carries it to an adjacent well, where the petroleum is then recovered and the detergent recycled.

In the United States, more than 300 billion barrels of oil remain in oil fields that were once thought to be depleted. As the technology for extracting this oil has become available, the USGS has revised earlier estimates of the oil available in North America.[2] According to the new report, improved extraction techniques have raised oil reserves by 41 percent between 1989 and 1995.

Because an oil well occupies only a few hundred square meters of land, most cause relatively little environmental damage. However, oil companies have begun to extract petroleum from fragile environments such as the ocean floor and the Arctic tundra. To obtain oil from the sea floor, engineers build platforms on pilings driven into the ocean floor and mount drill rigs on these steel islands (Fig. 21–20). Despite great care, accidents occur during drilling and extraction of oil. When accidents occur at sea, millions of barrels of oil can spread throughout the waters, poisoning marine life and disrupting marine ecosystems. Significant

---

[2] R.C. Burruss, "Petroleum Reserves in North America." *Geotimes*, July 1995, p. 14.

oil spills have occurred in virtually all offshore drilling areas.

Natural Gas

Natural gas, or methane (CH_4), forms naturally when crude oil is heated above $100°C$ during burial. Many oil wells contain natural gas floating above the heavier liquid petroleum. In other instances, the lighter, more mobile gas escaped into the atmosphere, or was trapped in a separate underground reservoir.

Natural gas is nearly pure methane and is used without refining for home heating, cooking, and to fuel large electrical generating plants. Because natural gas contains few impurities, it releases no sulfur and other pollutants when it burns. This fuel has a higher net energy yield, produces fewer pollutants, and is less expensive to produce than petroleum. At current consumption rates, global natural gas supplies will last for 80 to 200 years.

Tar Sands

In some regions, large sand deposits are permeated with heavy oil and an oil-like substance called **bitumen,** which are too thick to be pumped. The richest **tar sands** exist in Alberta (Canada), Utah, and Venezuela.

In Alberta alone, tar sands contain an estimated 1 trillion barrels of petroleum. About 10 percent of this fuel is shallow enough to be surface mined (Fig. 21–21). Tar sands are dug up and heated with steam to make the bitumen fluid enough to separate from

Figure 21–20 An offshore oil drilling platform extracts oil from the continental shelf beneath a shallow sea. *(Sun Oil)*

Figure 21–21 Tar sands are abundant in Alberta, Canada. *(Syncrude Canada, Limited)*

the sand. The bitumen is then treated chemically and heated to convert it to crude oil. At present, several companies mine tar sands profitably, producing 11 percent of Canada's petroleum. Deeper deposits, comprising the remaining 90 percent of the reserve, can be extracted using subsurface techniques similar to those discussed for secondary and tertiary recovery.

Oil Shale

Some shales and other sedimentary rocks contain a waxy, solid organic substance called **kerogen**. Kerogen is organic material that has not yet converted to oil. Kerogen-bearing rock is called **oil shale**. If oil shale is mined and heated in the presence of water, the kerogen converts to petroleum. In the United States, oil shales contain the energy equivalent of 2 to 5 trillion barrels of petroleum, enough to fuel the nation for 300 to 700 years at the 1997 consumption rate (Fig. 21–22). However, many oil shales are of such low grade that they require more energy to mine and convert the kerogen to petroleum than is generated by burning the oil, so they will probably never be used for fuel. Oil from higher-grade oil shales in the United States would supply this country for nearly 70 years if consumption rates remained at 1997 levels. Oil shale deposits in most other nations are not as rich, so oil shale is less promising as a global energy source.

Water consumption is a serious problem in oil shale development. Approximately two barrels of water are needed to produce each barrel of oil from shale. Oil shale occurs most abundantly in the semi-arid western United States. In this region, scarce water is also needed for agriculture, domestic use, and industry.

Figure 21–22 Secondary recovery, tar sands, and oil shale increase our petroleum reserves significantly.

When oil prices rose to $45 per barrel in 1981, major oil companies built experimental oil shale recovery plants. However, when prices plummeted a few years later, most of this activity came to a halt. Today, no large-scale oil shale mining is taking place in the United States.

21.7 Nuclear Fuels and Reactors

A modern nuclear power plant uses **nuclear fission** to produce heat and generate electricity. One isotope of uranium, U-235, is the major fuel. When a U-235 nucleus is bombarded with a neutron, it breaks apart (the word *fission* means "splitting"). The initial reaction releases two or three neutrons. Each of these neutrons can trigger the fission of additional nuclei; hence, this type of nuclear reaction is called a **branching chain reaction** (Fig. 21–23). Because this fission is initiated by neutron bombardment, it is not a spontaneous process and is different from natural radioactivity.

To fuel a nuclear reactor, uranium concentrated with U-235 is compressed into small pellets. Each pellet could easily fit into your hand but contains the energy equivalent of 1 ton of coal. A column of pellets is encased in a 2-meter-long pipe, called a **fuel rod** (Fig. 21–24). A typical nuclear power plant contains about 50,000 fuel rods bundled into assemblies of 200

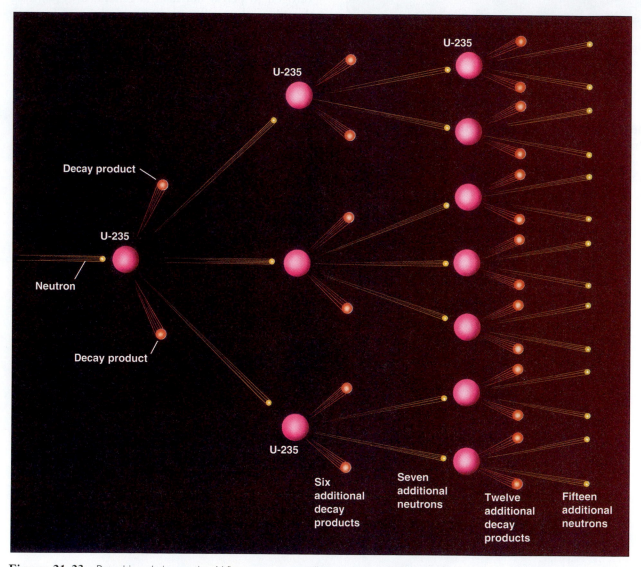

Figure 21–23 Branching chain reaction. When a neutron strikes a uranium–235 nucleus, the nucleus splits into two roughly equal fragments and emits two or three neutrons. These neutrons can then initiate additional reactions, which produce more neutrons. A branching chain reaction accelerates rapidly through a sample of concentrated uranium–235.

A   B

Figure 21–24   (A) Fuel pellets contain enriched uranium–235. Each pellet contains the energy equivalent of I ton of coal. (B) Fuel pellets are encased into narrow rods that are bundled together and lowered into the reactor core.   *(Courtesy Westinghouse Electric Corp., Commercial Nuclear Fuel Division)*

rods each. **Control rods** made of neutron-absorbing alloys are spaced among the fuel rods. The control rods fine-tune the reactor. If the reaction speeds up because too many neutrons are striking other uranium atoms, then the power plant operator lowers the control rods to absorb more neutrons and slow down the reaction. If fission slows down because too many neutrons are absorbed, the operator raises the control rods. If an accident occurs and all internal power systems fail, the control rods fall into the reactor core and quench the fission.

The reactor core produces tremendous amounts of heat. A fluid, usually water, is pumped through the reactor core to cool it. The cooling water (which is now radioactive from exposure to the core) is then passed through a radiator, where it heats another source of water to produce steam. The steam drives a turbine, which in turn generates electricity (Fig. 21–25).

The Nuclear Power Industry

Every step in the mining, processing, and use of nuclear fuel produces radioactive wastes. The mine waste

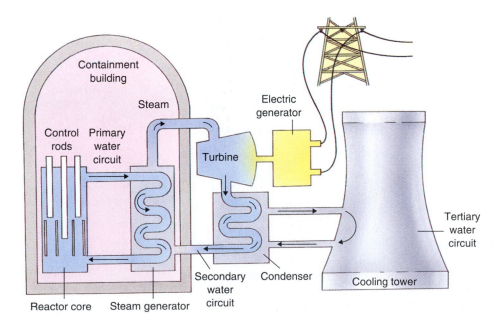

Figure 21–25   In a nuclear power plant, fission energy creates heat, which is used to produce steam. The steam drives a turbine, which generates electricity.

discarded during mining is radioactive. Enrichment of the ore produces additional radioactive waste. When a U-235 nucleus undergoes fission in a reactor, it splits into two useless radioactive nuclei that must be discarded. Finally, after several months in a reactor, the U-235 concentration in the fuel rods drops until the fuel pellets are no longer useful. In some countries, these pellets are reprocessed to recover U-235, but in the United States this process is not economical and the pellets are discarded as radioactive waste. In Chapter 20 we discussed proposals for storing radioactive wastes. To date, no satisfactory solution has been found.

In recent years, construction of new reactors has become so costly that electricity generated by nuclear power is more expensive than that generated by coal-fired power plants. Public concern about accidents and radioactive waste disposal has become acute. The demand for electricity has risen less than expected during the past two decades. As a result, growth of the nuclear power industry has halted. After 1974, many planned nuclear power plants were canceled; and after 1981, no new orders were placed for nuclear power plants in the United States.

In 1997, 109 commercial reactors were operating in the United States. These generators produced 23 percent of the total electricity consumed that year. Those numbers will decline in the coming decade because no new plants have been started and old plants must be decommissioned. *Forbes* business magazine called the United States nuclear power program "the largest managerial disaster in U.S. business history, involving $1 trillion in wasted investment and $10 billion in direct losses to stockholders."

21.8 Energy Strategies for the United States

Figure 21–26 summarizes contemporary energy production and use in the United States. Fossil fuels are the sources of 82 percent of U.S. energy production. The most abundant fossil fuel in the United States is coal. In 1994, we exported $2.4 billion worth of coal.

Before 1957, the United States was also an oil-exporting nation. However, oil consumption grew dramatically in the 1950s and 1960s. Domestic oil production peaked in 1970 and has declined since then. By 1993, the United States' oil production had declined to pre-1960 levels. These trends have led to a steady increase in oil imports (Fig. 21–27). In 1996, the United States was the second largest oil-producing nation (behind Saudi Arabia), but our consumption was so high that we imported 54 percent of our petroleum. The cash outflow has a negative impact on our national economy.

Oil dependence has also made the United States vulnerable to political instability. Twice in the 1970s, the major oil-producing nations reduced oil production and raised the price, causing economic problems in oil-importing nations. In 1990, Iraq invaded Kuwait

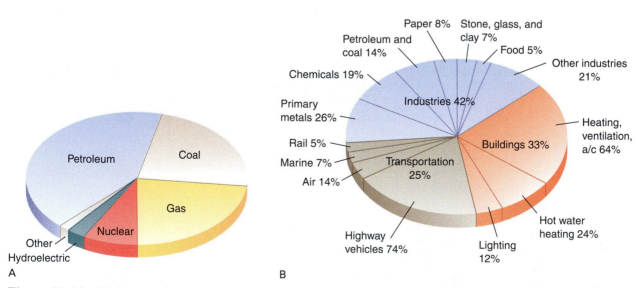

Figure 21–26 (A) Sources of energy production. (B) Energy use in the United States, with values outside of the pie diagram showing the proportional energy consumption within each of the three main sectors. Thus, highway vehicles use 74 percent of the energy consumed by all forms of transportation combined. *(Annual Energy Review 1996, Energy Information Administration)*

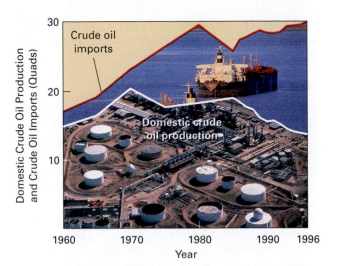

Figure 21–27 The graph shows that domestic oil production peaked in 1970 and had declined to pre-1960 levels by 1993. Total consumption increased rapidly until 1980 and has fluctuated since that time. As a result of these trends, our dependency on imported oil has risen steadily. In the background, Exxon supertankers bring Alaskan crude oil into Fidalgo Bay, Washington, where it will be refined. *(Vince Streano/Tony Stone Images)*

and threw global petroleum markets into turmoil. Within a week, the price of petroleum skyrocketed from below $20 per barrel to $40 per barrel; it then fell again after the brief Gulf War.

If war does not impede the free trade of oil, the energy future of the United States is closely linked with the global energy future. As mentioned earlier, coal reserves are expected to last for a few hundred years. However, experts are sharply divided on their predictions of when our oil supply will run out. In a commentary published in *Nature* in May 1997, the author argued that "a permanent decline in global oil production is virtually certain to begin within 20 years."[3] He based this conclusion on three observations and predictions:

- Between 1985 and 1995, global oil consumption increased by 16 percent.

- During the same time period, economic growth and per capita energy consumption was more rapid in developing countries than in the wealthier nations. Because more people live in developing nations than in the developed ones, the increase in per capita energy consumption in these areas will have a large effect on global demand. Thus, global oil consumption should continue to increase.

- Geological data indicate that there simply is not enough oil to meet this demand for longer than about 20 years.

In a rebuttal to this article, another expert argues that when oil supplies become scarce, the price will rise. As a result, secondary and tertiary reserves and oil shale will become profitable. He argues that "a gradual transition (from conventional to nonconventional sources) should be well within the power of market forces."[4]

No one knows exactly when oil supplies will run low. But we do know that we can decrease our dependence on imported oil either by using alternative energy resources or by conserving energy.

21.9 Alternative Energy Resources

Unlike fossil and nuclear fuels, alternative energy resources are renewable; natural processes replenish them as we use them. These energy resources include solar, wind, geothermal, and hydroelectric energy, wood and other biomass fuels, and ocean waves and currents. Although the amount of energy produced today by alternative sources is small compared to that provided by fossil and nuclear fuels, alternative resources have the potential to supply all of our energy needs. Excluding biomass fuels, alternative energy sources are also desirable because they emit no carbon dioxide and therefore do not contribute to global warming.

Solar Energy

Current technologies allow us to use solar energy in three ways: passive solar heating, active solar heating, and electricity production by solar cells.

A passive solar house is built to absorb and store the Sun's heat directly. In active solar heating systems, solar collectors absorb the Sun's energy and use it to heat water. Pumps then circulate the hot water through radiators to heat a building, or the inhabitants use the hot water directly for washing and bathing.

A **solar cell** produces electricity directly from sunlight. A modern solar cell is a semiconductor. Sunlight energizes electrons in the semiconductor so that they travel through the crystal, producing an electric current (Fig. 21–28). Some scientists have suggested that solar cells have the potential to supply between 50 and 100 percent of all electricity used in the United States.

At present, the cost of purchasing and operating solar cells to supply electricity to a home is approximately five times greater than purchasing the same

---

[3] Craig Bond Hatfield, "Oil Back on the Global Agenda." *Nature*, Vol. 387, May 8, 1997, p. 121.

[4] Hendrik S. Houthakker, *Nature*, Vol. 388, Aug. 14, 1998, p. 618.

Figure 21–28 This bank of solar cells generates electricity for the visitor center at Natural Bridges National Monument, Utah.

amount of electricity from a local power company over a 25-year period. Consequently, most domestic solar cell systems are sold to homeowners living in remote places that are not served by commercial power companies. Although solar cells cannot compete with other commercial systems at present, their potential for producing cheap, pollution-free energy is so great that a few research breakthroughs that lower their cost could alter the way the world produces its electricity.

Wind Energy

About 5000 clusters of wind generators now produce about 1745 megawatts in the United States, as much as that of 1.5 nuclear power plants (Fig. 21–29). The largest single wind farm in the United States is located on Hawaii's island of Oahu. A huge, untapped potential for wind generation exists in several midwestern and western states, where winds blow strongly and almost continuously.

Wind energy is cost efficient. The average price of electricity from conventional sources in 1996 varied from about 3.5 cents to 7 cents per kilowatt hour, and the cost of electricity generated by wind turbines varied from 3 to 5 cents per kilowatt hour. Wind energy is also clean and virtually limitless; however, wind farms are often unattractive, noisy, and they may interfere with migratory birds and birds of prey.

Geothermal Energy

Energy extracted from the Earth's internal heat is called geothermal energy. Hot ground water can be pumped to the surface to generate electricity, or it can be used directly to heat homes and other buildings. The United States is the largest producer of geothermal electricity in the world, with a production capacity of 2500 megawatts, an amount equivalent to the

power output of 2.5 large nuclear reactors and enough to provide over 1 million people with all of their electrical needs.

Hydroelectric Energy

If a river is dammed, the energy of water dropping downward through the dam can be harnessed to produce hydroelectric electricity. Hydroelectric generators supply between 15 and 20 percent of the world's electricity. They provide about 3.5 percent of all energy consumed in the United States, but about 15 percent of our electricity.

The United States is unlikely to increase its production of hydroelectric energy. Large dams are expensive to build, and few suitable sites remain. Environmentalists commonly oppose dam construction because reservoirs flood large areas, destroying wildlife habitat, agricultural land, towns, and fish harvests. Wild rivers and their canyons are often prized for their aesthetic and recreational value.

Biomass Energy

In 1996, biomass (plant) fuels produced 2.8 quads (quadrillion BTU) of energy. For comparison, nuclear power plants generated 6.83 quads. Wood was the most productive of all biofuels, generating 2.2 quads; controlled garbage incineration produced 0.5 quads;

Figure 21–29 Wind turbines generate electricity at Cowley Ridge, Alberta, Canada. *(Kenetech Windpower, Inc.)*

and alcohol fuels produced 0.1 quad. Wood-burning stoves produce so much smoke and particulate pollution that their use is banned in some cities and restricted in others. Recently, wood stoves have been developed that are 65 percent efficient and produce only 10 percent as much pollution as older wood stoves. But they are expensive, and wood will probably be a significant energy source only in rural areas.

Some biofuel experts suggest that trees grow too slowly and, instead, rapidly growing crops such as sugar cane, corn, or sunflowers should be grown for the energy content in their sugars, oils, or starches. In this case, the plant products are fermented to make ethanol. Ethanol burns more cleanly in automobile engines than conventional gasoline and is used as an additive to reduce urban air pollution. However, more energy from fossil fuels is required to grow the plants and distill the ethanol than is obtained from burning the ethanol. Thus, production and use of ethanol as a fuel involves an overall energy loss. In addition, the large-scale cultivation of ethanol-producing crops would require large amounts of agricultural land, putting energy production in direct competition with food production.

Energy from the Seas

Twice a day, seawater flows into a bay or estuary with the rising tide, and twice a day it rushes outward. The energy of these currents can be harnessed by a tidal dam and turbine. However, tidal dams are costly and create environmental problems. Tidal bays are commonly productive estuaries where fish breed, and they are also popular places for recreation. If these areas are dammed, some of the natural qualities and resources are lost.

Waves strike every coastline in the world, and their energy can be harnessed with a wave generator. In one type, water from an incoming wave rises in a concrete chamber, where it forces air through a narrow valve. The air spins a rotor that drives an electrical generator. At present, wave generators are too expensive to be practical in most places.

Hydrogen Fuel

When hydrogen gas burns, it combines with atmospheric oxygen to produce water vapor and no other products. Thus, it is a nonpolluting fuel. In addition, it is energy rich. Combustion of 1 kilogram of hydrogen produces two and one-half times as much energy as that produced by 1 kilogram of gasoline.

Hydrogen gas can be extracted from water by passing an electric current through the water. Because water is so abundant and the burning of hydrogen produces water to replenish the amount consumed, the hydrogen supply is limitless. Hydrogen can be stored under high pressure as a gas, or can be combined with metals to form solid compounds called **hydrides**. The hydrides can be heated to release the hydrogen gas to fuel autos, furnaces, or an electricity-generating fuel cell. In its gaseous form, the hydrogen can be shipped like natural gas. Gaseous hydrogen explodes easily, but the hydrides neither explode nor burn. Thus, they are safe and portable energy sources.

The extraction of hydrogen from water uses more energy than can be obtained by burning the hydrogen. It seems that such an energy budget would prevent its use as a fuel. However, hydrogen is energy rich, portable, and clean and thus is suited for use as a fuel for automobiles and other forms of transportation. In one plausible scenario, heavy, land-based power plants produce electricity from wind, solar, geothermal, and other renewable energy resources. Some of that electricity is used to extract hydrogen from seawater and convert it to hydrides. The hydrides, in turn, fuel autos, trucks, trains, and even airplanes.

Conservation as an Alternative Energy Resource

The single most effective way to increase the amount of energy available now, and to prolong the availability of cheap fossil fuels, is to conserve energy (Fig. 21–30). Some energy experts have suggested that as much as half of the energy consumed by industrialized nations could be conserved by changing to more efficient equipment and human habits. Two kinds of conservation strategies exist. Social solutions involve decisions to use existing energy systems more efficiently, although they may involve personal sacrifices. For example, if you choose to carpool rather than drive your own car, you save fuel but inconvenience yourself by coordinating your schedule with your carpool companions.

Technical solutions involve switching to more efficient implements. For example, a fluorescent lightbulb uses one fourth as much electricity as an incandescent bulb. A car that runs 40 miles on a gallon of gasoline can carry you as quickly and comfortably as one that gets 20 miles per gallon.

Technical solutions have saved large amounts of fuel and energy. In 1995, people in the United States used 15 million barrels of oil per day less than they would have if they had retained the technology of 1973.

Conservation Strategies Some social and technical solutions can be put into practice in the United States immediately. Many involve little or no cost; others have initial costs but pay for themselves within a few years. Consider the three major categories of

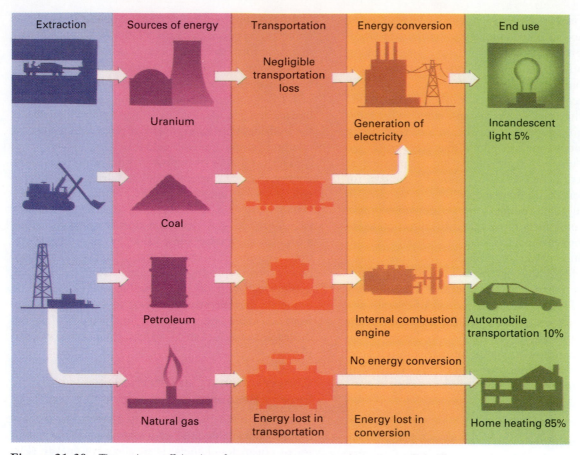

| Extraction | Sources of energy | Transportation | Energy conversion | End use |

Uranium — Negligible transportation loss — Generation of electricity — Incandescent light 5%

Coal

Petroleum — Internal combustion engine — Automobile transportation 10%

Natural gas — No energy conversion — Home heating 85%

Energy lost in transportation — Energy lost in conversion

Figure 21–30 The end use efficiencies of common energy-consuming systems. Only 15 percent of the energy generated in the United States performs useful tasks; 85 percent is wasted. Half of the waste is unavoidable but half could be avoided with improved technology and conservation. • Interactive Question: Use the data in this figure to list ways in which you could conserve energy this week. Order the list from most effective (greatest energy savings) to least effective.

energy use: energy for buildings and homes, energy for industry, and energy use in transportation.

Residential and commercial buildings consume about 36 percent of all the energy produced in the United States. Most of that energy is used for heating, air conditioning, and lighting. A super-insulated home or commercial building costs about 5 percent more to construct than a conventional one, but the energy savings usually compensate for the extra cost within about 5 years. After that time, the super-insulated construction saves money. About 25 percent of the electricity generated in the United States is used for lighting. A fluorescent bulb uses one fourth of the energy consumed by a comparable incandescent bulb. Fluorescent bulbs cost about $15 each, but they last 10 times longer and save three times as much money as they cost. As a result, switching to more efficient light sources would conserve the electricity generated by about 120 large generating plants and would save $30 billion in fuel and maintenance expenses.

Industry consumes another 36 percent of the energy used in the United States. About 70 percent of the electricity consumed by industry—half of all electricity produced in the United States—drives electric motors. Most motors are inefficient because they run only at full speed and are slowed by brakes to operate at the proper speeds to perform their tasks. This approach is like driving your car with the gas pedal on the floor and controlling your speed with the brakes. Replacing older electric motors with variable speed motors would save an amount of electricity equal to that generated by 150 large power plants. The reduced rate of electricity consumption would pay for the new motors in about a year.

Recognizing that money can be saved by conserving energy, U.S. industry reduced its energy consumption by about 70 percent per unit of production between 1973 and 1983. Since 1983, industrial efficiency has been relatively static. As a result, industry still wastes great amounts of energy.

About 28 percent of all energy and two thirds of all the oil consumed in the United States are used to transport people and goods. Americans own one third of all automobiles in the world and drive as much as the rest of the world's population combined. Ten percent of all oil consumed in the world is used by Americans on their way to and from work, about two thirds of them driving alone. The efficiency of auto and truck engines is about 10 percent. Thus, much energy can be saved by changing to more efficient cars and trucks and by using them more efficiently.

Several auto companies, including Volvo, Toyota, and Volkswagen, have produced prototypes of full-sized cars with fuel efficiencies of 30 to 60 kilometers per liter (70 to 140 miles per gallon). These cars, manufactured with strong and light materials, meet current safety and pollution standards, carry four or five passengers, and accelerate as rapidly as many current models. If mass produced, they would cost somewhat more than conventional autos, but the fuel savings would more than pay for the higher initial cost. A 40 percent increase in average automobile fuel economy over the current level would save 2.8 million barrels of oil a day, about one third of the current oil imports.

SUMMARY

Geologic resources fall into two major categories. (1) Useful rocks and minerals are called **mineral resources;** they include both **nonmetallic resources** and **metals.** All mineral resources are **nonrenewable.** (2) **Energy resources** include **fossil fuels, nuclear fuels,** and **alternative energy resources.**

Ore is a rock or other material that can be mined profitably. **Mineral reserves** are the estimated supply of ore in the ground. Five types of geologic processes concentrate elements to form ore: (1) **Magmatic processes,** such as **crystal settling,** form ore as magma solidifies. (2) **Hydrothermal processes** transport and precipitate metals from hot water. (3) Two types of **sedimentary processes** concentrate minerals. Flowing water deposits dense minerals to form **placer deposits. Evaporite deposits** and **banded iron formations** precipitate from lakes or seawater. (4) Weathering removes easily dissolved elements and minerals, leaving behind **residual deposits** such as **bauxite.** Metal ores and coal are extracted from **underground mines** and **surface mines.**

Fossil fuels include oil, gas, and coal. If oxygen and flowing water are excluded by rapid burial, plant matter decays partially to form **peat.** Peat converts to **coal** when it is buried further and subjected to elevated temperature and pressure. **Petroleum** forms from the remains of organisms that settle to the ocean floor or lake bed and are incorporated into **source rock.** The organic matter converts to liquid oil when it is buried and heated. The petroleum then migrates to a **reservoir,** where it is retained by an **oil trap.** Additional supplies of petroleum can be recovered by **secondary and tertiary extraction** from old wells and from **tar sands** and **oil shale.**

Nuclear power is expensive, and questions about the safety and disposal of nuclear wastes have diminished its future. Inexpensive uranium ore will be available for a century or more. Alternative energy resources currently supply a small fraction of our energy needs but have the potential to provide abundant renewable energy.

KEY TERMS

FOR REVIEW

1. Describe the two major categories of geologic resources.

2. Describe the differences between nonrenewable and renewable resources. List one example of each.

3. What is ore? What are mineral reserves? Describe three factors that can change estimates of mineral reserves.

4. If most elements are widely distributed in ordinary rocks, why should we worry about running short?

5. Explain crystal settling.

6. Discuss the formation of hydrothermal ore deposits.

7. Discuss the formation of marine evaporites.

8. Explain how coal forms. Why does it form in some environments but not in others?

9. Explain the importance of source rock, reservoir rock, and oil traps in the formation of petroleum reserves.

10. Discuss two sources of petroleum that will be available after conventional wells go dry.

11. List the relative advantages and disadvantages of using coal, petroleum, and natural gas as fuels.

12. Explain how a nuclear reactor works. Discuss the behavior of neutrons, the importance of control rods, and how the heat from the reaction is harnessed to produce useful energy.

13. Discuss the status of the nuclear power industry in the United States.

14. List the alternative energy resources described in this chapter. Can you think of others?

15. How does conservation act as an alternative energy resource?

16. Describe each of the ways in which solar energy can be used.

17. Describe the unique advantages of hydrogen fuel.

FOR DISCUSSION

1. What factors can make our metal reserves last longer? What factors can deplete them rapidly?

2. It is common for a single mine to contain ores of two or more metals. Discuss how geologic processes might favor concentration of two metals in a single deposit.

3. List ten objects that you own. What resources are they made of? How long will each of the objects be used before it is discarded? Will the materials eventually be recycled or deposited in the trash? Discuss ways of conserving resources in your own life.

4. If you were searching for petroleum, would you search primarily in sedimentary rock, metamorphic rock, or igneous rock? Explain.

5. If you were a space traveler abandoned on an unknown planet in a distant solar system, what clues would you look for if you were searching for fossil fuels?

6. Is an impermeable cap rock necessary to preserve coal deposits? Why or why not?

7. Discuss problems in predicting the future availability of fossil fuel reserves. What is the value of the predictions?

8. Compare the depletion of mineral reserves with the depletion of fossil fuels. How are the two problems similar, and how are they different?

9. Discuss the environmental, economic, and political implications of the development of a solar cell that produces electricity more cheaply than conventional sources.

10. Wind energy, hydroelectric power, and energy from sea waves are sometimes called secondary forms of solar energy. Discuss how the Sun is the primary source of those alternative energy resources.

Astronomy

This is part of the deepest image ever taken of the universe of galaxies. Two hundred seventy-six exposures, taken with the Hubble Space Telescope over 150 consecutive orbits, were combined to make an image that shows fainter galaxies than have ever been glimpsed before, including some that are a record-breaking 4 billion times fainter than the human eye can see. *(R. Williams, the Hubble Deep Field Team, & NASA)*

Birth of the Universe

In our search for the origin and history of the Earth, we have looked back more than 4.6 billion years to the time when a diffuse cloud of dust and gas coalesced to form the Solar System. But now, as we look into our galaxy and beyond, we ask, "How did that cloud of dust and gas form?" As our search for answers deepens, we finally ask, "How, when, and where did the Universe begin?"

Before we explore the origin of the Universe, we must ask an even more fundamental question: "Did it begin at all?" One possibility is that the Universe has always existed and there was no beginning, no start of time. An alternative hypothesis is that the Universe began at a specific time and has been evolving ever since.

In 1929 Edwin Hubble observed that all galaxies are moving away from each other. By projecting the galactic motion backward in time, he also discovered that they must have started moving outward from a common center, and at the same time. Therefore, scientists calculate that in the beginning, the entire Universe was compressed into a single infinitely dense point. This point was so small that we can't compare it with anything we know or can even imagine. According to modern theory this point exploded. But it was no ordinary explosion. It cannot even be compared with a hydrogen bomb or supernova explosion. This explosion, called the **big bang,** instantaneously created the Universe. Matter, energy, and space came into existence with this single event. It was the start of time.

Astronomers calculate the timing of the big bang by measuring speeds of galaxies and the distances among them. They then calculate backward in time to determine when they were all joined together. Estimates of the age of the Universe, starting at the big bang, vary from about 10 billion to 15 billion years. However, astronomers have found that the oldest stars appear to be older than the age of the Universe calculated from the rate of expansion. Scientists are working to resolve this discrepancy.

Even though our estimates of the time of the big bang vary by 5 billion years, astronomers have reconstructed a picture of the first few seconds after the origin of the Universe. This reconstruction is based on studies of how particles behave when they collide at very high velocities in modern particle accelerators. Other evidence comes from studying particles and radiant energy in space.

Scientists generally start their discussions of the Universe when it was a trillionth of a trillionth of a billionth of a second old. At that time, the Universe was about as big as a grapefruit and the temperature was about 100 billion degrees Celsius. During the first second, it cooled to about 10 billion degrees, 1000 times the temperature in the center of the modern Sun (Fig. 1). At such high temperatures, atoms do not exist. Most of the Universe consisted of a plasma of radiant energy, electrons, and extremely light particles called neutrinos. Protons and neutrons also began to form. After about 1.5 minutes, the temperature

fell to 1 billion degrees and a few simple atomic nuclei formed, although the temperature was still too hot for atoms. During the next million years the Universe continued to cool as it expanded. When the temperature dropped to a few thousand degrees, atoms formed and, in a sense, the modern Universe was born.

When you look up into the sky on a clear, moonless night, you see bands of stars sparkling in the blackness. If you owned a powerful telescope, you could peer deeper into space and detect distant galaxies, or even curious structures called quasars. All of these objects emit what astronomers call second generation energy. They use the term second generation because this light is emitted by concentrated matter—stars, galaxies, or quasars. But in the beginning, before these objects evolved, there was only a massive, hot plasma of elemental particles. In order to understand the early Universe, astronomers search for first generation energy, originating from that primordial plasma.

In 1964, Arno Penzias and Robert Wilson at Bell Laboratories calculated that the intensely hot first generation energy cooled over time as the Universe expanded. Today, it has cooled to a frigid 2.7°C above absolute zero. At that temperature, the remaining traces of first generation energy exist as microwave radiation. Penzias and Wilson built an antenna to search for the microwave signals from the early Universe. Scanning the sky, they detected a very faint radiation that appeared to be uniformly dis-

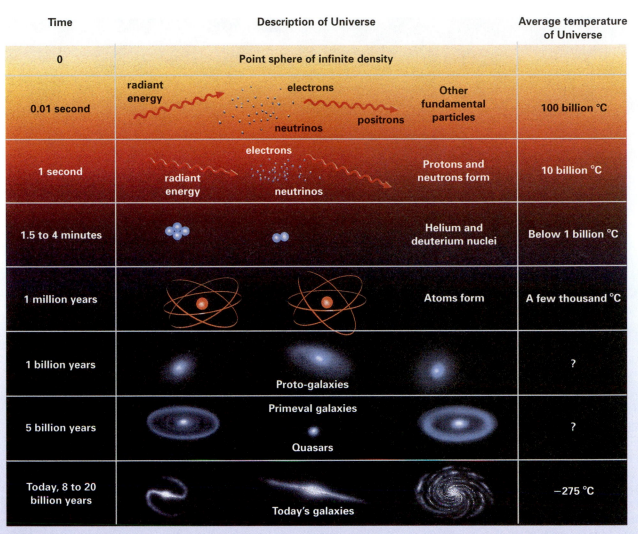

Time	Description of Universe			Average temperature of Universe
0	Point sphere of infinite density			
0.01 second	radiant energy	electrons neutrinos positrons	Other fundamental particles	100 billion °C
1 second	radiant energy	electrons neutrinos	Protons and neutrons form	10 billion °C
1.5 to 4 minutes			Helium and deuterium nuclei	Below 1 billion °C
1 million years			Atoms form	A few thousand °C
1 billion years		Proto-galaxies		?
5 billion years		Primeval galaxies Quasars		?
Today, 8 to 20 billion years		Today's galaxies		−275 °C

Figure 1 A brief pictorial outline of the evolution of the Universe.

tributed throughout space. This **cosmic background radiation** began traveling through space when the Universe was only a few thousand years old and has been called the echo of the big bang. They interpreted the apparently uniform distribution of this cosmic background radiation to mean that the Universe was homogeneous in its infancy.

However, the modern Universe is clearly not homogeneous. Matter is concentrated into stars, stars are clumped into galaxies, and galaxies are grouped into clusters containing tens of thousands of galaxies. Even the clusters group into su-

perclusters. Most of the space between the clusters and superclusters contains no galaxies at all (Fig. 2).

The question of how an initially uniform, homogeneous Universe could concentrate into stars, galaxies, and clusters disturbed cosmologists for nearly three decades. The validity of the big bang theory hinged on finding when, and how, the Universe became nonhomogeneous.

According to one hypothesis, the original grapefruit-sized Universe had to obey laws of quantum mechanics, the same laws that describe modern atoms. According to

quantum mechanics, the earliest Universe contained tiny waves of energy, space, and time. These waves, like sound waves, contained alternating regions of higher and lower densities. If this model were correct, then these bands of varying energy densities would have created tiny temperature differences in the cosmic background radiation.

In the 1980s, physicists calculated that the temperature differences would be about one ten thousandth of a degree Celsius. In order to search for these temperature differences, they had to build a much more precise radio tele-

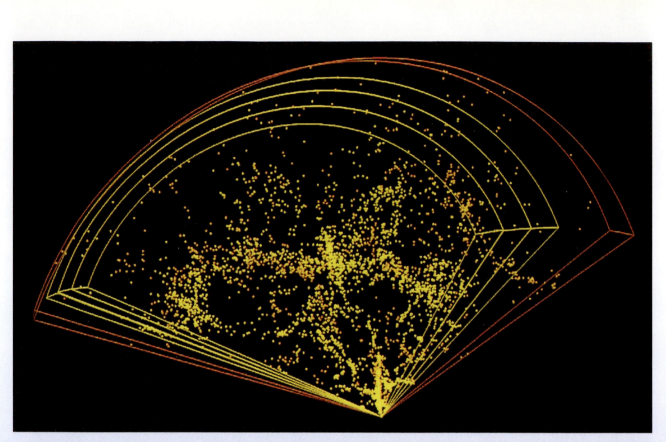

Figure 2 A three-dimensional computer-generated drawing of a portion of the Universe. Notice that the galaxies are distributed unevenly. *(Harvard Smithsonian Center for Astrophysics)*

scope than that used by Penzias and Wilson. Because the atmosphere interferes with microwave transmissions, they needed to mount this antenna on a satellite.

In 1989, astronomers launched the Cosmic Background Explorer (COBE) satellite which was capable of measuring 0.0001°C temperature differences in the cosmic background radiation. As it slowly scanned the Universe, COBE registered numerous fluctuations in the background temperature. After mapping the temperature of the Universe for three years, COBE scientists reported that cosmic background radiation varies by tiny amounts from one region to another. This data suggested that the primordial Universe was not homogeneous as Penzias and Wilson had inferred. Matter and energy had concentrated into clumps during the earliest infancy of the Universe, perhaps in the first bil-

lionth of a second. George Smoot, a researcher on the project, called these variations "the imprints of tiny ripples in the fabric of space-time put there by the primeval explosion."

From the first instant of the big bang, the Universe has been affected by two opposing forces. Matter is expanding as a result of the big bang, but at the same time gravity is pulling all matter back inward toward a common center.

If the gravitational force of the Universe is sufficient, all the galaxies will eventually slow down, reverse direction, and fall back to the center, forming another point of infinite density. This hypothesis is called a **closed Universe.** Some scientists have speculated that this point may then explode again to form a new Universe. In turn the new Universe will expand and then collapse, creating a continuous chain of universes. However, no

mathematical model explains how another big bang would occur. Instead, current models predict that in a closed system, the Universe would collapse into a mammoth black hole.

Another possibility is that the gravitational force of the Universe is not sufficient to stop the expansion, and the galaxies will continue to fly apart forever. Within each galaxy, stars will eventually consume all their nuclear fuel and stop producing energy. As the stars fade and cool, the galaxies will continue to separate into the cold void. This scenario is called the **open Universe.** Astronomers are attempting to measure the mass of the Universe to determine which of the two possibilities is correct (Fig. 3).

You can study the effect of gravity on an object by studying its motion. Thus, you can calculate the Earth's mass by observing the

518

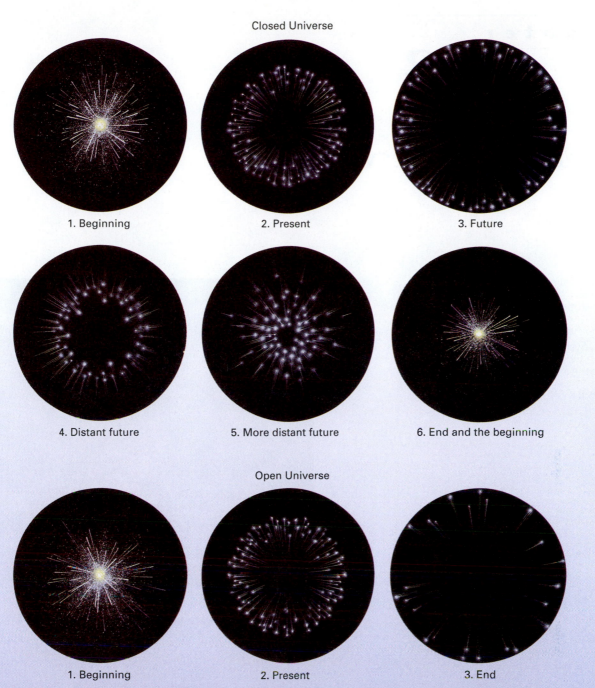

Closed Universe

1. Beginning   2. Present   3. Future

4. Distant future   5. More distant future   6. End and the beginning

Open Universe

1. Beginning   2. Present   3. End

Figure 3 A schematic representation of two possible cosmologies, a closed Universe cosmology and an open Universe cosmology.

flight of a ball thrown into the air, and you can calculate the relative masses of the Sun and the Earth by measuring the Earth's orbit. Observations of stellar and galactic motions yield a surprising result. Gravitational forces acting on the Universe are much greater than can be accounted for by all the known stars, dust clouds, and galaxies. By some estimates, matter that we see is only 10 percent of the mass of the Universe. The other 90 percent is invisible and is called **dark matter.** What is it? Suggestions include planet-like objects that do not radiate energy and are therefore invisible, and black holes that cannot radiate energy. One suggestion is that dark matter is made up of a type of particle that we have not yet detected. Because of these uncertainties about the total mass of the Universe, astronomers have been unable to agree whether the future of the Universe is a slow, cold death or an instantaneous, explosive one.

Motions in the Heavens

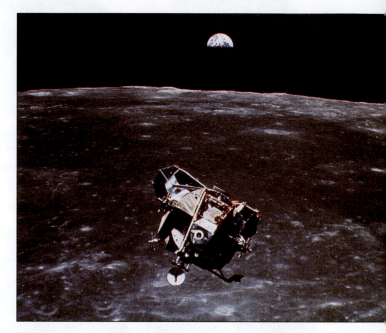

The *Apollo 11 Lunar Module* returning to the Command Module after the first landing of humans on the Moon. Smith's Sea is in the background with the Earth rising behind it.   *(NASA)*

Ancient astronomers used cycles of the Sun, Moon, and stars to tell time and to mark the seasons. However, these objects are so far away that, until recently, astronomers knew very little about them. The ancient Greeks believed that the Earth is stationary and that the Sun rises and sets because it orbits our planet daily. Today we realize that the Earth's rotation causes the cycle of night and day.

The Sun shines brightly in the sky and warms the Earth, whereas the stars twinkle feebly in the darkness. Modern astronomers understand that the Sun and stars are similar and that the Sun is merely much closer. However, the ancients had little concept of astronomical distances or sizes, and as a result, they didn't realize that the Sun is just an average-sized star.

In this and the following two chapters we will study the motions, compositions, and structures of the celestial bodies.   •

22.1 The Motions of the Heavenly Bodies

Even a casual observer notices that both the Sun and Moon rise in the east and set in the west. In the mid-latitudes, summer days are long and the Sun rises high in the sky. Winter days are shorter and, at mid-latitudes in the Northern Hemisphere, the Sun never rises very high above the southern horizon, even at noon. In contrast to the yearly cycle of seasons, the Moon completes its cycle once a month. It is full and round one night, then darkens slowly until it is a thin crescent. It disappears completely after two weeks, then reappears as a sliver that grows again, completing the entire cycle in about 29.5 days.

If you had been a cowboy in the 1800s, the foreman might have told you to guard the herd at night until the "dipper holds water." This phrase refers to two features of the night sky. First, stars remain in fixed positions relative to one another. This fact led the ancients to identify groups of stars, which they called **constellations** (Fig. 22–1). Second, the stars re-volve about a fixed point in the sky marked by the Pole Star, or North Star. In the Northern Hemisphere, the Pole Star remains almost motionless and all other stars appear to revolve around it (Fig. 22–2). This motion provided the clock for the cowboy on night watch because the dipper alternately holds and spills water as it revolves around the Pole Star.

Constellations appear and disappear with the seasons. For example, the Egyptians noted that Sirius, the brightest star in the sky, became visible at dawn just before the Nile began to flood. Farmers would therefore plant crops when this star first appeared, with the assurance that the high water soon would irrigate their fields.

Ancient astronomers noted several objects that appeared to be stars, but were different because they changed position with respect to the stars. The ancient Greeks called these objects **planets**, from the word meaning "wanderers." For most of the year, planets appear to drift eastward with respect to the stars, but sometimes they seem to reverse direction and drift westward. This apparent reverse movement is called **retrograde motion** (Fig. 22–3).

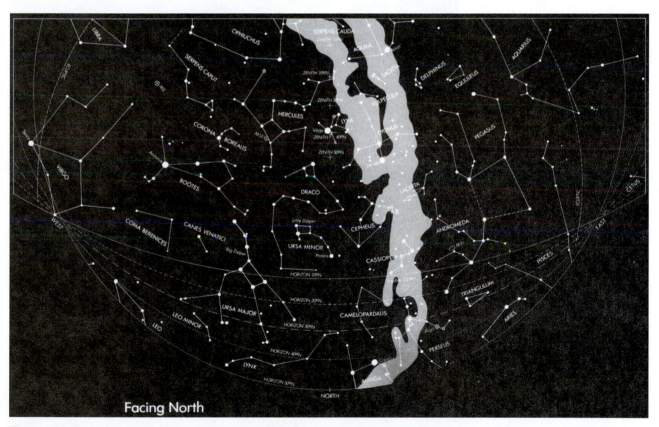

Figure 22–1 Different constellations appear at different seasons. This view shows the summer sky, as viewed by an observer facing north. The light shaded area is the Milky Way. *(Wil Tiron)*

Figure 22–2 A time exposure of the night sky shows the rotation of the stars around the Pole Star, which is nearly motionless. *(Yerkes Observatory)*

Figure 22–3 A planetarium simulation of the movement of Mars from August 1, 1990, through April 1, 1991. Mars appears to reverse direction, forming a retrograde loop. This motion was difficult to explain in the geocentric model. *(Geoff Chester, Albert Einstein Planetarium, Smithsonian Institution)*

The Constellations

The oldest known written record of the constellations appears in a 2000 B.C. Sumerian manuscript. Other accounts occur in old manuscripts and legends from numerous cultures. Ancient astronomers and religious leaders imagined that constellations represented animals such as Leo, the lion, or gods such as Aquarius, the Sumerian deity who pours the waters of immortality onto the Earth. The constellations have no particular significance to modern astronomy other than as a convenience to naming parts of the sky map. For example, Orion, the hunter, is easily recognized by the three bright stars of his sword belt (Fig. 1). The brightest star in Orion is Rigel, which forms his left kneecap. If you look at Rigel through a telescope or a good set of binoculars, you will note that it is bluish. In contrast, Betelgeuse (pronounced beetle juice), which defines Orion's right shoulder, is red. As we will learn in Chapter 24, color is an indication of the temperature of a star, and the temperature is an indication of its size and age. Blue stars such as Rigel are quite different from red ones such as Betelgeuse. In addition, the two stars are not even close to one another: Rigel is almost twice as far from Earth as Betelgeuse is. The only relationship between the two is that they appear close together in the night sky.

Each star in a constellation is independent from the others, and travels in its own path. These relative motions aren't apparent in a single lifetime, but they become significant over geologic time. As a result, constellations change shape over geologic time.

A   **B**

Figure 1 (A) A 17th century star atlas depicts the constellation Orion. *(J.M. Pasachoff and the Chapin Library)* (B) As seen in the night sky, the three stars in Orion's belt form a line in the center of the photograph. The bright stars below it represent his sword. *(Harvard College Observatory)*

22.2 Aristotle and the Earth-Centered Universe

The Greek philosopher and scientist Aristotle proposed a **geocentric,** or Earth-centered, Universe. In this model, the Earth is stationary and positioned at the center of the Universe. A series of concentric **celestial spheres** made of transparent crystal surrounds the Earth. The Sun, Moon, planets, and stars are imbedded in the spheres like jewels (Fig. 22–4). At any one time a person can see only a portion of each sphere, but as the sphere revolves around the Earth, objects appear and disappear.

Aristotle based his conclusions on two observations. First, he reasoned that the Earth must be stationary because people have no sensation of motion. People

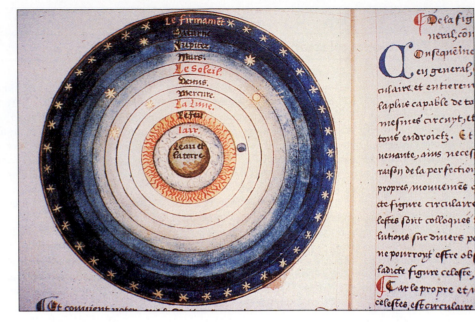

Figure 22–4 In Aristotle's cosmology, water and the Earth lie in the center of the Universe, surrounded by air and fire. Beyond these four basic elements, The Moon, Mercury, Venus, the Sun, Mars, Jupiter, Saturn, and the stars lie in concentric celestial spheres. *(Houghton Library, Harvard University)*

A

B

Figure 22–5 Parallax is illustrated by two photographs of a fence on the Montana prairie. (A) The photographer is nearly in line with the fence so the distant posts appear to be in line. (B) When the photographer moved, the posts appear to have shifted position. Now we can see spaces between the more distant posts. Of course, the posts haven't moved, only the photographer has. The same effect has been observed in astronomical studies. As the Earth revolves about the Sun, the stars appear to shift position relative to one another. ● Interactive Question: Imagine that there are two poles in the ground set 10 meters apart. You are directly in line with the two poles, 100 meters from the closest one. Now you step one meter to the left. Draw a scale diagram of your position and the positions of the two poles. With a protractor, estimate the parallax angle between you and the poles. Discuss the magnitude of the parallax angle if the units in your diagram were changed from meters to light-years.

tend to fall off the back of a chariot when horses start to gallop, but they do not fall off the Earth, so Aristotle reasoned that the Earth must not be moving.

Aristotle's second observation was based on **parallax,** the apparent change in position of an object due to the change in position of the observer. To understand parallax, consider the fence posts in Figure 22–5. They appear to stand one in front of the other when the photographer is in position (A) and seem offset when the photographer steps to the side (B). Thus the relative positions of the stationary posts appear to change with the movement of the observer. The ancients correctly reasoned that if the Earth moved around the Sun, the stars should change position relative to one another (Fig. 22–6). Since they observed no parallax shift, they concluded that the Earth must be stationary. The mistake arose not out of faulty reasoning, but because stars are so far away that their parallax shift is too small to be detected with the naked eye.

Aristotle incorporated philosophical concepts, in addition to observation, into scientific theory. He argued that the gods would create only perfection in the heavens and that a sphere is a perfectly symmetrical shape. Therefore, the celestial spheres must be a natural expression of the will of the gods. In addition, the Sun, Moon, planets, and stars must also be unblemished spheres. This incorporation of philosophy into science retarded debate and allowed dogma to rule over logic.

Aristotle's theory, although incorrect, did explain the motions of the Sun, Moon, and stars (Fig. 22–7). However, it failed to explain the retrograde motion of the planets. In about A.D. 150, Claudius Ptolemy modified the celestial sphere model to incorporate retrograde motion. In Ptolemy's model, each planet moves in small circles as it follows its larger orbit around

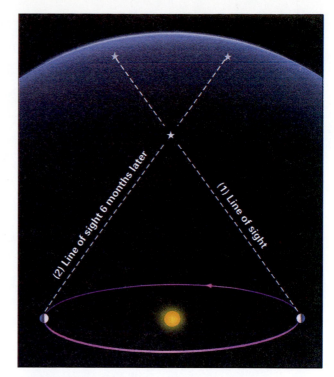

Figure 22–6 A nearby star appears to change position with respect to the distant stars as the Earth orbits around the Sun. This drawing is greatly exaggerated; in reality, the distance to the nearest stars is so much greater than the diameter of the Earth's orbit that the parallax angle is only a small fraction of a degree.

the Earth (Fig. 22–8). Ptolemy's sophisticated mathematics accurately described planetary motion and therefore his model was accepted. Still, he retained Aristotle's erroneous idea that the Sun and the planets orbit a stationary Earth.

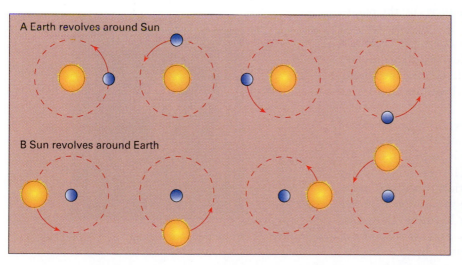

Figure 22–7 Series (A) shows the Earth revolving around the Sun, and series (B) shows the Sun revolving around the Earth. Now lay some thin paper over series (A) and trace the outlines of the Sun and Earth, but not the arrows or orbits. Lay this tracing over series (B) and note that they match exactly, after you shift the paper for each sketch to make sure the Sun and the Earth superimpose. Conclusion: There is no apparent difference between the Earth revolving around the Sun and vice versa, provided you do not refer to anything else, such as the outline of these diagrams or another star.

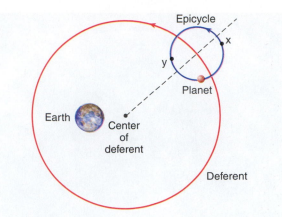

Figure 22–8 Ptolemy's explanation of retrograde motion. Each planet revolves in a small orbit (the epicycle) around the larger orbit (the deferent). When the planet is in position X, it appears to be moving eastward. When it is in position Y it appears to have reversed direction and moves westward. Ptolemy did not realize that the planets moved in elliptical orbits. To compensate for this error, he placed the Earth away from the center of the deferent.

22.3 The Renaissance and the Heliocentric Solar System

Aristotle's and Ptolemy's ideas remained essentially unchallenged for 1400 years. Then, in the 120 years from 1530 to 1650, several Renaissance scholars changed our understanding of motion in the Solar System and, in the process, revolutionized scientific thought.

Copernicus

In 1530 a Polish astronomer and cleric, Nicolaus Copernicus, proposed that the Sun, not the Earth, is the center of the Solar System and that the Earth is a planet, like the other "wanderers" in the sky (Fig. 22–9). Copernicus based his hypothesis on the philosophical premise that the Universe must operate by the simplest possible laws. Copernicus believed Ptolemy's model was too complex and that the motions of the heavenly bodies could be explained more concisely by a **heliocentric** model with the Sun at the center of the Solar System.

Figure 22–10 shows how the heliocentric model explains retrograde motion. Assume that initially the Earth is in position 1 and Mars is in position 1′. An observer on Earth looks past Mars (as shown by the white line) and records its position relative to more distant stars. Mars appears to be in position 1″ in the night sky. After a few weeks the Earth has moved to position 2, Mars has moved to 2′, and Mars's position relative to the stars is indicated by 2″. In the

Copernican model, Earth moves faster than Mars because Earth is closer to the Sun. Therefore, Earth catches up to and eventually passes Mars, as shown by positions 3, 3′ and 4, 4′. During this passage, Mars appears to turn around and move backward, although, of course, this appearance is merely an illusion against the backdrop of the stars. Mars appears to reverse direction again through positions 5 and 6, thus completing one cycle of retrograde motion. In reality, Mars never reverses direction, it just appears to behave in this way as Earth catches up to it and then passes it.

Brahe and Kepler

In the late 1500s, Tycho Brahe, a Danish astronomer, accurately mapped the positions and motions of all known bodies in the Solar System. His maps enabled him to predict where any planet would be seen at any time in the near future, but he never explained their motions. Brahe drank himself to death at a party in 1601, and his student Johannes Kepler inherited the vast amount of data Brahe had collected. Kepler calculated that the planets moved in elliptical orbits, not circular ones, and he derived a set of mathematical formulas to describe their paths. However, Kepler never answered the important question "Why do the planets move in orbits around the Sun rather than flying off into space in straight lines?"

Galileo

Galileo was an Italian mathematician, astronomer, and physicist who made so many contributions to science that he is often called the father of modern science (Fig. 22–11). Perhaps most significant was his realization that the laws of nature must be understood through observation, experimentation, and mathematical analysis. This concept freed scientists from the confines of Aristotelian thought.

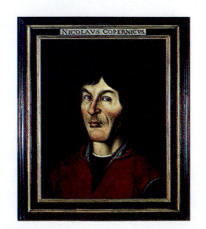

Figure 22–9 Nicolaus Copernicus. *(Marek Demianski, Torun Museum)*

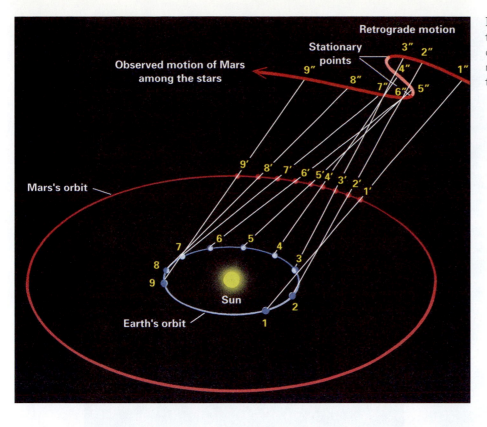

Figure 22–10 Retrograde motion is explained within the heliocentric model by changes in the relative positions of the Earth and the planet we are observing.

Recall that Aristotle's geocentric theory was based on two observations and one philosophical premise: (1) If the Earth were moving, people should fall off, but they don't. (2) Aristotle was unable to observe parallax shift of the stars. (3) The gods would create only an unblemished, symmetrical Universe. Galileo's experiments and observations showed that Aristotle's model was incorrect and led him to support Copernicus's heliocentric model.

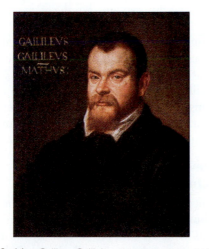

Figure 22–11 Galileo Galilei. *(Tintoretto; National Maritime Museum)*

Galileo studied the motion of balls rolling across a smooth marble floor, and legend tells us he also dropped objects from the Leaning Tower of Pisa. He organized the results of these experiments into laws of motion that were later expanded and quantified by Isaac Newton. One of these laws states that "an object at rest remains at rest and an object in uniform motion remains in uniform motion until forced to change." This corresponds to Newton's First Law of Motion, the law of **inertia.** Inertia is the tendency of an object to resist a change in motion.

According to the law of inertia, if the Earth were in uniform motion, a person on its surface would be in uniform motion along with it. The person therefore would travel with the Earth and would not fall off and be left behind, as Aristotle had assumed. In fact, the person could not even feel the motion.

Galileo built his first telescope in 1609 and turned it to the heavens soon afterwards. He learned that the hazy white line across the sky called the Milky Way was not a cloud of light as Aristotle had proposed but a vast collection of individual stars. Next, he trained his telescope at the Moon and saw hills, mountains, giant craters, and broad, flat regions on its surface, which he thought were seas and therefore named **maria** (Latin for "seas") (Fig. 22–12). Looking at the Sun, Galileo recorded dark regions, called sunspots, that appeared and then vanished.

Figure 22–12 A comparison of Galileo's drawing of the Moon with a modern photograph shows how accurately he drew topographical features, such as the maria A, B, and C. Although Galileo thought that the maria were seas, today we know that they are flat lava flows. *(J.M. Pasachoff after Ewen Whitaker, Lunar and Planetary Laboratory, University of Arizona; Photo, NASA)*

Although these discoveries had no direct bearing on the controversy of a geocentric versus a heliocentric Universe, they were important because they led Galileo to question Aristotle's views. The prevailing scientific opinion at the time was that if the Milky Way were a collection of stars, Aristotle would have known about it. Furthermore, Galileo's observations of the Sun and the Moon did not agree with Aristotle's philosophical assumptions that the heavenly bodies were perfectly homogeneous and unblemished. Galileo reasoned that if Aristotle was wrong about the structures of the Milky Way, the Sun, and the Moon, perhaps he was also wrong about the motions of these celestial bodies.

When Galileo studied Jupiter, he saw four moons orbiting the giant planet. (Today we know that Jupiter has 16 moons, but only four are large enough to have been seen with Galileo's telescope.) According to the geocentric model, *every* celestial body orbits the Earth. However, Jupiter's moons clearly orbited Jupiter, not the Earth. The contradiction increased Galileo's doubt about Aristotelian theories.

Finally, he observed that the planet Venus passes through phases as the Moon does. Such cyclical phases could not be readily explained in the geocentric model, even with Ptolemy's modifications (Fig. 22–13). As a result of his observations, Galileo proposed that the Sun is the center of the Solar System and that the planets orbit around it. The Roman Catholic Church rejected Galileo's model and ordered him to recant his conclusions. He bowed to the pressure rather than undergo imprisonment and torture. Despite his recantations, he died blind, poor, and under arrest.

Isaac Newton and the Glue of the Universe

Galileo successfully described the motions in the Solar System, but he never addressed the question "Why do the planets orbit the Sun rather than fly off in straight lines into space?" Aristotle had observed that an ar-

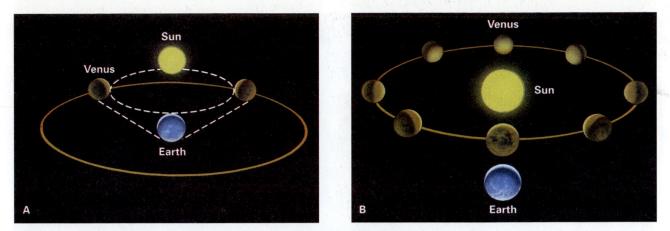

Figure 22–13 (A) In Ptolemy's theory, Venus could never move farther from the Sun than is shown by the dotted lines. Therefore it would always appear as a crescent. (B) In the heliocentric theory, Venus passes through phases like the moon. Galileo observed phases of Venus through a telescope, and concluded that the planets must orbit the Sun.

row shot from a bow flies in a straight line, but celestial bodies move in curved paths. He reasoned that arrows have an essence that compels them to move in straight lines, and planets and stars have an essence that compels them to move in circles. In the Renaissance, however, this answer was no longer acceptable.

Isaac Newton was born in 1643, the year Galileo died (Fig. 22–14). During his lifetime Newton made important contributions to physics and developed calculus. A popular legend tells that Newton was sitting under an apple tree one day when an apple fell on his head and—presto—he discovered gravity. Of course, people knew that unsupported objects fall to the ground long before Newton was born. But he was the

first to recognize that gravity is a universal force that governs all objects, including a falling apple, a flying arrow, and an orbiting planet.

According to the laws of motion introduced by Galileo and expanded by Newton, a moving body travels in a straight path unless it is acted on by an outside force. Just as a ball rolls in a straight line unless it is forced to change direction, a planet also moves in a straight line unless a force is exerted on it. The gravitational attraction between the Sun and a planet forces the planet to change direction and move in an elliptical orbit. Gravity is the glue of the Universe and affects the motions of all celestial bodies.

22.4 The Motions of the Earth and Moon

By about 1700 astronomers knew that the Sun is the center of our Solar System and that planets revolve around it in elliptical orbits. In addition to revolving around the Sun, the planets simultaneously spin on their axes. The Earth spins approximately 365 times for each complete orbit around the Sun. Each complete rotation of the Earth represents one day. As the Earth rotates about its axis, the Sun, Moon, and stars appear to move across the sky from east to west. We explained in Chapter 15 that the Earth's axis is tilted and that this tilt combined with the Earth's orbit around the Sun produces the seasons.

Recall from Section 22.1 that different stars and constellations are visible during different seasons. The Earth's revolution around the Sun causes this seasonal

Figure 22–14 Sir Isaac Newton. *(National Portrait Gallery, London)*

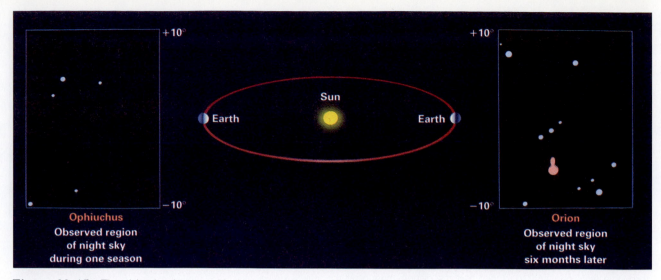

Figure 22–15 The night sky changes with the seasons because the Earth is continuously changing position as it orbits the Sun.

change in our view of the night sky (Fig. 22–15). Part of the sky is visible on a winter night when the Earth is on one side of the Sun, and a different part is visible on a summer night six months later.

More recent measurements have revealed several additional types of planetary motion. As the Earth rotates, its axis wobbles like a spinning top. This motion is called **precession** (Fig. 22–16A). At the present time, the axis points toward Polaris, the North Star (Fig. 22–16B). In 12,000 years, the axis will point toward Vega, and Vega will become the North Star. Because precession cycles on a 26,000-year period, the Earth's axis will point toward Polaris again by the year 28,000. Recall from Chapter 11 that precession is one factor that contributed to climatic fluctuations during the Pleistocene Ice Age.

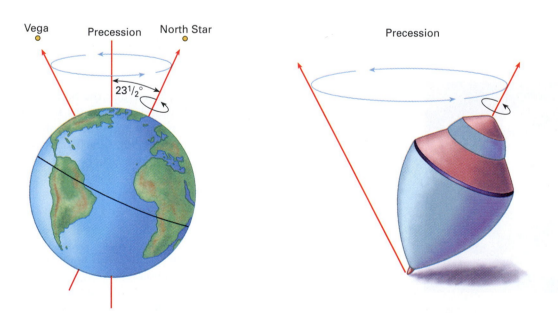

Figure 22–16 The Earth's axis wobbles or precesses like a top, completing one precession cycle every 26,000 years.

In addition, the Moon's gravity pulls the Earth slightly out of its orbit. Although we normally draw the Earth's orbit around the Sun as a line, the Moon's effect actually causes the Earth to spiral slightly as it circles the Sun. At the same time, the entire Solar System is moving toward the star Vega. Our Sun also orbits the center of the Milky Way galaxy. Traveling at a speed of 220 kilometers per second, it completes its orbit in about 200 million years. The entire Milky Way galaxy speeds along, carrying our Sun and planets through intergalactic space.

Motion of the Moon

The gravitational attraction between the Moon and the Earth not only holds the two in orbit around each other but it also affects the surfaces and interiors of both. We learned that the Moon's gravitation causes tides on Earth. In turn, the Earth's gravitation pulls on the Moon sufficiently to cause it to bulge, despite the fact that it is solid rock. The Earth's gravity attracts the bulge, so the side of the Moon with the bulge always faces the Earth. As a result, the Moon rotates on its axis at the same rate at which it orbits the Earth. Thus, we always see the same side of the Moon, and the other side was invisible to us until the Space Age.

Almost everyone has observed that the Moon appears to change shape over a month's time. When the moon is dark, it is called the **new moon.** Four days after the new moon, a thin **crescent moon** appears. The crescent grows and the Moon is **waxing** until it forms the first quarter 7 days after the new moon. Ten days after the new moon we see the bright waxing **gibbous moon**, with only a sliver of dark. About 14 to 15 days after the new moon, the moon appears circular and is a **full moon** (Fig. 22–17). A few evenings later, part of the disk is darkened. As the days progress, the visible portion shrinks and the Moon is said to be **waning.** Eventually, only a tiny curved sliver, the waning crescent, is left. After a total cycle of about 29.5 Earth days the Moon is dark because it has returned to the new moon position.

When the Moon is full, it rises approximately at sunset. On each successive evening it rises about 53 minutes later, so that in 7 days it rises in the middle of the night. In about a month the cycle is complete, and the next full moon rises in the early evening again.

To understand the phases of the Moon, we must first realize that the Moon does not emit its own light but reflects light from the Sun. The half of the Moon facing the Sun is always bathed in sunlight, while the other half is always dark. The phases of the Moon depend on how much of the sunlit area is visible from Earth. In turn, this visible area depends on the relative positions of the Sun, Moon, and Earth. The Earth's orbit around the Sun forms an elliptical plane,

with the Sun in the same plane and near its center. In a similar way, the Moon's orbit around the Earth describes another plane with the Earth at its center. But the plane of the Moon's orbit is tilted 5.2° with respect to the Earth's orbital plane. As a result, the Moon is not usually in the same plane as that of the Earth and Sun (Fig. 22–18). Therefore the Earth's shadow does not normally fall on the Moon and the Moon's shadow does not normally fall on Earth.

As the Moon orbits Earth, the side of the Moon facing the Sun is fully illuminated. When the Moon is on the opposite side of the Earth from the Sun the entire sunlit area is visible, and the Moon appears full (Fig. 22–17). However, if the Moon moves to a position between the Earth and the Sun, the Moon's sunlit side faces away from us. In this position the Moon appears new because the surface that faces the Earth is not illuminated. Midway between these two extremes, when the Moon is located 90° from the line between the Earth and the Sun, half of the illuminated side is visible, and the Moon is quartered. With each complete revolution around the Earth, the Moon passes through one complete cycle, from new to first quarter to full to third quarter and back to new again.

But why does the Moon rise at a different time each day? It travels in a complete elliptical orbit about the Earth about once a month. Imagine that today the Moon rises at 6:00 P.M. Twenty-four hours later the Earth will have rotated once, but in the meantime the Moon will have moved a short distance across the sky; therefore, the Earth must rotate an additional 13.2° in order to catch up with the Moon (Fig. 22–19). Thus, the timing of the moonrise depends on the Moon's position above a point on the Earth, and this also changes in a cyclical monthly pattern.

Eclipses of the Sun and the Moon

Recall that the plane of the Moon's orbit around the Earth is tilted with respect to that of the Earth's orbit about the Sun, so normally the Moon lies slightly out of the plane of the Earth-Sun orbit (Fig. 22–20A). As a result, during a new moon the Moon's shadow misses the Earth (Fig. 22–20B) and at a full moon the Earth's shadow misses the Moon (Fig. 22–20C).

However, on rare occasions, the new Moon passes through the Earth-Sun orbital plane. At these times, the Moon passes directly between the Earth and Sun (Fig. 22–20D). When this happens, the Moon's shadow falls on the Earth, producing a **solar eclipse** (Fig. 22–20E). As the Moon slides in front of the Sun, an eerie darkness descends, and the Earth becomes still and quiet. Birds return to their nests and stop singing. While the eclipse is total, the Moon blocks out the entire surface of the Sun, but the outer solar atmosphere, or **corona,** normally invisible owing to the Sun's

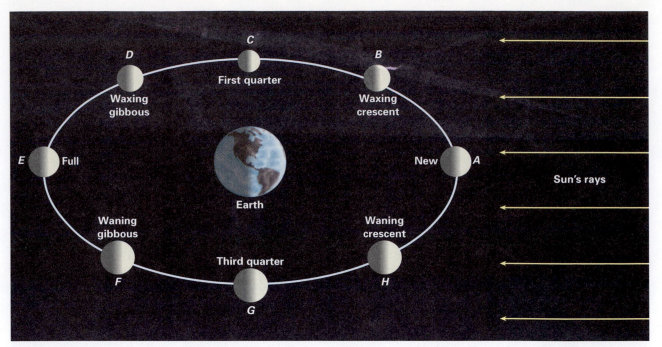

A

B Waxing crescent

C First quarter

D Waxing gibbons

E Full moon

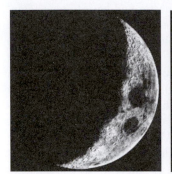

F Waning gibbous

G Third quarter

H Waning crescent

Figure 22–17 Approximately every 29.5 Earth days, the Moon passes through a complete cycle of phases. The upper part of the drawing shows the Earth-Moon orbital system viewed over the course of a month. The lower portion shows what the Moon looks like from Earth at the various phases. *(The Observatories of the Carnegie Institution)* • **Interactive Question:** How would the phases of the Moon differ if the Moon orbited the Earth within the Earth-Sun orbital plane?

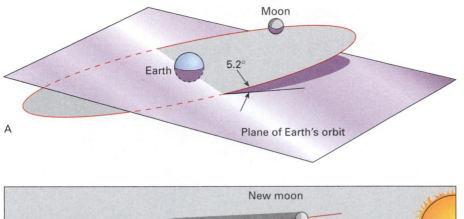

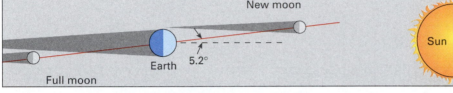

Figure 22–18 (A) The Sun and the Earth lie in one plane, while the Moon's orbit around the Earth lies in another. (B) A sideways look at the Sun, Moon, and Earth shows that most of the time the Moon's shadow misses the Earth and the Earth's shadow misses the Moon. Scales are exaggerated for emphasis.

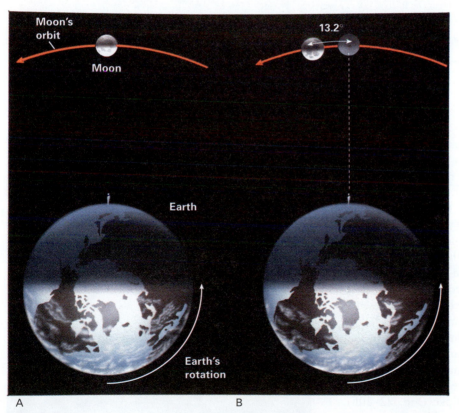

Figure 22–19 The Moon moves 13.2 degrees every day. (A) The Moon is directly above an observer on Earth. (B) One day later, the Earth has completed one complete rotation, but the Moon has traveled 13.2 degrees. The Earth must now travel for another 53 minutes before the observer is directly under it again.

Figure 22–20 Eclipses of the Sun and Moon. The Sun and the Earth lie in one plane, while the Moon's orbit around the Earth lies in another. Scales are exaggerated for emphasis. (A) Normally the Moon lies out of the plane of the Earth-Sun orbit. (B) During the new Moon the Moon's shadow misses the Earth. (C) During the full Moon the Earth's shadow misses the Moon. (D) However, if the Moon passes through the Earth-Sun plane when the three bodies are aligned properly, then an eclipse will occur. (E) An eclipse of the Sun occurs when the Moon is directly between the Sun and the Earth, and the Moon's shadow is cast on the Earth. (F) An eclipse of the Moon occurs when the Earth's shadow is cast on the Moon.

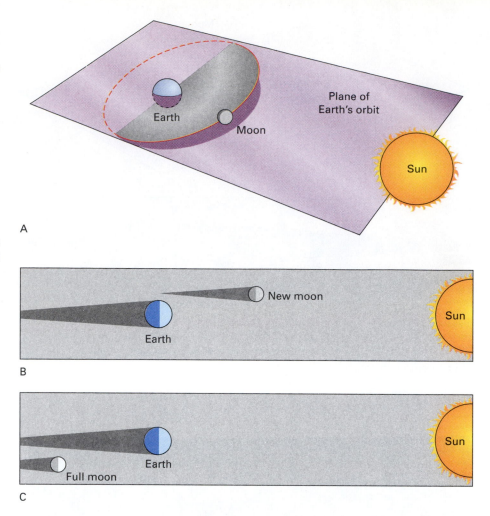

brilliance, appears as a halo around the black Moon (Fig. 22–21). Due to the relative distances between Sun, Moon, and Earth and their respective sizes, the Moon's shadow is only a narrow band on the Earth (Fig. 22–22). The band where the Sun is totally eclipsed, called the **umbra,** is never wider than 275 kilometers. In the **penumbra,** a wider band outside of the umbra, only a portion of the Sun is eclipsed. During a partial eclipse of the Sun, the sky loses some of its brilliance but does not become dark. Viewed through a dark filter such as a welder's mask, a semi-circular shadow cuts across the Sun (Fig. 22–23).

If the Moon passes through the Earth-Sun orbital plane when it is full, then the Earth lies directly between the Sun and the Moon. At these times the Earth's shadow falls on the Moon and the Moon temporarily darkens to produce a **lunar eclipse** (Fig. 22–20F). Lunar eclipses are more common and last longer than solar eclipses because the Earth is larger than the Moon, and therefore its shadow is more likely to cover the entire lunar surface. A lunar eclipse can last a few hours.

Figure 22–21 The solar corona appears as a bright halo around the eclipsed sun. This photograph was taken during the July 11, 1991 total eclipse of the Sun, in La Paz, Baja California. *(Stephen J. Edberg)*

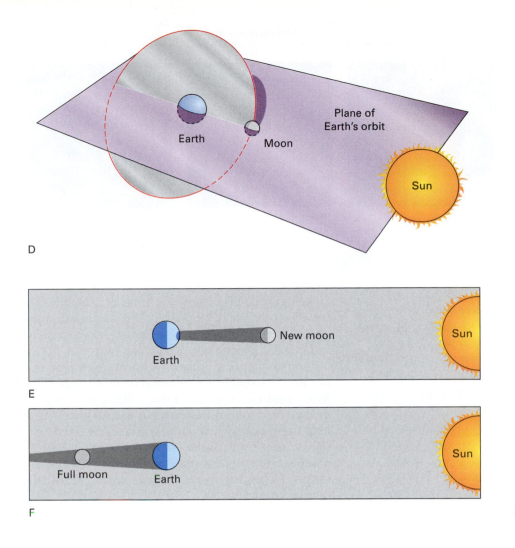

D

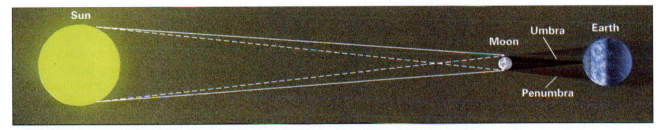

E

F

Figure 22–22 A total solar eclipse is viewed in the narrow band, called the umbra, formed by the projection of the Moon's shadow on the Earth. The Penumbra is the wider band where a partial eclipse is visible. (Drawing is not to scale.)

Figure 22–23 During a partial eclipse, a dark shadow obliterates a portion of the Sun. *(Jay Pasachoff)*

Figure 22–24 Resolution is the degree to which details are distinguishable in an image. The top photograph has poor resolution and appears to show a single object. With increasing resolution (center and right photos) we clearly see two distinct objects. *(Chris Jones, Union College)*

22.5 Modern Astronomy

Once astronomers understood the relative motions of the Sun, Moon, and planets, they began to ask questions about the nature and composition of these bodies, and to probe more deeply into space to study stars and other objects in the Universe. How do we gather data about such distant objects?

If you stand outside at night and look at a speck of light in the sky, the information you receive is limited by several factors. For one, your eye detects only visible light, which is only one millionth of 1 percent of the electromagnetic spectrum. Thus, more than 99.99 percent of the spectrum is invisible to the naked eye. In addition, the naked eye collects little light, and you may not see faint or distant objects at all. Your eye also has poor **resolution;** it may see one dot when two exist (Fig. 22–24). Finally, the light you see has been distorted by Earth's atmosphere. Modern astronomers attempt to overcome these difficulties with telescopes and other instruments.

Optical telescopes

A telescope is a device that collects light from a wide area and then focuses it where it can be detected. Detection devices may be simple, such as your eye or a photographic plate. In other instances, complex instruments analyze the images. In order to create a detectable image from a weak signal, a telescope may collect light for several hours.

Galileo and many other early astronomers used **refracting telescopes,** which employ two lenses. The first, called the **objective lens,** collects light from a distant object, and the **eyepiece** is a small magnifying lens. Light bends, or refracts, when it passes through the curved surface of the objective lens (Fig. 22–25). The bent light rays converge on the focus, forming an image of a distant object. The eyepiece then magnifies the image.

The problem with refracting telescopes is that different colors in the spectrum refract by different amounts. Therefore, if you focus the telescope to collect blue light sharply, red light will be fuzzy. As a result, most modern optical telescopes are **reflecting telescopes.** They collect light with a large curved mirror and reflect it to an eyepiece (Fig. 22–26). The reflecting mirror at the Palomar Observatory is 508 centimeters (200 inches) in diameter. It has 2.8 million times as much surface area as your pupil and thus collects 2.8 million times as much light. Modern telescopes are outfitted with both a camera and electronic detectors at the eyepiece.

Figure 22–25 In a refracting telescope, light is collected and focused by a large objective lens. A second lens, called the eyepiece, magnifies the image produced by the objective lens.

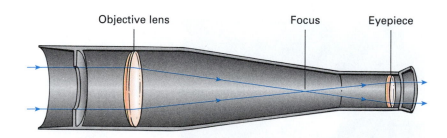

Objective lens   Focus   Eyepiece

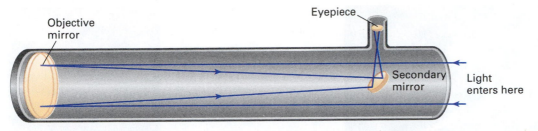

Figure 22–26 A Newtonian reflecting telescope. Incoming light (right) is collected and focused by a curved objective mirror. The light is then reflected to the eyepiece by a secondary mirror.

Astronomers have made many great discoveries with telescopes such as the one on Palomar Mountain, but telescopes of this type are limited. A mirror much larger than 600 centimeters sags under its own weight. In order to collect more light, the most recent telescopes use an array of smaller mirrors. The Keck telescopes on Mauna Kea in Hawaii contain 36 individual mirrors that are aligned and focused by computer (Fig. 22–27). The total mirror area is four times larger than the one at Palomar.

Problems caused by interference of the Earth's atmosphere can be solved by lifting a telescope into space. In 1990 the Hubble Space Telescope was launched into orbit around the Earth (Fig. 22–28). The Hubble has been repaired twice while in space and its high-resolution images have altered our understanding of many celestial bodies (Fig. 22–29). The Hubble telescope is outfitted with sensors to collect visible light and radiation in other portions of the spectrum.

Telescopes Using Other Wavelengths

Visible light is only a small portion of the electromagnetic spectrum. The wavelengths of electromagnetic radiation emitted by a star are determined by several factors, including the types of nuclear reactions that occur in the star, its chemical composition, and its temperature. In recent years, astronomers have enhanced our knowledge of stars and other objects in

A   B

Figure 22–27 (A) The 10-meter Keck Telescope is housed in the white dome on the left, located on Mount Mauna Kea in Hawaii. In this photograph, a twin telescope, named Keck II, is under construction on the right. *(Leonard Nakahashi, Keck Observatory)* (B) The 10-meter mirror of the Keck telescope is composed of 36 separate hexagonal sections, each adjusted by computer. This view shows the mirror under construction, with 18 of the mirrors installed. Note how small the worker appears compared with the overall size of the mirror. *(California Association for Research in Astronomy)*

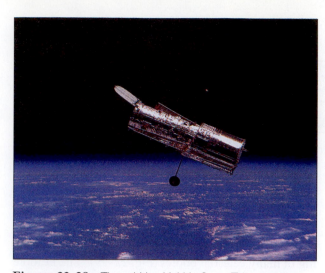

Figure 22–28 The orbiting Hubble Space Telescope has enabled astronomers to make many new discoveries of the Solar System, our galaxy, and intergalactic space. *(NASA/Johnson Space Center)*

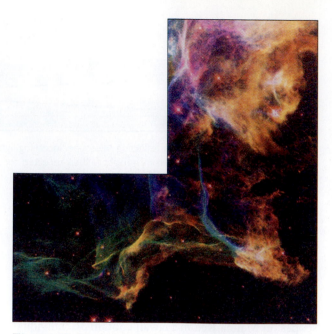

Figure 22–29 When a massive star dies, it explodes, blasting filaments of gas into space. The Hubble Space Telescope has photographed these stellar shreds in much more detail than is possible from ground-based instruments. *(Joachim Trümper, Max-Planck-Institut für Extraterrestrische Physik)*

space by studying many different wavelengths, from low-energy radio and infrared signals to high-energy gamma and X-rays (Fig. 22–30). Each wavelength provides specific information not available from other wavelengths. In Chapters 23 and 24, we will discuss some of the measurements made at various wavelengths.

Emission and Absorption Spectra

If light passes through a prism, it separates into a **spectrum,** an ordered array of colors[1] (Fig. 22–31). A rain-

---

[1] In modern instruments light is dispersed by a diffraction grating, not a prism, but the effect is the same.

Figure 22–30 The very large array (VLA) is a radio telescope consisting of 27 mobile antennas spread out over 36 kilometers. *(National Radio Astronomy Observatory)*

bow is such an array, with white sunlight separated into its individual colors. Each color is formed by a band of wavelengths.

As light passes from the hot interior of a star through the cooler outer layers, some wavelengths are selectively absorbed by atoms in the star's outer atmosphere. Therefore, in a spectrum of starlight, dark lines cross the band of colors. This is called an **absorption spectrum** (Fig. 22–32). Each dark line represents a wavelength that is absorbed by atoms of a particular element. Thus, an absorption spectrum enables us to determine the chemical composition of a star. This type of analysis is so effective that astronomers discovered helium in the Sun 27 years before chemists detected it on Earth. Because an atom's spectrum changes with temperature and pressure, spectra can also be used to determine surface temperatures and pressures of stars.

Whenever an atom absorbs radiation, it must eventually re-emit it. **Emission spectra** are often hard to detect against the bright background of starlight, but they can be seen along the outer edges of some stars and in large clouds of dust and gas in space.

Doppler Measurements

Have you ever stood by a train track and listened to a train speed by, blowing its whistle? As it approaches, the pitch of the whistle sounds higher than usual, and

Figure 22–31 A prism disperses a beam of sunlight into a spectrum of component colors. Each color represents a different band of wavelengths.

after it passes, the pitch lowers. The same effect can be duplicated with an electric razor. If you turn on an electric razor, hold it still, and listen to it, it produces a constant sound. Now move it quickly past your ear, and listen to the change in pitch. This change, called the **Doppler effect,** was first explained for both light and sound waves by the Austrian physicist Johann Christian Doppler in 1842. Three years later, a Dutch meteorologist, Christopher Heinrich Buys-Ballot, mounted an orchestra of trumpeters on an open railroad car and measured the frequency shifts as the musicians rode past him through the Dutch countryside.

A stationary object remains in the center of the circular waves it generates. The waves from a moving object crowd each other in the direction of the object's motion. The object, in effect, is catching up with its own waves (Fig. 22–33). If the object is moving toward you, you receive more waves per second (higher frequency) than you would if it were stationary, and if it is moving away from you, you receive fewer waves per second (lower frequency).

In the same way, the frequency of light waves changes with relative motion. The Doppler effect causes light from an object moving away from Earth to reach us at a lower frequency than it had when it was emitted. Lower-frequency light is closer to the red end of the spectrum. Thus, a Doppler shift to lower frequency is called a **red shift.** Alternatively, light from an object traveling toward us reaches us at higher frequency and is **blue shifted.** Using these principles, astronomers measure the relative velocities of stars, galaxies, and other celestial objects millions or billions of light-years away.

Spacecraft

On October 4, 1957, the U.S.S.R. launched the first spacecraft, Sputnik I. A few months later the Americans launched their first space probe. In the 1960s and 1970s hundreds of spacecraft were launched to study Earth, the Solar System, and distant galaxies. Some of the most notable missions include 6 manned lunar landings, 25 orbital or landing missions to Venus, 2 successful landings on Mars, and 4 probes to the outer planets. From 1978 to 1989 there was a sub-

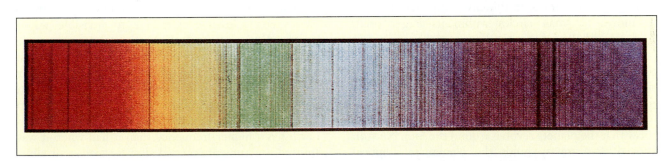

Figure 22–32 A copy of the first solar absorption spectrum taken in 1811. The dark lines are the absorption lines. *(Deutsches Museum)*

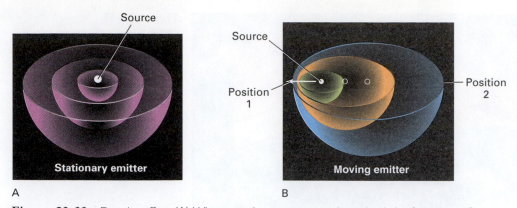

Figure 22–33 Doppler effect. (A) When a stationary source emits a signal, the frequency of the signal is unaffected by the source, and observers in any direction detect the same frequency. (B) The Doppler effect is the change in the frequency of a signal when it moves. If the source is moving toward an observer (Position 1), the observer detects waves squeezed close together and therefore with higher frequency than those detected from the same source when it is stationary. If the source is moving away from an observer (Position 2), the observer detects waves stretched farther apart and therefore with lower frequency.

stantial gap in space exploration. In the United States, most of the budget was concentrated on the Space Shuttle, and in Russia, exploration was reduced due to funding cutbacks.

However, a wave of spacecraft launched in the 1990s has provided new and fascinating data. As we will learn in Chapter 23, new planetary missions have increased our knowledge of the Solar System. In ad-dition, astronauts in the Russian Mir space station have performed numerous experiments and demonstrated that people can live in space for long periods of time. Future missions will alter some of the information in this book by the time you read it. That is one reason why science is so exciting; we are constantly probing deeper, with better instruments and new experiments.

SUMMARY

Aristotle proposed a **geocentric,** or Earth-centered, Universe in which a stationary, central Earth is surrounded by **celestial spheres** that contain the Sun, the Moon, the planets, and the stars. Ptolemy modified the geocentric model to explain the **retrograde motion** of the planets. In Ptolemy's model, each planet moves in small circles within its larger orbit. Copernicus believed that the Universe should operate in the simplest manner possible and showed that a **heliocentric** Solar System best explains planetary motion. Kepler described the elliptical orbits of the planets mathematically. Galileo used observation and experimentation to discredit the geocentric model and show that the planets revolve around the Sun. Newton proved that gravity holds the planets in elliptical orbits.

The revolution of the Moon around the Earth causes the phases of the Moon. A **lunar eclipse** occurs when the Earth lies directly between the Sun and the Moon. A **solar eclipse** occurs when the Moon lies directly between the Sun and Earth.

Objects in space are studied with both optical telescopes and telescopes sensitive to invisible wavelengths. **Emission and absorption spectra** provide information about the chemical compositions and temperatures of stars and other objects. The **Doppler effect** causes light from an object moving away from Earth to reach us at a lower frequency than it had when it was emitted. Light from an object moving toward Earth has a higher frequency than it had when it was emitted. Instruments carried aloft by spacecraft eliminate interference by the Earth's atmosphere and allow close inspection of objects in the Solar System.

constellation 521
planet 521
retrograde motion 521
geocentric 523
celestial sphere 523
parallax 525
heliocentric 526
maria 527

inertia 527
precession 530
new moon 531
crescent moon 531
waxing moon 531
gibbous moon 531
full moon 531
waning moon 531

solar eclipse 531
corona 531
umbra 534
penumbra 534
lunar eclipse 534
resolution 536
refracting telescope 536
objective lens 536

eyepiece 536
reflecting telescope 536
spectrum 538
absorption spectrum 538
emission spectra 538
doppler effect 539
red shift 539
blue shift 539

FOR REVIEW

1. Describe the apparent motion of the stars, as seen by an observer on Earth, over the course of a single night and over the course of a year.

2. List and explain Aristotle's observations and the reasoning he used to support his geocentric model of the Solar System.

3. What is retrograde motion? How is it explained in Ptolemy's and Copernicus' models?

4. Explain how Galileo's studies of physics contributed to his rejection of Aristotle's geocentric model.

5. Explain how Galileo's observations of the Milky Way, Sun, and Moon led him to question Aristotle's geocentric model.

6. What evidence convinced Galileo that the Earth revolves around the Sun?

7. What was the astronomical significance of Newton's studies of gravity?

8. How is the Moon positioned with respect to the Earth and the Sun when it is (a) full? (b) new? (c) gibbous? (d) crescent?

9. Why does the Moon rise approximately 53 minutes later each day than it did the previous day?

10. Draw a picture of the Sun, Earth, and Moon as they will be during an eclipse of the Sun. Draw a picture of the Sun, Earth, and Moon as they are aligned during an eclipse of the Moon. Explain how these positions produce eclipses.

11. Explain how optical telescopes work. How is a refracting telescope different from a reflecting telescope?

12. Describe the differences between absorption and emission spectra.

13. Explain how the frequency of radiation changes with the motion of an object that emits radiation.

FOR DISCUSSION

1. The Earth simultaneously rotates on its axis and revolves around the Sun. Which of these two motions is responsible for each of the following observations? (a) The Sun rises in the east and sets in the west. (b) Different constellations appear in the summer and winter. (c) On any given night the stars in the Northern Hemisphere appear to revolve around the Pole Star.

2. The Moon revolves around the Earth while the Earth rotates on its axis. Which of these two motions is responsible for each of the following observations? (a) The Moon rises in the east and sets in the west. (b) The Moon rises approximately 53 minutes later each day. (c) The Moon passes through monthly phases.

3. If you were the editor of a modern scientific journal, how would you respond to Aristotle's arguments for the geocentric model of the Universe? Defend your position.

4. Suppose that you are driving a car and the speedometer reads 80 km/hr. A person sitting next to you and looking at the speedometer at the same time may read a different value, perhaps 70 km/hr. Explain how two people can look at the same instrument at the same time and read different values.

5. With modern telescopes, astronomers can determine how far away some stars are by observing the parallactic shift as the Earth travels around the Sun. Using this technique, would it be easier to estimate distances to nearby stars or to distant stars, or would it be equally difficult for all stars? Defend your answer.

6. Describe the lunar cycle if the Sun, Earth, and Moon lay in a single plane.

7. Many societies use a lunar calendar instead of a solar one. In a lunar calendar, each month represents one full cycle of the Moon. How many days are there in a lunar month? If there are 12 months in a lunar year, is a lunar year longer or shorter than a solar year? Explain.

Planets and Their Moons

Spacecraft give us the best view of other planets in the Solar System and each space mission increases our understanding of our neighbors in space. This view of the surface of Mars was taken by the Mars *Pathfinder* lander on July 4, 1997. *(Lunar & Planetary Laboratory, University of Arizona; JPL/NASA)*

Recall from Chapter 1 that the entire Solar System formed from a single, homogeneous cloud of dust and gas. Today, however, each of the planets is quite different from the others. Some differences are easy to explain. Pluto, the farthest planet from the Sun, is frigid; Mercury, the closest, is torridly hot. However, neighboring planets, such as Venus, Earth, and Mars, have greater differences from one another than can be explained by their different distances from the Sun. Astronomers have learned that small initial differences in planetary environments have been amplified to create vastly dissimilar worlds. As explained in the opener to Unit IV, a runaway greenhouse effect on Venus created a hot, inhospitable environment. The climate on Mars was once temperate, and water flowed across the surface, but today this planet is cold and dry. Meanwhile on nearby Earth, flowers bloom, forests and prairies thrive, and animals, including humans, scurry across the surface. •

23.1 The Solar System: A Brief Overview

The Solar System formed about 4.6 billion years ago from a cold, diffuse cloud of dust and gas rotating slowly in space. The cloud was composed of about 92 percent hydrogen and 7.8 percent helium, about the same elemental composition as the Universe.[1] All of the other elements comprised only 0.2 percent of the Solar System.

A portion of the cloud gravitated toward its center to form the Sun. Here the pressure became so intense that hydrogen fused, producing energy as some of the hydrogen converted to helium. Hydrogen fusion is still the source of the Sun's energy. The remaining matter in the original cloud formed a disk-shaped rotating nebula that eventually coalesced into separate spheres to produce the planets. The inner planets were so close to the Sun, and their gravitational fields so weak, that

most of their earliest atmospheres of hydrogen, helium, and other light gases, boiled off into space or were blown away by the **solar wind,** a stream of positive ions and electrons radiating outward from the Sun at high speed. Rock and metal were left behind. Thus, Mercury, Venus, Earth, and Mars are mostly solid rock with metallic cores, and are called the **terrestrial planets.**

In contrast, the protoplanets in the outer reaches of the Solar System were so far from the Sun that they remained cool. They were also more massive and had stronger gravitational fields. As a result, the outer **Jovian planets,** Jupiter, Saturn, Uranus, and Neptune, have retained large amounts of hydrogen, helium, and other light elements. They may even have grown as they captured gases that escaped from the terrestrial planets. The Jovian planets all have relatively small rocky or metal cores surrounded by swirling liquid and gaseous atmospheres. The outermost planet, Pluto, is anomalous; it is relatively small and may be an escaped moon of Neptune.

Table 23–1 provides an overview of the nine major planets. Due to the differences in composition, the ter-

---

[1]Composition is given here in percentage of number of atoms. Percentage by mass is significantly different.

TABLE 23–1

Comparison of the Nine Major Planets

Planet	Distance from Sun (Millions of Kilometers)	Radius (Compared to Radius of Earth = 1)	Mass (Compared to Mass of Earth = 1)	Density (Compared to Density of Water = 1)	Composition of Planet	Density of Atmosphere (Compared to Earth's Atmosphere = 1)	Number of Satellites
Terrestrial Planets							
Mercury	58	0.38	0.06	5.4	Rocky with metallic core	One billionth	0
Venus	108	0.95	0.82	5.2		90	0
Earth	150	1	1	5.5		1	1
Mars	229	0.53	0.11	3.9		0.01	2
Jovian Planets							
Jupiter	778	11.2	318	1.3	Liquid hydrogen surface with liquid metallic mantle and solid core	Dense and turbulent	16
Saturn	1420	9.4	94	0.7			19
Uranus	2860	4.0	15	1.3	Hydrogen and helium outer layers with solid core	Similar to Jupiter except that some compounds that are gases on Jupiter are frozen on the outer planets	17
Neptune	4490	3.9	17	1.7			8
Most Distant Planet							
Pluto	5910	0.17	0.0025	2.0	Rock and ice	0.00001	1

restrial planets are much denser than the Jovian planets. However, the Jovian planets are much larger and more massive than the Earth and its neighbors.

23.2 Mercury: A Planet the Size of the Moon

Mercury has a radius of 2400 kilometers, less than four tenths that of Earth. It is the closest planet to the Sun and makes a complete circuit around the Sun faster than any other planet; each Mercurial year is only about 88 Earth days long. Mercury rotates slowly on its axis, so there are only three Mercurial days every two Mercurial years. Because Mercury is so close to the Sun and its days are so long, the temperature on its sunny side reaches 400°C, hot enough to melt lead. In contrast, the temperature on its dark side drops to −175°C. The lack of an atmosphere is partly responsible for these extremes of temperature because there is no wind to carry heat from one region to another.

Little was known about the surface of Mercury before the spring of 1974, when the spacecraft *Mariner 10* passed within a few hundred kilometers of the planet. Images relayed to Earth revealed a cratered surface remarkably similar to that of our Moon (Fig. 23–1). The craters on both Mercury and the Moon formed during intense meteorite bombardment early in the history of the Solar System. Earth was also pockmarked by the same episode of meteorite impacts; however, tectonic activity and erosion have erased the Earth's early meteorite craters. Craters have been preserved on Mercury and the Moon because they have not experienced tectonic activity or erosion.

Flat plains on Mercury are lava flows that formed early in its history, when the planet's interior was hot enough to produce magma. However, Mercury is so small that its interior cooled quickly and igneous activity ceased when the planet was still young. It is so close to the Sun that its water and atmosphere have boiled off into space, so no wind, rain, or rivers have eroded its surface. Today, 4-billion-year-old craters look as fresh as if they formed yesterday.

In 1991, radar images of Mercury revealed highly reflective regions at the planet's poles. Data indicated that these regions were composed of ice, but how can ice survive on a planet with a daytime surface temperature of 400°C? Mercury's spin axis is almost perpendicular to its orbital plane around the Sun, so the Sun never rises or sets at the poles but remains low on the horizon throughout the year. Because the Sun is so low in the sky, regions inside meteorite craters are perpetually in the shade. With virtually no atmosphere to transport heat, the shaded regions have remained

Figure 23–1 Mercury has a cratered surface. The craters are remarkably well preserved because there is no erosion or tectonic activity on Mercury. The photograph shows an area 580 km from side to side. *(NASA)*

below the freezing point of water for billions of years. According to one hypothesis, the ice was transported to Mercury by comets early in the history of the planet.

23.3 Venus: The Greenhouse Planet

Recall from the opening essay to Unit IV that Venus closely resembles Earth in size, density, and distance from the Sun. As a result, Venus and Earth probably had similar atmospheres early in their histories. However, Venus is closer to the Sun than is Earth, and therefore it was initially hotter. One hypothesis suggests that because of the higher temperature, water never condensed, or if it did, it quickly evaporated again. Because there were no seas for carbon dioxide to dissolve into, most of the carbon dioxide also remained in the atmosphere. Water and carbon dioxide combined to produce a runaway greenhouse effect. The molecular weight of water is less than half that of carbon dioxide. Because water is so light, the heat and the solar wind swept most of the water into space. Today, the Venusian atmosphere is 90 times denser than that of Earth. Thus, atmospheric pressure at the surface of Venus is equal to the pressure 1000 meters beneath the sea on our planet. The Venusian atmosphere is more than 97 percent carbon dioxide, with small amounts of nitrogen, helium, neon, sulfur dioxide, and other gases. Corrosive sulfuric acid aerosols float in a dense cloud layer that perpetually obscures the surface (Fig. 23–2). Due to greenhouse warming,

Figure 23–2 The solid surface of Venus is obscured by a dense, corrosive atmosphere and a turbulent cloud cover. *(NASA/JPL)*

the Venusian surface is hotter than that of Mercury, hot enough to destroy the complex organic molecules necessary for life.

Venusian Tectonics

Astronomers use spacecraft-based radar to penetrate the Venusian atmosphere and produce photo-like images of its surface. Gravity studies, also conducted by spacecraft, are used to infer the density of rocks near the Venusian surface. These density measurements provide information about the planet's mineralogy and internal structure. The most spectacular data were obtained by the orbiting *Magellan* spacecraft, which was launched in May 1989. In October 1994, its mission 99 percent accomplished and federal funding running out, scientists sent *Magellan* on one final suicide mission into the Venusian atmosphere to provide additional information about its density and composition.

During the early history of the Solar System, dense swarms of meteorites bombarded all the planets. However, *Magellan*'s detailed maps show few meteorite craters on Venus. This observation indicates that the Venusian surface was reshaped after the major meteorite bombardments. More detailed analysis shows that most of the landforms on Venus are 300 to 500 million years old.

James Head, *Magellan* investigator from Brown University, suggests that a catastrophic series of volcanic eruptions occurred 300 to 500 million years ago,

creating volcanic mountains and covering much of the surface with basalt flows (Fig. 23–3). According to this model, rising mantle plumes generated magma that repaved the planet in a short time with a rapid series of cataclysmic volcanic eruptions. Head concludes, "The planet's entire crust and lithosphere turned itself over. It certainly makes a strong case that catastrophic events are in the geological records of planets and it ought to make us think about the possibilities of such catastrophic events in Earth's past." An alternative hypothesis contends that the resurfacing occurred more slowly, in a piecemeal process.

Crater abundances indicate that most tectonic activity on Venus stopped after the intense volcanic activity 300 to 500 million years ago. Some evidence indicates that the volcanic activity has ceased permanently because the planet's interior cooled. This cooling may have occurred because radioactive elements floated toward the surface removing the mantle heat source. However, other data imply that Venus remains volcanically active and that another repaving event will occur in the future.[2]

---

[2]William Kaula, "Venus Reconsidered," *Science*, Vol. 270, Dec. 1, 1995, p. 1460.

Figure 23–3 The volcano Maat Mons, on Venus, has produced large lava flows, shown in the foreground. According to one hypothesis, a catastrophic series of volcanic eruptions altered the surface of Venus 300 to 500 million years ago. This image was produced from radar data recorded by *Magellan* spacecraft. Simulated color is based on color images supplied by Soviet spacecraft. *(NASA/JPL)*

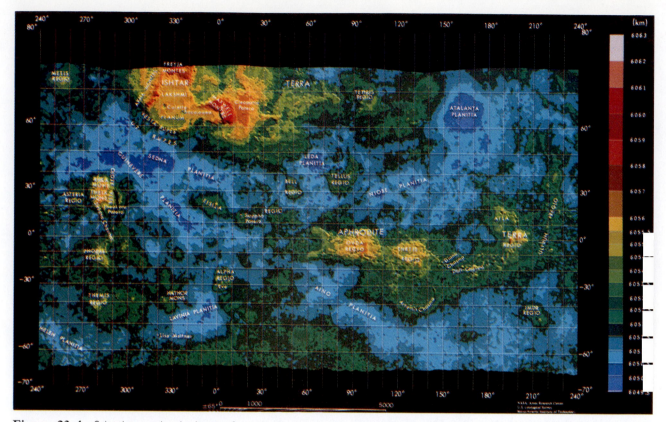

Figure 23–4 Scientists used radar images from the *Pioneer* Venus orbiter to produce this map of Venus. The lowland plains are shown in blue and the highlands are shown in yellow and red-brown. The colors are assigned arbitrarily. *(NASA/Ames Research Center)*

Most Earth volcanoes form at tectonic plate boundaries or over mantle plumes, so the discovery of volcanoes on Venus led planetary geologists to look for evidence of plate tectonic activity there. Radar images from *Magellan* and earlier *Pioneer* spacecraft show that 60 percent of Venus's surface consists of a flat plain. Two large and several smaller mountain chains rise from the plain (Fig. 23–4). The tallest mountain is 11 kilometers high—2 kilometers higher than Mount Everest. The images also show large crustal fractures and deep canyons. If Earth-like horizontal motion of tectonic plates caused these features, then spreading centers and subduction zones should exist on Venus. However, the *Magellan* data showed no features like a mid-ocean ridge, transform faults, or other evidence of lithospheric spreading. When the *Magellan* data were first received, planetary geologists couldn't identify any subduction zones, but more recently, two geophysicists located 10,000 kilometers of trench-like structures that they believe to be subduction zones. Despite this report, the most popular current model suggests that Venusian tectonics has been dominated by rising mantle plumes. In some regions the rock has

melted and erupted from volcanoes by processes similar to those that formed the Hawaiian Islands. In other regions, the hot Venusian mantle plumes have lifted the crust to form nonvolcanic mountain ranges. Just as the upward movement of mantle material on Earth must be balanced by subduction of material back into the mantle, upwelling of mantle plumes on Venus must somewhere be balanced by downwelling. Geologists continue to analyze data in search of landforms created by downwelling.

Some geologists have suggested the term **blob tectonics** to characterize Venusian tectonic activity because it is dominated by rising and sinking of the mantle and crust. In contrast, horizontal motion of lithospheric plates dominates tectonics on Earth.

Mantle plumes on Earth may initiate rifting of the lithosphere and formation of a spreading center. Why have spreading centers not developed over mantle plumes on Venus? Perhaps surface temperature on Venus is so high that the surface rocks are more plastic than those on Earth. Therefore, rock stretches and lithosphere-deep cracks do not form. It is also possible that Venus has a thicker lithosphere, which can move vertically but does not fracture and slide horizontally.

23.4 The Moon: Our Nearest Neighbor

Most planets have small orbiting satellites called moons. The Earth's Moon is close enough so that we can see some of its surface features with the naked eye. In the early 1600s, Galileo studied the Moon with the aid of a telescope and mapped its mountain ranges, craters, and plains. Galileo thought that the plains were oceans and called them seas, or maria. The word *maria* is still used today, although we now know that these regions are dry, barren, flat expanses of volcanic rock (Fig. 23–5). Much of the lunar surface is heavily cratered, similar to that of Mercury (Fig. 23–6).

The first close-up photographs of the Moon were taken by a Soviet orbiter in 1959. A decade later the United States landed the first of six manned *Apollo* spacecraft on the lunar surface (Fig. 23–7). The Apollo program was designed to answer several questions about the Moon. How did it form? What is its geologic history? Was it once hot and molten like the Earth? If so, does it still have a molten core, and is it tectonically active?

Formation of the Moon

According to the most popular current hypothesis, the Moon was created when a huge object—the size of

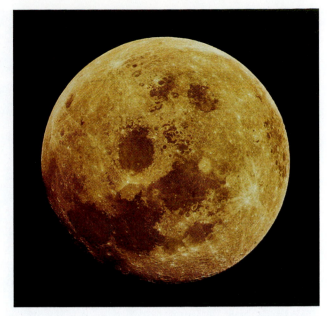

Figure 23–5 The Moon as photographed from the *Apollo* spacecraft from a distance of 18,000 kilometers. Even from this distance, we see cratered regions and the maria, which are smooth lava flows. *(NASA)* ● Interactive Question: Use or modify the principles of relative age dating (Chapter 4) to deduce the relative ages of different regions of the lunar surface in this photograph.

Mars or even larger—smashed into Earth shortly after our planet formed. This massive bolide plowed through the Earth's mantle, and silica-rich rocks from the mantles of both bodies vaporized and created a

Figure 23–6 Most of the lunar surface is heavily cratered. In places where smaller craters lie within the larger ones scientists deduce that the larger craters formed first. *(NASA)*

Figure 23–7 The six Apollo missions answered many scientific questions about the origin, structure, and history of the Moon. *(NASA)*

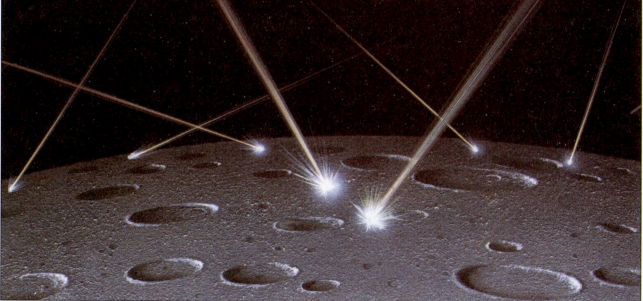

◀ **Figure 23–8** (A) According to the most widely held hypothesis, the Moon was formed about 4.5 billion years ago when a Mars-sized object struck the Earth, blasting a cloud of vaporized rock into orbit. The vaporized rock rapidly coalesced to form the Moon. (B) During its first 0.6 billion years, intense meteorite bombardment cratered the Moon's surface. (C) The dark, flat maria formed about 3.8 billion years ago as lava flows spread across portions of the surface. Today, all volcanic activity has ceased.

cloud around the Earth. The vaporized rock condensed and aggregated to form the Moon (Fig. 23–8A).

Perhaps the single most significant discovery of the Apollo program was that much of the Moon's surface consists of igneous rocks. The maria are mainly basalt flows. The highland rocks are predominantly anorthosite, a feldspar-rich igneous rock not common on Earth. Additionally, the rock of both the maria and the highlands has been crushed by meteorite impacts and then welded together by lava. Since igneous rocks form from magma, it is clear that portions of the Moon were once hot and liquid.

How did the Moon become hot enough to melt? The Earth was heated initially by gravitational coalescence, energy released by collisions among particles as they collapsed under the influence of gravity. Later, radioactive decay and intense meteorite bombardment heated the Earth further. But what about the Moon? Radiometric dating shows that the oldest lunar igneous rocks formed before there was enough time for radioactive decay to have melted a significant amount of lunar rock. Thus, gravitational coalescence and meteorite bombardment must have been the main causes of early melting of the Moon.

History of the Moon

The Moon formed about 4.45 to 4.5 billion years ago, shortly after the Earth. So much energy was released during the Moon's rapid accretion that, as the lunar sphere grew, it melted to a depth of a few hundred kilometers, forming a magma ocean. Meteorite bombardment kept the Moon's outermost layer molten. Eventually meteorite bombardment diminished enough for the Moon's surface to cool. The igneous rocks of the lunar highlands are about 4.4 billion years old, indicating that the highlands were solid by that time.

Swarms of meteorites, some as large as Rhode Island, bombarded the Moon again between 4.2 and 3.9 billion years ago (Fig. 23–8B). In the meantime, radioactive decay was also heating the lunar interior. As a result, by 3.8 billion years ago, most of the Moon's interior was molten and magma erupted onto the lunar surface. The maria formed when lava filled circular meteorite craters (Fig. 23–8C). This episode of volcanic activity lasted approximately 700 million years.

The Moon and Earth shared a similar history until this time, but the Moon is so much smaller that it soon cooled and has remained geologically inactive for the past 3.1 billion years. Seismographs left on the lunar surface by *Apollo* astronauts indicate that the energy released by moonquakes is only one billionth to one trillionth that released by earthquakes on our own planet. Because the Moon essentially froze geologically 3 billion years ago, scientists have learned much about Earth's early history by studying the Moon.

Data from the lunar *Prospector* spacecraft, released in March 1998, indicate that water is present in shadowed craters at the lunar poles. This discovery complements the earlier conclusion that ice exists on Mercury. The water on the Moon could be used to support human colonies or to synthesize hydrogen fuel from a Moon base for exploration of more distant planets.

23.5 Mars: A Search for Lost Water

The Geology of Mars

Mars's surface consists of old, heavily cratered plains and younger regions that have been altered by tectonic activity (Fig. 23–9). Lava flows much like those on Venus and the Moon cover the plains. The Tharsis bulge is the largest plain, crowned by Olympus Mons, the largest volcano in the Solar System (Fig. 23–10). Olympus Mons is nearly three times higher than Mount Everest, with a height of 25 kilometers and a diameter of 500 kilometers.

One hypothesis suggests that the geology of Mars is similar to that of Venus and is dominated by blob tectonics. Supporters of this concept point out that tremendous parallel cracks split the crust adjacent to the Tharsis bulge (Fig. 23–11). If this bulge lay near a tectonic plate boundary, there would be folding or offsetting of the cracks. However, the cracks are neither folded nor offset. Therefore, they suggest that a rising mantle plume formed the Tharsis bulge and its volcanoes. The parallel cracks may be the result of stretching as the crust uplifted.

An alternative hypothesis suggests that Earth-like, horizontal tectonics is occurring on Mars. As evidence, researchers have identified what they believe to be

Figure 23–9 The cratered plain on the left in this photo of the Mars surface is a geologically old surface, whereas the mountains on the lower right are evidence of more recent tectonic activity. *(NASA/JPL)*

Figure 23–10 Olympus Mons is the largest volcano on Mars and probably the largest in the Solar System. *(NASA/JPL)*

strike-slip faults. In addition, a linear distribution of volcanoes on and near the Tharsis bulge is similar to linear chains of volcanoes along terrestrial subduction zones.

The Mars *Pathfinder* spacecraft landed on Mars on July 4, 1997, and deployed a small rover, called *Sojourner* (Fig. 23–12). Scientists learned from one of

its analyses that rock near the landing site had high concentrations of silica and potassium whereas the soil was richer in magnesium and iron. Recall from Chapter 3 that on Earth, granite is a common silica- and potassium-rich rock that forms by partial melting of deep continental crust. Basalt is a mafic volcanic rock that contains a higher concentration of magne-

Figure 23–11 The huge parallel cracks near the Tharsis bulge on Mars are neither folded nor offset. Scientists speculate that the bulge and the cracks were formed by a rising mantle plume. However, the absence of folding or offset movement provides evidence that horizontal tectonics is not active in this region. Two large volcanoes appear at right. *(NASA/JPL)*

Figure 23–12 In July 1997, the Mars *Pathfinder* lander released the robot rover, *Sojourner*. In this photograph, *Sojourner* is extracting samples from a rock nicknamed Yogi. *(JPL/NASA)*

sium and iron, a composition closer to that of the mantle. Scientists concluded that the mafic soil near the Martian lander was transported from a remote site by wind. Furthermore, the samples of Martian rock were formed by different igneous processes, indicating Earth-like differentiation of the crust.

The Martian Atmosphere and Climate

Today, the Martian surface is frigid and dry. The surface temperature averages −60°C at the equator and never warms up enough to melt ice. At the poles, the temperature can dip to −120°C, freezing carbon dioxide to form "dry ice." The atmosphere contains about 0.007 bars of carbon dioxide with smaller amounts of water and other gases. (The Earth's atmospheric pressure is about one bar, about 140 times as dense as that on Mars.)

However, abundant evidence indicates that the Martian climate was once much warmer and that water flowed across the surface. Photographs from *Mariner* and *Viking* spacecraft show eroded crater walls and extinct stream and lake beds. One giant canyon, Valles Marineris, is approximately ten times longer and six times wider than Grand Canyon in the American southwest (Fig. 23–13). Massive alluvial fans at the mouths of Martian canyons indicate that floods prob-

ably raced across the land at speeds up to 270 kilometers per hour.

Given these observations, we ask two related questions: How long did water exist on Mars? How did Mars lose its water?

After studying the *Mariner* and *Viking* photographs, planetary scientists recognized two possibilities. One scenario is that Mars had a temperate climate, abundant water, and a relatively dense atmosphere for a long period of time, perhaps a few hundred million years. Another possibility is that occasional catastrophic events such as volcanic eruptions or cometary impacts produced transitory outbursts of water that flowed across the surface and then quickly evaporated. The difference between these two scenarios has important implications for our search for life on Mars. If the Martian climate was wet and temperate for a long time, we would expect a higher probability that life evolved than if water flowed for only brief episodes. "Focus on Extraterrestrial Life" discusses the search for life on Mars and on other planets and moons in the Solar System.

Scientists reasoned that if water had been abundant for a long time, then old seas, streams, and lake beds would be common over the entire surface. On the other hand, catastrophic episodes would produce flooding in more isolated regions. In 1997, photographs from the *Mars Pathfinder* showed layered rocks near the landing site. The news media immediately reported that these photos were evidence of sedimentary bedding in a watery environment. However, volcanic eruptions and windblown dust also produce layered rock and sediment, so the photographs by

(Continued on page 554)

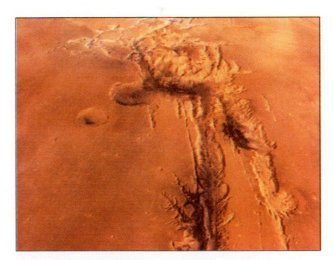

Figure 23–13 The giant canyon, Valles Marineris, was eroded by flowing water and is many times larger than Grand Canyon in the American southwest. *(NASA/JPL)*

Focus On

Extraterrestrial Life

Few scientific inquiries excite both scientists and nonscientists as much as the search for extraterrestrial life. Close scrutiny of our Solar System reveals no densely vegetated regions or advanced civilizations except for those on Earth. But even the presence of a few microscopic bacteria would be exciting proof that we are not alone in the Universe.

Temperatures have always been too hot to support life on Mercury. However, Mars once had a temperate climate and life could have evolved there.

The Search for Martian Life

Although Mars is inhospitable today, some scientists speculate that life may have evolved during its more temperate past. Supporters of this hypothesis point out that the primordial Martian atmosphere probably contained water, carbon dioxide, carbon monoxide, oxygen, hydrogen, nitrogen, and sulfur. On Earth these gases combined to form organic molecules—the building blocks of life. Moreover, the record of surface water on Mars tells us that at one time the Martian temperature was neither too hot nor too cold to support life. This evidence does not prove that life evolved on Mars, just that the necessary components and conditions were once present.

Spacecraft that landed on Mars in the 1970s searched for microorganisms in the soil and found none. However, they analyzed only a minuscule portion of the Martian surface.

In 1983, scientists found a meteorite, called ALH84001, on an Antarctic glacier (Fig. 1). The minerals and the oxygen isotope ratios indicated that the rock originated on Mars. Most scientists agree that about 16 million years ago, a powerful bolide impact blasted the Martian surface with enough force to eject rock into space. At least one chunk drifted through the Solar System until it crashed onto Earth recently enough to be found on the Antarctic ice.

In 1996, a team of NASA scientists announced that meteorite ALH84001 may contain fossil and chemical traces of Martian life (Fig. 2). The observations, the interpretation of the NASA team, and the critical response, are summarized in Table 1.

All scientists agree that each piece of evidence could be explained by either a biological or a nonbiological mechanism. However, the NASA meteorite research team argues that it is improbable that four independent, unlikely, nonbiological processes left their mark on a single rock. Since a bacterial colony would leave all of these traces, the most probable explanation is that life existed on Mars.

Critics argue that the evidence is not strong enough to make such a bold statement. The debate will continue until the next generation of spacecraft land on Mars and study its rock and sediment in more detail.

Vast communities of bacteria on Earth live adjacent to undersea volcanic vents or in crustal rock. These bacteria obtain their energy by chemical reactions with minerals rather than by photosynthesis or by eating products of photosynthesis. Scientists have speculated that if life evolved on Mars during its ancient past when it had a more benign climate, then some organisms may have migrated into the subsurface rock and survived even when surface conditions became too cold and dry to support life. However, remember that this scenario is speculation only.

Some biologists have also speculated that primitive life may have evolved in the atmosphere of Jupiter. The

Figure 1 Some scientists have hypothesized that the Martian meteorite ALH84001 harbors signs of early life on Mars. *(NASA)*

Figure 2 Microscopic analysis of meteorite ALH84001 reveals small bumps and tube-like structures. Some scientists have speculated that these structures are fossils of early life. *(NASA)*

TABLE 1

Data	Interpretation Favoring the Conclusion That Meteorite ALH84001 Contains Evidence of Martian Life	Critical Rebuttal
The meteorite contains numerous, 3.6-million-year-old carbonate globules.	Carbonates precipitate from water solutions, so these inclusions indicate that water once percolated through the rock. Furthermore, on Earth, similar carbonate globules are formed by bacteria.	Many carbonate globules form on Earth by purely chemical processes, unaided by living organisms.
Concentric rings of iron oxides and iron sulfides surround the largest carbonate globules.	On Earth, these minerals are formed commonly by bacteria, and rarely by nonbiological processes.	While it is true that formation of these minerals are uncommon by nonbiological processes, it does occur.
Researchers also found small amounts of complex organic molecules called polycyclic aromatic hydrocarbons (PAHs) in the rocks.	PAHs are formed abundantly by all living organisms but very rarely by nonbiological processes.	PAHs have been detected on comets, meteorites, and other extraterrestrial debris.
Meteorite ALH84001 contains tiny, tube-like structures, about 1/1000 the width of a human hair (Fig. 2).	These structures resemble microfossils of ancient bacteria found on Earth, even though they are much smaller.	Many different chemical and geological mechanisms can form tube-like crystal structures.

outermost region of the Jovian atmosphere is too cold for organisms to survive, and the interior is too hot. But in between, the temperature is just right. The atmosphere is composed primarily of hydrogen, helium, ammonia, water, methane, and hydrogen sulfide, the elements needed to build living tissue. In short, all the ingredients are present, and it would be thrilling, but not totally surprising, to find microorganisms floating about in the clouds of the giant planet.

When data from the *Galileo* spacecraft implied that water lies beneath the crust on Europa, scientists postulated that life could exist in the dark, warm, subterranean oceans or lakes. Other scientists suggest that biological evolution could have occurred within the methane seas of Titan.

One can only guess about life in other solar systems. Scientists have observed organic molecules on meteorites and comets and in dust clouds in interstellar space. Many scientists infer that, under the right conditions and with enough time, these molecules can combine to form living organisms. An average galaxy contains roughly 100 billion stars, and astronomers estimate that there are 10 to 50 billion galaxies in the Universe. In recent years, researchers found evidence for several planets orbiting stars outside the Solar System. One is estimated to have a surface temperature of about 80°C, roughly the temperature of warm tea, and a perfect temperature for life to evolve. Thus, evidence indicates that planets may be common stellar companions, and there are many, many stars in the Universe. Almost certainly conditions are favorable for life on some of the planets. Therefore, one argument concludes that the emergence of living organisms—and even their evolution into intelligent beings—is so probable that it almost surely must have happened in many parts of the Universe.

On the other hand, we cannot rule out the possibility that life is unique to Earth and exists nowhere else. Remember, there is no unequivocal proof of living entities anywhere but on our planet.

Focus Question:

Discuss the difficulties in detecting life outside the Solar System, even if it does exist.

themselves are not proof that the landing site was once wet.

Pathfinder's instruments analyzed soils for carbonates and evaporite minerals that normally precipitate from lakes and seas. The results showed some carbonate and salt-rich minerals, but not enough to prove that ancient seas existed at the landing site.

A few months after *Pathfinder* landed, a second spacecraft, the Mars *Global Surveyor*, went into orbit. *Surveyor* carries three primary instruments, a high resolution camera, a laser altimeter, and an infrared spectrometer. Every mineral emits radiation at a characteristic infrared frequency. If *Surveyor* found that evaporite deposits were common across the Martian landscape, then we could infer that water was abundant over a broad area. Unfortunately, preliminary results show that windblown sediment covers most topographic depressions, where evaporite deposits are most likely. Because the spacecraft's instruments only detect surface minerals, we can't determine whether or not evaporites are common. In summary, at this time we don't know how abundant water was on Mars or how long it flowed across the surface.

The second question, How did Mars lose its water, is discussed in the opening essay to Unit IV. Recall from that discussion that scientists suspect that Mars lost its water through a reverse greenhouse effect, perhaps augmented by the effects of the planet's weak magnetic field.

23.6   Jupiter: A Star That Failed

Structure and Composition of Jupiter

Jupiter is the largest planet in the Solar System, 71,000 kilometers in radius. It is composed mainly of hydrogen and helium, similar to the composition of the Sun. However, Jupiter isn't quite massive enough to generate fusion temperatures, so it never became a star.

Jupiter has no hard, solid, rocky crust on which an astronaut could land or walk. Instead, its surface is a vast sea of cold, liquid, molecular hydrogen, H_2, and atomic helium, He, 12,000 kilometers deep. Beneath the hydrogen/helium sea is a layer where temperatures are as high as 30,000°C and pressures are as great as 100 million times the Earth's atmospheric pressure at sea level (Fig. 23–14). Under these extreme conditions, hydrogen molecules dissociate to form atoms. Pressure forces the atoms together so tightly that the electrons move freely throughout the packed nuclei much as electrons travel freely among metal atoms. As a result, the hydrogen conducts electricity and is called **liquid metallic hydrogen.** Flow patterns in this fluid conductor generate a magnetic field ten times stronger than that of Earth.

Beneath the layer of metallic hydrogen, Jupiter's core is a sphere about 10 to 20 times as massive as Earth. It is probably composed of metals and rock surrounded by lighter elements such as carbon, nitrogen, and oxygen.

The *Galileo* spacecraft was launched in October 1989. In a six-year mission it passed Venus, then spiraled away from the Sun to rendezvous with Jupiter in December 1995. Once in orbit around the gas giant, the spacecraft launched a suicide probe to parachute through the outer atmosphere and collect data until it heated up and eventually vaporized. As expected, the atmosphere was primarily hydrogen and helium, with smaller concentrations of ammonia, water, and methane.

More than 300 years ago, two European astronomers reported a **Great Red Spot** on the surface

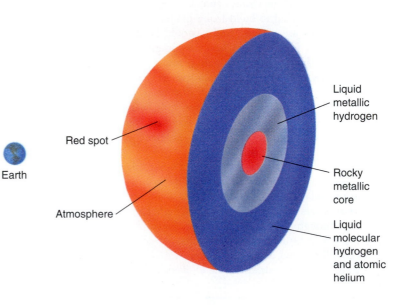

Earth

Figure 23–14   Jupiter consists of four main layers: a turbulent atmosphere, a sea consisting mainly of liquid molecular hydrogen with smaller amounts of atomic helium, a layer of liquid metallic hydrogen, and a rocky metallic core. The Earth is drawn to scale on the left.

Figure 23–15 Jupiter's Great Red Spot dwarfs the superimposed image of the Earth (to scale) shown on the bottom right. The vivid colors were generated by computer enhancement. (NASA)

of Jupiter. Although its size, shape, and color have changed from year to year, the spot remains intact to this day (Fig. 23–15). If the Earth's crust were peeled off like an orange rind and laid flat on the Jovian surface, it would fit entirely within the Great Red Spot. Measurements show that the Great Red Spot is a giant hurricane-like storm. Hurricanes on Earth dissipate after a week, yet this storm on Jupiter has existed for centuries. Other wind systems, rotating in linear bands around the planet, have also persisted for centuries (Fig. 23–16). One important mission of the *Galileo* suicide probe was to measure atmospheric pressures, temperatures, and wind speeds below the visible outer layers.

In the outer atmosphere, where the pressure was 0.4 bar, the *Galileo* probe measured a wind speed of 360 kilometers per hour with a temperature of −140°C. After falling 130 kilometers, the probe reported that the wind speed had increased to 650 kilometers per hour, the pressure was 22 bar, and the temperature was about +150°C. Buffeted by winds, squeezed by intense pressure, and heated beyond the

Figure 23–16 This colorful, turbulent complex cloud system of Jupiter was photographed by the *Voyager* spacecraft. The sphere on the left, in front of the Great Red Spot, is Io; Europa lies to the right against a white oval. (NASA/JPL)

tolerance of its electronics, the spacecraft stopped transmitting. Scientists calculate that 40 minutes later, when the spacecraft had sunk to deeper levels of the Jovian atmosphere, the temperature had increased to 650°C, the pressure increased to 260 bar, and the aluminum shell liquified. As the probe continued to fall, pressure and temperature rose until the titanium hull melted, streaming metallic droplets that flashed into vapor as the probe vanished.

On Earth, winds are driven by the Sun's heat. But Jupiter receives only about 4 percent of the solar energy that Earth receives. Moreover, strong winds rip through the Jovian atmosphere far below the deepest penetration of solar light and heat. For these reasons, scientists deduce that the Jovian weather is not driven by solar heat, but by heat from deep within the planet. This heat accumulated when Jupiter first formed. As the heat slowly rises, it warms the lower atmosphere, which then transmits heat by convection to higher levels. The weather systems are stable because the heat flux changes only over hundreds to thousands of years.

The Moons of Jupiter

In 1610 Galileo discovered four tiny specks of light orbiting Jupiter. He reasoned that they must be satellites of the giant planet. By mid-year 1997, astronomers had found 16 moons orbiting Jupiter. The four discovered by Galileo are the largest and most widely studied. In addition to the satellites, a rocky ring, similar to the rings of Saturn, lies inside the orbit of Jupiter's closest moon.

Io The innermost moon of Jupiter, Io, is about the size of the Earth's Moon and is slightly denser. Since it is too small to have retained heat generated during its formation or by radioactive decay, many astronomers expected that it would have a cold, lifeless, cratered, Moon-like surface. However, images beamed to Earth from the *Voyager* spacecraft showed huge masses of gas and rock erupting to a height of 200 kilometers above the satellite's surface. This was the first evidence of active extraterrestrial volcanism (Fig. 23–17).

Recall that the gravitational field of our Moon causes the rise and fall of ocean tides on Earth. At the same time, the Earth's gravity distorts lunar rock. Thus, the Earth's gravitation is responsible for deep-focus Moon quakes. Jupiter is 300 times more massive than Earth, so its gravitational effects on Io are correspondingly greater. In addition, the three nearby satellites, Europa, Ganymede, and Callisto, are large enough to exert significant gravitational forces on Io, but these forces pull in different directions from that of Jupiter. A few weeks before *Voyager* transmitted images of Io erupting, two scientists suggested that these gravitational forces might heat Io's interior enough to make the satellite tectonically active. We now know that this combination of oscillating and opposing tidal effects causes so much rock distortion and frictional heating that volcanic activity is nearly continuous on Io. Astronomers assume that meteorites bombarded Io and the other moons of Jupiter, as they did all other bodies in the Solar System. Yet the frequent lava flows have obliterated all ancient landforms, giving Io a smooth and nearly crater-free surface.

Europa The second closest of Jupiter's moons, Europa, is similar to Earth in that much of its interior

Figure 23–17 *Voyager I* captured an image of a volcanic explosion on Io (shown on the horizon). The eruption is ejecting solid material to an altitude of about 200 kilometers. *(NASA)* ● **Interactive Question:** The radius of Io (1815 km) is similar to that of Earth's Moon (1738 km). Discuss the differences between these moons and the reasons for these differences.

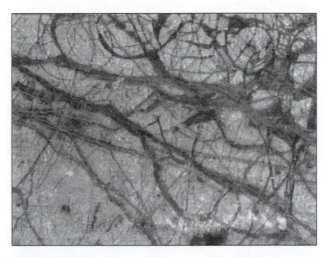

Figure 23–18 This jumbled terrain on Europa resembles Arctic pack ice as it breaks up in the spring. Scientists estimate that in this region, the ice crust is a few kilometers thick and is floating on subsurface water. *(NASA/JPL)*

is composed of rock, and much of its surface is covered with water. One major difference is that, on Europa, the water is frozen into a vast planetary ice crust. Early observations showed no craters on Europa, indicating that the surface had reformed after the intense meteor bombardment early in the history of the Solar System. In 1996 and 1997, the *Galileo* spacecraft transmitted images showing a fractured, jumbled, chaotic terrain resembling patterns created by Arctic ice on Earth during spring break-up (Fig. 23–18). Figure 23–19 shows a smooth region overlying an older wrinkled surface. Data suggest that this smooth region is young ice formed when liquid water erupted to the surface and froze. Thus pools or oceans of liquid water or a water/ice slurry probably lies beneath the surface ice. Astronomers estimate that this surface crust is ten kilometers thick, although it may be only a few kilometers thick in some places. Calculations show that the subterranean oceans are warmed by tidal effects similar to, though weaker than, those that cause Io's volcanism.

Scientists speculate that the chemical and physical environment in these subterranean oceans is favorable for life. However, there is no evidence whatsoever that life actually exists there.

Ganymede and Callisto The *Galileo* spacecraft measured a magnetic field on Ganymede, indicating that this moon has a convecting metallic core. Other measurements imply that the core is surrounded by a silicate mantle covered by a water/ice crust (Fig. 23–20). The surface ice is so cold that it is brittle and behaves much like rock. Photographs show two different types of terrain on Ganymede: One is densely cratered, and the other contains fewer craters (Fig. 23–21). The cratered regions were formed by ancient meteorite storms. The smooth regions probably developed when the crust cracked and water from the warm interior flowed over the surface and froze, much as lava flowed over the surfaces of the terrestrial planets

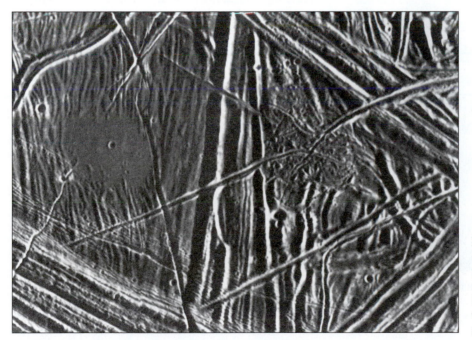

Figure 23–19 The smooth, circular region in the center left of this photograph was formed when subsurface water rose to the surface of Europa and froze, covering older wrinkles and fractures in the crust. *(NASA/JPL)*

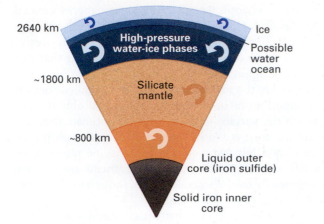

Figure 23–20 Recent data suggest that Ganymede has a conducting, convecting core, a silicate mantle, and surface layers consisting of ice and water. (Source: Stevenson, D.J., 1996, "When Galileo Met Ganymede," *Nature*, Vol. 384, Dec. 12, 1996, pp. 511–512)

Figure 23–21 Some areas of Ganymede are pockmarked by dense concentrations of impact craters (white spots), while other, younger regions contain fewer impact craters. *(NASA/JPL)* • **Interactive Question: Deduce the relative ages of different regions of the surface of Ganymede shown here.**

Figure 23–22 A close-up of a young terrain on Ganymede shows numerous grooves less than a kilometer wide. One likely explanation is that these grooves were formed by recent tectonic activity. *(NASA/JPL)*

and the Moon. Grooves and ridges in both types of terrain may be tectonic plate boundaries and mountain ranges formed by tectonic activity (Fig. 23–22).

Callisto, the outermost Galilean moon, is heavily cratered, indicating that its surface is very old. Its craters are shaped differently from those on either Ganymede or the Earth's Moon. Perhaps they have been modified by ice flowing slowly across its surface.

23.7 Saturn: The Ringed Giant

Saturn, the second largest planet, is similar to Jupiter. It has the lowest density of all the planets, so low that the entire planet would float on water if there were a basin large enough to hold it. Such a low density implies that it, like Jupiter, must be composed primarily of hydrogen and helium with a relatively small core of rock and metal. In fact, in many ways, Saturn and Jupiter are alike. For example, Saturn's atmosphere is similar to that of Jupiter. Dense clouds and great storm systems envelop the planet.

The Rings of Saturn

Saturn's most distinctive feature is its spectacular array of rings, which are visible from Earth even through

Figure 23–23 An image of Saturn shows its spectacular ring system. *(NASA/JPL)*

a small telescope. Photographs from space probes show six major rings, each containing thousands of smaller ringlets (Figs. 23–23 and 23–24). The entire ring system is only 10 to 25 meters thick, less than the length of a football field. However, the ring system is extremely wide. The innermost ring is only 7000 kilometers from Saturn's surface, whereas the outer edge of the most distant ring is 432,000 kilometers from the planet. Thus, the ring system measures 425,000 kilometers from its inner to its outer edge. A scale model of the ring system with the thickness of a CD (compact disk) would be 30 kilometers in diameter.

Saturn's rings are composed of dust, rock, and ice. The particles in the outer rings are only a few ten-thousandths of a centimeter in diameter (about the size of a clay particle), but those in the innermost rings are chunks of at least a few centimeters and possibly a few meters across. Each piece orbits the planet independently.

Saturn's rings may be fragments of a moon that never coalesced. Alternatively, they may be the remnants of a moon that formed and was then ripped apart by Saturn's gravitational field. If a moon were close enough to its planet, the tidal effects would be greater than the gravitational attraction holding the moon together, and it would break up. Thus, a solid moon cannot exist too close to a planet.

Titan

In addition to the rings, 18 moons orbit Saturn. Titan is the largest and is unique because it is the only moon in the Solar System with an appreciable atmosphere. This atmosphere has been retained because Titan is relatively massive and extremely cold. The atmosphere is composed primarily of nitrogen and a few percent methane, CH_4. The average temperature on the surface of Titan is $-180°C$, and the atmospheric pressure is 1.5 times greater than that on the Earth's surface. These conditions are close to the tempera-

Figure 23–24 *Voyager I* took this color-enhanced close-up view of Saturn's rings and ringlets. *(NASA/JPL)*

tures and pressures at which methane can exist as a solid, liquid, or vapor. Therefore, small changes in temperature or pressure on Titan could cause the methane in its atmosphere to freeze, melt, vaporize, or condense. Thus, methane clouds float in Titan's atmosphere, and methane seas, lakes, rivers, and ice caps may also exist. This situation is analogous to the Earth's environment, where water can exist as liquid, gas, or solid and frequently changes among those three states.

Methane, the simplest organic compound, reacts with other materials in Titan's environment to form more complex molecules. These organic compounds do not decompose at low temperature, so the satellite's surface is likely to be covered by a thick, tar-like organic goo. It is possible that a similar layer collected on the early Earth and later underwent chemical reactions to form life.

23.8 Uranus and Neptune: Distant Giants

Uranus and Neptune are so distant and faint that they were unknown to ancient astronomers. In 1977, the *Voyager II* spacecraft was launched to study the Jovian planets. It flew by Jupiter in 1979 and Saturn in 1981. It encountered Uranus by 1986 and Neptune in 1989. The journey from Earth to Neptune covered 7.1 billion kilometers and took 12 years. The craft passed within 4800 kilometers of Neptune's cloud tops, only 33 kilometers from the planned path. The strength of the radio signals received from *Voyager* measured one ten-quadrillionth of a watt $(1/10^{16})$. It took 38 radio antennas on four continents to absorb enough radio energy to interpret the signals (Fig. 23–25).

Composition of Uranus and Neptune

Both Uranus and Neptune are enveloped by thick atmospheres composed primarily of hydrogen and helium, with smaller amounts of carbon, nitrogen, and oxygen compounds. Beneath the atmosphere, their outer layers are molecular hydrogen, but neither body is massive enough to generate liquid metallic hydrogen. Their interiors are composed of methane, ammonia, and water, and the cores are probably a mixture of rock and metals. Uranus and Neptune are denser than Jupiter and Saturn, because these outermost giants contain relatively larger, solid cores.

Magnetic Fields of Uranus and Neptune

Voyager recorded that the magnetic field of Uranus is tilted 58° from its axis. This was unexpected, as current explanations suggest that the magnetic fields of all planets should be roughly aligned with the spin axis. At first, scientists thought that *Voyager* just happened to pass Uranus during a magnetic field reversal. However, *Voyager* later recorded that the magnetic field on Neptune is tilted 50° from its axis. Since the probability of catching two planets during magnetic reversals is extremely low, there must be another explanation. However, at present no satisfactory hypothesis has been developed.

Figure 23–25 The *Voyager I* and *II* planetary spacecraft provided valuable data on Jupiter, Saturn, Uranus, and Neptune. *(NASA/JPL)*

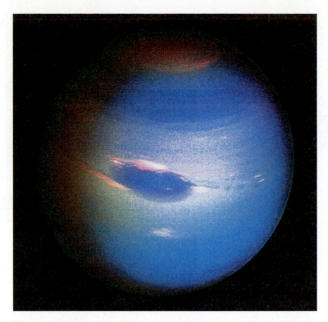

Figure 23–26 The Great Dark Spot shows clearly near the equator in this image of Neptune taken by *Voyager II.* *(NASA/JPL)*

Uranus and Neptune's Violent Weather

Spacecraft and the Hubble Space Telescope have revealed rapidly changing weather on Uranus and Neptune. On Neptune, winds of at least 1100 kilometers per hour rip through the atmosphere, clouds rise and fall, and one region is marked by a cyclonic storm system called the Great Dark Spot, similar to Jupiter's Great Red Spot (Fig. 23–26). According to one controversial hypothesis, under the intense pressure near the core, methane decomposes into carbon and hydrogen and the carbon then crystallizes into diamond. Convection currents carry the heat released during the formation of diamond to the planet's surface to power the winds.

Rings and Moons

A ring system and 17 moons orbit Uranus. Several of the moons are small and irregularly shaped, indicat-ing that they may be debris from a collision with a smaller planet or moon.

Neptune has a ring system and eight moons. The largest moon is Triton, which is about 75 percent rock and 25 percent ice. Like many other planets and moons, its surface is covered by impact craters, mountains, and flat, crater-free plains. While the maria on the Earth's Moon are blanketed by lava, those on Triton are filled with ice or frozen methane.

23.9 Pluto: The Ice Dwarf

Pluto is the outermost of the known planets and has never been visited by spacecraft. Our highest-resolution photographs of Pluto are of poor quality compared with those of other planets (Fig. 23–27). Yet astronomers have deduced the properties of this planet from a variety of data.

In 1978, astronomers discovered a satellite, Charon, orbiting Pluto. When Pluto and Charon orbit each other, they periodically block one another from view. By precisely measuring these appearances and disappearances, astronomers measured Charon's orbit. Using mathematical laws derived by Kepler and Newton in the 1600s, they calculated the relative masses of Pluto and Charon from the radius of Charon's orbit and the time required for one complete orbital revolution. From these calculations, astronomers learned that Pluto is the smallest planet in the Solar System, smaller than the Earth's Moon. It has a mass only 1/500 that of Earth.

From the diameter and mass, scientists calculated Pluto's density to be 2 g/cm^3. The density of granite is about 2.6 g/cm^3. The Earth, composed of a metallic core and rocky mantle and crust, has a density of 5.5 g/cm^3, and ice has a density of a little less than 1 g/cm^3. Since Pluto's density is greater than ice but less than rock, planetary scientists infer that Pluto is a mixture of rock and ice. Infrared measurements show that its surface temperature is about −220°C. Spectral analysis of Pluto's bright surface shows that it contains

A Ground-based Telescope

B Hubble Space Telescope

Pluto

Charon

Figure 23–27 (A) A ground-based image of Pluto and Charon and (B) a similar view from the Hubble Space Telescope. *(Space Telescope Science Institute)*

frozen methane. Its atmosphere is extremely thin and composed mainly of carbon monoxide, nitrogen, and some methane.

Pluto's orbit is more elliptical than that of any other planet. Pluto is also remarkably similar in size and density to Neptune's moon, Triton. The size and density similarities and orbital anomalies have led many astronomers to postulate that both Pluto and Charon were once moons of Neptune and were pulled out of their orbits by a close encounter with another object. Other astronomers have suggested that many ice dwarfs similar to Pluto exist in the outer Solar System, and they are now searching for more of these bodies.

23.10 Asteroids, Comets, and Meteoroids

Asteroids

Astronomers have discovered a wide ring containing tens of thousands of small orbiting bodies, called **asteroids.** This asteroid belt lies between the orbits of Mars and Jupiter. The largest asteroid, Ceres, has a diameter of 930 kilometers. Three others are about half that size, and most are far smaller. The orbit of an asteroid is not permanent like that of a planet. If an asteroid passes near a planet without getting too close, the planet's gravity pulls the asteroid out of its current orbit and deflects it into a new orbit around the Sun. Thus, an asteroid may change its orbit frequently and erratically. Many asteroids orbit near the Earth, and some even cross Earth's orbit. If an asteroid passes too close to a planet, it will crash into its surface. As discussed in the opening essay to Unit I, asteroid impacts may have caused mass extinctions on Earth.

Limited telescope observations and spacecraft data give an incomplete picture of asteroid structure and chemistry. One series of images of the asteroid Mathilde show an impact crater larger than the asteroid's mean radius. How could an object sustain such an impact without breaking apart? According to one hypothesis, the asteroid isn't solid rock as the Earth is. Instead that asteroid is a compressed mass of fractured rock and rubble. Other observations indicate that some asteroids are mostly metallic, whereas some may be composed largely of frozen volatiles. In 1999, a spacecraft called the *Near Earth Asteroid Rendezvous (NEAR)* will orbit an asteroid and attempt to answer some of these questions.

Comets

Occasionally, a glowing object appears in the sky, travels slowly around the Sun in an elongated elliptical or-

Figure 23–28 Comet Hale-Bopp was the brightest comet seen from Earth in decades. It was brightest in March and April of 1997. *(Jonathan Kern)*

bit, and then disappears into space (Fig. 23–28). Such an object is called a **comet,** after the Greek word for "long haired." Despite their fiery appearance, comets are cold, and their light is reflected sunlight.

Comets originate in the outer reaches of the Solar System, and much of the time they travel through the cold void beyond Pluto's orbit. A comet is composed of ice mixed with bits of silicate rock, metals, and frozen crystals of methane, ammonia, carbon dioxide, carbon monoxide, and other compounds. As explained in the opening essay of Unit III, comets may have transported most of the water and other volatiles to Earth.

When a comet is millions of kilometers from the Sun, it is a ball without a tail. As the comet approaches the Sun and is heated, some of its surface vaporizes. Solar wind blows some of the lighter particles away from the comet's head to form a long tail. At this time, the comet consists of a dense, solid **nucleus,** a bright outer sheath called a **coma,** and a long **tail** (Fig. 23–29). Some comet tails are more than 140 million kilometers long, almost as long as the distance from the Earth to the Sun. As a comet orbits the Sun, the solar wind constantly blows the tail so that it always extends away from the Sun. By terrestrial standards, a comet tail would represent a good, cold laboratory vacuum, yet viewed from a celestial perspective it looks like a hot, dense, fiery arrow.

Halley's comet passed so close to the Earth in 1910 that its visit was a momentous event. When the comet returned to the inner Solar System between 1985 and 1987, it was studied by six spacecraft as well as by several ground-based observatories. Its nucleus is a peanut-shaped mass approximately 16 by 8 by 8 kilometers, about the same size and shape as Manhattan Island. The cold, relatively dense coma of Halley's

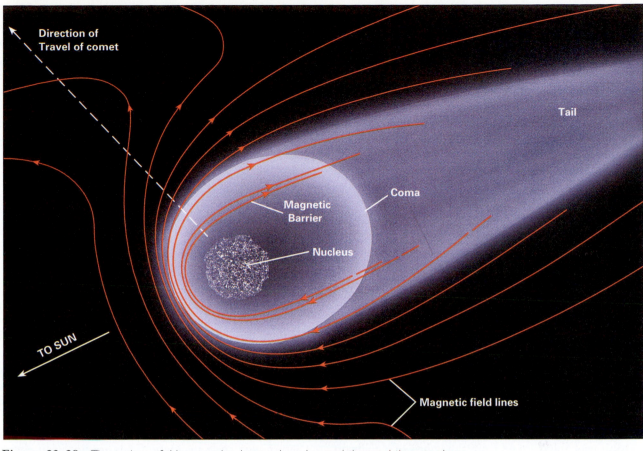

Figure 23–29 The nucleus of this comet has been enlarged several thousand times to show detail. When the comet interacts with the solar wind, magnetic field lines are generated as shown. Ions produced from gases streaming away from the nucleus are trapped within the field, creating the characteristically shaped tail.

comet had a radius of about 4500 kilometers when it passed by.

Meteoroids

As tens of thousands of asteroids race through the Solar System in changing paths, many collide and break apart, forming smaller fragments and pieces of dust. A **meteoroid** is an asteroid or a fragment of a comet that orbits through the inner Solar System. If a meteoroid travels too close to the Earth's gravitational field, it falls. Friction with the atmosphere heats it until it glows. To our eyes it is a fiery streak in the sky, which we call a **meteor** or, colloquially, a shooting star. Most meteors are barely larger than a grain of sand when they enter the atmosphere and vaporize completely during their descent. Larger ones, however, may reach the Earth's surface. A meteor that strikes the Earth's surface is called a **meteorite** (Fig. 23–30).

Figure 23–30 Scientists believe that this meteorite is a fragment from the asteroid Vesta. *(NASA)*

Most meteorites are **stony meteorites** and are composed of 90 percent silicate rock and 10 percent iron and nickel. The 90:10 mass ratio of rock to metal is similar to the mass ratio of the mantle to the core in the Earth. Therefore, geologists think that meteorites reflect the primordial composition of the Solar System and are windows into our Solar System's past. Most stony meteorites contain small grains about 1 millimeter in diameter called **chondrules,** which are composed largely of olivine and pyroxene. Many chondrules contain organic molecules, including amino acids, the building blocks of proteins.

Some meteorites are metallic and consist mainly of iron and nickel, the elements that make up the Earth's core, while the remainder are stony-iron, containing roughly equal quantities of silicates and iron-nickel. Some of our knowledge of the Earth's mantle and core comes from studying meteorites, which may be similar to the mantles and cores of other planetary bodies. With the exception of rocks returned from the *Apollo* moon missions, meteorites are the only physical samples we have from space.

SUMMARY

Mercury is the closest planet to the Sun. It has virtually no atmosphere and it rotates slowly on its axis; therefore, it experiences extremes of temperature. Mercury's surface is heavily cratered from meteorite bombardment that occurred early in the history of the Solar System. **Venus** has a hot, dense atmosphere as a result of greenhouse warming. Its surface shows signs of recent tectonic activity, probably resulting from vertical upwelling, called **blob tectonics.** The Earth's **Moon** probably formed from the debris of a collision between a Mars-sized body and the Earth. The Moon was heated by the energy released during condensation of the debris, by radioactive decay, and by meteorite bombardment. Evidence of ancient volcanism exists, but the Moon is cold and inactive today. **Mars** is a dry, cold planet with a thin atmosphere, but its surface bears signs of tectonic activity and ancient water erosion.

Jupiter, Saturn, Uranus, and **Neptune** are all large planets with low densities. Jupiter and Saturn have dense atmospheres, surfaces of liquid hydrogen, inner zones of liquid metallic hydrogen, and cores of rock and metal. The largest moons of Jupiter are **Io,** which is heated by gravitational forces and exhibits extensive volcanic activity; **Europa,** which is ice covered; and **Ganymede** and **Callisto,** which are large spheres of rock and ice. The rings of Saturn are made up of many small particles of dust, rock, and ice. They formed from a moon that was fragmented or from rock and ice that never coalesced to form a moon. **Titan,** the largest moon of Saturn, has an atmosphere and may be tectonically active. **Uranus** and **Neptune** have higher proportions of rock and ice than Jupiter and Saturn, and their magnetic fields are not in line with their axes of rotation. **Pluto,** the most distant planet, has a low density and is a small planet composed of ice and rock.

Asteroids are small planet-like bodies in solar orbits between those of Mars and Jupiter. When a **comet** is in the inner part of the Solar System, it consists of a small, dense **nucleus** composed of ice and rock; an outer sheath or **coma** composed of gases, water vapor, and dust; and a long **tail** made up of particles blown outward by the solar wind. A **meteorite** is a fallen **meteoroid,** a piece of matter from the inner Solar System. Most meteorites are stony, and some contain organic molecules. About 10 percent of all meteorites are metallic.

KEY TERMS

1. List the nine planets in order of distance from the Sun, and distinguish between the terrestrial and Jovian planets.

2. Give a brief description of Mercury. Include its atmosphere, surface temperature, surface features, and speed of rotation about its axis.

3. Why are there fewer meteorite craters visible on Venus than on Mercury?

4. Explain how Venusian tectonics differs from tectonics on Earth.

5. Compare and contrast the surface topography of Mercury, Venus, the Moon, Earth, and Mars.

6. Compare and contrast the atmospheres of Mercury, Venus, the Moon, Earth, and Mars.

7. Venus boiled, life evolved on Earth under moderate temperatures, and Mars froze. Discuss the evolution of the atmospheres and climates on these three planets.

8. What leads us to believe that the Moon was hot at one time in its history? How was the Moon heated?

9. Discuss the evidence that the Martian atmosphere was once considerably different than it is today.

10. Discuss evidence for blob tectonics and for horizontal tectonics on Mars.

11. Describe the composition of the planet Jupiter. How does it differ from that of Earth?

12. Discuss Jovian weather and what heat source drives it. Compare Jovian weather with that on Earth.

13. Compare and contrast the four Galilean moons of Jupiter.

14. Compare and contrast Saturn with Jupiter.

15. Compare and contrast Titan with the Earth.

16. Compare and contrast Pluto with Earth and Jupiter.

17. Compare and contrast Pluto with Triton.

18. Is a comet hot, dense, and fiery? If so, what is the energy source? If not, why do comets look like burning masses of gas?

1. If Mercury rotated once every 24 hours as the Earth does, would you expect daytime temperatures on that planet to be higher or lower than they are? Defend your answer.

2. Explain how we can learn about Earth's early history by studying the Moon.

3. At one time, Venus and Earth probably had similar climates, except that Venus was about 20°C warmer. If you could somehow cool the surface of Venus by 20°C, would conditions on that planet likely become similar to those on Earth? Explain.

4. Scientists speculate that even though the surface of Mercury is mostly hot, ice may exist in polar craters. The surface of Venus is also hot. Would you expect to find ice anywhere on Venus? If your answer is yes, where would you look? If your answer is no, why not?

5. Speculate on how a mantle plume could have generated faults on the surface of Venus even though it did not lead to rifting of tectonic plates.

6. Stephen Saunders, project scientist for NASA's *Magellan* mission, wrote, "Venus has been shaped by processes fundamentally similar to those that have taken place on Earth, but often with dramatically different results." Give examples to support or refute this statement.

7. Different minerals reflect radar differently, so radar data can be used to infer mineral structure. When a mineral weathers, it changes to a new mineral, and the radar reflectivity changes. Discuss how scientists could use radar reflectivity to age-date Venusian landforms.

8. Refer to Figures 23–12 and 23–13. Imagine that you know nothing about Mars except that it is a planet and that these two photographs were taken of its surface. What information can you deduce from these pictures alone? Defend your conclusions.

9. Imagine that three new planets were found between Earth and Mars. What could you tell about the geologic history of each, given the following limited data? (a) Planet X: The entire surface of this planet is covered with sedimentary rock. (b) Planet Y: This planet's atmosphere contains large quantities of water, ammonia, methane, and hydrogen sulfide. (c) Planet Z: About one third of this planet is covered with numerous impact craters. Smaller craters can be seen within the largest ones. Another third of the surface is much smoother and is scattered with a few small craters. The remainder of the planet has no visible craters but is marked by great topographic relief, including mountain ranges and smooth plains, but no canyons or river channels.

10. In his novel *2010: Odyssey II*, Arthur C. Clarke tells of a group of astronauts who traveled to the vicinity of Jupiter to retrieve a damaged spacecraft. At the end of the novel, Jupiter undergoes rapid changes and becomes a star. Is such a scenario plausible? Why did Mr. Clarke choose Jupiter as the planet to undergo such a change? Would any other planet have been as believable?

11. About 4 billion years from now, the Sun will probably grow significantly larger and hotter. How will this affect the composition and structure of Jupiter?

12. Write a short science fiction story about space travel within the Solar System. You may make the plot fantastic and fictitious, but place the characters in scientifically plausible settings.

Chapter

24

Stars, Space, and Galaxies

Massive stars produce so much energy at some stages of their life cycle that they eject huge outbursts of gas. The outer red region of this hot supergiant, Eta Carinae, is flying away from the center at about 1000 kilometers per second. The inner white region is moving slower and is therefore closer to the center. *(J. Hester and NASA)*

Our Sun is only one star among hundreds of billions in our Milky Way galaxy and the Milky Way, in turn, is only one galaxy of billions in the Universe. Looking beyond the Solar System into galactic and intergalactic space, we must stretch our minds to nearly unimaginable distances, look backward in time to events that occurred billions of years before the Earth formed, and attempt to fathom energy sources powerful enough to create an entire Universe. ●

24.1 Measuring Distances in Space

Distance on Earth is measured in millimeters, meters, or kilometers. However, even a kilometer is too small to express distance in space. Light travels in a vacuum at a constant rate—3.0×10^8 m/sec—never faster and never slower. One **light-year** is the distance traveled by light in a year, 9.5 trillion (9.5×10^{12}) kilometers. The closest star to our Solar System is 38×10^{12} (38 trillion) kilometers, or 4 light-years, away. But an object 4 light-years away is a close neighbor in space. Some objects are more than 12 billion light-years from Earth.

Recall from Chapter 22 that astronomers detected the Earth's motion around the Sun by observing the parallax shift of distant stars. The same approach can be used to calculate the distance to a star. The parallax angle is measured by observing the apparent shift in position of a star at a six-month interval, after the Earth has completed half of a revolution around the Sun (Fig. 24–1). The angles even to the closest stars are small and are expressed in arc seconds; one arc second is equal to 1/3600 of a degree. If you were to view a coin the size of a quarter from a distance of 5 kilometers, it would appear to have a diameter of 1 arc second. A common unit of distance, the **parsec**, is the distance to a star with a parallax angle of 1 arc second. One parsec is equal to 3.2 light-years or 3.1×10^{13} kilometers.

24.2 The Birth of a Star

A star exists for millions, billions, or tens of billions of years and we can never hope to observe one long enough to watch its birth, life, and death. We can, however, observe young, middle-aged, and old stars and thus piece together the story of stellar evolution. It is as if an alien came to Earth for one day to observe the life of a human being. It would be impossible to observe the birth, growth, and death of a single person, but the alien could observe babies, children, middle-aged people, and old people and thus infer the course of a human life.

Large clouds of dust and gas, called **nebulae,** exist within a galaxy. Even though a nebula looks spectacular when viewed through a powerful telescope, the densest nebulae are only 10^{-13} as dense as the Earth's atmosphere at sea level, and less dense ones are only 10^{-18} as dense as our atmosphere! A nebula consists mainly of hydrogen and helium, with traces of heavier elements.

One giant nebula lies in the constellation Orion. The entire structure is 100 light-years in length from north to south and even though it is extremely diffuse, the nebula contains a mass equivalent to 200,000 of our Suns (Fig. 24–2). Scientists subdivide this huge, diffuse cloud into several subregions that include the Horsehead Nebula, the Orion B cloud, and the Great Nebula, as shown in Figure 24–2. Figure 24–3 shows

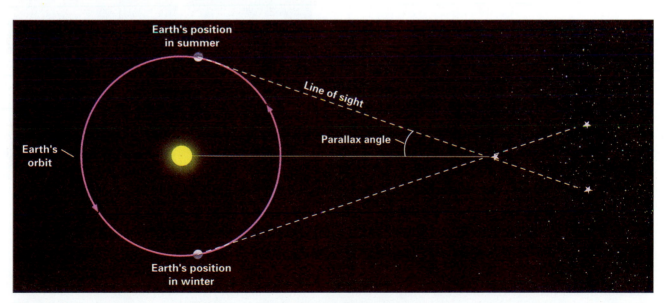

Figure 24–1   Scientists use simple trigonometry to calculate the distance to an object in space from a knowledge of the parallax angle.

A

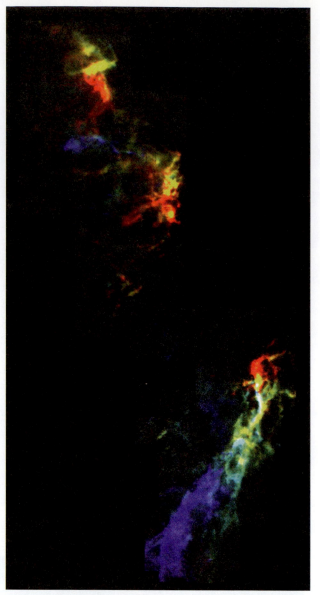

B

Figure 24–2 (A) The constellation Orion was named after a mythical Greek hunter. The three stars across the center of the constellation show the hunter's belt, and the points of light angling down from the belt show his sword. One of the objects in his belt is an emission nebula. *(Andrea Dupree, Harvard Smithsonian CA, and Ronald Gilliland, NASA)* (B) A close-up schematic view of Orion's belt region shows a huge system of nebulae. Young stars are evolving within this giant stellar nursery. (Bally, J. 1996, "Filaments in Creation's Heart," *Nature*, Vol. 382. July 11, 1996. p. 114.)

a close-up of the Horsehead Nebula, which appears near the easternmost star of Orion's belt. The brightly colored region surrounding the horsehead is an emission nebula. This portion is so hot that its atoms and molecules glow like a neon light. We will discuss emission nebulae later in the chapter. A dark region, shaped like a horse's head in Figure 24–3, occupies the middle of one of the red emission nebulae. This dark region is an absorption nebula. The dust and gases are frigidly cold, about −260°C, and absorb light from the emission nebula behind it. At these extremely low temperatures, most elements condense to form molecules. Although the two lightest elements, hydrogen and he-

lium, constitute over 99 percent of the cloud's total mass, the Orion absorption nebula contains all or most of the natural elements. Astronomers have detected 100 different molecules in the Orion absorption nebula. Most are simple, such as hydrogen, water, carbon monoxide, and ammonia, but several complex organic molecules have also been detected.

The stars in the Orion B (Fig. 24–2) cloud are at least 12 million years old. Stellar ages decrease progressively toward the southwest until the stars are less than one million years old in the Great Nebula region near Orion's sword. This age progression implies that star formation started in the Orion B portion of the

Figure 24–3 The Horsehead Nebula appears close to the bright star on Orion's belt shown on the center of this photograph. The bright white lights in the photographs are other stars. The reddish glows are emission nebulae, and the horsehead-shaped dark area is an absorption nebula in front of an emission nebula. *(Royal Observatory, Edinburgh/Anglo-Australian Telescope Board)*

cloud and moved steadily toward the current stellar nursery in the Great Nebula.

Astronomers are uncertain how the first stars formed in a large nebula. However, once the first stars began to generate energy, they heated the surrounding gas in the nebula. The heated gas expanded so that its pressure and density increased. From the rate of stellar evolution, astronomers calculated that one or more of the initial stars were massive and short lived. These stars quickly exhausted their supply of fuel and exploded. Hot gases burst into the surrounding nebula, increasing the pressure and density even more. When the pressure and gas density reached a critical value, gravitational forces between molecules were strong enough to pull the particles together. Matter separated into discrete regions falling toward common centers. Recent observations by the Hubble telescope show that these regions of varying density appear as towering pillars of light and shadow (Fig. 24–4). Hundreds of young stars, only 8 million to 300,000 years old, are forming along the edges of the cloud.

As the dust and gas in a nebula coalesce, particles accelerate under the force of gravity and collide rapidly with one another. Thus a small region of the cloud be-

comes very dense and hot. Under the intense heat, molecules break apart and electrons are stripped away from their atoms, leaving a plasma of positively charged nuclei and negatively charged electrons. Positive nuclei repel one another. However, if they are hot enough, they collide with sufficient energy to overcome the repulsion and fuse. In this way, hydrogen nuclei fuse to form nuclei of the next heavier element, helium. This nuclear fusion is similar to reactions that occur in a hydrogen bomb. If a sphere of hydrogen the size of a pinhead were to fuse completely, it would release as much energy as is released by burning several thousand tons of coal.

When fusion starts, the collapsing portion of the original nebula becomes a star. Fusion generates photons and heats atomic particles. Both the photons and the hot particles accelerate outward against the force of gravity. Thus, two opposing processes occur in a star. Gravity pulls particles inward, but at the same time fusion energy drives them outward (Fig. 24–5). The balance between these two processes determines the diameter and density of a star of a given mass. At equilibrium, a star with an average mass has a dense core surrounded by a less dense shell.

Figure 24–4 The Hubble Space telescope has provided a close-up view of stellar nurseries near Orion's belt. The pillar-like structures are about three light-years tall and are composed of cold gas and dust. Energy from nearby stars and shock waves from exploding stars have concentrated the gas in the pillars and initiated star formation. *(J. Hester and P. Scowen [ASU] and NASA)*

24.3 The Sun

The Sun is a star of average mass. However, since it is so much closer to Earth than any other star, we can observe it in more detail. The Sun's diameter is 1.4 million kilometers (109 Earth diameters). Hydrogen accounts for 92 percent of the Sun's atoms, and helium is second in abundance at 7.8 percent. All the remaining elements make up only 0.2 percent. The Sun's structure is shown in Figure 24–6.

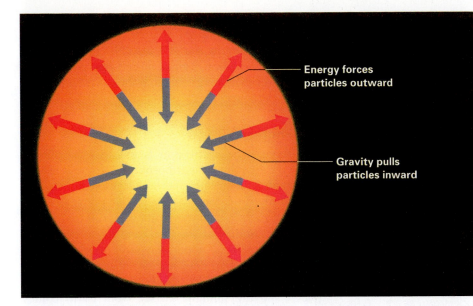

Figure 24–5 The diameter and density of a star are determined by two opposing forces. Gravity pulls particles inward while energy from fusion reactions in the core forces particles outward.

Energy forces particles outward

Gravity pulls particles inward

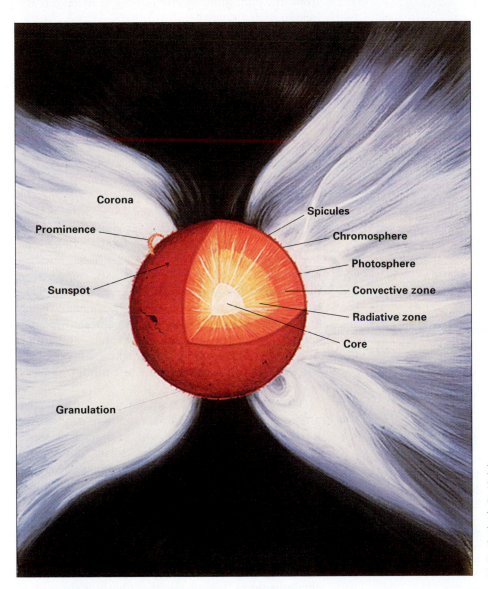

Corona

Prominence

Sunspot

Granulation

Spicules

Chromosphere

Photosphere

Convective zone

Radiative zone

Core

Figure 24–6 Structure of the Sun. Fusion occurs in the Sun's core. Energy escapes first by radiation, then by convection. The photosphere is the thin layer visible from Earth. Surface features are shown in more detail in Figures 24–7 through 24–9. *(Jay Pasachoff)*

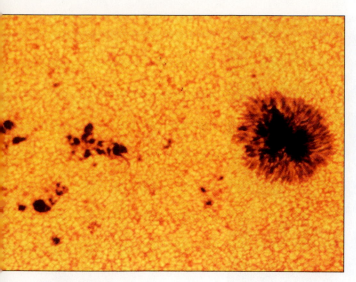

Figure 24–7 A close-up of the Sun's surface shows granular structures and sunspots. The large black structures are sunspots. Each small, bright yellow dot is a rising column of hot gas about 1000 km in diameter. The darker, reddish regions between the dots are cooler, descending gases. *(National Solar Observatory/National Optical Astronomy Observatory)*

The Core

The **core** of the Sun is extremely hot, more than 15,000,000°C, and its density is 150 times that of water.[1] The Sun's core is 140,000 kilometers in diameter and contains enough hydrogen to fuel its fusion reaction for another 5 billion years. When the hydrogen in the core is used up, the Sun will change drastically, as described in Section 24.4.

Most of the energy generated by fusion in the Sun's core is emitted as photons carrying high energy radiation. When a photon flies out of the core, it is almost immediately absorbed by atoms in a broad region surrounding the core called the **radiative zone.** Atoms re-emit photons soon after they absorb them, but the gas density in the region is so high that a photon is quickly reabsorbed again. This process of absorption and emission occurs repeatedly until the photon reaches the **convective zone.** The convective zone is much cooler on its surface than in its interior. As a result, hot gas rises, cools, and then sinks in huge convection cells. This convection transports heat to the outer layer that we see, called the **photosphere.**

---

[1] Astronomers generally report temperature on the Kelvin scale, which is based on fundamental thermodynamic properties. We use the Celsius scale here because it is more familiar. 0°C = 273°K. At high temperatures the 273° difference between the two is negligible.

If photons traveled directly from the core to the photosphere at the speed of light, they would complete the journey in about 4 seconds. However, the process of absorption, emission, and convection is so much slower that energy requires about a million years to make the journey.

The Photosphere

Even though the entire Sun is gaseous, the photosphere is called the solar atmosphere because it is cool and diffuse compared to the core. It is a thin surface veneer, only 400 kilometers deep. The pressure at the middle of the photosphere is about 1/100 that of the Earth's atmosphere at sea level, and its average temperature is 6000°C, about the same as that in the Earth's core. Fusion does not occur at this relatively low temperature. Thus, the core heats the photosphere, and the sunlight we see from Earth comes from this thin, glowing atmosphere of hydrogen and helium.

The photosphere has a granular structure with each grain about 1000 kilometers across, or about the size of Texas. The granules are caused by convection currents that carry energy to the surface. If you could watch the Sun's surface at close range, the granules would appear and disappear like bubbles in a pot of boiling water. The bright yellow granules in Figure 24–7 are formed by hot rising gas, and the darker, reddish regions are cooler areas of descending gas.

Large dark spots called **sunspots** also appear regularly on the Sun's surface. A single sunspot may be small and last for only a few days, or it may be as large as 150,000 kilometers in diameter and remain visible for months. Because a sunspot is 1000°C cooler than the surrounding area, it radiates only about one half as much energy and hence appears dark compared to the rest of the photosphere. Since heat normally flows from a hot body to a cooler one, a large, cool region on the Sun's surface should quickly be heated and disappear. Yet some sunspots persist for long times. Astronomers believe that the Sun's magnetic fields restrict mixing of the photosphere and inhibit warming of sunspots.

The magnetic fields associated with sunspots also produce flares of hot gas by accelerating gases to velocities in excess of 900 km/hr, and sending shock waves smashing through the solar atmosphere. Charged particles emitted by the flares interrupt radio communication and cause aurorae (northern and southern lights) on Earth.

The Sun's Outer Layers

The photosphere is not bounded by a sharp outer surface. A turbulent, diffuse, gaseous layer called the **chromosphere** lies above the photosphere and is

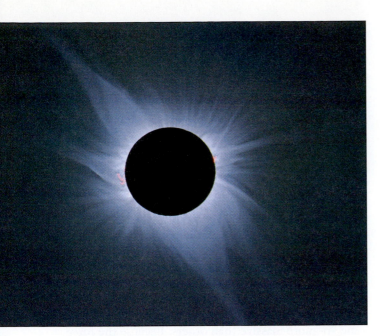

Figure 24–8 During a solar eclipse the photosphere is blocked by the Moon. The thin red streaks beyond the Moon's outline are portions of the chromosphere, and the large white zone is the corona. *(Kazuo Shiota)*

about 2000 to 3000 kilometers thick. The chromosphere can be seen with the naked eye only when the Moon blocks out the photosphere during a solar eclipse. Then it appears as a narrow red fringe at the edge of the blocked-out Sun. Jets of gas called **spicules** shoot upward from the chromosphere, looking like flames from a burning log. An average spicule is about 700 kilometers across and 7000 kilometers high, and lasts about 5 to 15 minutes.

An even more diffuse region called the **corona** lies beyond the chromosphere. During a full solar eclipse the corona appears as a beautiful halo around the Sun (Fig. 24–8). Its density is one-billionth that of the atmosphere at the Earth's surface. In a physics laboratory, that would be a good vacuum.

The corona is extremely hot, about 2,000,000°C. How does the photosphere, at 6000°C, heat the corona to 2,000,000°C? According to one hypothesis, twisting magnetic fields accelerate particles in the corona. When particles are moving quickly, their temperature is high. **Prominences** are red flame-like jets of gas that rise out of the corona and travel as much as 1 million kilometers into space (Fig. 24–9). Some promi-

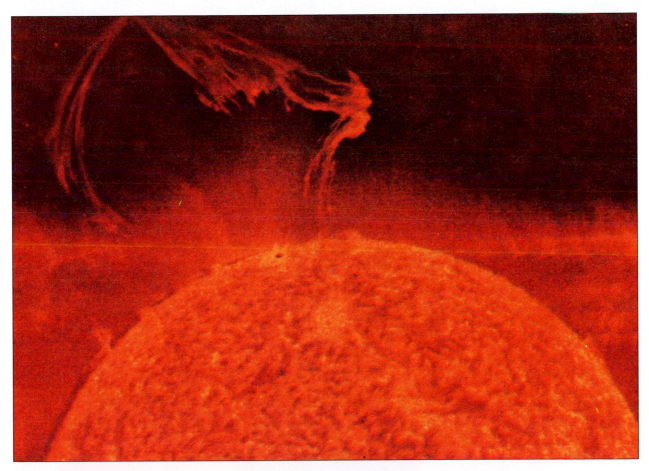

Figure 24–9 This prominence rises approximately 505,000 km above the photosphere.
(National Solar Observatory/National Optical Astronomy Observatory)

nences are held aloft for weeks or months by the Sun's magnetic fields.

The high temperature in the corona strips electrons from their atoms, reducing hydrogen and helium to bare nuclei in a sea of electrons. These nuclei and electrons are moving so rapidly that some fly off into space, forming the solar wind that extends outward toward the far reaches of the Solar System.

24.4 Stars: The Main Sequence

From our view on Earth, the Sun is the biggest and brightest object in the sky. However, the Sun looks so big and bright only because it is the closest star. When astronomers began to study stars in the early 1900s, one of their first efforts was to catalog them by their brightness. The **apparent magnitude** of a star is its brightness as seen from Earth. A star can appear luminous either because it really is bright or because it is close. The **absolute magnitude** is how bright a star would appear if it were 32 light-years (10 parsecs) away. Absolute magnitude is sometimes expressed in solar luminosities, a relative scale in which the luminosity of the Sun equals one.

A second visible property of a star is its color. Different stars have different colors; some are reddish, others white or blue. As explained in Chapter 22, the color of a star is a measure of its temperature. Between 1911 and 1913, Ejnar Hertzsprung and Henry Russell compared the absolute magnitudes of stars with their temperatures. Figure 24–10 is a **Hertzsprung-Russell** or **H-R diagram.**

In an H-R diagram luminosity increases toward the top and temperature increases toward the left. About 90 percent of all stars fall along a sinuous band from upper left (bright and hot) to lower right (dim and cool). This band is called the **main sequence.** Thus, luminosity increases with temperature in main-sequence stars.

All main-sequence stars are composed primarily of hydrogen and helium and are fueled by hydrogen fusion. The major reason for differences in temperature and luminosity among main-sequence stars is that some are more massive than others. Because the force of gravity is stronger in massive stars than in less massive stars, hydrogen nuclei are packed more tightly and move more rapidly. As a result, fusion is more rapid and intense in massive stars, so they are hotter and more luminous (upper left of the H-R diagram). The

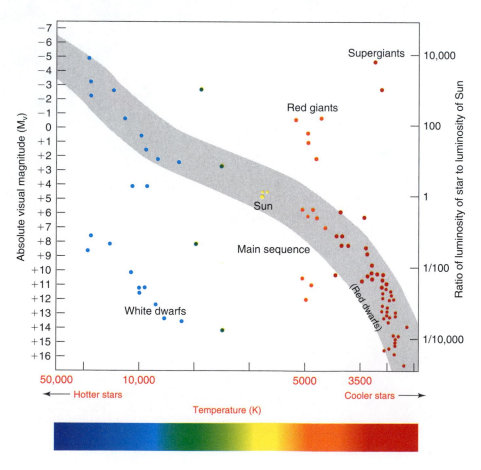

Figure 24–10 A Hertzsprung-Russell diagram. About 90 percent of all stars fall along the main sequence which is shaded in gray in this H-R diagram. The color bands below the graph illustrate changes in temperature. Our Sun is an average sized star with a surface temperature of about 6000°K, and glows yellow.

● Interactive Question: (A) What is the temperature range of blue main-sequence stars? (B) What is the characteristic luminosity and temperature of a red giant? A white dwarf?

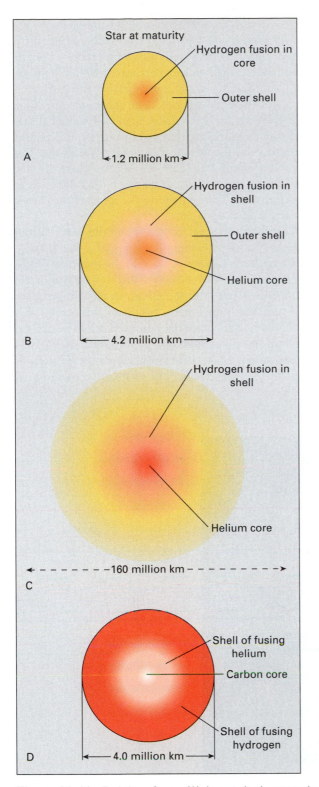

Star at maturity

Hydrogen fusion in core

Outer shell

A

|← 1.2 million km →|

Hydrogen fusion in shell

Outer shell

Helium core

B

|← 4.2 million km →|

Hydrogen fusion in shell

Helium core

C

←--------- 160 million km ---------→

Shell of fusing helium

Carbon core

Shell of fusing hydrogen

D

|← 4.0 million km →|

Figure 24–11 Evolution of a star. (A) At maturity, the energy in a star the size of our Sun is derived from hydrogen fusion in the core. (B) When most of the hydrogen in the core is consumed, the star first contracts, then expands as hydrogen fusion starts in the outer shell. (C) In the red giant phase, gravitational coalescence in the core and hydrogen fusion in the outer shell produce hundreds of times as much energy as was produced when the star was mature. (D) The star contracts again when helium fusion initiates in the core. The four parts cannot be drawn to scale as the Sun's diameter varies from 1.2 million kilometers to 160 million kilometers. Also, in all cases the core has been drawn larger than scale to show detail. (*After Jay Pasachoff,* Journey Through the Universe, *Saunders College Publishing, 1992*)

least massive stars, shown on the lower right, are cool and less luminous. Notice that our Sun is about halfway along the main sequence, in the yellow band.

To explain the stars that do not lie on the main sequence, we must consider the life and death of a star.

24.5   The Life and Death of a Star

Stars About the Same Mass As Our Sun

In a mature star such as our Sun, hydrogen nuclei fuse to form helium but the helium nuclei don't fuse to form heavier elements. For fusion to occur, two nuclei must collide so energetically that they overcome their nuclear repulsion. Hydrogen nuclei contain one proton, but helium contains two. Therefore the repulsion between two helium nuclei is much greater than that between two hydrogen nuclei. This stronger repulsion can be overcome if the nuclei are moving very rapidly. Since higher temperature causes nuclei to move more rapidly, helium fusion occurs when a star becomes much hotter than the temperature at which hydrogen fusion occurs.

The life cycle of a star with a mass similar to our Sun is shown in Figure 24–11. The Sun evolved from a cloud of gas and dust and is now midway through its mature phase as a main-sequence star. It has been shining for about 5 billion years and will continue to shine much as it is today for another 5 billion years (Fig. 24–11A). During this entire period, the Sun produces energy by hydrogen fusion and remains on the main sequence.

After about 10 billion years, the outer shell of a star such as our Sun still contains large quantities of hydrogen, but most of the hydrogen in the core has fused to helium (Fig. 24–11B). The star's behavior now changes drastically. Since the hydrogen in the core is nearly used up, hydrogen fusion slows down, less nuclear energy is produced, and the core cools. As it cools, the outward pressure of particles and energy decreases. Then the core starts to contract under the force of gravity. This gravitational contraction causes the core to grow hotter. It seems a paradox that when the nuclear reactions decrease, the core becomes hotter, but that is what happens.

As the core heats up, the rising temperature initiates hydrogen fusion in the outer shell. The star is now heated by both the gravitational coalescence in the core and the hydrogen fusion in the outer shell. As a result, the star releases hundreds of times as much energy as it did when it was mature. This intense energy output now causes the outer parts of the star to expand and become brighter. The star has become a **red giant** (Fig. 24–11C). A red giant is hundreds of times larger than an ordinary star. Its core is hotter, but its surface is so large that heat escapes and the surface cools. This cool surface emits red light (recall that red wavelengths have the lowest energy of the visible spectrum). These sudden changes in energy production and diameter move the star off the main sequence (Fig. 24–12).

Five billion years from now, the hydrogen in our Sun's core will be exhausted and the Sun will expand into a red giant. It will engulf Mercury, Venus, and Earth (Fig. 24–13). Perhaps the heat will blow much of Jupiter's atmosphere away, exposing a rocky surface.

The core of a red giant condenses under the influence of gravity and gets hotter until its temperature reaches 100,000,000°C. At this temperature, helium nuclei begin to fuse to form carbon nuclei. When helium fusion starts, radiant energy pushes outward once again, and the core expands. The star cools, its outer layers contract, and it enters a second stable phase. Gradually, as more helium fuses to carbon, the carbon accumulates in the core just as helium did during the earlier life of the star (Fig. 24–11D). When the helium is used up, fusion ceases again and the carbon core contracts. This gravitational contraction causes the core to heat up again.

What happens next depends on the star's initial mass. Astronomers express the mass of a star relative to that of the Sun. One **solar mass** is the mass of the Sun. In a star as massive as our Sun, contraction of the carbon core is not intense enough to raise its temperature sufficiently to initiate fusion of the carbon nuclei. However, gravitational contraction of the carbon core does release enough energy to blow a shell of gas out into space. This shell is called a **planetary nebula** (Fig. 24–14). Meanwhile, the material remaining in

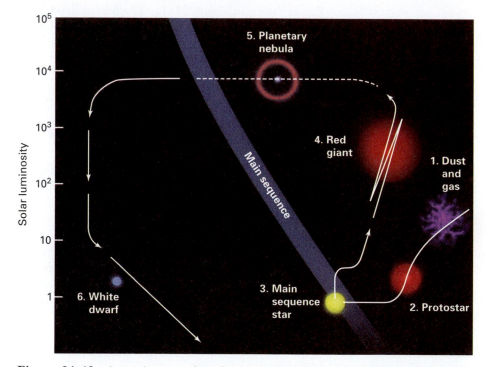

Figure 24–12 A star the mass of our Sun passes through six major stages in its life cycle. (1) After the original nebula condenses (2) the star glows from the heat of gravitational coalescence. (3) The star enters the main sequence when hydrogen fusion starts in the core. (4) After hydrogen fusion ends in the core, the star leaves the main sequence and passes through the red giant stage. (5) Finally it explodes to produce a planetary nebula and then (6) the remaining core glows as a white dwarf. • Interactive Question: List the solar luminosities and surface temperatures of the six major stages in the life of a star with the same mass as our Sun.

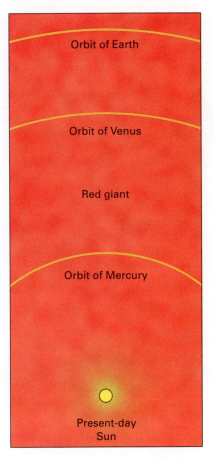

Figure 24–13 The Sun will engulf the Earth when it expands to its red giant stage. *(After Jay Pasachoff,* Journey Through the Universe, *Saunders College Publishing, 1992)*

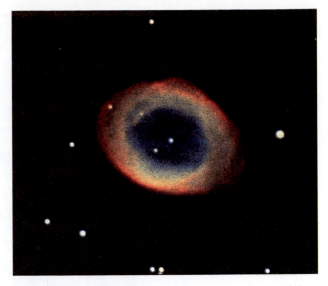

Figure 24–14 The Ring Nebula is a sphere of gas and dust expelled as a dying star exploded. *(National Optical Astronomy Observatories)*

the star contracts until atoms are squeezed so tightly together that only the pressure exerted by the electrons prevents further compression. A dying star as massive as our Sun will eventually shrink until its diameter is approximately that of the Earth. Such a shrunken star no longer produces energy, and it glows solely from its residual heat produced during past eras. The star has become a **white dwarf** (Fig. 24–15). It will continue to cool slowly over tens of billions of years, but it will never change diameter again. No further nuclear reactions will occur. Its gravitational force is not strong enough to overcome the strength of the electrons, so it will never contract further.

Stars with a Large Mass

Some stars do not die so gently (Fig. 24–15). If the star is larger than 1.44 solar masses, a white dwarf does not form. Instead, as helium fusion ends, gravitational contraction produces enough heat to fuse carbon. Renewed fusion produces increasingly heavier elements in a stepwise process until iron forms. Iron is

different from the lighter elements. When hydrogen or helium nuclei fuse, energy is released. In contrast, iron fusion absorbs energy, and thus cools a star. When this happens, the thermal pressure that forced the stellar gases outward diminishes, and the star collapses under the influence of gravity. This collapse releases large amounts of heat. Within a few seconds—a fantastically short time in the life of a star—the star's temperature reaches trillions of degrees and the star explodes to create a **supernova.** For a brief period, a supernova shines as brightly as hundreds of billions of normal stars and may even emit as much energy as an entire galaxy. To observers on Earth, it appears as though a new brilliant star suddenly materialized in the sky, only to become dim and disappear to the naked eye within a few months.

On February 24, 1987, an astronomer named Ian Shelton was carrying out research unrelated to supernovas. When he developed one of his photographic plates, he saw a bright star where previously there was only a dim one (Fig. 24–16). He walked outside, looked into the sky, and saw the star with his naked eye. This was the first supernova explosion visible to the naked eye since 1604, five years before the invention of the telescope.

A supernova explosion is violent enough to send shock waves racing through the atmosphere of the star, fragmenting atomic nuclei and shooting subatomic particles in all directions. Many of the nuclear particles collide with sufficient energy both to fuse and to split apart. These processes form all the known ele-

Figure 24–15 A star with a mass about the same as our Sun passes through its life cycle until it becomes a white dwarf. A more massive star follows a different route, ending as a neutron star or a black hole, depending on its initial mass. *(After Jay Pasachoff, Journey Through the Universe, Saunders College Publishing, 1992)*

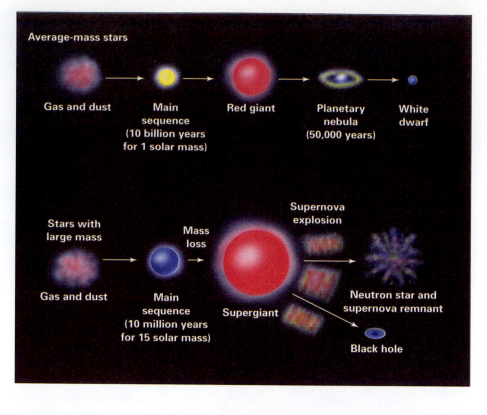

ments heavier than iron. Thus, in studying the evolution of stars, scientists learned how the heavy elements originated.

The high-temperature gas, streaming out of a supernova at thousands or even tens of thousands of kilometers per second, may collide with atoms and molecules in nearby cold nebulae, where collisions may raise the temperature of the cold nebula to a mil-

lion degrees. The hot gases then glow to create emission nebulae visible from Earth.

First- and Second-Generation Stars

Hydrogen was the first element to form when our Universe was young. Originally, all the stars in the Universe were composed of nearly pure hydrogen. These old stars are called **population II stars,** and a

A

B

Figure 24–16 During the supernova explosion of 1987, a relatively insignificant star (A) (see arrow) became the brightest object in that portion of the sky (B). *(Anglo-Australian Telescope Board)*

Chapter 24 Stars, Space, and Galaxies

few still exist. Within the cores of population II stars, hydrogen fused to helium, helium to carbon, and carbon to heavier elements, up to iron. Elements heavier than iron formed during supernova explosions of massive population II stars. The abundance of heavy elements in a population II star is determined by its mass and age. However, population II stars have a much lower proportion of heavy elements than population I stars.

When a star dies, it blasts gas and dust into space to form a nebula. Eventually, the nebulae condense once again into new stars, called **population I stars.** Population I stars begin life with a mixture of primordial hydrogen and heavy elements inherited from population II stars and from supernova explosions. The Sun is a population I star that condensed from a nebula containing many elements. Thus, our Solar System was born from the debris of one or more dying stars. Solar systems containing Earth-like planets and living organisms could never form around population II stars because these stars have virtually no heavy elements. As you can see, in studying the life cycle of stars we also learn about the origins of the elements that make up the Earth and its life-forms.

24.6 Neutron Stars, Pulsars, and Black Holes

Neutron Stars and Pulsars

In a supernova explosion, most of the matter in a star is blasted into a nebula, but a substantial fraction remains behind, compressed into a tight sphere. In the 1930s, scientists developed a hypothesis to explain what happens within this sphere. If it is between 2 and 3 solar masses, the gravitational force is so intense that the star cannot resist further compression the way a white dwarf does. Instead, the electrons and protons are squeezed together to form neutrons:

$$\text{Electrons} + \text{protons} \rightarrow \text{neutrons}$$

The neutrons then resist further compression and remain tightly packed. This ball of compressed neutrons is called a **neutron star.** A neutron star is extremely dense—approximately 10^{13} kg/cm^3 (Fig. 24–17). If the entire Earth were as dense, it would fit inside a football stadium.

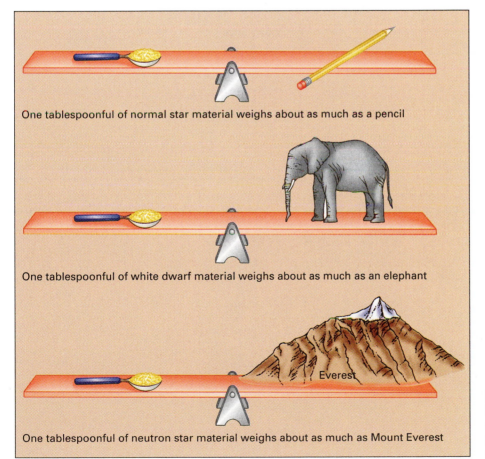

One tablespoonful of normal star material weighs about as much as a pencil

One tablespoonful of white dwarf material weighs about as much as an elephant

Everest

One tablespoonful of neutron star material weighs about as much as Mount Everest

Figure 24–17 A normal star has a relatively low density; a white dwarf is more dense, and a neutron star has even higher density.

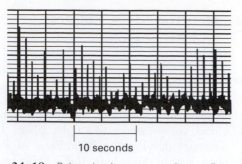

Figure 24–18 Pulsar signals appear as sharp spikes on the recording of a radio telescope.

10 seconds

The first neutron star was discovered by accident in 1967, thirty years after scientists postulated their existence. Jocelyn Bell Burnell, a graduate student at the time, was studying radio emissions from distant galaxies. In one part of the sky she detected a radio signal that switched on and off with a frequency of about one pulse every 1.33 seconds (Fig. 24–18). If such a signal were fed into the speaker of a conventional radio, you would hear a click, click, click, evenly spaced, with one click every 1.33 seconds. Many radio signals arrive at Earth from outer space, but the emissions Burnell heard were unusual because they were sharp, regular, and spaced a little over a second apart. At first, astronomers considered the possibility that they might be a signal from intelligent life, so they called the signals LGM, for "Little Green Men." But when Burnell found a similar pulsating source in a different region of the sky, scientists ruled out the possibility that two life-forms in different parts of the Universe would send similar signals. Once it was established that the signals did not originate from intelligent beings, their sources were called **pulsars.** But naming the source did not explain it.

The first step toward identifying pulsars was to estimate their sizes. Not all parts of an object in space are equidistant from Earth (Fig. 24–19). If a large

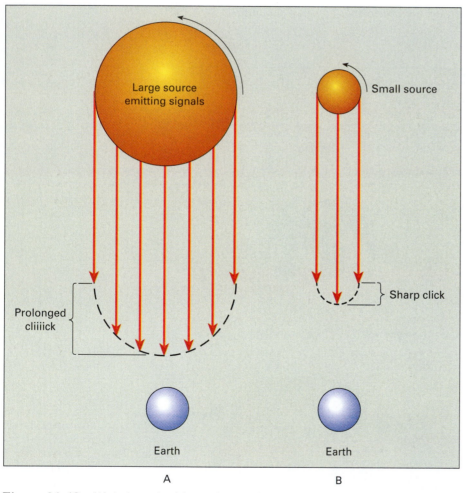

Figure 24–19 (A) A sharp signal from a larger sphere arrives over a longer time interval than (B) a sharp signal from a smaller sphere.

sphere emits a sharp burst of energy from its entire surface, some of the photons start their journey closer to Earth than others and therefore arrive sooner. A person on Earth listening to the radio noise hears not a sharp click but a more prolonged cliiiiiick, because it takes a while for all the radio waves to arrive. Alternatively, a signal from a small source is much sharper. Pulsar signals are sharp, indicating that the source must be unusually small for an energetic object in space—30 kilometers in diameter or less. The smallest star previously recorded was a white dwarf 16,000 kilometers in diameter. Scientists then reasoned that the pulsar detected by Burnell might be the long-searched-for neutron star.

Astronomers suggested that the radio signals are emitted by an electromagnetic storm on the surface of a neutron star or pulsar. According to this hypothesis, as the star rotates, so does the storm center. A receiver on Earth detects one click per revolution of the star, just as a lookout on a ship sees a lighthouse beam flash periodically as the beacon rotates (Fig. 24–20). Since a pulse is received once every 1.33 seconds, the pulsar must rotate that rapidly. White dwarfs can again be ruled out since they are too big to rotate so rapidly. But neutron stars are small enough to rotate that fast.

Astronomers searched for an example to test the hypothesis. They focused a radio telescope on the Crab Nebula, where a supernova explosion occurred during the Middle Ages. A pulsar signal was found exactly where the supernova had occurred about 950 years ago—precisely where a neutron star should be.

Black Holes

If a star with more than about 5 solar masses explodes, the core remnant remaining after the supernova explosion is thought to be more massive than 2 to 3 solar masses. When a sphere this massive contracts, the neutrons are not sufficiently strong to resist the gravitational force. Then the star shrinks to a diameter much smaller than a neutron star and becomes a **black hole.** Such a collapse is impossible to imagine in earthly terms. A tremendous mass, perhaps a trillion, trillion, trillion kilograms, shrinks to the size of a pinhead, then continues to shrink to the size of an atom, and then even smaller. Eventually it collapses to an infinitesimally small point of infinitely high density.

Such a small point of mass creates an extremely intense gravitational field. According to Einstein's theory of relativity, gravity affects photons. Thus, starlight bends as it passes the Sun. If an object is massive and dense enough, its gravitational field becomes so intense that light and other radiant energy cannot escape. So, just as you cannot throw a ball from the Earth to space because it falls back down, light cannot escape from a black hole because it is pulled back

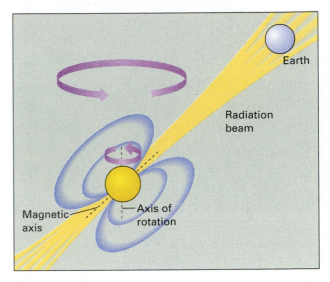

Figure 24–20 The radiation beam of a pulsar is detected only when it sweeps across the Earth.

downward. Since no light can escape such an object, it is invisible; hence the name black hole. If you were to shine a flashlight beam, a radar beam, or any kind of radiation at a black hole, the energy would be absorbed. The beam could never be reflected back to your eyes; therefore, you would never see it again. It would be as though the beam just vanished into space. Similarly, if a spaceship flew too close to a black hole, it would be sucked in. No engine could possibly be powerful enough to accelerate the rocket back out, for no object can travel faster than the speed of light.

The search for a black hole is therefore even more difficult than the search for a neutron star. How do you find an object that is invisible and can neither emit nor reflect any energy? In short, how do you find a hole in space? Although it is theoretically impossible to see a black hole, astronomers can observe the effects of its gravitational field. Many stars exist in pairs or small clusters. If two stars are close together, they orbit around each other. Even if one becomes a black hole, the two still orbit about each other, but one is visible and the other invisible. The visible one appears to be orbiting an imaginary partner. Astronomers have studied several stars that seem to orbit in this unusual manner. In several cases, the invisible member of the pair has a mass equal to or greater than 3 solar masses. Since a normal star of 3 solar masses would be visible, the invisible partner may be a black hole. However, the simple observation that a star moves around an invisible companion does not prove that a black hole exists.

If a star were orbiting a black hole, great masses of gas from the star would be sucked into the black hole,

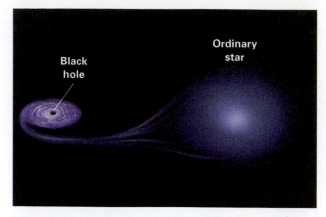

Figure 24–21 If a black hole and an ordinary star are in orbit around one another, gases from the star will eventually be sucked into the black hole.

to disappear forever (Fig. 24–21). As this matter started to fall into the hole, it would accelerate, just as a meteorite accelerates as it falls toward Earth. The gravitational field of a black hole is so intense that particles drawn into it would collide against each other with enough energy to emit X-rays. Thus, just as a falling meteorite glows white-hot as it enters the Earth's atmosphere, anything tumbling into a black hole would glow even more energetically; that is, it would emit X-rays. These X-rays might then be detected here on Earth. But, you may ask, if light cannot escape from a black hole, how can the X-rays escape? The answer is that the X-rays are produced and escape from just outside the black hole. Thus, matter being sucked into a black hole sends off one final message before being pulled into the void from which no message can ever be sent.

An important experiment, therefore, was to focus an X-ray telescope on portions of the sky where a star appeared to orbit an invisible partner. Such telescopes must be located aboard space satellites because X-rays do not penetrate the Earth's atmosphere. In the 1980s, an orbiting X-ray telescope detected X-ray sources adjacent to stars that appeared to orbit an unseen partner. More recent observations by ground-based observatories and the Hubble Space Telescope convince most astronomers that black holes exist.

24.7   Galaxies

In the late 1700s, a French astronomer, Charles Messier, was studying comets. He recorded more than 100 fuzzy objects in the sky that clearly were not stars. When these objects were studied in the 1850s with more powerful telescopes, many were observed to have

spiral structures like pinwheels. They were not comets, but what were they, and how far away were they? These questions were not answered until 1924 when Edwin Hubble determined that they were farther away than even the most distant known stars. In order for us to see them at all, they must be much more luminous than a star. Hubble concluded that each object is a **galaxy** composed of billions of stars. Today we recognize that galaxies and clusters of galaxies form the basic structure of the Universe.

About one third of the galaxies in our region of the Universe are elliptical (Fig. 24–22). Giant **elliptical galaxies** contain up to 10^{13} solar masses and may have a diameter larger than our Milky Way galaxy (which has a spiral shape, as discussed next). However most ellipticals are dwarf and contain only a few million solar masses. Recent observations indicate that dwarf ellipticals may be the most common type of galaxy far from our immediate galactic neighborhood.

Spiral galaxies are the most common bright galaxies in our region of the Universe. The stars in a spiral are arranged in a thin disk with arms radiating outward from a spherical center or nucleus (Fig. 24–23). The stars in the outer arms rotate around the nucleus like a giant pinwheel. A typical spiral galaxy contains hundreds of billions of stars.

About 30 percent of the spiral galaxies are **barred spirals,** having a straight bar of stars, gas, and dust extending across the nucleus (Fig. 24–24). A few percent of all galaxies are irregular and show no obvious pattern.

Figure 24–22 The giant elliptical galaxy M87 is nearly spherical, but other elliptical galaxies are more elongate and look like flying saucers. *(Anglo-Australian Telescope Board)*

Figure 24–23 The dense nucleus of the spiral galaxy NGC 2997 is surrounded by spiraling arms composed of billions of stars. *(Anglo-Australian Telescope Board)*

Figure 24–24 A barred spiral galaxy, NGC 1365, displays a conspicuous bar running through its nucleus. *(European Southern Observatory)*

The Milky Way

Our Sun lies in a spiral galaxy called the **Milky Way** (Fig. 24–25). The Milky Way's disk is 2000 light-years thick, nearly 100,000 light-years in diameter, and contains hundreds of billions of stars. Because the disk is thin, like a phonograph record, an observer on Earth sees relatively few stars perpendicular to its plane. Thus, most of the night sky contains a diffuse scattering of stars with large expanses of black space between them. However, if you look into the plane of the disk, you see a dense band of stars from horizon to horizon (Fig. 24–26). This band is commonly called the Milky Way, although astronomers use the term to describe the entire galaxy. The galactic disk rotates about its center once every 200 million years, so in the 4.6-billion-year history of the Earth we have completed about 23 rotations.

A spherical **galactic halo** of dust and gas surrounds the Milky Way's galactic disk. This halo is so large that

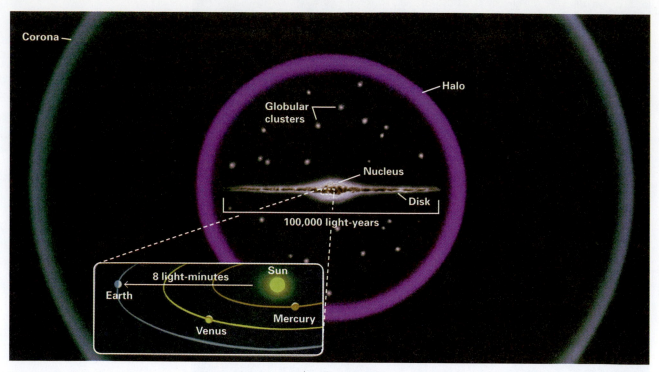

Figure 24–25 An artist's drawing of the Milky Way galaxy shows the galactic disk which is nearly 100,000 light-years in diameter and 2000 light-years thick. The distance from the Sun to Earth is only 8 light-minutes. *(After Jay Pasachoff,* Journey Through the Universe, *Saunders College Publishing, 1992)*

Figure 24–26 The Milky Way appears as a solid band of light. However, this light is produced by billions of stars in the plane of the galactic disk. *(Lund Observatory)*

even though it is extremely diffuse, it contains as much as 90 percent of the mass of the galaxy. Many dim and relatively old stars exist within this halo. Some are concentrated in groups of 10,000 to 1,000,000 stars. Each group is called a **globular cluster**. The galactic halo and globular clusters are probably remnants of one or more protogalaxies that condensed to form the Milky Way. The spherical structure of the halo suggests that the entire galaxy was once spherical.

Photographs of other spiral galaxies show that the galactic nucleus shines much more brightly than the disk. The concentration of stars in the galactic nucleus is perhaps one million times greater than in the outer disk. If you could visit a planet orbiting one of these stars, you would never experience night, for the stars would light the planet from all directions. However, stable solar systems could not exist in this region because gravitational forces among nearby stars would rip planets from their orbits.

It is impossible to see into the nucleus of our own galaxy because visible light doesn't penetrate through the interstellar nebulae that lie in the way. Looking into the nucleus is like trying to see a ship on a foggy day. However, just as sailors use radar to penetrate the fog, astronomers study the center of the galaxy by analyzing radio, infrared, and X-ray emissions that travel through the clouds. These studies provide a picture of the nucleus of the Milky Way.

A cloud of dust and gas orbits the galactic nucleus. Infrared measurements show that this cloud is so hot that it must be heated by an energy source with 10 to 80 million times the output of our Sun. The cloud is ring-shaped with a hole in its center, like a doughnut (Fig. 24–27). Relatively recently, 10,000 to 100,000 years ago, a giant explosion blew out the center of a larger cloud and created the hole. X-ray and gamma-ray emissions tell us that matter is now accelerating inward, back into the center at a rate of 1 solar mass every 1000 years (Fig. 24–28).

Astronomers hypothesize that the center of the galaxy is a massive black hole or a group of smaller black holes. According to this hypothesis, early in the history of the galaxy, one huge star or many large ones formed in the galactic center. These stars were large enough to pass through the main sequence quickly and then collapse to become one or many black holes that now form the center of our galaxy.

Galactic Motion

Recall from Chapter 22 that astronomers measure the relative velocities of distant objects by measuring the frequency of light they emit. In 1929, five years after he had described galaxies, Hubble noted that frequency of emitted light from almost every galaxy is shifted toward the red end of the spectrum. Hubble interpreted this red shift to mean that all the galaxies

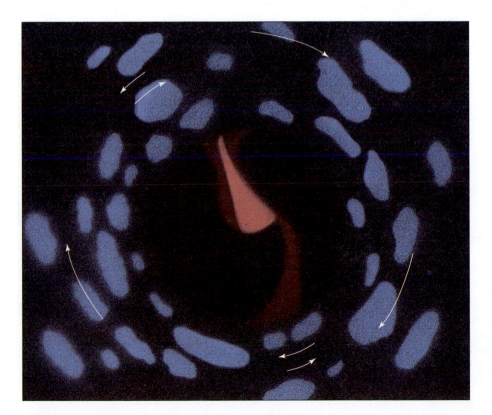

Figure 24–27 A schematic drawing of the center of the Milky Way galaxy. The orange represents streams of ionized gas falling into the center. Surrounding this region is a ring of dust and clouds of gas shown in blue. Arrows indicate direction of movement.

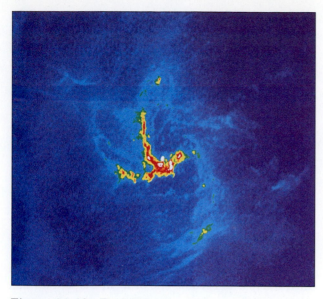

Figure 24–28 The bright region in the center of this image is a small but powerful radio emission from the center of the galaxy. The red, orange, and blue streamers are produced by gases falling toward the center. *(K.Y. Lo, University of Illinois)*

Figure 24–29 Hubble's Law states that the velocity of a galaxy is directly proportional to its distance from Earth.

are flying away from us and from each other; the Universe is expanding. Moreover, he observed that the most distant galaxies are moving outward at the greatest speeds, whereas the closer ones are receding more slowly. This relationship is known as **Hubble's Law** (Fig. 24–29). Using Hubble's Law, the distance from Earth to a galaxy can be calculated by measuring the galaxy's red shift.

24.8 Quasars

In the 1960s, astronomers studied many objects that look like stars but emit extremely large amounts of energy as radio waves. These objects were perplexing because normal stars, such as our Sun, emit mostly visible and ultraviolet light. During the following decade, astronomers discovered that the spectra of many of these objects coincide with that of hydrogen with a very large red shift. Such objects are now called **quasars** (Fig. 24–30).

Recall that a large red shift indicates that an object is moving away from Earth at a high speed. If quasars obey Hubble's Law, then they are very far away. In order to be visible when they are so far away, quasars must emit tremendous amounts of energy, 10 to 100 times more than an entire galaxy!

Quasars emit erratic bursts of energy. Recall from our discussion of pulsars that astronomers can estimate the diameter of a pulsar by the sharpness of a short burst of energy emitted from it. Similar techniques indicate that some quasars are several light-months in

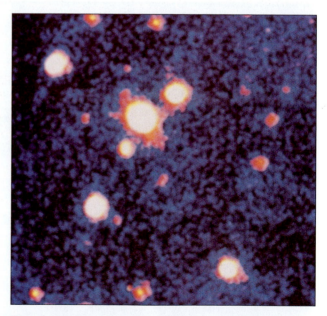

Figure 24–30 Quasar 3C 275.1 is the brightest object near the center of this photograph. The nucleus is surrounded by an elliptical gas cloud. This quasar is about two billion light-years from Earth. The smaller objects are normal galaxies. *(National Optical Astronomy Observatory)*

diameter. By comparison the distance from Earth to the nearest star is 4 light-years.

Thus, a quasar is much smaller than a galaxy but emits much more energy. Furthermore, it emits energy over a wide range of wavelengths. Most quasars are very distant from Earth. One common hypothesis for their origin is that a massive black hole, perhaps of a billion solar masses, lies at the center of a quasar, and the quasar emits energy as gas and dust accelerate into the hole.

Looking Backward into Time

Many quasars are 8 to 12 billion light-years away. If an object is 10 billion light-years away, the light we see today started its journey 10 billion years ago. Therefore, we see what was happening in the past, but not what is happening today. If the object blew up and disappeared 9 billion years ago, we will not know about it for another billion years! Thus, when we look at close objects we see what happened recently, but when we look at distant objects we see what happened in the distant past. One goal of building more powerful telescopes is to study more distant objects and therefore probe farther back in time.

The oldest quasars must have formed when the Universe was young. Moreover, quasars contain heavy elements, and heavy elements form only in the explosions of dying stars. Therefore, quasars are second-generation structures; they formed after earlier stars were born, evolved, exploded, and died. Hence, stars must have formed shortly after the Universe itself formed.

24.9 The Evolution of Galaxies: A Window into the Early Universe

In the opening essay to this unit, we explained that the Cosmic Background Explorer satellite (COBE) detected tiny temperature differences in the cosmic background radiation. This observation told us that matter and energy concentrated into clumps during the earliest infancy of the Universe. However, the COBE data did not reveal how and when these clumps coalesced into galaxies.

Shortly after the COBE data were released, the best available evidence implied that matter congregated into galaxies in one massive event early in the history of the Universe. This "single-burst model" suggested that galaxy formation and stellar evolution occurred simultaneously throughout the Universe in one relatively short period of time.

Astronomers tested this model in December 1995 by training the Hubble Space Telescope on one point

in the sky for ten days. The accumulated light was collected and combined by computer to produce the most accurate view of the distant Universe ever taken. Recall that light from distant objects is old, so a look into deep space is a view back into time. Astronomers analyzed the colors, and therefore the temperatures, of very distant, old galaxies. After comparing these data with the temperatures of young galaxies close to us, cosmologists concluded that the early Universe contained few elliptical galaxies. However, as mentioned in section 24.7, about one third of modern galaxies in our region of space are elliptical. This discrepancy implies that galaxies have evolved over time.

To explain these data, astronomers developed a new hypothesis stating that small protogalaxies evolved early in the history of the Universe. Gradually, many of these coalesced. During this process they formed new shapes, creating the large complex galaxies that currently populate our region in space. To support this hypothesis, recent Hubble photographs show two galaxies colliding and coalescing (Fig. 24–31).

As scientists continue to accumulate data, new models may emerge or old ones will be strengthened. Step by step, we increase our understanding of the evolution and current structure of the Universe.

Figure 24–31 This image from the Hubble Space Telescope shows two colliding spiral galaxies. The blue light comes from young, hot stars forming in the collision zone. *(Brad Whitmore, STScI, and NASA)*

Our study of Earth and space science started with a look at minerals, which we commonly observe as pinhead sized crystals imbedded in rock. Gradually, we expanded our view to look at the entire planet and its interacting systems—the geosphere, atmosphere, hydrosphere, and biosphere. In the final unit we probed into space to view other planetary systems and distant galaxies. Scientists strive to uncover the mysteries of the natural world. Over the centuries, scientific development has improved the lives of most people on Earth. As we enter a new century, we all hope that science will not only uncover new secrets, but also help us live in harmony with ourselves and our planet.

SUMMARY

Distances beyond the Solar System are measured in **light-years** and **parsecs.**

Stars form from condensing **nebulae,** clouds of gas and dust in interstellar space. Some are hot enough to emit light; others are cooler and absorb light. Two opposing forces occur in a star. Gravity pulls particles inward, but at the same time fusion energy generates gas pressure and drives them outward. The balance between these two forces determines the diameter and density of a star.

The central **core** of the Sun has a temperature of about 15,000,000°C and a density 150 times that of water. Hydrogen fusion occurs in this region. The visible surface of the Sun, called the **photosphere,** is only 6000°C and has a pressure of 1/10 of the Earth's atmosphere at sea level. **Sunspots** are magnetic storms on the surface of the Sun. The outer layers of the Sun's atmosphere, called the **chromosphere** and the **corona,** are turbulent and diffuse.

A **Hertzsprung Russell,** or **H-R, diagram** is a plot of luminosity versus stellar temperature. Most stars fall within the **main sequence.** All main-sequence stars are composed primarily of hydrogen and generate energy by hydrogen fusion.

Hydrogen nuclei in the core of a mature star fuse to form helium, with the release of large amounts of energy. When the hydrogen in the core is exhausted and hydrogen fusion ends in the core, gravity compresses the core and the temperature rises. Fusion then begins in the outer shell, and the star expands to become a **red giant.** In the following stage, helium nuclei fuse to form carbon. After helium fusion in an average-mass star ends, the star releases a **planetary nebula** and then shrinks to become a **white dwarf.** In a more massive star, enough heat is produced to fuse heavier elements; the iron-rich core then explodes to become a **supernova.** The remnant can contract to become a **neutron star** or, if the mass is great enough, a **black hole.**

Galaxies and clusters of galaxies form the basic structure of the Universe. Commonly, galaxies are **elliptical, spiral,** or **barred spiral,** although other shapes also exist. Our Sun lies in the **Milky Way,** a typical spiral galaxy. In addition to stars in the main disk, a galaxy contains diffuse clouds of gas and dust between the stars, and a **galactic halo** and **globular clusters** of stars surrounding the disk. The nucleus of the galaxy is dense and probably contains a massive black hole.

Hubble's Law states that the most distant galaxies are moving away from Earth at the greatest speeds, while closer ones are receding more slowly.

Quasars are very far away, emit as much as 100 times the energy of an entire galaxy, and are small compared with a galaxy. A black hole may lie in the center of a quasar. Modern models indicate that today's galaxies evolved through collisions of smaller protogalaxies.

KEY TERMS

light-year 567
parsec 567
nebulae 567
core 572
radiative zone 572
convective zone 572
photosphere 572
sunspot 572
chromosphere 573

spicules 573
corona 573
prominences 573
apparent magnitude 574
absolute magnitude 574
Hertzsprung-Russell, H-R
  diagram 574
main sequence 574
red giant 576

solar mass 576
planetary nebula 576
white dwarf 577
supernova 577
population II stars 578
population I stars 578
neutron star 579
pulsar 580
black hole 581

galaxy 582
elliptical galaxy 582
spiral galaxy 582
barred spirals 582
Milky Way 583
galactic halo 583
globular cluster 585
Hubble's Law 586
quasar 586

1. What is a light-year? How far is it in kilometers?

2. Briefly explain how a star is born.

3. Draw a cutaway diagram of the Sun, labeling the core, the photosphere, sunspots, granules, the chromosphere, and the corona. Label the temperature and the relative density of each region.

4. What is the fundamental source of energy within the Sun? Is the Sun's chemical composition constant, or is it continuously changing?

5. How does energy travel from the core of the Sun to the surface? How does it travel from the surface of the Sun to the Earth?

6. Compare and contrast the core of the Sun with its surface.

7. Compare and contrast the life cycles of stars the mass of our Sun, four times as massive as our Sun, and 20 times as massive as our Sun. At what points do the life cycles differ?

8. What is a supernova? Do all stars eventually explode as supernovas? Explain. What is a neutron star? Do all supernovas lead to the formation of neutron stars?

9. What is a black hole? How does it form, and how can we detect it?

10. Draw a picture of the Milky Way galaxy. Label the disk, the core, the halo, and globular clusters. Draw the approximate position of the Earth.

11. What evidence indicates that the center of our galaxy was once the scene of a violent explosion?

12. List four characteristics of quasars that distinguish them from ordinary stars.

FOR DISCUSSION

1. Could information be gained by building large microphones to detect the sounds of giant explosions in space?

2. What information (if any) could be gained by studying the absorption and emission spectra of the Moon and the planets?

3. Could a telescope be built to study a quasar as it exists today?

4. Hydrogen burns rapidly and explosively in air. Is this chemical combustion of hydrogen an important process within a star? Why or why not?

5. Explain why the density of the gases near the surface of the Sun is less than the density of the gases near the surface of the Earth, even though the gravitational force of the Sun is much greater.

6. Using Figure 24–10, what would be the color of a star whose surface temperature is 10,000°K? 3500°K?

7. What could you tell about the past history of a star if you knew that its core was composed primarily of carbon? Explain.

8. Compare and contrast a white dwarf with a red giant. Can a single star ever be both a red giant and a white dwarf during its lifetime?

9. Would you be likely to find life on planets that are orbiting around a star composed solely of hydrogen and helium? Explain.

10. Certain stars that lie above and below the plane of the Milky Way contain fewer heavy elements than does our own Sun. From this information alone, what can you tell about the history of these stars?

11. Explain why the pulsar signals detected by Jocelyn Bell Burnell could not have originated from (a) an ordinary star, (b) an unknown planet in our Solar System, (c) a distant galaxy, or (d) a large magnetic storm on a nearby star.

12. Could a black hole be hidden in our Solar System? Explain.

13. Would black holes represent a hazard to a rocket ship traveling to distant stars? Could the crew of such a rocket detect a black hole well in advance and avoid an encounter?

14. Is a quasar similar to a star, a galaxy, or neither? Discuss.

15. Arrange the following different environments in the order of increasing densities: (a) intergalactic space, (b) the core of the Sun, (c) the corona of the Sun, (d) the region of space between planets in our Solar System, (e) galactic space, (f) the Earth's atmosphere at sea level.

Identifying Common Minerals

About 3500 minerals exist in the Earth's crust. However, of this great number, only thirty or so are common. Therefore, when you pick up rocks and want to identify the minerals, you are most likely to be looking at the same small group of minerals over and over again. The following list includes the most common and abundant minerals in the Earth's crust. A few important ore minerals and other minerals of economic value, and a few popular precious and semi-precious gems, are included because they are of special interest.

The minerals in this table fall into four categories.

1. Rock-forming minerals are shown in *red*. They are the most abundant minerals in the crust, and make up the largest portions of all common rocks. The rock-forming minerals and mineral groups are feldspar, pyroxene, amphibole, mica, clay, olivine, quartz, calcite, and dolomite. If more than one mineral of a group is common, each mineral is listed under the group name. For example, three kinds of pyroxene are abundant: augite, diopside, and orthopyroxene. All three are described under pyroxene.

2. Accessory minerals are shown in *yellow*. They are minerals that are common, but that usually occur only in small amounts.

3. Ore minerals and other minerals of economic importance are shown in *green*. They are minerals from which metals or other elements can be profitably recovered.

4. Gems are shown in *blue*. If the common gem name(s) is different from the mineral name, the gem name is given in parentheses following the mineral name. For example, emerald is the gem variety of the mineral beryl, and is listed as Beryl (emerald).

Minerals are listed alphabetically within each of the four categories for quick reference. The physical properties most commonly used for identification of each mineral, and the kind(s) of rock in which each mineral is most often found, are listed to facilitate identification of these common minerals.

Rock-Forming Minerals

Mineral Group or Mineral		Chemical Composition	Habit, Cleavage, Fracture	Usual Color
Amphibole	Actinolite	$Ca_2(MgFe)_5Si_8O_{22}(OH)_2$	Slender crystals, radiating, fibrous	Blackish-green to black, dark green
	Hornblende	$(Ca,Na)_{2-3}(Mg,Fe,Al)_5Si_6(Si,Al)_2O_{22}(OH)_2$	Elongate crystals	Blackish-green to black, dark green
Calcite		$CaCO_3$	Perfect cleavage into rhombs	Usually white, but may be variously tinted
Clay Minerals	Illite	$K_{0.8}Al_2(Si_{3.2}Al_{0.8})O_{10}(OH)_2$		White
	Kaolinite	$Al_2Si_2O_5(OH)_4$		White
	Smectite	$Na_{0.3}Al_2(Si_{3.7}Al_{0.3})O_{10}(OH)_2$		White, buff
Dolomite		$CaMg(CO_3)_2$	Cleaves into rhombs; granular masses	White, pink, gray, brown
Feldspar	Albite (sodium feldspar)	$NaAlSi_3O_8$ (sodic plagioclase)	Good cleavage in two directions, nearly 90°	White, gray
	Orthoclase (potassium feldspar)	$KAlSi_3O_8$	Good cleavage in two directions at 90°	White, pink, red, yellow-green, gray
	Plagioclase (feldspar containing both sodium and calcium)	$(Na,Ca)(Al,Si)_4O_8$	Good cleavage in two directions at 90°	White, gray
Mica	Biotite	$K(Mg,Fe)_3AlSi_3O_{10}(OH)_2$	Perfect cleavage into thin sheets	Black, brown, green
	Muscovite	$KAl_2Si_3O_{10}(OH)_2$	Perfect cleavage into thin sheets	Colorless if thin
Olivine		$(MgFe)_2SiO_4$	Uneven fracture, often in granular masses	Various shades of green
Pyroxene	Augite	$Ca(Mg,Fe,Al)(Al,Si_2O_6)$	Short stubby crystals have 4 or 8 sides in cross-section	Blackish-green to light green
	Diopside	$CaMg(Si_2O_6)$	Usually short thick prisms; may be granular	White to light green
	Orthopyroxene	$MgSiO_3$	Cleavage good at 87° and 93°; usually massive	Pale green, brown, gray, or yellowish
Quartz		SiO_2	No cleavage, massive and as six-sided crystals	Colorless, white, or tinted any color by impurities

Hard-ness	Streak	Specific Gravity	Other Properties	Type(s) of Rock in Which the Mineral Is Most Commonly Found
5–6	Pale green	3.2–3.6	Vitreous luster.	Low- to medium-grade metamorphic rocks
5–6	Pale Green	3.2	Crystals six-sided with 124° between cleavage faces.	Common in many granitic to basaltic igneous rocks, and many metamorphic rocks
3	White	2.7	Transparent to opaque. Rapid effervescence with HCl.	Limestone, marble, cave deposits
}			The clay minerals are so fine-grained that most physical properties cannot be identified.	Shale Shale, weathered bedrock, and soil Shale, weathered bedrock, and soil
3.5–4	White to pale gray	3.9–4.2	Effervesces slightly in cold dilute HCl.	Dolomite
6–6.5	White	2.6	Many show fine striations (twinning lines) on cleavage faces.	Granite, rhyolite, low-grade metamorphic rocks
6	White	2.6	Vitreous to pearly luster.	Granite, rhyolite, metamorphic rocks
6	White	2.6–2.7	May show striations as in albite.	Basalt, andesite, medium- to high-grade metamorphic rocks
2.2–2.5	White, gray	2.7–3.1	Vitreous luster, divides readily into thin flexible sheets.	Granitic to intermediate igneous rocks, many metamorphic rocks
2–2.5	White	2.7–3	Vitreous or pearly; flexible and elastic; splits easily.	Many metamorphic rocks, granite
6.5–7	White	3.2–3.3	Vitreous, glassy luster.	Basalt, peridotite
5.6	Pale green	5–6	Vitreous, distinguished from hornblende by the 87° angle between cleavage faces.	Basalt, peridotite, andesite, high-grade metamorphic rocks
5–6	White to greenish	3.2–3.6	Vitreous luster.	Medium-grade metamorphic rocks
5.5	White	3.2–3.5	Vitreous luster.	Periodotite, basalt, high-grade metamorphic rocks
7	White	2.6	Includes rock crystal, rose and milky quartz, amethyst, smoky quartz, etc.	Granite rhyolite, metamorphic rocks of all grades, sandstone, siltstone

Mineral Group or Mineral		Chemical Composition	Habit, Cleavage, Fracture	Usual Color
Apatite		$Ca_5(OH,F,Cl)(PO_4)_3$	Massive, granular	Green, brown, red
Chlorite		$(Mg,Fe)_6(Si,Al)_4O_{10}(OH)_8$	Perfect cleavage as fine scales	Green
Corundum		Al_2O_3	Short, six-sided barrel-shaped crystals	Gray, light blue, and other colors
Epidote		$Ca_2(Al,Fe)Al_2O(SiO_4)(Si_2O_7)(OH)$	Usually granular masses; also as slender prisms	Yellow-green, olive-green, to nearly black
Fluorite		CaF_2	Octahedral and also cubic crystals	White, yellow, green, purple
Garnet	Almandine	$Fe_3Al_2(SiO_4)_3$	No cleavage, crystals 12- or 24-sided	Deep red
	Grossular	$Ca_3Al_2(SiO_4)_3$	No cleavage, crystals 12- or 24-sided	White, green, yellow, brown
Graphite		C	Foliated, scaly, or earthly masses	Steel gray to black
Hematite		Fe_2O_3	Granular, massive, or earthy	Brownish-red
Limonite		$2Fe_2O_3 \cdot 3H_2O$	Earthy fracture	Brown or yellow
Magnetite		Fe_3O_4	Uneven fracture, granular masses	Iron black
Pyrite		FeS_2	Uneven fracture cubes with striated faces, octahedrons	Pale brass yellow (lighter than chalcopyrite)
Serpentine		$Mg_3Si_2O_5(OH)_4$	Uneven, often splintery fracture	Light and dark green, yellow

Accessory Minerals

Hardness	Streak	Specific Gravity	Other Properties	Type(s) of Rock in Which the Mineral Is Most Commonly Found
4.5–5	Pale red-brown	3.1	Crystals may have a partly melted appearance, glassy.	Common in small amounts in many igneous, metamorphic, and sedimentary rocks
2.0–2.5	Gray, white, pale green	2.8	Pearly to vitreous luster.	Common in low-grade metamorphic rocks
9	None	3.9–4.1	Hardness is distinctive.	Metamorphic rocks, some igneous rocks
6.7	Pale yellow to white	3.3	Vitreous luster.	Low- to medium-grade metamorphic rocks
4	White	3.2	Cleaves easily, vitreous, transparent to translucent.	Hydrothermal veins
6.5–7.5	White	4.2	Vitreous to resinous luster.	The most common garnet in metamorphic rocks
6.5–7.5	White	3.6	Vitreous to resinous luster.	Metamorphosed sandy limestones
1–2	Gray or black	2.2	Feels greasy; marks paper.	Metamorphic rocks
2.5	Dark red	2.5–5	Often earthy, dull appearance.	Common in all types of rocks. It can form by weathering of iron minerals, and is the source of color in nearly all red rocks.
1.5–4	Brownish-yellow	3.6	Earthy masses that resemble clay.	Common in all types of rocks. It can form by weathering of iron minerals, and is the source of color in most yellow-brown rocks.
5.5	Iron black	5.2	Metallic luster. Strongly magnetic.	Common in small amounts in most igneous rocks
6–6.5	Greenish-black	5	Metallic luster, brittle, very common.	The most common sulfide mineral. Igneous, metamorphic, and sedimentary rocks; hydrothermal veins
2.5	White	2.5	Waxy luster, smooth feel, brittle.	Alteration or metamorphism of basalt, peridotite, and other magnesium-rich rocks

COMMON MINERALS AND THEIR PROPERTIES

Ore and Other Minerals of Economic Importance

Mineral Group or Mineral	Chemical Composition	Habit, Cleavage, Fracture	Usual Color
Anhydrite	$CaSO_4$	Granular masses, crystals with 2 good cleavage directions	White, gray, blue-gray
Asbestos	$Mg_3Si_2O_5(OH)_4$	Fibrous	White to pale olive-green
Azurite	$Cu_3(CO_3)_2(OH)_2$	Varied, may have fibrous crystals	Azure blue
Bauxite	$Al(OH)_3$	Earthy masses	Reddish to brown
Chalcopyrite	$CuFeS_2$	Uneven fracture	Brass yellow
Chromite	$FeCr_2O_4$	Massive, granular, compact	Black
Cinnabar	HgS	Compact, granular masses	Scarlet red to red-brown
Galena	PbS	Perfect cubic cleavage	Lead or silver gray
Gypsum	$CaSO_4 \cdot 2H_2O$	Tabular crystals, fibrous, or granular	White, pearly
Halite	$NaCl$	Granular masses, perfect cubic crystals	White, also pale colors and gray
Hematite	Fe_2O_3	Granular, massive, or earthy	Brownish-red, black
Malachite	$CuCO_3 \cdot Cu(OH)_2$	Uneven splintery fracture	Bright green, dark green
Native copper	Cu	Malleable and ductile	Copper red
Native gold	Au	Malleable and ductile	Yellow
Native silver	Ag	Malleable and ductile	Silver-white
Pyrolusite	MnO_2	Radiating or dendritic coatings on rocks	Black
Sphalerite	ZnS	Perfect cleavage in 6 directions at 120°	Shades of brown and red
Talc	$Mg_3Si_4O_{10}(OH)_2$	Perfect in one direction	Green, white, gray

Ore and Other Minerals of Economic Importance

Hard-ness	Streak	Specific Gravity	Other Properties	Type(s) of Rock in Which the Mineral Is Most Commonly Found
3–3.5	White	2.9–3	Brittle; resembles marble but acid has no effect.	Sedimentary evaporite deposits
1–2.5	White	2.6–2.8	Pearly to greasy luster. Flexible, easily separated fibers.	A variety of serpentine, found in the same rock types
4	Pale blue	3.8	Vitreous to earthy, effervesces with HCl.	Weathered copper deposits
1.5–3.5	Pale reddish-brown	2.5	Dull luster, claylike masses with small round concretions.	Weathering of many rock types
3.5–4.5	Greenish-black	4.2	Metallic luster, softer than pyrite.	The most common copper ore mineral; hydrothermal veins, porphyry copper deposits
5.5	Dark brown	4.4	Metallic to submetallic luster.	Peridotites and other ultramafic igneous rocks
2.5	Scarlet red	8	Color and streak distinctive.	The most important mercury ore mineral; hydrothermal veins in young volcanic rocks
2.5	Gray	7.6	Metallic luster.	The most important lead ore; commonly also contains silver; hydrothermal veins
1–2.5	White	2.2–2.4	Thin sheets (selenite), fibrous (satinspar), massive (alabaster).	Sedimentary evaporite deposits
2.5–3	White	2.2	Pearly luster, salty taste, soluble in water.	Sedimentary evaporite deposits
2.5	Dark red	2.5–5	Often earthy, dull appearance, sometimes metallic luster.	Huge concentrations occur as sedimentary iron ore; the most important source of iron
3.5–4	Emerald green	4	Effervesces with HCl. Associated with azurite.	Weathered copper deposits
2.5–3	Copper red	8.9	Metallic luster.	Basaltic lavas
2.5–3	Yellow	19.3	Metallic luster.	Hydrothermal quartz-gold veins, sedimentary placer deposits
2.5–3	Silver-white	10.5	Metallic luster.	Hydrothermal veins, weathered silver deposits
1–2	Black	4.7	Sooty appearance.	Black stains on weathered surfaces of many rocks, manganese nodules on the sea floor.
3.5	Reddish-brown	4	Resinous luster, may occur with galena, pyrite.	The most important ore mineral of zinc; hydrothermal veins
1–1.5	White	1–2.5	Greasy feel, occurs in foliated masses.	Low-grade metamorphic rocks

Mineral Group or Mineral	Chemical Composition	Habit, Cleavage, Fracture	Usual Color
Beryl (aquamarine, emerald)	$Be_3Al_2(SiO_3)_6$	Uneven fracture, hexagonal crystals	Green, yellow, blue, pink
Chrysoberyl (cat's eye, alexandrite)	$BeAl_2O_4$	Tabular crystals	Green, brown, yellow
Corundum (ruby, sapphire)	Al_2O_3	Short, six-sided barrel-shaped crystals	Gray, red (ruby), blue (sapphire)
Diamond	C	Octahedral crystals	Colorless or with pale tints
Garnet	$Fe_3Al_2(SiO_4)_3$	No cleavage, crystals 12- or 24-sided	Deep red
Jadite (jade) (a pyroxene)	$NaAl(Si_2O_6)$	Compact fibrous aggregates	Green
Olivine (peridot)	$(MgFe)_2SiO_4$	Uneven fracture, often in granular masses	Various shades of green
Opal	$SiO_2 \cdot NH_2O$	Conchoidal fracture, amorphous, massive	White and various colors
Quartz (rock crystal, amethyst, citrine, tiger eye, adventurine carneline, chrysoprase, agate, onyx, heliotrope, bloodstone, jasper)	SiO_2	No cleavage, massive and as six-sided crystals	Colorless, white, or tinted any color by impurities
Spinel	$MgAl_2O_4$	No cleavage, rare octahedral crystals	Black, dark green, or various colors
Topaz	$Al_2SiO_4(F,OH)_2$	Cleavage good in one direction; conchoidal fracture	Colorless, white, pale tints of blue, pink
Tourmaline	$(Na,Ca)(Li,Mg,Al)(Al,Fe,Mn)_6-(BO_3)_3(Si_6O_{18})(OH)_4$	Poor cleavage, uneven fracture; striated crystals	Black, brown, green, pink
Turquoise	$CuAl_6(PO_4)_4(OH)_8 \cdot 4H_2O$	Massive	Blue-green
Zircon	$Zr(SiO_4)$	Cleavage poor, but often well-formed tetragonal crystals	Colorless, gray, green, pink, bluish

Hard-ness	Streak	Specific Gravity	Other Properties	Type(s) of Rock in Which the Mineral Is Most Commonly Found
7.5–8.0	White	2.6–2.8	Vitreous luster.	Granite, granite pegmatite, mica schist
8.5	White	3.7–3.8	Vitreous luster.	Granite, granite pegmatite, mica schist
9	None	3.9–4.1	Ruby and sapphire are corundum varieties. Hardness is distinctive.	Metamorphic rocks, some igneous rocks
10	None	3.5	Adamantine luster. Hardness is distinctive.	Peridotite, kimberlite, sedimentary placer deposits
6.5–7.5	White	4.2	Vitreous to resinous luster.	Metamorphic rocks, igneous rocks, placer deposits
6.5–7	White, pale green	3.3	Vitreous luster.	High-pressure metamorphic rocks
6.5–7	White	3.2–3.3	Vitreous, glassy luster.	Basalt, peridotite
5.5–6.5	White	2.1	Vitreous, greasy, pearly luster. May show a play of colors.	Low-temperature hot springs and weathered near-surface deposits
7	White	2.6	Colors and other features differ among the varieties.	Quartz is found in nearly all rock types, although each of the gem varieties may form in special environments.
7.5–8.0	White	3.5–4.1	Vitreous luster. Hardness is distinctive.	High-grade metamorphic rocks, dark igneous rocks
8	Colorless	3.5–3.6	Vitreous luster.	Pegmatite, granite, rhyolite
7–7.5	White to gray	4.4–4.8	Vitreous, slightly resinous.	Pegmatite, granite, metamorphic rocks
6	Blue-green, white	2.6–2.8	Waxy luster.	Veins in weathered volcanic rocks in deserts
7.5	White	4.7	Adamantine luster.	Many types of igneous rocks

Systems of Measurement

I The SI System

In the past scientists from different parts of the world have used different systems of measurement. However, global cooperation and communication make it essential to adopt a standard system. The International System of Units (SI) defines various units of measurement as well as prefixes for multiplying or dividing the units by decimal factors. Some primary and derived units important to geologists are listed below.

Time

The SI unit is the **second,** s or sec, which used to be based on the rotation of the Earth but is now related to the vibration of atoms of cesium-133. SI prefixes are used for fractions of a second (such as milliseconds or microseconds), but the common words **minutes, hours,** and **days** are still used to express multiples of seconds.

Length

The SI unit is the **meter,** m, which used to be based on a standard platinum bar but is now defined in terms of wavelengths of light. The closest English equivalent is the **yard** (0.914 m). A **mile** is 1.61 kilometers (km). An inch is exactly 2.54 centimeters (cm).

Area

Area is length squared, as in **square meter, square foot,** and so on. The SI unit of area is the **are,** a, which is 100 sq m. More commonly used is the **hectare,** ha, which is 100 ares, or a square that is 100 m on each side. (The length of a U.S. football field plus one end zone is just about 100 m.) A hectare is 2.47 acres. An **acre** is 43,560 sq ft, which is a plot of 220 ft by 198 ft, for example.

Volume

Volume is length cubed, as in **cubic centimeter,** cm^3, **cubic foot,** ft^3, and so on. The SI unit is the **liter,** L, which is 1000 cm^3. A **quart** is 0.946 L; a U.S. liquid **gallon** (gal) is 3.785 L. A **barrel** of petroleum (U.S.) is 42 gal, or 159 L.

Mass

Mass is the amount of matter is an object. **Weight** is the force of gravity on an object. To illustrate the difference, an astronaut in space has no weight but still has mass. On Earth, the two terms are directly proportional and often used interchangeably. The SI unit of mass is the **kilogram,** kg, which is based on a standard platinum mass. A **pound** (avdp), lb, is a unit of weight. On the surface of the Earth, 1 lb is equal to 0.454 kg. A **metric ton,** also written as **tonne,** is 1000 kg, or about 2205 lb.

Temperature

The Celsius scale is used in most laboratories to measure temperature. On the Celsius scale the freezing point of water is 0°C and the boiling point of water is 100°C.

The SI unit of temperature is the **Kelvin.** The coldest possible temperature, which is −273°C, is zero on

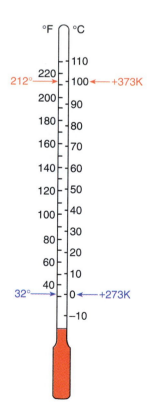

the Kelvin scale. The size of 1 degree Kelvin is equal to 1 degree Celsius.

Celsius temperature (°C) = Kelvin temperature (K) − 273 K

Fahrenheit temperature (°F) is not used in scientific writing, although it is still popular in English-speaking countries. Conversion between Fahrenheit and Celsius is shown below.

Energy

Energy is a measure of work or heat, which were once thought to be different quantities. Hence, two different sets of units were adopted and still persist, although we now know that work and heat are both forms of energy.

The SI unit of energy is the **joule,** J, the work required to exert a force of 1 newton through a distance of 1 m. In turn, a newton is the force that gives a mass of 1 kg an acceleration of 1 m/sec^2. In human terms, a joule is not much—it is about the amount of work required to lift a 100-g weight to a height of 1 m. Therefore, joule units are too small for discussions of machines, power plants, or energy policy. Larger units are

megajoule, MJ = 10^6 J (a day's work by one person)

gigajoule, GJ = 10^9 J (energy in half a tank of gasoline)

The energy unit used for heat is the **calorie,** cal, which is exactly 4.184 J. One calorie is just enough energy to warm 1 g of water 1°C. The more common unit used in measuring food energy is the **kilocalorie,** kcal, which is 1000 cal. When **Calorie** is spelled with a capital C, it means kcal. If a cookbook says that a jelly doughnut has 185 calories, that is an error—it should say 185 Calories (capital C), or 185 kcal. A value of 185 calories (small c) would be the energy in about one quarter of a thin slice of cucumber.

The unit of energy in the British system is the **British thermal unit,** or Btu, which is the energy needed to warm 1 lb of water 1°F.

1 Btu = 1054 J = 1.054 kJ = 252 cal

The unit often referred to in discussions of national energy policies is the **quad,** which is 1 quadrillion Btu, or 10^{15} Btu.

Some approximate energy values are

1 barrel (42 gal) of petroleum = 5900 MJ
1 ton of coal = 29,000 MJ
1 quad = 170 million barrels of oil, or 34 million tons of coal

TABLE II

Prefixes for Use with Basic Units of the Metric System

Prefix	Symbol†	Power		Equivalent
geo*		10^{20}		
tera	T	10^{12}	$= 1,000,000,000,000$	Trillion
giga	G	10^{9}	$=\ \ \ \ 1,000,000,000$	Billion
mega	M	10^{6}	$=\ \ \ \ \ \ \ \ 1,000,000$	Million
kilo	k	10^{3}	$=\ \ \ \ \ \ \ \ \ \ \ \ 1,000$	Thousand
hecto	h	10^{2}	$=\ \ \ \ \ \ \ \ \ \ \ \ \ \ 100$	Hundred
deca	da	10^{1}	$=\ \ \ \ \ \ \ \ \ \ \ \ \ \ \ \ 10$	Ten
—	—	10^{0}	$=\ \ \ \ \ \ \ \ \ \ \ \ \ \ \ \ \ 1$	One
deci	d	10^{-1}	$=\ \ \ \ \ \ \ \ \ \ \ \ \ \ \ .1$	Tenth
centi	c	10^{-2}	$=\ \ \ \ \ \ \ \ \ \ \ \ \ .01$	Hundredth
milli	m	10^{-3}	$=\ \ \ \ \ \ \ \ \ \ \ .001$	Thousandth
micro	μ	10^{-6}	$=\ \ \ \ \ \ \ .000001$	Millionth
nano	n	10^{-9}	$=\ \ \ .000000001$	Billionth
pico	p	10^{-12}	$=\ .000000000001$	Trillionth

*Not an official SI prefix but commonly used to describe very large quantities such as the mass of water in the oceans.

†The SI rules specify that its symbols are not followed by periods, nor are they changed in the plural. Thus, it is correct to write "The tree is 10 m high," not "10 m. high" or "10 ms high."

TABLE III

Handy Conversion Factors

To Convert From	To	Multiply By	
Centimeters	Feet	0.0328 ft/cm	
	Inches	0.394 in/cm	
	Meters	0.01 m/cm	(exactly)
	Micrometers (Microns)	1000 μm/cm	(")
	Miles (statute)	6.214×10^{-6} mi/cm	
	Millimeters	10 mm/cm	(exactly)
Feet	Centimeters	30.48 cm/ft	(exactly)
	Inches	12 in/ft	(")
	Meters	0.3048 m/ft	(")
	Micrometers (Microns)	304800 μm/ft	(")
	Miles (statute)	0.000189 mi/ft	
Grams	Kilograms	0.01 kg/g	(exactly)
	Micrograms	1×10^{6} μg/g	(")
	Ounces (avdp.)	0.03527 oz/g	
	Pounds (avdp.)	0.002205 lb/g	
Hectares	Acres	2.47 acres/ha	
Inches	Centimeters	2.54 cm/in	(exactly)
	Feet	0.0833 ft/in	
	Meters	0.0254 m/in	(exactly)
	Yards	0.0278 yd/in	
Kilograms	Ounces (avdp.)	35.27 oz/kg	
	Pounds (avdp.)	2.205 lb/kg	
Kilometers	Miles	0.6214 mi/km	

TABLE III

Handy Conversion Factors *(Continued)*

To Convert From	To	Multiply by	
Meters	Centimeters	100 cm/m	(exactly)
	Feet	3.2808 ft/m	
	Inches	39.37 in/m	
	Kilometers	0.001 km/m	(exactly)
	Miles (statute)	0.0006214 mi/m	
	Millimeters	1000 mm/m	(exactly)
	Yards	1.0936 yd/m	
Miles (statute)	Centimeters	160934 cm/mi	
	Feet	5280 ft/mi	(exactly)
	Inches	63360 in/mi	(exactly)
	Kilometers	1.609 km/mi	
	Meters	1609 m/mi	
	Yards	1760 yd/mi	(exactly)
Ounces (avdp.)	Grams	28.35 g/oz	
	Pounds (avdp.)	0.0625 lb/oz	(exactly)
Pounds (avdp.)	Grams	453.6 g/lb	
	Kilograms	0.454 kg/lb	
	Ounces (avdp.)	16 oz/lb	(exactly)

IV Exponential or Scientific Notation

Exponential or scientific notation is used by scientists all over the world. This system is based on exponents of 10, which are shorthand notations for repeated multiplications or divisions.

A positive exponent is a symbol for a number that is to be multiplied by itself a given number of times. Thus, the number 10^2 (read "ten squared" or "ten to the second power") is exponential notation for $10 \cdot 10 = 100$. Similarly, $3^4 = 3 \cdot 3 \cdot 3 \cdot 3 = 81$. The reciprocals of these numbers are expressed by negative exponents. Thus $10^{-2} = 1/10^2 = 1/(10 \cdot 10) = 1/100 = 0.01$.

To write 10^4 in longhand form you simply start with the number 1 and move the decimal four places to the right: 10000. Similarly, to write 10^{-4} you start with the number 1 and move the decimal four places to the left: 0.0001.

It is just as easy to go the other way—that is, to convert a number written in longhand form to an exponential expression. Thus, the decimal place of the number 1,000,000 is six places to the right of 1:

$$1\,000\,000 = 10^6$$
$$\text{6 places}$$

Similarly, the decimal place of the number 0.000001 is six places to the left of 1 and

$$0.000001 = 10^{-6}$$
$$\text{6 places}$$

What about a number like 3,000,000? If you write it $3 \cdot 1,000,000$, the exponential expression is simply $3 \cdot 10^6$. Thus, the mass of the Earth, which, expressed in long numerical form is 3,120,000,000,000,000,000,000,000 kg, can be written more conveniently as 3.12×10^{24} kg.

The Elements

State: S Solid · L Liquid · G Gas · X Not found in nature

Legend: Metals · Transition Metals · Nonmetals · Noble gases · Lanthanide series · Actinide series

Key example:
Atomic number 92 S · Symbol U · Uranium · Mass number 238.03

Group																	

1 G **H** Hydrogen 1.01

2 G **He** Helium 4.00

Group 1:
3 S **Li** Lithium 6.94 · 11 S **Na** Sodium 22.99 · 19 S **K** Potassium 39.10 · 37 S **Rb** Rubidium 85.47 · 55 S **Cs** Cesium 132.91 · 87 S **Fr** Francium 223

Group 2:
4 S **Be** Beryllium 9.01 · 12 S **Mg** Magnesium 24.31 · 20 S **Ca** Calcium 40.08 · 38 S **Sr** Strontium 87.62 · 56 S **Ba** Barium 137.34 · 88 S **Ra** Radium 226.03

21 S **Sc** Scandium 44.96 · 22 S **Ti** Titanium 47.90 · 23 S **V** Vanadium 50.94 · 24 S **Cr** Chromium 52.00 · 25 S **Mn** Manganese 54.94 · 26 S **Fe** Iron 55.85 · 27 S **Co** Cobalt 58.93 · 28 S **Ni** Nickel 58.71 · 29 S **Cu** Copper 63.55 · 30 S **Zn** Zinc 65.38

39 S **Y** Yttrium 88.91 · 40 S **Zr** Zirconium 91.22 · 41 S **Nb** Niobium 92.91 · 42 S **Mo** Molybdenum 95.94 · 43 X **Tc** Technetium 97 · 44 S **Ru** Ruthenium 101.07 · 45 S **Rh** Rhodium 102.91 · 46 S **Pd** Palladium 106.4 · 47 S **Ag** Silver 107.87 · 48 S **Cd** Cadmium 112.40

71 S **Lu** Lutinium 174.97 · 72 S **Hf** Hafnium 178.49 · 73 S **Ta** Tantalum 180.95 · 74 S **W** Tungsten 183.85 · 75 S **Re** Rhenium 186.21 · 76 S **Os** Osmium 190.2 · 77 S **Ir** Iridium 192.22 · 78 S **Pt** Platinum 195.09 · 79 S **Au** Gold 196.97 · 80 L **Hg** Mercury 200.59

103 S **Lr** Lawrencium 260 · 104 X **Rf** Rutherfordium 261 · 105 X **Db** Dubnium 262 · 106 X **Sg** Seaborgium 263 · 107 X **Bh** Bohrium 264 · 108 X **Hs** Hassium 265 · 109 X **Mt** Meitnerium 266 · 110 **Uun** 269 · 111 **Uuu** 272 · 112 **Uub** 277

Group 3:
5 S **B** Boron 10.81 · 13 S **Al** Aluminum 26.98 · 31 S **Ga** Gallium 69.72 · 49 S **In** Indium 114.82 · 81 S **Tl** Thallium 204.37

Group 4:
6 S **C** Carbon 12.01 · 14 S **Si** Silicon 28.09 · 32 S **Ge** Germanium 72.59 · 50 S **Sn** Tin 118.69 · 82 S **Pb** Lead 207.2

Group 5:
7 G **N** Nitrogen 14.01 · 15 S **P** Phosphorus 30.97 · 33 S **As** Arsenic 74.92 · 51 S **Sb** Antimony 121.75 · 83 S **Bi** Bismuth 208.98

Group 6:
8 G **O** Oxygen 16.00 · 16 S **S** Sulfur 32.06 · 34 S **Se** Selenium 78.96 · 52 S **Te** Tellurium 127.60 · 84 S **Po** Polonium 209

Group 7:
9 G **F** Fluorine 19.00 · 17 G **Cl** Chlorine 35.45 · 35 L **Br** Bromine 79.90 · 53 S **I** Iodine 126.90 · 85 S **At** Astatine 210

Group 8:
10 G **Ne** Neon 20.18 · 18 G **Ar** Argon 39.95 · 36 G **Kr** Krypton 83.80 · 54 G **Xe** Xenon 131.30 · 86 G **Rn** Radon 222

Lanthanide series:
57 S **La** Lanthanum 138.91 · 58 S **Ce** Cerium 140.12 · 59 S **Pr** Praseodymium 140.91 · 60 S **Nd** Neodymium 144.24 · 61 X **Pm** Promethium 145 · 62 S **Sm** Samarium 150.4 · 63 S **Eu** Europium 151.96 · 64 S **Gd** Gadolinium 157.25 · 65 S **Tb** Terbium 158.93 · 66 S **Dy** Dysprosium 162.50 · 67 S **Ho** Holmium 164.93 · 68 S **Er** Erbium 167.26 · 69 S **Tm** Thulium 168.93 · 70 S **Yb** Ytterbium 173.04

Actinide series:
89 S **Ac** Actinium 227 · 90 S **Th** Thorium 232.04 · 91 S **Pa** Protactinium 231.04 · 92 S **U** Uranium 238.03 · 93 X **Np** Neptunium 237.05 · 94 X **Pu** Plutonium 244 · 95 X **Am** Americium 243 · 96 X **Cm** Curium 247 · 97 X **Bk** Berkelium 247 · 98 X **Cf** Californium 251 · 99 X **Es** Einsteinium 254 · 100 X **Fm** Fermium 257 · 101 X **Md** Mendelevium 258 · 102 X **No** Nobilium 259

INTERNATIONAL TABLE OF ATOMIC WEIGHTS

Atomic Weights ($^{12}C = 12$)*†

Element	Symbol	Atomic Number	Atomic Weight	Element	Symbol	Atomic Number	Atomic Weight
Actinium	Ac	89	(227)	Mercury	Hg	80	200.59
Aluminum	Al	13	26.9815	Molybdenum	Mo	42	95.94
Americium	Am	95	(243)	Neodymium	Nd	60	144.24
Antimony	Sb	51	121.75	Neon	Ne	10	20.1797
Argon	Ar	18	39.948	Neptunium	Np	93	(237)
Arsenic	As	33	74.9216	Nickel	Ni	28	58.69
Astatine	At	85	(210)	Niobium	Nb	41	92.9064
Barium	Ba	56	137.327	Nitrogen	N	7	14.0067
Berkelium	Bk	97	(247)	Nobelium	No	102	(259)
Beryllium	Be	4	9.01218	Osmium	Os	76	190.2
Bohrium	Bh	107	(262)	Oxygen	O	8	15.9994
Bismuth	Bi	83	208.980	Palladium	Pd	46	106.42
Boron	B	5	10.811	Phosphorus	P	15	30.9738
Bromine	Br	35	79.904	Platinum	Pt	78	195.08
Cadmium	Cd	48	112.411	Plutonium	Pu	94	(244)
Calcium	Ca	20	40.078	Polonium	Po	84	(209)
Californium	Cf	98	(251)	Potassium	K	19	39.0983
Carbon	C	6	12.011	Praseodymium	Pr	59	140.908
Cerium	Ce	58	140.115	Promethium	Pm	61	(145)
Cesium	Cs	55	132.905	Protactinium	Pa	91	(231)
Chlorine	Cl	17	35.4527	Radium	Ra	88	(226)
Chromium	Cr	24	51.9961	Radon	Rn	86	(222)
Cobalt	Co	27	58.9332	Rhenium	Re	75	186.207
Copper	Cu	29	63.546	Rhodium	Rh	45	102.906
Curium	Cm	96	(247)	Rubidium	Rb	37	85.4678
Dubnium	Db	105	(262)	Ruthenium	Ru	44	101.07
Dysprosium	Dy	66	162.50	Rutherfordium	Rf	104	(261)
Einsteinium	Es	99	(252)	Samarium	Sm	62	150.36
Erbium	Er	68	167.26	Scandium	Sc	21	44.9559
Europium	Eu	63	151.965	Seaborgium	Sg	106	(263)
Fermium	Fm	100	(257)	Selenium	Se	34	78.96
Fluorine	F	9	18.9984	Silicon	Si	14	28.0855
Francium	Fr	87	(223)	Silver	Ag	47	107.868
Gadolinium	Gd	64	157.25	Sodium	Na	11	22.9898
Gallium	Ga	31	69.723	Strontium	Sr	38	87.62
Germanium	Ge	32	72.61	Sulfur	S	16	32.066
Gold	Au	79	196.967	Tantalum	Ta	73	180.948
Hafnium	Hf	72	178.49	Technetium	Tc	43	(98)
Hassium	Hs	108	(265)	Tellurium	Te	52	127.60
Helium	He	2	4.00260	Terbium	Tb	65	158.925
Holmium	Ho	67	164.930	Thallium	Tl	81	204.383
Hydrogen	H	1	1.00794	Thorium	Th	90	232.038
Indium	In	49	114.82	Thulium	Tm	69	168.934
Iodine	I	53	126.904	Tin	Sn	50	118.710
Iridium	Ir	77	192.22	Titanium	Ti	22	47.88
Iron	Fe	26	55.847	Tungsten	W	74	183.85
Krypton	Kr	36	83.80	Ununbium	Uub	112	(277)
Lanthanum	La	57	139.906	Ununnilium	Uun	110	(269)
Lawrencium	Lr	103	(262)	Unununium	Uuu	111	(272)
Lead	Pb	82	207.2	Uranium	U	92	238.029
Lithium	Li	3	6.941	Vanadium	V	23	50.9415
Lutetium	Lu	71	174.967	Xenon	Xe	54	131.29
Magnesium	Mg	12	24.3050	Ytterbium	Yb	70	173.04
Manganese	Mn	25	54.9381	Yttrium	Y	39	88.9059
Meitnerium	Mt	109	(266)	Zinc	Zn	30	65.39
Mendelevium	Md	101	(258)	Zirconium	Zr	40	91.224

*The atomic weights given are from the 1987 IUPAC Atomic Weight table, to a maximum of six significant figures.

†Value in parentheses is the mass number of the isotope of longest known half-life.

Rock Symbols

The symbols used in this book for types of rocks are shown below:

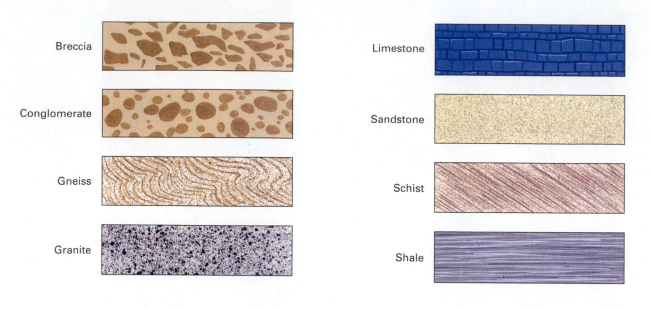

Breccia

Conglomerate

Gneiss

Granite

Limestone

Sandstone

Schist

Shale

In this book we have adopted consistent colors and style for depicting magma and layers in the upper mantle and crust.

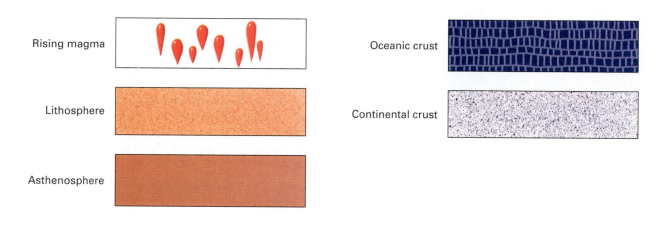

Rising magma

Lithosphere

Asthenosphere

Oceanic crust

Continental crust

SPRING SKY

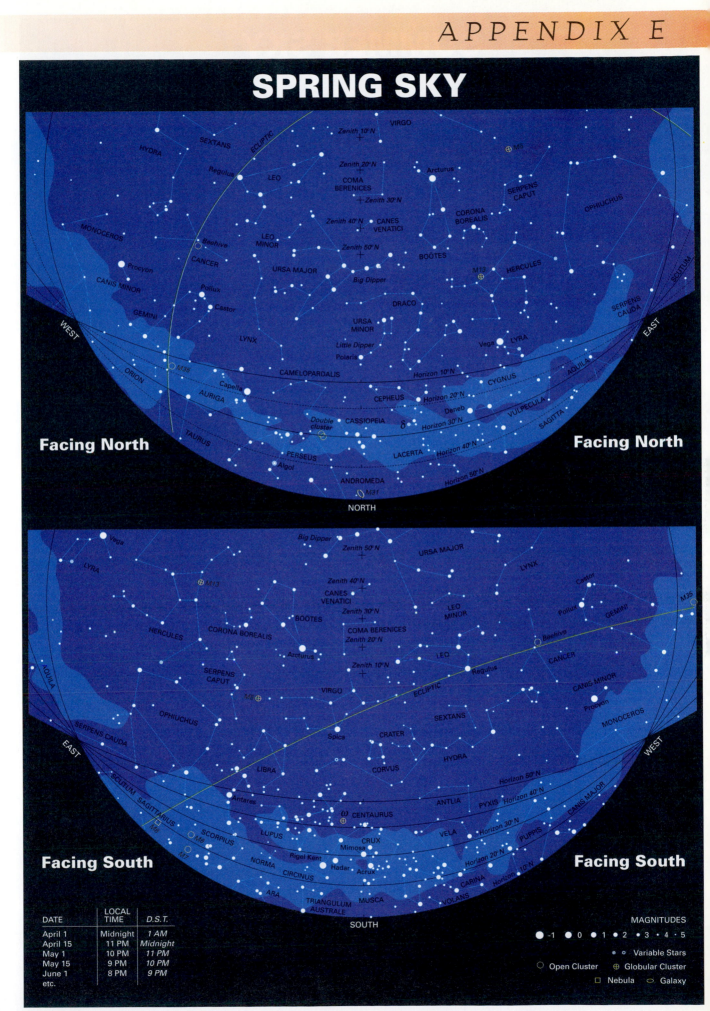

MAP BY WIL TIRION; FOR JAY M. PASACHOFF

SUMMER SKY

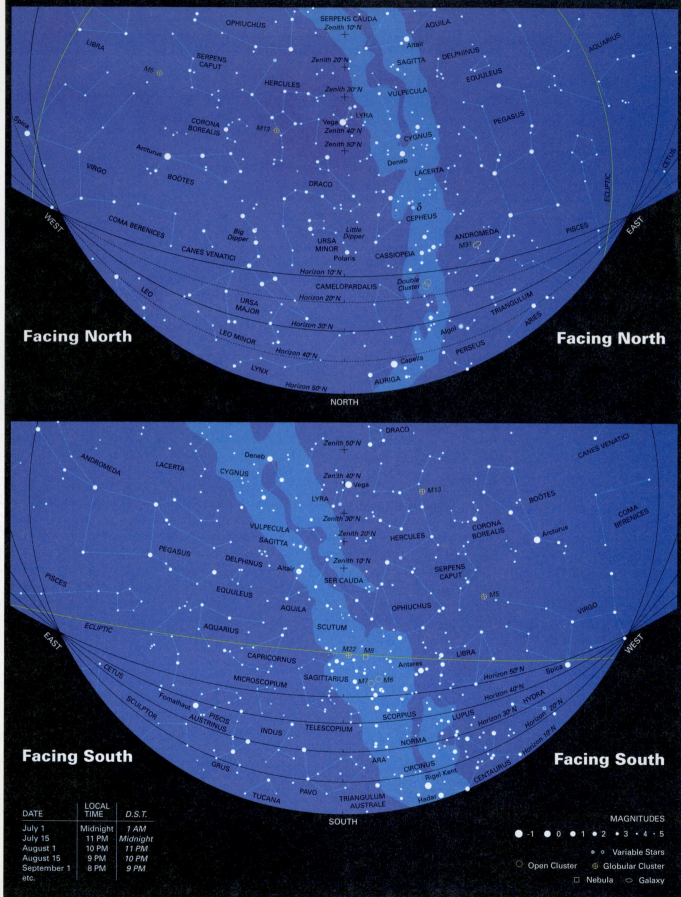

Facing North

Facing North

OPHIUCHUS
SERPENS CAUDA
Zenith 10° N
AQUILA
LIBRA
SERPENS CAPUT
Altair
SAGITTA
DELPHINUS
Zenith 20° N
M5 ⊕
HERCULES
Zenith 30° N
VULPECULA
EQUULEUS
Spica
CORONA BOREALIS
M13 ⊕
Vega
Zenith 40° N
LYRA
PEGASUS
Arcturus
CYGNUS
Zenith 50° N
VIRGO
BOÖTES
Deneb
LACERTA
CETUS
DRACO
ECLIPTIC
CEPHEUS
δ
WEST
ANDROMEDA
EAST
COMA BERENICES
Big Dipper
Little Dipper
M31
PISCES
CANES VENATICI
URSA MINOR
CASSIOPEIA
TRIANGULUM
Polaris
LEO
Horizon 10° N
Double Cluster
ARIES
URSA MAJOR
CAMELOPARDALIS
Horizon 20° N
LEO MINOR
Horizon 30° N
Algol ◉
PERSEUS
Horizon 40° N
LYNX
Capella
Horizon 50° N
AURIGA

NORTH

DRACO
Zenith 50° N
CANES VENATICI
Deneb
ANDROMEDA
LACERTA
CYGNUS
Vega
Zenith 40° N
M13 ⊕
BOÖTES
LYRA
COMA BERENICES
Zenith 30° N
VULPECULA
HERCULES
CORONA BOREALIS
Arcturus
SAGITTA
Zenith 20° N
PEGASUS
DELPHINUS
Altair
Zenith 10° N
SERPENS CAPUT
PISCES
EQUULEUS
SER CAUDA
M5 ⊕
VIRGO
AQUILA
OPHIUCHUS
EAST
ECLIPTIC
AQUARIUS
SCUTUM
WEST
CETUS
CAPRICORNUS
M22 ⊕ M8 □
LIBRA
Horizon 50° N
Spica
Fomalhaut
MICROSCOPIUM
SAGITTARIUS
Antares
Horizon 40° N
HYDRA
SCULPTOR
PISCIS AUSTRINUS
M7 ○ ○ M6
SCORPIUS
Horizon 30° N
Horizon 20° N
INDUS
TELESCOPIUM
NORMA
LUPUS
Horizon 10° N
GRUS
ARA
CIRCINUS
CENTAURUS
Rigel Kent
TUCANA
PAVO
TRIANGULUM AUSTRALE
Hadar

Facing South

Facing South

SOUTH

DATE	LOCAL TIME	D.S.T.
July 1	Midnight	*1 AM*
July 15	11 PM	*Midnight*
August 1	10 PM	*11 PM*
August 15	9 PM	*10 PM*
September 1	8 PM	*9 PM*
etc.		

MAGNITUDES

−1 0 1 2 3 · 4 · 5

Variable Stars

Open Cluster ⊕ Globular Cluster

□ Nebula ◯ Galaxy

AUTUMN SKY

Facing North (top chart)

AQUARIUS · PISCES · CETUS · ERIDANUS · LEPUS · CAPRICORNUS · PEGASUS · EQUULEUS · ARIES · TAURUS · ORION · Rigel · M42 · ANDROMEDA · TRIANGULUM · Pleiades · Hyades · Aldebaran · DELPHINUS · D_Zenith 10° N · Zenith 20° N · Zenith 30° N · Zenith 40° N · Zenith 50° N · M31 · Algol · PERSEUS · Betelgeuse · MONOCEROS · LACERTA · AQUILA · Altair · CYGNUS · Deneb · δ · CASSIOPEIA · Double Cluster · Capella · SAGITTA · VULPECULA · CEPHEUS · CAMELOPARDALIS · AURIGA · GEMINI · SERPENS CAUDA · LYRA · Vega · Little Dipper · Polaris · LYNX · Horizon 10° N · Castor · Horizon 20° N · Pollux · Horizon 30° N · DRACO · URSA MINOR · Horizon 40° N · CANCER · HERCULES · Horizon 50° N · M13 · Big Dipper · URSA MAJOR · BOÖTES · CANES VENATICI · LEO MINOR · ECLIPTIC · WEST · EAST

Facing North (left) · **Facing North** (right)

NORTH

Facing South (bottom chart)

Capella · CASSIOPEIA · Deneb · Vega · AURIGA · PERSEUS · Zenith 50° N · LACERTA · CYGNUS · LYRA · Algol · M31 · ANDROMEDA · Zenith 40° N · M35 · ECLIPTIC · TRIANGULUM · Zenith 30° N · PEGASUS · VULPECULA · GEMINI · Pleiades · ARIES · Zenith 20° N · SAGITTA · Hyades · Zenith 10° N · DELPHINUS · Altair · SERPENS CAUDA · ORION · Aldebaran · PISCES · EQUULEUS · AQUILA · Betelgeuse · TAURUS · Mira · MONOCEROS · CETUS · AQUARIUS · M42 · Rigel · ERIDANUS · CAPRICORNUS · Horizon 50° N · LEPUS · FORNAX · SCULPTOR · PISCIS AUSTRINUS · Fomalhaut · Horizon 40° N · CANIS MAJOR · COLUMBA · CAELUM · HOROLOGIUM · Achernar · PHOENIX · GRUS · MIC · Horizon 30° N · DORADO · PIScis AUSTRINUS · INDUS · Horizon 20° N · SAGITTARIUS · RETICULUM · HYDRUS · TUCANA · PAVO · Horizon 10° N · 47 Tuc · EAST · WEST

Facing South (left) · **Facing South** (right)

SOUTH

MAP BY WIL TIRION; FOR JAY M. PASACHOFF

DATE	LOCAL TIME	D.S.T.
October 1	Midnight	1 AM
October 15	11 PM	Midnight
November 1	10 PM	11 PM
November 15	9 PM	10 PM
December 1	8 PM	9 PM
etc.		

MAGNITUDES
-1 · 0 · 1 · 2 · 3 · 4 · 5

Variable Stars
Open Cluster · Globular Cluster
Nebula · Galaxy

WINTER SKY

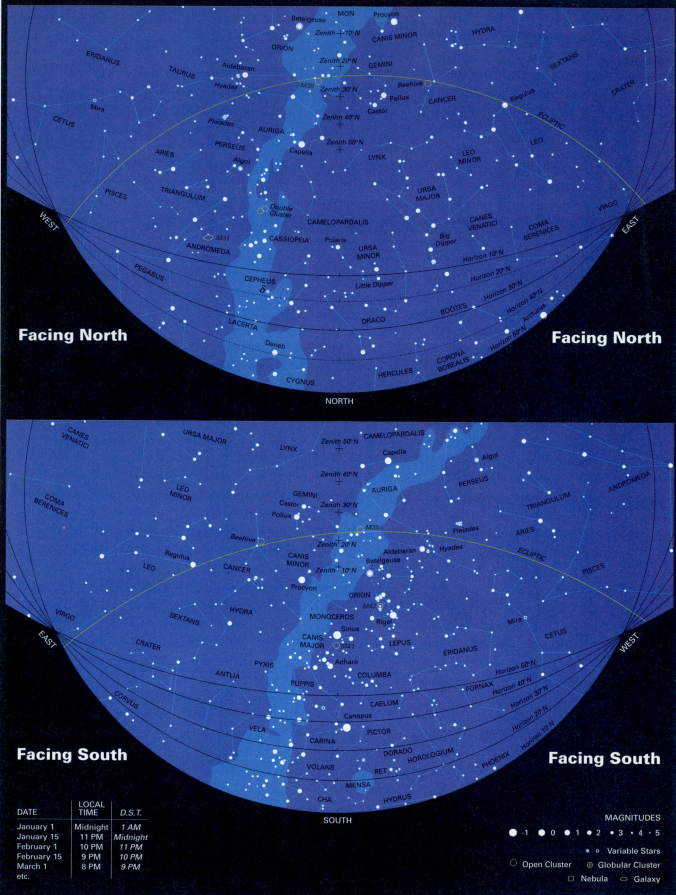

Facing North

Facing North

NORTH

Facing South

Facing South

SOUTH

DATE	LOCAL TIME	D.S.T.
January 1	Midnight	1 AM
January 15	11 PM	Midnight
February 1	10 PM	11 PM
February 15	9 PM	10 PM
March 1	8 PM	9 PM
etc.		

MAGNITUDES

-1 0 1 2 3 4 5

Variable Stars
Open Cluster   Globular Cluster
Nebula   Galaxy

MAP BY WIL TIRION; FOR JAY M. PASACHOFF

A horizon The uppermost layer of soil composed of a mixture of organic matter and leached and weathered minerals. (*syn:* topsoil)

aa A lava flow that has a jagged, rubbly, broken surface.

ablation area The lower portion of a glacier where more snow melts in summer than accumulates in winter so there is a net loss of glacial ice. (*syn:* zone of wastage)

abrasion The mechanical wearing and grinding of rock surfaces by friction and impact.

absolute age Time measured in years.

absolute humidity *See* humidity.

absolute magnitude The brightness that a star would appear to have if it were 32 light-years (10 parsecs) away.

absorption nebula A cold nebula that absorbs light.

absorption spectrum *See* spectrum.

abyssal fan A large, fan-shaped accumulation of sediment deposited at the bases of many submarine canyons adjacent to the deep sea floor. (*syn:* submarine fan)

abyssal plain A flat, level, largely featureless part of the ocean floor between the mid-oceanic ridge and the continental rise.

accreted terrain A land mass that originated as an island arc or a microcontinent that was later added onto a continent.

accumulation area The upper part of a glacier where accumulation of snow during the winter exceeds melting during the summer, causing a net gain of glacial ice.

acid precipitation (often called acid rain) A condition in which natural precipitation becomes acidic after reacting with air pollutants.

active continental margin A continental margin characterized by subduction of an oceanic lithospheric plate beneath a continental plate. (*syn:* Andean margin)

adiabatic temperature changes Temperature changes that occur without gain or loss of heat.

advection In meteorology, the horizontal component of a convection current in air, i.e., the surface movement that is commonly called wind.

advection fog Fog that forms when warm moist air from the sea blows onto cooler land, where the air cools and water vapor condenses at ground level.

aerosol Any small particle that is larger than a molecule and suspended in air.

air mass A large body of air that has approximately the same temperature and humidity throughout.

albedo The reflectivity of a surface. A mirror or bright snowy surface reflects most of the incoming light and has a high albedo, whereas a rough flat road surface has a low albedo.

alluvial fan A fan-like accumulation of sediment created where a steep stream slows down rapidly as it reaches a relatively flat valley floor.

alpine glacier A glacier that forms in mountainous terrain.

altimeter A barometer with a scale calibrated in units of elevation rather than pressure.

alternative energy resources All energy resources other than fossil fuels and nuclear fission; including solar energy; hydroelectric power; geothermal energy; wind energy; biomass energy; tidal, wave, and heat energy from the seas; and nuclear fusion.

altostratus cloud A high-level stratus cloud.

amphibole A group of double-chain silicate minerals. Hornblende is a common amphibole.

anaerobic Without oxygen; anaerobic bacteria are bacteria that live without oxygen.

Andean margin A continental margin characterized by subduction of an oceanic lithospheric plate beneath a continental plate. (*syn:* active continental margin)

andesite A fine-grained gray or green volcanic rock intermediate in composition between basalt and granite, consisting of about equal amounts of plagioclase feldspar and mafic minerals.

angle of repose The maximum slope or angle at which loose material remains stable.

angular unconformity An unconformity in which younger sediment or sedimentary rocks rest on the eroded surface of tilted or folded older rocks.

anion An ion that has a negative charge.

antecedent stream A stream that was established before local uplift started and cut its channel at the same rate the land was rising.

anticline A fold in rock that resembles an arch; the fold is convex upward and the oldest rocks are in the middle.

anticyclone A system of rotating winds that develop where descending air spreads out over the Earth's surface. In the Northern Hemisphere, the Coriolis effect deflects the diverging winds to the right, forming a pinwheel pattern with the air spiraling clockwise. In the Southern Hemisphere, the Coriolis effect deflects winds leftward, and creates a counterclockwise spiral.

apparent magnitude The brightness of a star as seen from Earth.

aquifer A porous and permeable body of rock that can yield economically significant quantities of ground water.

Archean Eon A division of geologic time 3.8 to 2.5 billion years ago. The oldest known rocks formed at the beginning of or just prior to the start of the Archean Eon.

arête A sharp, narrow ridge between adjacent valleys formed by glacial erosion.

artesian aquifer An inclined aquifer that is bounded top and bottom by layers of impermeable rock so the water is under pressure.

artesian well A well drilled into an artesian aquifer in which the water rises without pumping and in some cases spurts to the surface.

artificial channel Any channel dredged to modify the natural channel or to alter the course of a stream.

artificial levee   A wall built along the banks of a stream to prevent rising flood water from spilling out of the stream channel onto the flood plain.

aseismic ridge   A submarine mountain chain with little or no earthquake activity.

ash (volcanic)   Fine pyroclastic material less than 2 millimeters in diameter.

ash flow   A mixture of volcanic ash, larger pyroclastic particles, and gas that flows rapidly along the Earth's surface as a result of an explosive volcanic eruption. (*syn:* nuée ardente)

ash-flow tuff   A pyroclastic rock formed when an ash flow solidifies.

aspect   The orientation of a slope with respect to the Sun; the geographic orientation or exposure of a slope.

asteroid   One of the many small celestial bodies in orbit around the Sun. Most asteroids orbit between Mars and Jupiter.

asthenosphere   The portion of the upper mantle beneath the lithosphere. It consists of weak, plastic rock where magma may form and extends from a depth of about 100 kilometers to about 350 kilometers below the surface of the Earth.

atmosphere   A mixture of gases, mostly nitrogen and oxygen, with smaller amounts of argon, carbon dioxide, and other gases. The atmosphere is held to the Earth by gravity and thins rapidly with altitude.

atmospheric inversion   *See* inversion (atmospheric).

atoll   A circular reef that surrounds a lagoon and is bounded on the outside by deep water of the open sea.

atom   The fundamental unit of elements consisting of a small, dense, positively charged center called a *nucleus* surrounded by a diffuse cloud of negatively charged *electrons*.

B horizon   The soil layer just below the A horizon where ions leached from the A horizon accumulate.

back arc basin   A sedimentary basin on the opposite side of the magmatic arc from the trench, either in an island arc or in an Andean continental margin.

backshore   The upper zone of a beach that is usually dry but is washed by waves during storms.

bajada   A broad depositional surface extending outward from a mountain front formed by the merging of alluvial fans.

banded iron formation   Iron-rich layered sedimentary rocks precipitated from the seas mostly between 2.6 and 1.9 billion years ago, as a result of rising atmospheric oxygen concentrations.

bank   The rising slope bordering the side of a stream channel.

bar   An elongate mound of sediment, usually composed of sand or gravel, in a stream channel or along a coastline.

barchan dune   A crescent-shaped dune, highest in the center, with the tips facing downwind.

barometer   A device used to measure barometric pressure.

barometric pressure   The pressure exerted by the Earth's atmosphere.

barrier island   A long, narrow, low-lying island that extends parallel to the shoreline.

barrier reef   A reef separated from the coast by a deep, wide lagoon.

basal slip   Movement of the entire mass of a glacier along the bedrock.

basalt   A dark-colored, very fine grained, mafic, volcanic rock composed of about half calcium-rich plagioclase feldspar and half pyroxene.

base level   The deepest level to which a stream can erode its bed. The ultimate base level is usually sea level, but this is seldom attained.

basement rock   The older granitic and related metamorphic rocks of the Earth's crust that make up the foundations of continents.

basin   A low area of the Earth's crust of tectonic origin, commonly filled with sediment.

batholith   A large plutonic mass of intrusive rock with more than 100 square kilometers of surface exposed.

bauxite   A gray, yellow, or reddish brown rock composed of a mixture of aluminum oxides and hydroxides. It is the principal ore of aluminum.

baymouth bar   A bar that extends partially or completely across the entrance to a bay.

beach   Any strip of shoreline washed by waves or tides.

beach terrace   A level portion of old beach elevated above the modern beach by uplift of the shoreline or fall of sea level.

bed   The floor of a stream channel. Also the thinnest layer in sedimentary rocks, commonly ranging in thickness from a centimeter to a meter or two.

bed load   That portion of a stream's load that is transported on or immediately above the streambed.

bedding   Layering that develops as sediments are deposited.

bedrock   The solid rock that underlies soil or regolith.

Benioff zone   An inclined zone of earthquake activity that traces the upper portion of a subducting plane in a subduction zone.

big bang   An event thought to mark the beginning of our Universe. The big bang theory postulates that 10 to 20 billion years ago, all matter exploded from an infinitely compressed state.

bioclastic sedimentary rock   Sedimentary rocks such as most limestone, that are composed of broken shell fragments and similar remains of living organisms. The fragments are clastic, but they are of biological origin.

biome   A community of plants living in a large geographic area characterized by a particular climate.

bioremediation   The use of microorganisms to decompose an environmental contaminant.

biosphere   The zone inhabited by life.

biotite   Black, rock-forming mineral of the mica group.

bitumen   A general name for solid and semisolid hydrocarbons that are fusible and soluble in carbon bisulfide. Petroleum, asphalt, natural mineral wax, and asphaltites are all bitumens.

black hole   A small region of space that contains matter packed so densely that light cannot escape from its intense gravitational field.

black smoker   A jet of black water spouting from a fracture in the sea floor, commonly near the mid-oceanic ridge. The black color is caused by precipitation of very fine grained metal sulfide minerals as hydrothermal solutions cool by contact with seawater.

blowout   A small depression created by wind erosion.

body wave   A seismic wave that travels through the interior of the Earth.

braided stream   A stream that divides into a network of branching and reuniting shallow channels separated by mid-channel bars.

branching chain reaction   A nuclear fission reaction in which the initial reaction releases two or three neutrons, each of which triggers the fission of additional nuclei.

breccia   A coarse-grained sedimentary rock composed of angular, broken fragments cemented together.

brittle fracture   Rupture that occurs when a rock breaks sharply.

burial metamorphism   Metamorphism that results from deep burial of rocks in a sedimentary basin. Rocks metamorphosed in this way are usually unfoliated.

butte   A flat-topped mountain with several steep cliff faces. A butte is smaller and more tower-like than a mesa.

C horizon   The lowest soil layer composed of partly weathered bedrock grading downward into unweathered parent rock.

calcite   A common rock-forming mineral, $CaCO_3$.

caldera   A large circular depression caused by an explosive volcanic eruption.

caliche   A hard soil layer formed when calcium carbonate precipitates and cements the soil.

calving   A process in which large chunks of ice break off from tidewater glaciers to form icebergs.

cap rock   An impermeable rock, usually shale, that prevents oil or gas from escaping upward from a reservoir.

capacity   The maximum quantity of sediment that a stream can carry.

capillary action   The action by which water is pulled upward through small pores by electrical attraction to the pore walls.

carbonate platform   An extensive accumulation of limestone such as the Florida Keys and the Bahamas, formed on a continental shelf in warm regions where sediment does not muddy the water and reef-building organisms thrive.

carbonate rocks   Rocks such as limestone and dolomite made up primarily of carbonate minerals.

catastrophism   The model that Earth change occurs through a series of rare but cataclysmic events.

cation   A positively charged ion.

cave   *See* cavern.

cavern   An underground cavity or series of chambers created when ground water dissolves large amounts of rock, usually limestone. (*syn:* cave)

celestial spheres   A hypothetical series of concentric spheres centered at the center of the Earth. Aristotle postulated that the Sun, Moon, planets, and stars are imbedded in the spheres.

cementation   The process by which clastic sediment is lithified by precipitation of a mineral cement among the grains of the sediment.

Cenozoic era   The latest of the four eras into which geologic time is subdivided; 65 million years ago to the present.

chalk   A very fine grained, soft, earthy, white to gray bioclastic limestone made of the shells and skeletons of marine microorganisms.

chemical bond   The linkage between atoms in molecules and between molecules and ions in crystals.

chemical sedimentary rock   Rocks such as rock salt, that form by direct precipitation of minerals from solution.

chemical weathering   The chemical decomposition of rocks and minerals by exposure to air, water, and other chemicals in the environment.

chert   A hard, dense, sedimentary rock composed of microcrystalline quartz. (*syn:* flint)

chromosphere   A turbulent diffuse gaseous layer of the Sun that lies above the photosphere.

cinder cone   A small volcano, as high as 300 meters, made up of loose pyroclastic fragments blasted out of a central vent.

cinders (volcanic)   Glassy pyroclastic volcanic fragments 4 to 32 millimeters in size.

cirque   A steep-walled semicircular depression eroded into a mountain peak by a glacier.

cirrus cloud   A wispy, high-level cloud.

clastic sediment   Sediment composed of fragments of weathered rock that have been transported some distance from their place of origin.

clastic sedimentary rock   Rock composed of lithified clastic sediment.

clay   Any clastic mineral particle less than 1/256 millimeter in diameter. Also a group of layer silicate minerals.

claystone   A fine-grained clastic sedimentary rock composed predominantly of clay minerals and small amounts of quartz and other minerals of clay size.

Clean Water Act   A federal law mandating the cleaning of the nation's rivers, lakes, and wetlands, and forbidding the discharge of pollutants into waterways.

cleavage   The tendency of some minerals to break along certain crystallographic planes.

climate   The composite pattern of long-term weather conditions that can be expected in a given region. Climate refers to yearly cycles of temperature, wind, rainfall, etc., and not to daily variations. *See* weather.

cloud   A collection of minute water droplets or ice crystals in air.

coal   A flammable organic sedimentary rock formed from partially decomposed plant material and composed mainly of carbon.

cobbles   Rounded rock fragments in the 64 to 256 millimeter size range, larger than pebbles and smaller than boulders.

cold front   A front that forms when moving cold air collides with stationary or slower moving warm air. The dense, cold air distorts into a blunt wedge and pushes under the warmer air, creating a narrow band of violent weather commonly accompanied by cumulus and cumulonimbus clouds.

column   A dripstone or speleothem formed when a stalactite and a stalagmite meet and fuse together.

columnar joint A regularly spaced crack that commonly develops in lava flows forming five- or six-sided columns.

coma The bright outer sheath of a comet.

comet An interplanetary body composed of loosely bound rock and ice that forms a bright head and extended fuzzy tail when it enters the inner portion of the Solar System.

compaction Tighter packing of sedimentary grains causing weak lithification and a decrease in porosity, usually resulting from the weight of overlying sediment.

competence A measure of the largest particles that a stream can transport.

composite cone A volcano that consists of alternate layers of unconsolidated pyroclastic material and lava flows. (*syn:* stratovolcano)

compound A pure substance composed of two or more elements whose composition is constant.

compressive stress Stress that acts to shorten an object by squeezing it.

conduction The transport of heat by atomic or molecular motion.

cone of depression A cone-like depression in the water table formed when water is pumped out of a well more rapidly than it can flow through the aquifer.

confining pressure *See* confining stress.

confining stress A form of stress that develops when rock or sediment is buried. Confining stress compresses rocks but does not distort them because the compressive force acts equally in all directions.

conformable The condition in which sedimentary layers were deposited continuously without interruption.

conglomerate A coarse-grained clastic sedimentary rock, composed of rounded fragments larger than 2 millimeters in diameter cemented in a fine-grained matrix of sand or silt.

constellation One of 88 groups of stars that astronomers refer to for convenience in referring to the positions of objects in the night sky.

consumption Any process that uses water, and then returns it to the Earth far from its source.

contact A boundary between two different rock types or between rocks of different ages.

contact metamorphic ore deposit An ore deposit formed by contact metamorphism.

contact metamorphism Metamorphism caused by heating of country rock, and/or addition of fluids, from a nearby igneous intrusion.

continental crust The predominantly granitic portion of the crust, 20 to 70 kilometers thick, that makes up the continents.

continental drift The theory proposed by Alfred Wegener that continents were once joined together and later split and drifted apart. The continental drift theory has been replaced by the more complete plate tectonics theory.

continental glacier A glacier that forms a continuous cover of ice over areas of 50,000 square kilometers or more and spreads outward in all directions under the influence of its own weight. (*syn:* ice sheet)

continental margin The region between the shoreline of a continent and the deep ocean basins including the continental shelf, continental slope, and continental rise. Also

the region where thick, granitic continental crust joins thinner, basaltic oceanic crust.

continental rifting The process by which a continent is pulled apart at a divergent plate boundary.

continental rise An apron of sediment between the continental slope and the deep sea floor.

continental shelf A shallow, nearly level area of continental crust covered by sediment and sedimentary rocks that is submerged below sea level at the edge of a continent between the shoreline and the continental slope.

continental slope The relatively steep (3° to 6°) underwater slope between the continental shelf and the continental rise.

control rod A column of neutron-absorbing alloys that are spaced among fuel rods to fine-tune nuclear fission in a reactor.

convection The transport of heat by the movement of currents. In meteorology, horizontal air flow is called advection, whereas *convection* is reserved for vertical air flow.

convective zone The subsurface zone in a star where energy is transmitted primarily by convection.

convergent boundary A boundary where two lithospheric plates collide head-on.

coquina A bioclastic limestone consisting of coarse shell fragments cemented together.

core The innermost region of the Earth probably consisting of iron and nickel.

Coriolis effect The deflection of air or water currents caused by the rotation of the Earth.

corona The luminous irregular envelope of highly ionized gas outside the chromosphere of the Sun.

correlation Demonstration of the equivalence of rocks or geologic features age from different locations.

cost-benefit analysis A system of analysis that attempts to weigh the cost of an act or policy, such as pollution control, directly against the economic benefits.

country rock The older rock intruded by a younger igneous intrusion or mineral deposit.

crater A bowl-like depression at the summit of the volcano.

craton A segment of continental crust, usually in the interior of a continent, that has been tectonically stable for a long time, commonly a billion years or longer.

creep The slow movement of unconsolidated material downslope under the influence of gravity.

crest (of a wave) The highest part of a wave.

crevasse A fracture or crack in the upper 40 to 50 meters of a glacier.

cross-bedding An arrangement of small beds at an angle to the main sedimentary layering.

cross-cutting relationship Any relationship in which younger rocks or geological structures interrupt or cut across older rocks or structures.

crust The Earth's outermost layer about 7 to 70 kilometers thick, composed of relatively low-density silicate rocks.

crystal A solid element or compound whose atoms are arranged in a regular, orderly, periodically repeated array.

crystal face A planar surface that develops if a crystal grows freely in an uncrowded environment.

crystal habit The shape in which individual crystals grow and the manner in which crystals grow together in aggregates.

crystal settling A process in which the crystals that solidify first from a cooling magma settle to the bottom of a magma chamber because the solid minerals are more dense than liquid magma.

crystalline structure The orderly, repetitive arrangement of atoms in a crystal.

cumulonimbus cloud A cumulus cloud from which precipitation is falling.

cumulus cloud A column-like cloud with a flat bottom and a billowy top.

Curie point The temperature below which rocks can retain magnetism.

current A continuous flow of water in a concerted direction.

cyclone A low-pressure region with its accompanying surface wind. (*syn:* tropical cyclone, hurricane)

daughter isotope An isotope formed by radioactive decay of another isotope.

debris flow A type of mass wasting in which particles move as a fluid and more than half of the particles are larger than sand.

deflation Erosion by wind.

deformation Folding, faulting, and other changes in shape of rocks or minerals in response to mechanical forces, such as those that occur in tectonically active regions.

degeneracy pressure The strength of the atomic particles that holds a white dwarf star from further collapse.

delta The nearly flat, alluvial, fan-shaped tract of land at the mouth of a stream.

deposition The laying down of rock-forming materials by any natural agent.

desert A region with less than 25 centimeters of rainfall a year. Also defined as a region that supports only a sparse plant cover.

desert pavement A continuous cover of stones created as wind erodes fine sediment, leaving larger rocks behind.

desertification A process by which semi-arid land is converted to desert, often by improper farming or by climate change.

dew Moisture condensed onto objects from the atmosphere, usually during the night, when the ground and leaf surfaces become cooler than the surrounding air.

dew point The temperature at which the relative humidity of air reaches 100 percent, and the air becomes saturated. If saturated air cools below the dew point, some of the water vapor generally condenses into liquid droplets (although sometimes the relative humidity can rise above 100 percent).

differential weathering The process by which certain rocks weather more rapidly than adjacent rocks, usually resulting in an uneven surface.

dike A sheet-like igneous rock that cuts across the structure of country rock.

directed stress Stress that acts most strongly in one direction.

discharge The volume of water flowing downstream per unit time. It is measured in units of cubic meters per second (m^3/sec).

disconformity A type of unconformity in which the sedimentary layers above and below the unconformity are parallel.

discordant Pertaining to a dike or other feature that cuts across sedimentary layers or other kinds of layering in country rock.

disseminated ore deposit A large low-grade ore deposit in which generally fine-grained metal-bearing minerals are widely scattered throughout a rock body in sufficient concentration to make the deposit economical to mine.

dissolved load The portion of a stream's sediment load that is carried in solution.

distributary A channel that flows outward from the main stream channel, such as is commonly found in deltas.

divergent boundary The boundary or zone where lithospheric plates separate from each other. (*syn:* spreading center)

diversion All processes that transfer ground or surface water from its natural place and path in the hydrologic cycle to a new place and path to serve human needs.

docking The accretion of island arcs or microcontinents onto a continental margin.

doldrums A region of the Earth near the equator in which hot, humid air moves vertically upward, forming a vast low-pressure region. Local squalls and rainstorms are common, and steady winds are rare.

dolomite Common rock-forming mineral similar to calcite, $CaMg(CO_3)_2$.

dome A circular or elliptical anticlinal structure.

Doppler effect The observed change in frequency of light or sound that occurs when the source of the wave is moving relative to the observer.

downcutting Downward erosion by a stream.

drainage basin The region that is ultimately drained by a single river.

drift (glacial) Any rock or sediment transported and deposited by a glacier or by glacial meltwater.

drumlin An elongate hill formed when a glacier flows over and reshapes a mound of till or stratified drift.

dry adiabatic lapse rate The rate of cooling that occurs when dry air rises without gain or loss of heat.

dune A mound or ridge of wind-deposited sand.

earthflow A flowing mass of fine-grained soil particles mixed with water. Earthflows are less fluid than mudflows.

earthquake A sudden motion or trembling of the Earth caused by the abrupt release of slowly accumulated elastic energy in rocks.

echo sounder An instrument that emits sound waves and then records them after they reflect off the sea floor. The data is then used to record the topography of the sea floor.

eccentricity The shape of the ellipse that constitutes the Earth's orbit around the Sun.

eclipse A phenomenon that occurs when a heavenly body is shadowed by another and therefore rendered invisible. When the Moon lies directly between the Earth and the

Sun, the Moon blocks our view of the Sun and we observe a *solar eclipse*. When the Earth lies directly between the Sun and the Moon, the Earth's shadow falls on the Moon and we observe a *lunar eclipse*.

effluent stream   A stream that receives water from ground water because its channel lies below the water table. (*syn:* gaining stream)

elastic deformation   A deformation such that if the stress is removed, the material springs back to its original size and shape.

elastic limit   The maximum stress that an object can withstand without permanent deformation.

electromagnetic radiation   The transfer of energy by an oscillating electric and magnetic field; it travels as a wave and also behaves as a stream of particles.

electromagnetic spectrum   The entire range of electromagnetic radiation from very long wavelength (low frequency) radiation to very short wavelength (high frequency) radiation.

electron   A fundamental particle which forms a diffuse cloud of negative charge around an atom.

element   A substance that cannot be broken down into other substances by ordinary chemical means. An element is made up of the same kind of atoms.

elliptical galaxy   A galaxy with an elliptical appearance.

emergent coastline   A coastline that was recently under water but has been exposed because either the land has risen or sea level has fallen.

emission nebula   A glowing cloud of interstellar gas.

emission spectrum   *See* spectrum.

end moraine   A moraine that forms at the end, or terminus, of a glacier.

energy resources   Geologic resources, including petroleum, coal, natural gas, and nuclear fuels, used for heat, light, work, and communication.

eon   The longest unit of geologic time. The most recent eon, the Phanerozoic Eon, is further subdivided into eras.

epicenter   The point on the Earth's surface directly above the focus of an earthquake.

epidemiology   The study of the distribution of sickness in a population.

epoch   The smallest unit of geologic time. Periods are divided into epochs.

equinox   Either of two times during a year when the Sun shines directly overhead at the equator. The equinoxes are the beginnings of spring and fall, when every portion of the Earth receives 12 hours of daylight and 12 hours of darkness.

era   A geologic time unit. Eons are divided into eras and in turn eras are subdivided into periods.

erosion   The removal of weathered rocks by moving water, wind, ice, or gravity.

erratic   A boulder that was transported to its present location by a glacier. Usually different from the bedrock in its immediate vicinity.

esker   A long snake-like ridge formed by deposition in a stream that flowed on, within, or beneath a glacier.

estuary   A bay that formed when a broad river valley was submerged by rising sea level or a sinking coast.

eutrophic lake   A lake characterized by abundant dissolved nitrates, phosphates, and other plant nutrients and by a seasonal deficiency of oxygen in bottom water. Such lakes are commonly shallow.

evaporation   The transformation of a liquid to a gas.

evaporation fog   Fog that forms when air is cooled by evaporation from a body of water, commonly a lake or river. The water evaporates, but the vapor cools and condenses to fog.

evaporite deposit   A chemically precipitated sedimentary rock that formed when dissolved ions were concentrated by evaporation of water.

evolution   The change in the physical and genetic characteristics of a species over time.

exfoliation   Fracturing in which concentric plates or shells split from the main rock mass like the layers of an onion.

extensional stress   Tectonic stress in which rocks are pulled apart.

externalities   The cost of living in a degraded environment, including direct costs such as medical bills, lost work, and damage to waterways, crops, and livestock. It also includes indirect costs of environmental degradation such as reduction in tourism and lowered land values.

extrusive igneous rock   An igneous rock formed from material that has erupted onto the surface of the Earth.

fall   A type of mass wasting in which unconsolidated material falls freely or bounces down the face of a cliff.

fault   A fracture in rock along which displacement has occurred.

fault creep   A continuous, slow movement of solid rock along a fault resulting from a constant stress acting over a long time.

fault zone   An area of numerous, closely spaced faults.

faunal succession (principle of)   *See* principle of faunal succession.

feedback mechanism   A feedback mechanism occurs when a small initial perturbation affects another component of the Earth systems, which amplifies the original effect, which perturbs the system even more, which leads to an even greater effect, and so on.

feldspar   A common group of aluminum silicate rock-forming minerals that contain potassium, sodium, or calcium.

felsic   A term describing any light-colored igneous rock containing large amounts of *fel*dspar and *si*lica.

fetch   The distance that the wind has traveled over the ocean without interruption.

fiord   A long, deep, narrow arm of the sea bounded by steep walls, generally formed by submergence of a glacially eroded valley.

firn   Hard, dense snow that has survived through one summer melt season. Firn is transitional between snow and glacial ice.

fissility   Fine layering along which a rock splits easily.

fissure   An extensive break, crack, or fracture in rocks. Fissure eruptions are usually gentle volcanic eruptions in which lava flows from fissures in the Earth's crust.

flash flood   A rapid, intense, local flood of short duration.

flint   *See* chert.

flood Any relatively high stream flow that overtops the stream banks in any part of its course, covering land that is not usually under water.

flood basalt Basaltic lava that erupts gently in great volume from cracks at the Earth's surface to cover large areas of land and form basalt plateaus. (*syns:* basalt plateau, lava plateau)

flood plain That portion of a river valley adjacent to the channel; it is built by sediment deposited during floods and is covered by water during a flood.

flow Mass wasting in which individual particles move downslope as a semifluid not as a consolidated mass.

fly ash Minerals that escape into the atmosphere, usually when coal burns and eventually settles as gritty dust.

focus The initial rupture point of an earthquake.

fog A cloud that forms at or very close to ground level.

fold A bend in rock.

foliation Layering in rock created by metamorphism.

footwall The rock beneath an inclined fault.

forearc basin A sedimentary basin between the trench and the magmatic arc in either an island arc or an Andean continental margin.

foreshock Small earthquakes that precede a large quake by an interval ranging from a few seconds to a few weeks.

foreshore The zone that lies between the high and low tides; the intertidal region.

formation A lithologically distinct body of sedimentary, igneous, or metamorphic rock that can be recognized in the field and can be mapped.

fossil The preserved trace, imprint, or remains of a plant or animal.

fossil fuel Fuels formed from the partially decayed remains of plants and animals. The most commonly used fossil fuels are petroleum, coal, and natural gas.

fracture (1) The manner in which minerals break other than along planes of cleavage. (2) A crack, joint, or fault in bedrock.

front In meteorology, a line separating air masses of different temperature or density.

frontal weather system A weather system that develops when air masses collide.

frontal wedging A condition in which a moving mass of cool, dense air encounters a mass of warm, less dense air; the cool, denser air slides under the warm air mass, forcing the warm air upward to create a weather front.

frost Ice crystals formed directly from vapor.

frost wedging A process in which water freezes in a crack in rock and the expansion wedges the rock apart.

fuel rod A 2-meter-long column of fuel-grade uranium pellets used to fuel a nuclear reactor.

fusion (of atomic nuclei) The combination of nuclei of light elements (particularly hydrogen) to form heavier nuclei.

gabbro Igneous rock that is mineralogically identical to basalt, but that has a medium- to coarse-grained texture because of its plutonic origin.

gaining stream A stream that receives water from ground water because its channel lies below the water table. (*syn:* effluent stream)

galactic halo A spherical cloud of dust and gas that surrounds the Milky Way's galactic disk.

galaxy A large volume of space containing many billions of stars, held together by mutual gravitational attraction.

gem A mineral that is prized for its rarity and beauty rather than for industrial use.

geocentric A model that places the Earth at the center of the celestial bodies.

geologic column A composite columnar diagram that shows the sequence of rocks at a given place or region arranged to show their position in the geologic time scale.

geologic structure Any feature formed by deformation or movement of rocks, such as a fold or a fault. Also, the combination of all such features of an area or region.

geologic time scale A chronological arrangement of geologic time subdivided into units.

geology The study of the Earth, the materials that it is made of, the physical and chemical changes that occur on its surface and in its interior, and the history of the planet and its life forms.

geosphere The solid Earth, consisting of the entire planet from the center of the core to the outer crust.

geothermal energy Energy derived from the heat of the Earth.

geothermal gradient The rate at which temperature increases with depth in the Earth.

geyser A type of hot spring that intermittently erupts jets of hot water and steam. Geysers occur when ground water comes in contact with hot rock.

glacial polish A smooth polish on bedrock created when fine particles transported at the base of a glacier abrade the bedrock.

glacial striation Parallel grooves and scratches in bedrock that form as rocks are dragged along at the base of a glacier.

glacier A massive, long-lasting accumulation of compacted snow and ice that forms on land and moves downslope or outward under its own weight.

glaze An ice coating that forms when rain falls on surfaces that are below freezing.

globular cluster A spherically symmetrical collection of stars that shared a common origin. Numerous globular clusters lie within the Milky Way's galactic halo.

gneiss A foliated rock with banded appearance formed by regional metamorphism.

graben A wedge-shaped block of rock that has dropped downward between two normal faults.

graded bedding A type of bedding in which larger particles are at the bottom of each bed, and the particle size decreases toward the top.

graded stream A stream with a smooth concave profile. A graded stream is in equilibrium with its sediment supply; it transports all the sediment supplied to it with neither erosion nor deposition in the streambed.

gradient The vertical drop of a stream over a specific distance.

gradualism A theory of evolution that proposes that species change gradually in small increments.

granite A medium- to coarse-grained, sialic, plutonic rock made predominately of potassium feldspar and quartz.

gravel Unconsolidated sediment consisting mainly of rounded particles with a diameter greater than 2 millimeters.

greenhouse effect An increase in the temperature of a planet's atmosphere caused when infrared-absorbing gases are introduced into the atmosphere.

groin A narrow wall built perpendicular to the shore to trap sand transported by currents and waves.

ground moraine A moraine formed when a melting glacier deposits till in a relatively thin layer over a broad area.

ground water Water contained in soil and bedrock. All subsurface water.

guyot A flat-topped seamount.

gypsum A mineral with the formula, $(CaSO_4 \cdot 2H_2O)$. It commonly forms in evaporite deposits.

gyre A closed, circular current in either water or air.

Hadean Eon The earliest time in the Earth's history.

hail Ice globules varying from 5 millimeters to a record 14 centimeters in diameter that fall from cumulonimbus clouds.

half-life The time it takes for half of the nuclei of a radioactive isotope in a sample to decompose.

hanging valley A tributary glacial valley whose mouth lies high above the floor of the main valley.

hanging wall The rock above an inclined fault.

hardness The resistance of the surface of a mineral to scratching.

heliocentric A model that places the Sun at the center of the Solar System.

Hertzsprung-Russell diagram (H-R diagram) A plot of absolute stellar magnitude (or luminosity) against temperature.

horn A sharp, pyramid-shaped rock summit where three or more cirques intersect near the summit.

hornblende A rock-forming mineral. The most common member of the amphibole group.

hornfels A fine-grained rock formed by contact metamorphism.

horse latitudes A region of the Earth lying at about 30° north and south latitudes, in which air is falling, forming a vast high-pressure region. Generally dry conditions prevail, and steady winds are rare.

horst A block of rock that has moved relatively upward and is bounded by two faults.

hot spot A persistent volcanic center thought to be located directly above a rising plume of hot mantle rock.

hot spring A spring formed where hot ground water flows to the surface.

Hubble's Law A law that states that the velocity of a galaxy is proportional to its distance from Earth. Thus the most distant galaxies are traveling at the highest velocities.

humidity A measure of the amount of water vapor in the air. *Absolute humidity* is the amount of water vapor in a given volume of air. *Relative humidity* is the ratio of the amount of water vapor in a given volume of air divided by the maximum amount of water vapor that can be held by that air at a given temperature.

humus The dark organic component of soil composed of litter that has decomposed sufficiently so that the origin of the individual pieces cannot be determined.

hurricane *See* tropical cyclone.

hydration The chemical combination of water with another substance.

hydraulic action The mechanical loosening and removal of material by flowing water.

hydride A compound of hydrogen and one or more metals. Hydrides can be heated to release hydrogen gas for use as a fuel.

hydrologic cycle The constant circulation of water among the sea, the atmosphere, and the land.

hydrolysis A decomposition reaction involving water.

hydrosphere All of the Earth's water, which circulates among oceans, continents, and the atmosphere.

hydrothermal metamorphism Changes in rock that are primarily caused by migrating hot water and by ions dissolved in the hot water.

hydrothermal ore deposit An ore deposit formed by precipitation of dissolved metals from hot water solutions.

hydrothermal vein A sheet-like mineral deposit that fills a fault or other fracture, precipitated from hot water solutions.

ice age A time of extensive glacial activity, when alpine glaciers descended into lowland valleys and continental glaciers spread over the higher latitudes.

ice sheet A glacier that forms a continuous cover of ice over areas of 50,000 square kilometers or more and spreads outward under the influence of its own weight. (*syn:* continental glacier, ice cap)

iceberg A large chunk of ice that breaks from a glacier into a body of water.

icefall A section of a glacier that flows down a steep gradient so that the ice fractures into numerous crevasses and towering ice pinnacles.

igneous rock Rock that solidified from magma.

index fossil A fossil that identifies and dates the layers in which it is found. Index fossils are abundantly preserved in rocks, widespread geographically, and existed as a species or genus for only a relatively short time.

industrial mineral Any rock or mineral of economic value exclusive of metal ores, fuels, and gems.

influent stream A stream that lies above the water table. Water percolates from the stream channel downward into the saturated zone. (*syn:* losing stream)

intensity (of an earthquake) A measure of an earthquake's force by its effects on buildings and people in a particular place.

intermediate rock Igneous rock with chemical and mineral composition between those of granite and basalt.

internal processes Earth processes that are initiated by movements within the Earth and those internal movements themselves. For example, formation of magma, earthquakes, mountain building, and tectonic plate movement.

intertidal zone The part of a beach that lies between the high and low tide lines.

intracratonic basin A sedimentary basin located within a craton.

intrusive igneous rock A rock formed when magma solidifies within bodies of preexisting rock.

inversion (atmospheric) A meteorological condition in which the lower layers of air are cooler than those at higher altitudes. This cool air can remain relatively stagnant and allows air pollutants to concentrate in urban areas.

ion An atom with an electrical charge.

ionic substitution The replacement of one or more kinds of ions in a mineral by other kinds of ions of similar size and charge.

island arc A gently curving chain of volcanic islands in the ocean formed by convergence of two plates each bearing ocean crust, and the resulting subduction of one plate beneath the other.

isobar Line on a weather map connecting points of equal pressure.

isostasy The condition in which the lithosphere floats on the asthenosphere as an iceberg floats on water.

isostatic adjustment The rising and settling of portions of the lithosphere to maintain equilibrium as they float on the plastic asthenosphere.

isotherm A line on a weather map connecting points of equal temperature.

isotopes Atoms of the same element that have the same number of protons but different numbers of neutrons.

jet stream A relatively narrow, high altitude, fast-moving air current.

joint A fracture that occurs without movement of rock on either side of the break.

Jovian planets The outer planets—Jupiter, Saturn, Uranus, and Neptune—which are massive with a high proportion of the lighter elements.

kame A small mound, or ridge of layered sediment deposited by a stream that flows on top of, within, or beneath a glacier.

karst topography A type of topography formed over limestone or other soluble rock and characterized by caverns, sinkholes, and underground drainage.

kerogen The solid bitumen in oil shales that yields oil when the shales are heated and distilled; the precursor of liquid petroleum.

kettle A depression in glacial drift created by melting of a large chunk of ice left buried in the drift by a receding glacier. The ice prevents sediment from collecting; when the ice melts a lake or swamp may fill the depression.

key bed A thin, widespread, easily recognized sedimentary layer that can be used for correlation.

lagoon A protected body of water separated from the sea by a reef or barrier island.

lake A large, inland body of standing water that occupies a depression in the land surface.

landslide A general term for the downslope movement of rock and regolith under the influence of gravity.

latent heat The heat released or absorbed by a substance during a change in state, i.e., melting, freezing, vaporization, condensation, or sublimation.

lateral erosion The action of a stream as it cuts into and erodes its banks. A meandering stream commonly swings from one side of its channel to the other, causing lateral erosion.

lateral moraine A ridge-like moraine that forms on or adjacent to the sides of a mountain glacier.

laterite A highly weathered soil rich in oxides of iron and aluminum that usually develops in warm, moist tropical or temperate regions.

lava Fluid magma that flows onto the Earth's surface from a volcano or fissure. Also, the rock formed by solidification of the same material.

lava plateau A sequence of horizontal basalt lava flows that were extruded rapidly to cover a large region of the Earth's surface. (*syns:* flood basalt, basalt plateau)

layers 1, 2, 3 The three layers of oceanic crust. The uppermost layer, layer 1, consists of mud. Layer 2 consists of pillow basalt. Layer 3 directly overlies the mantle and consists of basalt dikes and gabbro.

leaching The dissolution and downward movement of soluble components of rock and soil by percolating water.

light year The distance traveled by light in one year, approximately 9.5×10^{12} kilometers.

limb The side of a fold in rock.

limestone A sedimentary rock consisting chiefly of calcium carbonate.

liquefaction A process in which a soil loses its shear strength during an earthquake and becomes a fluid.

lithification The conversion of loose sediment to solid rock.

lithologic correlation Correlation based on the lithologic continuity of a rock unit.

lithosphere The cool, rigid, outer layer of the Earth, about 100 kilometers thick, which includes the crust and part of the upper mantle.

litter Leaves, twigs, and other plant or animal material that have fallen to the surface of the soil but have not decomposed.

loam Soil that contains a mixture of sand, clay, and silt and a generous amount of organic matter.

loess A homogenous, unlayered deposit of windblown silt, usually of glacial origin.

longitudinal dune A long symmetrical dune oriented parallel with the direction of the prevailing wind.

longshore current A current flowing parallel and close to the coast that is generated when waves strike a shore at an angle.

longshore drift Sediment carried by longshore currents.

losing stream A stream that lies above the water table. Water percolates from the stream channel downward into the saturated zone. (*syn:* influent stream)

low velocity layer A region of the upper mantle where seismic waves travel relatively slowly, approximately the same as the asthenosphere.

lunar eclipse   *See* eclipse.

luster   The quality and intensity of light reflected from the surface of a mineral.

mafic rock   Dark-colored igneous rock with high magnesium and iron content, and composed chiefly of iron- and magnesium-rich minerals.

magma   Molten rock generated within the Earth.

magmatic arc   A narrow elongate band of intrusive and volcanic activity associated with subduction.

magnetic reversal   A change in the Earth's magnetic field in which the north magnetic pole becomes the south magnetic pole, and vice versa.

magnetometer   An instrument that measures the Earth's magnetic field.

magnitude (of an earthquake)   A measure of the strength of an earthquake determined from seismic recordings. *See also* Richter scale.

main sequence   A band running across a Hertzsprung-Russell diagram that contains most of the stars. Hydrogen fusion generates the energy in a main sequence star.

manganese nodule   A manganese-rich, potato-shaped rock found on the ocean floor.

mantle   A mostly solid layer of the Earth lying beneath the crust and above the core. The mantle extends from the base of the crust to a depth of about 2900 kilometers.

mantle convection   The convective flow of solid rock in the mantle.

mantle plume   A rising vertical column of mantle rock.

marble   A metamorphic rock consisting of fine- to coarse-grained recrystallized calcite and/or dolomite.

maria   Dry, barren, flat expanses of volcanic rock on the Moon first thought to be seas.

mass wasting   The movement of earth material downslope primarily under the influence of gravity.

meander   One of a series of sinuous curves or loops in the course of a stream.

mechanical weathering (physical)   The disintegration of rock into smaller pieces by physical processes.

medial moraine   A moraine formed in or on the middle of a glacier by the merging of lateral moraines as two glaciers flow together.

Mercalli scale   A scale of earthquake intensity that measures the strength of an earthquake in a particular place by its effect on buildings and people. It has been replaced by the Richter scale.

mesa   A flat-topped mountain or a tableland that is smaller than a plateau and larger than a butte.

mesosphere   The layer of air that lies above the stratopause and extends from about 55 kilometers upward to 80 kilometers above the Earth's surface. Temperature decreases with elevation in the mesosphere.

Mesozoic era   The part of geologic time roughly 245 to 65 million years ago. Dinosaurs rose to prominence and became extinct during this era.

metallic mineral resources   Ore; valuable concentrations of metals.

metamorphic facies   A set of all metamorphic rock types that formed under similar temperature and pressure conditions.

metamorphic grade   The intensity of metamorphism that formed a rock; the maximum temperature and pressure attained during metamorphism.

metamorphic rock   A rock formed when igneous, sedimentary, or other metamorphic rocks recrystallize in response to elevated temperature, increased pressure, chemical change, and/or deformation.

metamorphism   The process by which rocks and minerals change in response to changes in temperature, pressure, chemical conditions, and/or deformation.

meteorite   A fallen meteoroid.

meteoroid   A small interplanetary body in an irregular orbit. Many meteoroids are fragments of asteroids formed during collisions.

mica   A layer silicate mineral with a distinctive platy crystal habit and perfect cleavage. Muscovite and biotite are common micas.

microcontinent   A small mass of continental crust, or an island arc, similar to Japan, New Zealand, and the modern islands of the southwest Pacific Ocean. The first supercontinent is thought to have formed when many microcontinents joined together about 2 billion years ago. Modern continents consist of many microcontinents that welded together over geologic time.

microwave radar   Instruments and processes using reflections of electromagnetic radiation in the microwave range. The data measure subtle swells and depressions on the sea surface, which reflect sea-floor topography.

mid-channel bar   An elongate lobe of sand and gravel formed in a stream channel.

mid-oceanic ridge   A continuous submarine mountain chain that forms at the boundary between divergent tectonic plates within oceanic crust.

migmatite   A rock composed of both igneous and metamorphic-looking materials. It forms at very high metamorphic grades when rock begins to partially melt to form magma.

mineral   A naturally occurring inorganic solid with a definite chemical composition and a crystalline structure.

mineral deposit   A local enrichment of one or more minerals.

mineral reserve   The known supply of ore in the ground.

mineral resources   All useful rocks and minerals.

mineralization   A process of fossilization in which the organic components of an organism are replaced by minerals.

Mohorovičić discontinuity (Moho)   The boundary between the crust and the mantle identified by a change in the velocity of seismic waves.

Mohs hardness scale   A standard, numbered from 1 to 10, to measure and express the hardness of minerals based on a series of ten fairly common minerals, each of which is harder than those lower on the scale.

moment magnitude   An earthquake scale based on the amount of movement and the surface area of a fault. The moment magnitude scale closely reflects the total amount of energy released during an earthquake.

monocline   A fold with only one limb.

monsoon   A continental wind system caused by uneven heating of land and sea. Monsoons generally blow from the sea to the land in the summer, when the continents are warmer than the ocean, and bring rain. In winter, when the ocean is warmer than the land, the monsoon winds reverse.

moraine   A mound or ridge of till deposited directly by glacial ice.

mud   Wet silt and clay.

mud cracks   Irregular, usually polygonal fractures that develop when mud dries. The patterns may be preserved when the mud is lithified.

mudflow   Mass wasting of fine-grained soil particles mixed with a large amount of water.

mudstone   A non-fissile rock composed of a mixture of clay and silt.

natural gas   A mixture of naturally occurring light hydrocarbons composed mainly of methane, CH_4.

natural levee   A ridge or embankment of flood-deposited sediment along both banks of a stream channel.

neap tide   A relatively small tide that forms when the Moon is 90° out of alignment with the Sun and the Earth.

nebula   An interstellar cloud of gas and dust. A *planetary nebula* is created when a star the size of our Sun explodes.

neutron   A subatomic particle with the mass of a proton but no electrical charge.

neutron star   A small, extremely dense star composed almost entirely of neutrons. *See also* pulsar.

nimbo   a prefix or suffix added to cloud types to indicate precipitation.

nimbostratus cloud   A stratus cloud from which precipitation is falling.

nonconformity   A type of unconformity in which layered sedimentary rocks lie on an erosion surface cut into igneous or metamorphic rocks.

nonfoliated   The lack of layering in metamorphic rock.

nonpoint source pollution   Pollution generated over a broad area, such as fertilizers and pesticides spread over agricultural fields.

nonrenewable resource   A resource in which formation of new deposits occurs much more slowly than consumption.

normal fault   A fault in which the hanging wall has moved downward relative to the footwall.

normal lapse rate   The decrease in temperature with elevation in air that is neither rising nor falling.

normal polarity   A magnetic orientation the same as that of the Earth's modern magnetic field.

nuclear fission   The breakdown of an atomic nucleus of an element of relatively high atomic number into two or more nuclei of lower atomic number, with conversion of part of its mass into energy.

nuclear fuel   Radioactive isotopes, such as those of uranium, used to generate electricity in nuclear reactors.

nucleus   The small, dense, central portion of an atom composed of protons and neutrons. Nearly all of the mass of an atom is concentrated in the nucleus.

nuée ardente   A swiftly flowing, often red-hot cloud of gas, volcanic ash, and other pyroclastics formed by an explosive volcanic eruption. (*syn:* ash flow)

obsidian   A black or dark-colored glassy volcanic rock usually of rhyolitic composition.

occluded front   A front that forms when a faster moving cold air mass traps a warm air mass against a second mass of cold air. The faster moving cold air mass slides beneath the warm air, lifting it off the ground. Precipitation occurs along both frontal boundaries, resulting in a large zone of inclement weather.

oceanic crust   The 7- to 10-kilometer-thick layer of sediment and basalt that underlies the ocean basins.

oceanic island   A volcanic island formed at a hot spot above a mantle plume.

oceanic trench   A long, narrow, steep-sided depression of the sea floor formed where a subducting oceanic plate sinks into the mantle.

Ogallala aquifer   The aquifer that extends for almost 1000 kilometers from the Rocky Mountains eastward beneath portions of the Great Plains.

oil   A naturally occurring liquid or gas composed of a complex mixture of hydrocarbons. (*syn:* petroleum)

oil shale   A kerogen-bearing sedimentary rock that yields liquid or gaseous hydrocarbons when heated.

oil trap   Any rock barrier that accumulates oil or gas by preventing its upward movement.

oligotrophic lake   A lake characterized by nearly pure water with low concentrations of nitrates, phosphates, and other plant nutrients. Oligotrophic lakes have low productivities, meaning that they sustain relatively few living organisms, although lakes of this type typically contain a few huge trout or similar game fish, and are commonly deep.

olivine   A common rock-forming mineral in mafic and ultramafic rocks with a composition that varies between Mg_2SiO_4 and Fe_2SiO_4.

ore   A natural material that is sufficiently enriched in one or more minerals to be mined profitably.

original horizontality (principle of)   *See* principle of original horizontality.

organic sedimentary rock   Sedimentary rocks such as coal, that consist of the lithified organic remains of plants or animals.

orogeny   The process of mountain building; all tectonic processes associated with mountain building.

orographic lifting   Lifting of air that occurs when air flows over a mountain.

orthoclase   A common rock-forming mineral; a variety of potassium feldspar, $KAlSi_3O_8$.

outwash plain   A broad level surface formed where glacial sediment is deposited in front of or beyond a glacier.

oxbow lake   A crescent-shaped lake formed where a meander is cut off from a stream and the ends of the cut-off meander become plugged with sediment.

oxidation   The loss of electrons from a compound or element during a chemical reaction. In the weathering of common minerals, oxidation usually occurs when a mineral reacts with molecular oxygen.

P **wave**   *See* primary (P) wave.

pahoehoe   A basaltic lava flow with a smooth, billowy, or "ropy" surface.

paleoclimatology The study of ancient climates.

paleontology The study of life that existed in the past.

Paleozoic era The part of geologic time 570 to 245 million years ago. During this era invertebrates, fishes, amphibians, reptiles, ferns, and cone-bearing trees were dominant.

parabolic dune A crescent-shaped dune with tips pointing into the wind.

parallax The apparent displacement of an object caused by the movement of the observer.

parent isotope A radioactive isotope that decays to produce a daughter isotope.

parent rock Any original rock before it is changed by weathering, metamorphism, or other geological processes.

parsec The distance to an object that would have a stellar parallax angle of 1 arc second. One parsec is about 3.2 light years or 3×10^{13} kilometers.

partial melting The process in which a silicate rock only partly melts as it is heated, to form magma that is more silica rich than the original rock.

particles In air pollution terminology, any pollutant larger than a molecule.

passive continental margin A margin characterized by a firm connection between continental and oceanic crust where little tectonic activity occurs.

paternoster lake One of a series of lakes, strung out like beads and connected by short streams and waterfalls, created by glacial erosion.

peat A loose, unconsolidated, brownish mass of partially decayed plant matter; a precursor to coal.

pebble A sedimentary particle between 2 and 64 millimeters in diameter, larger than sand and smaller than a cobble.

pedalfer A common soil type that forms in humid environments, characterized by abundant iron and aluminum oxides and a concentration of clay in the B horizon.

pediment A gently sloping erosional surface, that forms along a mountain front uphill from a bajada, usually covered by a patchy veneer of gravel only a few meters thick.

pedocal A soil formed in arid and semi-arid climates characterized by an accumulation of calcium carbonate.

pegmatite An exceptionally coarse-grained igneous rock, usually with the same mineral content as granite.

pelagic sediment Muddy ocean sediment that consists of a mixture of clay and the skeletons of microscopic marine organisms.

penumbra A wide band outside of the umbra, where only a portion of the Sun is eclipsed during a solar eclipse.

perched water table The top of a localized lens of ground water that lies above the main water table, formed by a layer of impermeable rock or clay.

peridotite A coarse-grained plutonic rock composed mainly of olivine; it may also contain pyroxene, amphibole, or mica but little or no feldspar. The mantle is thought to be made of peridotite.

period A geologic-time unit longer than an epoch and shorter than an era.

permafrost A layer of permanently frozen soil or subsoil which lies from about a half meter to a few meters beneath the surface in arctic environments.

permeability A measure of the speed at which fluid can travel through a porous material.

petroleum A naturally occurring liquid composed of a complex mixture of hydrocarbons. (*syn:* oil)

Phanerozoic Eon The most recent 570 million years of geologic time represented by rocks that contain evident and abundant fossils.

phenocryst A large, early formed crystal in a finer matrix in igneous rock.

photon The smallest particle or packet of electromagnetic energy, such as light, infrared radiation, etc.

photosphere The surface of the Sun visible from Earth.

phyllite A metamorphic rock with a silky appearance and commonly wrinkled surface, intermediate in grade between slate and schist.

pillow basalt Lava that solidified under water, forming spheroidal lumps like a stack of pillows.

placer deposit A surface mineral deposit formed by the mechanical concentration of mineral particles (usually by water) from weathered debris.

planetary nebula A nebula created when a star the size of our Sun explodes.

plastic deformation A permanent change in the original shape of a solid that occurs without fracture.

plastic flow *See* plastic deformation.

plate A relatively rigid independent segment of the lithosphere that can move independently of other plates.

plate boundary A boundary between two lithospheric plates.

plate tectonics theory A theory of global tectonics in which the lithosphere is segmented into several plates that move about relative to one another by floating on and gliding over the plastic asthenosphere. Seismic and tectonic activity occur mainly at the plate boundaries.

plateau A large elevated area of comparatively flat land.

platform The part of a continent covered by a thin layer of nearly horizontal sedimentary rocks overlying older igneous and metamorphic rocks of the craton.

playa A dry desert lake bed.

playa lake An intermittent desert lake.

Pleistocene Ice Age A span of time from roughly 2 or 3 million to 8,000 years ago characterized by several advances and retreats of glaciers.

plucking A process in which glacial ice erodes rock by loosening particles and then lifting and carrying them downslope.

plume (of pollution) The three-dimensional zone of an aquifer or of surface water affected by a dispersing pollutant.

pluton An igneous intrusion.

plutonic rock An igneous rock that forms deep (a kilometer or more) beneath the Earth's surface.

pluvial lake A lake formed during a time of abundant precipitation. Many pluvial lakes formed as continental ice sheets melted.

point bar A stream deposit located on the inside of a growing meander.

point source pollution Pollution which arises from a specific site such as a septic tank or a factory.

polarity The magnetically positive (north) or negative (south) character of a magnetic pole.

pollution   The reduction of the quality of a resource by the introduction of impurities.

population I star   A relatively young star, formed from material ejected by an older, dying star, composed mainly of hydrogen and helium, with 1 percent heavier elements. Our Sun is a population I star.

population II star   An old star with lower concentration of heavy elements than a population I star.

pore space   The open space between grains in rock, sediment, or soil.

porosity   The proportion of the volume of a material that consists of open spaces.

porphyry   Any igneous rock containing larger crystals (phenocrysts) in a relatively fine-grained matrix.

porphyry copper   A large body of porphyritic igneous rock that contains disseminated copper sulfide minerals, and is usually mined by open pit methods.

Precambrian   All of geologic time before the Paleozoic era, encompassing approximately the first 4 billion years of Earth's history. Also, all rocks formed during that time.

precession   The circling or wobbling of the tilt of the Earth's orbit.

precipitation   (1) A chemical reaction that produces a solid salt, or precipitate, from a solution. (2) Any form in which atmospheric moisture returns to the Earth's surface—rain, snow, hail, and sleet.

pressure gradient   The change in air pressure over distance.

pressure-relief melting   The melting of rock and the resulting formation of magma caused by a drop in pressure at constant temperature.

primary air pollutant   A pollutant released directly into the atmosphere.

primary (P) wave   A seismic wave formed by alternate compression and expansion of rock. P waves travel faster than any other seismic waves.

principle of cross-cutting relationships   The principle that a rock or feature must first exist before anything can happen to it or before another rock can cut across it.

principle of faunal succession   The principle that fossil organisms succeed one another in a definite and recognizable sequence, so that sedimentary rocks of different ages contain different fossils, and rocks of the same age contain identical fossils. Therefore, the relative ages of rocks can be identified from their fossils.

principle of original horizontality   The principle that most sediment is deposited as nearly horizontal beds, and therefore most sedimentary rocks started out with nearly horizontal layering.

principle of superposition   The principle that states that in any undisturbed sequence of sediment or sedimentary rocks, the age becomes progressively younger from bottom to top.

prominence   A flame-like jet of gas rising from the Sun's corona.

Proterozoic Eon   The portion of geological time from 2.5 billion to 570 million years ago.

proton   A dense, massive, positively charged particle found in the nucleus of an atom.

protoplanets   The planets in their earliest, incipient stage of formation.

protosun   The Sun in its earliest, incipient stage of formation. The protosun was a condensing agglomeration of dust and gas.

pulsar   A neutron star that emits a pulsating radio signal.

pumice   Frothy, usually rhyolitic magma solidified into a rock so full of gas bubbles that it can float on water.

pyroclastic rock   Any rock made up of material ejected explosively from a volcanic vent.

pyroxene   A rock-forming silicate mineral group that consists of many similar minerals. Members of the pyroxene group are major constituents of basalt and gabbro.

quartz   A rock-forming silicate mineral, SiO_2. Quartz is a widespread and abundant component of continental rocks but is rare in the oceanic crust and mantle.

quartzite   A metamorphic rock composed mostly of quartz formed by recrystallization of sandstone.

quasar   An object, less than one light year in diameter and very distant from Earth, that emits an extremely large quantity of energy.

radiative zone   The zone of a star surrounding the core where energy is transmitted by absorption and radiation.

radioactivity   The natural spontaneous decay of unstable nuclei.

radiometric dating   The process of measuring the absolute age of geologic material by measuring the concentrations of radioactive isotopes and their decay products.

rain-shadow desert   A desert formed on the lee side of a mountain range.

recessional moraine   A moraine that forms at the terminus of a glacier as the glacier stabilizes temporarily during retreat.

recharge   The replenishment of an aquifer by the addition of water.

red giant   A stage in the life of a star when its core is composed of helium that is not undergoing fusion. A hot shell of hydrogen around this core is fusing at a rapid rate, producing enough energy to cause the star to expand.

red shift   The frequency shift of light waves observed in the spectrum of an object traveling away from an observer. This shift is caused by the Doppler effect.

reef   A wave-resistant ridge or mound built by corals or other marine organisms.

reflection   The return of a wave that strikes a surface.

refraction   The bending of a wave that occurs when the wave changes velocity as it passes from one medium to another.

regional metamorphism   Metamorphism that is broadly regional in extent, involving very large areas and volumes of rock. Includes both regional dynamothermal and regional burial metamorphism.

regolith   The loose, unconsolidated, weathered material that overlies bedrock.

reflecting telescope   A type of telescope that uses a mirror or mirrors to form the primary image.

refracting telescope   A type of telescope where the primary image is formed by a lens or lenses.

relative age Time expressed as the order in which rocks formed and geological events occurred, but not measured in years.

relative humidity *See* humidity.

remediation The treatment of a contaminated substance to remove or decompose a pollutant.

remote sensing The collection of information about an object by instruments that are not in direct contact with it.

reserves Known geological deposits that can be extracted profitably under current conditions.

reservoir Porous and permeable rock in which liquid petroleum or gas accumulates.

residual deposit A mineral deposit formed when water dissolves and removes most of the soluble ions from soil and rock near the Earth's surface, leaving only relatively insoluble ions in the soil. Bauxite, the principal source of aluminum, forms as a residual deposit, and in some instances iron also concentrates by this process to become ore.

resolution The ability of an optical telescope to distinguish details.

retrograde motion The apparent motion of the planets where they appear to move backward (westward) with respect to the stars.

reverse fault A fault in which the hanging wall has moved up relative to the footwall.

reversed polarity Magnetic orientations in rock which are opposite to the present orientation of the Earth's field. Also, the condition in which the Earth's magnetic field is opposite to its present orientation.

revolve To orbit a central point. A satellite revolves around the Earth. *See* rotate.

rhyolite A fine-grained extrusive igneous rock compositionally equivalent to granite.

Richter scale A numerical scale of earthquake magnitude measured by the amplitude of the largest wave on a standardized seismograph.

rift valley An elongate depression that develops at a divergent plate boundary. Examples include continental rift valleys and the rift valley along the center of the mid-oceanic ridge system.

rift zone A zone of separation of tectonic plates at a divergent plate boundary.

ring of fire The belt of subduction zones and major tectonic activity including extensive volcanism that borders the Pacific Ocean along the continental margins of Asia and the Americas.

rip current A current created when water flows back toward the sea after a wave breaks against the shore. (*syn:* undertow)

rock A naturally formed solid that is an aggregate of one or more different minerals.

rock avalanche A type of mass wasting in which a segment of bedrock slides over a tilted bedding plane or fracture. The moving mass usually breaks into fragments. (*syn:* rockslide)

rock cycle The sequence of events in which rocks are formed, destroyed, altered, and reformed by geological processes.

rock dredge An open-mouthed steel net dragged along the sea floor behind a research ship for the purpose of sampling rocks from submarine outcrops.

rock flour Finely ground, silt-sized rock fragments formed by glacial abrasion.

rock-forming minerals The most abundant minerals in the Earth's crust; the nine minerals or mineral "groups" that make up most rocks of the Earth's crust. They are olivine, pyroxene, amphibole, mica, the clay minerals, quartz, feldspar, calcite, and dolomite.

rockslide A type of slide in which a segment of bedrock slides along a tilted bedding plane or fracture. The moving mass usually breaks into fragments. (*syn:* rock avalanche)

rotate To turn or spin on an axis. Tops and planets rotate on their axes. *See* revolve.

rubble Angular particles with a diameter greater than 2 millimeters.

runoff Water that flows back to the oceans in surface streams.

S wave A seismic wave consisting of a shearing motion in which the oscillation is perpendicular to the direction of wave travel. S waves travel slower than P waves. (*syn:* secondary wave)

salinity The saltiness of seawater.

salinization A process whereby salts accumulate in soil that is irrigated heavily.

salt cracking A weathering process in which salts that are dissolved in water found in the pores of rock crystallize. This process widens cracks and pushes grains apart.

saltation Sediment transport in which particles bounce and hop along the surface.

salt-water intrusion A condition in coastal regions in which excessive pumping of fresh ground water causes salty ground water to invade an aquifer.

sand Sedimentary grains that range from 1/16 to 2 millimeters in diameter.

sandstone Clastic sedimentary rock comprised primarily of lithified sand.

saturated zone The region below the water table where all the pores in rock or regolith are filled with water.

saturation The maximum amount of water vapor that air can hold; the maximum amount of a substance that can dissolve in another substance.

schist A strongly foliated metamorphic rock that has a well-developed parallelism of minerals such as micas.

sea arch An opening created when a cave is eroded all the way through a narrow headland.

sea stack A pillar of rock left when a sea arch collapses or when the inshore portion of a headland erodes faster than the tip.

sea-floor drilling A process in which drill rigs mounted on offshore platforms or on research vessels cut cylindrical cores from both sediment and rock of the sea floor, which are then brought to the surface for study.

sea-floor spreading The hypothesis that segments of oceanic crust are separating at the mid-oceanic ridge.

seamount A submarine mountain, usually of volcanic origin, that rises 1 kilometer or more above the surrounding sea floor.

secondary air pollutant A pollutant generated by chemical reactions within the atmosphere.

secondary and tertiary recovery Production of oil or gas by artificially augmenting the reservoir energy, as by injection of water, detergent, or other fluid.

secondary wave A seismic wave consisting of a shearing motion in which the oscillation is perpendicular to the direction of wave travel. S waves travel slower than P waves. (*syn*: S wave)

sediment Solid rock or mineral fragments transported and deposited by wind, water, gravity, or ice, precipitated by chemical reactions, or secreted by organisms. Sediment accumulates as layers in loose, unconsolidated form.

sedimentary rock A rock formed when sediment is lithified.

sedimentary structure Any structure formed in sedimentary rock during deposition or by later sedimentary processes; for example, bedding.

seismic profiler A device used to construct a topographic profile of the ocean floor and to reveal layering in sediment and rock beneath the sea floor.

seismic tomography A technique whereby seismic data from many earthquakes and recording stations are analyzed to provide a three-dimensional view of the interior of the Earth.

seismic wave All elastic waves that travel through rock, produced by an earthquake or explosion.

seismogram The record made by a seismograph.

seismograph An instrument that records seismic waves.

seismology The study of earthquake waves and the interpretation of this data to elucidate the structure of the interior of the Earth.

shale A fine-grained clastic sedimentary rock with finely layered structure composed predominantly of clay minerals.

shear strength The resistance of materials to being pulled apart along their cross-section.

shear stress Stress that acts in parallel but opposite directions.

sheet flood A broad, thin sheet of flowing water that is not concentrated into channels, typically in arid regions.

shield volcano A large, gently sloping volcanic mountain formed by successive flows of basaltic magma.

sialic rock A rock such as granite and rhyolite that contains large proportions of silicon and aluminum.

silica Silicon dioxide, SiO_2. Includes quartz, opal, chert, and many other varieties.

silicate All minerals whose crystal structures contain silica tetrahedra. All rocks composed principally of silicate minerals.

silicate tetrahedron A pyramid-shaped structure of a silicon ion bonded to four oxygen ions, $(SiO_4)^{4-}$.

sill A tabular or sheet-like igneous intrusion parallel to the grain or layering of country rock.

silt All sedimentary particles from 1/256 to 1/16 millimeter in size.

siltstone Lithified silt.

sinkhole A circular depression in karst topography caused by the collapse of a cavern roof or by dissolution of surface rocks.

slate A compact, fine-grained, low-grade metamorphic rock with slaty cleavage that can be split into slabs and thin plates. Intermediate in grade between shale and phyllite.

slaty cleavage A parallel metamorphic foliation in a plane perpendicular to the direction of tectonic compression.

sleet Small spheres of ice that develop when raindrops form in a warm cloud and freeze as they fall through a layer of cold air at lower elevation.

slide All types of mass wasting in which the rock or regolith initially moves as a consolidated unit over a fracture surface.

slip The distance that rocks on opposite sides of a fault have moved.

slip face The steep lee side of a dune that is at the angle of repose for loose sand so that the sand slides or slips.

slump A type of mass wasting in which the rock and regolith move as a consolidated unit with a backward rotation along a concave fracture.

smog Smoky fog. The word is used loosely to define visible air pollution.

snowline The altitude above which there is permanent snow.

soil Soil scientists define soil as the upper layers of regolith that support plant growth. Engineers define soil synonymous with regolith.

soil horizon A layer of soil that is distinguishable from other horizons because of differences in appearance and in physical and chemical properties.

soil-moisture belt The relatively thin, moist surface layer of soil above the unsaturated zone beneath it.

solar cell A device that produces electricity directly from sunlight.

solar eclipse *See* eclipse.

solar mass The mass of our Sun.

solar wind A stream of atomic particles shot into space by violent storms occurring in the outer regions of the Sun's atmosphere.

solifluction A type of mass wasting that occurs when water-saturated soil moves slowly over permafrost.

solstice Either of two times per year when the Sun shines directly overhead furthest from the equator. The solstices mark the beginnings of summer and winter. One solstice occurs on or about June 21 and marks the longest day of the year in the Northern Hemisphere and the shortest day in the Southern Hemisphere. The other solstice occurs on or about December 22, marking the longest day in the Southern Hemisphere and the shortest day in the Northern Hemisphere.

source rock The geologic formation in which oil or gas originates.

specific gravity The weight of a substance relative to the weight of an equal volume of water.

specific heat The heat required to raise the temperature of one gram of a substance 1° Celsius.

spectrum (electromagnetic) A pattern of wavelengths into which a beam of light or other electromagnetic radiation is separated. The spectrum is seen as colors, or is photographed, or is detected by an electronic device. An **emission spectrum** is obtained from radiation emitted from a source. An **absorption spectrum** is obtained after radiation from a source has passed through a substance that absorbs some of the wavelengths.

speleothems Any mineral deposit formed in caves by the action of water.

spheroidal weathering Weathering in which the edges and corners of a rock weather more rapidly than the flat faces, giving rise to a rounded shape.

spicule A small jet of gas at the edge of the Sun. An average spicule is about 700 kilometers across and 7000 kilometers high, and lasts about 5 to 15 minutes.

spiral galaxy A type of galaxy characterized by arms that radiate out from the center like a pinwheel.

spit A small point of sand or gravel extending from shore into a body of water.

spreading center The boundary or zone where lithospheric plates rift or separate from each other. (*syn:* divergent plate boundary)

spring A place where ground water flows out of the Earth to form a small stream or pool.

spring tide A relatively large tide that forms when the Sun, Moon, and Earth are aligned.

stalactite An icicle-like dripstone, deposited from drops of water, that hangs from the ceiling of a cavern.

stalagmite A deposit of mineral matter that forms on the floor of a cavern by the action of dripping water.

stationary front A boundary between two stationary air masses of different temperatures.

stock An igneous intrusion with an exposed surface area of less than 100 square kilometers.

storm surge An abnormally high coastal water level created by a combination of strong onshore winds and the low atmospheric pressure of a storm.

storm track A path repeatedly followed by storms.

strain The deformation (change in size or shape) that results from stress.

stratified drift Sediment that was transported by a glacier and then sorted, deposited, and layered by glacial meltwater.

stratocumulus Low sheet-like clouds with some vertical structure.

stratopause The boundary between the stratosphere and the mesosphere.

stratosphere The layer of air extending from the tropopause to about 55 kilometers. Temperature remains constant and then increases with elevation in the stratosphere.

stratovolcano A steep-sided volcano formed by an alternating series of lava flows and pyroclastic eruptions. (*syn:* composite cone)

stratus cloud A horizontally layered, sheet-like cloud.

streak The color of a fine powder of a mineral usually obtained by rubbing the mineral on an unglazed porcelain streak plate.

stream A moving body of water, confined in a channel, and flowing downslope.

stress The force per unit area exerted against an object.

striations Parallel scratches in bedrock caused by rocks embedded in the base of a flowing glacier.

strike-slip fault A fault on which the motion is parallel with its strike and is primarily horizontal.

subduction The process in which a lithospheric plate descends beneath another plate and dives into the asthenosphere.

subduction zone (or subduction boundary) The region or boundary where a lithospheric plate descends into the asthenosphere.

sublimation The process by which a solid transforms directly to a vapor or a vapor transforms directly to a solid without passing through the liquid phase.

submarine canyon A deep, V-shaped, steep-walled trough eroded into a continental shelf and slope.

submarine fan A large, fan-shaped accumulation of sediment deposited at the bases of many submarine canyons adjacent to the deep sea floor. (*syn:* abyssal fan)

submergent coastline A coastline that was recently above sea level but has been drowned because either the land has sunk or sea level has risen.

subsidence Settling of the Earth's surface that can occur as either petroleum or ground water is removed by natural processes.

sunspot A cool dark region on the Sun's surface formed by an intense magnetic disturbance.

supercontinent A continent, such as Alfred Wegener's Pangaea, consisting of all or most of the Earth's continental crust joined together to form a single large landmass. At least three supercontinents are thought to have existed during the past 2 billion years, and each broke apart after a few hundred million years.

supercooling A condition in which water droplets in air do not freeze even when the air cools below the freezing point.

Superfund The Comprehensive Environmental Response, Compensation, and Liability Act (CERCLA). An act that provides an emergency fund to clean up chemical hazards and imposes fines for polluting the environment.

supernova An exploding star that is releasing massive amounts of energy.

superposition (principle of) *See* principle of superposition.

supersaturation A condition in which the concentration of a solution exceeds the saturation point; a condition in which the relative humidity of air exceeds 100 percent.

surf The chaotic turbulence created when a wave breaks near the beach.

surface mine A hole excavated into the Earth's surface for the purpose of recovering mineral or fuel resources.

surface processes All processes that sculpt the Earth's surface, such as erosion, transport, and deposition.

surface wave An earthquake wave that travels along the surface of the Earth, or along a boundary between layers within the Earth. (*syn:* L wave)

suspended load That portion of a stream's load that is carried for a considerable time in suspension, free from contact with the streambed.

suture The junction created when two continents or other masses of crust collide and weld into a single mass of continental crust.

syncline A fold that arches downward and whose center contains the youngest rocks.

system Any combination of interrelated, interacting components.

talus An accumulation of loose angular rocks at the base of a cliff from which they have been cleared by mass wasting.

tarn A small lake at the base of a cirque.

tectonic plate A large segment of the Earth's outermost, cool, rigid shell, consisting of the lithosphere and overlying crust. Tectonic plates float on the weak, plastic asthenosphere.

tectonics A branch of geology dealing with the broad architecture of the outer part of the Earth; specifically the relationships, origins, and histories of major structural and deformational features.

tensional stress Stress that pulls rock apart and is the opposite of tectonic compression.

terminal moraine An end moraine that forms when a glacier is at its greatest advance.

terminus The end, or foot, of a glacier.

terrestrial planets The four Earth-like planets closest to the Sun—Mercury, Venus, Earth, and Mars—which are composed primarily of rocky and metallic substances.

terrigenous sediment Sea-floor sediment derived directly from land.

texture The size, shape, and arrangement of mineral grains, or crystals in a rock.

thermocline A layer of ocean water between 0.5 and 2 kilometers deep where the temperature drops rapidly with depth.

thermosphere An extremely high and diffuse region of the atmosphere lying above the mesosphere. Temperature remains constant and then rises rapidly with elevation in the thermosphere.

threshold effect A threshold effect occurs when the environment initially changes slowly (or not at all) in response to a small perturbation, but after the threshold is crossed, an additional small perturbation causes rapid change.

thrust fault A type of reverse fault with a dip of 45° or less over most of its extent.

tidal current A current caused by the tides.

tide The cyclic rise and fall of ocean water caused by the gravitational force of the Moon and, to a lesser extent, of the Sun.

tidewater glacier A glacier that flows directly into the sea.

till Sediment deposited directly by glacial ice and that has not been resorted by a stream.

time correlation Correlation based on age equivalence of rock units.

time-travel curve A curve that records arrival times of P and S earthquake waves, used to measure the distance from a recording station to an earthquake epicenter.

topsoil The fertile, dark-colored surface soil, or A horizon.

tornado A small, intense, short-lived, funnel-shaped storm that protrudes from the base of a cumulonimbus cloud.

trade winds The winds that blow steadily from the northeast in the Northern Hemisphere and southeast in the Southern Hemisphere between 5° and 30° north and south latitudes.

transform boundary A boundary between two lithospheric plates where the plates are sliding horizontally past one another.

transform fault A strike-slip fault between two offset segments of a mid-oceanic ridge.

transpiration Direct evaporation from the leaf surfaces of plants.

transport The movement of sediment by flowing water, ice, wind, or gravity.

transverse dune A relatively long, straight dune with a gently sloping windward side and a steep lee face that is orientated perpendicular to the prevailing wind.

trench A long, narrow depression of the sea floor formed where a subducting plate sinks into the mantle.

tributary Any stream that contributes water to another stream.

Tropic of Cancer The latitude 23.5° north of the equator. On June 21, the summer solstice in the Northern Hemisphere, sunlight strikes the Earth from directly overhead at noon at this latitude.

Tropic of Capricorn The latitude 23.5° south of the equator. On December 22, the summer solstice in the Southern Hemisphere, sunlight strikes the Earth from directly overhead at noon at this latitude.

tropical cyclone A broad, circular storm with intense low pressure that forms over warm oceans (also called a *hurricane*, *typhoon*, or *cyclone*).

tropopause The boundary between the troposphere and the stratosphere.

troposphere The layer of air that lies closest to the Earth's surface and extends upward to about 17 kilometers. Temperature generally decreases with elevation in the troposphere.

trough The lowest part of a wave.

truncated spur A triangular-shaped rock face that forms when a valley glacier cuts off the lower portion of an arête.

tsunami A large sea wave, produced by a submarine earthquake or a volcano, characterized by long wavelength and great speed.

turbidity current A rapidly flowing submarine current laden with suspended sediment, that results from mass wasting on the continental shelf or slope.

turnover A process, usually occurring in fall and spring in temperate-climate lakes, in which convection mixes lake water so that it becomes of uniform temperature from top to bottom of the lake.

typhoon *See also* tropical cyclone, hurricane, cyclone.

U-shaped valley A glacially eroded valley with a characteristic U-shaped cross-section.

ultimate base level The lowest possible level of downcutting of a stream, usually sea level.

ultramafic rock Rock composed mostly of minerals containing iron and magnesium, for example, peridotite.

umbra The part of an eclipse shadow where the Sun is completely eclipsed.

unconformity A gap in the geological record, such as an interruption of deposition of sediments, or a break between eroded igneous and overlying sedimentary strata, usually of long duration.

underground mine A mine consisting of subterranean passages that commonly follow ore veins or coal seams.

underthrusting The process by which one continent may be forced beneath the other during a continent-continent collision.

uniformitarianism The principle that states that geological processes and scientific laws operating today also operated in the past and that past geologic events can be explained by forces observable today. "The present is the key to the past." The principle does not imply that geologic change goes on at a constant rate, nor does it exclude catastrophes such as impacts of large meteorites.

unit cell The smallest group of atoms that perfectly describes the arrangement of all atoms in a crystal, and repeats itself to form the crystal structure.

unsaturated zone A subsurface zone above the water table that may be moist but is not saturated; it lies above the zone of saturation. (*syn:* zone of aeration)

upper mantle The part of the mantle that extends from the base of the crust downward to about 670 kilometers beneath the surface.

upslope fog Fog that forms when air cools as it rises along a land surface.

upwelling A rising ocean current that transports water from the depths to the surface.

urban heat island effect A local change in climate caused by a city.

valley train A long and relatively narrow strip of outwash deposited in a mountain valley by the streams flowing from an alpine glacier.

vent A volcanic opening through which lava and rock fragments erupt.

vesicle A bubble formed by expanding gases in volcanic rocks.

volatile A compound that evaporates rapidly and therefore easily escapes into the atmosphere.

volcanic neck A vertical pipe-like intrusion formed by the solidification of magma in the vent of a volcano.

volcanic rock A rock that formed when magma erupted, cooled, and solidified within a kilometer or less of the Earth's surface.

volcano A hill or mountain formed from lava and rock fragments ejected through a volcanic vent.

warm front A front that forms when moving warm air collides with a stationary or slower moving cold air mass. The moving warm air rises over the denser cold air as the two masses collide, cools adiabatically, and the cooling generates clouds and precipitation.

wash An intermittent stream channel found in a desert.

water table The upper surface of a body of ground water at the top of the zone of saturation and below the zone of aeration.

wave height The vertical distance from the crest to the trough of a wave.

wave period The time interval between two crests (or two troughs) as a wave passes a stationary observer.

wave-cut cliff A cliff created when a rocky coast is eroded by waves.

wave-cut platform A flat or gently sloping platform created by erosion of a rocky shoreline.

wavelength The distance between successive wave crests (or troughs).

weather The state of the atmosphere on a particular day as characterized by temperature, wind, cloudiness, humidity, and precipitation. *See* climate.

weathering The decomposition and disintegration of rocks and minerals at the Earth's surface by mechanical and chemical processes.

welded tuff A hard, tough, glass-rich pyroclastic rock formed by cooling of an ash flow that was hot enough to deform plastically and partly melt after it stopped moving; it often appears layered or streaky.

well A hole dug or drilled into the Earth, generally for the production of water, petroleum, natural gas, brine, sulfur, or for exploration.

wet adiabatic lapse rate The rate of cooling that occurs when moist air rises and condensation occurs, without gain or loss of heat.

wetlands Areas known variously as swamps, bogs, marshes, sloughs, mud flats, flood plains, and many other names, wetlands are the boundary between land and water. Some are water soaked or flooded throughout the entire year; others are dry for much of the year and wet only during times of high water. Some are wet only during exceptionally wet years and may be dry for several years at a time.

white dwarf A stage in the life of a star when fusion has halted and the star glows solely from the residual heat produced during past eras. White dwarfs are very small stars.

wind Horizontal air flow caused by the pressure differences resulting from unequal heating of the Earth's atmosphere. Winds near the Earth's surface always flow from a region of high pressure toward a low-pressure region.

wind shear A condition in which air currents rise and fall simultaneously within the same cloud.

withdrawal Any process that uses water, and then returns it to the Earth locally.

zone of aeration A subsurface zone above the water table that may be moist but is not saturated; it lies above the zone of saturation. (*syn:* unsaturated zone)

zone of saturation A subsurface zone below the water table in which the soil and bedrock are completely saturated with water.

zone of wastage The lower portion of a glacier where more snow melts in summer than accumulates in winter so there is a net loss of glacial ice. (*syn:* ablation area)

Note: Page numbers in **boldface** type indicate pages on which terms are defined. *Italic* page numbers indicate pages with illustrations, and t indicates table.

East African rift, 93, *95*
Easterlies, polar, **403**
East Pacific Rise, 291, *293, 495*
Eccentricity, **259**, *260*
Echo sounder, **292**, *292*
Eclipses
    of moon, 531–535, *534*
    of sun, 531–535, *534, 573*
Ecosystems
    destruction of, 419–420
    global, 5
    human impact on, 5, *5*
    study of, 194
    of wetlands, 239
Effluent stream, **233**
Eiffel Tower, lightning on, *390*
Elastic deformation, **108**, *109*
Electricity
    generating plant, *448*
    in lightning, 390, *391*
    from nuclear power, *507*
    from wind, *510*
Electromagnetic radiation, **350**, 537
Electromagnetic spectrum, **351**
Electron, in radioactive decay, 75
Electron microscope photo, *300*
Electrons, **24**, *26*
Elements, **23**
    in crust, 23, *492*
    important, *491*
    periodic table of, 23, A.14
    radioactive decay, 74–75
    uses of, *491*
Elevation
    and atmospheric pressure, 349, *350*
    temperature and, 355–356
Elliptical galaxies, **582**, *582*
El Niño, 384–387
Emergent coastline, **328**–330, *329, 331*
Emission nebula, in Orion, *568*
Emission spectra, **538**
Emissions, of air pollution, 444
Emperor seamount chain, *308*
End moraine, **254**, *255*
Energy
    from biomass, 510–511
    geothermal, **237**, 510
    hydroelectric, 510
    measurement of, A.11
    from oceans, 511
    solar, 509–510
    in a system, 10
Energy-consuming systems, efficiency
    of, *512*
Energy resources, **492**
    alternative, **492**, 509–510
    of United States, 508–509, *508*
Environmental Protection Agency
    (EPA), 37, 57, 420, 444, 453,
    456, 477

Eons, **77**
    Archean, **79**
    earliest, 78–79
    Hadeon, 78–79
    Phanerozoic, **79**–80
    Proterozoic, **79**
EPA, *see* Environmental Protection
    Agency
Epicenter, **109**, *110*, 113–114
Epidemiology, **448**, 479
Epochs, **77**
Equant garnet, *28*
Equator
    moisture at, 266–267
    sunshine at, 357
Equinoxes, **359**
Eras, **77**
    Cenozoic, **80**
    Mesozoic, **79**
    Paleozoic, **79**, *79*
Erosion, **181**–182, *181*
    along coastlines, 323–324
    of desert mountains, 273
    due to dams, 466
    by glaciers, 248–253
    by humankind, 195–197
    lateral, **216**–217
    on Mars, 285, *286*
    natural, 195
    by streams, 212–214
    by wind, 275–280
Erratics, **254**
Escalante River, sandstones along, *6,*
    162
Esker, **257**, *257*
Estuary, **332**
Eta Carinae, *566*
Ethiopia, soil loss in, 170
Eurasian plate, *91*
Europa, 556–557
Eustatic, **328**
Eutrophic lake, **228**, *229*, 485
Eutrophic waterway, *478*
Evaporation, 361, *362*
Evaporation fog, **374**
Evaporites, **52**, *52*, **496**–497
Everglades, *239*
Evolution, **72**
    of life on Earth, 6–7, *7, 17,* 79
    of landforms, 221–222
    of oceans, 287, 288–289
    of stars, 575–577, *575–578*
Exfoliation, **187**–188, *187*
Expansion, thermal, 183–184
Exponential notation, A.13
Extensional stress, **155**, *155*
Externalities, **479**
Extinction, mass, 16–20
Extraterrestrial impacts, and mass ex-
    tinctions, 18

Extraterrestrial life, 552–553
Extrusive igneous rocks, **45**, *45*
Eyepiece, **536**, *536*

F

Face, of crystal, **27**
Fall, **199**, *200,* 204, *205*
    ice, *247*
Fan
    abyssal, *303,* **304**
    alluvial, **219**–220, *219, 273*
    submarine, *303,* **304**
Fault creep, **116**
Faults, **108**, *109,* **110,** 157–161
    of California, *115*
    normal, **158**, *159*
    reverse, **160**, *161*
    San Andreas, 86, *110,* **114,** *115*
    strike-slip, *114, 115,* **160,** *161*
    thrust, **116,** *160, 161*
    transform, **296,** *299*
    in Yucca Mountain repository, *483*
Fault zone, *159,* 161
Faunal succession, principle of, 71–**72**
Feedback mechanism, **12**
    in Earth systems, 11–13, 19
Feldspar, 22, 33, **34**–35
    cleavage planes of, *29*
    in granite, 46
    hydrolysis of, 186
    as phenocrysts, *46*
Felsic, **46**
Fern, fossil, *429*
Fertilizer, water pollution by, 477–478
Fetch, 318
Fibrous crystals, *28*
Finger Lakes, in New York, *253*
Fire, from earthquakes, 122
Firn, **244**, *244*
First-generation stars, 578–579
First Law of Motion, 527
Fish harvests
    of Chesapeake Bay, 333
    global, 322
Fissility, **51**
Fission, nuclear, *506*
Fissures, **141**
Fjords, **252**, *253,* **332**
Flash flood, **270**
Flathead River valley, *218, 227*
Flint, **51**
Flood basalt, **141**, *142*
Flood plain, **212**, *222*
    flooding of, *224*
    management of, 226
Floods, **212**
    catastrophic, 11
    on coastlines, 336–338
    control of, 223

Harcourt
College Publishers

Where Learning Comes to Life

TECHNOLOGY

Technology is changing the learning experience, by increasing the power of your textbook and other learning materials; by allowing you to access more information, more quickly; and by bringing a wider array of choices in your course and content information sources.

Harcourt College Publishers has developed the most comprehensive Web sites, e-books, and electronic learning materials on the market to help you use technology to achieve your goals.

PARTNERS IN LEARNING

Harcourt partners with other companies to make technology work for you and to supply the learning resources you want and need. More importantly, Harcourt and its partners provide avenues to help you reduce your research time of numerous information sources.

Harcourt College Publishers and its partners offer increased opportunities to enhance your learning resources and address your learning style. With quick access to chapter-specific Web sites and e-books . . . from interactive study materials to quizzing, testing, and career advice . . . Harcourt and its partners bring learning to life.

Harcourt's partnership with Digital:Convergence™ brings :CRQ™ technology and the :CueCat™ reader to you and allows Harcourt to provide you with a complete and dynamic list of resources designed to help you achieve your learning goals. Just swipe the cue to view a list of Harcourt's partners and Harcourt's print and electronic learning solutions.

http://www.harcourtcollege.com/partners/

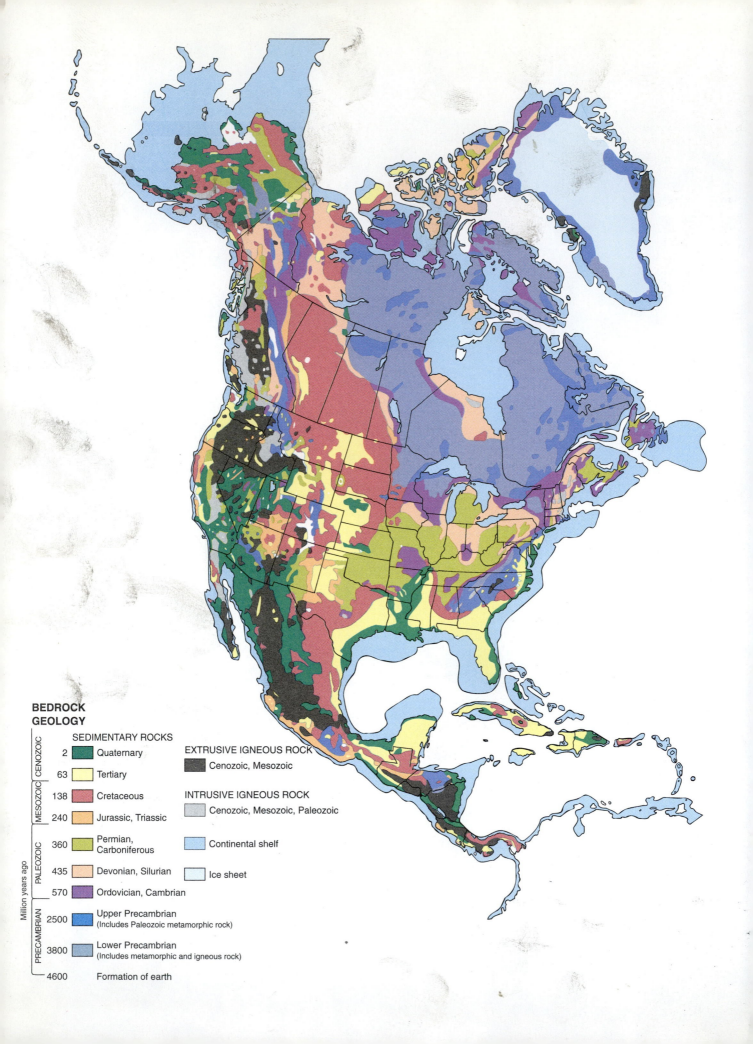

BEDROCK GEOLOGY

Million years ago

SEDIMENTARY ROCKS

CENOZOIC		
2	🟢	Quaternary
63	🟡	Tertiary

MESOZOIC		
138	🔴	Cretaceous
240	🟧	Jurassic, Triassic

PALEOZOIC		
360	🟩	Permian, Carboniferous
435	🟧	Devonian, Silurian
570	🟪	Ordovician, Cambrian

PRECAMBRIAN		
2500	🔵	Upper Precambrian (Includes Paleozoic metamorphic rock)
3800	🔵	Lower Precambrian (Includes metamorphic and igneous rock)
4600		Formation of earth

EXTRUSIVE IGNEOUS ROCK

⬛ Cenozoic, Mesozoic

INTRUSIVE IGNEOUS ROCK

⬜ Cenozoic, Mesozoic, Paleozoic

🟦 Continental shelf

🟦 Ice sheet